Marine Algae of California

T0177214

With Contributions by

H. William Johansen   James N. Norris

Susan Loiseaux   George F. Papenfuss

Nancy L. Nicholson   Joan G. Stewart

Elise M. Wollaston

Isabella A. Abbott
and George J. Hollenberg

Marine Algae of California

STANFORD UNIVERSITY PRESS

Stanford, California

Stanford University Press
Stanford, California
© 1976 by the Board of Trustees of the
Leland Stanford Junior University
Printed in the United States of America
Cloth ISBN 0-8047-0867-3
Paper ISBN 0-8047-2152-1
Original printing 1976
Last figure below indicates year of this printing:
15 14 13 12 11 10 09 08 07

Dedicated in Admiration and Affection
to
GILBERT MORGAN SMITH (1885–1959)
and
ELMER YALE DAWSON (1918–1966)

Preface

THE MARINE ALGAE of North America's Pacific shores, though forming
a distinct and complex flora, have never been comprehensively studied as
a geographic unit or treated in a single volume. Excellent monographs on
particular groups do exist, such as Setchell and Gardner's *Chlorophyceae*
(1920b) and *Melanophyceae* (1925). And there are a number of good
though geographically limited studies, such as Scagel's *Marine Algae of
British Columbia and Northern Washington* (1966), Doty's *Marine Algae
of Oregon* (1947), Smith's *Marine Algae of the Monterey Peninsula, California* (1944; with *Supplement*, 1969), and Dawson's series *Marine Red
Algae of Pacific Mexico*. Of these, Smith's work has been the most widely
used, partly because of its superb descriptions and luxuriant illustration,
but chiefly because the species it covers range beyond the Peninsula itself
to Alaska and Mexico; Dr. Smith himself, however, scrupulously limited
his actual descriptions of species to Monterey Peninsula specimens.

What we have sought to do in this book, *Marine Algae of California*, is
to extend the descriptions offered by Smith to embrace the variations shown
by the species, as we understand them, throughout their California distribution. This treatment may be unsatisfactory for those working with small
and perhaps little-changing populations, but it is not for these specialists
that we have written. Rather, we have written for those who, like most of
our students, have come from elsewhere to the Monterey Peninsula and
have found on returning home that Smith's descriptions, *sine qua non* for
the Monterey area, do not "fit" the specimens collected elsewhere. This
book, then, is intended to replace Smith's study, but only in the sense that
it describes the marine algae of the Monterey Peninsula and a good deal of
the coastline beyond. It cannot be expected to replace the detailed specific
descriptions given by Smith, for in treating the more varied populations of
a much longer coastline, and half again the number of species, we have
necessarily prepared descriptions that suffice for a species as a whole rather
than for a few small or limited populations included within that species.

The book is thus intended to be used as a manual—a combined laboratory and field reference for identifying the marine algae of California. It

differs from its predecessors (Smith 1944; Hollenberg & Abbott 1966) in placing less emphasis on a previous knowledge of lower plants ("crypto-gams"), a decision that should render the material a bit more accessible. It is evident that fewer college students are trained in the lower plants today, owing in part to the consolidation of botany departments into biol-ogy departments and in part to the need to forgo some of the more classically oriented courses in favor of molecular biology. Moreover, a good many zoologists, ecologists, environmental-impact analysts, sanitary engi-neers, and the like, all of them lacking the classical lower-plant training, are taking a professional interest in algae. And an increasing number of nonprofessionals are finding marine algae both attractive and useful: the plants are interesting and often beautiful, and many are valuable sources of food or industrial chemicals.

The late E. Yale Dawson, realizing this shift in emphasis, had asked the authors to join him in writing a book of the present scope, and a contract was signed in 1966. When Dr. Dawson died in a drowning accident, just three months later, the authors invited Dr. Paul C. Silva of the University of California, Berkeley, to join them. In time, Dr. Silva found other com-mitments too pressing and was unable to meet deadlines. His text assign-ments were accordingly shared by the authors or taken up by a number of colleagues, three of them specialists: Dr. Susan Loiseaux of France (the Myrionemataceae of the brown algae), Dr. H. William Johansen of Clark University, Massachusetts (the Corallinaceae of the red algae), and Dr. Elise Wollaston of Australia (*Antithamnion, Antithamnionella, Hollen-bergia, Scagelia,* and *Platythamnion* of the red algae). Three other sections were written by former students: Dr. James N. Norris (*Blidingia* and *Enteromorpha* of the green algae), Dr. Nancy L. Nicholson (the Lami-nariales and Fucales of the brown algae), and Dr. Joan G. Stewart (the Gelidiaceae of the red algae). A very valuable contribution on the history of West Coast algal exploration and study has been added by Professor George F. Papenfuss of the University of California, Berkeley.

In 1970, Dr. Nicholson offered us a large and fascinating collection of intertidal and subtidal algae from the Channel Islands of southern Cali-fornia. Working this material into the descriptions and records, which had been drawn mainly from mainland material, took a good deal of time.

In late 1971, Dr. Peter Dixon was asked to read the entire manuscript for accuracy and consistency. This reading in itself prolonged the final writing, for it became necessary to determine once again whether the species of many of the older authors had been validly published. Many, especially those first appearing in the collections of dried algae known as the Phyco-theca Boreali-Americana, or familiarly as the P.B.-A., were not. Each of these cases necessitated a search for a valid name. These searches prompted us, in fact, to decide that the nomenclatural information provided (the

data following the heading for a species name and author) would be kept
to the minimum consonant with good taxonomic practice.

In the Smith, Hollenberg & Abbott (1969), which combines the 1944
first edition of Smith and the 1966 *Supplement*, there are 448 species. This
new volume describes 669 species. The gain in number of species is due
partly to our subsuming the previous California records of species not occurring on the Monterey Peninsula, and thus not treated in Smith; to new
additions to the California marine flora in the form of species newly described since 1969; to a few geographical extensions (into the Monterey
Peninsula) of species previously known only to the north or to the south;
and to a few reassignments of former varietal epithets as species.

Space-saving was an important concern throughout this process, for we
hoped to put into a book the size of the Smith (1944)—or at least not much
larger than the Smith, Hollenberg & Abbott (1969)—roughly half again as
many species. This meant writing each description as tightly as possible,
emphasizing the more diagnostic characters. It also meant carefully con-
trolling the content of the illustrations, to ensure that each drawing expresses not only the way the species appears in nature but also contains
the characters mentioned in the description. Extending the coverage from
the Monterey Peninsula to the entire California coastline abetted our efforts
to condense, since a description that suffices for a great many populations
necessarily entails more generality in individual characters than does a
description based on one or a few highly homogeneous local populations.

We know that we have not included all species of large marine algae
occurring in California: we ourselves have collected specimens of perhaps
a dozen that are yet undescribed; and others, perhaps already described
for other floras, we cannot identify with certainty because we lack fertile
material. (Except for *Neoagardhiella baileyi*, in fact, we have not added
any change of name made since 1972.) Even so, we believe that we have
included at least 98 percent of the species that have been collected along
California's long coastline prior to 1976. It will be clear to the reader that
in this project we have depended heavily on the previous major studies
of Pacific Coast algae, especially those of Setchell and Gardner, Smith,
Dawson, and Scagel, as well as our own. Some descriptions (e.g. those
for *Ceramium* species) are literally those of Dawson, generally because
he had had more material than we upon which to base his descriptions.

In carrying through both condensation of detail and expansion of coverage, we have tried to retain some of the more valuable and successful
features of the Smith (1944). Thus there is a greatly expanded Master Key
to the Genera, which holds the use of specialized technical terms to a
minimum and can easily be used by those with only a meager grounding
in biology. The Literature Cited section has been rigorously assembled as
a reference list for species only (more general references are discussed in

the Introduction). And we have continued Smith's strong emphasis on excellence in illustration. In our combined 28 years of teaching courses in marine algae at the Hopkins Marine Station, we have become very much aware that a great many people who have studied California algae have relied on Smith's own figures as the basis for taxonomic judgment.

Each of the 669 species described in this work is illustrated by one or more line drawings, as are all but a handful of the infraspecific taxa. In all, the book employs 891 separate drawings, reproduced at substantially larger scale than are those in the Smith. About half of these have been newly drawn for this book, but wherever possible we have made use of the superb original drawings from the Smith (1944), all of which are by Jeanne Russell Janish. A few of these, passed through many hands over the years, were lost and have been redrawn from new specimens. A few have since been shown to be inadequate representations of the species they were intended to illustrate and have been replaced. And of course, a number of California species and varieties do not occur near Monterey or have been described in the years following the original publication of Smith's work. For many of these last we have used drawings from the *Supplement* to the Smith (Hollenberg & Abbott 1966); for others, new drawings have been prepared; and in a few cases, drawings from other sources have been borrowed (as indicated in the legends) or copied (so signified by "after").

For all of the new drawings, we have tried to maintain the quality of illustration set by Mrs. Janish, and wherever possible we have prepared them using fresh material. In many cases a pencil drawing—typically a microscopic cross section—was prepared by one of the authors, then inked by an artist. This is indicated in the legends by combining the initials of the two: for example, "(DBP/H)" means that Dana Bean Pierce inked a drawing by George Hollenberg; "(SM/A)" means that Susan Manchester inked a drawing by Isabella Abbott. Mrs. Janish's initials are routinely combined with Smith's, since it cannot be determined, at this remove, where Dr. Smith initiated an illustration and where he did not—thus "(JRJ/S)," in all cases. Those whose illustrations appear in this book are the following:

IA	Isabella Abbott	NLN	Nancy L. Nicholson
EYD	E. Yale Dawson	DBP	Dana Bean Pierce
LH	Lisa Haderlie	WAS	William A. Setchell
GJH	George J. Hollenberg	CS	Cathy Short
JRJ	Jeanne Russell Janish	JGS	Joan G. Stewart
HWJ	H. William Johansen	FT	Frances Thompson
SM	Susan Manchester	EW	Elise Wollaston

No two people, no matter how persistent, could have written this book without the help of contributors and of other colleagues, friends, and students. We are grateful first of all to Peter S. Dixon for his time, effort, and advice on all aspects of the book. Michael J. Wynne provided much useful advice on the brown algae. And we wish once again to thank those who have contributed text material: George F. Papenfuss, Elise Wollaston, H. William Johansen, Joan G. Stewart, Susan Loiseaux, James N. Norris, and Nancy L. Nicholson.

We express our appreciation for the continued loan of herbarium materials, so essential to the preparation of an extensive flora, to the keepers and curators of collections at many institutions: Agardh Herbarium, Lund University, Sweden (Dr. Ove Almborn); Muséum National d'Histoire Naturelle, Paris (Dr. Pierre Bourrelly); Farlow Herbarium, Harvard University (Dr. I. MacKenzie Lamb); New York Botanical Garden (Dr. Clark Rogerson); Smithsonian Institution (Dr. Arthur Dahl); University of British Columbia (Dr. Robert F. Scagel); University of Washington (Dr. Richard E. Norris); University of California, Berkeley (Dr. G. F. Papenfuss, Dr. Paul C. Silva); University of California, Santa Barbara (Dr. Michael Neushul); Allan Hancock Foundation, University of Southern California (Dr. Nancy L. Nicholson, Robert Setzer); University of Hawaii (Dr. Maxwell S. Doty); Hokkaido University (the late Prof. Yukio Yamada, Prof. Munenao Kurogi); and Tokyo University (Prof. H. Hara). We record our thanks, as well, to Dr. Richard S. Cowan of the Smithsonian Institution for allowing us to use the notes and illustrations of the late E. Yale Dawson.

We thank also Stanford University Press, for permitting us to use as many of the Smith illustrations as we needed; the University of Redlands, for some of the secretarial help; the secretaries at Hopkins Marine Station; and especially Faylla Chapman, at Hopkins, for much technical, secretarial, and editorial help. James Norris, Nancy Nicholson, Lester Hair, and Robert Setzer are among the numerous students who contributed specimens. Dr. Wheeler North contributed outstanding subtidal collections from throughout California.

The first-named author acknowledges with pleasure her indebtedness to the U.S. Office of Naval Research (Contract N-00014-67-A-0112) and the U.S.-Japan Cooperative Science Program (Grant GF-219). Their support provided most of the funds needed to illustrate this book, allowed the examination of related Japanese specimens in the herbarium and in the field, and sustained much curatorial assistance. We gratefully acknowledge also a gift from The David and Lucile Packard Foundation toward production costs of the book.

Five artists, among those mentioned earlier, spent many tedious hours during the past eight years in making the algae we have added to this volume come alive on paper. It has been an instructive and altogether pleasant experience working with Dana Bean Pierce, Frances Thompson, Cathy Short, Susan Manchester, and Lisa Haderlie.

In the 10 years that we have worked with him in the course of writing the *Supplement* (1966) and this volume, we have come to know William W. Carver, Executive Editor of Stanford University Press, as a friend, and we thank him for his cheerfulness and his patience. We must also thank Barbara E. Mnookin, Elizabeth Spurr, and James R. Trosper, of the Press, who contributed in many ways to the editorial task, and Albert P. Burkhardt, who designed this volume and prepared the maps for the Introduction.

Finally, to Naomi Hollenberg and Donald Abbott, we say in public, thank you for your patience and support while we devoted so much time to what proved to be both an intellectual challenge and a physical strain.

IAA   GJH

Contents

Marine Algae of
California

Introduction

Following the precedent set by Smith (1944), this book treats only benthic (sea-floor) marine algae of multicellular structure and macroscopic size. Within these limits, we have tried to include all previously described species found along the California coast, with the exception of the blue-green algae (Cyanophyta). The taxonomy of the blue-greens is extraordinarily complex, and can be adequately dealt with only by specialists. There are few published accounts of California blue-greens; and in any case very few taxa in this division are marine. One might also question our omission of the unicellular greens; and Smith in point of fact did include a few taxa from this group (e.g. Volvocales). Smith, however, was perhaps the foremost specialist of his day in the freshwater algae, many of which are unicellular greens. In general, these microscopic greens, like the blue-greens, should be left to those trained in the elaborate electron-microscopy and culture techniques their study requires.[*]

CLASSIFICATION, FORM, AND PHYSIOLOGY

Following a classification that has not been challenged in its major categories for more than 125 years, we recognize three phyla (sometimes called divisions by botanists) among the macroscopic marine algae: the green algae (Chlorophyta), the brown algae (Phaeophyta), and the red algae (Rhodophyta). As the names indicate, these groups were originally set off by their colors, which are apparent to the eye; the validity of these colors as a distinguishing characteristic has since been demonstrated by chemical means. In limiting our coverage (with one exception) to these three divisions, we have excluded several other algal divisions in the ocean whose total biomass and diversity (species numbers) surpass those of the macroscopic algae. These are the microscopic Bacillariophyta (diatoms) and

---

[*] See J. D. Pickett-Heaps. 1975. *Algae: Structure, reproduction and evolution in selected genera.* Sunderland, Mass.: Sinauer Assoc. 606 pp. Along the California coast, marine unicellular algae often coat granite rock in thin sheets or form darker patches in shady crevices. These have been collectively named GATGORE ("green algae that grow on rocks everywhere") by Stanford University undergraduates, who know that most may be identified if cultured but are extraordinarily difficult to study in field-collected material.

Pyrrophyta (dinoflagellates, including the luminescent *Noctiluca* and some "red-tide" organisms). Groups with smaller numbers of species (e.g. the euglenoid flagellates) also occur in the ocean. Some of the yellow-brown algae (Chrysophyta) are marine, but the group is far more numerous and diverse in freshwater (P. Bourrelly. 1968. *Les Algues d eau douce, algues jaunes et brunes*, Vol. 2. Paris: N. Boubée. 438 pp.). From this phylum, however, we have decided to include the single genus *Vaucheria*, a macroscopic alga (most other Chrysophyta are unicellular). In the form of thin patches on the substratum, *Vaucheria* occurs at many points in California; moreover, it is familiar to many students owing to years of laboratory use as a "representative" alga (though incorrectly considered one of the green algae for most of this time).

For Californian waters, our treatment of the algae embraces the following:

Phylum	Classes	Orders	Families	Genera	Species
Yellow-browns (Chrysophyta)	1	1	1	1	1
Greens (Chlorophyta)	2	6	10	27	72
Browns (Phaeophyta)	1	10	20	69	137
Reds (Rhodophyta)	2	7	35	186	459
TOTAL	6	24	66	283	669

The book also treats two subspecies (beyond one per species), 41 varieties (beyond one per species), and three formae (beyond one per species or variety), as well as nine entities of questionable status (these set off by quotation marks). All 669 species and all but 23 of the 55 infraspecific taxa (701 taxa in all) are illustrated.

All major groups of algae occur in both marine and freshwater habitats, though the browns and reds are predominantly marine. Certain orders of green algae are predominantly or entirely marine (Siphonocladales, Codiales), and some are wholly tropical (Caulerpales, Dasycladales). Among the brown algae the order Laminariales is of temperate seas, its members occurring almost entirely in water temperatures lower than 24°C.

As regards structure, the macroscopic algae may in general be expected to have a holdfast, a stipe, and a frond or blade. Most may be characterized in this fashion, but many will lack one or more of these structures, owing to morphological modification and adaptation. The three structures of a typical alga may appear to be equivalent to the root, stem, and leaf of a flowering plant. However, the resemblance is entirely superficial: algal tissues are not modified internally to assume the specialized structural and translocatory functions of the "higher plants," even those of the mosses (Bryophyta). The larger brown algae, or kelps (e.g. *Macrocystis*), some of which are taller than most trees, do not have the highly modified internal tissues of any tree, for they are supported by the water in which they live. The roots of a tree, to take another example, contain cells and tissues specialized for their purpose and quite unlike those in most other parts of the plant; by contrast, the rootlike holdfast of a large kelp, though modified on the external, macroscopic level, is composed of vegetative cells largely similar to those of the stipe or blade.

All macroscopic algae have plastids containing one or more chlorophylls (and usually other pigments). The paler epiphytic species sometimes show the remnants of plastids in thin sections prepared for the electron microscope, and this finding brings into question the parasitic or nonparasitic nature of the epiphyte. Moreover, one may encounter very small, colorless gall-like structures that contain no plastids and are now suspected of being fungal, bacterial, or viral infections: several of these continue to be accorded taxonomic status as algae.

It is generally known that the free-floating phytoplankton of the oceans fix carbon in respectable amounts through the photosynthetic process. Many benthic algae and sea grasses also fix large amounts, even though these plants are restricted to the relatively narrow continental shelves and shallower waters. *Macrocystis*, for example, can fix from 1 to 4.8 kg of carbon per square meter of plant surface each year (W. J. North, personal communication). Other species may display even higher productivities, offering abundant food and shelter for a variety of marine animals, and forming the base of an elaborate food web and an important part of the biogeochemical cycle.

The depth at which effective photosynthesis can take place is determined by a number of factors, of which temperature and penetration of light are the most important. On the California coast very few benthic algae are found below 40 meters. Some of the larger kelps (e.g. *Pelagophycus*) may be attached to the substratum near or below this limit; but these species have generally developed floatlike pneumatocysts that keep stipes and fronds much nearer the surface.

The high productivity and potential food value of many of the macroscopic algae have been recognized most notably by the Japanese, who depend heavily on seaweeds for food. Hundreds of metric tons of dried seaweed are imported by the United States each year; very little of it is used directly as food, though much is added to foodstuffs. Except for the giant kelp, *Macrocystis*, few native algae are used commercially in the United States. *Macrocystis* yields alginic acid, used for dozens of purposes as diverse as preparing dental molds, brewing, and making candy and fancy pastry. Industrial and sewage pollution have taken their toll of the kelp beds, apparently causing conditions incompatible with balanced growth and the maintenance of the species. The effort to restore and manage these kelp beds has been a valiant and at times disheartening struggle over the last 15 years (W. J. North, ed. 1971. Biology of the giant kelp beds. *Nova Hedwigia*, 32. 600 pp.). Regulated by the State of California, and leased for private exploitation, the kelp beds are worth several hundred thousand dollars annually to the State, and many times that to the kelp harvesters. American investment in *Macrocystis* and other seaweeds totals many millions of dollars annually.

It is not feasible, of course, in a book like this, to furnish detailed information on all aspects of algal classification, morphology, cytology, and related topics. *Marine algae of California* is oriented toward the needs of the field and laboratory worker, and it is essentially a guide for the identification of the California algal taxa as we presently understand them. Obviously, there is more to phycology than this; and the interested reader will find much useful material on the biology and classification of marine algae in the following general works:

Chapman, V. J. 1970. *Seaweeds and their uses*, 2d ed. London: Methuen. Bot. Monogr. 10. 304 pp. A very readable account of the folk and industrial uses of seaweeds, covering both nutritional and economic value.

Dawson, E. Y. 1966. *Marine botany, an introduction*. New York: Holt, Rinehart & Winston. 371 pp. An extremely useful general book, requiring very little advance technical knowledge, and containing the best life-history diagrams given by the books listed here. It also reflects the good nature and interests of the late Dr. Dawson.

Fritsch, F. E. 1935, 1945. *The structure and reproduction of the algae*. 2 vols. Cambridge: Cambridge University Press. 791 and 939 pp. These two books summarize in great detail what was known of the morphology and reproduction of algae up to 1945. Cell development and structure are explained on the basis of an example in each taxonomic group, and differences or similarities in close relatives are discussed. The time and expense necessary to bring these volumes up to date would be prohibitive—and it is possible that no one person could do the job, inasmuch as knowledge of the algae has mushroomed in the interval that has passed.

Smith, G. M. 1955. *Cryptogamic botany* (2d ed.), Vol. 1. New York: McGraw-Hill. 546 pp. A generation of American phycologists, including most of the more established teachers of phycology, were brought up on this work. Like Fritsch, it is out of date not so much because the basic information is old, but because ultrastructural, physiological, and cytological discussions are completely lacking. Nonetheless, the data included are essential to an understanding of the place of algae in the plant kingdom.

Stewart, W. D. P., ed. 1974. *Algal physiology and biochemistry*. Berkeley, Calif.: Univ. Calif. Bot. Monogr. 10. 989 pp. Current information on the physiology of the algae is masterfully summarized in this volume. It should be read with current research papers on ultrastructure in hand.

Proceedings of the I–VIII International Seaweed Symposia. Published separately following each Symposium, at about five-year intervals. The last to appear was VII, edited by K. Nisizawa and published by the University of Tokyo Press in 1972. Offering a wide variety of papers, both classical and technological, these volumes are excellent places to learn in what direction some research is leading, and what some of the new approaches are.

Besides these general works, there are a number of useful monographs on the different algal groups, both for the Pacific coast and for other marine provinces:

Cupp, E. E. 1943. Marine plankton diatoms of the west coast of North America. *Bull. Scripps Inst. Oceanography*, 5. 238 pp. Although this volume treats chiefly the southern California diatoms, it is as useful in Monterey Bay, Puget Sound, or even Hawaii. Until a similarly useful volume is written to replace it, it will probably continue to be used internationally.

Dawson, E. Y. 1953–63. Marine red algae of Pacific Mexico. *Allan Hancock Pac. Exped.*, 17 (398 pp.) and 26 (208 pp.). *Pac. Naturalist*, 2: 1–126; 189–375. *Nova Hedwigia*, 5: 437–76; 6: 401–81. Baja California, Mexico, the region chiefly served by these publications, may be defined as "warm-temperate" or "subtropical." These publications will continue to be essential reading for southern California phycologists, since new records for the flora are likely to be made from the south rather than from the north.

Dixon, P. 1970. The Rhodophyta: Some aspects of their biology, II. *Ann. Rev. Oceanogr. & Marine Biol.*, 8: 307–52. A review of the contributions of recent research to our knowledge of the red algae, particularly in the areas of life history, morphology, and physiology.

―――. 1973. *Biology of the Rhodophyta*. New York: Hafner. 285 pp. Most recent information on the red algae appears in scattered research papers, or is published in German or French. This short review of the biology of red algae is sufficiently detailed for the serious student.

Kylin, H. 1956. *Die Gattungen der Rhodophyceen*. Lund, Sweden: Gleerup. 673 pp. Well-illustrated and relatively expensive, this is the one book a serious student of the red algae must have. The genera are characterized and described; the structure and reproduction of orders and families are delimited; and distribution is given at the species level. Although more and more studies of the red algae are under way, much of this research simply fills in the chinks in Kylin's classification, rather than disturbing its overall arrangement.

Scagel, R. F. 1966. Marine algae of British Columbia and northern Washington, Part I: Chlorophyceae (green algae). *Nat. Mus. Canada Bull.*, 207. 257 pp. The thoroughgoing scholarship evident in the pages of this work makes it one of the most reliable handbooks for the Pacific coast. It accords the green algae, which tend to be overlooked in favor of the larger brown algae and the more attractive red algae, a firm place in the flora.

―――. 1966. The Phaeophyceae in perspective. *Ann. Rev. Oceanogr. & Marine Biol.*, 4: 123–94. Although ten years old, this review of the brown algae is so well done that it will be used for years to come. It shows excellent balance and judgment in its selection and explanation of topics.

THE CALIFORNIA COAST

As is well known, the California Current* brings cold water southward from the North Pacific; indeed, summer surface temperatures show a difference of only 5–10° C from north to south along the California coast. In the winter the northward-flowing Davidson Current hugs the coastline north of Point Conception, bringing from the south oceanic water with different temperature and salinity characteristics.† Upwelling, most intense

---

* The California Current is the eastward extremity of the Japanese Current, or Kuroshio. After warming southern Japan with tropical waters, the Kuroshio, flowing northward and mixing with the arctic Oyashio Current, turns east across the North Pacific, bringing cool waters to the California coast.
† The effect of this current on the phytoplankton community structure in Monterey

in late winter and spring, also occurs along the coast, and altogether a cool temperature prevails, accounting for a fairly uniform algal flora. The commonest species (perhaps 20 percent of those described for the flora) may be found all along the California coast, with more distinctive species or forms occurring in the northern or southern reaches,* on the outer coasts, or in protected bays, particularly where local wind and bottom configurations allow water temperatures to rise higher than those of the adjacent coast (Newport Bay is a good example). The more uniform subtidal temperatures at 20 m appear to encourage the development of a flora largely different from the intertidal flora.

The most dramatic temperature change along the coast, affecting plant but especially animal species composition, occurs at Point Conception, near Santa Barbara; here the California coast turns sharply eastward, and the California Current continues south and west. The effects are especially noticeable in the California Channel Islands: the northern islands support a fauna and flora similar to those of, for example, Pacific Grove; the southern islands, especially Santa Catalina, support a subtropical flora more closely allied to that of Baja California and the Pacific coast of Mexico. Many species of algae found in the southern islands, and occasionally at La Jolla as well, are the northern distribution records of essentially more southern species.

East and south of Point Conception, seawater temperatures usually remain above 18°C at the surface, whereas north of the point they are usually below 18°C. This difference is reflected both in kinds of algal species and in the diversity of their forms. Northward, the Laminariales are the most conspicuous brown algae; southward, it is the Fucales that are common. Northward, the large, fleshy, often foliose red algae, such as *Iridaea*, are common; southward, the shores are dominated by shorter and more densely branched species, such as *Laurencia* and *Pterocladia*. In the north there is frequently a greater biomass comprising fewer species per square meter (lower diversity); in the south it is usual to find a lesser biomass but more species (higher diversity).†

Throughout California each year, tons of shifting sand move onshore from deep water, offshore from the intertidal, or laterally along the beaches, uncovering rocks not previously seen or covering over rocks with existing

Bay is especially noteworthy. See R. L. Bolin and D. P. Abbott. 1962. Studies on the marine climate and phytoplankton of the central coastal area of California, 1954–60. *Calif. Coop. Oceanic Fish. Invest. Bull.*, 9: 23–45.

* I. A. Abbott and W. J. North. 1972. Temperature influences on floral composition in California coastal waters. *Proc. VII International Seaweed Symposium*, pp. 72–79.

† M. M. Littler and S. N. Murray. 1974. The primary productivity of marine macrophytes from a rocky intertidal community. *Marine Biol.*, 27: 131–35, presenting the first report on primary productivity for algae in southern California, showed a clear relationship between effective productivity and structure. In general, the sheetlike or finely branched species were found to have greater production rates.

algal populations. Most algae do poorly in such unstable situations; but some, notably *Phaeostrophion irregulare, Ahnfeltia plicata,* and *Gigartina volans,* actually appear to favor these habitats. In general, however, sandy beaches may be considered unfavorable habitats for a marine flora, rocky substrata being preferred.

The winter and spring storms responsible for much of the substratum movement along the coast also dislodge many old, senescent plants. Great quantities of drift algae are to be found on certain beaches at these times (see C. Zobell, pp. 269–314 in W. J. North. 1971. The biology of giant kelp beds, *Nova Hedwigia,* 32). The late spring and summer storms also bring plants ashore, and these midseason drifts tend to stay on the beaches longer than the winter and spring drifts, which are quickly swept back into the sea by the higher winter waves. The largest kelp plants are usually dislodged in winter. The effect of all this dislodgment on the habitat appears to be that of weeding a garden; the plants that remain have more space and more light in which to grow.

The combination of these various factors has produced along the west coast of North America a marine algal flora that is second in number of unique species only to the floras of Australia and New Zealand, and on a par with the rich algal populations of Japan and South Africa. Many of the larger and more spectacular taxa from this region (such as most of the kelps) occur nowhere else. And the California coastline proper is representative to an unusual degree of the Pacific coast as a whole: a high percentage of all the taxa are found there, as well as some of the densest and most diverse populations. These circumstances, combined with the accessibility of so many California beaches to the phycologist, have rendered the California flora one of the most studied and best understood in the world—though our knowledge of it is still barely beyond the rudiments.

The geographic area whose algal population is most like that of California is Japan. The marine flora of Northern Honshu and Hokkaido would seem very familiar to a Californian; at the same time, because of the warm Kuroshio, which turns north from the equatorial regions south of Japan, the flora of southern Japan is unlike that of southern California. It is estimated that about 30–40 percent of the species occurring in California may also be found in Japan. This is a larger percentage than a comparison of California with the North Atlantic would realize, though temperature barriers in the Pacific are no less formidable as obstacles than the land barriers separating California from the Atlantic.

Although California and Japan share many genera, there are usually larger numbers of species in given genera on one side of the Pacific than on the other, implying a genetic if not an ecological diversity. One genus, *Laurencia,* has 12 eastern Pacific species and 23 western Pacific species.

By contrast, 12 *Gigartina* species, surely among the most common algae in terms of both occurrence and biomass in California, are matched by only six sparsely occurring species in Japan. *Chondrus* is common in Japan, with seven species and dozens of varieties and forms, but is evidently absent from California.

COLLECTING AND PREPARING SPECIMENS

The five coastline maps on pp. 10–13 carry every California place name cited in the species descriptions in this book. For obvious reasons, the names given tend to be points of transition between populations; or, just as often, they are localities that are rich in algae and easily reached from university research centers or marine laboratories. Other outlying or isolated localities will recall field trips when certain (rare) algae were collected, or someone fell in, or we got lost. The names on the maps do not reflect how many times a locality was visited, or how many important collections were made there. We and our colleagues all have favorite places to which we return again and again, and in order to protect those places— a busload of people can leave a stretch of intertidal devastated—we suggest only that field trips be taken to the outer coast, where one can obtain the best cross-section of an algal flora with the least overall disturbance.

But a note of caution is in order for would-be algae collectors in California: you must have a permit from the State Department of Parks and Recreation to collect in a state park. And *in any locality*, should you remove any animals with the seaweeds—a happenstance almost impossible to avoid—you are expected to have both a fishing license and a collecting permit from the State Department of Fish and Game.

A good low tide is the collector's best ally in obtaining specimens that adequately represent a range of species. For a given collecting trip, 15–20 specimens presumably representing about the same number of species is a good, workable quantity. Anything beyond this leads to a group that is unwieldy to study, identify, and press—though additional specimens will be needed for species requiring a demonstration of variation. Entries in a field notebook, made at the time of collection, should log the specimens and should note any ecological data that might be helpful in describing variation. Since tides are cyclical, any three or four consecutive days will yield that many low-tide periods, during which one may either collect repeatedly in one area or sample localities expressing ecological similarities or differences. With either approach, the collections should be made as part of a single project. With experience, it will become clear which species are the most common, or are the most likely to be found in a variety of situations; and similarly, which species are rare, or are to be found under a particular set of conditions.

The rarest species tend to be subtidal; and for that habitat, reached only by diving or by dredge, we still know too little to predict the occurrence of the rare entities in a given region. However, deeper waters tend to offer more consistent habitats; thus the subtidal species tend to be consistent in morphology, and their identification is usually far easier than the identification or characterization of upper-midtidal inhabitants. The greatest variation is shown by the more common midtidal or low-intertidal plants, and large numbers of specimens may be necessary to appreciate the morphological variation shown by some species. This is especially true of the common intertidal genera *Ulva, Enteromorpha, Pelvetia, Fucus, Gelidium, Prionitis, Callophyllis, Gigartina,* and *Rhodymenia.* Many species of these genera are difficult to characterize, and the eventual collection of large numbers of specimens may show that the measurements, color ranges, or degrees of branching given in a description are unreliable. Nevertheless, one must have some basis for judgment, and this is what our descriptions attempt to provide.

An adequate specimen of a macroscopic marine alga should include a holdfast (or basal portion), as well as a stipe and branches or one or more blades. It should also bear reproductive structures, for these are the only means by which genera (or, for that matter, families, orders, or classes) can be definitively identified. The small, sterile specimens frequently collected by marine ecologists simply because they were present in a quadrat can be identified (even by specialists) only with difficulty. Small, outlying (atypical) specimens, often taken when tides are not low, are also difficult to identify. Collecting algae in drift is much to be preferred; here there may be a choice of fully formed, mature, and reproductive thalli.

Deepwater (subtidal) algae have been traditionally collected by dredging, but more and more, scuba diving is being employed for detailed habitat studies and systematic collections. Years of dredging by Dr. Smith and beachcombing by the authors of this book (using the waves as our dredges) have yielded all but one or two subtidal species known from the Monterey Peninsula. However, specimens collected *in situ* at the depths reached by divers are usually intact, and tell us more precisely what a species is like than scraps and pieces pulled from drift can. Underwater collecting has allowed us to describe the subtidal habitats and flora more completely; and in some cases it has led to a revision in our descriptions of species that can also be collected in the intertidal areas.

Wherever a particular plant may be collected, every effort should be made to detach it carefully, disentangling it from associated plants and animals. All parts of the plant should be included, and the plant body should be intact, though this is not always possible in the case of encrusting species or very large kelps. Specimens are best transported by placing each in a separate tagged or labeled plastic bag, along with a small amount of seawater to prevent crushing, matting, or desiccation. One or more bags can then be easily carried in the shore collector's old standby, a metal or plastic bucket.

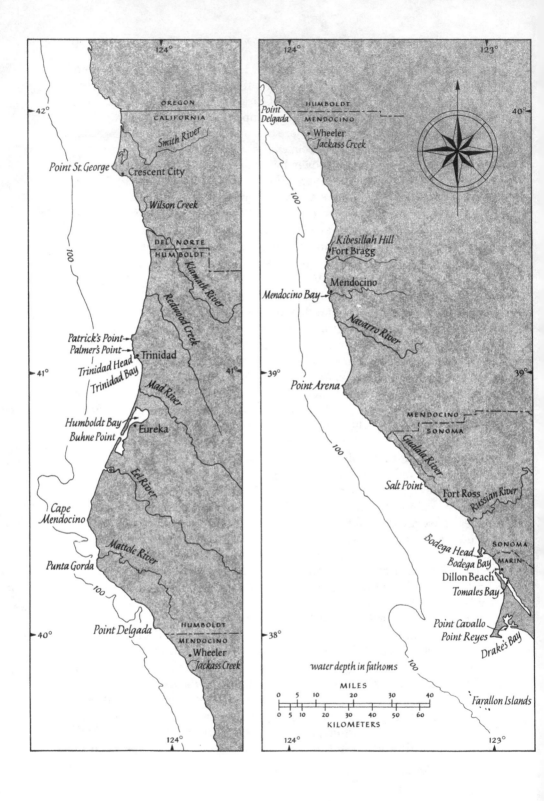

Left map:

124°

42°

OREGON
CALIFORNIA

Smith River

Point St. George
• Crescent City

Wilson Creek

DEL NORTE
HUMBOLDT

41°

Klamath River

Redwood Creek

Patrick's Point→
Palmer's Point→
• Trinidad
Trinidad Head
Trinidad Bay

Mad River

Humboldt Bay
Buhne Point
• Eureka

Eel River

Cape
Mendocino

Mattole River

Punta Gorda

100

40°

HUMBOLDT
MENDOCINO
• Wheeler
Jackass Creek

124°

Right map:

124°   123°

Point
Delgada

HUMBOLDT
MENDOCINO
• Wheeler
Jackass Creek

40°

100

Kibesillah Hill
Fort Bragg

Mendocino
Mendocino Bay •

Navarro River

39°   39°

Point Arena

MENDOCINO
SONOMA

Gualala River

Salt Point   Fort Ross   *Russian River*

100

Bodega Head
Bodega Bay   SONOMA
Dillon Beach   MARIN
Tomales Bay

Point Cavallo
Point Reyes
Drake's Bay

38°

water depth in fathoms

100

MILES
0   5   10   20   30   40

0 5 10 20 30 40 50 60
KILOMETERS

Farallon Islands

124°   123°

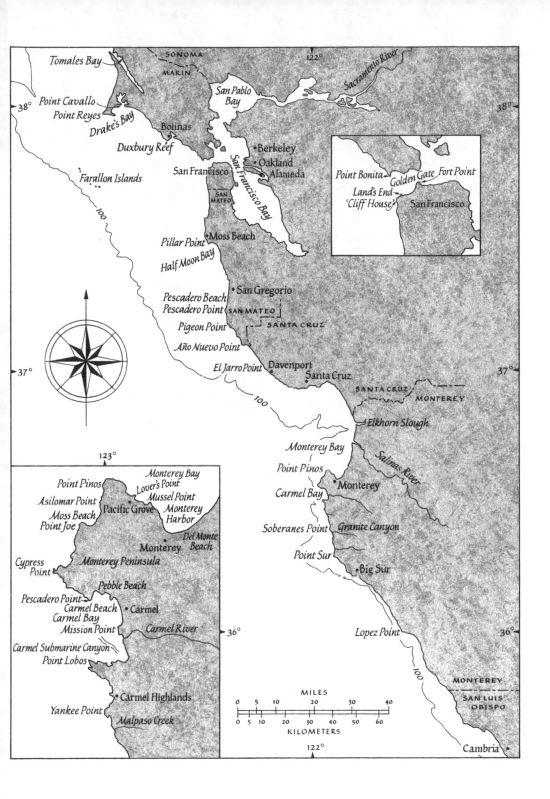

Tomales Bay

Point Cavallo
Point Reyes

Drake's Bay

Bolinas

Duxbury Reef

Farallon Islands

SONOMA
MARIN

San Pablo
Bay

122°

Sacramento River

38°

38°

Berkeley
Oakland
Alameda

San Francisco

SAN
MATEO

San Francisco Bay

Point Bonita
Golden Gate Fort Point

Land's End
'Cliff House'

San Francisco

100

Pillar Point
Moss Beach

Half Moon Bay

Pescadero Beach
Pescadero Point

Pigeon Point

Año Nuevo Point

El Jarro Point

San Gregorio

SAN MATEO

SANTA CRUZ

Davenport

Santa Cruz

37°

SANTA CRUZ

MONTEREY

37°

100

Elkhorn Slough

Monterey Bay

Point Pinos

Carmel Bay

Soberanes Point

Point Sur

Salinas River

Monterey

Granite Canyon

Big Sur

123°

Point Pinos

Asilomar Point
Moss Beach
Point Joe

Cypress
Point

Monterey Bay
Lover's Point
Mussel Point
Pacific Grove

Monterey Peninsula

Pescadero Point
Carmel Beach
Carmel Bay
Mission Point

Carmel Submarine Canyon
Point Lobos

Yankee Point

Monterey
Harbor

Del Monte
Beach

Monterey

Pebble Beach

Carmel

Carmel River

36°

Carmel Highlands

Malpaso Creek

Lopez Point

MONTEREY

SAN LUIS
OBISPO

100

36°

MILES

0   5   10   20   30   40

0 5 10   20   30   40   50   60

KILOMETERS

122°

Cambria

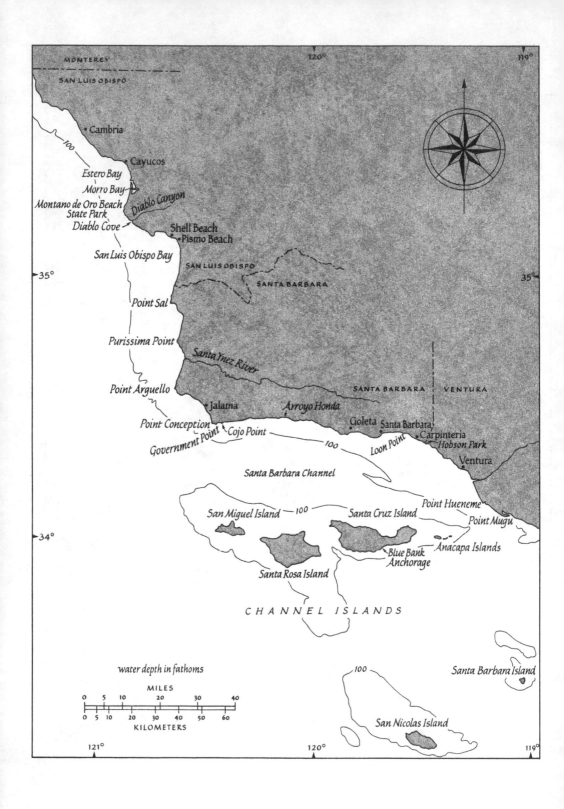

MONTEREY
SAN LUIS OBISPO

120°   119°

• Cambria

100

• Cayucos

Estero Bay
Morro Bay
Montano de Oro Beach
State Park
Diablo Cove
Diablo Canyon

Shell Beach
• Pismo Beach

San Luis Obispo Bay

SAN LUIS OBISPO

SANTA BARBARA

35°   35°

Point Sal

Purissima Point

Santa Ynez River

Point Arguello

SANTA BARBARA   VENTURA

• Jalama   *Arroyo Honda*

Point Conception   Goleta   Santa Barbara
Government Point *Cojo Point*   Carpinteria
                                      100   *Loon Point*   *Hobson Park*
                                                              Ventura

Santa Barbara Channel

Point Hueneme

San Miguel Island   100   Santa Cruz Island   Point Mugu

34°   Anacapa Islands

Blue Bank
Anchorage

Santa Rosa Island

C H A N N E L   I S L A N D S

water depth in fathoms   100   Santa Barbara Island

MILES
0   5 10   20   30   40

0 5 10   20   30   40   50   60
KILOMETERS

San Nicolas Island

121°   120°   119°

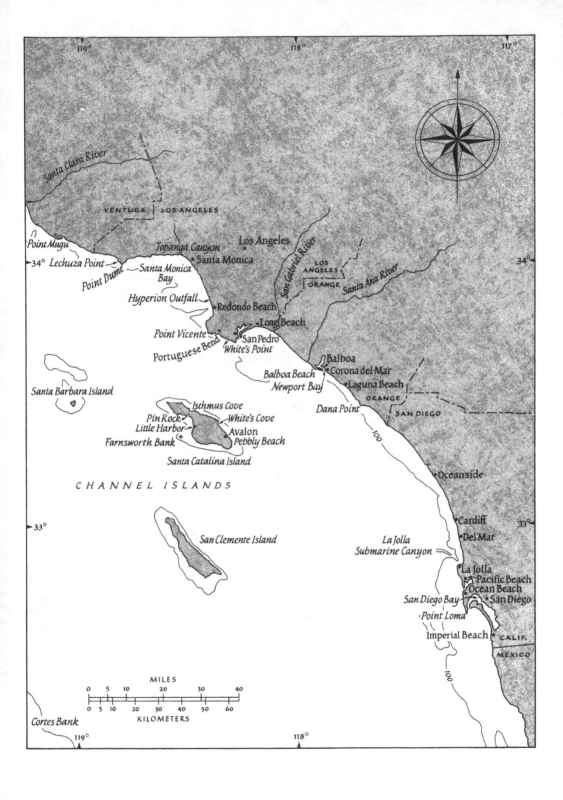

MILES
0   5 10   20   30   40
0 5 10   20   30   40   50   60
KILOMETERS

Santa Clara River
VENTURA | LOS ANGELES
Point Mugu
34° Lechuza Point
Point Dume
Topanga Canyon
Los Angeles
Santa Monica
Santa Monica Bay
San Gabriel River
LOS ANGELES
ORANGE
Santa Ana River
34°
Hyperion Outfall
Redondo Beach
Long Beach
Point Vicente
San Pedro
Portuguese Bend
White's Point
Balboa
Corona del Mar
Balboa Beach
Newport Bay
Laguna Beach
ORANGE
Santa Barbara Island
SAN DIEGO
Dana Point
Isthmus Cove
Pin Rock
White's Cove
Little Harbor
Avalon
Farnsworth Bank
Pebbly Beach
Santa Catalina Island
100
CHANNEL ISLANDS
Oceanside
San Clemente Island
Cardiff
33°
Del Mar
La Jolla
Submarine Canyon
La Jolla
Pacific Beach
Ocean Beach
San Diego Bay
San Diego
Point Loma
Imperial Beach
CALIF.
MEXICO
100
Cortes Bank
119°
118°

Each specimen should be logged separately in the fieldbook, with species identified, where known. The exact collection information to be recorded will depend on the intended use of the specimen: if it is suspected that a plant may become the type of a new taxon, for example, as many circumstances as possible should be entered; again, the coastal ecologist may be intensely interested in information of only marginal use to the systematic phycologist. The descriptions in this book will give the reader some idea of the information we consider essential or desirable. But there are many conditions that might be noted down where there is opportunity. To name only a few of the more important: date of collection; locality, as closely as possible; exposure to wave action, and in the case of intertidal plants, duration of exposure to air by the local tides; type of substratum; water depth, bottom type (e.g. muddy), and perhaps temperature; and associated plants and animals, perhaps including others of the same apparent species that were not collected. Any of these data could easily be of use to a particular researcher. It is also helpful to record information about the plant itself that may not be apparent when the specimen reaches the laboratory— for example, growth habit or visual appearance before collection (a great many algae show quite different coloring under water than they do on the beach or the laboratory table).

Preparing the specimen also warrants great care. Again, it is important to begin with a properly collected and recorded specimen, for working on freshly collected material is by far the best way to learn about structure. First float the plants out in a shallow pan of seawater (fresh water can have disastrous effects on some algae) and study the branching patterns and relationships of parts. If the plant is fertile, a section should be made, using a single-edged razor blade on a glass slide. The arrangement of cells and tissues will become apparent under the microscope or even under a strong glass, as will the kinds or locations of reproductive material.

If all or part of the plant is to be saved for later study, it must be preserved by wet-fixing or drying. Wet preservation for ordinary purposes can be done by placing the specimen in a 3- to 4-percent solution of formalin in seawater, buffered with 2 to 3 tablespoons of borax per gallon. More elaborate fixatives are necessary for cytological studies. Specimens may also be frozen after excess water has been removed.

A dried specimen is easily prepared for study and storage. First, slide a sheet or card of good-quality paper under the specimen while it floats in a pan of seawater. Then arrange the specimen on the sheet so that all parts are flat and lying in a single plane, removing excess material in the process if necessary, and carefully lift the paper, with specimen atop, from the water. Drain off the excess water, and place the paper with its specimen between newspapers or blotters, or between ventilators made of cardboard. For gelatinous specimens, a piece of waxed paper is often used to cover the specimen; for delicate specimens, a piece of unbleached muslin (preferably) or cheesecloth may be placed over the specimen. Stack up several specimens thus prepared, and place a weight over them. Exchange the cloths or newspapers for dry ones in about 12 hours, and change them again at 24-hour intervals until the plants are dry to the touch. Material prepared in this fashion should be labeled, and may be kept in three-ring notebooks or filed in an herbarium. Properly prepared material is quite durable: Linnaean specimens of marine algae are still usable and recognizable.

MAKING EFFECTIVE USE OF THIS BOOK

The various elements of our text have been assembled with a view to their practical utility in identifying California marine algae. With this in mind, we have not always adopted policies or formats identical to those used by other writers on the algae. The reader, and especially the beginning phycologist, can use this book most efficiently if he first understands the advantages and limitations of its contents as he might use them in identifying a plant he has collected.*

To identify either a freshly collected or a dried alga, one must first decide whether it is a green, a brown, or a red. The green algae are usually grass-green; a paler shade of green generally indicates that the specimen belongs to another phylum. The brown algae are usually olive to golden brown, but some are more black than brown. The red algae are usually dark rose in color, but may be brown, yellowish-green, or blackish. An old trick of Dr. Smith's is useful when a fresh alga is suspected of being a red, but does not look rose-colored or violet or purple: throw the plant into hot (not boiling) water for about 5 minutes, then dip it in alcohol; it should turn reddish-orange if it is a red alga (the dominant green pigment is removed by this treatment). A microscopic section may be necessary to establish structural details, such as the presence of pit connections in most red algae, and in general the kinds of vegetative cells present and their arrangement (see p. 747 for other considerations in making the initial choice of phylum).

Once it has been decided that a given alga is a green, a brown, or a red, the Master Key to Genera (pp. 747–66) should be consulted. This key (as also the keys to species in the body of text) offers a series of paired choices (dichotomies) leading by stages to the identification of material in hand. If you are unfamiliar with the use of a particular key, it is wise

---

* Assuming, of course, that what has been collected is in fact an alga, as understood in this book, or even a plant. Viewed macroscopically in nature, some algae (e.g. encrusting forms) may bear a resemblance to invertebrate animals common in the intertidal, such as bryozoans or hydroids. And there are marine representatives of other plant groups, such as marine lichens. Under a lens or microscope, however, any doubtful specimen can usually be pigeonholed at least at this level.

to write down, at each stage of the procedure, the lead number you select as being more appropriate to your specimen; in this way you may retrace your path and determine where a wrong choice was made. When your series of choices delivers you in due course to a lead leg ending in a genus name, turn to the page number given and carefully read the description of the genus. If you have made your choices accurately all along the way, most of the description will directly apply to your specimen. If the genus description does indeed fit the specimen, use the key to species following the genus text (a key is given wherever there are three or more species in a genus). This will lead you in the same manner to a lead leg ending in a species name, and thence to a description and an illustration (the species descriptions following a genus key are arranged alphabetically).

If you are not satisfied with a species identification, but retracing your steps offers nothing better, consulting some of the literature cited under the species name may be helpful. A good herbarium might also be consulted. Additionally, though we believe that the Master Key offers the easiest approach to identification, it is supplemented in the text by keys at the intermediate levels of order and family. These may be useful when one is able to match a specimen to a description given at one of these levels; and for those more familiar with the flora, the keys in the text offer more rapid access to species identity and a concise listing of the members of a class, order, family, or genus.

Within the text, descriptions follow the accepted ordering: phylum (or division), class, order, family, and genus. Some of the higher taxa, especially some classes and orders, are described in what may seem to be burdensome detail. Our intention here has been to assemble information that is otherwise contained only in scattered references, some published quite recently. For example, the Class Prasinophyceae of the green algae is a taxon that was established only when the electron microscope revealed the fine structure now associated with this group. For taxa whose definitions have not changed appreciably in the course of decades (for example, the Class Chlorophyceae of the green algae), the descriptions are less lengthy.

In two groups, the Fucales of the brown algae and the Corallinaceae of the red algae, we use the additional categories subspecies (ssp.) and variety. As used by some phycologists, a subspecies reflects the opinion of the author of a binomial that certain specimens are distinct from others assigned to the same species because of consistent differences in *geographically* distinct populations, whereas the rank of variety conveys, for these authors, a supposed inherent *genetic* difference, possibly occurring in geographically overlapping populations. In actual practice, and within the intent of modern botanical classification, these two terms merely imply differences that occur with sufficient constancy to be recognized, but that are not considered to warrant specific discrimination; understood in this fashion, subspecies and variety are no more than successively lower ranks in the hierarchy of infraspecific taxa.

All the taxa given in keys and headings may be taken as reflecting the best opinion of a good many phycologists. But they should not be considered sacrosanct, and the integrity of various species, in particular, has been called into question by our accumulating knowledge of the physiology, life history, and cytology of certain algae. Two recurring problems of this kind are evident in these pages. First, specimens once thought to represent quite distinct taxa—and classified, perhaps, in separate families or orders—have sometimes been found to be merely different stages in the life history of a single entity. We have retained the original names and separate descriptions for both entities in such cases, while making clear the taxonomic reallocations these situations entail (or anticipate). In some cases a definite relationship has been demonstrated in culture studies; in others the relationship is suspected; and in still others a given taxon is evidently superfluous but cannot as yet be associated with another. A second problem concerns certain supposed epiphytic or endophytic species that are at once host-specific and host-related. These entities are generally paired with their hosts in present classifications: e.g. *Rhodymenia* and *Rhodymeniocolax*. They are increasingly suspected to be no more than malignancies of the host plant, but the problem cannot be resolved within present knowledge.

Each heading at the species level (a binomial, or genus name plus species epithet) is followed by a line giving the minimum of nomenclatural information that we consider necessary. The first reference under the species name and author cites the first valid publication of the epithet, under whatever generic name the species was first assigned to; a second reference, if present, cites the first combination of the epithet with what we consider to be the most acceptable generic assignment. In many cases additional names, or synonyms, are also supplied: some of these names reflect the most important or most widespread usage of the species epithet (sometimes incorrectly applied) in the past; other names, with different epithets, are also, in our judgment, important synonyms. In many references we have made use of the parenthetical phrase "(incl. synonymy)" to avoid repeating all of the more secondary information there contained. We have also eliminated all references that did not add substantive information or employ representative illustrations; and we have eliminated all references to illustrations. In many cases brevity in the lists of synonyms has been made possible by prior publication elsewhere by one or both of us.

The index includes all algal names appearing in this book, whatever their taxonomic rank or status. A note to the index explains the typographic distinction of valid names from those we hold to be synonyms.

All sources cited in the names and synonyms at species level and below are included in the Literature Cited (pp. 783–807); sources cited only in the names of genera are omitted. One method of citation that we have employed may very well offend librarians, but it is not for them that our list has been assembled. In adhering to the International Code of Botanical Nomenclature,* under which we all work, we refer the reader directly to the author of a binomial even if the binomial is published in a paper or book by a second author. This avoids the poorly understood phrases "Abbott *in* Hollenberg 1965," or "Abbott *apud* Hollenberg 1965," or "Abbott *ex* Hollenberg 1965." In the text in this book, an author is cited simply as "Abbott 1965." Reference to the Literature Cited under "Abbott 1965" will disclose whether the name in question appeared in a paper by Abbott alone or, as in the example above, in material by Abbott contained in a paper by Hollenberg.

The morphological descriptions in the text follow a standardized sequence, with some information omitted in cases where it has already been given in the description of a subsuming higher taxon. All measurements used are metric. The range cited for a given measurement (e.g. "8–15 cm") is that encountered in the majority of known specimens of the species. Where a few specimens have shown extreme measurements falling significantly outside this range, the variation is given in parentheses (e.g. "(6)8–15(20) cm").

All terms employed in the text will be found in the Glossary (pp. 769–81), which has been through several drafts and the hands of several colleagues. In nearly all cases we have tried to use terms as others might use them; a few are of our own coinage. Concerning some words or phrases we have reservations. For the terms "sex," "carpospores," and "spermatia," as employed in the Bangiophyceae, there remains some question whether this condition and these structures operate as in other red algae (for the red algae as a whole the sexual process is still unclear, being cytologically demonstrated in only a few cases). Also, we do not use the term "parasite" in its strictest sense—that of an organism deriving its nourishment from another living organism, often to the detriment of the host. Recent ultrastructural studies have shown that chloroplasts or reduced chloroplasts are present in the cells of many species of red algae formerly believed to be parasites. Thus in our usage the term (frequently in quotation marks) often implies only that the "parasite" is epiphytic (i.e., it simply gains a foothold on the host), though in some cases it may also derive some or all of its nourishment from the host.

Each species description is followed by a paragraph giving habitat (on rocks, epiphytic, etc.), tidal distribution (intertidal, subtidal, etc.), and geographic distribution from north to south. Geographically discontinuous

* F. A. Stafleu, ed. 1972. International Code of Botanical Nomenclature. Utrecht. 426 pp.

distribution is indicated by specific localities, and continuous distribution by broad ranges. The type locality, or point of collection specified in the original published description of the species, is also given.

The book thus offers to phycologists—we prefer the term to "algologists"—a good deal of information and illustration on a great variety of plants. And today there are many more phycologists than at any previous time, possibly more than during *all* previous time. The recent literature suggests that their research problems lean toward culture studies and ultrastructure research, perhaps ultimately biochemical. Because few of these workers will have the time to understand systematics and ontogenetic development, instruction in classical taxonomy and morphology must be both simplified and clarified. However, the new breed of diver / marine phycologist, collecting in localities and at depths previously unsampled, has uncovered organisms whose characteristics call into question the traditional system of classification, forcing the more classically trained to reexamine both materials and criteria. Perhaps the most unusual discoveries in recent years have been two green algae, wholly unexpected in their structure and in their life histories: H. B. S. Womersley. 1971. *Palmoclathrus*, a new deepwater genus of Chlorophyta. *Phycologia*, 10: 229–33; G. N. MacRaild and H. B. S. Womersley. 1974. The morphology and reproduction of *Derbesia clavaeformis* (J. Agardh) DeToni (Chlorophyta). *Phycologia*, 13: 83–93. The challenge posed by such discoveries is healthy, and the efforts that lead to them should be encouraged. Furthermore, algal population biology, or ecology, is still a young science: life histories are inadequately known for most species; taxa are poorly known on a comparative basis; and reproductive structures and strategies are not understood (for many red algae, even sexual reproduction remains an article of faith!). Each year brings new discoveries that resolve old problems, as well as new facts that shake previously held ideas. It is an exciting time.

This book has been many years in the making; the Preface offers a chronology and philosophy of its preparation. The Preface does not convey, however, the excitement and challenge of seeking greater order in our understanding of the marine algae, or the sense of participation in the history of phycology. We are proud to succeed Dr. Smith; and one of us (the second-named) was in fact Dr. Smith's first Ph.D. student. The other (the first-named) was the first Ph.D. student trained by Dr. Papenfuss; and we are equally proud to offer, on the following pages, his unique and fascinating account of West Coast algal exploration and study.

Landmarks in Pacific North American Marine Phycology

GEORGE F. PAPENFUSS

KNOWLEDGE of the marine algae of the Pacific coast of North America begins with the 1791–95 expedition of Captain George Vancouver. (See Anderson, 1960, for an excellent account of this expedition.) On the recommendation of the botanist Sir Joseph Banks (who as a young man had been a member of the scientific staff on Cook's first voyage, 1768–71), Archibald Menzies, a surgeon, was appointed botanist of the Vancouver expedition. Menzies had earlier served on a fur-trading vessel plying the northeastern Pacific and had collected plants from the Bering Strait to Nootka Sound, on the west coast of Vancouver Island, in the years 1787 and 1788 (Jepson, 1929b; Scagel, 1957, p. 4), but I have come across no records of algae collected by him at that time.

As a young midshipman Vancouver had been to the northeastern Pacific with Cook's third voyage in 1778. Now, in 1791, his expedition consisted of two ships, the sloop *Discovery* and the armed tender *Chatham*. The ships came to the north Pacific by way of the Cape of Good Hope, Australia, New Zealand, Tahiti, and the Sandwich Islands (Hawaii). They sailed from Hawaii on March 16, 1792, sighted the Mendocino coast of California (or Nova Albion [New Britain], the name given to northern California and Oregon by Drake and the name by which this region was still known among English navigators in Vancouver's time) on April 18, and proceeded north to explore the coast. Returning south in autumn, the ships called at San Francisco and Monterey in November. After a stay of nearly two months at the Spanish capital city of Monterey, they wintered in the Hawaiian Islands.

On their return in the spring, the ships parted company, with the *Chatham* going direct to Nootka Sound, and the *Discovery* making land on

This historical review is an expanded version of a paper presented in a series of lectures in botany honoring Gilbert Morgan Smith (1885–1959) at Stanford University in February 1960. I am indebted to my colleague, the late Professor Johannes Proskauer, for his suggestion that this might be an appropriate topic for my lecture.

George F. Papenfuss is professor of botany, emeritus, at the University of California, Berkeley. Sources cited in this essay are given on pp. 41–46.

the Humboldt coast of California. After a stay of three days, including a shore excursion at Trinidad Bay on May 3, 1793, the *Discovery* proceeded northward for summer surveying. Reunited, the ships moved south in the autumn, and the *Chatham*, with Menzies on board, stopped for one day (October 20) in Bodega Bay, where a shore trip was made. The expedition then called again at San Francisco and Monterey, but the reception this time was unfriendly, and Vancouver did not allow Menzies ashore at either place. Moving southward, the ships reached Santa Barbara and San Diego toward the end of November. There the reception was cordial, and shore excursions were made at both points. From San Diego, the ships surveyed the southern coast as far as Rosarío, in Baja California, and then made for Hawaii, which was reached in early January 1794.

In mid-March 1794 the ships sailed again, for Cook Inlet in Alaska. In the fall they headed south once more, stopping first at Nootka Sound and then, in early November, at Monterey. With the reception once again hospitable, they stayed on to the end of November, taking on stores for the long voyage home, before setting sail at last to make for Cape Horn—and England.

The specimens of marine algae collected by Menzies were entrusted to the foremost marine phycologist of the time, Dawson Turner, a banker in Yarmouth and the father-in-law of the distinguished botanist Sir William Jackson Hooker. Turner described and illustrated them, together with species from many other parts of the world, in his classic four-volume work, *Fuci*, published in the years 1808–19. Like most algologists of his day, Turner followed Linnaeus's system of classification, according to which the macroscopic marine algae were divided into three genera: *Conferva*, containing the filamentous species; *Ulva*, containing the membranous species; and *Fucus*, containing all the fleshy or bulky species. Turner dealt only with taxa referable to *Fucus*, and described just seven species from the Menzies material from North America: *F. menziesii*, *F. herbaceus*, *F. osmundaceus*, *F. cordatus*, *F. larix*, *F. linearis*, and *F. costatus*. So few species suggests either that Menzies collected only a few algae or that he gave only a small part of his collection to Turner. Three additional species from the Menzies collection—*F. asplenioides*, *F. floccosus*, and *F. nootkanus* (see Silva, 1953)—were described by Professor Eugen J. C. Esper of Erlangen, on the basis of specimens sent to him by Turner (who did not, however, expect him to publish them). Three of the species Turner had described later became the types of new genera: *Fucus cordatus* became the lectotype of the red algal genus *Iridaea*, established by Baron Jean Baptiste Bory de Saint-Vincent in 1826; *Fucus costatus* became the type of the laminarialean genus *Costaria*, erected by Robert K. Greville in 1830; and John Erhard Areschoug (whom we shall take up later) in 1876 made *Fucus menziesii* the type of the new genus *Egregia*, a common member of the Laminariales along the Pacific coast from northern British Columbia to northern Baja California.

Two of the officers on the Vancouver expedition are indirectly commemorated by genera of red algae occurring on the Pacific coast: *Whidbeyella* Setchell & Gardner derives from Whidbey Island in the San Juan Archipelago, which is named for Joseph Whidbey; and *Pugetia* Kylin derives from Puget Sound, which is named for Peter Puget. Vancouver is indirectly honored by *Urospora vancouveriana* (Tilden) Scagel, *Corallina vancouveriensis* Yendo, and *Pleonosporium vancouverianum* (J. Agardh) J. Agardh, all derived from Vancouver Island. And Dawson Turner, for his part, is commemorated by the red algal genus *Turnerella* Schmitz (which is based on *Iridaea mertensiana* Postels & Ruprecht, a species from the northern Pacific, including British Columbia and northern Washington) and by *Gigartina turneri* Setchell & Gardner.

In September 1791, shortly before Menzies collected on the Pacific coast, Dr. Thaddaeus Haenke, a native of Bohemia who was botanist on a Spanish expedition under the command of Captain Alejandro Malaspina (Galbraith, 1924; Cutter, 1960; Kühnel, 1960; Barneby, 1963), had also collected a few algae at Monterey. These specimens were not described, however, until Carl A. Agardh (and later his son Jacob) published them in 1822, 1824, and (in K. B. Presl) 1825. Moreover, the source of Haenke's algae was not altogether clear. Since his specimens had been labeled as obtained *in mari australi*, botanists for many years assumed they had been collected somewhere in the South Pacific, which had been visited by the Malaspina expedition. It was not appreciated that much of the Pacific Ocean—and perhaps all of it, but including at least the northeastern reaches as far north as Nootka Sound and the western Pacific to at least Guam—for a long time was referred to as *mare australe*, the name (as *Mar del Sur*) given to the Pacific Ocean by its discoverer Vasco Núñez de Balboa in 1513.° The identity of at least four of the species obtained by Haenke—*Cystoseira australis* C. Agardh (in Presl, 1825), *C. caudata* C. Agardh (in Presl, 1825), *Fucus compressus* C. Agardh (1824), and *Grateloupia hystrix* C. Agardh (1822)—is still uncertain; and a fifth, *Cystoseira tuberculata* C. Agardh (1824), was suspected by Ruprecht (1852, p. 70) of being the same as *C. osmundacea* (Turner) C. Agardh. The voucher material of these taxa presumably is preserved in the National Museum of Prague. (Concerning four other West Coast taxa based on Haenke material from *mare australe*, see Kylin, 1941, pp. 10, 12, 16, and 28.)

Carl Adolph Agardh was professor of botany at the University of Lund from 1812 until 1835, when he became Bishop of the Karlstad Diocese

---

° When Balboa saw the Pacific Ocean, after crossing the Isthmus of Panama, he was looking southward.

(Areschoug, 1870; Krok, 1925). His main botanical interest was the marine algae. He is best known for his two works, the *Species algarum* (1820–28) and the *Systema algarum* (1824), in which he described taxa from many parts of the world. Even more famous was his son, Jacob Georg Agardh, who was appointed professor of botany at Lund in 1839 (Eriksson, 1916; Krok, 1925). Jacob continued in his father's footsteps in specializing in the marine algae. One of the most distinguished phycologists of all time, he earned international renown, and for half a century was the man to whom collectors worldwide sent material for determination and often for herbarium deposit. On his death in 1901, at the age of 88, he had a publishing record extending back over 68 years. In consequence of the work and fame of the Agardhs, the most important algal herbarium in the world was assembled at Lund. Because of its richness in type specimens and other published specimens, the Agardh Herbarium is perhaps of greater importance to the algal taxonomist than the Linnean Herbarium is to the phanerogamic taxonomist.

In the course of his productive career J. G. Agardh described a number of genera and species of algae from the Pacific coast. He received material from Mrs. R. F. Bingham of Santa Barbara (after whom he named *Binghamia, Dictyota binghamiae, Endarachne binghamiae, Leptocladia binghamiae,* and *Gigartina binghamiae*), Dr. L. N. Dimmick of Santa Barbara, and Dr. Charles L. Anderson of Santa Cruz; all three later contributed specimens to W. G. Farlow (see p. 30) and Anderson was later to publish on the California algae himself (see p. 31). In 1847 J. Agardh published on algae from Mexico collected by the botanist Professor Frederik Michael Liebmann of the University of Copenhagen, after whom he named *Sargassum liebmannii.* Among the algae of this coast, there are several that commemorate J. Agardh: *Sargassum agardhianum* Farlow *ex* J. Agardh, *Iridaea agardhiana* (Setchell & Gardner) Kylin, and *Gigartina agardhii* Setchell & Gardner. The tradition of phycology at Lund remained strong for more than 125 years, ending only in 1949 with the death of Harald Kylin (to whom we shall return).

In October 1816 the *Rurik,* a Russian ship with the Romanzoff expedition under the command of Captain Otto von Kotzebue, visited San Francisco for a month (Mahr, 1932). Adelbert von Chamisso, poet and keeper of the Berlin Herbarium, was the naturalist of the expedition; Dr. Ivan Eschscholtz (for whom the California state flower, the California poppy, *Eschscholzia californica* Chamisso, is named) was the ship's surgeon; and Login Choris was the painter and chronicler. It was in Choris's (1822) account of the journey that Chamisso published his description of the remarkable southern fucoid *Fucus antarcticus* (*Durvillea antarctica*).

Chamisso collected algae at a number of the places the *Rurik* visited,

William Henry Harvey (1811–1866)   Jacob Georg Agardh (1813–1901)

but confusion persists concerning the origin of some of his specimens. At least three West Coast species—*Fucus furcatus* C. Agardh, *Gigartina papillata* (C. Agardh) J. Agardh and *G. volans* (C. Agardh) J. Agardh—are based on material he collected, but their sources were given as Unalaska, Hawaii, and the Cape of Good Hope, respectively. In 1954 E. Yale Dawson produced convincing evidence that another species, *Sphaerococcus salicornia* C. Agardh, reported as having been collected in Unalaska, had almost certainly come from the Philippines.

Alexander Collie and George Tradescant Lay, surgeon and naturalist, respectively, on Captain Frederick William Beechey's (1831) exploring expedition in H.M.S. *Blossom*, collected algae at San Francisco and especially at Monterey between November 1825 and January 1826 and from October to December 1827. Their specimens were determined by the Irish phycologist William Henry Harvey, who in 1833 published a list of the species in *The Botany of Captain Beechey's Voyage*, by W. J. Hooker and G. A. W. Arnott. This list of 24 species constitutes the first unified account of the algae of Pacific North America.

The botanical explorer David Douglas* (Hooker, 1836; A. Harvey, 1947)

---

* No algae were named in honor of Douglas, but the most important lumber tree of North America, the Douglas Fir, will always bear his name, even though its proper botanical name is *Pseudotsuga menziesii* (Mirbel) Franco.

collected algae at Monterey (and at least one specimen at San Francisco) in 1831. Harvey published these specimens together with some collected at San Francisco in October and November 1837 and September 1839 by Dr. Andrew Sinclair, the surgeon on H.M.S. *Sulphur* with Captain Edward Belcher's expedition (Pierce & Winslow, 1969), in a second list in 1841, when Hooker and Arnott issued a supplementary fascicle of *The Botany of Captain Beechey's Voyage.* All told, 39 species were published in the two lists. Dr. Sinclair later settled in New Zealand and contributed much to the botanical exploration of that country (Glenn, 1950). He is commemorated by *Laminaria sinclairii* (Harvey) Farlow, Anderson & Eaton of the Pacific coast and by several species of New Zealand algae.

While Douglas was in Monterey, another important collector of western American plants established headquarters there. This was Dr. Thomas Coulter (Coville, 1895), who came to Monterey from Mexico, where he had been company physician for a mining crew. Coulter, who later became curator of the herbarium of Trinity College in Dublin, stayed in Monterey off and on during 1831–33. In that period he collected many marine algae, which were published by Harvey in *Nereis Boreali-Americana* (1852–58). Coulter is commemorated by *Gelidium coulteri* Harvey, *Gastroclonium coulteri* (Harvey) Kylin, and *Microcladia coulteri* Harvey.

William Henry Harvey (L. Fisher, 1869; Oliver, 1913) was one of the most distinguished students of marine algae in the nineteenth century, and an extraordinary man. He contributed immensely to North American phycology during the formative period of this branch of botany on this continent. Harvey was born at Limerick, Ireland, in 1811, the youngest of 11 children. At the age of 15 he had already established algae as his overriding interest—"To be useless, various, and abstruse is a sufficient recommendation of a science to make it pleasing to me" (L. Fisher, 1869, p. 4). A few years later in writing to his brother Jacob, who had emigrated to America, he remarked: "All I have taste for is natural history, and that might possibly lead in days to come to a genus called *Harveya*, and the letters F.L.S. after my name, . . ." (L. Fisher, 1869, p. 6). His discovery in 1831 of the moss *Hookeria laetevirens* at Killarney, new to Ireland, led to a lifelong friendship with Sir William Jackson Hooker, who was then Regius Professor of Botany at Glasgow University. "Hooker recognized at once the extraordinary talent of the shy young man of twenty, lent him books, asked him to visit . . . and predicted for him a rapid advance to the top of [phycology]" (Oliver, 1913, pp. 207–8). Soon afterward Hooker invited him to contribute the section on algae for his *British Flora* (1833) as well as the section on algae for *The Botany of Captain Beechey's Voyage.*

The death of Harvey's father in 1834 broke up his home life. A great urge to visit distant lands led to inquiries after an appointment in the

Colonies. He obtained the post of Colonial Treasurer (or so he probably thought) at the Cape, and in July 1835 left for South Africa.

Asa Gray (1866), in his biographical sketch of Harvey, tells of the curious circumstances connected with this appointment. In error, the appointment was made out in the name of an elder brother, Joseph, and an inopportune change of government at the time frustrated all attempts at rectification. However, Joseph took his brother with him as assistant. In Cape Town, William at once settled down to collecting. Within a few weeks he was engaged in the description of new genera and species of vascular plants, and in three months his herbarium contained 800 species. Shortly after their arrival in South Africa his brother fell ill, and in April 1836 he died. William was appointed to fill the vacancy of Colonial Treasurer. He now spent his days in official duties and the early mornings and nights at botany.

During this appointment, Harvey published his *Genera of South African Plants* (1838), the forerunner of his *Flora Capensis* (in collaboration with Dr. Wilhelm Sonder of Hamburg) and his *Thesaurus Capensis*, a series of plates designed to supplement and illustrate this unillustrated flora. Both of these are classic works dealing with the vascular plants of South Africa. Harvey's first major work on the algae, his *Manual of British Algae* (1841), also appeared while he was in South Africa. In 1842, a complete physical breakdown brought on by overwork forced Harvey to return to Ireland. After two years of attending to the uncongenial duties of the family business, he obtained, in 1844, the appointment of Keeper of the Herbarium of Trinity College.* The position had fallen vacant owing to the death of Dr. Thomas Coulter, the botanical explorer of Central Mexico and California. College vacations were now usually spent at Kew, Harvey staying with his friend Sir William Hooker, who now was Director of Kew.

In 1849, Harvey accepted an invitation from Harvard University and the Smithsonian Institution to deliver a series of lectures in Boston and Washington. During this visit, he made extensive collections of seaweeds along the Atlantic seaboard as far south as the Florida Keys and also arranged with people at various points to send him specimens. The product of all this was Harvey's *Nereis Boreali-Americana*, which appeared in three parts between 1852 and 1858 in the *Smithsonian Contributions to Knowledge*. Harvey himself lithographed the colored plates. This work, the first—and to this day the only—marine algal flora of North America, included many taxa from the Pacific coast. These were based upon material collected by Menzies, Collie, Lay, Sinclair, Douglas, Coulter, Voznesenski, Wilkes, and

---

* Harvey had just previously become a candidate for the post of Professor of Botany at Trinity College, but a difficulty arose because an act prescribed that a candidate hold a medical degree or license. Though the degree of M.D. was quickly conferred on him *honoris causa*, this solution was held to be inadmissible.

several others. A number of Pacific coast specimens were credited to Colonel Nicolas Pike (Howe, 1904), who served as United States Consul at both Oporto, Portugal, and Mauritius; Harvey named two taxa after him: *Pikea* and *Callithamnion pikeanum*. However, it later came to Harvey's attention (1858, p. 5) that these specimens had actually been collected by one A. D. Frye (not to be confused with T. C. Frye), who is remembered by *Nitophyllum fryeanum* Harvey. So far as we know, Colonel Pike did not himself collect on the Pacific coast of America.

In 1851, the year in which the last volume of Harvey's three-volume *Phycologia Britannica* appeared, he and his old friend Professor J. W. Bailey of West Point published an account of the algae collected by the United States Exploring Expedition of 1838–42, under the command of Charles Wilkes. The expedition's Pacific Coast collections were made in the Puget Sound region. Professor Bailey is remembered in the West Coast algae by *Lomentaria baileyana* (Harvey) Farlow, *Pterosiphonia baileyi* (Harvey) Falkenberg, and *Neoagardhiella baileyi* (Kützing) Wynne & Taylor.

In 1862 Harvey published on a collection of algae made in British Columbia and northern Washington by Dr. David Lyall, an English surgeon and botanist attached to the British Columbia Boundary Commission from 1858 to 1861. During Sir James Clark Ross's Antarctic expedition (1839–43), Lyall had worked alongside Joseph D. Hooker, Sir William's son, Lyall as assistant surgeon on the *Terror*, Hooker as assistant surgeon and botanist on the *Erebus*, the *Terror*'s companion ship (Glenn, 1950). A few years later Lyall, as surgeon and naturalist on H.M.S. *Acheron*, made important botanical collections in New Zealand and other parts of the southern hemisphere. He is commemorated by *Prionitis lyallii* Harvey and *Odonthalia lyallii* (Harvey) J. Agardh of the Pacific coast and by several New Zealand species of algae, among other taxa.

In 1853 Harvey had set out on a three-year collecting trip to Ceylon, Australia, New Zealand, and various islands in the South Pacific. He returned from this journey in October 1856, and soon thereafter was appointed Professor of Botany at Trinity College, Dublin. No objections to his appointment were raised at this time, although the law remained the same. Within two years he began publishing his magnificent five-volume *Phycologia Australica*, which was issued in parts between 1858 and 1863. This was to be his last great work. Though his marriage to Miss Phelps of Limerick in 1861 gave him a few brief years of happiness, in the winter of 1865 he became seriously ill. An immediate change to a mild climate was ordered, and he and his wife took quarters in Torquay with Lady Hooker, widow of Sir William, who had himself died only months before. But the change was of no avail—William Henry Harvey died that spring, on May 15, 1866.

The many important works Harvey published in his short 55 years are lasting monuments to his industry. In the marine algae alone, he studied the species of Great Britain, North America, South Africa, Australia (including Tasmania), New Zealand, the Antarctic (with J. D. Hooker), and several other regions. He was the first to segregate, in 1836, the algae into major groups (phyla), of which he recognized four: Melanospermeae (brown algae), Rhodospermeae (red algae), Chlorospermeae (green algae), and Diatomaceae (in which he placed, in error, the desmids as well as the diatoms).

Asa Gray of Harvard, one of Harvey's hosts in 1849, eloquently summed up Harvey's work and character (1866, p. 277): "He was a keen observer and a capital describer. He investigated accurately, worked readily and easily with microscope, pencil, and pen, wrote perspicuously and, where the subject permitted, with captivating grace; affording in his lighter productions mere glimpses of the warm and poetical imagination, delicate humor, refined feeling, and sincere goodness which were charmingly revealed in intimate intercourse and correspondence, and which won the admiration and the love of all who knew him well. Handsome in person, gentle and fascinating in manners, genial and warm-hearted but of very retiring disposition, simple in his tastes and unaffectedly devout, it is not surprising that he attracted friends wherever he went, so that his death will be sensibly felt on every continent and in the islands of the sea."

Harvey was elected not only Fellow of the Linnean Society but also Fellow of the Royal Society. In his honor a flowering plant genus *Harveya* W. J. Hooker and a red algal genus *Harveyella* Schmitz & Reinke were erected; the latter is represented on the Pacific Coast. Among the algae of the Pacific Coast he is immortalized by *Hesperophycus harveyanus* (Decaisne) Setchell & Gardner and *Gigartina harveyana* (Kützing) Setchell & Gardner.

Working in roughly the same period as Harvey was Franz Joseph Ruprecht (Maximowicz, 1871), director of the Botanical Museum of the St. Petersburg Imperial Academy of Sciences. It was in 1840 that he and Alexander F. Postels published their magnificent elephant folio volume (*Illustrationes algarum . . .*) on the marine algae collected on the round-the-world voyage of the *Seniavin* (1826–29). Postels, a professor of geology at the Academy (see Hultén, 1940) and a member of the Russian expedition, made the illustrations from living specimens (see Maximowicz, 1871, p. 3); Ruprecht prepared the text.[*]

A number of our common algae, such as *Endocladia muricata* (Postels

[*] The commander of the expedition, Captain Friedrich Benjamin von Lütke (see Komarov, 1926) was very much interested in science himself, and later became president of the St. Petersburg Academy (Maximowicz, 1871, p. 3).

& Ruprecht) J. Agardh and the genus *Nereocystis*, were first described in this work. The species for which *Nereocystis* was erected commemorates Captain Lütke and was described as *Fucus luetkeanus* by the naturalist of the expedition, Dr. Carl Heinrich Mertens, in a letter to his father, Professor Franz Carl Mertens of Bremen, a friend of Dawson Turner's. In this letter, Mertens (1829, p. 48; for an English translation, see Mertens, 1833) notes that the Russians called this alga sea otter cabbage, reflecting the sea otter's tendency to find refuge in beds of this kelp and to rock and sleep on the terminal part of the stipe. He also noted that the Aleuts used the stipes for fishing lines. Mertens is commemorated by *Spongomorpha mertensii* (Ruprecht) Setchell & Gardner, *Costaria mertensii* J. Agardh, and *Turnerella mertensiana* (Postels & Ruprecht) Schmitz of the Pacific coast.

In 1852 Ruprecht published a beautifully illustrated paper on several brown algae from California. Here were first described the sea palm, *Postelsia palmaeformis* (type locality, Bodega Head), *Pterygophora californica* (type locality, Fort Ross), and *Dictyoneurum californicum* (type locality, Fort Ross), all three representing monotypic kelp genera. The material of these and other taxa was collected by Ilya Gavrilovic Voznesenski, a preparator of the St. Petersburg Zoological Museum who was sent by the Academy to the Russian American colonies in 1839 to make collections for the Museum (Schmidt, 1926; Eastwood, 1939). It was Ruprecht who proposed the name Rhodophyceae, by which the red algae have generally been known. He is commemorated by *Botryoglossum ruprechtianum* (J. Agardh) J. DeToni and *Hypophyllum ruprechtianum* Zinova among the algae of the eastern Pacific.

Frans Reinhold Kjellman, docent and later professor of botany at Uppsala University, was botanist on the Swedish ship *Vega*, which on its 1878 to 1880 cruise under the command of Nils Adolph Erik Nordenskiöld was the first ship to circumnavigate Europe and Asia (Nordenskiöld, 1882). In 1889 Kjellman, the man under whom Kylin (see below, p. 37) received his doctorate, published a monograph on the algae of the Bering Sea, many of which were obtained in Alaska—at Port Clarence on the mainland and on St. Lawrence Island.

In 1874 one of Asa Gray's bright young students, William Gilson Farlow (Setchell, 1926; Taylor, 1945), returned to Harvard University from two years of study in Sweden (especially with J. Agardh at Lund), Germany, and France. In 1879 he was appointed professor of cryptogamic botany at Harvard. The first holder of a professorship in that branch of botany in the United States, Farlow exerted enormous influence on its development and growth in this country. During the early part of his career Farlow devoted much of his research to the algae, whereas in later years he concentrated on the fungi, although not to the exclusion of the algae. In the voting for names to be included in the first edition of *American Men of Science*, Farlow ranked first among the botanists (Taylor, 1945, p. 63).

Farlow described a number of algae from California, and in the company of Asa Gray visited California in 1885. He received many of his specimens from Dr. C. L. Anderson of Santa Cruz and Daniel Cleveland of San Diego (Jepson, 1929b). Other specimens were sent by Dr. Edward Palmer (Thomas, 1969) of San Diego; Mrs. Mary S. Snyder of San Diego and Santa Cruz (Taylor, 1964); Mrs. A. E. Bush of San Pedro and San Jose (Dawson, 1966, p. 295); Dr. L. N. Dimmick, Mrs. Ellwood Cooper, and Mrs. R. F. Bingham, all of Santa Barbara; and Miss Helen Lennebacker of Santa Barbara and Santa Cruz. Farlow is commemorated by *Farlowia* J. Agardh, *Laminaria farlowii* Setchell, *Zonaria farlowii* Setchell & Gardner, *Pseudolithophyllum neofarlowii* (Setchell & L. Mason) Adey, and *Botryoglossum farlowianum* (J. Agardh) J. DeToni; Cleveland by *Ozophora clevelandii* (Farlow) Abbott, *Platysiphonia clevelandii* (Farlow) Papenfuss, and *Pterosiphonia clevelandii* (Farlow) Hollenberg; Palmer by *Microdictyon palmeri* Setchell, *Codium palmeri* Dawson, and *Sargassum palmeri* Grunow; Mrs. Snyder by *Pseudoscinaia snyderiae* Setchell, *Tiffaniella snyderiae* (Farlow) Abbott, and *Laurencia snyderiae* Dawson; Mrs. Bush by *Carpopeltis bushiae* (Farlow) Kylin; and Miss Lennebacker by *Taonia lennebackeriae* Farlow *ex* J. Agardh.[*]

In 1876 the Swedish phycologist John Erhard Areschoug (Krok, 1925), professor of botany at Uppsala University, reported on a few brown algae that were collected in California by Dr. Gustavus A. Eisen (Ewan, 1955, pp. 28, 48), a docent in zoology at Uppsala and later a curator with the California Academy of Sciences in San Francisco. In this paper Areschoug erected the genus *Egregia* to receive *Fucus menziesii* Turner, and also named the new kelp genus *Eisenia*, with the species *E. arborea* Areschoug, for a plant obtained at Santa Catalina Island (Areschoug, 1881). In addition, he described as new the remarkable elk kelp, under the name *Nereocystis gigantea* (not realizing that the plant had been described by Leman in 1822 as *Laminaria porra*), and in 1881 he erected the genus *Pelagophycus* for *N. gigantea* (Setchell corrected the name to *P. porra*, in 1908).

Dr. Charles Lewis Anderson, who lived in Santa Cruz from 1866 until his death in 1910 (Jepson, 1929a), was the first resident of California to publish on the algae of that state. Several of his papers appeared during the 1890's. Between 1877 and 1889 he collaborated with Farlow and Professor Daniel Cady Eaton (Setchell, 1895) of Yale University on the first exsiccatae of North American algae, the *Algae exsiccatae Americae Borealis*, Anderson contributing the bulk of the Pacific coast material. He is

---

[*] Names commemorating Dr. Anderson and Mrs. Bingham are given on pp. 32 and 24, respectively.

Charles Lewis Anderson (1827–1910)   William Albert Setchell (1864–1943)

commemorated by *Haplogloia andersonii* (Farlow) Levring, *Cumagloia andersonii* (Farlow) Setchell & Gardner, *Prionitis andersonii* Eaton *ex* J. Agardh, *Gracilaria andersonii* (Grunow) Kylin, and *Nienburgia andersoniana* (J. Agardh) Kylin. Eaton, though a specialist on ferns, was very much interested in the algae, as is evidenced by his work on the exsiccatae. He is remembered by *Gigartina eatoniana* J. Agardh and *Ceramium eatonianum* (Farlow) J. DeToni from the Pacific coast.

In 1893 Marshall Avery Howe (Setchell, 1938), curator and later director of the New York Botanical Garden, published a list of 105 species of algae that he had collected at Monterey the previous summer. Until the appearance of Gilbert M. Smith's *Marine Algae of the Monterey Peninsula, California* in 1944, Howe's was the only published account dealing exclusively with the rich algal flora at Monterey. In 1911 Howe published a paper on some algae from La Paz and San Felipe Bay in the Gulf of California. In it, he named *Dictyota vivesii* and *Gracilaria vivesii* after Señor G. J. Vives, who collected most of the La Paz specimens, and *Cladophora macdougalii* after Dr. Daniel T. MacDougal, director of the Carnegie Desert Laboratory at Tucson, Arizona (Humphrey, 1961, pp. 155–59), who collected the San Felipe material. MacDougal is also remembered by *Codium macdougalii* Dawson, *Sargassum macdougalii* Dawson, and *Gigartina macdougalii* Dawson. Among the Pacific coast algae, Howe is commemorated by *Archaeolithothamnium howeii* Lemoine and *Grateloupia howeii* Setchell & Gardner.

The French phycologist Paul Hariot in 1895 reported on a small collection, including three new species, made in the Gulf of California by Léon Diguet, a student of Mexican cacti. Diguet is remembered by *Lithophyllum diguetii* (Hariot) Heydrich.

De Alton Saunders collected on the Monterey Peninsula during the summers of 1894 and 1896 and published several papers on the algae of the California coast. A number of new genera and species, especially of brown algae, were described in these papers (1895, 1898, 1899, 1901b). Professor of botany at the South Dakota Agricultural College, Brookings, Saunders was a member of the Harriman Alaska Expedition of 1899, and in 1901 he wrote an account of the algae obtained. He is remembered by *Saundersella* Kylin, a genus of brown algae, by *Giffordia saundersii* (Setchell & Gardner) Hollenberg & Abbott, and by *Helminthora saundersii* Gardner.

An event of major importance for phycology on this coast occurred in 1895, when William Albert Setchell (Goodspeed, 1936; Campbell, 1945; Papenfuss, 1973a) accepted the invitation of the University of California to become professor of botany and chairman of the department at Berkeley, a position he held for 39 years, until his retirement in 1934. Setchell did his undergraduate work at Yale, where he received much encouragement and stimulus from D. C. Eaton.

Before moving on to Harvard to work for a doctorate under Farlow, Setchell had collected and studied algae with Isaac Holden of Bridgeport, Connecticut (Collins, 1903), who was vice-president and business manager of the Wheeler and Wilson Sewing Machine Company, and an enthusiastic collector of seaweeds. Between 1895 and 1919, Setchell, Holden, and another amateur phycologist, Frank Shipley Collins of Malden, Massachusetts (Setchell, 1925; see below, p. 37), an accountant in the Malden Rubber Shoe Company and a student of both marine and freshwater algae, issued a series of mounted and named specimens of North American algae. All told, 51 fascicles of 25 to 50 numbers each of this famous exsiccata, the *Phycotheca Boreali-Americana*, were distributed to subscribers in many parts of the world. (Dawson, 1966, p. 296, estimated that during these 25 years Collins handled over 200,000 specimens and in his spare time wrote some 86 botanical papers.) The *P.B.-A.* included a large number of specimens from the West Coast, some of them supplied by two of Setchell's colleagues at Berkeley, Professors Nathaniel L. Gardner and Winthrop J. V. Osterhout (later professor at Harvard University).

By the time Setchell came to Berkeley many of the more striking algal genera of Pacific North America (e.g. *Postelsia, Nereocystis, Pelagophycus, Eisenia*) had been described by European botanists, and a few short lists of algae occurring on this coast had been published. Yet no one had fully appreciated the true nature of the Pacific coast marine flora, one of the richest in the world. It is no exaggeration to say that Setchell found still largely virgin territory in the West. The diversity of the flora and the work confronting him must have seemed overwhelming. Fortunately, the man with whom he was to collaborate so fruitfully for the next 40-odd years—Nathaniel Lyon Gardner (Setchell, 1937)—contacted Setchell early in his tenure at Berkeley. In 1897 Gardner, a schoolteacher at Coupeville on Whidbey Island in the state of Washington, wrote Setchell to ask whether he would be willing to identify some specimens of marine algae from that region. Soon after, with Setchell's encouragement, Gardner decided to continue his education. He took a B.Sc. degree at the University of Washington in 1900, then did graduate work at Berkeley, taking his doctorate in 1906. Three years later, in 1909, he was appointed to the Botany faculty at Berkeley.*

Setchell and Gardner complemented each other admirably—Gardner as an outstanding field man and microscopist who, as Setchell himself admitted (1937, p. 128), did "the greater part of the detailed work," Setchell as the "broad-stroke" man, who excelled at putting the pieces together and wrote with great ease. As a team—most of their publications relating to the algae of Pacific North America were done jointly—they put knowledge of the marine algae of the Pacific coast on a sound footing. On a few occasions Gardner or Setchell also published on algae from other parts of the world, and Setchell frequently allowed himself the pleasure of pursuing some of his many other interests, which included work on fungi, bryophytes, marine and terrestrial flowering plant taxonomy, plant geography, coral reef structure, and ethnobotany. Setchell and Gardner's greatest contributions are their three-part study of the blue-green, green, and brown algae of the Pacific coast (their work on the red algae was never completed), their 1903 report on the algae of northwestern North America, and their 1924 paper on the algae of the Gulf of California. Some of these works were truly comprehensive in the sense of dealing not only with the composition of the flora but also with the overall systems of classification of the groups concerned. They served as reference works throughout the world for many years (as they do today, to some extent) for the classification of orders families, genera, and species, and also for bibliographic information.

In 1899 Setchell published a paper on the algae of the Pribilof Islands in the Bering Sea, based on material received from various sources, and in

---

* In 1915 Gardner established one of the many fruitful links between Stanford University and the University of California by marrying the daughter of David Starr Jordan, the distinguished ichthyologist and President of Stanford.

1930 he and Gardner published on the algae of the Revillagigedo Islands, on the basis of specimens collected in 1925 by members, especially Herbert L. Mason, of a California Academy of Sciences expedition. Mason, now Emeritus Professor of Botany at Berkeley, is honored by the brown algal genus *Masonophycus* and by specific names in *Sphacelaria, Dictyota, Eisenia, Predaea, Polysiphonia,* and *Laurencia,* all described by Setchell and Gardner. The material for their paper on the algae of the Gulf of California was obtained by Townshend S. Brandegee, Walter E. Bryant, and Ivan M. Johnston on expeditions of the California Academy of Sciences in 1890 and 1921, and by Dr. and Mrs. Marchant, who collected especially in the La Paz area in 1917 (Setchell & Gardner, 1924, p. 696). Brandegee (Setchell, 1928) was associated in his later years with the Department of Botany at the University of California; Bryant (W. K. Fisher, 1905) was an ornithologist and for a time a curator with the California Academy of Sciences; and Johnston was a graduate student at Berkeley and later on the staff of the Arnold Arboretum of Harvard University (Thomas, 1969, pp. 21–23). (Biographical data on Dr. and Mrs. Marchant are not known to me.) Brandegee, Bryant, Mrs. Marchant, and Johnston have been commemorated by names of species in the flora of the Gulf of California, as follows: *Nemacystus brandegeei* (Setchell & Gardner) Kylin, *Platythalia brandegeei* (Setchell & Gardner) Womersley, *Sargassum brandegeei* Setchell & Gardner, *Ectocarpus bryantii* Setchell & Gardner, *Pringsheimiella marchantiae* (Setchell & Gardner) Schmidt, *Myriactula marchantiae* (Setchell & Gardner) J. Feldmann, *Hypnea marchantiae* Setchell & Gardner, *Polysiphonia marchantiae* Setchell & Gardner, *Gigartina johnstonii* Dawson, and *Myriactula johnstonii, Sargassum johnstonii, Gelidium johnstonii, Grateloupia johnstonii, Hypnea johnstonii, Gymnogongrus johnstonii, Polysiphonia johnstonii,* and *Laurencia johnstonii,* the last eight species described by Setchell and Gardner.

Many other persons furnished Setchell and Gardner with specimens from California, notably the Misses A. and A. Bayles of Pacific Grove; Mrs. Elizabeth E. Johnston of San Pedro (*Streblonema johnstoniae* Setchell & Gardner, *Scinaia johnstoniae* Setchell); Professor George R. Johnstone (Silva, 1975) of the University of Southern California, Los Angeles (*Codium johnstonei* Silva, *Dictyopteris johnstonei* Gardner); Harold E. Parks (Bonar, 1970), who at one time was collector for the Department of Botany at Berkeley and later resided in Trinidad, California (*Ectocarpus parksii* Setchell & Gardner, *Fucus parksii* Setchell & Gardner, *Erythrotrichia parksii* Gardner); and Mrs. J. M. Weeks of Pacific Grove and elsewhere (*Weeksia* Setchell, Weeksiaceae Abbott, and *Membranoptera weeksiae* Setchell & Gardner). Mikael Heggelund Foslie, curator of the Royal Norwegian Scientific Society museum in Trondheim (Printz, 1929), was

Nathaniel Lyon Gardner (1864–1937)   Harald Johann Kylin (1879–1949)

especially helpful to Setchell in the determination of the crustose corallines of the Pacific coast; he is remembered by *Fosliella* Howe.

Setchell and Gardner both reached retirement age in 1934; Gardner died in 1937, Setchell in 1943. Setchell is commemorated by specific names in *Ulvella, Codium, Laminaria, Cystoseira, Gelidium, Neogoniolithon, Grateloupia, Membranoptera, Phycodrys*, and *Hymenena*; and Gardner by *Gardneriella* Kylin, by specific names in *Pilayella, Pleurophycus, Porphyrella, Gelidium, Gracilariophila, Fryeella, Antithamnion, Ceramium, Asterocolax, Levringiella, Pterosiphonia, Polysiphonia, Laurencia*, and *Janczewskia*, and by varieties in *Polysiphonia* and *Hypnea*.

At the turn of the century Professor Josephine Tilden of the University of Minnesota established a laboratory known as the Minnesota Seaside Station at Port Renfrew, on the west coast of Vancouver Island. During its five or six years of existence (it was abandoned in 1906), Professor Tilden and her students and associates made important contributions to the knowledge of the algae of that region. Miss Tilden published some of the work done at the Station in a yearbook, *Postelsia*. She also issued six volumes of dried algae, each containing 100 specimens, under the general title *American Algae* (1894–1902), which included species from British Columbia and Washington. One of her visitors at the Station was the distinguished Japanese phycologist Professor Kichisaburo Yendo, who in 1902 published a paper on some coralline algae from the Port Renfrew area; he is remembered by *Yendonia* Kylin, a monotypic genus based on material from St. Paul Island in the Bering Sea.

In 1902 Dr. Lorenzo Gordin Yates (Britten & Boulger, 1931) of Santa Barbara published a list of the marine algae of Santa Barbara County. And in 1913 Frank S. Collins, collaborator on the *Phycotheca Boreali-Americana*, published a paper on the marine algae of Vancouver Island, based largely on material supplied by the naturalist John Macoun (Britten & Boulger, 1931; Humphrey, 1961, pp. 160–62). Collins's memory is perpetuated among the algae of the Pacific coast by *Collinsiella* Setchell & Gardner; Macoun is remembered by *Acrochaetium macounii* (Collins) Hamel and *Polysiphonia macounii* Hollenberg.

In 1914 Professor Theodore C. Frye of the Department of Botany of the University of Washington was appointed director of the University's Puget Sound Biological Laboratory (now the Friday Harbor Laboratories), established a few years earlier on San Juan Island (Phifer & Phifer, 1930). Frye and his colleague Professor George B. Rigg and their students and visiting investigators (e.g., Professor Walter C. Muenscher of Cornell University) published a number of papers on the algae of that region, some of which appeared in a short-lived journal called *Publications of the Puget Sound Biological Station*. Frye is commemorated by *Fryeella* Kylin, *Weeksia fryeana* Setchell, *Fauchea fryeana* Setchell, and *Internoretia fryeana* Setchell & Gardner; and Rigg by *Phycodrys riggii* Gardner.

A man who merits special recognition in the advancement of phycology on the Pacific coast is Harald Kylin (Tuneld, 1950; Papenfuss, 1973b), who like the Agardhs was professor of botany at the University of Lund. As Kylin, who was first and foremost a morphologist, unraveled the step-by-step development of the vegetative and reproductive structures of the algae and the details of their life histories, he unearthed so much that was wrong with their taxonomy that he was always deeply involved in their systematics, as well. The systems of classification of the brown and red algae used today are largely his. He was the world specialist on the red algae, the reproductive morphology of which is the most complex of all the algae.

In 1923, in an important monograph on the morphology of 25 genera of Swedish red algae, Kylin elaborated on the system of classification of the red algae ushered in by Friedrich Schmitz in 1883 and emended by Friedrich Oltmanns in 1904. In framing this paper, Kylin realized that the ontogeny of practically every genus of red algae required thorough investigation, and to obtain properly prepared material he had to visit other parts of the world. In the summer of 1922, while his paper was still in press, he visited the United States, collecting at the Monterey Peninsula, La Jolla, Friday Harbor, and Woods Hole. He returned to Friday Harbor in 1924 to teach the Laboratory's summer course on the algae, and in the following year published an important paper on the red algae of that region. In 1923, 1927, 1928, and 1929 he collected at a number of places in Europe.

Using material from all these various places and from the Agardh Herbarium, Kylin published a series of significant morphological and taxonomic monographs, culminating in one on the Gigartinales in 1932. Through these studies he immensely advanced our knowledge of the morphology and interrelationships of members of the large and diversified class Florideophyceae. In 1941 he published an important paper titled "Californische Rhodophyceen," which was based in large part on material from the Monterey Peninsula supplied by Professor Gilbert M. Smith of Stanford. Kylin's last great work, *Die Gattungen der Rhodophyceen*, which appeared posthumously in 1956, contains descriptions of several new genera from the Pacific coast. Obviously, Kylin could study only a fair cross section of the genera of red algae in detail, but he acquired an almost uncanny feel for these algae, and the majority of them are probably correctly placed by him in relation to their nearest allies. It will take many years and the efforts of many phycologists to complete the work that he started. Three Pacific coast taxa (*Antithamnion kylinii* Gardner, *Erythrotrichia kylinii* Gardner, and *Hymenena kylinii* Gardner) are named in his honor, as is *Kylinia* Rosenvinge, which is represented on the Pacific coast.

Dr. Kathleen Drew Baker of the University of Manchester (Lund *et al.*, 1958) spent some time at Berkeley in 1925–27, studying with Gardner, and in 1928 she published an important monograph on the *Acrochaetium-Rhodochorton* complex of the red algae of the Pacific coast.

In 1933 Dr. Kintaro Okamura, director of the Imperial Fisheries Institute in Tokyo, published on a collection of algae made in the Aleutian Islands by Y. Kobayashi. The latter is remembered by *Pleonosporium kobayashii* Okamura.

The man who did more than anyone else in introducing the marine algae of Pacific North America to the world at large was Professor Gilbert Morgan Smith of Stanford University (Page, 1959; Silva, 1960; Wiggins, 1962). Smith came to Stanford from the University of Wisconsin in 1925. His first love was the freshwater algae, and he retained a lifelong interest in them, as his excellent book on the freshwater algae of the United States attests. Among marine phycologists he is best known for his *Marine Algae of the Monterey Peninsula, California*, published in 1944. Setchell and Gardner had not lived to complete their work on the red algae of the Pacific coast, and Smith's volume (Papenfuss, 1944) has therefore served an especially useful purpose in treating a large percentage of that group. But there are other features of this book that mark it as the best marine algal flora ever

Gilbert Morgan Smith (1885–1959)   Elmer Yale Dawson (1918–1966)

published: the descriptions and illustrations are superior, references are given for the original place of publication of each genus and species, and notes on intertidal distribution and on the morphology and life history of species are included wherever these are known. The volume has served as a reference work and guide throughout the world for more than 30 years. Smith retired in 1950 and died in 1959; he is commemorated by *Smithora* Hollenberg, *Hymenena smithii* Kylin, *Porphyra smithii* Hollenberg & Abbott, and *Polysiphonia flaccidissima* var. *smithii* Hollenberg on the Pacific Coast.

One of the major contributors to knowledge of the marine algae of Pacific North America, from Alaska to Panama but especially Mexico and Central America, was Dr. Elmer Yale Dawson (Abbott, 1967; Silva, 1967). While an undergraduate at Berkeley, Dawson came under the influence of Setchell. In his senior year he participated in the marine biological expedition to the Gulf of California sponsored by Captain G. Allan Hancock of Los Angeles. The algae collected formed the subject of his Ph.D. dissertation at Berkeley, which appeared in the same year (1944) in *Allan Hancock Pacific Expeditions*. Dawson published over 60 papers on the algae of southern California, Mexico, and the Central American countries and many other papers and books (see, for example, 1966) on algae and other subjects. One of his major works was an eight-part study on the red algae of Mexico, completed in 1963, just three years before his death by drowning in the Red Sea. The original plan for *Marine Algae of California* was also Dawson's; he was to have been its senior author.

For the greater part of his professional life, Dawson was associated with the Allan Hancock Foundation of the University of Southern California. During this period he participated in the Foundation's expeditions along the coasts of southern California and Mexico in the *Velero IV*. The Hancock Foundation thus is the repository of the voucher material for many of Dawson's collections, and also for many of those of Professor William Randolph Taylor of the University of Michigan, who made extensive collections on two Hancock expeditions. Taxa in the following genera were named by Dawson after Captain Hancock: *Ectocarpus, Sphacelaria, Cutleria, Ralfsia, Acrochaetium, Gelidiella, Grateloupia, Cruoriella, Lithophyllum, Rhodoglossum, Rhodymenia, Botryocladia, Polyneurella, Polysiphonia,* and *Laurencia.* The *Velero* is commemorated by *Veleroa, Cryptonemia veleroae, Dermatolithon veleroae, Lithophyllum veleroae,* and *Gracilaria veleroae,* all described by Dawson.

Dawson's career spanned a period of only 25 years, including graduate study and military service. At the time of his death, in June 1966, he was Curator of Algal Collections in the United States National Museum of the Smithsonian Institution, Washington. With his untimely death at the age of 48, modern phycology was deprived of its most energetic and one of its most brilliant and productive taxonomists. He is commemorated by *Percursaria dawsonii* Hollenberg & Abbott, *Porphyropsis coccinea* var. *dawsonii* Hollenberg & Abbott, *Hildenbrandia dawsonii* (Ardré) Hollenberg, *Murrayellopsis dawsonii* Post, *Rhodymenia dawsonii* Taylor, and *Schizymenia dawsonii* Abbott. Many more Pacific coast species would have borne his name but for the fact that it was he who had first discovered and described them.

Gilbert Smith wrote in the Preface to the Monterey volume that "the flora of the Monterey Peninsula is far richer than that of any other portion of the West Coast of this country," and that the volume would "enable one to identify at least 80 percent of the species to be found anywhere between Puget Sound and southern California." Research in the years since has shown Smith to have been right about the richness of the Peninsula's flora: the *Supplement* to the Smith, published in 1966 (Hollenberg & Abbott, 1966; combined edition, Smith, Hollenberg & Abbott, 1969) added some 55 species to the 393 treated in Smith. But Smith appears to have been a bit sanguine about percentages: Abbott and Hollenberg in *Marine Algae of California* (1976), basically an extension of Smith's treatment to the entire California coast, describe 667 species and do not, of course, treat Oregon, Washington, or British Columbia species that do not extend into California. Indeed, of the 313 species and varieties given in Doty's (1947) Oregon list, 83 do not occur in California; and of the 478 species and varieties given in Scagel's (1957) British Columbia and Washington list, 155 do not occur in California. There remains much to study!

Doty has been commemorated by *Scytosiphon dotyi* Wynne; Scagel by *Scagelia* Wollaston, *Scagelonema* Norris & Wynne, and *Ulva scagelii* Chihara; Hollenberg by *Hollenbergia* Wollaston, *Derbesia hollenbergii* Taylor, *Porphyra hollenbergii* Dawson, *Herposiphonia hollenbergii* Dawson, *Nitophyllum hollenbergii* (Kylin) Abbott, and *Halymenia hollenbergii* Abbott; Abbott by *Porphyra abbottiae* Krishnamurthy, *Phycodrys isabelliae* Norris & Wynne, *Liagora abbottiae* Dawson, and, very recently, *Isabbottia* Balakrishnan.

A perusal of the pages of the present volume will reveal that in the tradition of Setchell and Gardner, Kylin, Smith, Dawson, and the earlier workers, Professors Hollenberg and Abbott find the algae of Pacific North America a continually rewarding subject of study. The world of phycology is indeed fortunate that these two, who know the plants so well, have undertaken to provide us with the first full-fledged flora of the benthic algae of California.

REFERENCES

Abbott, Isabella A. 1967. Elmer Yale Dawson, 1918–1966. *Jour. Phycol.*, 2: 129–32. Portrait.

Abbott, Isabella A., and G. J. Hollenberg. 1976. *Marine Algae of California*. Stanford, Calif. 800 pp. 702 figs.

Agardh, C. A. 1822. *Species algarum*. . . . Lund. Vol. 1(2). Pp. [8], 169–531.

———. 1824. *Systema algarum*. Lund. xxxviii + 312 pp.

Agardh, J. G. 1847. Nya alger från Mexico. *Öfvers. Kgl.* [*Svensk.*] *Vetensk.-Ak. Förh.*, 4: 5–17.

Anderson, B. 1960. *Surveyor of the Sea: The Life and Voyages of Captain George Vancouver*. Seattle, Wash. vii + 274 pp. Illustrated.

Areschoug, J. E. 1870. Carl Adolph Agardh: Professor, Biskop. *Kgl. Svensk. Vetensk.-Ak. Lefnadsteck.*, 1: 251–96.

———. 1876. De tribus Laminarieis et de Stephanocystide osmundaceae (Turn.) Trev. observationes praecursorias. *Bot. Notiser*, 1876: 65–73.

———. 1881. Beskrifning på ett nytt algslägte Pelagophycus, hörande till Laminarieernas familj. *Ibid.*, 1881: 49–50.

Barneby, R. C. 1963. Treasures of the Garden's herbarium: Reliquiae Haenkeanae. *Jour. N.Y. Bot. Gard.*, 13: 139–40, 142. 2 figs.

Beechey, F. W. 1831. *Narrative of a Voyage to the Pacific and Beering's Strait . . . in His Majesty's Ship "Blossom" . . . in the Years 1825, 26, 27, 28*. London. 2 vols.

Bonar, L. 1970. Harold Ernest Parks. *Madroño*, 20: 373–77. Portrait.

Bory de Saint-Vincent, J. B. 1826. Iridée. Iridea, *in* J. B. Bory de Saint-Vincent (ed.), *Dict. Class. Hist. Nat.* Vol. 9. Paris. Pp. 15–16.

Britten, J., and G. S. Boulger. 1931. *A Biographical Index of Deceased British and Irish Botanists*. London. 2d ed. Rev. and comp. A. B. Rendle. xxii + 342 pp.

Campbell, D. H. 1945. William Albert Setchell. *Biogr. Memoirs Nat. Acad. Sc. U.S.A.*, 23: 127–47.

Choris, L. 1822. *Voyage pittoresque autour du monde* . . . [Part I]. Paris. vi + 17 pp. 12 pls.

Collins, F. S. 1903. Isaac Holden. *Rhodora*, 5: 219–20.

———. 1913. The marine algae of Vancouver Island. *Victoria Mem. Mus. Bull.*, 1: 99–137.

Coville, F. V. 1895. The botanical explorations of Thomas Coulter in Mexico and California. *Bot. Gaz.*, 20: 519–31. 1 pl.

Cutter, D. C. 1960. *Malaspina in California*. San Francisco, Calif. viii + 96 pp. Illustrated.

Dawson, E. Y. 1944. The marine algae of the Gulf of California. *Allan Hancock Pacific Expeditions*, 3: 189–453. 47 pls.

———. 1954. Notes on tropical Pacific marine algae. *Bull. S. Calif. Acad. Sc.*, 53: 1–7. 4 figs.

———. 1963. Marine red algae of Pacific Mexico. VIII. Ceramiales: Dasyaceae, Rhodomelaceae. *Nova Hedwigia*, 6: 401–81. 46 pls.

———. 1966. *Marine Botany: An Introduction*. New York. xii + 371 pp. Illustrated.

Doty, M. S. 1947a. The marine algae of Oregon. I. Chlorophyta and Phaeophyta. *Farlowia*, 3: 1–65. 10 pls.

———. 1947b. The marine algae of Oregon. II. Rhodophyta. *Ibid.*, 3: 159–215. 4 pls.

Drew, Kathleen M. 1928. A revision of the genera Chantransia, Rhodochorton, and Acrochaetium, with descriptions of the marine species of Rhodochorton (Naeg.) gen. emend. on the Pacific coast of North America. *Univ. Calif. Publ. Bot.*, 14: 139–224. 12 pls.

Eastwood, Alice. 1939. Early botanical explorers on the Pacific coast and the trees they found there. *Calif. Hist. Soc. Quart.* 18: 335–46.

Eriksson, J. 1916. Jacob Georg Agardh. *Kgl. Svensk. Vetensk.-Ak. Lefnadsteck.*, 5: 1–136. Portrait.

Esper, E. J. C. 1797–1808. *Icones fucorum* . . . Nürnberg. 139 pp. 177 pls.

Ewan, J. 1955. San Francisco as a Mecca for nineteenth century naturalists . . . , in *A Century of Progress in the Natural Sciences*. San Francisco, Calif. (Calif. Acad. Sc.). Pp. 1–63. 4 portraits.

[Fisher, Lydia, ed.]. 1869. *Memoir of W. H. Harvey, M.D., F.R.S., . . . with Selections from His Journal and Correspondence*. London. xi + 372 pp. Portrait. [The information that this memoir was edited by Mrs. Lydia Fisher, a cousin of Harvey's, and published after his death is furnished by R. Lloyd Praeger, in Oliver, 1913, p. 205 (q.v.).]

Fisher, W. K. 1905. In memoriam: Walter E. Bryant . . . *The Condor*, 7: 129–31. Portrait.

Galbraith, Edith C. 1924. Malaspina's voyage around the world. *Calif. Hist. Soc. Quart.*, 3: 215–37.

Glenn, Rewa. 1950. *The Botanical Explorers of New Zealand*. Wellington. 176 pp. Illustrated.

Goodspeed, T. H. 1936. William Albert Setchell: a biographical sketch, *in* T. H. Goodspeed, ed., *Essays in Geobotany in Honor of William Albert Setchell.* Berkeley, Calif. Pp. xi–xxv. Portrait.

Gray, A. 1866. William Henry Harvey. *Am. Jour. Sc. & Arts*, 92: 273–77.

Greville, R. K. 1830. *Algae britannicae* . . . Edinburgh. lxxxviii + 218 pp. 19 pls.

Hariot, P. 1895. Algues du Golfe de Californie recueillies par M. Diguet. *Jour. de Bot.*, 9: 167–70.

Harvey, A. G. 1947. *Douglas of the Fir* . . . Cambridge, Mass. x + 290 pp. Illustrated, frontispiece.

Harvey, W. H. 1833. Algae, in W. J. Hooker and G. A. W. Arnott, *The Botany of Captain Beechey's Voyage* . . . London. Pp. 163–65.

———. 1840. Algae, in *ibid.* (Suppl.). Pp. 406–9.

———. 1852. Nereis boreali-americana . . . I. Melanospermeae. *Smithsonian Contr. to Knowledge*, 3 (art. 4). 150 pp. 12 pls.

———. 1853. Nereis boreali-americana . . . II. Rhodospermeae. *Smithsonian Contr. to Knowledge*, 5 (art. 5). 258 pp. 24 pls.

———. 1858. Nereis boreali-americana . . . III. Chlorospermeae. *Smithsonian Contr. to Knowledge*, 10 (art. 2). 140 pp. 14 pls.

———. 1862. Notice of a collection of algae made on the northwest coast of North America, chiefly at Vancouver's Island, by David Lyall, Esq., M.D., R.N., in the years 1859–61. *Jour. Linn. Soc., Bot.*, 6: 157–77.

Harvey, W. H., and J. W. Bailey. 1851. Descriptions of seventeen new species of algae, collected by the United States Exploring Expedition. *Proc. Boston Soc. Nat. Hist.*, 3: 370–73.

Hollenberg, G. J., and Isabella A. Abbott. 1966. *Supplement to Smith's Marine Algae of the Monterey Peninsula.* Stanford, Calif. ix + 130 pp. 53 figs. Map.

Hooker, W. J. 1836. A brief memoir of the life of Mr. David Douglas, with extracts from his letters. *Companion to Bot. Mag.*, 2: 79–182.

Howe, M. A. 1893. A month on the shores of Monterey Bay. *Erythea*, 1: 63–68.

———. 1904. The Pike collection of algae. *Jour. N.Y. Bot. Gard.*, 5: 86–87.

———. 1911. Phycological studies. V. Some marine algae of Lower California, Mexico. *Bull. Torrey Bot. Club*, 38: 489–514. 1 fig. 8 pls.

Hultén, E. 1940. History of botanical exploration in Alaska and Yukon territories from the time of their discovery to 1940. *Bot. Notiser*, 1940: 289–346. Map.

Humphrey, H. B. 1961. *Makers of North American Botany.* New York. xi + 265 pp.

Jepson, W. L. 1929a. The botanical explorers of California, V. *Madroño*, 1: 214–16. Portrait of Anderson.

———. 1929b. The botanical explorers of California, VI. *Ibid.*, 1: 262–70. 3 portraits.

Kjellman, F. R. 1889. Om Beringhafvets Algflora. *Kgl. Svensk. Vetensk.-Ak. Handl.*, 23(8). 58 pp. 7 pls.

Komarov, V. 1926. Botany, in Acad. Sc. USSR, ed., *The Pacific Russian Scientific Investigations.* Leningrad. pp. 121–36. Portrait.

Krok, T. O. B. N. 1925. *Bibliotheca Botanica Suecana* . . . Uppsala. xvi + 799 pp. Portrait.

Kühnel, J. 1960. *Thaddaeus Haenke, Leben und Wirken eines Forschers.* Munich. 276 + [2] pp. Illustrated.

Kylin, H. 1923. Studien über die Entwicklungsgeschichte der Florideen. *Kgl. Svensk. Vetensk.-Ak. Handl.* 63(11). 139 pp. 82 figs.

————. 1925. The marine red algae in the vicinity of the biological station at Friday Harbor, Washington. *Lunds Univ. Årsskr.*, n.f., avd. 2, 21(9). 87 pp. 47 figs.

————. 1932. Die Florideenordnung Gigartinales. *Lunds Univ. Årsskr.*, n.f., avd. 2, 28(8). 88 pp. 22 figs. 28 pls.

————. 1941. Californische Rhodophyceen. *Ibid.*, 37(2). 51 pp. 7 figs. 13 pls.

————. 1956. *Die Gattungen der Rhodophyceen.* Lund. xv + 673 pp. 458 figs.

Leman, D. S. 1822. Laminaria, Laminaire, *in* F. G. Levrault (ed.). *Dict. Sc. Nat.* Vol. 25. Strasbourg and Paris. Pp. 185–90.

Lund, J. W. G., Margaret T. Martin, H. T. Powell, G. F. Papenfuss, J. Feldmann, Mary Calder, Lily Newton, C. W. Wardlaw, Edna M. Lind, Dorothy Brittain, and Editha Jackson. 1958. Kathleen M. Drew: [expressions of appreciation]. *Phycolog. Bull.*, No. 6: iv, 1–12. Portrait.

Mahr, A. C. 1932. The visit of the *Rurik* to San Francisco in 1816. *Stanford Univ. Publ. Hist., Econ., Poli. Sc.*, 2: 267–460. Illustrated, frontispiece.

Maximowicz, C. J. 1871. Dr. Franz Joseph Ruprecht. *Bull. Acad. Imp. Sc. St.-Pétersb.*, 16 (Suppl.). 21 pp. Portrait.

Mertens, H. 1829. Zwei botanisch-wissenschaftliche Berichte vom Dr. Heinrich Mertens . . . *Linnaea*, 4: 43–58.

————. 1833. Two scientific botanical notices . . . . Extracted from the *Linnaea*, Vol. IV, January 1829. *Bot. Misc.*, 3: 1–23.

Nordenskiöld, [N.] A. E. 1882. *The Voyage of the "Vega" Round Asia and Europe* . . . Tr. A. Leslie. New York. xxvi + 756 pp. Illustrated.

Okamura, K. 1933. On the algae from Alaska collected by Y. Kobayashi. *Records Oceanogr. Works Japan*, 5: 85–97. 2 pls.

Oliver, F. W., ed. 1913. *Makers of British Botany* . . . Cambridge, Eng. [6] + 332 pp. 1 fig. 26 pls. Frontispiece.

Page, R. M. 1959. Gilbert Morgan Smith, Botanist. *Science*, 130: 1693–95. Portrait.

Papenfuss, G. F. 1944. [Review of] G. M. Smith: Marine algae of the Monterey Peninsula, California. *Madroño*, 7: 226–31.

————. 1973a. Setchell, William Albert. *Dictionary of American Biography*, Suppl. 3, 1941–45. New York. Pp. 703–4.

————. 1973b. Kylin, Johann Harald. *Dictionary of Scientific Biography*, Vol. 7. New York. Pp. 534–36.

Phifer, L. D., and Margaret W. Phifer. 1930. The Puget Sound Biological Station. *The Biologist*, 12: 5–20.

Pierce, R. A., and J. W. Winslow, eds. 1969. *H.M.S. "Sulphur" at California, 1837 and 1839. Being the Accounts of Midshipman Francis Guillemard Simpkinson and Captain Edward Belcher.* San Francisco. xvi + 70 pp. Illustrated.

Postels, A., and F. J. Ruprecht. 1840. *Illustrationes algarum* . . . St. Petersburg. iv + 22 + [2] pp. 40 pls.

Presl, K. B. 1825. *Reliquiae Haenkeanae* . . . Prague. Vol. 1(1). xv + 84 pp.

Printz, H., ed. 1929. M. Foslie, *Contributions to a Monograph of the Lithothamnia*. Collected and edited after the author's death by H. Printz. Trondheim. 60 pp. 75 pls. Portrait.

Ruprecht, F. J. 1852. Neue oder unvollständig bekannte Pflanzen aus dem nördlichen Theile des Stillen Oceans. *Mém. Acad. St.-Pétersb. Sc. Nat. Bot.*, sér. 6, 7: 55–82. 8 pls.

Saunders, De A. 1895. A preliminary paper on Costaria with description of a new species. *Bot. Gaz.*, 20: 54–58. 1 pl.

———. 1898. Phycological memoirs. *Proc. Calif. Acad. Sc.*, ser. 3, Botany, 1: 147–68, 21 pls.

———. 1899. New and little-known brown algae of the Pacific coast. *Erythea*, 7: 37–40. 1 pl.

———. 1901a. Papers from the Harriman Alaska Expedition. XXV. The algae. *Proc. Wash. Acad. Sc.*, 3: 391–486. 20 pls.

———. 1901b. A new species of Alaria. *Minn. Bot. Stud.*, 2: 561–62. 1 pl.

Scagel, R. F. 1957. An annotated list of the marine algae of British Columbia and northern Washington (including keys to genera). *Bull. Natl. Mus. Canada*, 150. vi + 289 pp. 1 fig.

Schmidt, P. 1926. Zoology, in Acad. Sc. USSR, ed., *The Pacific Russian Scientific Investigations*. Leningrad. Pp. 137–60. 1 fig. 2 portraits.

Setchell, W. A. 1895. Daniel Cady Eaton, 1834–1895. *Bull. Torrey Bot. Club*, 22: 341–51. Portrait.

———. 1899. Algae of the Pribilof Islands, in D. S. Jordan, ed., *The Fur Seals and Fur-Seal Islands of the North Pacific Ocean*. Washington, D.C. III: 589–96. 1 pl.

———. 1908. Nereocystis and Pelagophycus. *Bot. Gaz.*, 45: 125–34.

———. 1925. Frank Shipley Collins, 1848–1920. *Amer. Jour. Bot.*, 12: 54–62. Portrait.

———. 1926. William Gilson Farlow, 1844–1919. *Mem. Nat. Acad. Sc.*, 21(4). 22 pp. Portrait.

———. 1928. Townshend Stith Brandegee and Mary Katharine (Layne) (Curran) Brandegee. *Univ. Calif. Publ. Bot.*, 13: 155–78. 2 pls.

———. 1937. Nathaniel Lyon Gardner, 1864–1937. *Madroño*, 4: 126–28. Portrait.

———. 1938. Biographical memoir of Marshall Avery Howe, 1867–1936. (With bibliography by John Hendley Barnhart.) *Biogr. Mem. Nat. Acad. Sc.*, 19: 243–69. Portrait.

Setchell, W. A., and N. L. Gardner. 1903. Algae of northwestern America. *Univ. Calif. Publ. Bot.*, 1: 165–419. 11 pls. Map.

———. 1924. New marine algae from the Gulf of California. *Proc. Calif. Acad. Sc.*, 4th ser., 12: 695–949. 77 pls.

———. 1930. Marine algae of the Revillagigedo Islands Expedition in 1925. *Ibid.*, 19: 109–215. 12 pls.

Silva, P. C. 1953. The identity of certain Fuci of Esper. *Wasmann Jour. Biol.*, 11: 221–32.

———. 1960. Gilbert Morgan Smith, 1885–1959. *Rev. Algologique*, sér. 2, 5: 97–102. Portrait.

———. 1967. E. Yale Dawson, 1918–1966. *Phycologia*, 6: 218–36. Portrait.

———. 1975. George Rufus Johnstone, 1888–1971. *Phycologia*, 14: 49–51. Portrait.

Smith, G. M. 1944. *Marine Algae of the Monterey Peninsula, California.* Stanford, Calif. vii + 622 pp. 98 pls.

Smith, G. M., G. J. Hollenberg, and Isabella A. Abbott. 1969. *Marine Algae of the Monterey Peninsula, California.* 2d ed., incorporating the 1966 *Supplement.* Stanford, Calif. xi + 752 pp. 53 figs. 98 pls.

Taylor, W. R. 1945. William Gilson Farlow, promoter of phycological research in America, 1844–1919. *Farlowia*, 2: 53–70.

———. 1964. A valuable old collection of Florida marine algae. *Quart. Jour. Florida Acad. Sc.*, 27: 1–8.

Thomas, J. H. 1969. Botanical explorations in Washington, Oregon, California and adjacent regions. *Huntia*, 3: 5–66. 27 portraits.

Tuneld, J. 1950. Bibliografi över Harald Kylins tryckta skrifter. *Bot. Notiser*, 1950: 106–16.

Turner, D. 1808. *Fuci* ... Vol. 1. London. [2] + 164 + [2] pp. 71 pls.

———. 1809. *Fuci* ... Vol. 2. London. 162 + [2] pp. 63 pls.

———. 1811. *Fuci* ... Vol. 3. London. 148 + [2] pp. 62 pls.

———. 1819. *Fuci* ... Vol. 4. London. [2] + 153 + [2] + 7 pp. 62 pls.

Wiggins, I. L. 1962. Gilbert Morgan Smith, January 6, 1885–July 11, 1959. *Biogr. Mem. Nat. Acad. Sc.*, 36: 289–313. Portrait.

Yates, L. G. 1902. The marine algae of Santa Barbara County. *Bull. Santa Barbara Soc. Nat. Hist.*, 1(3): 3–20.

Yendo, K. 1902. Corallinae verae of Port Renfrew. *Minn. Bot. Studies*, 2: 711–22. 6 pls.

THE ALGAE

Division Chrysophyta

The Yellow-Brown Algae

THALLI OF VARIOUS vegetative forms, unicellular, flagellate, coccoid or filamentous with overlapping walls, more or less silicified. Cells mostly uninucleate, with 1 to many chloroplasts containing chlorophyll *a* and usually chlorophyll *c*, beta-carotene, and several characteristic xanthophyll pigments, with or without pyrenoids; reserve material consisting of oil droplets or chrysolaminarin (a polysaccharide). Asexual reproduction by cell division, akinetes, aplanospores, or zoospores; sexual reproduction (isogamous, anisogamous, or oogamous) rare. Endogenous cysts (statospores) characteristic of division, with walls nearly equal or highly unequal in size.

Current classification at class, order, family, and genus levels in the Chrysophyta is dependent on ultrastructural details that show relationship within the division more clearly than is possible with the use of a light microscope.

The majority of the Chrysophyta are unicellular and occur in freshwater habitats.

Class XANTHOPHYCEAE

Flagellate, coccoid or filamentous; usually containing several to many chloroplasts with semi-immersed pyrenoids not surrounded by food reserves; reserve material free in cell as oil droplets. Motile cells (zoospores) bilaterally symmetrical, with unequal flagella, the longer one directed anteriorly and bearing two rows of stiff hairs, the shorter one emerging laterally without hairs. Endogenous cysts with two overlapping walls approximately equal in size, more or less silicified.

Order VAUCHERIALES

Thalli coenocytic, tubular, nonseptate; chloroplasts many.

Family VAUCHERIACEAE

Thalli branched, tubular, nonseptate, oogamous. Chlorophyll *a* included in greater proportion than xanthophylls, giving a dark green color to thalli. Cell walls containing cellulose.

Vaucheria DeCandolle 1801

Thalli oogamous. Species distinctions dependent on location, shape, and size of oogonia and antheridia. Sperms laterally biflagellate, the anterior flagellum pantonematic, the posterior one acronematic.

Vaucheria longicaulis Hopp.

Hoppaugh 1930: 332; Taylor 1952: 274.

Thalli filamentous, 33–60 μm diam., forming thin patches on substratum; dioecious; antheridia cylindrical, 45–60 μm diam., 336–730 μm long, terminal on branches, separated from branch by empty supporting compartment, the antheridia with mostly 5–8 short lateral projections, each with terminal round pore; oogonia club-shaped, 110–165 μm diam., 275–440 μm long, terminal on lateral branches, becoming terminally enlarged, with apex gelatinizing at maturity; zygote subspherical, distal in oogonium.

Frequent on intertidal mudflats, Bodega Bay (Sonoma Co.) and Elkhorn Slough (Monterey Co.; type locality), Calif. Also Brazil.

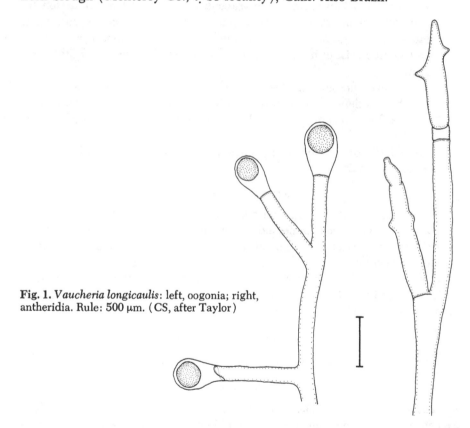

Fig. 1. *Vaucheria longicaulis*: left, oogonia; right, antheridia. Rule: 500 μm. (CS, after Taylor)

Division Chlorophyta

The Green Algae

THALLI GRASS-GREEN, with pigments in plastids; usually with 1 or more pyrenoids. Chlorophyll *a* and *b*, beta-carotene and oxy-carotenes present, xanthin varying with different groups; reserve carbohydrate alpha-1:4 linked starch. Asexual reproduction by zoospores, aplanospores, or akinetes; sexual reproduction isogamous, anisogamous, or oogamous; gametes usually flagellated. Motile cells with 2–4+ flagella of equal length, apically inserted. Life history various.

Two classes in the California flora, the Prasinophyceae (below) and the Chlorophyceae (p. 53).

Class PRASINOPHYCEAE

Thalli essentially unicellular, occasionally associated in dendroid colonies. Cell wall absent, the motile cells mostly with 2 or 4 flagella emerging from an apical pit or groove, the flagella usually bearing scales or mastigonemes or both, these also present on cell body of some genera. Chloroplast single, usually cup-shaped, sometimes lobed, with elaborate pyrenoid closely associated (by virtue of internal canals) with nucleus or cytoplasm; starch plates variable in formation and generally more complex than those of Class Chlorophyceae. Asexual reproduction by longitudinal division of protoplast into zoospores. Sexual reproduction uncertain or unknown.

The species *Stephanoptera gracilis* (Artari) Smith and *Platymonas subcordiformis* (Wille) Hazen listed by Smith (1944) for the Monterey Peninsula are now placed in the Class Prasinophyceae, but are unicellular forms and have therefore been excluded from treatment here.

Order PRASINOCLADALES

Cells tending to remain nonmotile, becoming motile only on formation of short-lived zoospores.

Family CHLORODENDRACEAE

Thalli sedentary, solitary or forming dendroid colonies connected by tough gelatinous sheaths arising from stipelike anterior extensions of indi-

vidual cell sheath (theca). Chloroplasts cup-shaped, entire, lobed or reticulate. Cells dividing longitudinally; daughter cells either remaining per-
manently ensheathed and contributing to stalk system or becoming flagellate and functioning as zoospores. Sexual reproduction unknown.

Prasinocladus Kuckuck 1894

Thalli of 2 different vegetative forms, 1 unicellular and motile, the other
stalked and sedentary. Unicellular motile stage (cells or zooids) with
single noncellulose theca and terminal row of 4 flagella arising in anterior
depression; flagella pantonematic and covered with 2 layers of scales;
nucleus closely appressed to yellowish-green chloroplast and to single
pyrenoid, the nucleus with 2-layered eyespot; starch sheath around pyrenoid continuous or interrupted at anterior pole. Stalked sedentary stage
arising by formation of a new theca inside nonmotile cell, the inner wall
elongating rapidly and the older rupturing, resulting in single stalked cell
or seemingly branched colonies owing to longitudinal divisions of protoplast and simultaneous development of stalked cells, but any cell capable
of producing new theca within old one with or without cell division; several successive thecae usually present; newly reconstituted cell or new cells
placed with anterior end toward stalk. Any cell potentially a motile zooid,
swimming for short time before settling and producing stalked stage again,
or becoming encysted. Sexual reproduction unknown.

Prasinocladus ascus Prosk.

Proskauer 1950: 65; Chihara 1963: 19.

Stalked stages mostly solitary but sometimes branched; stalks mostly
nonseptate, with empty chambers to 30 times as long as diam.; protoplasts
8.5–16 µm diam., 19–50 µm long; chloroplast cup-shaped, occasionally with
anterior lobing; pyrenoid axial near center or more posterior, with continuous starch plates; eyespot usually anterior, sometimes fragmented;
zooids 9–10 µm diam., 19–32 µm long.

Uncommon, on rocks in shallow upper tidepools, Wash. to Calif.; in
Calif., known from Bodega Head (Sonoma Co.), Santa Cruz (type locality), and Pt. Lobos (Monterey Co.). Also Japan.

Prasinocladus marinus (Cienk.) Waern

Chlorangium marinum Cienkowski 1881: 152. Prasinocladus marinus (Cienk.) Waern
1952: 85. P. lubricus Kuckuck 1894: 261; Smith 1944: 32.

Stalked stages to 250 µm tall, with repeatedly dichotomously branched
and transversely septate stalks; chambers of stalks barrel-shaped, 6–8 µm
diam., 6–12 µm long; cells of colonies 4–11 µm diam., 12–25 µm long;
zooids mostly 2–7 µm diam., 16–20 µm long, with single theca and with
anteriorly laciniate chloroplast closely appressed to nucleus; pyrenoid lateral; starch sheath cup-shaped, usually interrupted at anterior pole.

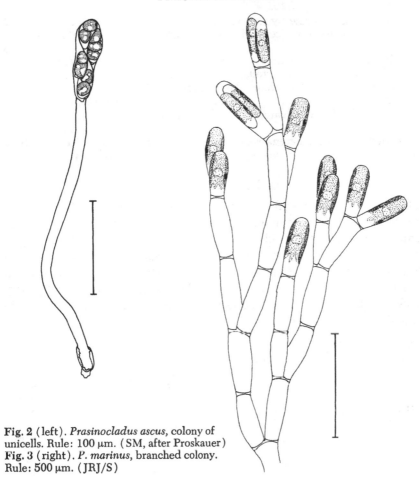

Fig. 2 (left). *Prasinocladus ascus*, colony of unicells. Rule: 100 μm. (SM, after Proskauer)
Fig. 3 (right). *P. marinus*, branched colony. Rule: 500 μm. (JRJ/S)

Motile stage common in upper tidepools, colonial stage occasional on upper intertidal rocks, Friday Harbor, Wash., to Monterey Co., Calif. Common in N. Europe and Japan. Type locality: White Sea, U.S.S.R.

Class Chlorophyceae

Thalli unicellular, colonial or multicellular. Growth intercalary or diffuse, rarely apical. Cell wall usually of cellulose. Motile cells with 2–4 similar apical acronematic flagella, emerging individually through holes in cell wall. Chloroplast 1 to many, cup-shaped, ribbonlike, band-shaped, stellate, disk-shaped, or reticulate, with 1 to many pyrenoids, with or without starch plates; starch plates arranged in several ways but never as elaborate as in Class Prasinophyceae. Asexual reproduction by fragmentation, zoospores, akinetes, or aplanospores; sexual reproduction isogamous, anisogamous, or oogamous.

1. Thalli increasing in size by division and growth of cells.............. 2
1. Thalli not increasing in size by division and growth of cells (thalli coenocytic) 4
   2. Chloroplast single, stellate, central.............. PRASIOLALES (p. 89)
   2. Chloroplast single, not stellate 3
3. Cells mostly uninucleate; chloroplast imperforate, laminate...........
.... ULOTRICHALES (below)
3. Cells typically multinucleate; chloroplast usually reticulate ....
.... CLADOPHORALES (p. 90)
   4. Growth involving segregative cell formation... SIPHONOCLADALES (p. 108)
   4. Growth not involving segregative cell formation..... CODIALES (p. 111)

Order ULOTRICHALES

Thalli filamentous, membranous, or parenchymatous, branched or unbranched. Cells mostly uninucleate. Chloroplast mostly single, parietal,
usually with several pyrenoids. Life history mostly with isomorphic stages.
Asexual reproduction by zoospores or aplanospores, the spores formed
mostly in unmodified vegetative cells, the zoospores biflagellate or quadriflagellate, usually many per sporangium. Sexual reproduction mostly isogamous or anisogamous, the gametes mostly biflagellate, usually formed in
unmodified vegetative cells.

1. Plants filamentous 2
1. Mature plants 2 to many cells broad 3
   2. Filaments unbranched .... ULOTRICHACEAE (below)
   2. Filaments branched, sometimes aggregated and forming a pseudoparenchyma .... CHAETOPHORACEAE (p. 56)
3. Thalli bladelike, monostromatic .... MONOSTROMATACEAE (p. 65)
3. Thalli mostly bladelike and distromatic or partly to wholly tubular.....
.... ULVACEAE (p. 69)

Family ULOTRICHACEAE

Thalli filamentous, mostly unbranched, usually basally attached. Cells
cylindrical, uninucleate. Chloroplast single, parietal, band-shaped, completely or incompletely encircling cell, with 1 to several pyrenoids.

Ulothrix Kützing 1833

Thalli filamentous, mostly unbranched, with basal attachment. Cells
uninucleate. Chloroplast single. Zoospores quadriflagellate, formed in mul-
tiples of 2 in a cell and liberated through a single pore. Gametes isogamous,
biflagellate. Zygote secreting a thick wall, liberating zoospores after dormant period.

1. Filaments often laterally adjoined, occasionally branched..... *U. laetevirens*
1. Filaments not laterally adjoined, unbranched 2
   2. Chloroplast a complete ring .... *U. flacca*
   2. Chloroplast an incomplete ring .... *U. pseudoflacca*

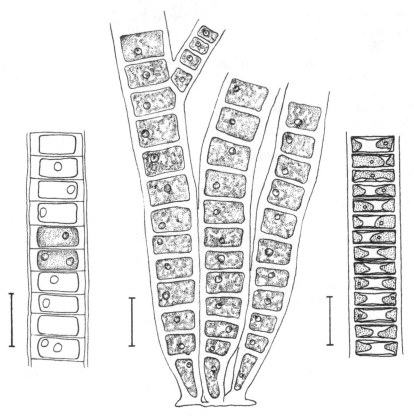

Fig. 4 (left). *Ulothrix flacca.* Rule: 20 μm. (LH, after Hazen) Fig. 5 (center).
U. laetevirens. Rule: 20 μm. (DBP/H) Fig. 6 (right). *U. pseudoflacca.* Rule: 20 μm.
(SM, after Setchell & Gardner)

Ulothrix flacca (Dillw.) Thur.

Conferva flacca Dillwyn 1802–9 (1805): pl. 49. *Ulothrix flacca* (Dillw.) Thuret 1863:
56; Setchell & Gardner 1903: 217. *U. speciosa sensu* Kornmann 1964: 29.

Filaments 10–25 μm diam., in bright or dark green masses, often much
entangled; the cells 0.25–0.75 times as long as diam., considerably swollen
when fertile; chloroplast lining entire cell wall, 1–3+ pyrenoids per cell.

Frequent on rocks, wood, or other algae, intertidal, Japan and Alaska to
Santa Catalina I., Calif. Widely distributed. Type locality: Wales.

Ulothrix laetevirens (Hook.) Coll.

Bangia laetevirens Hooker 1833: 317. *Ulothrix laetevirens* (Hook.) Collins 1909a: 186;
Setchell & Gardner 1920b: 286.

Filaments 10–25 μm diam., groups of 2 or 3 often laterally adjoined and
subparenchymatous below, with occasional to frequent branches, mostly

more slender than main filament; cells mostly 0.25–0.75 times as long as diam.; chloroplast usually lining entire cell wall, and having a single pyrenoid.

Rare, intertidal on wood or other algae, Alaska to Humboldt Co., Calif. Type locality: Ireland.

Ulothrix pseudoflacca Wille

Wille 1901: 22; Setchell & Gardner 1920b: 285; Hollenberg & Abbott 1966: 9. *U. implexa* Kützing 1849: 349; Smith 1944: 34.

Filaments free, 1–4 mm long, unbranched, (8)22–25(30) μm diam., the cells 0.3–1 times as long as diam.; filaments attached by elongate, basally narrowed cell; cell wall thin; chloroplast a broken ring encircling cell, the pyrenoid single; fertile cells not enlarged.

Frequent on rocks, wood, or other algae, upper intertidal, Sitka, Alaska, and from San Francisco to Redondo Beach, Calif. Type locality: Norway.

Family CHAETOPHORACEAE

Thalli filamentous, branched, free or adjoined, sometimes forming pseudoparenchyma; with or without colorless hairs (setae); chloroplasts mostly laminate and parietal, with 1 or more pyrenoids.

Pilinella Hollenberg 1971

Thalli minute epiphytes, with erect filaments from compact layer of spreading prostrate filaments, the filaments not penetrating host. Erect filaments mostly unbranched except near base, each bearing long, colorless, terminal unicellular seta. Sporangia (gametangia?) terminal on much shorter erect filaments.

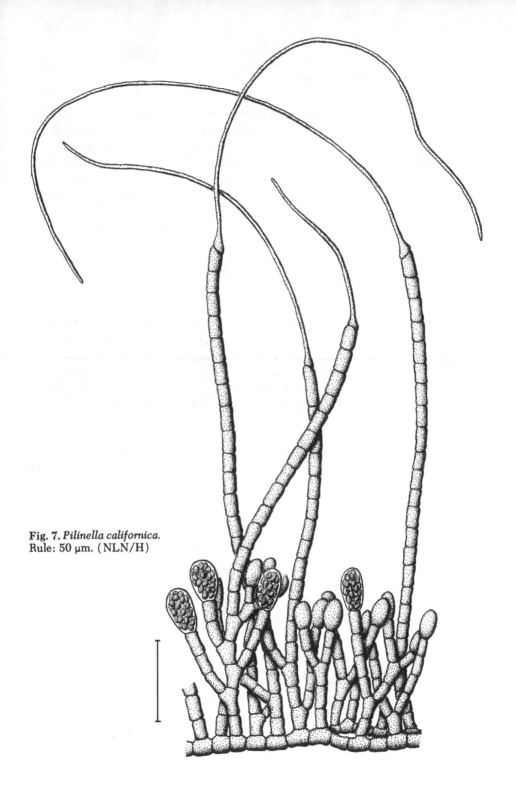

Fig. 7. *Pilinella californica.*
Rule: 50 μm. (NLN/H)

Pilinella californica Hollenb.

Hollenberg 1971a: 11.

Thalli bright green, with nontapering erect filaments 270–400 µm tall, unbranched except near base; erect filaments composed of cells 6–8.5 µm diam., 1.2–2.5 times as long, with little or no constriction at crosswalls; chloroplast single, parietal, almost completely lining cell wall, with single pyrenoid; cell walls exhibiting exfoliation; terminal setae 80–500 µm long, 2–4 µm diam. at broad, slightly bulbous base, tapering to very slender apex; fertile filaments 60–100 µm tall; sporangia (?) 9–10 µm diam., 15–20 µm long, terminal on fertile filaments; spores (gametes?) numerous, their structural features not observed.

Occasional, low intertidal on other algae, Santa Catalina I., Corona del Mar (Orange Co.), and La Jolla (type locality), Calif.

Bolbocoleon Pringsheim 1863

Thalli microscopic, creeping filaments, branched, usually epiphytic or endophytic. Cells irregularly shaped, bearing bulbous, unicellular branchlets each surmounted by long, nonseptate and colorless seta. Chloroplast in cells without seta parietal, fenestrate, with 5–10 pyrenoids; chloroplast in cells with seta an irregularly toothed plate, with 2 pyrenoids. Reproduction by numerous quadriflagellate zoospores formed in cells without seta.

Bolbocoleon piliferum Pringsh.

Pringsheim 1863: 2; Setchell & Gardner 1920b: 288; Hollenberg 1971b: 281.

Thalli making very dark green, minute patches; vegetative cells 12–16

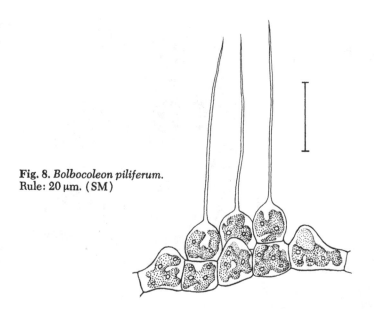

Fig. 8. *Bolbocoleon piliferum.*
Rule: 20 µm. (SM)

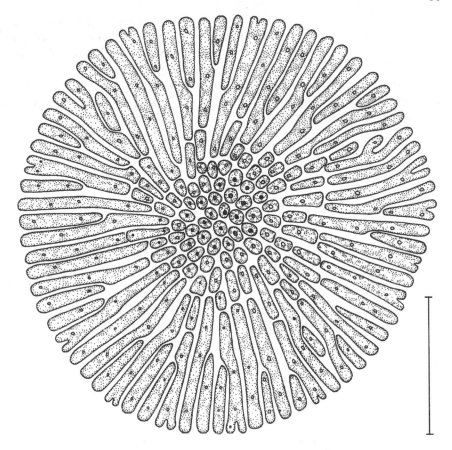

Fig. 9. *Ulvella setchellii*. Rule: 50 µm. (DBP/H)

µm diam., 2–3 times as long as diam., subcylindrical or somewhat conical.

Uncommonly reported, high to midtidal, on various algae, rarely on rocks, Br. Columbia to Santa Catalina I. and Laguna Beach, Calif. Widely distributed in N. Atlantic. Type locality: Helgoland, Germany.

Ulvella Crouan & Crouan 1859

Thalli epiphytic, endophytic, or saxicolous, minute, discoid, without rhizoids, primarily monostromatic. Cells in laterally adjoined dichotomously branched rows radiating from central region, this sometimes multistratose. Chloroplast single, laminate, with or without pyrenoids. Cells uninucleate (or multinucleate?). Zoospores with 2(4?) flagella, arising in surface cells of central region. Sexual reproduction uncertain.

Ulvella setchellii Dang.

Dangeard 1931: 318; 1969: 37; Smith 1944: 38. *Ulvella lens sensu* Setchell & Gardner 1920b: 295.

Thalli discoid, to 2 mm broad, the filaments commonly bifurcate, indicating manner of origin of branch; marginal cells 3–5 μm diam., 10–50 μm long; central region 2–4 cells thick at maturity, with rounded cells 5–8 μm diam.; cells uninucleate with several pyrenoids.

Epiphytic or endophytic in outer walls of various algae, especially *Amplisiphonia*, midtidal to subtidal, San Juan I., Wash., to Baja Calif.; probably common throughout Calif. Type locality: France.

Pseudulvella Wille 1909

Thalli mostly epiphytic or epizoic, discoid, basically filamentous, monostromatic at margin, the central portion several cells thick without rhizoids.

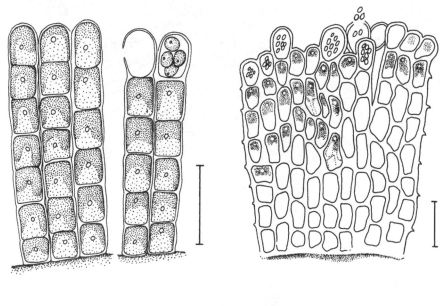

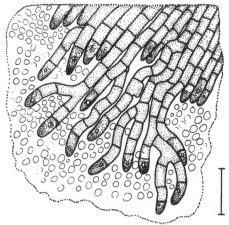

Fig. 10 (above left). *Pseudulvella applanata*. Rule: 20 μm. (DBP/H) Fig. 11 (above right). *P. consociata*. Rule: 20 μm. (SM, after Setchell & Gardner) Fig. 12 (below right). *P. prostrata*. Rule: 20 μm. (SM, after Setchell & Gardner)

Cells of monostromatic margin commonly in radial rows; all cells uninucleate. Chloroplast parietal, laminate to cup-shaped, with 1 pyrenoid. Zoospores quadriflagellate (at least in certain species), formed in surface cells of central region. Sexual reproduction unknown. Distinction from *Ulvella* somewhat doubtful.

1. Basal layer not evidently filamentous .... *P. applanata*
1. Basal layer of radiating filaments 2
    2. Radiating filaments distinct, loosely adjoined .... *P. prostrata*
    2. Radiating filaments relatively indistinct, closely adjoined...*P. consociata*

Pseudulvella applanata S. & G.

Setchell & Gardner 1920a: 295; 1920b: 298.

Thalli smooth and glossy, grass green; thin, discoid, pseudoparenchymatous, with marginal growth, to several mm broad, 45–55 μm thick; cells in erect rows, 6–7.5 μm diam., nearly isodiametric; chloroplast lining cell wall; sporangia(?) slightly modified surface cells.

Growing on shells of periwinkles (*Littorina planaxis* Nutt.), probably coextensive with the host, abundant on high rocks, Alaska to San Diego, Calif. Type locality: Carmel Bay, Calif.

Pseudulvella consociata S. & G.

Setchell & Gardner 1920a: 296; Doty 1947a: 7.

Thalli dark green, irregularly discoid, with closely adjoined radiating prostrate filaments evident only at margins; central region 100–140 μm thick, of erect filaments composed of cells 7–10 μm diam., 1–2 times as long, firmly adjoined, appearing parenchymatous; zoosporangia terminal, pyriform to spherical, producing 8 zoospores.

Growing on shells of various mollusks, midtidal; known only from South Bay, Ore., and Alameda, Calif. (type locality).

Pseudulvella prostrata (Gardn.) S. & G.

Ulvella prostrata Gardner 1909: 373. *Pseudulvella prostrata* (Gardn.) Setchell & Gardner 1920a: 295; 1920b: 297.

Thalli very dark green, discoid, epiphytic, firmly adherent, 2–3 mm broad, with 2 or 3 cell layers in center; monostromatic margins with free to loosely adjoined radiating filaments with cells 6–7 μm diam., 1.5–2.5 times as long, the cells of central region mostly quadrate; reproduction unknown.

On *Iridaea* sp., San Francisco, Calif. (type and only known locality).

Pseudopringsheimia Wille 1909

Thalli minute, pulvinate, nonpiliferous, epiphytic. Basal layer pseudoparenchymatous, monostromatic, with laterally adjoined radiating filaments and with short, rhizoidal branches penetrating host cells and bear-

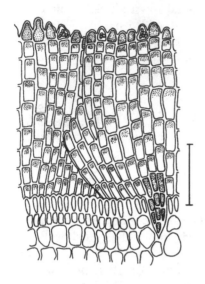

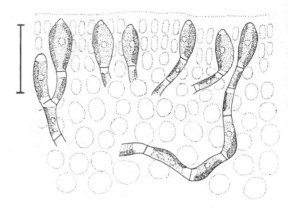

Fig. 13 (left). *Pseudopringsheimia apiculata.*
Rule: 50 μm. (SM, after Setchell & Gardner)
Fig. 14 (above). *Endophyton ramosum.*
Rule: 250 μm. (JRJ/S)

ing simple or sparsely branched erect filaments. Erect filaments laterally adjoined. Chloroplast single, at upper end of cell, with 1 pyrenoid. Zoosporangia(?) mostly terminal on erect filaments.

Pseudopringsheimia apiculata S. & G.

Setchell & Gardner 1920a: 297; Smith 1944: 37.

Thalli forming low to hemispherical cushions; erect filaments 145–160 μm tall, 8–12 μm diam., of 9–12 cylindrical to slightly inflated cells 0.5–2.5 times as long as diam.; terminal cells of erect filaments conical to apiculate; rhizoids in short conical clusters; zoospores quadriflagellate, ±8 per cell.

Frequent on *Egregia menziesii* or *Dictyoneurum californicum*, San Francisco (type locality) to Monterey Co., Calif.

Endophyton Gardner 1909

Thalli with filaments endophytic, irregularly branched, nonpiliferous, penetrating deeply perpendicular to host surface. Cells cylindrical, those terminal on anticlinal branches ovoid, with apiculate apex. Chloroplast single, laminate, parietal, with 1 pyrenoid. Sporangia enlarged, terminal-apiculate, intracortical. Zoospores biflagellate; gametes similar but distinctly smaller.

Endophyton ramosum Gardn.

Gardner 1909: 372; Setchell & Gardner 1920b: 292.

Thalli forming irregular dark-colored areas in lower part of host; cells

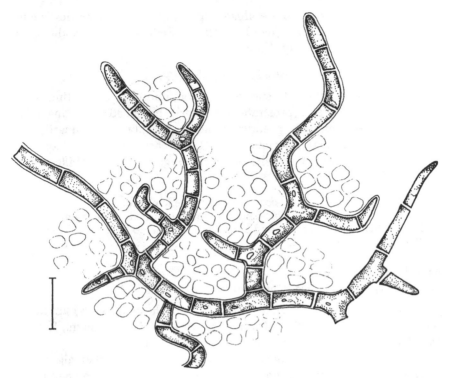

Fig. 15. *Pseudodictyon geniculatum*. Rule: 20 μm. (SM, after Setchell & Gardner)

3–6 μm diam., 2–6 times as long; sporangia about twice diam. of vegetative cells; zoospores 3–4 μm diam., with stigma.

Frequent endophyte in various red or brown algae, low intertidal, Strait of Juan de Fuca to Redondo Beach, Calif. Type locality: San Francisco, Calif.

Pseudodictyon Gardner 1909

Thalli endophytic, much branched, tortuous, with filaments creeping among outer cortical cells of host, not penetrating medullary tissue. Filaments branching at right angles in plant parallel to host surface, resulting in netlike appearance. Short erect branches of 2 or 3 cells, with enlarged terminal cells, projecting toward host surface. Chloroplast single, parietal, with 1 pyrenoid. Reproduction unkown.

Pseudodictyon geniculatum Gardn.

Gardner 1909: 374; Setchell & Gardner 1920b: 293.

Thalli forming greenish areas on host; cells of creeping filaments 3–6 μm diam., 2–3 times as long; terminal cells of erect branches 8–12 μm diam.

Occasional in various foliose algae, especially the Laminariales, sometimes in *Gigartina* spp., San Juan I., Wash., to Redondo Beach, Calif. Type locality: San Francisco Bay, Calif.

Entocladia Reinke 1879

Thalli epiphytic or mostly endophytic, nonpiliferous, penetrating outer walls of host. Filaments prostrate, irregularly branched, radiating from common center, sometimes pseudoparenchymatous, but not primarily discoid. Chloroplast single, parietal, with 1 or more pyrenoids. Zoospores quadriflagellate, formed in cells of central region. Gametes biflagellate, similarly formed.

1. Filaments free, not forming central pseudoparenchyma...........*E. viridis*
1. Filaments in center of thallus compact, forming pseudoparenchyma..... 2
   2. Free filaments numerous and long ....*E. codicola*
   2. Free filaments few, short ....*E. cingens*

Entocladia cingens S. & G.

Setchell & Gardner 1920a: 292; 1920b: 291.

Thalli creeping, free marginal filaments short, the cells 2–4(8) µm diam., 2–3 times as long; cells of central region crowded, 5–8 µm diam., forming pseudoparenchyma; reproduction not definitely known.

Within outer walls of *Chaetomorpha, Ceramium,* or other algae, and on chiton bristles, Monterey and Santa Monica to Ocean Beach (San Diego Co.; type locality), Calif.; probably common in Calif.

Entocladia codicola S. & G.

Setchell & Gardner 1920a: 293; 1920b: 290; Smith 1944: 35.

Thalli making irregular disks of numerous radiating and branched filaments with pseudoparenchymatous central region; cells of central region polygonal, those of marginal branches cylindrical, 3–4 µm diam., 2.5–6 times as long; chloroplast with or without 1 pyrenoid; reproduction unknown.

Within apical walls of utricles of *Codium fragile,* Monterey Co. to Laguna Beach, Calif.; probably common in C. and S. Calif. Type locality: Redondo Beach, Calif.

Entocladia viridis Reinke

Reinke 1879: 476; Setchell & Gardner 1920b: 289; Smith 1944: 35.

Thalli freely branched, without central pseudoparenchymatous region; cells irregular to cylindrical; cylindrical peripheral cells 3–8 µm diam., 2–6 times as long.

In outer walls of various algae; reported from S. Br. Columbia to Laguna Beach, Calif.; Galápagos Is.; probably common throughout Calif. Widely distributed. Type locality: Italy.

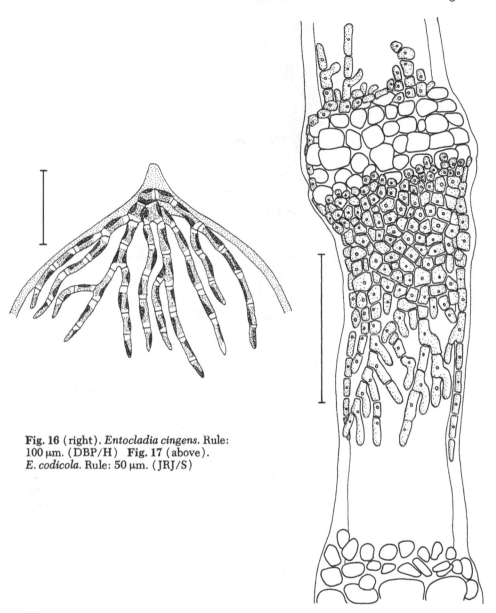

Fig. 16 (right). *Entocladia cingens*. Rule: 100 μm. (DBP/H) **Fig. 17** (above). *E. codicola*. Rule: 50 μm. (JRJ/S)

Family Monostromataceae

Thalli usually with discoid developmental stage, which is more or less microscopic and which by upheaval develops an erect tubular to saccate stage, this splitting to form the macroscopic monostromatic blades associated with species of this family. Cells uninucleate, the chloroplast single with 1 to several pyrenoids or without pyrenoids. Lower cells of blades

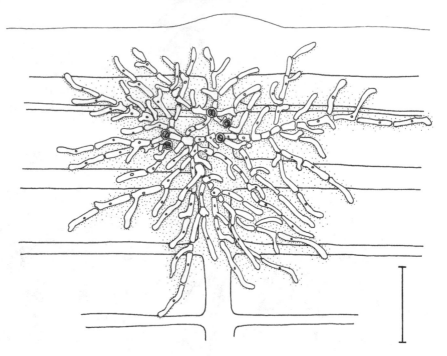

Fig. 18. *Entocladia viridis.* Rule: 50 μm. (DBP/H)

mostly without rhizoidal extensions. Reproduction variable; life histories variable. Blades producing quadriflagellate zoospores in some species, biflagellate gametes in other species; zoospores and zygotes sometimes producing discoid stage. Sporangial and gametangial stages similar in at least 1 species; sporangial stage cystlike in certain species.

Bliding (1968) and certain other phycologists recognize several genera in this family. Hirose and Yoshida (1964, fig. 3) illustrated ten types of life histories in *Monostroma*. For the present, however, it seems best to recognize only the genus *Monostroma*.

Monostroma Thuret 1854

Characters of the family. Traditionally, flat blades that are monostromatic and whose life history shows dissimilar generations are placed in *Monostroma*, in contrast to species of *Ulva*, which are distromatic and with similar generations. *"Collinsiella tuberculata"* and *"Gomontia polyrhiza,"* either of which may be a stage in the life history of a species of *Monostroma*, are described on pp. 119, 120.

1. Thalli epiphytic on *Phyllospadix* or *Zostera*; blades producing quadriflagellate zoospores ....*M. zostericola*
1. Thalli not epiphytic on *Phyllospadix* or *Zostera*; blades producing biflagellate zoospores or biflagellate gametes......................... 2

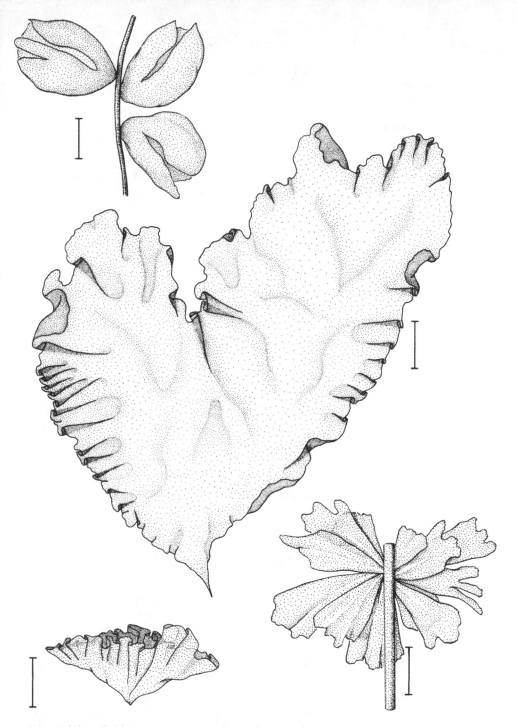

Fig. 19 (above left). *Monostroma grevillei*. Rule: 1 cm. (DBP/H) **Fig. 20** (center & below left). *M. oxyspermum*. Rule: 1 cm. (CS) **Fig. 21** (below right). *M. zostericola*. Rule: 1 cm. (DBP/H)

2. Blades producing biflagellate gametes and cystlike zygotes...*M. grevillei*
2. Blades producing biflagellate zoospores; sexual reproduction reported
to be lacking ....*M. oxyspermum*

Monostroma grevillei (Thur.) Wittr.

Enteromorpha grevillei Thuret 1854: 25. *Monostroma grevillei* (Thur.) Wittrock 1866: 57; Setchell & Gardner 1920b: 236; Bliding 1968: 602.

Macroscopic thalli with globular saccate stage to 2–3 cm diam., larger bladelike stage 5–12 cm tall, these yellowish-green, soft and lubricous, 21–30 μm thick; cells in groups or rows, rectangular to polygonal, ±16 μm diam., to 240 μm long (to 100 μm long near base); chloroplast single, parietal, with 1–3 pyrenoids; thalli dioecious and anisogamous; gametes biflagellate; zygote cystlike, free-living or producing a *"Codiolum"*-like stage in shells, releasing quadriflagellate zoospores after dormant period.

Infrequent, epiphytic or saxicolous, low intertidal, Alaska to Santa Cruz and Monterey Co., Calif. Widely distributed in cold waters of N. Hemisphere. Type locality: France.

Monostroma oxyspermum (Kütz.) Doty

Ulva oxyspermum Kützing 1843: 296. *Monostroma oxyspermum* (Kütz.) Doty 1947a: 12 (incl. synonymy); Hollenberg & Abbott 1966: 11. *Ulvaria oxyspermum* (Kütz.) Bliding 1968: 585.

Macroscopic thallus saccate, globular, 1–2 cm diam., the membrane 15–20 μm thick, the thallus later membranous, free-floating, soft and delicate, irregularly lobed and folded, often radially plicate, with undulate margin, pale green; cells quadrate to round in surface view, ultimately relatively remote, sometimes in groups of 2–6 commonly slightly elongate in section; chloroplast single, parietal, with 1 pyrenoid; zoospores biflagellate; sexual reproduction reported to be lacking.

Epiphytic or saxicolous, or free-floating, mostly in quiet brackish or fresh water, S. Br. Columbia and Ore. to S. Calif.; common in northern part of range. Also from N. Europe and Japan. Type locality: Germany.

Monostroma zostericola Tild.

Tilden 1894–1909 (1900): no. 388; Setchell & Gardner 1920b: 238; Smith 1944: 43; Tatewaki 1969: 26. *Kornmannia zostericola* (Tild.) Bliding 1968: 620.

Thalli saccate or bladelike, the saccate stage mostly less than 3 mm diam., soon forming cluster of medium green cuneate or fan-shaped blades mostly 1.5–2.5 cm tall (to 4 cm tall in northern part of range); membrane 7–10(14) μm thick; cells rounded-angular, small, mostly 4–6 μm diam. in surface view and oblong in anticlinal section, sometimes in more or less distinct rows; chloroplast single, parietal, apparently without pyrenoid; zoospores quadriflagellate, giving rise to discoid stage, this producing biflagellate gametes; zygote also forming discoid stage, this becoming saccate and splitting to form blades, these producing zoospores; gametes under certain conditions may parthenogenetically repeat discoid stage.

Spring annual common in certain localities on *Zostera* and *Phyllospadix*, January to June in Calif.; S. Br. Columbia to San Luis Obispo Co., Calif. Also N. Japan. Type locality: San Juan I., Wash.

Family ULVACEAE

Thalli multicellular, nonfilamentous, mostly tubular or membranous, usually attached. Cells mostly uninucleate. Chloroplast single, laminate or cup-shaped, mostly with 1 or more pyrenoids. Sporangial and gametangial stages similar; or thalli with only sporangial or only gametangial stages during life history. Zoospores quadriflagellate.

1. Thalli wholly or partly tubular 2
1. Thalli normally neither wholly nor partly tubular.................... 3
    2. Tubular parts arising from prostrate discoid base; lacking rhizoidal processes .... *Blidingia* (p. 70)
    2. Tubular parts not arising from prostrate base; with rhizoidal processes from lower cells......................*Enteromorpha* (p. 73)
3. Thalli slender, unbranched, mostly of 2 longitudinal rows of symmetrically placed cells, the cells lacking rhizoidal processes...*Percursaria* (below)
3. Thalli bladelike, distromatic, many cells broad, with rhizoidal processes from lower cells ....*Ulva* (p. 77)

Percursaria Bory 1828

Thalli slender, unbranched, mostly of 2 longitudinal rows of cells symmetrically placed throughout, arising from discoid base without distinct rhizoidal cells. Chloroplast single, parietal, with or without pyrenoids. Sporangial and gametangial thalli morphologically similar. Reproduction by zoospores and anisogametes.

Percursaria dawsonii Hollenb. & Abb.

Hollenberg & Abbott 1968: 1235.

Thalli with dark green, densely congested erect fronds, arising from monostromatic discoid base; fronds mostly simple, to 3 mm tall, at first with single row of cells, but soon 2 cells wide and 1 cell thick, with cells of the 2 rows in staggered arrangement; cells 5–6 μm long, somewhat narrower than long; upper parts sometimes with 4+ cells in cross section, rarely tubular above, with mostly isodiametric cells; chloroplast parietal, without pyrenoid; erect parts without rhizoidal filaments; reproduction by zoospores (gametes?).

Occasional to locally frequent, forming light-green velvety stratum on midtidal limpets and chitons, San Juan I., Wash., and Pacific Grove, Calif. (type locality).

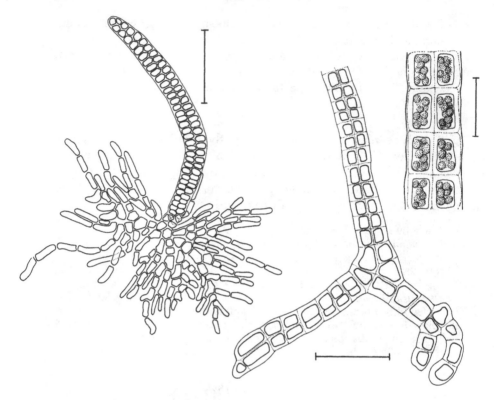

Fig. 22 (left). *Percursaria dawsonii*. Rule: 50 μm. (DBP/H)   **Fig. 23** (two at right). *P. percursa*. Rules: habits, 50 μm; detail, 20 μm. (both DBP/H)

Percursaria percursa (C. Ag.) Rosenv.

Conferva percursa C. Agardh 1817: 87. *Percursaria percursa* (C. Ag.) Rosenvinge 1893: 963; Setchell & Gardner 1920b: 274; Kornmann 1956: 259.

Thalli usually unbranched, to 3+ cm long, flexuous and contorted, pale green, with 2 cell rows placed symmetrically side by side throughout, or sometimes with 1 cell row in part; cells 10–12(15) μm diam., mostly 1–2 times as long; chloroplast single, laminate, with 2–3+ pyrenoids; life stages isomorphic; zoospores 4–5 μm diam., quadriflagellate; gametes reportedly biflagellate, slightly anisogamous.

Along with other algae frequently forming floating entangled masses, proliferating vegetatively, in upper tidepools and salt marshes, Alaska to San Francisco Bay, Calif.; in Calif., known only from San Francisco Bay. Also N. Atlantic and Japan. Type locality: Denmark.

Blidingia Kylin 1947
(Contributed by James N. Norris)

Thalli tubular, gregarious, branched or unbranched, with several erect tubes arising from polystromatic, parenchymatous cushions by upheaval

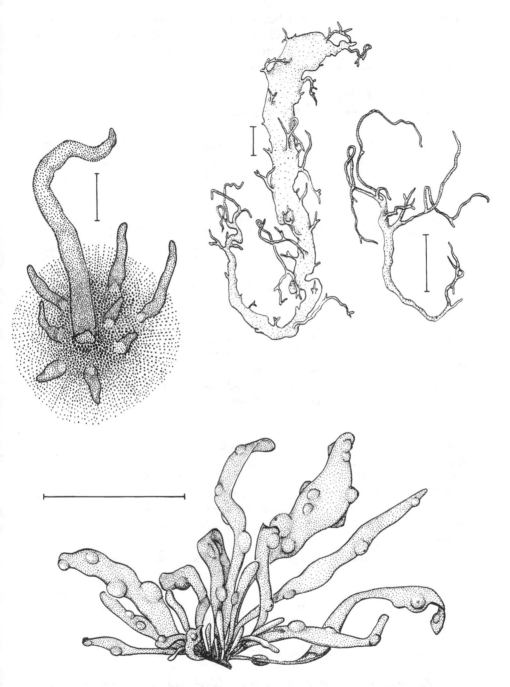

Fig. 24 (above left). *Blidingia minima* var. *minima*. Rule: 1 cm. (DBP) **Fig. 25** (two above right). *B. minima* var. *subsalsa*. Rules: both 1 cm. (both DBP) **Fig. 26** (below). *B. minima* var. *vexata*. Rule: 1 cm. (DBP)

of cells of the cushion, forming hollow areas continuous with center of the tubes, without basal rhizoids. Cells angular, irregularly arranged, with single parietal chloroplast containing 1 pyrenoid. Reproduction by quadriflagellate zoospores lacking an eyespot; sexual reproduction unknown.

Blidingia minima (Kütz.) Kyl.

Enteromorpha minima Kützing 1849: 482. *Blidingia minima* (Kütz.) Kylin 1947b: 8; J. Norris 1971: 146 (incl. synonymy).

Thalli tubular, simple or branched, cylindrical to flattened or compressed, 1 to several thalli arising from common discoid cushion. Cells irregularly arranged, angular, sometimes with rounded corners, under 10 μm diam., with single parietal chloroplast and 1 centrally located pyrenoid.

1. Branches many, often of several orders.....................var. *subsalsa*
1. Branches lacking, though occasionally a few short proliferations near base 2
   2. Surface cells dark olive green, with black, wartlike perithecia of fungus infestation ....var. *vexata*
   2. Surface cells bright green, lacking fungus infestation.......var. *minima*

Blidingia minima var. minima

Thalli green, gregarious, unbranched, or occasionally with short proliferations at base, (1)5–50(120) mm tall, 0.5–3(5) mm wide, broadening distally and narrowing toward a common base of nonrhizoidal cells.

On rocks and old wood, high to midtidal, Dutch Harbor, Alaska, to Mexico, and in Chile; common in Calif. Type locality: Helgoland, Germany.

Blidingia minima var. subsalsa (Kjellm.) Scag.

Enteromorpha micrococca f. *subsalsa* Kjellman 1883: 292. *Blidingia minima* var. *subsalsa* (Kjellm.) Scagel 1957: 37; J. Norris 1971: 146 (incl. synonymy).

Thalli light green, often entangled, usually convoluted and distorted, with few to numerous branches and proliferations, 1–18 cm long, branches with basal portion broad, 0.1–2 cm broad when intact and to 4 cm broad when torn, tapering distally, often to a point.

On rocks or mudflats, epiphytic on other aquatic vegetation, or free-floating, commonly found in estuaries, river mouths, or other protected brackish habitats, Skagway, Alaska, to Ventura Co., Calif. Frequent in Calif. Probable type locality: Arctic Ocean.

Blidingia minima var. vexata (S. & G.) J. Norr.

Ulva vexata Setchell & Gardner 1920a: 282. *Blidingia minima* var. *vexata* (S. & G.) J. Norris 1971: 148 (incl. synonymy).

Thalli deep green, dark olive, or nearly black, tubular, flattened to compressed, linear to spatulate, usually simple, occasionally branched, 1–3 cm long, the surface corrugated or distorted, roughened by few to numerous black or brown wartlike perithecia; thickness of fronds dependent on amount of fungal infestation.

On rocks in splash to high intertidal, Pine Is., Br. Columbia, to Monterey, Co., Calif.; occasional in Calif. Type locality: Fort Point (San Francisco), Calif.

Enteromorpha Link 1820
(Contributed by James N. Norris)

Thalli hollow and tubular, the wall 1 cell thick, the thallus either cylindrical throughout, or basally hollow with hollow portion often extending beyond base, the terminal portions sometimes expanded and bladelike or sometimes flattened or compressed with only the margins hollow; thalli either branching, with proliferations, or not branched, basally attached by system of nonseptate filaments, frequently unattached and floating. Cells uninucleate, embedded in homogeneous mucilaginous matrix. Chloroplast single, laminate or cup-shaped, with 1 to several pyrenoids. Reproduction by fragmentation, the zoospores motile, quadriflagellate; sexual reproduction by terminally biflagellate gametes, these equal or unequal in size and function. In some species only zoospores produced.

"*Collinsiella tuberculata*," which may be a stage in the life history of one or more species of *Enteromorpha*, is described on p. 119.

1. Thallus unbranched, tubular near base and broadening distally; with 2 cell layers completely joined or with hollow margins..............*E. linza*
1. Thallus simple or branched, either tubular throughout, cylindrical throughout or in part, compressed throughout, or only distally flattened; cell layers never completely joined 2
  2. Cells in surface view not arranged in longitudinal rows in any part of plant 3
  2. Cells in surface view in longitudinal rows, at least in younger or narrower parts 4
3. Thallus usually not branched, cylindrical throughout or flattened distally; in section, outer cell walls thin ....*E. intestinalis*
3. Thallus usually branched, with branches constricted at origin, the cylindrical basal portion becoming compressed and usually broadening distally; outer cell walls thick ....*E. compressa*
  4. Thallus profusely and repeatedly branched...............*E. clathrata*
  4. Thallus unbranched, or forming main axis with lateral branches...... 5
5. Thallus branched, rarely unbranched; cells in surface view not in transverse rows ....*E. prolifera*
5. Thallus unbranched, rarely branched; cells in surface view often in transverse rows (though not always distinct).................*E. flexuosa*

Enteromorpha clathrata (Roth) Grev.

Conferva clathrata Roth 1806: 175. *Enteromorpha clathrata* (Roth) Greville 1830: 181; Doty 1947a: 16 (incl. synonymy); Bliding 1963: 107; Scagel 1966: 46.

Thalli 0.5 to 5 mm diam., filiform, cylindrical or compressed, to 40 cm long, profusely and repeatedly branched, with numerous radially arranged proliferations; thallus usually attached, becoming free-floating; cells in surface view in distinct longitudinal rows in younger, narrower portions of thallus and more or less in distinct order or not in order in older, broader portions; cells angular, often rounded, sometimes quadrangular or rectangular, 8–17(25) μm wide, (8)17–35(50) μm long; chloroplast single, small, with 1 to several pyrenoids; transection 18–30 μm thick; cells 8–25 μm wide, 14–25 μm long.

Enteromorpha clathrata var. clathrata

Branch endings multiseriate (rarely uniseriate); chloroplasts small, not filling cell and close to outer wall.

On rocks or mudflats, or epiphytic on other algae, or free-floating, high to midtidal, Valdez, Alaska, to Mexico; frequent in Calif. Type locality: Baltic Sea.

Although occasionally found in unprotected coastal areas, this variety is most commonly found in protected brackish habitats.

Enteromorpha clathrata var. crinita (Roth) Hauck

Conferva crinita Roth 1797: 162. *Enteromorpha clathrata* var. *crinita* (Roth) Hauck 1885: 429. *E. crinita* J. Agardh 1883: 145; Scagel 1966: 48 (incl. synonymy).

Branch endings uniseriate (rarely multiseriate); chloroplasts nearly filling cell.

On rocks or mudflats, or epiphytic on other aquatic vegetation, or free-floating, in sloughs, estuaries, or unprotected bays, high to midtidal, Alaska to Mexico and in Galápagos Is.; frequent in Calif. Type locality: E. Germany.

Of the free-floating species and varieties of *Enteromorpha*, *E. clathrata* var. *crinita* is the most common in estuaries.

Enteromorpha compressa (L.) Grev.

Ulva compressa Linnaeus 1755: 433. *Enteromorpha compressa* (L.) Greville 1830: 180; Doty 1947a: 14; Bliding 1963: 132; Scagel 1966: 47 (incl. synonymy).

Thalli tubular, the upper portion of main axis compressed, tapering to base, to 40 cm tall; thallus usually branched, rarely unbranched; branches compressed, narrowed at base and broadening distally, often with hollow margins; cells in surface view irregularly arranged, polygonal, usually with rounded corners, (5)8.5–12(18) μm diam.; chloroplast laminate, with 1 pyrenoid. Transection 20–40 μm thick in basal portion, with cells 5–12 μm wide and 15–30 μm long; thallus 25–50 μm thick in distromatic compressed portion of upper thallus, the cells 5–12 μm wide and 23.5 μm long; protoplasts usually centrally located.

On rocks or rarely epiphytic on other algae, high to low intertidal, in semiprotected or occasionally unprotected coastal areas or estuaries, Ber-

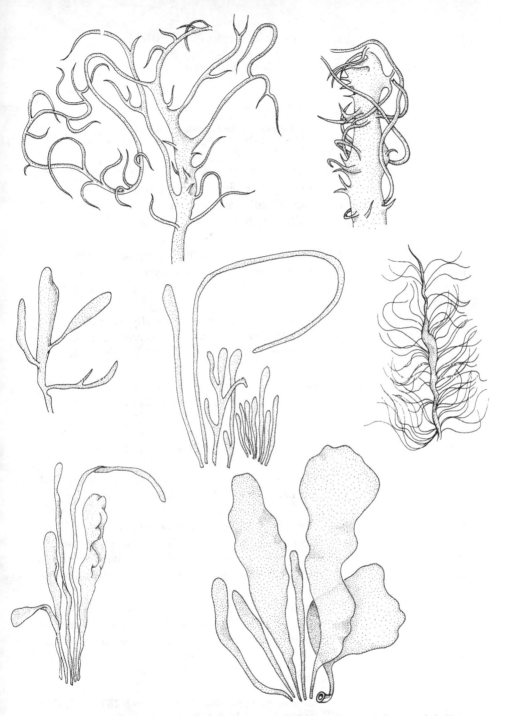

Fig. 27 (above left). *Enteromorpha clathrata* var. *clathrata*.   Fig. 28 (above right). *E. clathrata* var. *crinita*.   Fig. 29 (middle left). *E. compressa*.   Fig. 30 (center). *E. flexuosa:*   Fig. 31 (below left). *E. intestinalis*.   Fig. 32 (below right). *E. linza*.   Fig. 33 (middle right). *E. prolifera*. (all SM, diagrammatic)

ing Sea to Costa Rica, and in Chile; infrequent in Calif. Type locality: Sweden.

Enteromorpha flexuosa (Roth) J. Ag.

Conferva flexuosa Roth 1800: 188. *Enteromorpha flexuosa* (Roth) J. Agardh 1883: 126; Bliding 1963: 73. *E. tubulosa* Kützing 1856: 11; Scagel 1966: 56 (incl. synonymy). *E. prolifera* var. *flexuosa* (Roth) Doty 1947a: 15.

Thalli usually unbranched, 4–8(30) cm tall, occasionally sparsely branched, rarely with secondary branches, without proliferations, cylindrical throughout, sometimes slightly compressed; cells in surface view in longitudinal series and often in transverse series in narrow portions, usually not in distinct order in broader portions, angular, often squarish to elongate, (8.5)12–15(20) μm wide and 12–26(34) μm long; chloroplast parietal, almost filling cell and with 1 or 2(6) pyrenoids; transection (15)20–25(34) μm thick, with cells 8.5–20 μm wide and (10)12–20(30) μm long.

On rocks or epiphytic on other algae or occasionally on mudflats or free-floating, in estuaries, high to low intertidal, Vancouver I., Br. Columbia, to C. America and Galápagos Is.; common in Calif. Type locality: Yugoslavia, near Trieste.

As described from Europe, thalli of this species are often found to be branching; branching forms have not been observed in Calif. specimens.

Enteromorpha intestinalis (L.) Link

Ulva intestinalis Linnaeus 1753: 1163. *Enteromorpha intestinalis* (L.) Link 1820: 5; Doty 1947a: 14; Bliding 1963: 139; Scagel 1966: 49 (incl. synonymy).

Thalli unbranched, 4–20(200) cm long, tubular, cylindrical throughout, or broadening distally, often becoming slightly compressed and expanded above, or occasionally with a few small basal proliferations; cells in surface view irregularly arranged, angular, usually with rounded corners (8)10–18 μm diam.; chloroplast almost filling cell and with 1(2 or 3) pyrenoids; transection 20–55 μm thick, thicker near base; cells near base vertically elongated, (8)12–18 μm wide, 12–30 μm long; protoplast located closer to outer wall in lower portion of thallus.

On rocks, epiphytic on other algae or often free-floating, high to midtidal, in protected areas of bays or estuaries, Kukak Bay, Alaska, to Mexico, and in Chile; common throughout Calif. Type locality: probably N. Europe.

Enteromorpha linza (L.) J. Ag.

Ulva linza Linnaeus 1753: 1163. *Enteromorpha linza* (L.) J. Agardh 1883: 134; Doty 1947a: 18; Bliding 1963: 127; Scagel 1966: 52 (incl. synonymy).

Thalli unbranched, silky, often gregarious, to 50(175) cm tall, 1–10(45) cm wide; distromatic flattened upper portion with hollow monostromatic margins, tapering to tubular hollow monostromatic cylindrical or compressed stipitate basal portion; cells in surface view above stipe portion angular, not ordered, 10–20 μm diam.; cells of stipe region in longitudinal rows, usually elongate, (8.5)12–25(34) μm wide, 17–44 μm long; chloroplast laminate to cup-shaped, with 1(2 or 3) pyrenoids; transection of distromatic upper portion of thallus (25)35–60(80) μm thick, with cells 8.5–25.5 μm wide and 15–35 μm long; monostromatic region of stipe to 45 μm thick, with cells 8.5–25.5 μm wide and to 40 μm long.

On rocks or rarely epiphytic on other algae, midtidal to low intertidal in bays, estuaries, or other semiprotected localities, Orce, Alaska, to Mexico, and Chile; common in Calif. Type locality: probably N. Europe.

Enteromorpha prolifera (Müll.) J. Ag.

Ulva prolifera Müller 1778: 7. *Enteromorpha prolifera* (Müll.) J. Agardh 1883: 129; Doty 1947a: 14; Bliding 1963: 45; Scagel 1966: 54 (incl. synonymy).

Thalli richly branched, rarely unbranched, varying from long slender filaments to broad, expanded, sometimes convoluted masses; distinct main axis to several meters long, with few to numerous proliferations and lateral branches; main axis tubular, cylindrical to flattened, the upper portion sometimes compressed, distromatic with hollow monostromatic margins, tapering to narrow hollow monostromatic stipe; branches and proliferations tubular; cells in surface view in longitudinal series in younger, narrower portions, more or less in distinct order or not in order in older, broader portions; cells angular, elongated near base, polygonal to squarish above, (6)10–12(18) μm wide, (10)12–18(34) μm long; chloroplast single, usually filling cell, with 1(2–4) pyrenoids; transection 14–22 μm thick; protoplast centrally located, 8.5–15 μm wide, to 20 μm long.

On rocks or free-floating in bays, estuaries, or similar sheltered habitats, high to midtidal, Golovnin Bay, Alaska, to C. Mexico, and in Peru; common in Monterey Co., but less frequent north and south in Calif. Type locality: Germany.

Ulva Linnaeus 1753

Thalli membranous blades, broadly expanded, distromatic, mostly without hollow margins. Blades annual, mostly without stipe; rhizoidal processes from multinucleate lower cells extending downward between blade margins forming usually perennial holdfast. Cells of blade mostly uninucleate. Chloroplast single, laminate or cup-shaped, usually on outer face of cell, with 1 to several pyrenoids. Sporangial and gametangial thalli usually morphologically similar; fertile areas marginal or terminal; zoospores quadriflagellate. Gametes biflagellate, isogamous or anisogamous. Zygote germinating without dormant period.

"*Gomontia polyrhiza*," which may be a stage in the life history of species of *Ulva*, is described on p. 120.

1. Blades mostly less than 2 cm tall, usually in dense groups.....*U. californica*
1. Blades mostly over 6 cm tall at maturity; usually not in dense groups.... 2
   2. Cells in middle of blade quadrate or periclinally elongate.....*U. lactuca*
   2. Cells in middle of blade anticlinally elongate 3
3. Blades digitate, with several divisions tapering from basal portion.....
   .... *U. dactylifera*
3. Blades not digitate; if divided, with portions narrowed toward base...... 4
   4. Blades or blade divisions 6 or more times as long as broad.......... 5
   4. Blades orbicular or less than 3 times as long as broad............. 8
5. Blades with much-thickened median strip visibly resembling midrib...
   .... *U. costata*
5. Blades lacking median strip 6
   6. Blades usually deeply divided ....*U. taeniata*
   6. Blades simple, with deeply ruffled margins 7
7. Blades lanceolate to elliptical, 50–100 mm broad, 60–110 µm thick;
   cells lacking pyrenoid ....*U. stenophylla*
7. Blades linear to oblanceolate, mostly less than 50 mm broad, 35–45 µm
   thick; cells with 1 pyrenoid ....*U. angusta*
   8. Blades relatively stiff, usually deeply divided, plane to slightly ruffled,
      dark green ....*U. rigida*
   8. Blades neither stiff nor deeply divided, often much ruffled.......... 9
9. Blades entire, orbicular or irregularly expanded, commonly free-floating
   in quiet water ....*U. expansa*
9. Blades usually deeply lobed, broadly obovate, usually attached intertidally ....*U. lobata*

Ulva angusta S. & G.

Setchell & Gardner 1920a: 283; Smith 1944: 45. *Enteromorpha angusta* (S. & G.) Doty 1947a: 20.

Blades pale grass-green, mostly simple, linear to oblanceolate, often spirally twisted, to 1 m tall, to 5 cm broad, mostly 35–45 µm thick; margins
usually strongly ruffled; cells in anticlinal section mostly subquadrate, 5–12
µm diam., to 5–18 µm long; chloroplast laminate, covering outer wall, with
1 pyrenoid; gametes anisogamous.

Saxicolous or epiphytic, intertidal and probably subtidal, Ore. to Ventura Co., Calif., and to Baja Calif. Type locality: San Francisco, Calif.

Ulva californica Wille

Wille, P.B.-A., 1895–1919 [1899]: no. 611; Setchell & Gardner 1920b: 264; Doty 1947a: 11.

Blades mostly 1.5–2 cm tall; cuneate to ovate or reniform, commonly
much crisped, with broad attachment or slender stipe; membrane 30–35(60)
µm thick, commonly much thicker near base; cells in surface view irregularly polygonal with rounded corners and irregular arrangement or in distinct rows; (5)10–15 µm diam., 1–1.5 times as long as diam. anticlinally;
chloroplast with 1 or 2(4) pyrenoids.

Usually in dense, turflike stands atop rocks, occasionally epiphytic, midtidal to upper intertidal, Ore. to I. Magdalena, Baja Calif.; formerly com-

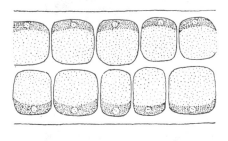

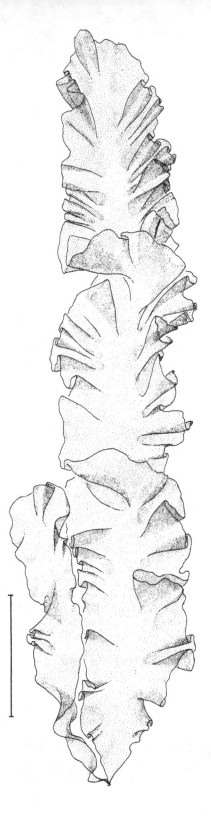

Fig. 34. *Ulva angusta.* Rules: above &
below, 50 μm; right, 5 cm. (all JRJ/S)

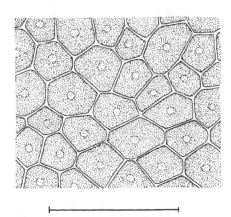

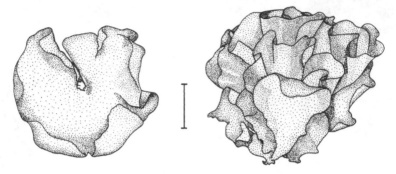

Fig. 35. *Ulva californica.* Rule: 1 cm. (DBP)

mon in S. Calif., seemingly rare or absent in C. and N. Calif. Type locality: La Jolla, Calif.

Ulva costata (Howe) Hollenb.

Ulva fasciata f. *costata* Howe 1914: 20. *U. costata* (Howe) Hollenberg 1971b: 283.

Blades of Calif. specimens to 32 cm tall, simple or with 1 or 2 basal branches, with main divisions linear, to 2 cm broad; blades with strongly ruffled margins, sometimes slightly spirally twisted, with distinct midrib, this to 160 μm thick; cells of midrib 15–20 μm diam., 3–4 times as long anticlinally; lateral parts of blade 50–60 μm thick, with cells 12–16 μm diam., and only slightly longer anticlinally; basal portion subterete, thickened; fertile area a narrow marginal band; reproductive cells not observed.

Infrequent, on rocks or on jointed coralline algae, low intertidal to subtidal, Santa Barbara to Redondo Beach, Calif., and probably Baja Calif. Type locality: Peru.

Ulva dactylifera S. & G.

Setchell & Gardner 1920a: 285; 1920b: 272.

Blades sessile or with very short stipe, to 18 cm tall, with mostly 5 or 6 lanceolate divisions from an orbicular to reniform basal portion; divisions much crisped, 5–15 mm broad; membrane of basal portion 50 μm thick at margin, to 100 μm thick in middle, with cells in surface view 16–20 μm diam., quadrate to twice as long in anticlinal section; blades of divisions to 190 μm thick in middle, with cells 12–16 μm diam. in surface view, anticlinally quadrate in marginal portions to 5 times as long as diam. in central parts. Reproductive parts and number of pyrenoids unknown.

Infrequent, on exposed rocks, upper intertidal, near Redondo Beach, Calif., to Baja Calif. and Gulf of Calif. Type locality not specified.

Ulva expansa (Setch.) S. & G.

Ulva fasciata f. *expansa* Setchell, P.B.-A., 1895–1919 [1905]: no. 77. *U. expansa* (Setch.) Setchell & Gardner 1920a: 284; 1920b: 268; Smith 1944: 46.

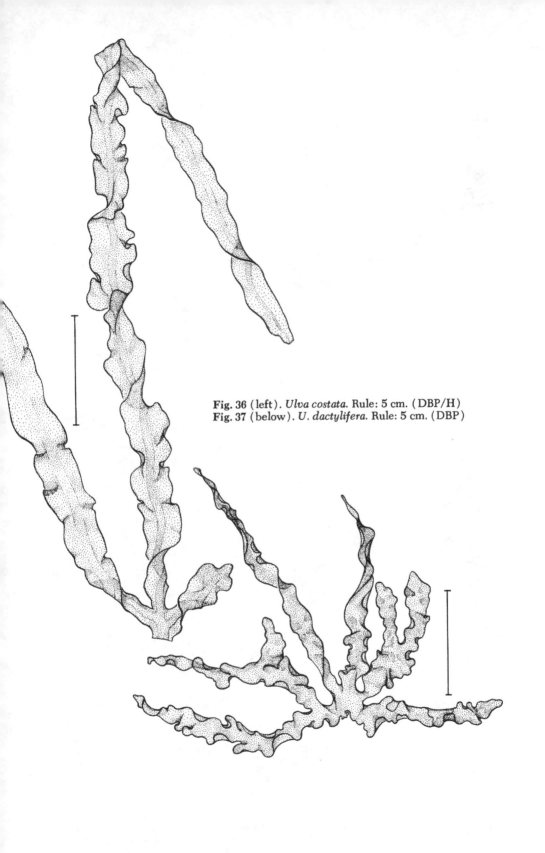

Fig. 36 (left). *Ulva costata*. Rule: 5 cm. (DBP/H)
Fig. 37 (below). *U. dactylifera*. Rule: 5 cm. (DBP)

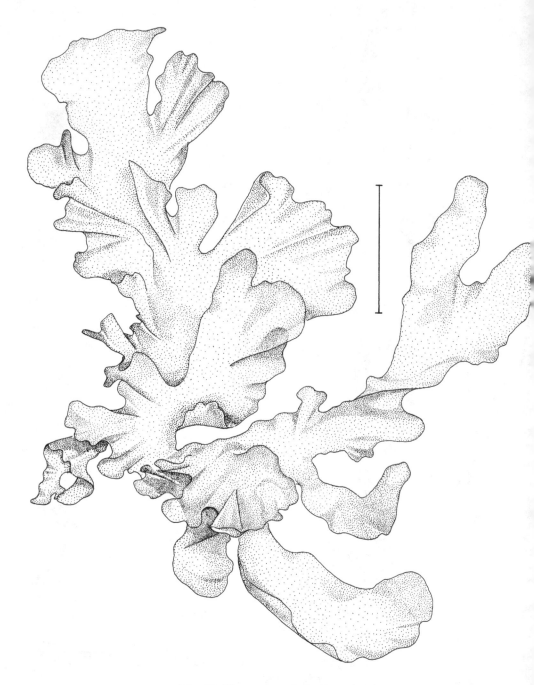

Fig. 38. *Ulva expansa.* Rule: 5 cm. (SM)

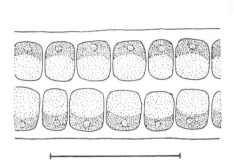

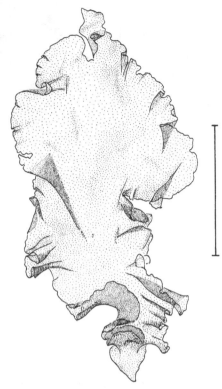

Fig. 39. *Ulva lactuca*. Rules: above, 50 μm; right, 5 cm. (both JRJ/S)

Blades medium to pale green, elongate or orbicular, sometimes expanded, to 1(3) m long, 20–30(75) cm broad, usually not lobed but frequently with deeply ruffled margins; cells of marginal parts 14–17 μm diam., quadrate in anticlinal section; cells of central and basal portion 13–18 μm diam., 25–30 μm long anticlinally; chloroplast usually on lateral walls, with 1–3 pyrenoids.

Weakly attached to rocks or epiphytic, lower intertidal to subtidal, in sheltered water common and usually free-floating, S. Br. Columbia to Baja Calif. Type locality: Monterey, Calif.

Ulva lactuca L.

Linnaeus 1753: 1163; Setchell & Gardner 1920b: 265; Smith 1944: 45; Papenfuss 1960: 304.

Blades light to medium green, with discoid attachment, soft and lubricous, sometimes deeply incised, the margins plane or ruffled, to 18 cm tall, sometimes broader than long, 35–40 μm thick at margins, 45–90 μm in middle; cells irregularly arranged, 10–20 μm diam., quadrate to slightly elongate anticlinally; chloroplast cup-shaped, filling outer third of cell, with 1–3 pyrenoids; gametes conspicuously anisogamous; thalli probably monoecious.

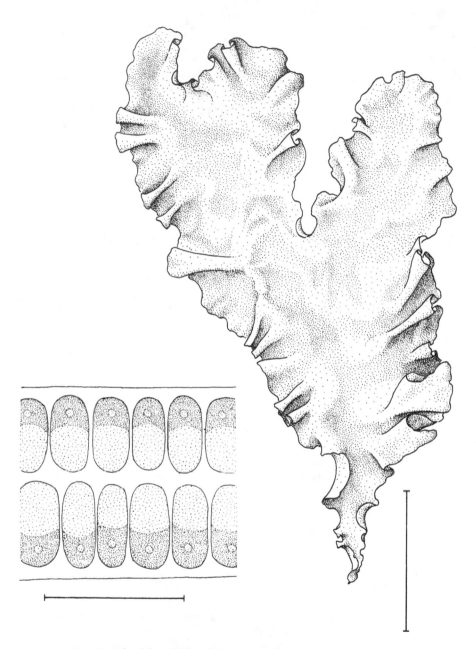

Fig. 40. *Ulva lobata.* Rules: left, 50 μm; right, 5 cm. (both JRJ/S)

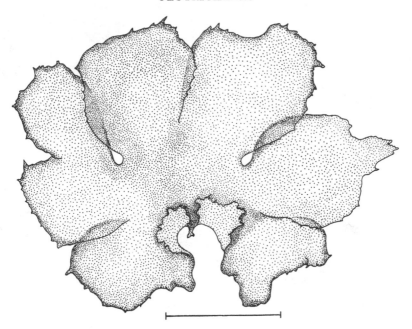

Fig. 41. *Ulva rigida.* Rule: 5 cm. (DBP)

Occasional on rocks or other algae, upper intertidal to subtidal, sometimes floating on mudflats in lagoons, Bering Sea to Chile; in Calif., known from Humboldt Co. to San Diego Co. Widely distributed. Type locality: N. Europe.

Field experience on the Monterey Peninsula has demonstrated that this species is far less abundant than herbarium material would indicate, and that *U. lactuca* is probably misidentified more frequently than any other species of *Ulva*. Size of cells in section, and shape of cells, are far more critical for the identification of this species than for that of other species in the flora. *Ulva expansa* and *U. lobata* are frequently identified as *U. lactuca*, but have cells often twice the thickness of those of this species, and are more vertically elongate. Both *U. expansa* and *U. lobata* are far more abundant than *U. lactuca*.

Ulva lobata (Kütz.) S. & G.

Phycoseris lobata Kützing 1849: 477. *Ulva lobata* (Kütz.) Setchell & Gardner 1920a: 284; 1920b: 268; Smith 1944: 46.

Blades rich grass-green, broadly obovate, often deeply lobed or divided, slightly to deeply ruffled, with smooth or undulate margins, 10–30 cm tall, to 15 cm broad, 40–50 µm thick at margins to 90 µm in middle, the blade base gradually narrowed and sometimes stipelike; cells irregularly arranged, in anticlinal section 10–12 µm diam., 28–30 µm long, the chloro-

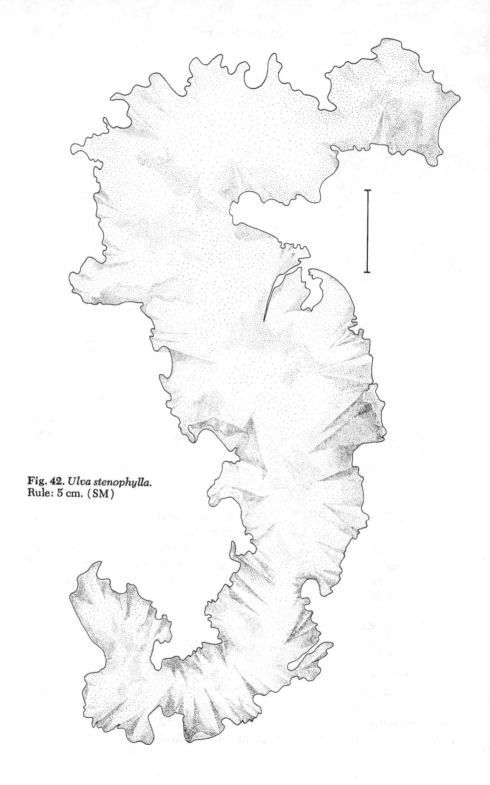

Fig. 42. *Ulva stenophylla.*
Rule: 5 cm. (SM)

plast filling outer third of cell, usually with 1 pyrenoid; gametes anisogamous.

On rocks or occasionally epiphytic, midtidal to subtidal, Ore. to Guerrero, Mexico, and to Ecuador and Chile (type locality); common intertidally in C. Calif.

Ulva rigida C. Ag.

C. Agardh 1822a: 410; Setchell & Gardner 1920b: 269; Smith 1944: 47; Papenfuss 1960: 305.

Blades dark green, drying to black, firm and relatively stiff, broadly orbicular, usually deeply lobed and somewhat ruffled, 8–10(15) cm tall, 10–30 cm broad, 40–60(110) µm thick, commonly with short, distinct, solid stipe; cells in surface view irregularly arranged, 10–15 µm diam., 1.5–3 times as long in anticlinal section; 2 layers of blade separated by pronounced accumulation of mucilaginous material; chloroplast filling outer half of cell, with 1 or 2(3) pyrenoids; gametes probably anisogamous.

Occasional on rocks, woodwork, or other algae, upper intertidal, Alaska to Baja Calif. and Chile. Widely distributed. Type locality: Cape of Good Hope, S. Africa.

Ulva stenophylla S. & G.

Setchell & Gardner 1920a: 282; 1920b: 271.

Blades dark green, linear-lanceolate, usually simple, with entire, broadly ruffled margins, 50–80 cm tall, 5–10 cm broad, tapering at base to short, flattened-cuneate stipe; membrane tough, uniformly 60–100 µm thick; cells squarish in surface view, 14–20 µm diam., 1.5–2 times as long in anticlinal section; chloroplast parietal, thin, covering most of cell wall, without pyrenoid; reproductive structures unknown.

Relatively rare, on rocks, lower intertidal, Ore. to Monterey (type locality) and Big Sur (Monterey Co.), Calif.

Ulva taeniata (Setch.) S. & G.

Ulva fasciata f. *taeniata* Setchell, P.B.-A., 1895–1919 [1901]: no. 862. *U. taeniata* (Setch.) Setchell & Gardner 1920a: 286; Smith 1944: 48.

Blades mostly grass-green, simple or with several long narrow divisions from discoid base; 20–50(150) cm tall, 3–4 cm broad, densely ruffled and commonly spirally twisted, the margins usually more or less dentate below, continuously dentate in quiet water, the membrane mostly 30–45 µm thick at margins, 75–140 µm thick in middle; cells often in rows, 7–10 µm diam., in anticlinal section subquadrate near margins and about twice as long as diam. in axial portions; chloroplast filling outer third of cell, mostly with 1 pyrenoid; gametes isogamous.

Common on rocks, midtidal and lower intertidal, Coos Bay, Ore. to Ventura, Calif. Type locality: Monterey, Calif.

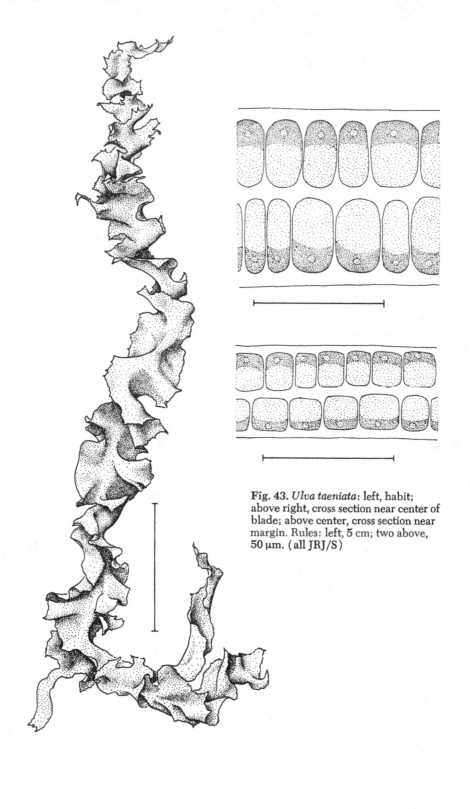

Fig. 43. *Ulva taeniata*: left, habit; above right, cross section near center of blade; above center, cross section near margin. Rules: left, 5 cm; two above, 50 μm. (all JRJ/S)

Order PRASIOLALES

Thalli mostly filamentous or foliaceous, with cells dividing in 1–3 planes. Cells uninucleate. Chloroplast single, axial, stellate, with 1 central pyrenoid. Vegetative reproduction frequent. Asexual reproduction chiefly by aplanospores. Sexual reproduction and zoospores in some species.

Family PRASIOLACEAE

With characters of the order.

Prasiola Meneghini 1838

Thalli foliaceous, monostromatic, the blades commonly stipitate, the plant frequently with rhizoidal attachment. Cells cubical or elongate to rounded, commonly in groups of 4, repeatedly dividing and forming longitudinal and transverse rows. Vegetative reproduction by abscission of marginal strips. Asexual reproduction by aplanospores or akinetes. Some species with zoospores and motile gametes, having morphologically similar gametangial and sporangial stages in the life history.

Prasiola meridionalis S. & G.

Setchell & Gardner 1920a: 291; 1920b: 278; Smith 1944: 53; Bravo 1965: 177. *Gayella constricta* S. & G. 1917: 384. *Rosenvingiella constricta* (S. & G.) Silva 1957b: 41.

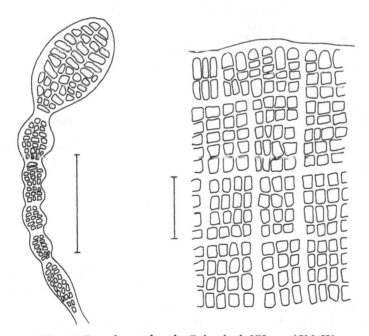

Fig. 44. *Prasiola meridionalis*. Rules: both 250 µm. (SM/H)

Blades deep green, broadly ovate, with crisped margins, to 8(10) mm tall, 40–45 μm thick; cells cubical or anticlinally elongate, mostly 5–8 μm diam., 14–20 μm long; cylindrical form (=*Rosenvingiella*) and diploid spores reported to be formed during summer and fall, haploid gametes during winter; sexual reproduction oogamous.

Frequent to locally abundant, mostly in spray zone on exposed rock coated with guano, Friday Harbor, Wash., to Carmel and Santa Cruz I., Calif. Type locality: entrance to Tomales Bay (Marin Co.), Calif.

Order CLADOPHORALES

Thalli composed of coenocytic cells united end to end in branched or unbranched filaments, these usually attached by rhizoids or modified rhizoid-like branches. Chloroplast parietal, reticulate, with few to many pyrenoids. Asexual reproduction frequently by fragmentation, or with quadriflagellate zoospores formed in undifferentiated vegetative cells functioning as sporangia. Sexual reproduction by biflagellate gametes, similarly formed in undifferentiated cells. Alternating generations similar or dissimilar.

Family CLADOPHORACEAE

The Cladophorales are usually placed in a single family, the Cladophoraceae, with the characters of the order.

1. Filaments unbranched or with short rhizoidal branches 2
1. Filaments profusely branched 5
  2. Filaments usually free-floating, but if attached, with occasional short
     rhizoidal branches along entire length. . . . .*Rhizoclonium* (below)
  2. Filaments attached by rhizoids, these restricted to base 3
3. Rhizoids from several cells near base, these highly modified. . . . .
  . . . . *Urospora* (p. 93)
3. Rhizoids from basal cell only 4
  4. Basal cell long in comparison to width. . . . .*Chaetomorpha* (p. 100)
  4. Basal cell a normal vegetative cell. . . . .*Lola* (p. 92)
5. Branches laterally attached to one another by hooked or spinelike
  branchlets . . . . *Spongomorpha* (p. 96)
5. Branches not laterally attached by special branchlets. . .*Cladophora* (p. 103)

Rhizoclonium Kützing 1843

Thalli of uniseriate filaments with or without short, simple rhizoidal branches, the filaments forming small entangled masses on other algae, on wood, or on flowering plants, or forming larger masses on mud, or free-floating. Cells multinucleate; chloroplast reticulate, with few to many pyrenoids. Cell length/width ratio used as primary taxonomic character but unreliable. Vegetative reproduction by fragmentation. Asexual reproduction by quadriflagellate zoospores, akinetes, and biflagellate zoospores from unmodified vegetative cells. Sexual reproduction by biflagellate isogametes. Zygotes developing directly into sporangial plant. Life stages isomorphic.

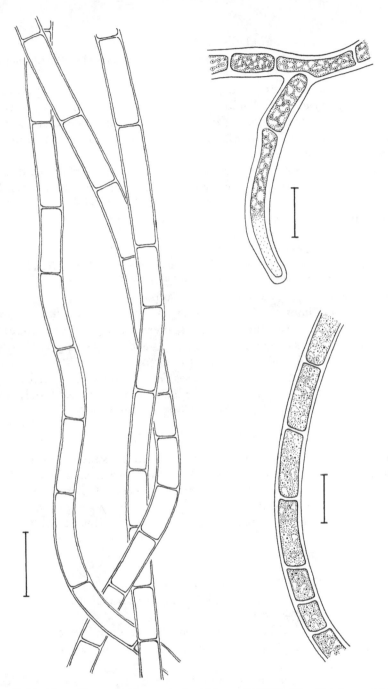

Fig. 45 (left). *Rhizoclonium implexum*. Rule: 100 µm. (JRJ/S) **Fig. 46** (above right). *R. riparium*. Rule: 100 µm. (SM, after Scagel) **Fig. 47** (below right). *Lola lubrica*. Rule: 100 µm. (SM, after Scagel)

Rhizoclonium implexum (Dillw.) Kütz.

Conferva implexa Dillwyn 1809: 46. *Rhizoclonium implexum* (Dillw.) Kützing 1845: 206; Smith 1944: 62; Scagel 1966: 73.

Thalli in small floating mats, entangled with other algae (particularly *Enteromorpha*) or mud-dwelling; filaments unbranched, very rarely with unicellular rhizoidal branches; cells cylindrical, slightly wider terminally than basally, 10–21(30) μm diam., 2–8 times longer; chloroplast reticulate, with 6–8 pyrenoids.

Locally frequent in quiet areas such as sloughs or marshes, floating or in mud, Alaska to Mexico; in Calif., definitely known from Dillon Beach (Marin Co.), San Francisco Bay, Elkhorn Slough (Monterey Co.), Pacific Grove, and Carmel Bay. Type locality: Ireland.

Rhizoclonium riparium (Roth) Harv.

Conferva riparia Roth 1806: 216. *Rhizoclonium riparium* (Roth) Harvey 1846–51 (1849): 238; Scagel 1966: 74. *R. tortuosum sensu* Smith 1944: 62.

Thalli in tangled masses, dark green to pale yellowish-green. Uniseriate filaments sometimes twisted or contorted, usually with numerous short, tapering, irregular rhizoidal branches, these nonseptate or composed of 2–5 cells; filaments 35–50(70) μm diam., 1–3(6) times longer; chloroplast coarsely reticulate, with 10–12(24) pyrenoids.

Locally abundant, in skeinlike masses on various substrata including pilings and clay banks, upper and midtidal, Alaska to Chile. Widely distributed in temperate regions. Type locality: Germany.

Lola A. & G. Hamel 1929

Filaments unbranched, uniseriate, attached by simple, nonseptate rhizoids, or free-floating in dense, soft, slippery mats. Cells multinucleate, cylindrical or elongate-oval, somewhat constricted. Chloroplast reticulate with numerous pyrenoids. Asexual reproduction by zoospores. Sexual reproduction anisogamous, the gametes biflagellate. Other life history details unknown.

The genus is monotypic.

Lola lubrica (S. & G.) Ham. & Ham.

Rhizoclonium lubricum Setchell & Gardner 1919: 492. *Lola lubrica* (S. & G.) Hamel & Hamel 1929: 1094; Scagel 1966: 76.

Filaments attached in common groups, dark emerald green; cells 35–48 (60) μm diam., 65–180 μm long, the upper cells frequently elongate-oval, somewhat constricted.

Infrequent, usually entangled with other algae, or on mud, N. Wash. to Costa Rica. Also from Atlantic coast of France. Type locality: Oakland, Calif.

Urospora Areschoug 1866

Thalli with filaments unbranched, conspicuous for modified basal rhizoidal portion, unlike any other unbranched green algae. Rhizoids from several consecutive cells at base of filament, either emerging directly from a cell or growing downward for some distance between wall layers. Cell division intercalary and occurring in all but lowermost cells. Cells uninucleate or multinucleate. Chloroplast band-shaped or reticulate, with few to many pyrenoids. Vegetative reproduction by fragmentation. Quadriflagellate zoospores distinguished by long-attenuate posterior portion, or by akinetes. Sexual reproduction by biflagellate gametes; zygote developing into free-living *"Codiolum"* stage (see p. 95), producing quadriflagellate zoosporangia. Life stages heteromorphic.

1. Filaments less than 100 μm diam. . . . .*U. penicilliformis*
1. Filaments 100+ μm diam. 2
  2. Rhizoids retained within basal cell walls . . . .*U. wormskioldii*
  2. Rhizoids emerging directly from (and lying external to) cell walls of lower cells . . . .*U. doliifera*

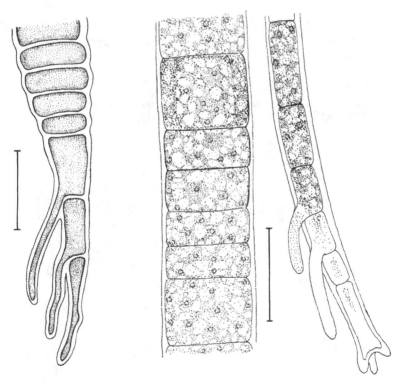

Fig. 48 (left). *Urospora doliifera*. Rule: 100 μm. (SM, after Doty)
Fig. 49 (two at right). *U. penicilliformis*. Rule: 50 μm. (JRJ/S)

Urospora doliifera (S. & G.) Doty

Hormiscia doliifera Setchell & Gardner 1920a: 279; 1920b: 193. *Urospora doliifera* (S. & G.) Doty 1947a: 26.

Filaments 3–4 cm tall, distinctly clavate, dark green; rhizoids emerging directly from (and lying external to) cell walls of lower cells; cells frequently 3–4 times broader than long, (25)40–60 μm diam. at base, to 200–250 μm diam. in terminal portions, variable in diameter along filament; cells barrel-shaped, multinucleate; chloroplast reticulate, with numerous small pyrenoids.

Rare, on boulders in upper intertidal, Chetco Cove (Curry Co.), Ore., and San Francisco, Calif. (type locality).

Urospora penicilliformis (Roth) Aresch.

Conferva penicilliformis Roth 1806: 271. *Urospora penicilliformis* (Roth) Areschoug 1866: 16; Smith 1944: 64. *Chaetomorpha catalinae* Dawson 1949b: 2.

Filaments 3–4 cm tall, rather uniform in width; rhizoids at base of filament growing downward external to cell walls; cells 32–55(100) μm diam., 0.5–2 times longer, multinucleate; chloroplast reticulate, with many pyrenoids.

Abundant when found, but discontinuous in distribution, usually on rounded boulders making dark green, slippery, hairlike coating, high intertidal, Alaska to S. Calif. Type locality: Germany.

Urospora wormskioldii (Mert.) Rosenv.

Conferva wormskioldia Mertens 1818: 6. *Urospora wormskioldii* (Mert.) Rosenvinge 1893: 920; Scagel 1966: 81.

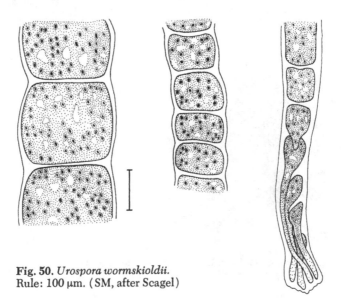

Fig. 50. *Urospora wormskioldii.*
Rule: 100 μm. (SM, after Scagel)

Filaments to 24 cm tall, bright green, attached by rhizoids retained within lateral walls of lower 12–15(30) cells; cells multinucleate, cylindrical or slightly constricted, 25–70 μm diam., gradually increasing in diameter distally to 120–300 μm in terminal portions, squarish to elongate, to 10 times longer than broad; chloroplast variable, ranging from reticulate to band-shaped.

Infrequent along Pacific Coast, but abundant where found, on rocks, shells, or pilings, or epiphytic on other algae, intertidal to upper subtidal, Alaska to S. Calif. Type locality: Greenland.

Recent collections from S. Calif. (Pt. Conception, Santa Barbara, Channel Is., Laguna Beach) show this to be the commonest *Urospora* in that region.

Sporophyte of *Urospora*

"Codiolum gregarium" Braun

Braun 1855: 20; Scagel 1966: 105.

Thalli unicellular, clavate, multinucleate, usually stipitate, although stipe may be short and almost sessile, the plant (45)55–80 μm diam., (280)375–425 μm long.

Free-living, S. Br. Columbia to N. Wash.; not reported from Calif. but probably similar in distribution to *Urospora*. Type locality: Helgoland.

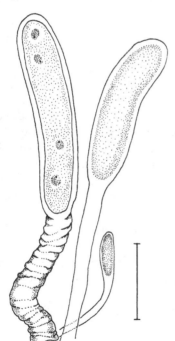

Fig. 51. "*Codiolum gregarium*."
Rule: 100 μm. (SM, after Scagel)

Spongomorpha Kützing 1843

Thallus of profusely branched, uniseriate filaments, the major axes entangled in ropelike strands, spongy in texture. Branches ropelike primarily because of hooklike lateral branchlets entangling adjacent branchlets, and because of lateral fusion of lower axes by descending rhizoids. Terminal branchlets usually rebranched, rarely unilateral in position. Terminal cells elongate, cell division usually intercalary. Cells multinucleate. Chloroplast single, parietal, reticulate, with numerous pyrenoids. Asexual reproduction by fragmentation, or by biflagellate zoospores from undifferentiated vegetative cells functioning as sporangia. Sexual reproduction isogamous, the gametes formed from undifferentiated vegetative cells; as far as is known these large gametangial plants alternating with unicellular "*Codiolum*" stage or "*Chlorochytrium*" stage, from which zoospores give rise to gametangial phase. Cytological details of unicellular stage unknown.

Two "species" of algae now known to be phases in the life histories of *Spongomorpha* species are described on pp. 98–100.

1. Terminal branches with dilated apical cells; hooklike lateral branches usually lacking ....*S. saxatilis*
1. Terminal branches lacking dilated apical cells; hooklike lateral branches abundant 2
   2. Cells usually shorter than diam.; ropelike lower portions repeatedly branched ....*S. coalita*
   2. Cells usually longer than diam.; ropelike lower portions usually not branched ....*S. mertensii*

Spongomorpha coalita (Rupr.) Coll.

Conferva coalita Ruprecht 1851: 404. *Spongomorpha coalita* (Rupr.) Collins 1909a: 361; Smith 1944: 65; Scagel 1966: 97.

Thallus tufted, bright grass-green to dark green, frequently with yellowish tips, 25–45 cm tall, having appearance of branched, much-frayed green rope; groups of branches 1.5–3 cm wide, more or less flabellate, joined laterally and matted basally; filaments dichotomously branched below, becoming irregularly alternate or unilateral above; cells shorter to nearly as wide as long, (100)250–300 μm from middle to lower axes; short, compound branchlets with recurved apices forming hooks abundant in lower part of thallus, these sometimes becoming rhizoidal and on entangling with other rhizoids producing attachment structure.

Seasonally abundant, saxicolous or occasionally epiphytic, low intertidal, Alaska to San Luis Obispo Co., Calif.; a summer annual in C. Calif. Type locality: N. Calif.

"*Codiolum petrocelidis*," now known to be a stage in the life history of *S. coalita*, is described on p. 99.

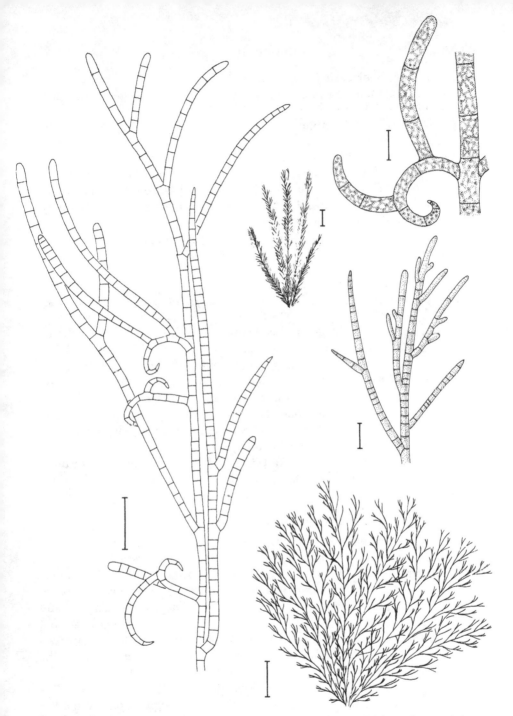

Fig. 52 (left). *Spongomorpha coalita*. Rule: 1 mm. (JRJ/S)   **Fig. 53** (two above right).
S. mertensii. Rules: left, 1 cm; right, 200 μm. (both SM, after Scagel)   **Fig. 54** (two
below right). *S. saxatilis*. Rules: habit, 20 μm; detail of apical portion, 200 μm. (CS &
SM; both after Scagel)

Spongomorpha mertensii (Rupr.) S. & G.

Conferva mertensii Ruprecht 1851: 403. *Spongomorpha mertensii* (Rupr.) Setchell & Gardner 1920b: 227; Doty 1947a: 24; Scagel 1966: 98.

Thallus 8–11 cm tall, bright green, the entire thallus ropelike but undivided; filaments fairly rigid, the cells of main axes toward distal end of thallus 95–125(160) µm diam., usually 0.5–3(10) times longer, those toward basal end smaller, 80–95 µm diam., to 2–5 times longer; filaments laterally held together, matted basally because of hooked branchlets and elongated intertwined rhizoidal filaments; branchlets not as elaborately rebranched and recurved as in *S. coalita*.

Infrequent, saxicolous, low intertidal, N. Japan and Alaska (type locality) to Carmel, Calif.

Spongomorpha saxatilis (Rupr.) Coll.

Conferva saxatilis Ruprecht 1851: 403. *Spongomorpha saxatilis* (Rupr.) Collins 1909a: 360; Setchell & Gardner 1920b: 226; Scagel 1966: 99.

Thallus at first bright green, becoming darker, forming dense tufts of erect branches 10–15 cm tall, not matted together; cells of filaments 80–120 (165) µm diam., 3–6 times as long, nearly uniformly wide throughout, the filaments free from one another except at base, there held together by branched rhizoids; branchlets short and straight, not curved or hooked; branching more or less alternate but sometimes opposite; filaments terminal on thallus having dilated apical cells, nonsporangial; filaments lower on thallus with acuminate apical cells, sporangial.

Infrequent, saxicolous, low intertidal, N. Japan and Kurile Is. through Alaska to N. Wash., rare in Ore. and Calif. (San Francisco). Type locality: Sea of Okhotsk.

Sporophytes of *Spongomorpha* and Possibly Other Algae

The genera "*Codiolum*" and "*Chlorochytrium*" have traditionally been assigned to the order Chlorococcales (chiefly of fresh water) because of their multinucleate, unicellular structure, but current studies of members of this order emphasize other characteristics that are believed to be of greater taxonomic significance, such as the tendency to lose flagella and assume a "coccoid" habit, and the ability to divide simultaneously so that many daughter cells are associated in immobile groups. These characteristics are not shown by either "*Codiolum*" or "*Chlorochytrium*," hence the doubtful systematic position of these taxa. Moreover, culture studies by several workers have found "*Codiolum*"-like and "*Chlorochytrium*"-like stages in the life histories of species of *Urospora*, *Spongomorpha*, and other green algae. On this coast, Hollenberg (1957, 1958b) and Fan (1959) found that "*Codiolum petrocelidis*" was clearly associated with *Spongomorpha coalita*. More recently, Chihara (1969) demonstrated conclusively that

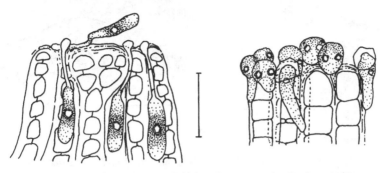

Fig. 55. *"Codiolum petrocelidis."* Rule: 10 μm. (LH, after Fan)

"Chlorochytrium inclusum" is associated with another species of *Spongomorpha*, possibly *S. spinescens* (not in our flora). It is thus clear that those taxa we have previously identified with either *"Codiolum"* or *"Chlorochytrium"* are actually stages in the life history of one or more other algae, and indeed the possibility exists that when growing under different conditions or in different hosts they have shown one or more characteristics that taxonomists have thought were of specific distinction.

Neither *"Codiolum"* nor *"Chlorochytrium,"* however, can be identified in the field as a phase in the life history of a particular macroscopic plant. Three of these "species" are described here as independent taxa, since they are recognizable in the field; two of them are known to be associated in the life history of other algae, and the third, while not yet implicated, is very doubtfully an independent species related to these.

"Codiolum petrocelidis" Kuck.

Kuckuck 1894: 259; 1896: 396; Smith 1944: 68; Hollenberg 1957: 76; 1958b: 249; Fan 1959: 1.

Thalli unicellular, multinucleate, endophytic within tissue of *Petrocelis* (an encrusting red alga), usually with a stout colorless stalk 25–50 μm tall; cell globose to ovoid, 25–50 μm diam., 125–175 μm long; chloroplast laminate to reticulate with 1 to many pyrenoids; reproduction by quadriflagellate zoospores, which on germination in Calif. give rise to the gamete-producing plants known as *Spongomorpha coalita* (Rupr.) Coll.

Uncommon, endophytic in *Petrocelis*, high to midtidal on rocks, N. Wash. to Carmel, Calif. Type locality: N. Europe.

"Chlorochytrium inclusum" Kjellm.

Kjellman 1883: 320; Smith 1944: 67; Chihara 1969: 127.

Thalli unicellular, multinucleate, endophytic within cortex of a variety of large foliose red algae, without a colorless stalk; cells spherical to subspherical, (75)80–100 μm diam., with thickened walls; chloroplast parietal,

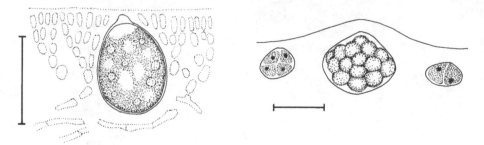

Fig. 56 (left). "*Chlorochytrium inclusum.*" Rule: 50 µm. (JRJ/S) **Fig. 57** (right). *Chlorochytrium porphyrae.* Rule: 50 µm. (LH, after Scagel)

containing many pyrenoids. Reproduction by quadriflagellate zoospores, which on germination (in Br. Columbia and Washington) give rise to gamete-producing plants known as a species of *Spongomorpha*.

Frequent in foliose red algae, low intertidal to subtidal, Alaska to Carmel Bay, Calif. Type locality: Arctic Ocean.

Chlorochytrium porphyrae S. & G.
Setchell & Gardner 1917i: 379; 1920b: 150; Smith 1944: 67.

Thalli unicellular, multinucleate, endophytic within cell wall of *Porphyra* spp., cells oblate-spherical, 40–60 µm diam., with thickened walls; chloroplast parietal, cup-shaped to stellate, with 1 pyrenoid; reproduction by isogametes; asexual reproduction unknown.

Locally abundant in high-intertidal species of *Porphyra*, Wash. to San Luis Obispo Co., Calif. Type locality: San Francisco, Calif.

This species has not been implicated in the life history of another alga.

Chaetomorpha Kützing 1845

Thalli uniseriate unbranched filaments, solitary or in tufts, attached or free-floating, entangled, epiphytic. Holdfast discoid or with nonseptate rhizoidal filaments growing from proximal end of elongate basal cell. Cells multinucleate, most cells fairly uniform in size in a given thallus; size and shape of basal cell diagnostic in taxonomy but probably not reliable. Chloroplast parietal, reticulate (discoid with age), with numerous pyrenoids. Asexual reproduction by fragmentation, or by quadriflagellate zoospores produced in large numbers from undifferentiated vegetative cells. Sexual reproduction by biflagellate gametes. Life stages isomorphic.

1. Filaments less than 40 µm diam. .... *C. californica*
1. Filaments more than 100 µm diam. 2
  2. Thalli usually attached, solitary or in tufts .... *C. linum*
  2. Thalli usually free-floating, but may be entangled 3
3. Filaments coiled and contorted .... *C. spiralis*
3. Filaments not coiled and contorted, the apices clavate........ *C. antennina*

Chaetomorpha antennina (Bory) Kütz.

Conferva antennina Bory 1804: 161. *Chaetomorpha antennina* (Bory) Kützing 1849: 379; Setchell & Gardner 1920b: 203.

Thalli growing in stiff, brushlike tufts, 4–12 cm tall; basal cell 10–20 times longer than wide, to 9 mm long; proximal end of basal cell giving off many branched, twisted rhizoids; lateral walls of lower cells (including basal cell) frequently with annular constrictions; upper cells 0.45–1 mm diam., 1–4 times as long.

Infrequent in S. Calif.; midtidal, on rocks or mollusks or entangled with other algae in warm pools, more common and more widely distributed in tropics. Type locality: Réunion I., Indian Ocean.

Chaetomorpha californica Coll.

Collins, P.B.-A., 1895–1919 [1900]: no. 664; Setchell & Gardner 1920b: 200; Scagel 1966: 83.

Thalli erect slender filaments, to 20 cm long, attached by elongate basal cell 100–150(200) µm long, this anchored at proximal end by means of nonseptate rhizoidal branches forming a disk; upper cells 20–25(40) µm diam., 1–5 times longer, nearly uniform diam. throughout; lateral walls distinctly thickened in layers.

Infrequent, saxicolous, intertidal, Laguna Beach and La Jolla (type locality), Calif.; also known subtidally (to 10 m) from Strait of Juan de Fuca and Hein Bank, Wash.

Chaetomorpha linum (Müll.) Kütz.

Conferva linum Müller 1775: 771. *Chaetomorpha linum* (Müll.) Kützing 1845: 204. *Chaetomorpha aerea* (Dillwyn) Kütz. 1849: 379; Smith 1944: 56. *Conferva aerea* Dillw. 1809: pl. 80.

Thalli grass-green to yellowish-green, 4–8(30) cm tall, the filaments straight, basally attached, gregarious; delicate rhizoids from proximal end of basal cell sometimes forming disk; basal cell tapering downward, 4–6 times longer than ordinary vegetative cell; vegetative cells short-cylindrical, 125–400 µm diam., 0.5–2 times as long, somewhat constricted; fertile cells barrel-shaped to subglobose, 400–600 µm diam.

Abundant, on shaded banks or flat rocks, lining sandy tidepools or floating, high to midtidal, Bolinas (Marin Co.) to San Diego, Calif. Type material from England and Wales.

Chaetomorpha spiralis Okam.

Okamura 1903: 3; 1907–42 (1912): 162; Abbott 1972b: 259. *Chaetomorpha clavata* var. *torta* Collins 1909a: 323. *C. torta* Yendo 1914: 264; Coll. 1918: 78.

Filaments rigid, attached when young, soon loosened and entangled among other algae, much coiled and contorted, 20–49(60) cm long, 0.75–1.25 mm diam.; cells moniliform to nearly cylindrical, 1–1.5 times as long

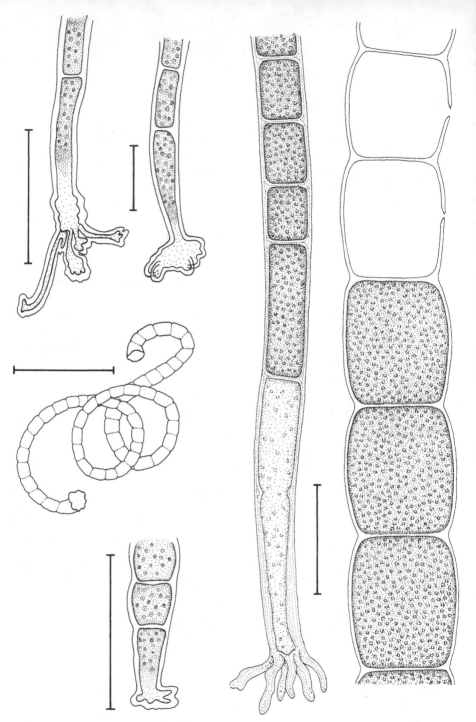

Fig. 58 (above left). *Chaetomorpha antennina*. Rule: 5 mm. (SM) **Fig. 59** (above center). *C. californica*. Rule: 5 mm. (SM, after Scagel) **Fig. 60** (two at right). *C. linum*. Rule: 250 μm. (both JRJ/S) **Fig. 61** (two below left). *C. spiralis*. Rules: both 5 mm. (both SM)

as diam.; thallus appearing iridescent bluish-green because of thick, opaque cell walls.

Infrequent, usually floating and entangled with other algae, low intertidal to subtidal (10 m), Channel Is., Redondo Beach, San Pedro, and La Jolla, Calif., to I. Magdalena, Baja Calif. Type locality: Japan.

Cladophora Kützing 1843

Thalli sparsely to profusely branched, erect, the upper branches usually conspicuously branching pectinately or unilaterally. Chloroplasts conspicuously reticulate or disklike, containing many pyrenoids. Lower branches dichotomously or many times branched. Lateral branches short, not branching further, or the lower laterals longer than the upper newly formed branches, occasionally branching again. Erect branches not entangled by hooks, but lower portions sometimes matted and difficult to separate because of adventitious rhizoids produced by lower cells and fusing with adjacent cells. Basal portion usually of short creeping rhizoids growing together and making firm mat. Asexual reproduction by quadriflagellate zoospores formed in terminal and subterminal cells of branches; pore of escape of zoospores thought to have some taxonomic utility. Sexual reproduction by biflagellate isogametes.

The species of *Cladophora* are highly variable with respect to morphology and ecology, as is clearly reflected in the nomenclatural confusion surrounding the taxa. Van den Hoek (1963) performed a herculean task in collecting, culturing, and clarifying the status of the European taxa. It is clear from a critical assessment of his studies that for the most part the application of names of European species to Pacific Coast *Cladophora* is in error, either because of the existence of earlier homonyms (e.g. *C. sericea*) or because as presently understood the West Coast specimens do not fit the newly drawn detailed descriptions (e.g. *C. columbiana*, previously known as *C. trichotoma*). Furthermore, the lack of understanding of Western Pacific (Asian) species of *Cladophora* has led to the exclusion of those taxa from the North American West Coast. Though the following classification is imperfect, it does represent a determined effort to apply correct names to the West Coast species. The lack of large numbers of collections with adequate locality data and the lack of culture studies hamper any effort to be definitive. Nonetheless, we believe that this classification will at least acknowledge the contributions of van den Hoek, as well as those of Sakai (1964) from Japan, and that further studies, so badly needed, will begin with a firmer grounding than if the attempt had not been made.

1. Basal portion of thallus procumbent; average diam. of apical cells more than 130 μm ....*C. columbiana*
1. Basal portion of thallus erect 2

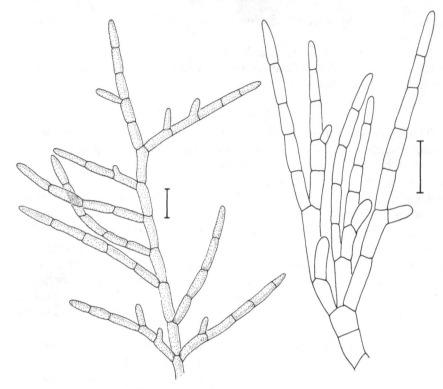

Fig. 62 (left). *Cladophora albida*. Rule: 5 mm. (SM, after Scagel) **Fig. 63** (right).
C. columbiana. Rule: 500 μm. (SM, after Scagel)

2. Thalli coarse, with few orders of branching 3
2. Thalli delicate, with many orders of branching 5
3. Cells from apex to midportions 8–12 times as long as diam. ..... *C. graminea*
3. Cells from apex to midportions seldom more than 4 times as long as
    diam. 4
   4. Ultimate branchlets curving away from axis, usually rebranched....
       .... *C. microcladioides*
   4. Ultimate branchlets close to axis, not rebranched............. *C. sakaii*
5. Ultimate branchlets usually one-half to one-third diam. of bearing axis
    .... *C. sericea*
5. Ultimate branchlets nearly same width as bearing axis 6
   6. Upper branches regular, arcuate, occurring on nearly every cell, the
       older rebranched .... *C. albida*
   6. Upper branches scattered, not occurring on every cell, rarely re-
       branched .... *C. stimpsonii*

Cladophora albida (Huds.) Kütz.

Conferva albida Hudson 1762: 595. *Cladophora albida* (Huds.) Kützing 1843: 267;
van den Hoek 1963: 94 (incl. synonymy); Scagel 1966: 86. *Cladophora bertoloni* var.

hamosa sensu Collins 1909a: 344; Smith 1944: 61. *Cladophora delicatula sensu* Coll. 1909a: 336; P.B.-A., 1895–1919 [1909]: no. 1582.

Thallus delicate, pale yellowish-green, densely tufted, often matted 5–10 cm thick; filaments 26–39(44) µm diam., 2–4 times as long from apex to midportions, longer below; branching more or less dichotomous below, sometimes trichotomous; ultimate branches with 1 to several cells, unilateral or irregular; base with irregular knobby rhizoids; plant with descending rhizoids from lower portions of erect branches, sometimes fusing laterally.

Infrequent, saxicolous, midtidal in tidepools, N. Wash. to S. Calif. Type locality: England.

Cladophora columbiana Coll.

Collins 1903: 226. *Cladophora trichotoma sensu* Setchell & Gardner 1920b: 210; Smith 1944: 58; Scagel 1966: 93. Not *C. trichotoma* (C. Agardh) Kützing (=*C. pellucida* (Hudson) Kütz., see van den Hoek 1963: 215). *C. hemisphaerica* Gardner 1918: 83; Smith 1944: 57.

Thalli bright green, in low matted tufts, 3–5(15) cm tall, 2–4(20) cm diam.; apical cells 90–150(170) µm diam., including lateral walls of 23.4–26 µm; cells of main axes 200–250 µm diam., 5–10 times as long; cells cylindrical; filaments unbranched or branched in various ways, the base a single, only slightly modified cell; or rhizoidal base giving rise to low, horizontally directed branches bearing on their upper surfaces rows of short, irregularly branched, erect filaments; or basal rhizoids mingled and occasionally fusing with adventitious rhizoids from lower cells of erect filaments.

Frequent to locally abundant, saxicolous, midtidal to low intertidal, Vancouver I., Br. Columbia (type locality), to Baja Calif. and Gulf of Calif.

This is the most abundant and characteristic species of *Cladophora* on the Pacific Coast. The common form grows on rocks, and as Smith (1944) has pointed out, sand accumulates between the branches. The less common form (=*C. hemisphaerica*), which makes very large loose tufts, grows in tidepools; the structure of the basal portions, the branching habit, and the sizes of the filaments are, however, extremely similar to those of the rock-dwelling form.

Cladophora graminea Coll.

Collins 1909b: 19; Smith 1944: 59 (incl. synonymy).

Thalli in hemispherical tufts 4–10 cm tall, grayish-green to dark green, sometimes appearing striped; branches stiff, mostly long and uninterrupted by branching for some distance; apical cells 100–150 µm diam.; basal cells 300–500 µm diam., 20–30 times as long; branching primarily dichotomous or trichotomous in lower portions, alternate above; branchlets with 1–3 cells, narrowing from base to apex.

Frequent in small caves or shaded overhangs, saxicolous and usually

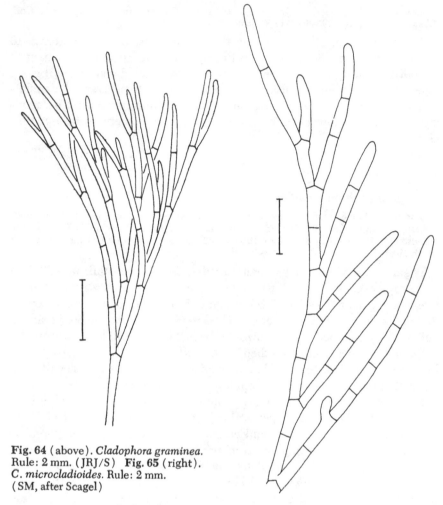

Fig. 64 (above). *Cladophora graminea.*
Rule: 2 mm. (JRJ/S) **Fig. 65** (right).
C. microcladioides. Rule: 2 mm.
(SM, after Scagel)

growing intermingled with sponges, midtidal to low intertidal, Santa Cruz
to San Pedro, Calif. Type locality: Monterey, Calif.

No other species of *Cladophora* on this coast has cells so long in relation
to diameter; nor does any other favor shaded overhangs.

Cladophora microcladioides Coll.

Collins 1909b: 17; Smith 1944: 59; Scagel 1966: 92.

Thalli of erect spreading tufts 5–10(20) cm tall, dull to glaucous green;
apical cells 35–50 μm diam.; cells in midportions 60–80 μm diam.; basal
cells 100–150 μm diam., 1.5–4 times as long throughout; ultimate branches
tapering slightly to rounded apices; branching in lower main axes occasionally dichotomous to trichotomous, alternate above, the oldest laterals

branched pectinately, the youngest generally simple and progressively shorter toward apex of axis.

Infrequent, saxicolous or on wood or plastic floats, low intertidal, Br. Columbia to Baja Calif. and Gulf of Calif. Type locality: San Pedro, Calif.

Cladophora sakaii Abb.

Abbott 1972b: 259. *Cladophora ovoidea sensu* Collins 1909a: 346; Smith 1944: 60. *C. elmorei* Dawson 1949b: 3.

Thalli growing in groups 5–10(30) cm long, dark green, coarse, the lower portions matted laterally, ropelike; the thalli repeatedly branched, with fascicles of branchlets unilateral or on either side of axes, or sparingly branched; branches di- or trichotomously divided, the ultimate branches straight, tapering slightly to apex, of 1–3 cells; apical cells 65–100 μm diam.; cells in midportions 120–180 μm diam.; cells toward base 200–220(500)

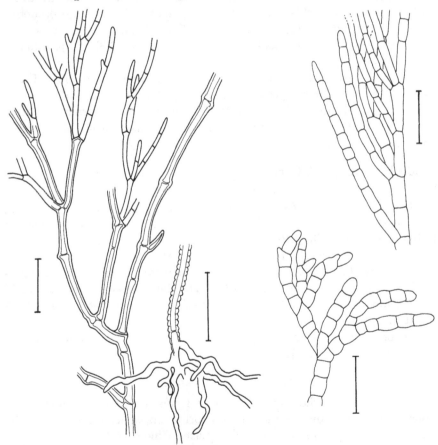

Fig. 66 (two at left). *Cladophora sakaii.* Rules: both 1 mm. (CS, after Sakai) **Fig. 67** (two at right). *C. sericea.* Rules: above, 200 μm; below, 500 μm. (CS, after Scagel)

μm diam., 3–8 times as long throughout; basal portions with annular constrictions along lateral walls; filaments laterally adjoined; basal rhizoids creeping, matted, firmly fastened to substratum.

Seasonally abundant, saxicolous in areas scoured by sand, low intertidal, Pigeon Pt. (San Mateo Co.) to San Pedro and Santa Catalina I., Calif. Type locality: Japan.

Cladophora sericea (Huds.) Kütz.

Conferva sericea Hudson 1762: 485. *Cladophora sericea* (Huds.) Kützing 1843: 264; van den Hoek 1963: 77 (incl. synonymy). West Coast records believed to be synonymous with these are the following: *Cladophora flexuosa sensu* Collins 1909a: 339; Smith 1944: 60; *Cladophora glaucescens sensu* Coll. 1909a: 336; *Cladophora rudolphiana sensu* Setchell & Gardner 1920b: 217.

Thalli 5–15(50) cm tall, yellowish-green to light green; main axes branching irregularly dichotomously, with branches unilateral or bilateral, these with or without branchlets; after producing spores the ultimate branches sometimes eroded to axes; apical cells 38–45 μm diam.; cells of upper axes 55–70 μm diam., 3–4 times as long; cells of lower axes 100–120 μm diam., 4–5 times as long.

Occasional, especially floating in quiet pools, Alaska to San Diego, Calif. Widely distributed in temperate regions. Type locality: England.

Most of the slender, rather delicate specimens from California that have been examined in herbaria are probably to be included in this variable species.

Cladophora stimpsonii Harv.

Harvey 1859: 333; Sakai 1964: 50; Scagel 1966: 93.

Thalli forming low tufts 1 to several cm diam., loosely tufted in spring, forming dark green, discrete clumps 5–8 cm tall, later in season becoming lighter green, spreading, 10–20 cm tall; rhizoids weak; lower branches entangled but not matted; basal cells frequently fused laterally owing to adventitious rhizoids; upper filaments somewhat straight, the branches close to bearing axes; apical cells 26–40 μm diam.; cells of upper axes 49–57(73) μm diam.; cells of lower axes 120–170 μm diam., 6–8 times as long as diam.

Frequent, on shells or rocks, protected intertidal, S. Br. Columbia to S. Calif. Type locality: N. Japan.

Order SIPHONOCLADALES

Thalli siphonous, the plants of various forms. Plants mostly noncalcareous, with growth by segregative cell formation, septate and coenocytic at maturity. Nuclei internal to chloroplasts. Chloroplasts reticulate, with or without pyrenoids. Zoospores quadriflagellate as far as known; life history involving isomorphic stages as far as known.

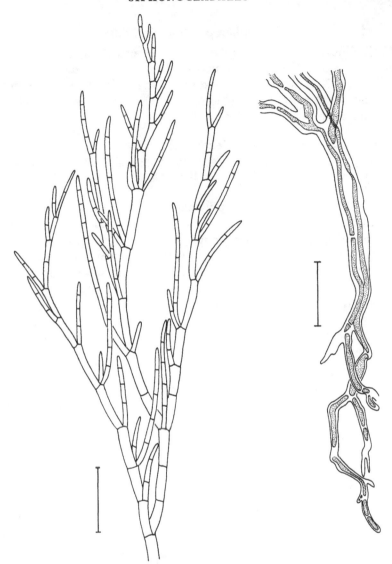

Fig. 68. *Cladophora stimpsonii*. Rules: both 1 cm. (both CS, after Sakai)

Family SIPHONOCLADACEAE

Thalli mostly erect, branching, with growth initiated by apical cells and additionally by segregative cell formation. Erect, freely branched axes commonly arising from septate rhizomelike portion with rhizoidal attachment. Erect axes composed of linear series of segments, these commonly developing branches from distal ends without formation of crosswall at base of branches.

Cladophoropsis Børgesen 1905

Thalli filiform, erect, *Cladophora*-like axes usually clumped, with rhizoidal attachment. Cells of axes multinucleate, with branches arising from distal ends at irregular intervals and remaining, at least for a time, in open connection with mother cell; chloroplasts reticulate to isolated with age, with numerous small pyrenoids.

Cladophoropsis fasciculatus (Kjellm.) Okam.

Siphonocladus fasciculatus Kjellman 1897: 36. *Cladophoropsis fasciculatus* (Kjellm.) Okamura 1907–42 (1921): 75; Dawson 1958: 65; Abbott 1972b: 259. *C. coriacea* Yendo 1920: 1.

Thalli filiform, caespitose to pulvinate, 2–3 cm tall, bluish-green, rigid when fresh; filaments 125–350 µm diam., profusely and variously branched and densely compacted; branches spreading, somewhat curved, the upper branches mostly longer than the lower, with frequent rhizoidal branches conjoining filaments; crosswalls sparse, not occurring at point of branch origin.

Reported in California, without habitat or depth, only from Lechuza

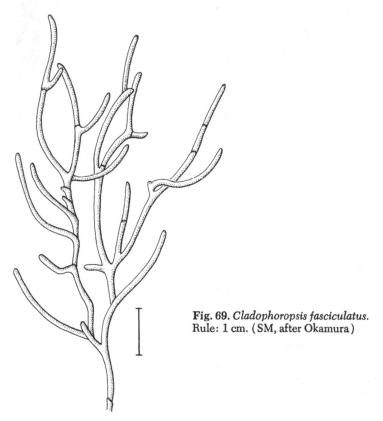

Fig. 69. *Cladophoropsis fasciculatus.*
Rule: 1 cm. (SM, after Okamura)

Point (Los Angeles Co.) and Dana Point (Orange Co.). Type locality: Japan.

Order CODIALES

Thalli coenocytic, mostly tubular and branched, septate or nonseptate. Chloroplasts numerous, discoid, with or without single pyrenoid; characteristic pigment, siphonoxanthin, present. Plants sometimes multiplying vegetatively by abscission of branches. Asexual reproduction by zoospores and aplanospores. Sexual reproduction mostly anisogamous.

1. Tubular branches closely grouped, forming compact, spongy thallus...
 .... CODIACEAE (116)
1. Tubular branches not forming compact, spongy thallus............... 2
 2. Branches with percurrent axes; zoospores lacking ....
 .... BRYOPSIDACEAE (below)
 2. Branches lacking percurrent axes; zoospores multiflagellate; sporangia globular to obovoid .... DERBESIACEAE (113)

Family BRYOPSIDACEAE

Thalli much branched, tubular, coenocytic, the branches nonseptate. Vegetative reproduction occasionally by abscission of branches. No asexual reproduction. Sexual reproduction by biflagellate anisogametes formed in slightly modified branches.

Bryopsis Lamouroux 1809b

Thalli erect, usually densely branched but with few orders of branches. Branches mostly with percurrent axes, pinnately or radially branched. Chloroplasts numerous, discoid, with conspicuous pyrenoid. Thalli monoecious or dioecious, the gametes liberated through several pores in wall of gametangium. Zygote developing directly, without sporangial phase in life history.

1. Erect branches radially branched ....*B. hypnoides*
1. Erect branches pinnately branched 2
 2. Plants mostly less than 4 cm long; branchlets limited to short terminal tuft on erect axes ....*B. pennatula*
 2. Plants mostly 8+ cm long; branchlets not limited to short terminal tuft ....*B. corticulans*

Bryopsis corticulans Setch.

Setchell, P.B.-A., 1895–1919 [1903]: no. 626; Smith 1944: 73 (incl. synonymy).

Thalli mostly tufted, 5–10(16) cm tall, blackish-green; main erect axes to 1 mm diam., naked below, abundantly pinnately branched above; older branches with coarse, descending rhizoidal branches from their bases; ultimate branches 150–300 µm diam., cylindrical, abruptly constricted at base.

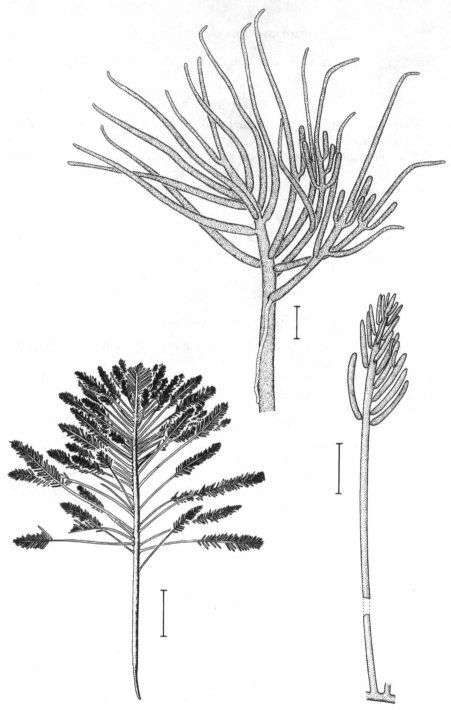

Fig. 70 (below left). *Bryopsis corticulans*. Rule: 1 cm. (JRJ/S) **Fig. 71** (above). *B. hypnoides*. Rule: 1 mm. (JRJ/S) **Fig. 72** (below right). *B. pennatula*. Rule: 1 mm. (SM)

Frequent, on vertical sides of rocks exposed to strong surf, very abundant locally in fall as epiphytes on denuded branches of *Egregia menziesii*, midtidal to lower intertidal, S. Br. Columbia to Is. San Benito, Baja Calif. Type locality: Pacific Grove, Calif.

Bryopsis hypnoides Lamour.

Lamouroux 1809: 135; Setchell & Gardner 1920b: 159; Smith 1944: 73.

Thalli tufted, to 4 cm tall, dull green, profusely radially branched; main erect axes to 430 μm diam.; long, downwardly directed rhizoids arising from base of main branches; ultimate branches 60–75 μm diam., 2–5 mm long, gradually tapering upward, abruptly constricted at bases.

Occasional to locally abundant, on sand-covered rocks, mostly midtidal, S. Br. Columbia to Panama; in Calif., from Humboldt Co. to San Pedro. Widely distributed. Type locality: Mediterranean.

Bryopsis pennatula J. Ag.

J. Agardh 1847: 6; Kützing 1856: 27; Setchell & Gardner 1920b: 158; Hollenberg & Abbott 1966: 14 (incl. synonymy).

Thalli tufted to matted, to 3 cm tall, with mostly simple erect axes, these to 250 μm diam., arising from matted basal branches; erect axes each bearing short terminal tuft of branchlets, these cylindrical, blunt, distichous to partly radial, mostly less than 1 mm long, 20–30 μm diam., limited to terminal 2 mm of erect axes.

Infrequent, chiefly atop large rocks, midtidal to upper intertidal, Pacific Grove, Santa Monica, and Laguna Beach, Calif., to St. Augustine (Oaxaca; type locality), Mexico.

On certain substrata (e.g. sponge tissue), attachment may involve a penetrating branch resembling a taproot, bearing a dense compound tuft of slender erect branches. This form is known only from Pacific Grove, Calif.

Family DERBESIACEAE

Thalli normally with 2 morphologically different stages in life history. Sporangial thalli of tufted tubular branches, with numerous disk-shaped chloroplasts, with or without pyrenoids, and producing relatively large sporangia and large, multiflagellate zoospores. Gametangial thalli globular, with many disk-shaped chloroplasts, lacking pyrenoids, and producing numerous small, anisogamous biflagellate gametes.

Derbesia Solier 1846

With features of the family. Certain species with pyrenoids, others without. Some species reported to lack gametangial stage under culture con-

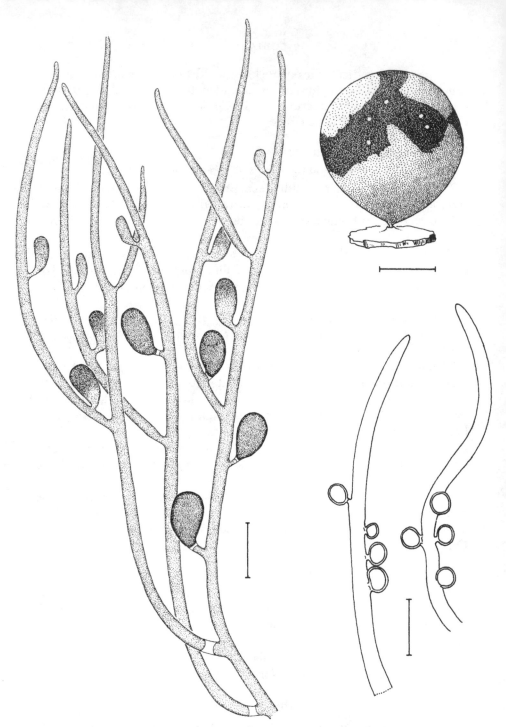

Fig. 73 (left). *Derbesia marina.* Rule: 200 μm. (JRJ/S)   **Fig. 74** (two below right).
D. prolifica. Rule: 200 μm. (SM, after Taylor)   **Fig. 75** (above right). *"Halicystis
ovalis."* Rule: 5 mm. (JRJ/S)

ditions; many species known in nature that have never been matched to a gametangial stage.

Derbesia marina (Lyngb.) Sol.

Vaucheria marina Lyngbye 1819: 79. *Derbesia marina* (Lyngb.) Solier 1846: 452; Setchell & Gardner 1920b: 165; Smith 1944: 71. *Gastridium ovale* Lyngb. 1819: 72. *Halicystis ovalis* (Lyngb.) Areschoug 1850: 447; Hollenberg 1935: 782; 1936: 1; Kornmann 1938: 464.

Sporangial stage (*Derbesia*): Thalli tufted, to 1.5 cm tall, attached by irregularly branched, prostrate and often penetrating branches; erect branches 50–65 μm diam., frequently branched without percurrent axes; chloroplasts without pyrenoids; sporangia lateral, obovoid to obpyriform, 100–160 μm diam., 150–200 μm long, on brief pedicels with a biconcave partition; zoospores mostly 20–30 per sporangium, with anterior crown of flagella, germinating directly and slowly developing into sexual phase.

Frequent on rocks or on the chiton *Cryptochiton stelleri*, or on encrusting coralline algae, low intertidal on exposed rocky shores to subtidal (23 m), Sitka, Alaska, to La Jolla, Calif. Also N. Europe. Type locality: Faeroes.

Gametangial stage (*"Halicystis ovalis"*): see below.

Derbesia prolifica Tayl.

Taylor 1945: 75.

Thalli to 15 cm tall, matted below, tufted above, frequently (and mostly laterally) branched; filaments flexuous or slightly arcuate, 120–250 μm diam.; sporangia very numerous and often unilateral, toward branch apices, 160–210 μm diam., subspherical, on mostly short pedicels 32–42 μm diam.

Rare, known from subtidal (15 m), Santa Catalina I., Calif., and from tidepools, Galápagos Is. (type locality). Also in tidepools, S. Africa.

Gametangial Stage of *Derbesia marina*

"Halicystis ovalis" (Lyngb.) Aresch.

Gastridium ovale Lyngb. 1819: 72. *Halicystis ovalis* (Lyngb.) Areschoug 1850: 447.

Thalli spherical, coenocytic, 2–10(15) mm diam., with rhizomelike basal part penetrating encrusting coralline algae; protoplasm forming thin layer next to wall, with chloroplasts next to the large vacuole and numerous nuclei next to wall; reproduction by anisogamous biflagellate gametes; gametangial areas irregularly band-shaped, formed in spherical portion, not cut off by cell wall from vegetative portion; numerous gametes liberated periodically and forcefully through 1 or more pores.

Frequent to locally abundant, low intertidal (on *Lithophyllum* spp. on exposed rocky shores) to subtidal (23 m), Silver Bay, Alaska, to Pta. Eugenio, Baja Calif.; in Calif., known chiefly south of Pt. Lobos (Monterey

Co.), to Santa Catalina I. Widely distributed in N. Hemisphere. Type locality: N. Europe.

Family CODIACEAE

Thalli much branched, tubular, coenocytic, composed of nonseptate branches, these commonly densely compacted, forming a plant body of macroscopic form. Thallus variously shaped, in some genera encrusted with lime. Surface branches commonly organized into palisade-like layers. Chloroplasts numerous, discoid, without pyrenoids. No asexual reproduction. Sexual reproduction anisogamous; gametes biflagellate, formed in gametangia of distinctive shape.

Codium Stackhouse 1797

Thalli much branched, the branches compacted and forming erect or prostrate, noncalcareous spongy thallus of definite form, basally attached by rhizoids. Surface layer composed of palisade-like, photosynthetic, enlarged branch tips (utricles). Interior of thallus composed of slender, colorless, intertwined tubular branches. Thalli mostly dioecious; gametangia fusiform to cylindrical, borne laterally on utricles and sealed off at base by annular thickenings. Gametes biflagellate, anisogamous, formed following meiosis. Zygote developing directly into diploid plant.

1. Thalli erect 2
1. Thalli prostrate, more or less pulvinate 4
   2. Thalli unbranched ....*C. johnstonei*
   2. Thalli branched 3
3. Branches cylindrical throughout; apices of utricles mucronate.....*C. fragile*
3. Branches mostly flattened; apices of utricles not mucronate....*C. cuneatum*
   4. Apical walls of utricles pitted or alveolate ....*C. hubbsii*
   4. Apical walls of utricles neither pitted nor alveolate.........*C. setchellii*

Codium cuneatum S. & G.

Setchell & Gardner 1924a: 708; Dawson 1944a: 218; Silva 1951: 99.

Thalli erect, to 35 cm tall, regularly or irregularly dichotomo-flabellate, with more or less cuneate segments and rounded axils; branches 2–15 mm broad, usually compressed throughout except for branch tips and stipe, these sometimes terete, or in occasional specimens the branches nearly terete throughout; utricles unbranched, cylindrical to broadly clavate, occasionally with hairs, (55)69–250(440) µm diam., 400–700(960) µm long, the apex truncate or sometimes rounded, to as much as 32 µm thick, lamellate.

Occasional, on rocks, intertidal to subtidal (45 m), Santa Cruz I., Calif., to Bahía de Los Angeles, Baja Calif., and I. Smith, Gulf of Calif. (type locality).

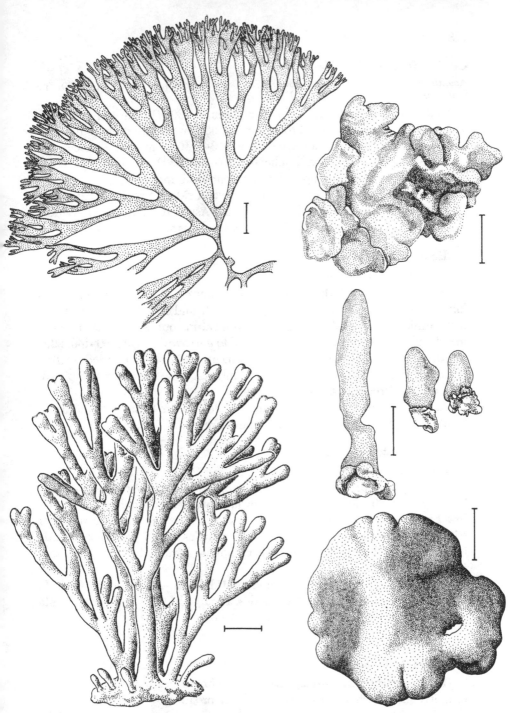

Fig. 76 (above left). *Codium cuneatum*. Rule: 2 cm. (SM) **Fig. 77** (below left).
C. fragile. Rule: 2 cm. (JRJ/S) **Fig. 78** (above right). *C. hubbsii*. Rule: 2 cm. (SM)
Fig. 79 (three at middle right). *C. johnstonei*. Rule: 2 cm. (SM) **Fig. 80** (below right).
C. setchellii. Rule: 2 cm. (JRJ/S)

Codium fragile (Sur.) Har.

Acanthocodium fragile Suringar 1867: 258. *Codium fragile* (Sur.) Hariot 1889: 32; Smith 1944: 75; Silva 1951: 96 (incl. synonymy).

Thalli erect, dark green to blackish-green, 10–30(41) cm tall, with 1 to several erect branches from broad basal disk, abundantly dichotomo-fastigiately branched; the branches cylindrical, 3–8(10) mm diam.; utricles unbranched, narrowly to broadly clavate, 150–350(630) μm diam., 600–1500 μm long, with mucronate apex, hairs absent to abundant.

Frequent to common, intertidal on tops and sides of rocks, occasionally subtidal (to 45 m), Sitka, Alaska, to Bahía Asunción, Baja Calif. Widely distributed in temperate seas. Type locality: Japan.

Codium hubbsii Daws.

Dawson 1950a: 151; Silva 1951: 85.

Thalli pulvinate, dark green, 3–8 mm thick, orbicular to marginally lobed, commonly joining with adjacent thalli and forming a more extensive cushion moderately adhering to substratum but with loose margins; utricles abundantly branched, cylindrical to narrowly clavate, 60–140(220) μm diam., 0.8–1.5(2) mm long; apex truncate or slightly rounded; wall at apex of utricle 5–32 μm thick, slightly umbonate, mostly internally pitted; hair scars in irregular whorls near apex.

Occasional, on rocks, midtidal to subtidal (25 m), Santa Catalina I., Calif., to Is. San Benito, Mexico (type locality).

Codium johnstonei Silva

Silva 1951: 94.

Mature plants dark green, erect, cylindrical to slightly compressed, unbranched, 15–20 mm diam., to 10 cm tall; utricles unbranched or with a few branches below, 70–110(140) μm diam., 2+ mm long, truncate or slightly rounded at apex; terminal wall of utricle to 65 μm thick, slightly laminate and slightly to markedly umbonate internally; utricular hairs infrequent.

Occasional, mostly saxicolous and subtidal (to 46 m), Santa Cruz I., Calif. (type locality), to Is. Coronados, Baja Calif.; in Calif., from Santa Cruz I., Redondo Beach, and Laguna Beach.

Codium setchellii Gardn.

Gardner 1919: 489; Setchell & Gardner 1920b: 168; Smith 1944: 75; Silva 1951: 83.

Thalli prostrate, pulvinate, forming indefinitely expanded, dark green, firm, irregular cushion to 25 cm broad (or broader by coalescence), 6–15 mm thick, often coarsely rugose, closely adherent to substratum; utricles abundantly branched, cylindrical to narrowly clavate, mostly 65–90 μm

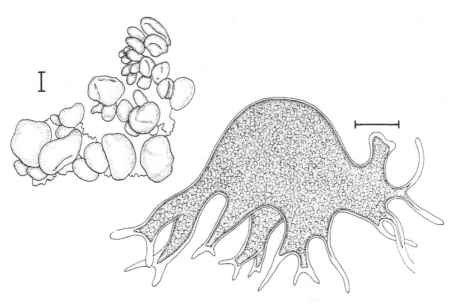

Fig. 81 (left). *"Collinsiella tuberculata."* Rule: 2 cm. (SM, after Setchell & Gardner)
Fig. 82 (right). *"Gomontia polyrhiza."* Rule: 1 mm. (JRJ/S)

diam., 0.65–1 mm long; utricle wall to 16 μm thick and lamellate at apex, but not internally sculptured.

Frequent on exposed rocks, low intertidal, from Sitka, Alaska, throughout Wash. and Calif. to Pta. Baja, Baja Calif. Type locality: Pacific Grove, Calif.

Algae of Uncertain Relationship

Chlorophyta of uncertain systematic position reported to represent stages in the life history of one or more species of green algae.

"Collinsiella tuberculata" S. & G.

Setchell & Gardner 1903: 204; Chihara 1958: 307; 1960: 181; Scagel 1960: 969.

Plant forming globose, gelatinous colonies, these commonly tuberculately lobed, at first solid, later becoming hollow; cells terminal on branched stalk, pyriform, 12–20 μm long, 5–12 μm diam., the narrower end directed toward interior of colony; chloroplast single, cup- or band-shaped, with 1 pyrenoid.

Common on rocks or old shells, midtidal, S. Br. Columbia to Pacific Grove, Calif. Type locality: Whidbey I., Wash.

Culture studies by Scagel (1960) and Chihara (1960) indicate that *"Collinsiella"* may be a stage in the life history of a species of *Enteromorpha* and/or *Monostroma*.

"Gomontia polyrhiza" (Lagerh.) Born. & Flah.

Codiolum polyrhizum Lagerheim 1885: 22. *Gomontia polyrhiza* (Lagerh.) Bornet & Flahault 1888: 163; Setchell & Gardner 1920b: 302; Smith 1944: 41.

Cells irregularly arranged, or arranged in filaments, irregularly branched, 4–10 μm wide, 2–6 times as long. Sporangia with 1 to several simple or forked rhizoidal processes.

Frequent, forming green patches within shells or wood, S. Br. Columbia to Monterey Peninsula, Calif. Widely distributed. Type locality: Kristineberg, Sweden.

Culture studies seem to indicate that "*Gomontia*" may be a stage in the life history of certain species of *Monostroma* and perhaps of *Ulva* and other genera.

Division Phaeophyta

The Brown Algae

PREDOMINANTLY MARINE algae; except for large kelps, mostly intertidal or in upper subtidal zone. Thalli all multicellular, the plants usually small to medium-sized, filamentous, cushion-shaped, bladelike, or tubular to sac-like, or large with complex vegetative structure. Cells mostly uninucleate; chloroplasts 1 or more, with an olive to gold-brown pigment, fucoxanthin, predominant; carbohydrate food reserve never starch. Growth intercalary, mostly trichothallic, or apical. Plants usually perennial. Separate sporangial and gametangial stages in life history; asexual reproduction typically by zoospores, less commonly by aplanospores; sexual stage microscopic or macroscopic. Sporangia with single compartment (unilocular) or with many compartments (plurilocular); gametangia mostly plurilocular. Sexual reproduction isogamous, anisogamous, or oogamous; motile spores and male gametes mostly with 2 laterally inserted, dissimilar flagella. Zygotes developing without dormant period.

Class PHAEOPHYCEAE

We recognize only one class, with characters of the division.

1. Thalli of uniseriate filaments, occasionally multiseriate, never pseudo-parenchymatous .... ECTOCARPALES (below)
1. Thalli with at least some part a pseudoparenchyma or parenchyma....... 2
  2. Growth trichothallic; hairs conspicuous, deciduous................ 3
  2. Growth apical, intercalary, or diffuse; hairs, if present, inconspicuous.. 5
3. Thalli foliose to finely branched, with acid taste, discoloring, producing acrid odor after death...................... DESMARESTIALES (p. 220)
3. Thalli crustose, globular, or erect, with subapical meristem often located at base of a tuft of filaments 4
  4. Thalli parenchymatous .... SPOROCHNALES (p. 184)
  4. Thalli of closely packed filaments .... CHORDARIALES (p. 156)
5. Growth apical 6
5. Growth intercalary or diffuse 8
  6. Plants finely branched, growing in stiff tufts, polysiphonous in structure, with deciduous branches (propagula) .... SPHACELARIALES (p. 216)
  6. Plants complanate, growing from an apical cell or row of apical cells... 7

7. Reproductive structures borne on surface of thallus...DICTYOTALES (p. 206)
7. Reproductive structures (gametangia) borne in conceptacles embedded
    in swollen receptacles ....FUCALES (p. 257)
    8. Thalli with macroscopically distinguishable blade, stipe, and holdfast
        (except *Hedophyllum*) and internally differentiated into epidermis,
        cortex, and medulla .......................LAMINARIALES (p. 228)
    8. Thalli not differentiated into separate tissues or structures 9
9. Plants macroscopic, bearing uniseriate plurangia only, these mostly not
    in sori ...................................SCYTOSIPHONALES (p. 198)
9. Plants macroscopic, bearing only unilocular or both unilocular and pluri-
    locular organs, these mostly in sori...........DICTYOSIPHONALES (p. 186)

Order ECTOCARPALES

Thalli filamentous, with growth mostly intercalary, occasionally becoming multiseriate but not forming a pseudoparenchyma. Chloroplasts
disk-shaped, or band-shaped. Typically with all plants of a species morphologically similar. Sporangial plants bearing only unangia (unilocular
sporangia), or bearing both unangia and plurangia (plurilocular reproductive structures). Gametangial plants bearing plurangia (gametangia) morphologically similar to plurangia on sporangial plants. Gametes isogamous
or anisogamous.

A group of great variability with a single family.

Family ECTOCARPACEAE

With the characters of the order.

1. Vegetative filaments wholly or largely endophytic....*Streblonema* (p. 149)
1. Vegetative filaments chiefly epiphytic or free-living 2
    2. Reproductive structures intercalary, mostly in series ....
        .... *Pilayella* (p. 146)
    2. Reproductive structures terminal or lateral 3
3. Erect filaments in ropelike strands held together by hooked branchlets..
    .... *Spongonema* (p. 148)
3. Erect filaments not in ropelike strands, without hooked branchlets 4
    4. Plurangia small, clustered, often several on a cell....*Sorocarpus* (p. 130)
    4. Plurangia not clustered, borne singly 5
5. Chloroplasts band-shaped, few per cell 6
5. Chloroplasts discoid, mostly many per cell 7
    6. Filaments with true sheathed hairs.............*Kuckuckia* (p. 140)
    6. Filaments without true hairs but sometimes with colorless extensions
        of filaments ....*Ectocarpus* (p. 123)
7. Thalli without clearly delimited growth zones; plurangia sessile........
    .... *Giffordia* (p. 140)
7. Thalli with 1 or more clearly delimited growth zones; plurangia com-
    monly pedicellate 8

* Culture studies by Ravanko (1970) lead her to conclude that *Kuckuckia* cannot
be retained as a genus distinct from *Ectocarpus*.

8. Thalli mostly with a single growth zone per branch, above which no lateral branches occur ....*Feldmannia* (p. 130)
8. Thalli with a number of distinct growth zones distributed in branches .... *Acinetospora** (p. 136)

Ectocarpus Lyngbye 1819

Thalli much branched, attached basally by prostrate or rhizoidal portion, the erect portion variously branched. Growth mostly intercalary. Chloroplasts parietal, band-shaped, 1 to several per cell, each with 1 to several pyrenoids. True hairs lacking. Reproductive structures sessile or pedicellate. Sporangial plants frequently bearing unangia and more commonly also plurangia. Sexual plants bearing plurilocular gametangia; plurangia and gametangia multiseriate.

1. Main filaments corticated below 2
1. Filaments not corticated 3
    2. Plurangia cylindroconical ....*E. acutus*
    2. Plurangia ovoid ....*E. corticulatus*
3. Growing on parasitic isopods ....*E. isopodicola*
3. Not growing on parasitic isopods 4
    4. Upper branches commonly in secund series; plurangia frequently 500 μm or more long, often with sterile apex ....*E. siliculosus*
    4. Upper branches mostly not in secund series; plurangia much shorter, without sterile apex 5
5. Plants saxicolous; branch growth apical ....*E. chantransioides*
5. Plants typically epiphytic; branch growth intercalary 6
    6. Less than 0.5 mm tall, partly endophytic ....*E. gonodioides*
    6. Mostly more than 1 mm tall, not endophytic 7
7. Erect branches simple, 1–2 mm tall; plurangia mostly sessile 8
7. Erect branches simple or branched, to 25 mm tall; plurangia sessile or pedicellate 9
    8. Erect branches piliferous, not tufted ....*E. taoniae*
    8. Erect branches not piliferous, tufted ....*E. simulans*
9. Erect filaments simple or with few branches ....*E. parvus*
9. Erect filaments profusely branched, the branches beset with numerous short ramuli ....*E. fructosus*

Ectocarpus acutus S. & G.

Setchell & Gardner 1922: 404; Smith 1944: 80.

Thalli epiphytic, 5–7 cm tall, feathery, medium brown to olive brown; prostrate filaments not penetrating deeply into host; main erect filaments corticated below, profusely branched; branching mostly alternate below, secund above, with gradually attenuated apices; chloroplasts few per cell, each with several pyrenoids.

* See footnote opposite. Ravanko also questions the taxonomic status of the genus *Acinetospora*.

Ectocarpus acutus var. acutus

Thalli more than 3 cm tall, cells barrel-shaped, 40–60 μm diam. in lower portion of filaments, 0.4–2 times as long; unangia unknown; plurangia on erect filaments only, very numerous, cylindroconical, regularly spaced, 20–35 μm diam., 100–150(230) μm long, on pedicels of mostly 1 or 2 cells.

Frequent on *Desmarestia*, especially *D. ligulata*, low intertidal to subtidal, Br. Columbia and Ore.; in Calif., from Humboldt Co. to San Pedro. Type locality: Carmel, Calif.

Ectocarpus acutus var. haplogloiae Doty

Doty 1947a: 32.

Thalli mostly less than 3 cm tall; cells of main axes 60 μm diam., mostly shorter than broad; plurangia 19–25 μm diam., 100–200 μm long, elliptical-lanceolate, on 1- or few-celled pedicels.

Occasional, on *Haplogloia andersonii*, midtidal on rocks, Wash. to San Mateo Co., Calif. Type locality: Middle Bay (Cape Arago) Ore.

Ectocarpus chantransioides S. & G.

Setchell & Gardner 1922: 406; 1925: 430.

Thalli forming hemispherical cushions; creeping filaments contorted, bearing numerous erect branches 4–8 mm tall; erect filaments profusely branched, with alternate branching below and mostly secund above; main filaments and branches not attenuated, with distinct apical growth; cells 8–10 μm diam., 2–3 times as long in middle parts of branches; unangia unknown; plurangia mostly sessile or on short pedicels, narrowly cylindroconical, 80–110 μm long, 16–20 μm diam. at base.

Saxicolous, lower intertidal on boulders; known only from near Santa Monica (type locality) and Corona del Mar (Orange Co.), Calif.

Ectocarpus corticulatus Saund.

Saunders 1898: 152; Setchell & Gardner 1925: 418. *Ectocarpus corticulans* of Smith 1944: 81.

Thalli usually epiphytic, 3–30 mm tall, tufted, arising from small, compact mass of creeping filaments; erect filaments 90–120 μm diam., usually densely corticated below, frequently and irregularly branched, with non-tapering branches of variable length; cells in lower part of erect filaments barrel-shaped, mostly 0.8–1.4 times as long as diam.; chloroplasts at first band-shaped, few per cell, later forming numerous divisions, each division with 1 pyrenoid; unangia unknown; plurangia on erect filaments only, narrowly to broadly ovoid, sessile or on short pedicels, often in secund series on upper branchlets, 30–70(90) μm long, 12–30 μm diam.

Occasional on larger brown algae or on *Zostera*, Alaska to San Pedro, Calif. Type locality not specified, probably Pacific Grove, Calif.

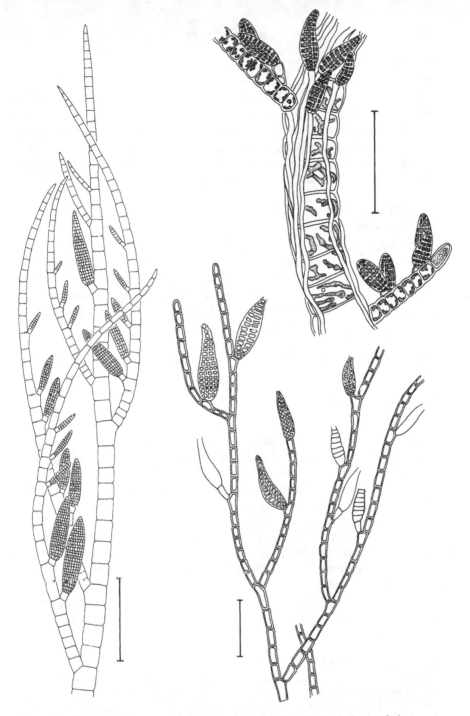

Fig. 83 (left). *Ectocarpus acutus*. Rule: 100 μm. (JRJ/S)   **Fig. 84** (below right). *E. chantransioides*. Rule: 100 μm. (SM, after Setchell & Gardner)   **Fig. 85** (above right). *E. corticulatus*. Rule: 100 μm. (DBP, after Saunders)

Ectocarpus fructosus S. & G.

Setchell & Gardner 1922: 410; 1925: 419.

Thalli tufted, profusely alternately branched, to 2.5 cm tall, arising from mass of short creeping filaments; erect filaments with many long branches provided throughout with numerous short ramuli, bearing numerous short, sessile or briefly pedicellate, broadly conical plurangia; unangia unknown.

Epiphytic on floating pneumatocysts of *Nereocystis*, Moss Beach (San Mateo Co.; type and only known locality), Calif.

Ectocarpus gonodioides S. & G.

Setchell & Gardner 1924a: 721.

Thalli epiphytic, densely tufted, less than 1 mm tall; prostrate filaments much branched, deeply penetrating host; erect filaments branched below, simple above, the apex attenuate to blunt; cells of lower parts of erect filaments 8–14(24) μm diam., mostly 1–1.5 times as long; chloroplasts few per cell; unangia unknown; plurangia restricted to base of erect branches, narrowly fusiform, 70–110 μm long, 15–25 μm diam., on pedicels of 1 to many cells.

Rare, on *Codium* spp., low intertidal, Carmel, Calif., and I. Smith (type locality), Gulf of Calif.

Ectocarpus isopodicola Daws.

Dawson 1945b: 81.

Thalli minute, epizoic, attached to isopods parasitizing fish, 2–4 mm tall, from low-pulvinate base of densely branched filaments, these not penetrating host tissues; erect filaments ecorticate, with attenuate apices, often piliferous, mostly simple below, sparingly branched above, with relatively long branches; cells of main erect filaments 45–55 μm diam., as long as diam. or longer, distinctly constricted at crosswalls; chloroplasts band-shaped, with several to many pyrenoids per cell; unangia unknown; plurangia numerous and closely spaced, borne on upper parts of erect branches, conical to fusiform, mostly 70–80 μm long, 25–30 μm diam., sessile, or on pedicels of 1–4 cells.

On isopods found in gill chambers of certain bottom fish; known only from Pacific Grove (sterile specimen) and Newport Bay (Orange Co.; type locality), Calif.

Ectocarpus parvus (Saund.) Hollenb.

Ectocarpus siliculosus var. *parvus* Saunders 1898: 153. *E. parvus* (Saund.) Hollenberg 1971b: 283. *E. confervoides* f. *parvus* (Saund.) Setchell & Gardner 1922: 414. *E. commensalis* S. & G. 1922: 407. *E. eramosus* S. & G. 1922: 407. *E. mesogloiae* S. & G. 1922: 411. *E. dimorphus* Silva 1957b: 42. *E. confervoides* f. *pygmaeus sensu* S. & G. 1925: 415. *E. confervoides* var. *pygmaeus sensu* Hollenberg & Abbott 1968: 1237.

Thalli epiphytic, forming more or less extended patches on host, usually

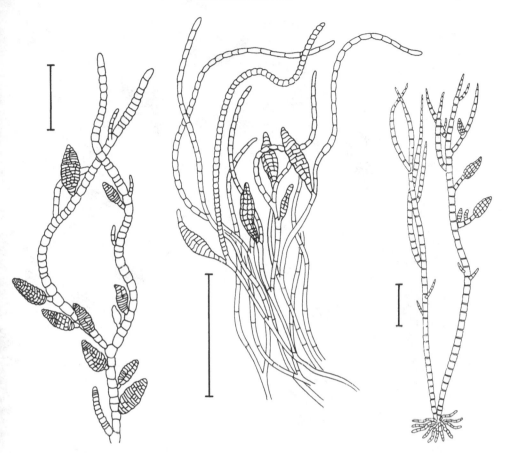

Fig. 86 (left). *Ectocarpus fructosus.* Rule: 250 µm. (SM, after Setchell & Gardner)
Fig. 87 (center). *E. gonodioides.* Rule: 250 µm. (SM, after Setchell & Gardner) **Fig. 88**
(right). *E. isopodicola.* Rule: 250 µm. (SM, after Dawson)

not penetrating host; erect filaments mostly 4–12 mm tall, mostly 14–25
µm diam., simple or sparingly branched, arising from spreading prostrate
filaments; cells of erect filaments mostly 1.5–2 times as long as diam. below,
shorter above, only slightly or not at all constricted at crosswalls, slightly
attenuated apically or sometimes ending in a pseudohair; chloroplasts
irregularly band-shaped, 1 to several per cell; unangia ovoid to globose,
30–35 µm diam., 40–50 µm long, mostly pedicellate, infrequent; plurangia
terminal or lateral on erect filaments, sessile or briefly pedicellate, conical
to fusiform, 60–125 µm long, 20–33 µm diam.; plurangia infrequently intercalary, seriate, to 400 µm long.

Common, epiphytic on Laminariales and Desmarestiales, low intertidal
to subtidal, Alaska to San Diego, Calif., and Baja Calif.; common throughout Calif. Type locality: San Pedro, Calif.

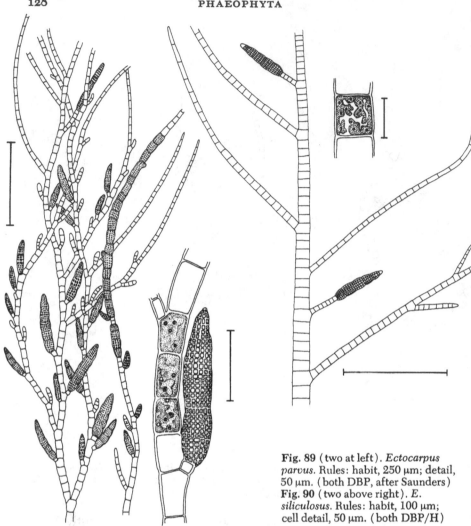

Fig. 89 (two at left). *Ectocarpus parvus.* Rules: habit, 250 μm; detail, 50 μm. (both DBP, after Saunders) **Fig. 90** (two above right). *E. siliculosus.* Rules: habit, 100 μm; cell detail, 50 μm. (both DBP/H)

Ectocarpus siliculosus (Dillw.) Lyngb.

Conferva siliculosa Dillwyn 1809: 69. *Ectocarpus siliculosus* (Dillw.) Lyngbye 1819: 131; Setchell & Gardner 1925: 410.

Thalli tufted, basically epiphytic, 3–30 mm long, light brown or yellowish, with branches pseudodichotomously branched below, alternately secund above; lesser branches erect at mostly narrow angles, never patent or fascicled; cells in major branches 50–60 μm diam., 1–2 times as long; chloroplasts irregularly ribbonlike, several per cell, each with several pyrenoids; unangia unknown for Calif.; plurangia subulate-conical, 12–25 μm diam., 100–600 μm long, sometimes ending in a long multicellular extension, on mostly short pedicels.

Infrequent, usually epiphytic, Coos Bay, Ore., and San Francisco Bay and Malpaso Creek (south of Carmel), Calif. Widely distributed in N. Atlantic. Type locality: England.

Free-floating forms are probably to be identified with *E. siliculosus* f. *subulatus* (Kütz.) S. & G., 1922: 416. A dwarf form only a few mm high is occasional as an epiphyte on Laminariales on exposed coasts of Central Calif.

Ectocarpus simulans S. & G.

Setchell & Gardner 1922: 412; 1925: 422. *Ectocarpus terminalis sensu* S. & G. 1925: 421.

Thalli epiphytic, tufted, 1–2 mm tall, arising from small cushion of tortuous prostrate filaments; erect filaments ecorticate, mostly unbranched, slightly attenuated to rounded apices; cells in middle portion of erect filaments 11–13(25) μm diam., 1–2.5 times as long; chloroplasts few in each

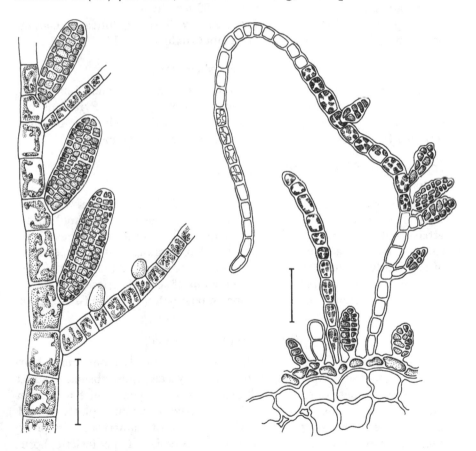

Fig. 91 (left). *Ectocarpus simulans*. Rule: 50 μm. (SM, after Setchell & Gardner)
Fig. 92 (right). *E. taoniae*. Rule: 50 μm. (SM, after Setchell & Gardner)

cell, pyrenoids unknown; unangia unknown; plurangia on erect filaments only, sessile, mostly lateral, ellipsoidal, with blunt apices, 55–65(90) μm long, 15–20(35) μm diam.

Rare, on other algae or on wood, high intertidal, Cypress Pt. (Monterey Co.; type locality) and Laguna Beach, Calif.

Ectocarpus taoniae S. & G.

Setchell & Gardner 1922: 413; 1925: 420.

Thalli minute, 0.5–1.5 mm tall, arising from profusely branched creeping filaments; erect filaments simple, slightly narrowed at base, gradually attenuated upward, piliferous; cells of erect filaments 8–10 μm diam., quadrate below, to 6 times as long as diam. above; chloroplasts short and relatively broad irregular bands, pyrenoids unknown; plurangia ellipsoidal, sometimes slightly asymmetrical, mostly sessile on creeping and erect filaments, 20–29(40) μm long, 15–20 μm diam.

Infrequent, epiphytic on *Taonia* and *Zonaria*, forming diffuse patches on host, San Pedro (type locality) and Santa Catalina I., Calif.

Sorocarpus Pringsheim 1863

Thalli erect, filamentous, with uniseriate ecorticate branches. Commonly with colorless hairs. Chloroplasts discoid, many per cell. Pyrenoids unknown. Plurangia small, multiseriate, mostly in dense clusters. Unangia reported for certain Japanese species. Gametes anisogamous.

Sorocarpus pacifica Hollenb.

Hollenberg 1971a: 14.

Filaments to 1 cm tall, repeatedly branched; main branches 35–45 μm diam., narrowing to slender apices and abruptly narrowed to simple basal attachment; cells at middle of main filaments 0.5–1 times as long as diam., to twice as long above; hairs not observed; plurangia ovoid to conical, often slightly asymmetrical, 14–20 μm long, 10–13 μm diam., mostly clustered on short lateral branchlets or in compact secund series.

Known only from a single collection on boat bottom, Pacific Grove, Calif. (type locality).

Feldmannia Hamel 1939

Thalli small, tufted, with branching in general mostly basal; sterile hair-like filaments usually present; true (phaeophycean) hairs absent. Distinct growth zone at base of elongate ultimate branches beyond which no further laterals occur, frequently terminating as sterile hair. Chloroplasts discoid, numerous, each with 1 pyrenoid. Unangia ovoid to spherical, infrequent; plurangia usually symmetrical, multiseriate, occasionally pedicellate, borne below growth regions.

The genus is not clearly delimited.

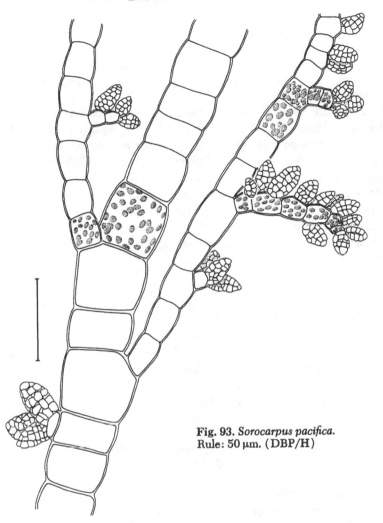

Fig. 93. *Sorocarpus pacifica.*
Rule: 50 µm. (DBP/H)

1. Plurangia subglobular ....*F. globifera*
1. Plurangia not subglobular 2
    2. Thalli growing in conceptacles of *Cystoseira* and *Halidrys*; plurangia
    long-acuminate ....*F. acuminata*
    2. Thalli not growing in fucoid conceptacles; plurangia not long-acumi-
    nate 3
3. Plurangia mostly pedicellate 4
3. Plurangia mostly sessile 5
    4. Branching mostly basal; plurangia mostly toward base of erect
    branches ....*F. cylindrica*
    4. Branching not mostly basal; plurangia distributed throughout thallus
    ....*F. hemispherica*
5. Plurangia conical, commonly in axillary groups ....*F. irregularis*
5. Plurangia not conical, not in axillary groups 6

6. Thalli epizoic, 1–2 mm tall . . . . *F. chitonicola*
6. Thalli epiphytic or saxicolous, to 15+ mm tall . . . . *F. rhizoidea*

Feldmannia acuminata (Saund.) Hollenb. & Abb.

Ectocarpus acuminatus Saunders 1898: 149; Setchell & Gardner 1925: 435. *Feldmannia acuminata* (Saund.) Hollenberg & Abbott 1966: 18.

Thalli minute, attached basally by a network of delicate rhizoidal filaments; erect filaments to 1 mm tall, simple or sparingly branched below; cells of erect filaments 12–14 μm diam., about as long as diam. below, to 5 times as long above; unangia unknown; plurangia 90–300 μm long, 20–30 μm diam. at base, cylindroconical to very long-acuminate, with 1 or more sterile cells at apex, sessile on creeping filaments or bases of erect filaments.

Occasional, upper midtidal, growing within conceptacles of *Cystoseira* and *Halidrys*, only the upper part of erect filaments projecting through ostiole of host conceptacle, Pacific Grove (type locality) and San Diego, Calif.

Feldmannia chitonicola (Saund.) Levr.

Ectocarpus chitonicola Saunders 1898: 150; Setchell & Gardner 1925: 436. *Feldmannia chitonicola* (Saund.) Levring 1960: 15.

Thalli tufted, 1–2 mm tall; prostrate filaments numerous, irregularly branched, not penetrating host; erect filaments mostly simple, sometimes sparingly branched near base, slightly attenuated upward; cells of erect filaments about 14 μm diam., 0.5–2 times as long at base, to 3 times as long above; unangia unknown; plurangia 90–175(250) μm long, 25–35 μm diam., the longer ones with many small loculi, the shorter ones with few large loculi.

Common on shells of chitons or limpets, midtidal, Pacific Grove (type locality) and Carmel Bay, Calif. Also Chile.

Feldmannia cylindrica (Saund.) Hollenb. & Abb.

Ectocarpus cylindricus Saunders 1898: 150. *Feldmannia cylindrica* (Saund.) Hollenberg & Abbott 1966: 19. *F. simplex sensu* R. Norris & Wynne 1968: 135 (in part). *E. cylindricus* var. *codiophilus* S. & G. 1922: 415. *E. flocculiformis* S. & G. 1922: 409. *E. socialis* S. & G. 1922: 412.

Thalli in dense tufts or patches, mostly epiphytic, 1–2(10) mm tall; prostrate filaments irregularly branched, often penetrating host tissues; erect filaments mostly not attenuated above, rarely piliferous, with few short lateral branches; cells 18–30 μm diam., nearly as long in midpart, 1.5–3 times as long near base and apex; reproductive structures mostly near base of erect filaments, sometimes variously distributed, frequently in opposite pairs; unangia ovoid to ellipsoid, sessile or pedicellate, 60–120 μm long, 30–40 μm diam.; plurangia cylindrical with broadly rounded apices, or sometimes ellipsoidal to somewhat conical, mostly pedicellate, 80–270 μm long, 30–44 μm diam.

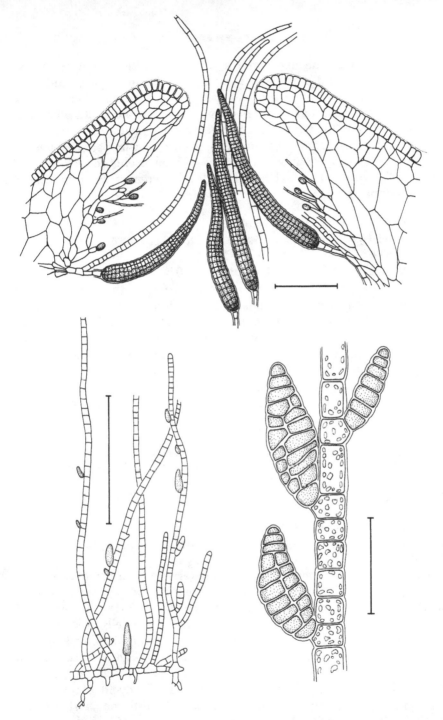

Fig. 94 (above). *Feldmannia acuminata*, in conceptacle of *Halidrys*. Rule: 60 μm. (DBP/H) **Fig. 95** (two below). *F. chitonicola*. Rules: below left, 500 μm; below right, 50 μm. (both DBP, after Saunders)

Frequent, on other algae or limpets, mid- to lower intertidal, N. Wash. to La Jolla, Calif., and Baja Calif. Type locality: Pacific Grove, Calif.

Found chiefly on Laminariales, this species includes a piliferous form on *Codium*, with rhizoidal filaments penetrating the host and with plurangia confined mostly to the basal part of erect filaments; and a rare form to 10 mm tall, with mostly opposite plurangia, growing on limpets (Setchell & Gardner 1922: 415). Culture studies would be useful in determining whether the relatively small morphological differences are a reflection of substratum preference.

Feldmannia globifera (Kütz.) Ham.

Ectocarpus globifer Kützing 1843: 289; Setchell & Gardner 1925: 438. *Feldmannia globifera* (Kütz.) Hamel 1931–39 (1939): xvii.

Thalli densely tufted, mostly epiphytic, 0.5–1(5) cm tall, sparingly branched, attached by numerous rhizoidal filaments, these in epiphytic specimens penetrating deeply into host; branches mostly basal, alternate or opposite, divaricate, constricted at base, slightly attenuated above to pseudohairs above a distinct growth zone; cells cylindrical, slightly or not at all constricted at crosswalls, 35–50 µm diam. in main filaments, 0.5–1.5 times as long in midpart, 2–4(7) times as long above; unangia spherical or subspherical, 30–40 µm diam., mostly on short 1-celled pedicels; plurangia broadly ovoid to subspherical, 70–100(240) µm long, 40–50(70) µm diam., mostly on short 1-celled pedicels, distributed along lower half of filaments.

Infrequent to rare epiphyte (mostly on *Pelvetia*, *Codium*, or other algae), occasionally on wood, mid- to lower intertidal, San Pedro to La Jolla, Calif.; very common on Santa Catalina I., Calif. Frequent in Europe. Type locality: Yugoslavia.

Feldmannia hemispherica (Saund.) Hollenb.

Ectocarpus hemisphericus Saunders 1898: 151. *Feldmannia hemispherica* (Saund.) Hollenberg 1971b: 285.

Thalli pulvinate epiphytes 2–10 mm tall, arising from compact mass of creeping filaments, these sometimes penetrating host deeply; erect branches sparsely pseudodichotomously branched below, with few to numerous short lateral branches; main branches long-piliferous; cells in lower parts 14–26 µm diam., 1–3 times as long as diam. or considerably longer in pseudohairs above distinct growth zone; filaments slightly constricted at crosswalls; unangia cylindrical, 30–35(50) µm long, 15–25 µm diam., on short, 1-celled pedicels; plurangia fusiform to broadly ovoid, obtuse, 14–20 (35) µm diam., 30–90(125) µm long, sessile or mostly on short, 1-celled pedicels, distributed mostly on middle or lower parts of branches.

Infrequent, mostly on *Pelvetia*, but sometimes on other algae, midtidal,

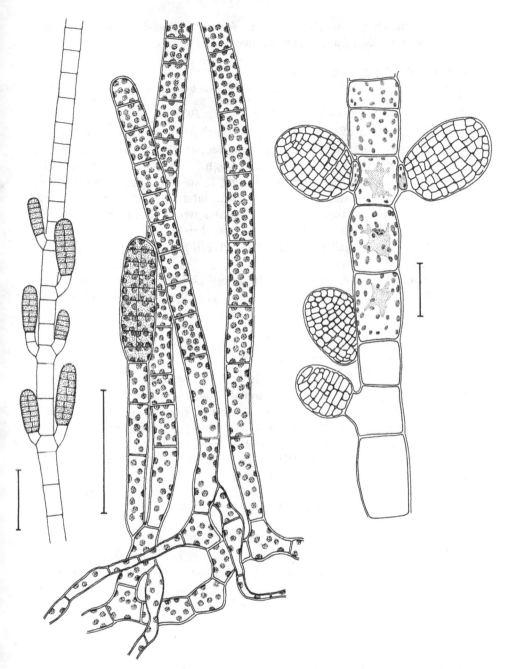

Fig. 96 (left & center). *Feldmannia cylindrica*. Rules: left, 50 μm; center, 100 μm. (both JRJ/S) Fig. 97 (right). *F. globifera*. Rule: 50 μm. (DBP/H)

Redondo Beach to San Diego, Calif.; formerly abundant at Corona del Mar (Orange Co.), Calif. Type locality: San Pedro, Calif.

Feldmannia irregularis (Kütz.) Ham.

Ectocarpus irregularis Kützing 1845: 234. *Feldmannia irregularis* (Kütz.) Hamel 1931–39 (1939): xi; Cardinal 1964: 54. *E. mucronatus* Saunders 1898: 152. *E. coniferus* Børgesen 1914: 8. *Giffordia conifera* (Børg.) Taylor 1960: 207.

Thalli tufted, to 1.2 cm tall, attached by short creeping filaments; erect axes numerous, sparingly to richly branched, ecorticate, 25–40 µm diam., with cells mostly 0.5–1 times as long, slightly constricted at crosswalls, mostly with distinct growth zones, ending in short or long colorless hairs, these with cells 2–4 times as long as diam.; lateral branches mostly short, tapering, sometimes recurved; unangia unknown for Calif.; plurangia cylindroconical, broadly sessile or sometimes briefly pedicellate, lateral or frequently in brief axillary series, (55)70–140 µm long, 25–40 µm diam. at base, mostly symmetrical.

Relatively rare, saxicolous or epiphytic, low intertidal, San Pedro and La Jolla, Calif., and Pta. Banda, Baja Calif. Widely distributed. Type locality: Adriatic.

Feldmannia rhizoidea Hollenb. & Abb.

Hollenberg & Abbott 1968: 1238.

Thalli tufted, 7–15 mm tall, basally attached by several short branching rhizoidal filaments; erect branches mostly 25–35 µm diam., simple or with occasional rhizoidal laterals above, narrowed toward apices but hardly piliferous, the cells mostly 1.5–2.5 times as long as diam., commonly with a single intercalary growth zone in lower part; unangia unknown; plurangia fusiform-conical, 100–150(210) µm long, 40–50(70) µm diam., mostly sessile, solitary or in short secund series.

Probably infrequent, epiphytic or on wood, midtidal, Pacific Grove, Calif. (type and only reported locality).

Acinetospora Bornet 1891

Thalli erect, delicate, the filaments profusely branched; branches with 1 to many distinct growth zones. Chloroplasts numerous, discoid, each with a lateral pyrenoid. Unangia infrequent, lateral, spherical or ovoid. Plurangia small, lateral, with relatively large loculi.

Acinetospora nicholsoniae Hollenb.

Hollenberg 1971a: 12.

Thalli with pale brown filaments, forming tangled, stringy to filmy masses 10–18 cm tall, arising from short-celled creeping branches supplemented by rhizoids; main filaments 30–45 µm diam., with cells 0.8–1.5

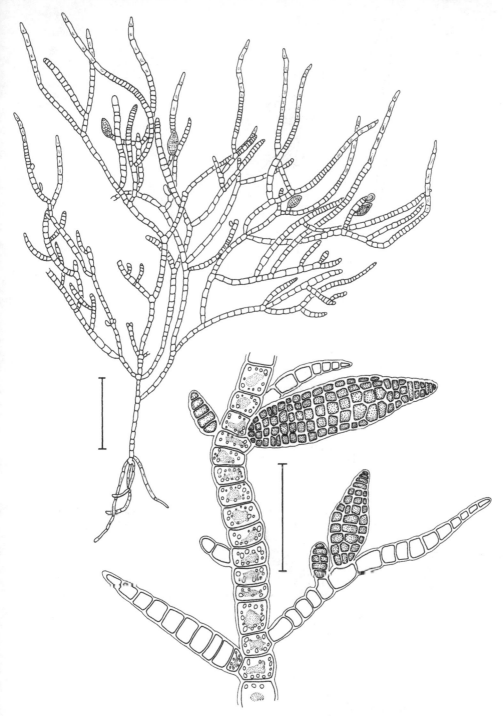

Fig. 98 (above). *Feldmannia hemispherica*. Rule: 250 μm. (DBP, after Saunders)
Fig. 99 (below). *F. irregularis*. Rule: 50 μm. (DBP/H)

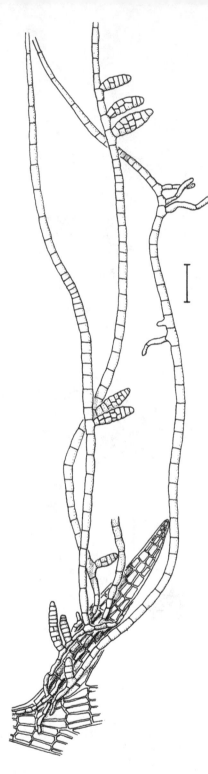

Fig. 100. *Feldmannia rhizoidea.*
Rule: 100 μm. (DBP/H)

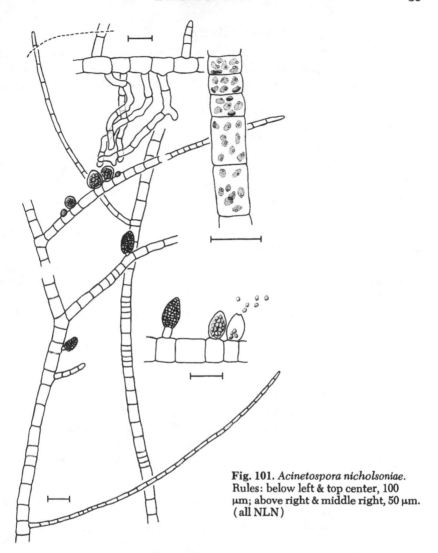

Fig. 101. *Acinetospora nicholsoniae.*
Rules: below left & top center, 100
μm; above right & middle right, 50 μm.
(all NLN)

times as long, narrowing gradually to 13–17 μm diam.; branches of several to many orders, pseudodichotomously branched below, spreading above; unangia globular to shortly elliptical, 24–33 μm diam., sessile; plurangia elliptical to conical, mostly 35–44 μm long, 17–24 μm diam., sessile or mostly on short, 1-celled pedicels.

Growing in abundance on *Sargassum* spp. or any of a variety of other algae; low intertidal, known only from type locality, Santa Catalina I., Calif.

Kuckuckia Hamel 1939

Thalli erect, frequently to profusely branched; endogenous hairs present. Chloroplasts few, ribbonlike, each with a number of small pyrenoids. Unangia ovoid. Plurangia conical to cylindrical, sometimes with an apical hair.

Kuckuckia catalinae Hollenb.

Hollenberg 1971a: 13.

Thalli pulvinate tufts 2–4 mm diam.; filaments mostly 9–12 µm diam., with rounded apices, arising from compact mass of short-celled creeping branches; cells of erect branches mostly 1.5–2.5 times as long as diam., only slightly narrowed at crosswalls; branches of 3 or 4 orders, alternate to unilateral; endogenous hairs 1–3 mm long, 6.5–8 µm diam., with basal meristem and sheath, and with distal cells to 10 times as long as diam.; chloroplasts parietal, irregularly ribbon-shaped; unangia unknown; plurangia solitary, sessile, lanceolate, in 2 or 3 series, with uniseriate terminal portion, borne near middle of erect branches.

Common, saxicolous, on upper intertidal rocks; known only from type locality, Santa Catalina I., Calif.

Giffordia Batters 1893

Thalli filamentous, of variable height, mostly much branched; growth zones not obvious. Chloroplasts mostly numerous, parietal, discoid, each with 1 pyrenoid. Unangia rare. Plurangia nearly always sessile, commonly asymmetrical and in secund series. At least some species with plurangia of 3 types: neutral sporangia on asexual plants, and male and female gametangia with small and large loculi respectively; gametes in such species anisogamous.

1. Plurangia in long, continuous secund series, 1 on each cell . . . . . G. hincksiae
1. Plurangia in short series or scattered 2
   2. Opposite branches frequent; plurangia prominently asymmetrical. . .
   . . . . G. granulosa
   2. Opposite branches lacking or rare 3
3. Plurangia subfusiform to broadly ovoid; slightly asymmetrical; main axes usually corticated below . . . . G. sandriana
3. Plurangia ellipsoidal to cylindrical, symmetrical; main axes without cortical filaments 4
   4. Plurangia pedicellate . . . . G. saundersii
   4. Plurangia sessile . . . . G. mitchelliae

Giffordia granulosa (J. E. Smith) Ham.

Conferva granulosa J. E. Smith 1814: pl. 2351. Giffordia granulosa (J. E. Smith) Hamel 1931–39 (1939): xv; Cardinal 1964: 39. Ectocarpus granulosus (J. E. Smith) C. Agardh 1828: 45; Setchell & Gardner 1925: 426. E. oviger Harvey 1862: 167.

Thalli profusely branched, 1–25 cm tall, attached by rhizoidal cushion;

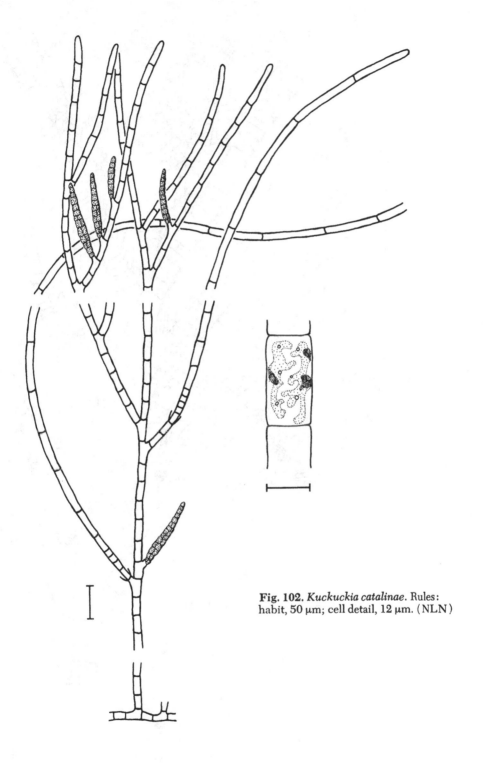

Fig. 102. *Kuckuckia catalinae.* Rules: habit, 50 μm; cell detail, 12 μm. (NLN)

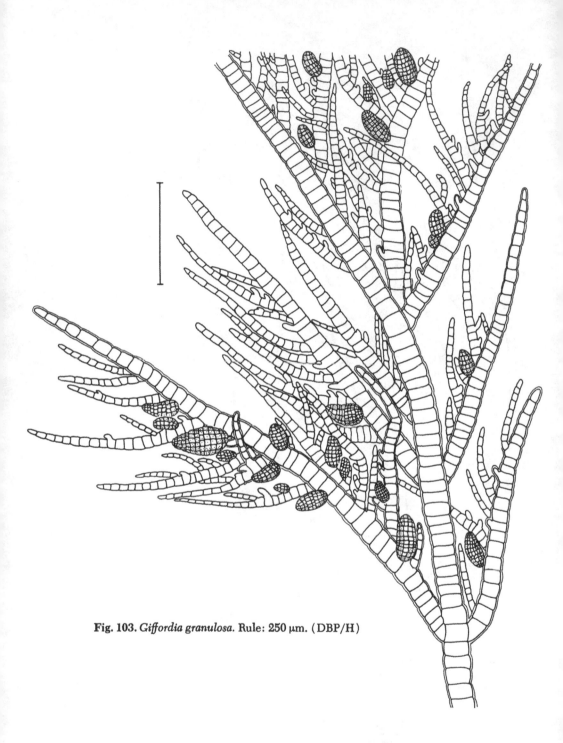

Fig. 103. *Giffordia granulosa.* Rule: 250 µm. (DBP/H)

major branches corticated below, lateral branches commonly opposite, ultimate branchlets secund and gradually attenuated to rounded apices, these sometimes piliferous; cells in major branches barrel-shaped, 50–100 μm diam., mostly shorter than diam.; unangia globose, not reported for Calif.; plurangia broadly ovoid, asymmetrical, broadly sessile, mostly secund, seriate and adaxial, 60–100(165) μm long, 30–60 μm diam.

Occasional on rocks or various algae, low intertidal, S. Br. Columbia to Pta. Thurloe, Baja Calif.; frequent in Calif., a common spring and summer annual in S. Calif. Also N. Atlantic. Type locality: England.

Giffordia hincksiae (Harv.) Ham.

Ectocarpus hincksiae Harvey 1841b: 40. *Giffordia hincksiae* (Harv.) Hamel 1931–39 (1939): xiv.

The type variety does not occur in Calif.

Giffordia hincksiae var. californica Hollenb. & Abb.

Hollenberg & Abbott 1968: 1238.

Thalli minute, epiphytic to 4.5 μm tall, arising from basal mass of rhizoidal filaments; main erect branches simple or with a few secund lateral branches, ecorticate, nonpiliferous; cells 130–145 μm diam., mostly slightly longer than diam. in lower vegetative parts, mostly slightly shorter in fertile branches; chloroplasts relatively few per cell; unangia unknown; plurangia sessile, broadly conical to ovoid, 150–200 μm long, 90–100 μm diam., in long continuous secund series, 1 on each cell.

Rare, on *Laminaria* stipes, low intertidal, C. Calif. (Santa Cruz and San Luis Obispo Cos.). Type locality: Davenport (Santa Cruz Co.), Calif.

This variety is much smaller than the type variety, with few lateral branches, much broader main branches, and larger plurangia.

Giffordia mitchelliae (Harv.) Ham.

Ectocarpus mitchelliae Harvey 1852: 142; Setchell & Gardner 1925: 428. *Giffordia mitchelliae* (Harv.) Hamel 1931–39 (1939): xiv; Cardinal 1964: 45.

Thalli densely tufted, feathery, 2–8 cm tall, attached by long creeping filaments, sometimes slightly corticated below, light brown to yellowish, drying to olive green; branching alternate; branches divaricate, attenuate, often ending in short pseudohairs; cells in main branches 25–40 μm diam., 1–3 times as long; unangia rare, oval, 70–100 μm long, 35–50 μm diam.; plurangia neutral, male, or female, the 3 types differing in size, all plurangia 50–150 μm long, 18–35 μm diam., sessile, mostly secund, ellipsoid to cylindrical.

Common, epiphytic or saxicolous, lower intertidal, Monterey, Redondo Beach and Santa Catalina I., Calif., to Baja Calif. Widely distributed in N. Hemisphere. Type locality: Massachusetts.

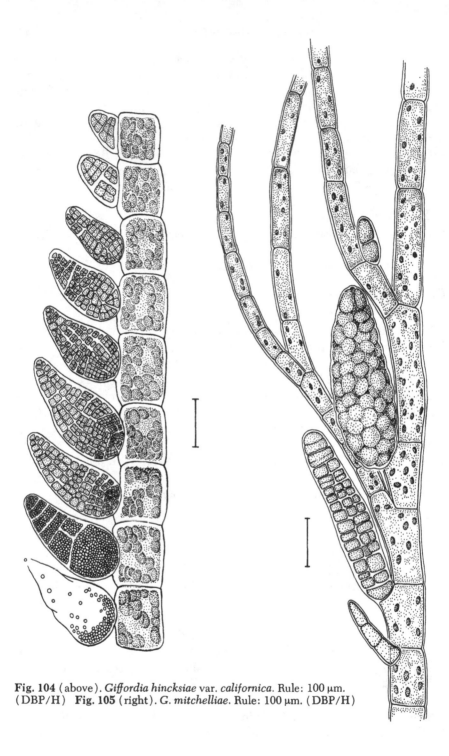

Fig. 104 (above). *Giffordia hincksiae* var. *californica*. Rule: 100 μm. (DBP/H) **Fig. 105** (right). *G. mitchelliae*. Rule: 100 μm. (DBP/H)

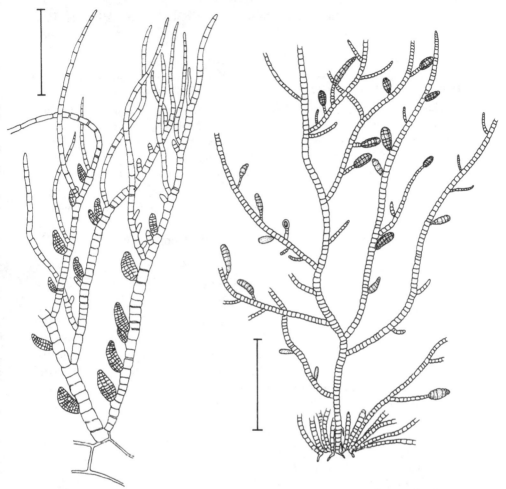

Fig. 106 (left). *Giffordia sandriana*. Rule: 250 μm. (DBP, after Setchell & Gardner)
Fig. 107 (right). *G. saundersii*. Rule: 250 μm. (DBP, after Saunders)

Giffordia sandriana (Zan.) Ham.

Ectocarpus sandrianus Zanardini 1843: 41. *Giffordia sandriana* (Zan.) Hamel 1931–39 (1939): xiv; Cardinal 1964: 37. *E. granulosoides* Setchell & Gardner 1922: 410.

Thalli epiphytic, 2–5 cm tall, profusely branched, the main branches sometimes corticated below, subdichotomously or alternately branched below, more or less secund above; ultimate ramuli acute, nonpiliferous; all branches abruptly narrowed at base; cells of main filaments 40–80 μm diam., 0.5–1 times as long; unangia not known for Calif.; plurangia secund on adaxial side of ultimate and subultimate ramuli, 40–60 μm long, 12–20 μm diam., slightly asymmetrical.

Occasional epiphyte on larger algae, intertidal to subtidal, Steamboat I.,

Puget Sd., to San Diego, Calif. Widely distributed in N. Atlantic. Type locality: Yugoslavia.

Giffordia saundersii (S. & G.) Hollenb. & Abb.

Ectocarpus saundersii Setchell & Gardner 1922: 411. *Giffordia saundersii* (S. & G.) Hollenberg & Abbott 1966: 17. *E. paradoxus* var. *pacificus* Saunders 1898: 152.

Thalli usually epiphytic, 2–5 mm tall; prostrate filaments branched, densely compacted, spreading; erect filaments much branched, the branching alternate, the branches divergent, tapering to acute apices; cells in lower part of erect branches 25–40 μm diam., about as long as diam.; chloroplasts many per cell; reproductive structures on erect filaments only; unangia globose, pedicellate, occasionally intercalary, about 30 μm diam.; plurangia cylindrical to ovoid, with obtuse or acuminate apices, pedicellate, 70–150 μm long, 25–50 μm diam.

On *Fucus* or piling; midtidal, known only from Monterey Co., Calif. Type locality: Pacific Grove, Calif.

Pilayella Bory 1823
(= *Pylaiella* Bory)

Thalli filamentous, with prostrate portion of irregularly branched filaments bearing simple or branched erect filaments. Growth intercalary, the growth zones not readily distinguishable. True (phaeophycean) hairs absent. Chloroplasts parietal, discoid to band-shaped, each with 1 pyrenoid. Unangia and plurangia intercalary or sometimes terminal, commonly in continuous series. Gametangial and sporangial plants similar in size and vegetative structure.

1. Erect filaments richly branched ....*P. littoralis*
1. Erect filaments simple or sparingly branched 2
  2. Thalli growing on *Postelsia* ....*P. gardneri*
  2. Thalli growing on *Pleurophycus* ....*P. tenella*

Pilayella gardneri Coll.

Collins, P.B.-A., 1895–1919 [1912]: no. 1384. *Leptonema fasciculatum sensu* Saunders 1899: 38.

Thalli tufted, epiphytic or slightly endophytic, 0.4–1.2(2) cm tall, arising from creeping filaments 6–10 μm diam.; erect filaments simple or occasionally with opposite branches, these to 500 μm long, clavate to fusiform, occasionally whorled; cells in lower part of erect filaments 6–10(15) μm diam., 1.5–2 times as long, increasing slightly in diam. upward; upper parts of erect filaments and laterals becoming fertile, forming continuous multiseriate plurangia; chloroplasts band-shaped, 1–3 per cell, later dividing into several fragments; unangia unknown; plurangia commonly 4–6 cells wide, cylindrical, to 60 μm diam., 2–4 times longer, interrupted at points of attachment of lateral branches, where axis is 1 or 2 cells wide.

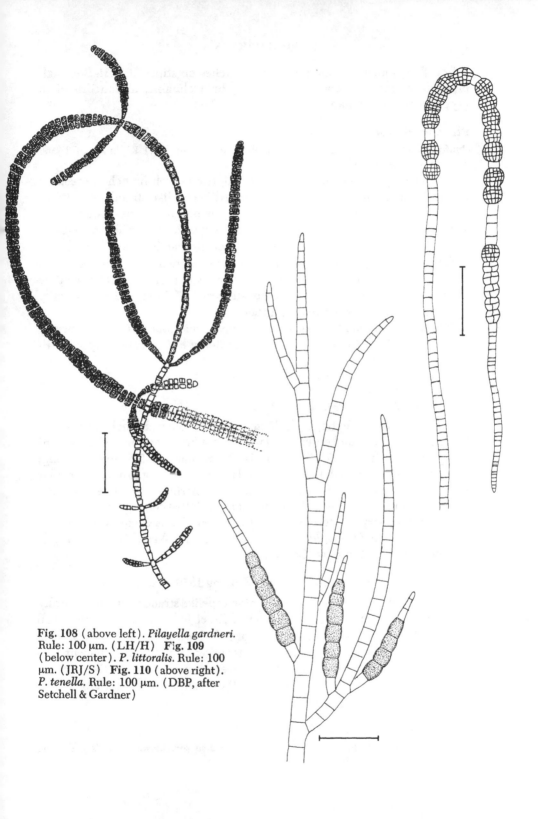

Fig. 108 (above left). *Pilayella gardneri.*
Rule: 100 μm. (LH/H) **Fig. 109**
(below center). *P. littoralis.* Rule: 100
μm. (JRJ/S) **Fig. 110** (above right).
P. tenella. Rule: 100 μm. (DBP, after
Setchell & Gardner)

Locally common, forming extensive patches on stipes of *Postelsia*, high intertidal exposed to surf, Vancouver I., Br. Columbia, to San Francisco (type locality) and San Luis Obispo Co., Calif.

Pilayella littoralis (L.) Kjellm.

Conferva littoralis Linnaeus 1753: 1165. *Pilayella littoralis* (L.) Kjellman 1872: 99; Setchell & Gardner 1925: 402.

Thalli richly branched, 2–5(60) cm tall, from much branched creeping filaments; erect filaments commonly twisted in cordlike strands, yellowish-brown to dark brown; branches opposite or alternate, attenuated, occasionally piliferous; cells of main erect filaments 25–60 µm diam., 1–3(6) times as long; chloroplasts numerous, approximately discoid, with 1+ pyrenoids; unangia chiefly in lateral branches, intercalary in catenate series of 5–15(25), barrel-shaped to subglobose, mostly broader than vegetative cells; plurangia multiseriate, in catenate series of 2–30 or more, usually appearing as a single intercalary group.

Infrequent on wood, rocks, or other algae, midtidal, Japan and Bering Sea to S. Calif.; in Calif., known from Monterey to San Pedro. Type locality: N. Europe.

Pilayella tenella S. & G.

Setchell & Gardner 1922: 385; 1925: 406.

Thalli tufted, epiphytic, 0.5–0.75 mm tall, with short, branched prostrate filaments and unbranched erect filaments gradually attenuated to a subacute apex; cells of erect filaments 7–10 µm diam., 1–2.5 times as long; chloroplast single, lobed, band-shaped, later fragmenting to several smaller chloroplasts; unangia intercalary, catenate, subterminal, at times biseriate, with individual cells longer than adjacent vegetative cells; plurangia multiseriate, intercalary, subterminal, mostly in continuous catenate series.

Infrequent, on *Pleurophycus*, low intertidal, N. Wash. (type locality), and on *Zostera*, Monterey, Calif.

Spongonema Kützing 1849

Thalli of tufted erect filaments, forming ropelike strands, attached basally by a mass of short, branching filaments. Erect filaments ecorticate, much branched, with numerous irregularly spaced, short, patent branchlets, some of which are hooked or prominently recurved, entangling adjacent filaments. Chloroplasts band-shaped, 1 to several per cell, each with 1–2 pyrenoids. Unangia infrequent, pedicellate. Plurangia multiseriate, lateral, patent, sessile or briefly pedicellate.

Spongonema tomentosum (Huds.) Kütz.

Conferva tomentosa Hudson 1762: 480. *Spongonema tomentosum* (Huds.) Kützing

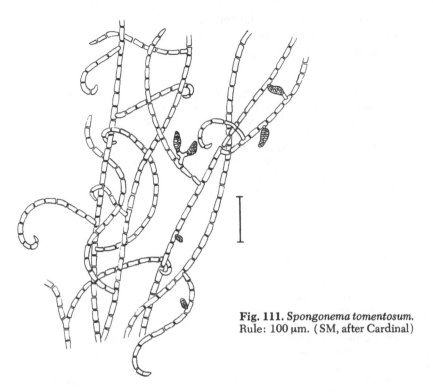

Fig. 111. *Spongonema tomentosum.*
Rule: 100 μm. (SM, after Cardinal)

1849: 461; Cardinal 1964: 34 (incl. synonymy); Saunders 1898: 155; Setchell & Gardner 1925: 417; Smith 1944: 83. *Ectocarpus ellipticus* Saund. 1898: 149.

Thalli epiphytic, 0.5–1.5(4) cm tall, yellowish-brown to dark brown; branching alternate; cells of erect filaments cylindrical, 8–12 μm diam., 1–2.5 times as long; unangia infrequent, pedicellate, ovoid to ellipsoid, terminal or lateral, 28–36(52) μm long, 20–30 μm diam.; plurangia mostly 40–100 μm long, 10–15 μm diam., lateral, patent, often secund, rarely intercalary, frequently recurved, cylindroconical, sessile or briefly pedicellate.

Infrequent to locally frequent, formerly abundant in certain areas, attached to *Fucus, Hesperophycus* or *Pelvetia,* mostly on rim of conceptacles and hairpits, midtidal, Alaska to Br. Columbia and Chile; in Calif., reported only from Monterey Co. and Laguna Beach. Type locality: England.

Pacific Coast specimens are much smaller than most of those from N. Atlantic.

Streblonema Derbès & Solier 1851

Thalli with vegetative filaments wholly or chiefly endophytic, irregularly branched, growing between cells of host; fertile branches mostly, or partly, extending above surface of host. Growth intercalary. Filaments with meristem basal to colorless multicellular unbranched hairs frequently present. Chloroplasts mostly band-shaped, 1 to several per cell, pyrenoids absent. Unangia known only in a few species. Plurangia mostly uniseriate. Life history mostly unknown.*

1. Thalli producing noticeable distortion of host surface ....S. scabiosum
1. Thalli not producing noticeable distortion of host surface 2
   2. Thalli producing noticeable colored patches on host 3
   2. Thalli not producing noticeable colored patches on host 8
3. Colored patches on host large and irregular in shape, to 20+ mm broad.... 4
3. Colored patches orbicular, mostly less than 8 mm broad 5
   4. Colored areas roughly orbicular, the plant growing on *Laminaria dentigera* ....S. evagatum
   4. Colored areas indefinite in outline and extent, the thalli growing on *Hesperophycus* ....S. penetrale
5. Thalli growing on *Desmarestia ligulata*; plurangia multiseriate........
   .... S. transfixum
5. Thalli not growing on *Desmarestia*; plurangia mostly uniseriate 6
   6. Colored areas 2–4 mm diam. ....S. pacificum
   6. Colored areas microscopic 7
7. Filaments penetrating only 2 or 3 cell layers of host.....S. myrionematoides
7. Filaments penetrating 5+ cell layers of host..............S. aecidioides
   8. Plurangia uniseriate in dense, corymbose clusters ....S. corymbiferum
   8. Plurangia partly biseriate or multiseriate, not densely corymbose...... 9
9. Plurangia fusiform; chloroplasts band-shaped; endophytic in *Smithora*..
   .... S. porphyrae
9. Plurangia cylindrical; chloroplasts discoid, several per cell; endophytic in *Helminthocladia* ....S. investiens

Streblonema aecidioides DeToni

DeToni 1895: 577. *Streblonema aecidioides* f. *pacificum* Setchell & Gardner 1922: 395. *S. aecidioides* Rosenvinge 1893: 894. *Entonema aecidioides* (Rosenv.) Kylin 1947a: 21.

Thalli microscopic, appearing as small surface elevations on host, 75–150 μm diam.; vegetative part filamentous, ultimately forming more or less parenchymatous layer immediately beneath surface cell layer of host, with a few deeply penetrating rhizoidal filaments; superficial hairs colorless, 4–5.5 μm diam.; chloroplasts small, several per cell; unangia(?) narrowly clavate, sessile, 22–28 μm long, 8–12 μm diam. at apex; plurangia numerous, closely crowded, cylindrical, sessile on the vegetative layer, uniseriate, 45–55 μm long, 5–6.5 μm diam.

Infrequent on blades of *Hedophyllum*, low intertidal, S. Br. Columbia to

* Species descriptions mostly after Setchell & Gardner. *Streblonema anomalum* S. & G. 1922: 392 has been reported by Loiseaux (1970a: 185) to be a stage in the life history of a tiny species of *Scytosiphon*. We suspect that *Streblonema johnstonae* S. & G. 1922: 394 is the same endophyte. *Streblonema luteolum* (Sauv.) DeToni has been reported to be a stage in the life history of *Spongonema tomentosum* (cf. Kornmann 1960). *Streblonema investiens* (Coll.) S. & G. (1922) is of questionable status.

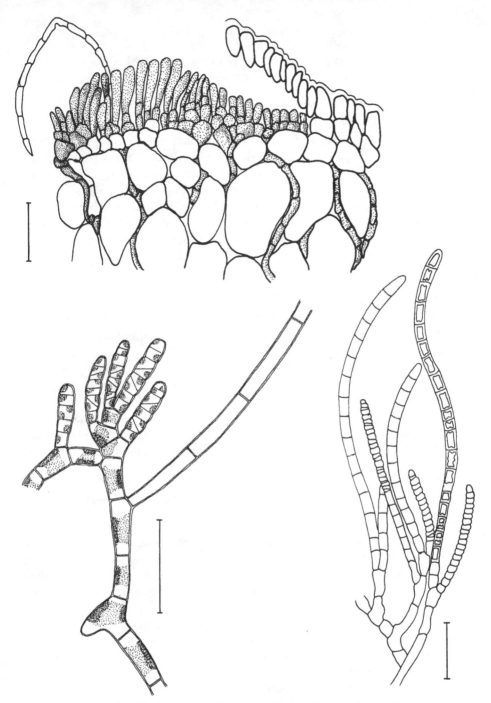

Fig. 112 (above). *Streblonema aecidioides*. Rule: 25 μm. (SM, after Setchell & Gardner)
Fig. 113 (below left). *S. corymbiferum*. Rule: 25 μm. (JRJ/S) Fig. 114 (below right).
S. evagatum. Rule: 25 μm. (SM, after Setchell & Gardner)

N. Calif.; in Calif., known only from Wilson Creek (Del Norte Co.). Also from N. Atlantic. Type locality: Greenland.

Streblonema corymbiferum S. & G.

Setchell & Gardner 1922: 391; 1925: 441.

Filaments irregularly and alternately branched, penetrating among filaments of host; fertile branches in dense, corymbose clusters near surface of host; cells mostly cylindrical, 4–5 µm diam., 1.5–4 times as long; hairs absent; chloroplasts band-shaped, 1 per cell, not fully covering cell walls; unangia unknown; plurangia cylindrical to slightly fusiform, blunt, short-pedicellate or sessile, in dense clusters near surface of host, 25–35 µm long, 4.5–5.5 µm diam., uniseriate, with dividing walls frequently oblique.

Infrequent, in *Cumagloia*, high intertidal, Monterey Co. and San Pedro (type locality), Calif.

Streblonema evagatum S. & .G.

Setchell & Gardner 1922: 390; 1925: 449; Smith 1944: 91.

Thalli forming circular patches 1–2 cm broad; creeping filaments irregular, much branched, extending along bases of host sporangia; erect filaments fasciculately branched, 190–230 µm long; cells of creeping filaments 3.5–4 µm diam., those of erect filaments at base about 4 µm diam., 2–3.5 times as long, those of widest parts 6.7–7.5 µm diam., 1–2 times as long, cylindrical, slightly constricted at crosswalls; hairs absent; chloroplasts band-shaped, 1 or 2 per cell; unangia unknown; plurangia numerous, cylindrical, sessile or short-pedicellate, uniseriate, 65–80 µm long, 5.5–6.5 µm diam.

Infrequent, in blades of *Laminaria dentigera*; known only from Patrick's Pt. (Humboldt Co.) and Cypress Pt. (Monterey Co.; type locality), Calif.

Streblonema investiens (Coll.) S. & G.

Strepsithalia investiens Collins, P.B.-A., 1895–1919 [1900]: no. 738. *Streblonema investiens* (Coll.) Setchell & Gardner 1922: 396; 1925: 452.

Thalli forming areas of indefinite form and extent on host; creeping filaments irregularly branched, with short, simple or sparingly branched erect filaments; superficial hairs sparse; cells of creeping filaments 5–8 µm diam., 1–2 times as long, swollen or cylindrical; cells of erect filaments about 6 µm diam., 1–2 times as long; chloroplasts discoid, several per cell; unangia ovoid, sessile or on 1-celled pedicels on both erect branches and creeping filaments, 15 µm diam., 20 µm long; plurangia cylindrical, 8–10 µm diam., 25–40 µm long, mostly uniseriate, on same plant as unangia.

Infrequent, among branching filaments of *Helminthocladia*, low intertidal, San Pedro (type locality) and La Jolla, Calif.

Of questionable status.

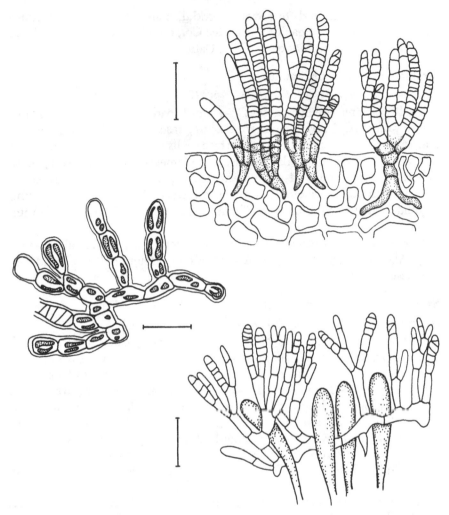

Fig. 115 (middle left). *Streblonema investiens*. Rule: 25 μm. (LH/A) **Fig. 116** (above right). *S. myrionematoides*. Rule: 25 μm. (SM, after Setchell & Gardner) **Fig. 117** (below right). *S. pacificum*. Rule: 25 μm. (SM, after Setchell & Gardner)

Streblonema myrionematoides S. & G.

Setchell & Gardner 1922: 387; 1925: 452.

Prostrate filaments penetrating slightly among 2 or 3 outer cell layers of host; erect filaments fasciculately branched at surface of host, 65–80 μm tall, mostly reproductive, with a few hairs; cells of penetrating filaments 4–5 μm diam., irregular in shape; chloroplasts unknown; unangia unknown; plurangia numerous, cylindrical, crowded, 50–65 μm long, 4.5–6.5 μm diam., uniseriate.

Locally common, mid- to lower intertidal, penetrating blades of *Laminaria dentigera*, Moss Beach (San Mateo Co.; type locality), and *Leathesia nana*, Pebble Beach (Monterey Co.), Calif.

Streblonema pacificum Saund.

Saunders 1901a: 417; Setchell & Gardner 1925: 448.

Thalli forming circular patches 2–4 mm broad; creeping filaments profusely branched, bearing numerous acute, fasciculately branched fertile filaments, these extending to host surface; cells of creeping filaments 3.5–4 μm diam., those of rhizoidal penetrating filaments 1–2.5 μm diam., those of erect filaments mostly 4–6 μm diam.; chloroplasts unknown; unangia unknown; plurangia numerous, terminal on erect filaments and projecting beyond host surface, bluntly fusiform, uniseriate, 20–28 μm long, 4–6 μm diam.

Infrequent, penetrating sporophylls of *Alaria*, low intertidal, Yakutat Bay, Alaska (type locality), and Patrick's Pt. (Humboldt Co.) and San Francisco, Calif.

Streblonema penetrale S. & G.

Setchell & Gardner 1922: 388; 1925: 446.

Thalli forming pulvinate colored areas on lower parts of host, of indefinite shape, to 20+ mm diam.; rhizoidal filaments penetrating deeply into and mostly perpendicular to host surface; hairs absent; cells of penetrating filaments 6.5–8 μm diam., 1.5–2.5 times as long in upper parts and to 5 times as long in deeper parts; erect filaments fasciculately branched at host surface, 70–125 μm long, tapering slightly upward, not piliferous; cells of erect filaments cylindrical to slightly doliform, 6.5–8 μm diam., 1.5–2.5 times as long; chloroplast single, band-shaped; unangia unknown; plurangia cylindrical to bluntly fusiform, uniseriate, 30–40 μm long, 8–11 μm diam., on pedicels of 2–4 cells at host surface.

On stipes of *Hesperophycus*, midtidal, known only from type locality, Pacific Grove, Calif.

Streblonema porphyrae S. & G.

Setchell & Gardner 1922: 387; 1925: 445; Smith 1944: 91.

Thalli endophytic in blades and basal cushions of *Smithora*; filaments very tortuous, branched; filaments growing in blades usually restricted to 1 side of blade in matrix external to host cells; filaments in basal cushion growing between host cells; cells 3–4 μm diam.; hairs absent; chloroplast band-shaped, 1 per cell; unangia unknown; plurangia limited to filaments growing in basal cushion, the plurangia there terminal on outermost filaments, fusiform in 1 or 2 series, 25–35 μm long, 5–8 μm diam. (After G. M. Smith.)

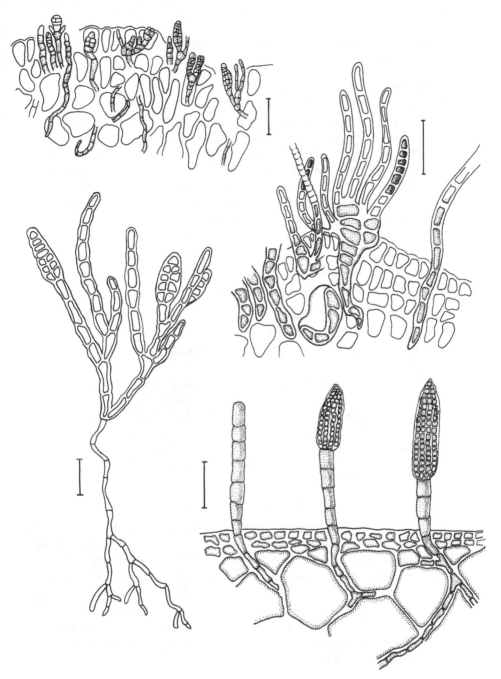

Fig. 118 (below left). *Streblonema penetrale*. Rule: 25 μm. (SM, after Setchell & Gardner) Fig. 119 (above left). *S. porphyrae*. Rule: 25 μm. (SM, after Setchell & Gardner) Fig. 120 (above right). *S. scabiosum*. Rule: 25 μm. (SM, after Setchell & Gardner) Fig. 121 (below right). *S. transfixum*. Rule: 25 μm. (DBP/H)

Common in *Smithora*, low intertidal, Pacific Grove (type locality) and San Luis Obispo Co., Calif.

Streblonema scabiosum S. & G.

Setchell & Gardner 1922: 389; 1925: 450.

Thalli forming circular to elliptical surface distortions of variable size on host; filaments penetrating and profusely branched among host cells, ultimately causing complete disintegration of some host cells; erect filaments partly external, unbranched or more or less fasciculately branched at base, 50–80 μm tall; cells of creeping filaments very irregular in shape and size; cells of erect filaments cylindrical, 4.5–5.5 μm diam., 1–2.5 times as long; hairs absent; chloroplast single, band-shaped; unangia unknown; plurangia cylindrical, sessile or short-pedicellate, 40–60 μm long, 4.5–6 μm diam., extending beyond host surface.

On stipe of *Nereocystis* cast ashore, "Cliff House," San Francisco, Calif. (type and only known locality).

Streblonema transfixum S. & G.

Setchell & Gardner 1922: 391; 1925: 446.

Thalli forming patches 5–8 mm broad; creeping filaments penetrating deeply among cells of host, distorted, irregularly branched, 4–5 μm diam.; erect filaments scattered, short, unbranched, protruding from host surface, mostly with terminal plurangia; cells of erect filaments cylindrical, 7–9 μm diam., 0.75–1.5 times as long; hairs absent; chloroplast band-shaped, single, nearly covering cell wall; unangia unknown; plurangia cylindroconical, blunt, 40–60 μm long, 8–12 μm diam., in 3 or 4 series.

Infrequent, on *Desmarestia ligulata*, low intertidal, Pacific Grove and San Pedro (type locality), Calif.

Order CHORDARIALES

Life history mostly heteromorphic, the sporangial stage macroscopic (some minute); gametangial stage mostly microscopic. Sporangial stage of variable form and mostly pseudoparenchymatous, with apical and intercalary growth, mostly in filaments of cortex or cortex-like tissue; some species clearly uniaxial or multiaxial. Chloroplasts 1 to many, plate-like, band-shaped or discoid, mostly with pyrenoids. Sporangial plants mostly with unangia only. Gametangial stage with branched filaments; unangia mostly not aggregated in sori; plurangia uniseriate or multiseriate; gametes mostly isogamous as far as known.

1. Thalli chiefly erect, pseudoparenchymatous . . . . . . CHORDARIACEAE (p. 179)
1. Thalli not erect (crustose, pulvinate, or globose), with or without free
   erect filaments 2

Family MYRIONEMATACEAE

(With the assistance of Susan Loiseaux)

Thalli minute, generally epiphytic, most thalli diploid, with monostromatic or distromatic basal layer bearing short erect structures, simple or sparingly branched filaments, also bearing endogenous hairs, single-celled paraphyses (ascocysts), plurangia, or unangia. Growth intercalary, marginal or terminal. Chloroplasts 1–8 per cell, parietal, discoid or irregular, each with 1 pyrenoid.

Myrionema Greville 1827

Thalli small disks, the cells of monostromatic basal layer of prostrate filaments bearing erect structures of equal height except for ascocysts and hairs. Plurangia cylindrical, usually uniseriate. Growth mostly apical. Life histories varied.*

1. Thalli with unangia only . . . .*M. strangulans*
1. Thalli with plurangia only or with both unangia and plurangia 2
   2. Plurangia 4–7 μm diam., with narrow loculi . . . .*M. corunnae*
   2. Plurangia 6–9 μm diam., with larger regular loculi 3
3. Thalli with ascocysts . . . .*M. magnusii*
3. Thalli without ascocysts . . . .*M. balticum*

Myrionema balticum (Reinke) Fosl.

Ascocyclus balticus Reinke 1889a: 45. *Myrionema balticum* (Reinke) Foslie 1894: 131; Setchell & Gardner 1922: 341; Loiseaux 1970b: 248. *M. attenuatum* S. & G. 1922: 344; 1925: 468; Smith 1944: 105. *M. attenuatum* var. *trichophora* Hollenberg 1970: 61.

Thalli microscopic disks generally less than 1 mm broad; basal layer

---

* The following species of *Myrionema*, described by Setchell & Gardner (1922: 334–52), do not in our opinion merit recognition: *M. globosum* f. *affine*, *M. obscurum*, *M. phyllophilum*, and *M. setiferum*.

composed of regularly radiating filaments, each cell bearing an erect filament, endogenous hair, or plurangium; chloroplasts discoid, 1 or 2 per cell; erect filaments of same length as plurangia when young, much longer on older plants, cylindrical or with barrel-shaped cells, slightly attenuated at base and apex, 70–275 μm long, 5.5–7.8(10) μm diam. (mostly ±6.7 μm); plurangia with few oblique divisions, pedicellate, or more rarely lateral at base of erect filaments, 30–150 μm long, 6–9 μm diam., with pedicel 1–15 cells long.

Frequent, on various algae or *Phyllospadix*, low intertidal, Bodega Head (Sonoma Co.) to Monterey Co., Calif. Type locality: Baltic Sea.

Myrionema corunnae Sauv.

Sauvageau 1897: 237; Jaasund 1957: 223; Loiseaux 1970b: 251.

Thalli small, dark circular disks, ±1 mm diam., epiphytic on large brown algae; cells of basal layer bearing short endogenous hairs 4–5.5 μm diam., sterile erect filaments, and plurangia; rhizoids more or less regularly present on basal layer; generally 1 discoid chloroplast per cell; ascocysts probably absent; unangia unknown; plurangia sessile or on pedicels of 1–4 cells, sometimes lateral or branched in older plants, 4–7 μm diam., 25–120 μm long.

Occasionally abundant, epiphytic on Laminariales, low intertidal, reported from Humboldt Co., San Francisco, and Monterey Co., Calif. Widely distributed in N. Europe. Type locality: Atlantic coast of France.

Setchell & Gardner (1922: 334–52) recognized five forms similar to the species, but differing in minor features.

Myrionema magnusii (Sauv.) Lois.

Ascocyclus magnusii Sauvageau 1927a: 14; J. Feldmann 1937: 116. *Myrionema magnusii* (Sauv.) Loiseaux 1967a: 338; 1967b: 567.

Thalli light brown disks, generally less than 1 mm diam.; endogenous hairs numerous, 10–12 μm diam.; ascocysts long and narrow, 10 μm diam., 80–100 μm long; chloroplasts discoid, yellowish, 2 or 3 per cell; unangia absent or exceedingly rare; plurangia sessile, uniseriate, 25–80 μm long, 8–9 μm diam.

Rare, low intertidal, epiphytic on leaves of *Zostera*, Bodega Head (Sonoma Co.) and on *Phyllospadix*, Pacific Grove, Calif. Type locality: North Sea.

Myrionema strangulans Grev.

Greville 1822–28 (1827): pl. 300; Setchell & Gardner 1925: 471; Smith 1944: 106.

Basal disk strictly monostromatic, 1–2 mm broad, without rhizoids, composed of cell rows 4–4.5 μm diam., with cells about twice as long; central erect filaments 5–7 cells long, shorter toward margin of disk, clavate, with

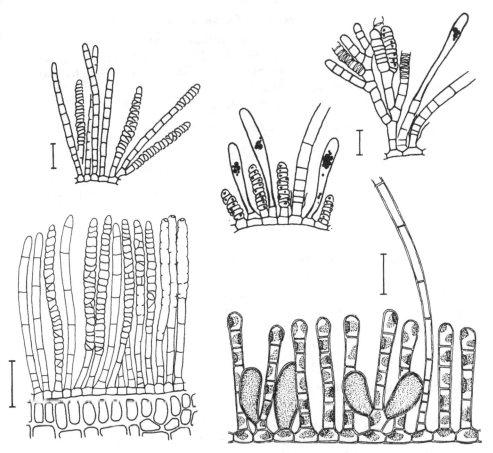

Fig. 122 (above left). *Myrionema balticum.* Rule: 20 µm. (DBP/H)   **Fig. 123** (below left). *M. corunnae.* Rule: 25 µm. (SM, after Setchell & Gardner)   **Fig. 124** (above center & above right). *M. magnusii*: above center, portion of maturing disk; right, portion of old disk. Rule: 20 µm. (both DBP, after Loiseaux)   **Fig. 125** (below right). *M. strangulans.* Rule: 20 µm. (JRJ/S)

uppermost cell 6–8 µm diam.; chloroplasts many, discoid; sheathed hairs 1–1.5 mm long, composed of cells to 100 µm long above; unangia sessile or on 1-celled pedicel, among erect filaments, occasionally lateral at base of erect filaments, 38–46(60) µm long, 18–24 µm diam.; plurangia unknown for Calif.

Generally rare, occasionally abundant, commonly on *Ulva* spp., intertidal, Alaska to Calif.; in Calif., known only from Monterey Co. Frequent in N. Europe. Type locality: Scotland.

The identity of the Pacific Coast material must be considered in need of confirmation.

Compsonema Kuckuck 1899

Thallus a disk with basal layer monostromatic, of irregular, contorted filaments. Erect filaments of different heights, simple or branched, each cell with 1 or more band-shaped chloroplasts. Plurangia multiseriate, borne on erect filaments, frequently in series.

A poorly delimited genus of questionable status.[*]

1. Erect filaments commonly with lateral branches, 1–4 mm tall . . . .
. . . . C. intricatum
1. Erect filaments mostly simple, usually less than 500 μm tall 2
2. Erect filaments 10–12 μm diam. in largest part, attenuated upward . . .
. . . . C. serpens
2. Erect filaments 8–9 μm diam., mostly not attenuated . . . . C. fructosum

Compsonema fructosum S. & G.

Setchell & Gardner 1922: 355; 1925: 476.

Thalli circular or irregular cushions 4–5 mm diam.; monostromatic basal disk composed of tortuous filaments; erect filaments unbranched, 190–230 μm tall, 8–9 μm diam.; hairs absent; unangia clavate, terminal, 55–65(100) μm long, 22–28 μm diam.; plurangia terminal, occupying definite zone on outer ends of erect filaments, 80–120 μm long, 12–16 μm diam.

Epiphytic on floating pneumatocysts of *Nereocystis*, Tomales Bay (Marin Co.; type locality), Calif.; no other collections known.

Compsonema intricatum S. & G.

Setchell & Gardner 1922: 354; 1925: 482; Smith 1944: 111.

Thalli more or less confluent, forming a velvety layer of indefinite extent; creeping filaments profusely branched, densely grouped, contorted; erect filaments 1.5–2.5 mm tall, unbranched or with a few short branches near base, gradually attenuated at apices, partly piliferous, 8–9 μm diam.; hairs absent; unangia(?) pedicellate on prostrate filaments, or lateral and sessile, or pedicellate on erect filaments, 18–22 μm diam., 25–33 μm long; plurangia terminal, lateral, or intercalary, 10–14 μm diam., 80–120(600) μm long.

Epiphytic on *Fucus distichus*; midtidal, known only from type locality, Carmel Bay, Calif.

Compsonema serpens S. & G.

Setchell & Gardner 1922: 363; 1925: 480; Smith 1944: 110.

Thallus an expanded irregular stratum on host surface; creeping filaments much contorted, irregularly branched, at times with a few endophytic

[*] Loiseaux (1970a: 185) is of the opinion that *Compsonema sporangiferum* S. & G. (1922: 357) is a stage in the life history of a minute species of *Scytosiphon*. *C. coniferum, dubium, fasciculatum, immixtum, sessile,* and *tenue,* all described by S. & G. (1922: 353–76), do not in our opinion merit recognition.

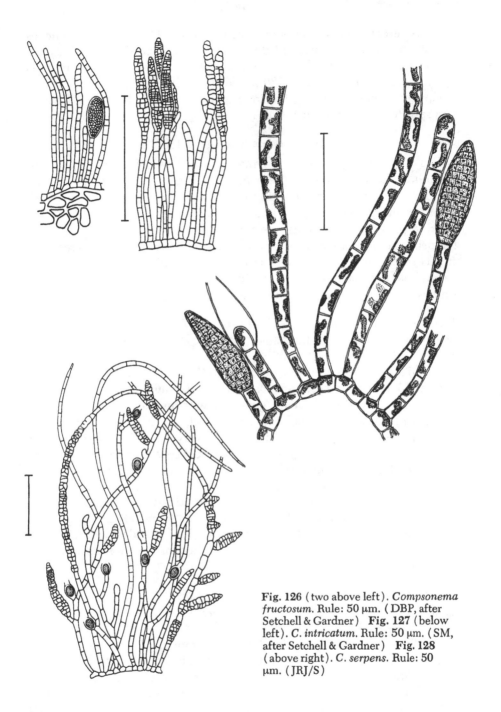

Fig. 126 (two above left). *Compsonema fructosum*. Rule: 50 μm. (DBP, after Setchell & Gardner) **Fig. 127** (below left). *C. intricatum*. Rule: 50 μm. (SM, after Setchell & Gardner) **Fig. 128** (above right). *C. serpens*. Rule: 50 μm. (JRJ/S)

rhizoids; erect filaments unbranched, or with a few subulate branches above, attenuated upward and downward, to 550 μm tall, 5.5–8(10) μm diam. at base, 10–20 μm diam. at widest part; unangia unknown; plurangia cylindroconical, terminal, 60–130 μm long, 15–28 μm diam.

Epiphytic on larger algae; low intertidal, known only from type locality, Cypress Pt. (Monterey Co.), Calif.

Hecatonema Sauvageau 1897

Thalli minute, crustose or pulvinate; basal layer of laterally adjoined radiating filaments, cells of these mostly dividing unequally in a plane parallel with substratum, resulting in a partly distromatic basal disk; cell divisions otherwise, marginal and intercalary; lower cells of disk often bearing penetrating rhizoids, the upper cells mostly producing an erect filament, endogenous hair, ascocyst, or reproductive structure; erect filaments simple or branched, of variable length; chloroplasts discoid, 1 to several per cell, each with 1 pyrenoid; unangia unknown or mostly sessile on basal layer; plurangia multiseriate, more or less fusiform.

Hecatonema primarium (S. & G.) Lois.

Myrionema primarium Setchell & Gardner 1922: 334. *Hecatonema primarium* (S. & G.) Loiseaux 1970b: 257 (incl. synonymy).

Thalli small circular disks, generally less than 2 mm diam., sometimes to 5–6 mm; basal layer mostly monostromatic, distromatic in part, composed of regularly radiating filaments, generally without rhizoids, though sometimes with short rhizoids; chloroplasts discoid, 1 or 2 per cell, each with 1 pyrenoid; erect filaments unbranched, of very variable length, 6–8 μm diam., 70–400 μm long, composed of cylindrical cells, often absent in young plants; endogenous hairs rarely absent, sessile, 4–7.5 μm diam.; ascocysts numerous, sometimes missing in young plants, sessile or pedicellate, 8–12 μm diam., 30–65 μm long; unangia unknown in nature; plurangia usually sessile, mostly multiseriate, 6–8(10) μm diam., 30–65(80) μm long.

Common, epiphytic on blades of various Laminariales, low intertidal, Alaska to Pt. Conception, Calif.; in Calif., known mostly from Monterey Co. Type locality: Coos Bay, Ore.

Hecatonema streblonematoides (S. & G.) Lois.

Compsonema streblonematoides Setchell & Gardner 1922: 353; 1925: 481. *Hecatonema streblonematoides* (S. & G.) Loiseaux 1970b: 253 (incl. synonymy).

Thalli circular or irregular cushions 2–10 mm diam., often more or less confluent, forming a velvety stratum of indefinite extent; basal layer partly distromatic, composed of regular or contorted filaments, with rhizoids; erect filaments unbranched, with secund branching in old specimens, 0.15–0.5(1) mm tall, 5–7 μm diam. at base, 10–12(15) μm diam. above; cells

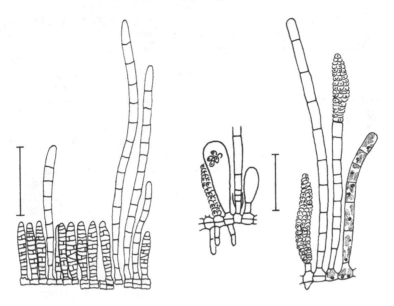

Fig. 129 (left). *Hecatonema primarium*. Rule: 50 μm. (SM, after Setchell & Gardner)
Fig. 130 (center & right). *H. streblonematoides*: center, thallus with ascocyst and hair; right, pedicellate plurangia. Rule: 50 μm. (both CS, after Loiseaux)

of erect filaments cylindrical, sometimes divided longitudinally; endogenous hairs rare or absent in young plants, abundant in some older specimens, 4–7 μm diam.; ascocysts rare, mostly lateral on erect filaments, 10–15 μm diam., 20–50 μm long; unangia on prostrate filaments, sessile or pedicellate, 15–30 μm diam., 40–80 μm long; plurangia 12–17 μm diam., 40–180 μm long, extremely variable in shape and position, mostly terminal or lateral and sessile on erect filaments, in older plants characteristically in long terminal rows, intercalary and unilaterally secund on upper part of reflexed filaments, taking the form of a cockscomb, to 400 μm long.

Infrequent, epiphytic or saxicolous, low intertidal, Alaska to Baja Calif.; in Calif., reported from Tomales Bay (Marin Co.; type locality) and Monterey Co.

Family RALFSIACEAE

Thalli crustose, saxicolous, pseudoparenchymatous, with horizontally extended basal layer, mostly without rhizoids, usually several cells thick, and generally with a thicker upper layer of assurgent to erect and usually firmly adjoined filaments. Growth marginal or intercalary. Chloroplasts laminate, 1 to many per cell, without pyrenoid. Reproductive structures mostly in ill-defined sori; unangia and plurangia mostly on separate but similar individuals, or with only 1 type of reproductive structure known.

1. Crust with overlapping lobes or divisions, loosely attached to substratum... 2
1. Crust undivided, firmly attached to substratum 3
  2. Divisions many, narrow, mostly 2–4 mm broad ....
  .... *Hapterophycus* (p. 172)
  2. Lobes relatively few, rounded, overlapping, to 60 mm broad.......
  .... *Ralfsia fungiformis* (p. 165)
3. Erect filaments of crust firmly adjoined, not readily separating under
  pressure 4
3. Erect filaments of crust readily separating under pressure 6
  4. Plurangia without sterile terminal cells......*Pseudolithoderma* (p. 174)
  4. Plurangia with 1 to several sterile terminal cells 5
5. Chloroplasts 1 per cell; plurangia with a single sterile terminal cell.....
  .... *Ralfsia* (below)
5. Chloroplasts several to many per cell; plurangia with 2–5 sterile terminal
  cells .... *Endoplura* (p. 173)
  6. Plurangia mostly in terminal pairs; unangia unknown...*Diplura* (p. 172)
  6. Plurangia not in terminal pairs; unangia terminal 7
7. Crust very gelatinous; erect filaments to 750 μm tall, more or less assurgent at base ....*Hapalospongidion* (p. 170)
7. Crust not gelatinous; erect filaments mostly less than 100 μm tall, not
  assurgent at base ....*Petroderma* (p. 174)

Ralfsia Berkeley 1831

Thalli macroscopic, crustose, mostly without rhizoids on lower surface; basal layer horizontally expanded, 1 to several cells thick, erect filaments firmly adjoined, mostly branching and commonly assurgent. Chloroplast single, parietal. Unangia borne at base of loosely associated multicellular paraphyses; plurangia on similar thalli, terminal on erect filaments, with single sterile terminal cell (1–3 cells in *R. fungiformis*).

1. Crusts with prominently overlapping broad lobes; attached by numerous
  rhizoids ....*R. fungiformis*
1. Crusts not with prominently overlapping lobes; firmly attached, mostly
  without rhizoids 2
  2. Crusts mostly less than 200 μm thick; filaments strictly erect...*R. confusa*
  2. Crusts mostly 300+ μm or more thick; erect filaments more or less
  assurgent 3
3. Crusts fleshy or coriaceous; cells of upper horizontal layer commonly 3–4
  times as long as diam. ....*R. hesperia*
3. Crusts firm, not fleshy or coriaceous; cells of horizontal to assurgent layers
  mostly not over twice as long as diam. 4
  4. Unangia in a single large and visibly evident sorus; plurangia unknown and probably lacking ....*R. integra*
  4. Unangia in small, inconspicuous sori on separate individuals....*R. pacifica*

Ralfsia confusa Hollenb.

Hollenberg 1969: 291.

Thalli light brown to medium brown, crustose, mostly 5–10 mm broad, gregarious and commonly confluent, 150–250 μm thick in fertile parts;

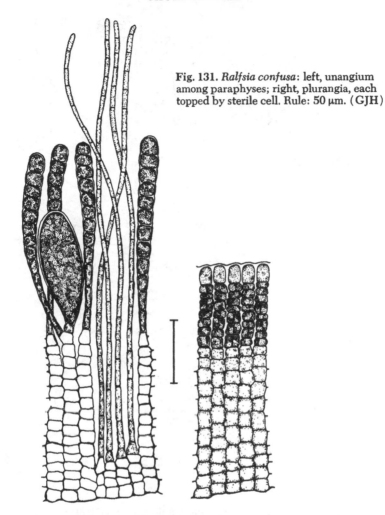

Fig. 131. *Ralfsia confusa*: left, unangium among paraphyses; right, plurangia, each topped by sterile cell. Rule: 50 µm. (GJH)

basal layer mostly 1 or 2 cells thick; erect filaments of 10–15 or more cells, 5–6 µm diam. and long; chloroplast single; hair pits frequent; sori mostly single and central on crust; unangia 70–90 µm long, 20–25 µm diam., sessile or on pedicels of 1–2(6) cells; paraphyses mostly 75–100 µm long, 6–8 µm diam. above, typically of 10–12 cells, occasionally of only 3 or 4 cells; plurangia mostly 30–45 µm long, about 5 µm diam., uniseriate, with single sterile terminal cell.

Frequent to common on rocks, upper to lower intertidal, Santa Monica, Calif., to Baja Calif. Type locality: Corona del Mar (Orange Co.), Calif.

Ralfsia fungiformis (Gunn.) S. & G.

Fucus fungiformis Gunnerus 1766: 107. *Ralfsia fungiformis* (Gunn.) Setchell & Gardner 1924b: 11.

Fig. 132. *Ralfsia fungiformis.* Rule: 50 µm. (GJH)

Thalli coriaceous, dark brown, crustose, 2–6 cm broad, loosely attached by numerous unicellular to multicellular rhizoids, free at margins, with free overlapping lobes, more or less circular in outline and with prominent concentric and radial growth lines; assurgent and downwardly curving cell rows from a central layer in radial anticlinal section; chloroplast single; unangia in sori, borne at base of loosely associated multicellular paraphyses; plurangia in sori, on separate crusts or on same crust, arising in upper part of erect filaments, uniseriate, with 3–5 sterile terminal cells; reproductive structures otherwise unknown for Calif. specimens.

Frequent, on rocks in midtidal to upper intertidal, Alaska to Trinidad (Humboldt Co.), Calif. Widely distributed in colder waters of N. Hemisphere. Type locality: Norway.

Ralfsia hesperia S. & G.

Setchell & Gardner 1924b: 2; 1925: 498; Smith 1944: 96.

Thalli fleshy to coriaceous, circular in outline, firmly adhering to substratum, occasionally with a few rhizoids, 3–4 cm broad, light brown, with some concentric ridges; cells in horizontal layer 16–18 µm diam., 3–4 times as long; assurgent cell rows grading into erect cell rows, with cells 8–15 µm diam., mostly 2–3 times as long; chloroplast single; unangia mostly 100–135 µm long, 23–26 µm diam.; paraphyses slightly clavate, 290–400 µm long, of 6–12 cells, these 8–9.5 µm diam., 1.5–2.5 times as long above, much narrower and 6–9 times as long below; plurangia unknown and probably lacking.

Infrequent to rare, upper intertidal rocks, Trinidad Bay (Humboldt Co.) and Carmel Bay (type locality) to Corona del Mar (Orange Co.), Calif.

Ralfsia integra Hollenb.

Hollenberg 1969: 295.

Thalli medium brown to dark brown, crustose, 1–8 cm broad, mostly 400–800 µm thick in mature fertile portions; sterile margins showing concentric and often radial growth ridges; cells of basal layer horizontally elongate, 12–14 µm diam., 1–2 times as long, merging upward into assurgent filaments 7–8 µm diam., with cells 1–1.5 times as long in erect upper parts; chloroplast single, parietal or terminal; hair pits frequent, relatively deep; sorus single, large, central, visibly darker than sterile portion and often rugose in older specimens, covering half or more of crust diam.; unangia subclavate, 18–25 µm diam., 60–140 µm long, sessile at base of paraphyses, or becoming pedicellate as new sporangia are proliferated; paraphyses 130–200 µm long, 8–12 µm diam. at apex, of mostly 8–12 cells about as long as diam. above, gradually narrower and longer below; plurangia unknown and probably lacking.

Locally abundant on upper intertidal rocks, usually in shallow pools, Corona del Mar (Orange Co.), Calif., to Baja Calif. Type locality: Laguna Beach, Calif.

Ralfsia pacifica Hollenb.

Hollenberg 1944b: 95; 1969: 296. *Ralfsia occidentalis* Hollenb. 1945b: 81.

Thalli crustose, often confluent, olive brown to dark brown, irregularly circular, (2)5–20 cm diam., 0.4–1 mm thick, firmly attached to substratum without rhizoids, frequently with prominent radial and concentric ridges; cells of basal stratum 10–16 µm diam., 1.5–2.5 times as long, merging assurgently into similar forking cell rows, these narrowing gradually to erect filaments; upper cells of erect filaments 7–10 µm diam., mostly 0.7–1.5 times as long; chloroplast single; hair pits frequent; sori small, 1 mm diam. or less; unangia oblong to subclavate, 70–90(140) µm long, 18–30 µm diam., sessile at base of paraphyses; paraphyses 90–120(180) µm long, of 7–12

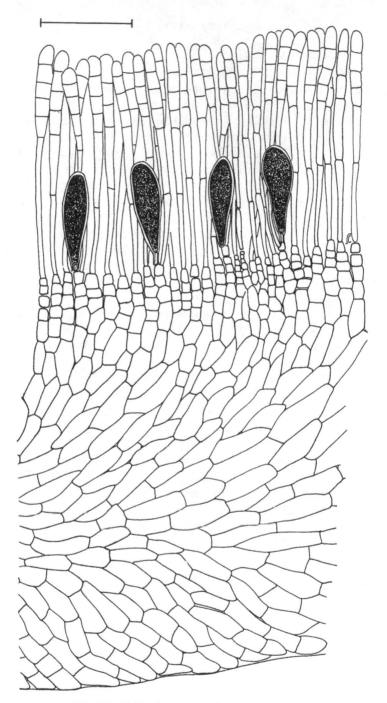

Fig. 133. *Ralfsia hesperia.* Rule: 100 μm. (GJH)

Fig. 134. *Ralfsia integra.* Rule: 50 μm. (GJH)

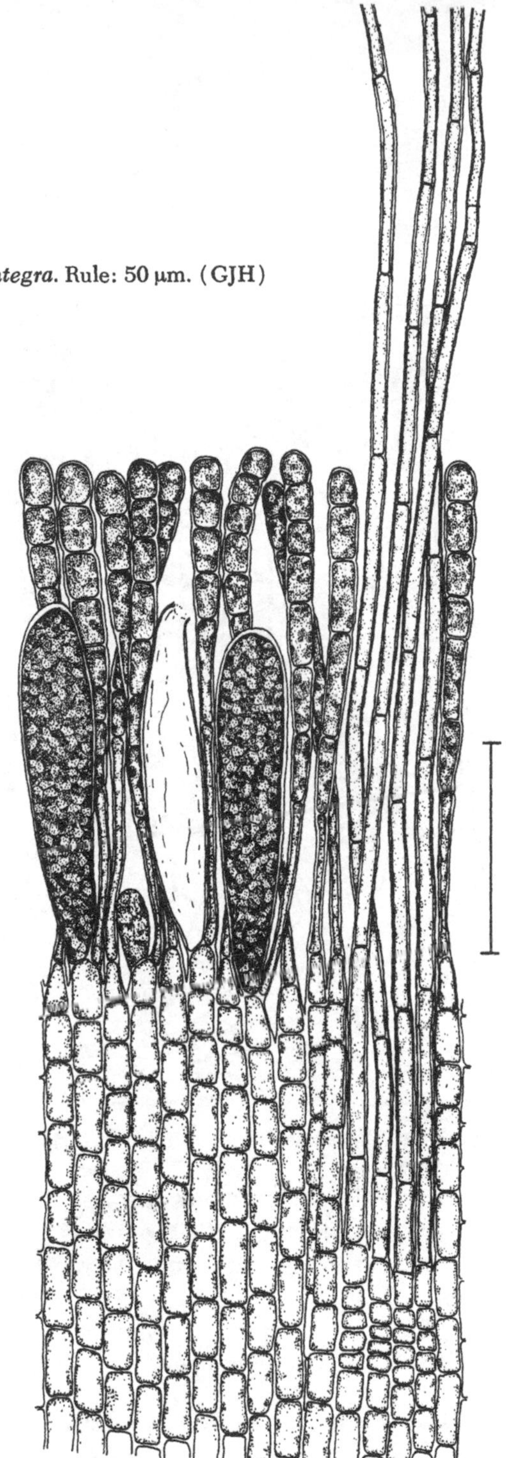

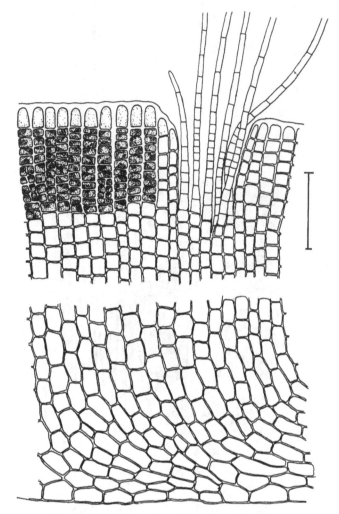

Fig. 135. *Ralfsia pacifica.* Rule: 50 μm. (GJH)

cells, these 6–9 μm diam., 1–1.5 times as long near apex, 2–2.5 times as long toward much narrower base; plurangia 40–90 μm long, 5–6 μm diam., uniseriate, terminal on erect filaments, with single sterile terminal cell ±1.5 times as long as diam.

Common on rocks, midtidal to upper intertidal, Alaska to Sinaloa, Mexico. Type locality: Corona del Mar (Orange Co.), Calif.

Hapalospongidion Saunders 1899

Thalli small, gelatinous, crustose, firmly attached to substratum without rhizoids. Horizontally expanded basal layer mostly 2 cells thick, bearing assurgent to erect, sparsely branched and loosely adjoined filaments, these readily separating under pressure. Colorless hairs frequent. Chloroplasts laminate, mostly 1 per cell. Unangia terminal on erect filaments; plurangia intercalary in upper part of erect filaments.

Hapalospongidion gelatinosum Saund.

Saunders 1899: 37; Hollenberg 1942a: 528. *Microspongium saundersii* Setchell & Gardner 1924b: 12; 1925: 492.

Thalli gelatinous, crustose, a few mm broad or by confluence to 10+ cm broad; erect filaments 250–750 μm tall, the lower cells cylindrical, 4–5 μm diam., the upper cells barrel-shaped, 7–10 μm diam., about twice as long;

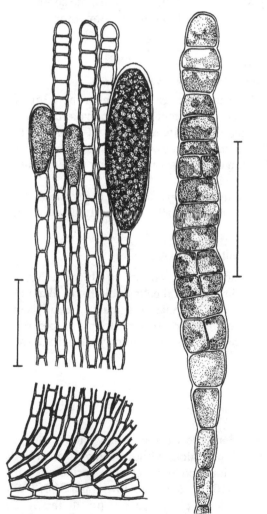

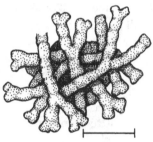

Fig. 136 (three at left).
Hapalospongidion gelatinosum.
Rules: both 50 μm. (all GJH)
Fig. 137 (above right).
Hapterophycus canaliculatus.
Rule: 1 cm. (DBP/H)

without evident sori; unangia terminal, 25–35 μm diam., 80–140 μm long, elliptical; plurangia 20–25 μm diam., 60–200 μm long, subterminal, in 1 or 2 series, on plants similar to but separate from those bearing unangia.

Occasional on rocks, mostly at or above high-tide level in areas of strong wave action, Pacific Grove, Calif. (type locality), to Guerrero, Mexico.

Hapterophycus Setchell & Gardner 1912

Thalli saxicolous, loosely crustose, deeply divided into linear branched divisions, these with middle layer of horizontally extended cells and with downwardly directed filaments and assurgent filaments terminating in firmly adjoined erect filaments. Chloroplast parietal, 1 per cell. Hair pits frequent. Unangia in sori nearly as broad as thallus divisions, the sporangia sessile at base of multicellular paraphyses; plurangia unknown for macroscopic phase of life history.

Hapterophycus canaliculatus S. & G.

Setchell & Gardner 1912a: 233; Hollenberg 1941: 676; Wynne 1969a: 6.

Thalli medium brown, somewhat cartilaginous, to 5+ cm or more broad; divisions hapteroid, 2.5–3 mm broad, 1–1.5 mm thick, concave on lower side, convex on upper side, with slightly crenate margins and blunt apices; unangia 65–75 μm long, 18–25 μm diam., briefly pedicellate; paraphyses of mostly 5–9 cylindrical cells 6–9 μm diam., 1–2 times as long above, longer and narrower below.

Frequent to common in shallow upper intertidal rock depressions, Redondo Beach, Calif., to Is. San Benito, Baja Calif. Also Japan. Type locality: San Pedro, Calif.

Diplura Hollenberg 1969

Thalli saxicolous, crustose, horizontally expanded, firmly adhering to substratum without rhizoids. Basal layer several cells thick. Erect filaments sparsely branched, loosely held together by gelatinous matrix, readily separating under pressure. Chloroplasts discoid, several to many per cell. Unangia unknown; plurangia uniseriate, single or mostly in pairs terminating erect filaments.

A monotypic genus.

Diplura simulans Hollenb.

Hollenberg 1969: 298.

Thalli individual or confluent, dark brown and somewhat mottled in living state, 1–8 cm broad, 150–250 μm thick, without rhizoids or growth lines; cells of basal layers slightly flattened parallel with surface of substratum; cells of erect filaments 7–10 μm diam. in lower parts, 5–7 μm diam. above, mostly 1–1.5 times as long; plurangia 4–5 μm diam., 30–55 μm long, uniseriate, of cells mostly shorter than diam., with sterile termi-

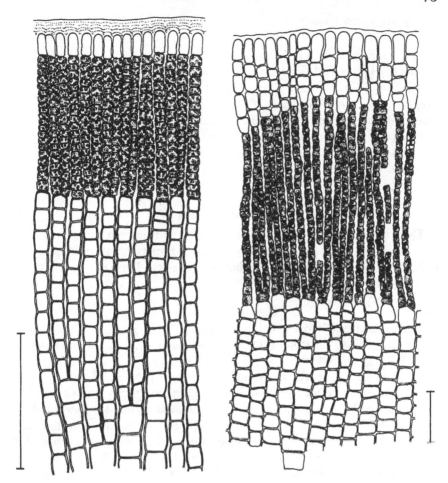

Fig. 138 (left). *Diplura simulans*. Rule: 30 μm. (GJH) Fig. 139 (right). *Endoplura aurea*. Rule: 50 μm. (GJH)

nal cell of similar diam. and 1.3–1.5 times as long as diam.; plurangia in irregular, extensive sori, not visibly evident.

Occasional on rocks in shallow upper intertidal pools, Corona del Mar (Orange Co.; type locality), Calif., to Santa Rosalía, Baja Calif.

Endoplura Hollenberg 1969

Thalli saxicolous, crustose, firmly adhering to substratum without rhizoids, with indistinct horizontal stratum and closely adjoined erect filaments. Chloroplasts discoid, small, several to many per cell. Unangia unknown, probably lacking; plurangia in sori below surface of crust, uniseriate or partly biseriate, with mostly 2–4 terminal sterile cells.

A monotypic genus.

Endoplura aurea Hollenb.

Hollenberg 1969: 300.

Thalli light yellow to golden brown, nongelatinous, to 8 cm broad, 300–600 μm thick, relatively smooth, without evident growth lines; cells of erect filaments 6–7.5 μm diam., mostly about same length; plurangia forming irregular, relatively large sori 100–130 μm deep anticlinally, each cell remaining undivided or dividing longitudinally into 2–4 small cells; sterile terminal cells 5–7 μm diam., about twice as long; hairs not observed.

Frequent, mostly in shallow upper intertidal pools on wave-swept rocks, Corona del Mar (Orange Co.), Calif., to Gulf of Calif. Type locality: Laguna Beach, Calif.

Petroderma Kuckuck 1896

Thalli thin, crustose, with free erect filaments from monostromatic basal layer; erect filaments not firmly adjoined. Chloroplast single, without pyrenoid. Unangia terminal on erect filaments, not associated with paraphyses; plurangia terminal, mostly biseriate.

A monotypic genus.

Petroderma maculiforme (Wollny) Kuck.

Lithoderma maculiforme Wollny 1881: 31. *Petroderma maculiforme* (Wollny) Kuckuck 1896: 382; Wynne 1969a: 9; Wilce *et al.* 1970: 190.

Thalli crustose, small to extensive, medium brown, 50–96 μm thick, firmly attached without rhizoids; erect filaments mostly unbranched, of 5–11 cells; cells of erect filaments 7.5–10 μm diam., 0.7–1.5 times as long; unangia terminal, 15–18(42) μm long, 10–14 μm diam.; plurangia and hairs not known for Pacific Coast specimens, and crusts much thinner than those reported by Edelstein & McLachlan (1969) from Nova Scotia.

On small rocks, lower intertidal; known on Pacific Coast only from Smith I., Wash., and Pillar Pt. (San Mateo Co.), Calif. Frequent in N. Atlantic. Type locality: Helgoland, Germany.

Pseudolithoderma Svedelius 1910

Thalli horizontally expanded, crustose, firmly adhering to substratum without rhizoids. Erect filaments firmly adjoined, sparsely branched, arising from relatively indistinct basal layer, this primarily 1 cell thick. Cells with several to many small, discoid chloroplasts, these lacking pyrenoids. Hair pits occasional. Unangia present in some species, terminal on erect filaments. Plurangia terminal on erect filaments, without sterile terminal cell.

Pseudolithoderma nigra Hollenb.

Hollenberg 1969: 297.

Thallus 2–10+ cm broad, nearly black, mostly 300–500 μm thick; growth

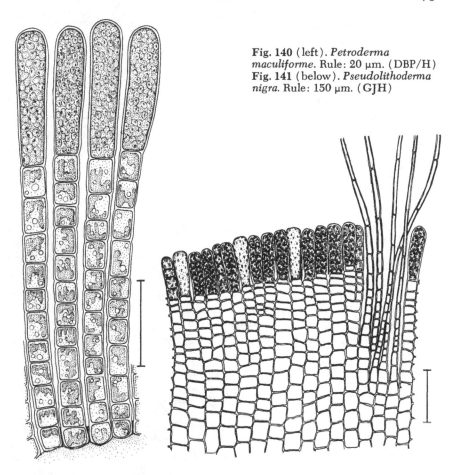

Fig. 140 (left). *Petroderma maculiforme.* Rule: 20 µm. (DBP/H)
Fig. 141 (below). *Pseudolithoderma nigra.* Rule: 150 µm. (GJH)

lines inconspicuous or absent; cells of erect filaments 9–10 µm diam., mostly shorter than broad; unangia unknown and probably absent; plurangia in extensive irregular sori, mostly 30–50 µm long, 8–14 µm diam., composed of tiers of 4 cells, without sterile terminal cells; nearly colorless paraphysis-like cells scattered among plurangia.

Common, often forming extensive areas on rocks, upper intertidal in *Pelvetia* zone, Shell Beach (San Luis Obispo Co.), Channel Is., and Corona del Mar (Orange Co.; type locality), Calif., to Oaxaca, Mexico.

Family CORYNOPHLAEACEAE

Thalli pulvinate or globose, fleshy, without distinct basal layer. Cortex of anticlinal filaments with intercalary growth. Chloroplasts several per cell, each with 1 pyrenoid. Sporangial plants with unangia, with or without plurangia. Gametangial plants microscopic and filamentous, with uniseriate gametangia.

With two genera in the California flora, *Leathesia* (below) and *Cylindrocarpus* (p. 177).

Leathesia S. F. Gray 1821

Thalli of sporangial plants globular and fleshy, mucilaginous; inner tissue of large, colorless cells in radiately branched filaments; cortical layer of pigmented smaller cells forming palisade-like filaments 3–6 cells long, usually with enlarged terminal cell and with or without fascicles of long, colorless hairs projecting beyond surface. Unangia and plurangia arising at base of cortical filaments. Gametangial plants microscopic, filamentous, with plurangia.

Leathesia difformis (L.) Aresch.

Tremella difformis Linnaeus 1755: 429. *Leathesia difformis* (L.) Areschoug 1847: 376; Smith 1944: 114 (incl. synonymy).

Thalli yellowish-brown, globular to irregularly expanded, to 12 cm diam., hollow and much convoluted at maturity, surface texture slippery, commonly with numerous hair fascicles; unangia ovoid, 40–50 μm long, 18–24 μm diam.; plurangia uniseriate, 30–45 μm long, 4–6 μm diam., of 7–10 short cells, usually in clusters among cortical filaments.

Annual, common early spring to October, on rocky or occasionally epi-

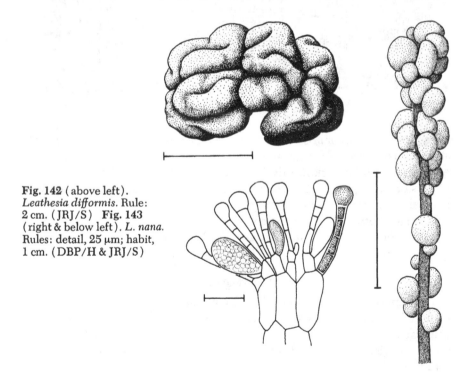

Fig. 142 (above left).
Leathesia difformis. Rule:
2 cm. (JRJ/S) Fig. 143
(right & below left). *L. nana.*
Rules: detail, 25 μm; habit,
1 cm. (DBP/H & JRJ/S)

phytic, very abundant in sheltered locations, midtidal to upper intertidal, Bering Sea to Baja Calif. and Chile; common throughout Calif. Also Europe. Type locality: Sweden.

Leathesia nana S. & G.

Setchell & Gardner 1924b: 3; 1925: 511; Smith 1944: 115.

Thalli epiphytic, globular, to 5 mm diam., not convoluted, usually not hollow, commonly densely aggregated, light to medium brown; cortical filaments 3–5 cells long, the terminal cell large and globular; unangia 25–40 μm long, 10–18 μm diam.; plurangia 20–30 μm long, 3.5–4.5 μm diam., of 4 or 5 short cells.

Occasional spring and summer annual, epiphytic on *Phyllospadix, Rhodomela,* or *Odonthalia,* intertidal to upper subtidal, N. Wash. to Carpinteria (Santa Barbara Co.), Calif. Type locality: Monterey, Calif.

Specimens on *Phyllospadix* bear plurangia only; those on *Rhodomela* and *Odonthalia,* unangia almost exclusively.

Cylindrocarpus Crouan & Crouan 1851

Thalli of sporangial plants pulvinate, spongy, convoluted-rugose. Internal tissue of branched colorless filaments of relatively large cells, with progressively smaller pigmented cells forming a cortical layer; colorless slender filaments extending horizontally among inner colorless large cells. Chloroplasts restricted largely to cortical filaments, there discoid, several per cell, each with pyrenoid. Unangia lateral at base of cortical filaments; plurangia none. Gametangial plants microscopic, filamentous.

Cylindrocarpus rugosus Okam.

Okamura 1907–42 (1907): 20. *Petrospongium rugosum* (Okam.) Setchell & Gardner 1924b: 12; Smith 1944: 116.

Thalli chestnut brown to dark brown, pulvinate, mostly circular, with upper surface irregularly convoluted, to 8 cm diam.; pigmented cortical filaments mostly unbranched, 8–11 μm diam., of 8–12 cells; long, colorless multicellular hairs frequent among cortical filaments; mature unangia reniform, 16–22 μm diam., 75–90 μm long, laterally attached by pedicels of 1 or 2 cells.

Annual, common late spring to October, on upper intertidal rocks, Sonoma Co., Calif., to Baja Calif. Type locality: Japan.

Family ELACHISTACEAE

Thalli pulvinate or tufted, primarily filamentous, with mostly compact pseudoparenchymatous basal portion. Erect filaments, commonly bearing short lateral branchlets below. Reproductive structures borne on free erect filaments or at base of lateral branchlets.

Fig. 144. *Cylindrocarpus rugosus*:
below, habit; right, vegetative
filaments, showing unangia.
Rules: below, 2 cm; right, 50 μm.
(both JRJ/S)

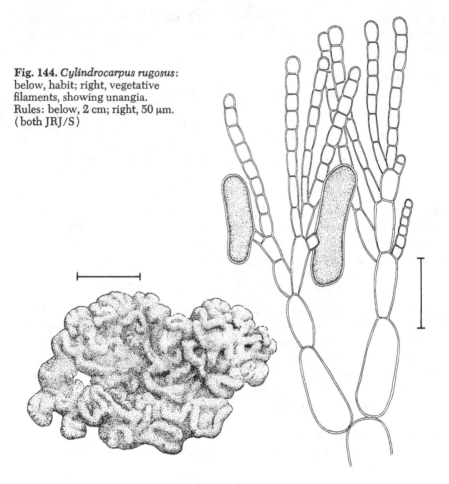

Elachista Duby 1830

Thalli minute, tufted, epiphytic with dense basal mass of closely intertwined branched filaments bearing numerous free erect filaments and short,
curved paraphyses. Chloroplasts in young cells reticulate, becoming fragmented with age; pyrenoids unknown. Unangia borne at base of paraphyses; plurangia unknown for California.

Elachista fucicola (Vell.) Aresch.

Conferva fucicola Velley 1795: fig. 4. *Elachista fucicola* (Vell.) Areschoug 1842: 235;
Setchell & Gardner 1925: 503.

Thalli tufted, 4–7 mm tall, the basal cushion hemispherical; erect filaments tapering briefly near base, gradually above, not piliferous, mostly

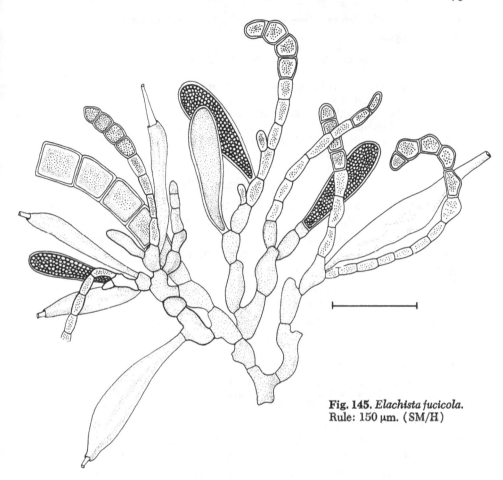

Fig. 145. *Elachista fucicola.*
Rule: 150 µm. (SM/H)

40–50 µm diam., with cells 0.5–2 times as long, thick-walled; paraphyses clavate, uncinate, moniliform, 170–200 µm long, 18–24 µm diam. above; unangia obpyriform, 90–110 µm long, 22–26 µm diam.

Rare, epiphytic on *Fucus* spp., midtidal, reported from Sitka, Alaska, Rosario Beach, Wash., Coos Bay, Ore., and Humboldt Bay, Calif. Common in N. Atlantic. Type locality: England.

Family CHORDARIACEAE

Thalli of sporangial plants erect, uniaxial or multiaxial. Growth mostly trichothallic, with subapical meristem at branch apex and intercalary divisions in pigmented anticlinal cortical filaments. Cortical filaments frequently terminated by a colorless hair. Chloroplasts discoid, usually many per cell. Unangia borne at base of cortical filaments, not grouped in sori;

plurangia usually absent. Gametangial plants of various forms, macroscopic or microscopic.

1. Main axes percurrent; branches numerous, short, nongelatinous.......
   ....*Analipus* (below)
1. Main axes not percurrent; branches gelatinous, with relatively few or no
   short lateral branches 2
   2. Branches uniaxial, of several to many orders; inner cortex without
      radial filaments ....*Haplogloia* (below)
   2. Branches multiaxial, of 1 or 2 orders; inner cortex with numerous
      slender radial filaments ....*Tinocladia* (p. 183)

Analipus Kjellman 1889

Thallus perennial, with a number of main erect axes arising from profusely branched, rhizomatous, persistent base. Erect axes percurrent, densely clothed on all sides with unbranched laterals of limited growth. Medulla of longitudinally elongate, colorless cells, surrounded by inner cortex of shorter colorless cells and outer cortex of short, anticlinal rows of pigmented cells. Chloroplasts discoid, many per cell, without pyrenoid. Unangia and plurangia on separate plants of similar size and structure; unangia at base of cortical filaments among scattered hairs; plurangia pluriseriate, intercalary in middle and lower parts of cortical filaments.

Analipus japonicus (Harv.) Wynne

Halosaccion (*Halocoelia*) *japonica* Harvey 1857: 331. *Analipus japonicus* (Harv.) Wynne 1971: 172. *Chordaria abietina* Farlow 1875: 357. *Heterochordaria abietina* (Farl.) Setchell & Gardner 1924b: 6; 1925: 550.

Thallus dark brown to tan; base crustose to 5 cm broad; erect branches to 35 cm tall; determinate lateral branches cylindrical to distinctly flattened, commonly curved, to 30 mm long; axis and branches at first solid, later becoming hollow; erect axes disappearing in late fall, new axes arising from perennial base the following spring.

Frequent to common on upper intertidal rocks in areas exposed to moderately heavy wave action, Alaska to Pt. Conception, Calif. Type locality: Japan.

Haplogloia Levring 1939

Thalli of sporangial plants irregularly branched; branches cylindrical, soft and gelatinous, with single axial filament surrounded by compact medullary layer of elongate, colorless cells forming longitudinal filaments. Cortex of unbranched, pigmented, anticlinal filaments and numerous, much longer, unbranched, evanescent colored hairs; true endogenous hairs absent. Chloroplasts band-shaped, 1 to many per cell, disappearing with age. Unangia ovoid, at base of cortical filaments; plurangia lacking. Gametangial plants compact microscopic disks.

Fig. 146. *Analipus japonicus.*
Rule: 1 cm. (JRJ/S)

Haplogloia andersonii (Farl.) Levr.

Mesogloia andersonii Farlow 1889: 9. *Haplogloia andersonii* (Farl.) Levring 1939: 50; Smith 1944: 117; Wynne 1969a: 13. *Myriogloia andersonii* Setchell & Gardner 1925: 556.

Thalli of sporangial plants to 25–40 cm tall, mostly light tan, frequently with several erect, irregularly branched axes from a common base; branches, including deciduous hairs, mostly 2–3 mm diam.; older axes commonly hollow; cortical filaments of 5–9 cells, outwardly larger and frequently curved at apex; unangia 24–28 µm long, 10–16 µm diam.

Spring annual on rocks, lower intertidal in tidepools to upper subtidal,

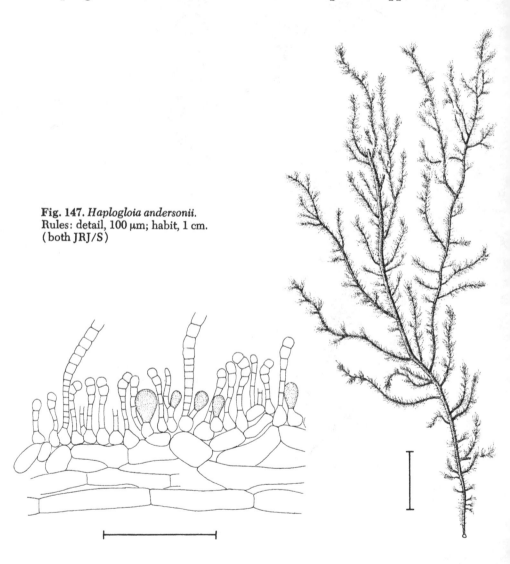

Fig. 147. *Haplogloia andersonii.*
Rules: detail, 100 µm; habit, 1 cm.
(both JRJ/S)

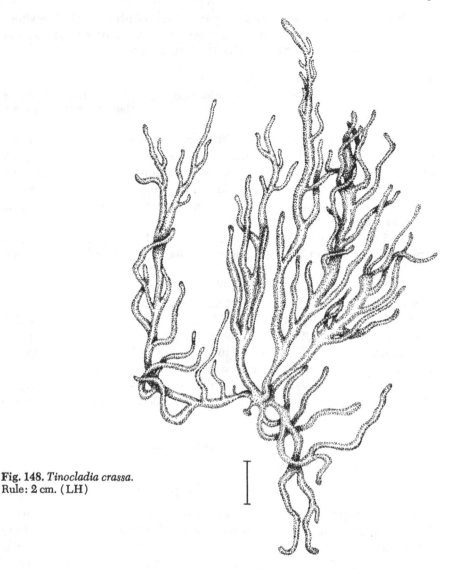

Fig. 148. *Tinocladia crassa.*
Rule: 2 cm. (LH)

Sitka, Alaska, to Baja Calif. and Gulf of Calif.; occasional to locally abundant in Calif. (Humboldt Co. to San Diego). Type locality: Santa Cruz, Calif.

Tinocladia Kylin 1940

Thalli erect, irregularly branched, gelatinous. Branches cylindrical, multiaxial. Central core surrounded by layer of thin rhizoidal filaments. Inner cortex of sparsely branched, radially directed, colorless filaments. Outer cortex composed of fascicles of moniliform, little-branched filaments with occasional phaeophycean hairs arising from base of filament cluster. Chloroplasts discoid, 1 to several per cell. Unangia borne at base of pigmented filaments. Plurangia and gametangial thalli unknown.

Tinocladia crassa (Suring.) Kyl.

Mesogloia crassa Suringar 1873: 85. *Tinocladia crassa* (Suring.) Kylin 1940: 34; Mower & Widdowson 1969: 72. *Aegira virescens sensu* Setchell & Gardner 1925: 547; *sensu* Dawson 1944b: 96.

Thalli lubricous, 15–25+ cm tall, irregularly branched; branches of 1 or 2 orders, cylindrical, 2–3 mm diam.; pigmented cortical filaments slightly arcuate, of 10–15 cells, 6–8 µm diam., 8–15 µm long; cells of inner cortical filaments 3–10 µm diam., 12–14 µm long; phaeophycean hairs on outside of cluster, 6–8 µm diam., unbranched; unangia nearly spherical to elongate, 31–50 µm diam., 30–42(75) µm long.

Occasional on rocks, low intertidal to subtidal (5 m), Pt. Dume (Los Angeles Co.), Santa Catalina I., and La Jolla, Calif. Type locality: Japan.

Order SPOROCHNALES

Spore-bearing thalli erect, branching, with terete or flattened branches, solid or hollow, parenchymatous. Meristem subterminal, cambium-like, basal to a terminal tuft of hairs, forming a plate, cutting off cells distally and contributing to the terminal tuft of colored filaments. Chloroplasts discoid to irregular, many per cell, without pyrenoids. Plurangia unknown; unangia clustered, several borne laterally on shortened branches. Gametangial thalli microscopic, uniseriate, with terminal oogonia and antheridia isolated or in groups.

Family SPOROCHNACEAE

With characters of the order.

Sporochnus C. Agardh 1817

Thallus alternately branched, cylindrical to slightly flattened, with inner layer of large, colorless cells and outer layer of pigmented cells. Meristem subterminal on terminally enlarged branchlets, with terminal tuft of filaments. Unangia lateral, on special branchlets at base of terminal filaments of lateral branches.

Sporochnus pedunculatus (Huds.) C. Ag.

Fucus pedunculatus Hudson 1778: 587; Turner 1811: 126. *Sporochnus pedunculatus* (Huds.) C. Agardh 1817: xii; Kützing 1859: 34; Taylor 1928: 114; 1960: 253. *S. apodus sensu* Mower & Widdowson 1969: 75.

Thallus erect, with axis to 50 cm tall, flexible, golden brown in life, drying to dark brown, bearing branches in an irregular spiral, these appearing

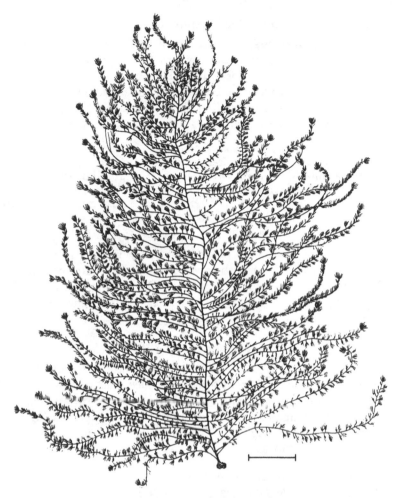

Fig. 149. *Sporochnus pedunculatus.* Rule: 2 cm. (SM, after Okamura)

alternate in pressed material; branches cylindrical, 0.5–1.2 mm diam., bearing numerous short-determinate branchlets terminating in tufts of dark brown, monosiphonous filaments; fertile branchlets at first subsessile in older specimens; with stipe 0.2 mm diam, and to 2.5 mm long; fertile portions cylindrical, 0.5 mm diam. to 2 mm long, composed of slender, branching paraphyses with globose end cells lateral and unangia to 58 μm long.

Frequent on rocks, reefs, or pebbles, subtidal (11–13 m), Santa Catalina I., Calif., to Scammon Lagoon, Baja Calif. Widely distributed in N. Atlantic; also reported from Brazil. Type locality: England.

Order Dictyosiphonales

Thalli of sporangial plants macroscopic, of various forms, parenchymatous throughout. Meristem intercalary or sometimes apical. Hairs frequent, scattered or grouped. Chloroplasts discoid, many per cell. Reproductive structures transformed vegetative cells, in sori or scattered; with unangia only, with plurangia only, or with both on the same individual, or on separate individuals. Where known, gametangial thalli microscopic.

1. Branches terminating in a uniseriate filament........Striariaceae (below)
1. Branches not terminating in a uniseriate filament....................2
   2. Sporangia embedded below surface layer..Dictyosiphonaceae (p. 187)
   2. Sporangia arising from transformed superficial cells ....
    .... Punctariaceae (p. 191)

Family Striariaceae

Thalli erect, cylindrical, with branch apices terminating in a uniseriate filament bearing a hair, and with lower parts multiseriate, parenchymatous. Chloroplasts discoid, pyrenoid unknown. Meristem intercalary. Unangia and plurangia developing from superficial cells. Gametangial thalli microscopic, bearing plurangia.

Stictyosiphon Kützing 1843

Thalli erect, mostly solid, arising from densely interwoven mass of rhizoids. Flaments uniseriate or multiseriate. Surface cells cubical, with discoid to band-shaped chloroplasts, several per cell. Subsurface cells rectangular, radially elongate, without chloroplasts. Plurangia arising singly or in small groups intermingled with vegetative cells, embedded or protruding, without paraphyses. Unangia rare.

Stictyosiphon tortilis (Rupr.) Reinke

Scytosiphon tortilis Ruprecht 1851: 373. *Stictyosiphon tortilis* (Rupr.) Reinke 1889b: 55; 1892: 47; Setchell & Gardner 1925: 529; Smith 1944: 133.

Thalli of Calif. specimens to 10 cm tall, medium brown to yellowish-brown, with hemispherical holdfast; erect branches freely to sparsely branched, often fistulous below; branches alternate, to 65 μm diam., mostly less than 12 surface cells in circumference; surface cells slightly longer than wide; cells of uniseriate parts about half as long as diam.; chloroplasts in multiseriate parts discoid to narrowly elliptical, those in uniseriate parts discoid. Plurangia 20–25 μm diam., and unangia restricted to multiseriate parts, mostly solitary, embedded, the free end bulging outward.

Infrequent on subtidal rocks (to 11 m), Port Clarence, Alaska, to Monterey, Calif. Type locality: Arctic Ocean.

According to Wynne (1972) the alga that has been recorded from Wash.

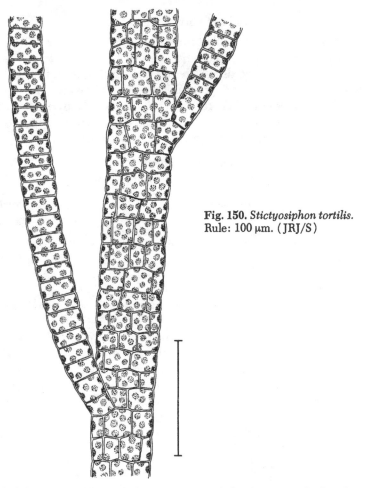

Fig. 150. *Stictyosiphon tortilis.*
Rule: 100 μm. (JRJ/S)

and Calif. is probably not this species and furthermore lacks the phaeophycean hairs on the branch endings that are characteristic of the genus.

Family DICTYOSIPHONACEAE

Thalli erect, cylindrical, filiform, solid or tubular-saccate, simple or
branched, epiphytic or saxicolous, attached by a discoid base or by penetrating rhizoids. Surface cells small and pigmented, with progressively
larger and less pigmented cells inward. Colorless hairs occasionally present. Chloroplasts discoid, many per cell. Growth apical, subapical, or intercalary and diffuse. Unangia embedded among outer cortical cells. Sporangial plants macroscopic, bearing unangia only or both unangia and plurangia. Gametangial plants microscopic, filamentous, with plurilocular
gametangia.

Coilodesme Strömfelt 1886

Thalli saccate, cylindrical to flattened, entire or sometimes with frayed divisions at maturity, mostly epiphytic, usually briefly stipitate, attached by a small disk or with rhizoids apparently penetrating host. Cortical cells mostly not in anticlinical rows, progressively larger inward. Chloroplasts discoid, without pyrenoid. Unangia narrowly to broadly pyriform or very irregular in form, scattered and singly embedded just below surface. Gametangial plants microscopic, bearing uniseriate plurangia.

1. Thallus saccate, not compressed, thin-walled. . . . .*C. californica*
1. Thallus basically saccate but distinctly compressed, relatively thick-walled 2
   2. Surface very closely wrinkled . . . .*C. corrugata*
   2. Surface relatively smooth 3
3. Thallus with ruffled margins, the blades linear, to 58 cm long. . . . . .*C. plana*
3. Thallus without ruffled margins, the blades broadly rounded, mostly less than 10 cm long . . . .*C. rigida*

Coilodesme californica (Rupr.) Kjellm.

Adenocystis californica Ruprecht 1851: 291. *Coilodesme californica* (Rupr.) Kjellman 1889b: 4; Setchell & Gardner 1925: 579.

Thalli mostly 25–35 cm long (occasionally to 1 m), to 12 cm broad, light olive tan, drying to greenish; thalli saccate from earliest stages, fragile, easily ruptured at maturity, mostly wrinkled with age; base tapering to short stipe 1–3 mm long; outer cortical cells in anticlinal rows of 2–4 cells, the inner cortical layers lacking longitudinal filaments; sporangia irregular in shape, 40–45 µm long anticlinally.

Common as epiphyte on *Cystoseira osmundacea*, low intertidal, mostly in sheltered water, Queen Charlotte Strait, Br. Columbia, to Baja Calif. Type locality: Fort Ross (Sonoma Co.), Calif.

Coilodesme corrugata S. & G.

Setchell & Gardner 1924b: 8; 1925: 582.

Thalli mostly less than 10 cm long and 2 cm broad, complanate, with finely wrinkled surface; walls 40–45 µm thick; outer cortical cells not in anticlinal rows; longitudinal filaments among inner cortical cells lacking; sporangia mostly broader than long.

Occasional epiphyte on *Cystoseira neglecta*, low intertidal to subtidal; known only from type locality, Santa Catalina I., Calif.

Coilodesme plana Hollenb. & Abb.

Hollenberg & Abbott 1965: 1177; Smith 1969: 635.

Thalli medium brown to dark olive brown, linear, strictly complanate

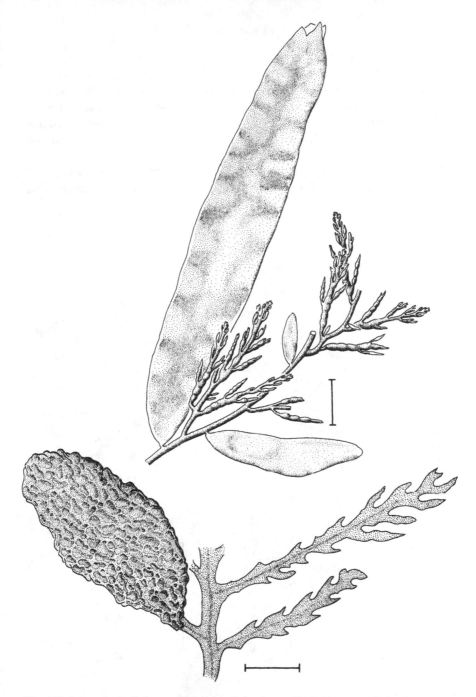

Fig. 151 (above). *Coilodesme californica.* Rule: 2 cm. (JRJ/S) **Fig. 152** (below).
C. corrugata. Rule: 2 cm. (CS)

even when very young, relatively firm but pliable, to 58 cm long, 4–6 cm broad; base abruptly narrowed, the margins commonly somewhat ruffled; older specimens often longitudinally laciniate in upper parts, the divisions frequently with finely dentate and crisped margins; inner tissue of relatively large cells intermixed with abundant interwoven filaments extending mostly longitudinally; sporangia ovoid, 15–20 µm diam., 20–26 µm long.

Occasional to common epiphyte on *Cystoseira osmundacea*, in exposed localities, low intertidal, Cape Mendocino to Soberanes Pt. (Monterey Co.; type locality), Calif.

Fig. 153. *Coilodesme plana.*
Rule: 10 cm. (DBP)

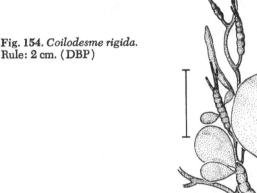

Fig. 154. *Coilodesme rigida*.
Rule: 2 cm. (DBP)

Coilodesme rigida S. & G.

Setchell & Gardner 1924b: 9; 1925: 584.

Thalli firm and leathery, dark brown to medium brown, strictly complanate, the surface and margins smooth, the apex broadly rounded, the stipe short and thick, the blades 5–10(25) cm long, 1–2.5 cm broad, 300–375 μm thick, with thicker margins; outer cortex of anticlinal cell rows, the inner cortex with slender filaments extending in all directions among larger cortical cells; sporangia irregular in shape, several times longer anticlinally than diam.

Frequent on *Halidrys dioica*, upper subtidal to lower intertidal, Redondo Beach (type locality), Calif., to Bahía Tortuga, Baja Calif.; common on Santa Catalina I., Calif.

Family PUNCTARIACEAE

Thalli bladelike, saccate, or cylindrical, usually unbranched; parenchymatous throughout. Meristems diffuse or intercalary. Chloroplasts discoid, many per cell, or lobed and single, mostly with pyrenoids. Sporangia transformed vegetative cells, embedded in thallus. With unangia only or with both unangia and plurangia on same plant. Microscopic plants (plethysmothalli) that repeat sporangial phase common. Where known, gametangial plants microscopic, filamentous or discoid, bearing uniseriate gametangia.

1. Plants globular, hollow .... *Soranthera* (p. 196)
1. Plants cylindrical or bladelike 2

Melanosiphon Wynne 1969a

Thalli cylindrical or slightly compressed, unbranched, erect, flaccid, loosely caespitose, solid when young, later tubular. Growth mostly diffuse. Medulla of large colorless cells, these soon breaking down; cortex of 2–4 layers of small cells. With unangia only or with plurangia only; unangia superficial, among multicellular paraphyses over major surface of plant; plurangia in sori formed in locally thickened cortical areas.

Melanosiphon intestinalis (Saund.) Wynne

Myelophycus intestinalis Saunders 1901a: 420; Setchell & Gardner 1925: 527. *Melanosiphon intestinalis* (Saund.) Wynne 1969a: 45; Edelstein, Wynne & McLachlan 1970: 5.

Thalli caespitose, 1–15 cm tall, 2–4 mm diam., usually twisted, dark reddish-brown; unangia ellipsoidal to obovate, scattered over surface of axes among superficial paraphyses composed of 4–8(18) cells; plurangia in sori on separate individuals, these probably not gametangial stages.

Occasional on rocks, mid- to upper intertidal, Alaska to Moss Beach (San Mateo Co.), Calif. Also Nova Scotia, Canada, and Japan. Type locality: Popoff I., Alaska.

Melanosiphon intestinalis f. tenuis (S. & G.) Wynne

Myelophycus intestinalis f. *tenuis* Setchell & Gardner 1917: 385. *Melanosiphon intestinalis* f. *tenuis* (S. & G.) Wynne 1969a: 45.

Distinguished chiefly by its smaller size, 1.5–2.5 cm tall, less than 1 mm diam.

Common on rocks, high intertidal, Coos Bay, Ore., Duxbury Reef (Marin Co.) and San Francisco (type locality), Calif.

Punctaria Greville 1830

Thalli of sporangial plants bladelike annuals, with 1 or more erect blades arising from discoid base. Blades linear to broadly elliptical, tapering to short slender stipe, generally with multicellular surface hairs borne singly or in tufts. Blades composed of cubical cells, 3–7 cells thick, the inner cells only slightly larger than surface cells. Chloroplasts discoid, few to many per cell, with pyrenoid. Unangia subcubical, scattered singly, partly exserted, formed by transformation of surface cells. Plurangia multiseriate,

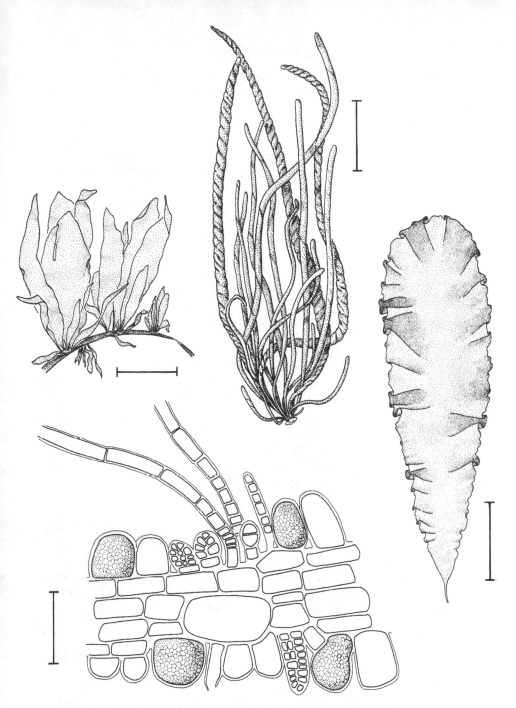

Fig. 155 (above center). *Melanosiphon intestinalis*. Rule: 2 cm. (DBP) **Fig. 156** (above left & below). *Punctaria hesperia*: above left, habit on *Phyllospadix*; below, cross section, showing unangia and plurangia. Rules: habit, 1 cm; section, 50 μm. (DBP after Saunders) **Fig. 157** (right). *P. occidentalis*. Rule: 2 cm. (JRJ/S)

usually grouped in small sori without paraphyses, frequently with projecting apex. Gametangial plants microscopic branching filaments; gametes isogamous.

Punctaria hesperia S. & G.

Setchell & Gardner 1924b: 3; 1925: 517.

Thalli 1.5–2.5 cm tall, 5–10 mm broad; blades 4–6 cells thick, solitary or clustered, linear to oblanceolate, the margin plane, the apex blunt, the stipe very short; blades with frequent small tufts of hairs; unangia numerous, intermingled with plurangia; plurangia very numerous, closely grouped or scattered, conical with upper half projecting beyond surface of blade.

Occasional epiphyte on *Phyllospadix* or *Codium*; low intertidal, known only from Vancouver I., Br. Columbia, and Monterey Peninsula, San Luis Obispo Co., and San Pedro, Calif. Type locality: Pacific Grove, Calif.

Punctaria occidentalis S. & G.

Setchell & Gardner 1924b: 4; 1925: 520.

Thalli light brown, to 20 cm tall; blades to 12 cm broad, linear-lanceolate to oblanceolate or broadly elliptical, with ruffled margins; blades 2–7 cells thick, thickest in median region; stipe short; base small, discoid; mature plants seemingly without hairs; unangia cubical to subspherical, 30–40(70) μm diam., slightly protuberant; plurangia cylindrical to conical, with upper half projecting above blade surface.

Rare, on leaves of the eelgrass *Zostera*, subtidal (5 m), Crescent City (Del Norte Co.) and Monterey (type locality), Calif.

Halorhipis Saunders 1898

Thalli annual, with 1 to several erect blades arising from discoid base. Blades bearing numerous small tufts of multicellular hairs. Medulla of several layers of large, colorless, cylindrical cells extending lengthwise in blades; cortex of 1 or 2 layers of smaller, cubical, pigmented cells. With unangia only, grouped in small sori surrounding hair tufts, projecting above surface of blade. Gametangial plants unknown.

Halorhipis winstonii (Anders.) Saund.

Punctaria winstonii Anderson 1894: 358. *Halorhipis winstonii* (Anders.) Saunders 1898: 161; Setchell & Gardner 1925: 524; Smith 1944: 125.

Thalli light olive tan, to 35 cm tall, 5 cm broad; blades lanceolate to spatulate; sori minute, numerous; unangia 30–45 μm long, 20–30 μm diam., obpyriform to ellipsoidal.

Locally abundant, on *Egregia menziesii* or rocks, low intertidal, near Mendocino (city) and Carmel Bay (type locality), Calif.

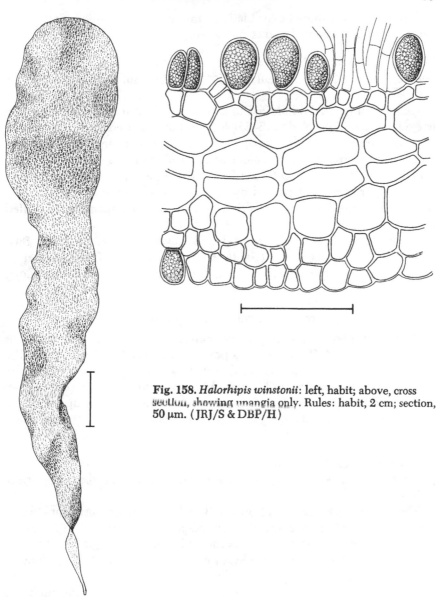

Fig. 158. *Halorhipis winstonii*: left, habit; above, cross section, showing unangia only. Rules: habit, 2 cm; section, 50 μm. (JRJ/S & DBP/H)

Phaeostrophion Setchell & Gardner 1924b

Thalli perennial, with 1 to several erect blades arising from discoid holdfast. Inner cells of blades large and elongate; surface cells cuboidal to cylindrical in cross section of blades. Chloroplasts small, several per cell, with pyrenoid only in germlings. Blades bearing unangia only, or unangia and plurangia distributed over blades. Plurangia accompanied by 2-celled paraphyses. Gametangial plants unknown.

Phaeostrophion irregulare S. & G.

Setchell & Gardner 1924b: 10; Mathieson 1967: 293. *Phaeostrophion australe* Dawson 1958: 65.

Thalli with perennial holdfast to 20 cm broad, bearing several to many irregularly spatulate blades 15–25(40) cm tall, 1.5–4 cm broad, attenuate to a slender meristematic stipe; medulla parenchymatous to somewhat filamentous; cortex 1–3 cells thick; unangia abundant, more or less ellipsoidal, 20–35 μm diam., 30–60 μm long, forming a more or less continuous surface layer, not accompanied by paraphyses; plurangia cylindrical, 10–25 μm diam., 35–75 μm long, multiseriate, usually accompanied by 2-celled paraphyses.

On rocks in sandy areas, midtidal to lower intertidal, Yakutat Bay, Alaska, to Government Pt. (Santa Barbara Co.), Calif.; occasional in Calif. to locally frequent farther north. Type locality: Bolinas (Marin Co.), Calif.

Soranthera Postels & Ruprecht 1840

Thalli epiphytic, annual, globose, hollow at maturity, with broad or narrow attachment. Wall of thallus parenchymatous, 5 or 6 cells thick, the cells progressively smaller outward. Unangia in conspicuous, slightly elevated sori, among multicellular paraphyses. Plurangia not present on sporangial plants. Zoospores from unangia giving rise to microscopic filamentous gametangial plants, the plants bearing uniseriate to biseriate plurangia.

Soranthera ulvoidea Post. & Rupr.

Postels & Ruprecht 1840: 19; Setchell & Gardner 1925: 545; Smith 1944: 127; Wynne 1969a: 39.

Thalli globose, hollow, olive brown, 3–5(10) cm diam.; sori 1 mm broad, distributed evenly over entire thallus, much darker than vegetative areas, each sorus surrounding a tuft of multicellular hairs; unangia ovoid to clavate, 78–100 μm long, the paraphyses clavate and nearly twice as long as unangia, containing mostly 6–14 cells.

Common as summer annual on *Rhodomela* and *Odonthalia*, midtidal, N. Japan and Bering Sea to Government Pt. (Santa Barbara Co.), Calif. Type locality: Sitka, Alaska.

Plants on *Rhodomela*, which are more spherical, thick-walled, and broadly attached, occur mostly in N. and C. Calif.; those on *Odonthalia*, which are mostly obovoid, thin-walled, and narrowly attached, occur from Alaska to C. Calif.

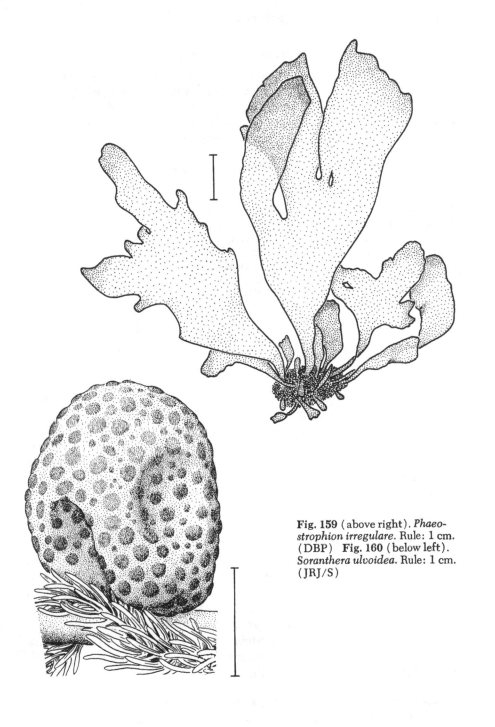

Fig. 159 (above right). *Phaeo-strophion irregulare*. Rule: 1 cm. (DBP) Fig. 160 (below left). *Soranthera ulvoidea*. Rule: 1 cm. (JRJ/S)

Order SCYTOSIPHONALES

Thalli unbranched, erect, bladelike to globular. Growth diffuse or inter-calary. Chloroplast large, parietal, 1 per cell, each with single large pyre-noid. Hairs frequent. Plurangia numerous, uniseriate, mostly not in well-defined sori. Unangia occurring on small, crustose stage in some genera. Fusion of motile cells from plurangia reported for *Petalonia*.

Family SCYTOSIPHONACEAE

With the characters of the order.

1. Thallus hollow, globular to saccate, at least in early stages............ 2
1. Thallus, if hollow, not globular nor saccate 3
  2. Thallus with numerous perforations; globular in early stages.......
  .... *Hydroclathrus* (p. 206)
  2. Thallus normally without perforations .... *Colpomenia* (p. 202)
3. Erect divisions bladelike 4
3. Erect divisions not bladelike, mostly cylindrical 5
  4. Medulla of thick-walled, intertwined filaments..... *Endarachne* (p. 200)
  4. Medulla parenchymatous, with few or no filaments.... *Petalonia* (p. 200)
5. Thallus branched, not tufted .... *Rosenvingea* (p. 202)
5. Thallus unbranched, tufted .... *Scytosiphon* (below)

Scytosiphon C. Agardh 1820

Thallus usually tufted, with several to numerous erect fronds arising from discoid base. Erect parts simple, solid or mostly tubular, cylindrical or flat-tened, commonly with frequent constrictions when tubular. Inner cells of thallus elongate, thick-walled; cortical layer with cells progressively smaller toward surface. Phaeophycean hairs commonly present. Unangia borne on crustose stage resembling *Ralfsia*. Plurangia forming on extensive surface areas.

Scytosiphon dotyi Wynne

Wynne 1969a: 34 (incl. synonymy).

Thalli tufted, cylindrical to slightly flattened, to 12 cm tall, ±1 mm diam., mostly not constricted, sometimes twisted, attenuated above and below, greenish to dark brown; hairs in dense tufts in small depressions on surface; plurangia 4–6 cells long, in extensive areas over most of surface; unicellular structures resembling paraphyses absent; unangia unknown.

Occasional to frequent, mostly a winter annual restricted to vertical faces of large boulders and cliffs, upper intertidal, Ore. to Baja Calif. Type local-ity: Pillar Pt. (San Mateo Co.), Calif.

Scytosiphon lomentaria (Lyngb.) J. Ag.

Chorda lomentaria Lyngbye 1819: 74. *Scytosiphon lomentaria* (Lyngb.) J. Agardh 1848: 126; Setchell & Gardner 1925: 531; Smith 1944: 129; Wynne 1969a: 31.

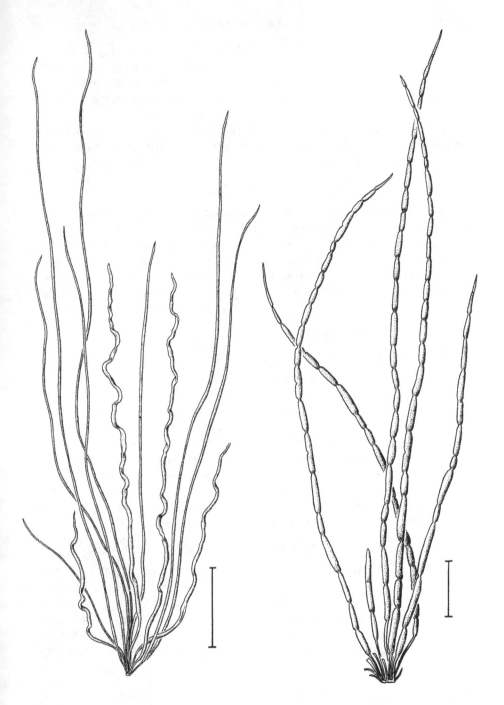

Fig. 161 (left). *Scytosiphon dotyi*. Rule: 2 cm. (JRJ/S) Fig. 162 (right). *S. lomentaria*.
Rule: 2 cm. (JRJ/S)

Thallus erect, parts 20–30(70) cm tall, mostly 4–6 mm diam., tubular, usually constricted; hairs arising singly on superficial cells; plurangia 4–8 μm diam., 40–65 μm long; colorless, unicellular structures resembling paraphyses (ascocysts?) present among plurangia; unangia occurring on crustose stage of plant.

Abundant in localized areas on sheltered rocks in lower intertidal, or in shallow upper-intertidal pools. Bering Sea to Baja Calif. Widely distributed. Type locality: Denmark.

Petalonia Derbès & Solier 1850

Thalli annuals, 1 or more erect blades arising from small, discoid base. Blades linear to broadly lanceolate, with small tufts of multicellular hairs; surface cells small, subcubical; medulla with several layers of much larger colorless cells, the plurangia uniseriate, ultimately covering both surfaces of blades and maturing progressively from apex to base of blades. Unangia borne at base of free, pigmented paraphyses on very small, thin, *Ralfsia*-like, crustose stage developing from plurangial motile cells (swarmers).

Petalonia fascia (Müll.) Kuntze

Fucus fascia Müller 1775–82 (1778): 7. *Petalonia fascia* (Müll.) Kuntze 1898: 419; Wynne 1969a: 17. *Ilea fascia* (Müll.) Fries 1835: 321 (in part); Setchell & Gardner 1925: 535; Smith 1944: 126. *P. debilis sensu* Hollenberg & Abbott 1966: 23 (incl. synonymy).

Blades to 35 cm tall, 30 cm broad, of variable width in a given tuft, linear-lanceolate, with tapering apex and cuneate base, greenish-brown to dark brown; plurangia 6–8 cells long, uniseriate.

Frequent on upper to midtidal rocks, or epiphytic on *Phyllospadix*, N. Japan and Alaska to Isla Magdalena, Baja Calif. Widely distributed; also in Europe and Chile. Type locality: Norway.

It seems best to consider the various forms listed by Setchell and Gardner (1925: 536) as representing a single variable species.

Endarachne J. Agardh 1896

Thalli bladed, without midrib, solid, several blades arising from discoid base. Surface cells cuboidal. Inner cortical layer of larger, thick-walled cells; medullary layer of very thick-walled, densely intertwined, branched filaments, extending mostly in longitudinal direction. Hairs in fascicles. Plurangia uniseriate and lacking paraphyses, forming continuous palisade-like layer covering most of surface area of blades, maturing from apex to base of blades. Unangia unknown.

Endarachne binghamiae J. Ag.

J. Agardh 1896: 26; Saunders 1898: 162; Setchell & Gardner 1925: 538.

Blades golden to dark brown, usually in clusters, 10–18(25) cm tall,

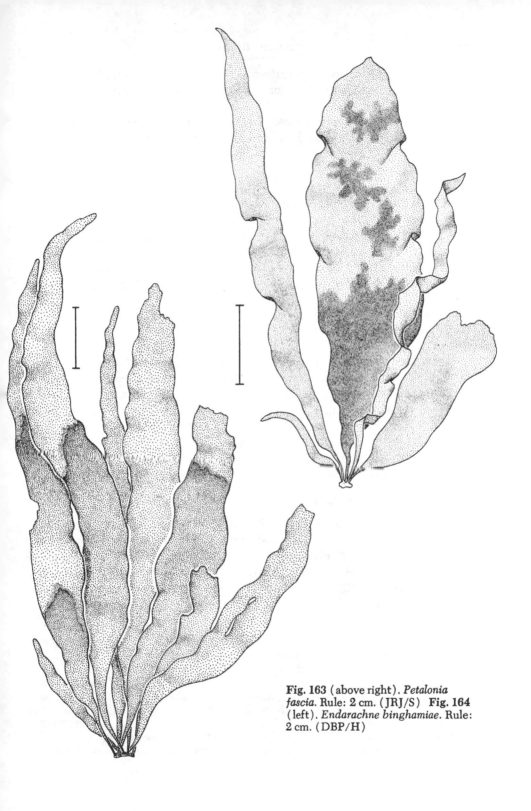

Fig. 163 (above right). *Petalonia fascia*. Rule: 2 cm. (JRJ/S) **Fig. 164** (left). *Endarachne binghamiae*. Rule: 2 cm. (DBP/H)

1–3(5) cm broad, linear to cuneate or broadly spatulate, usually eroding above fertile zone at maturity; plurangia slender, 4–5 μm diam., 40–50 μm long.

Occasional to frequent on rocks, midtidal to upper intertidal, S. Calif. to Bahía Asunción, Baja Calif. Also Japan. Type locality: Calif., possibly Santa Barbara.

Rosenvingea Børgesen 1914

Thalli tubular, branched, cylindrical or compressed, attached by basal disk. Branching sparse. Growth intercalary. Walls of axes composed of 3 or 4 layers of cells; outer cells close, angular, with 1 discoid chloroplast per cell, without pyrenoid; inner cells large, colorless. Hairs scattered singly or in groups. Plurangia subcylindrical to clavate, formed by division of outer cell layer; paraphyses lacking. Unangia unknown.

Rosenvingea floridana (Tayl.) Tayl.

Cladosiphon floridana Taylor 1928: 113. *Rosenvingea floridana* (Tayl.) Tayl. 1955: 72; 1960: 262.

Thalli erect, golden brown, to 4.5 cm tall, hollow except at solid base; main axes 0.5–2.2 mm diam.; branching irregular and frequent, often with 3 (rarely more) branches arising at 1 place; branches narrowed at base; apices acute and tapering; surface hairs 15–18 μm diam., often broken; surface cells 10–17 μm diam., inner cells 28–94 μm diam.; reproductive structures unknown for Calif. specimens.

Occasional, subtidal on sand (to 10 m), Santa Catalina I., Calif. Frequent to common, Gulf of Mexico, S. Fla. Type locality: Dry Tortugas, Fla.

Colpomenia Derbès & Solier 1851

Thalli globular or saccate, with broad basal attachment, at first solid, later becoming hollow. Walls several to many cells thick. Inner cells large and colorless, progressively smaller and more pigmented outwardly. Plurangia at first limited to areas around hair tufts, later forming extensive surface areas. Unicellular, colorless paraphyses frequent among plurangia. Unangia unknown.

1. Thallus with erect, more or less elongate and fingerlike projections . . . . .
. . . . *C. bullosa*
1. Thallus globular or irregularly expanded 2
   2. Walls relatively thick and rigid to coriaceous, with blunt points or
      tubercles at maturity . . . .*C. tuberculata*
   2. Walls mostly thin and membranous, without tubercles 3
3. Walls of 3–5 cell layers, smooth, drying to greenish; thallus not much
expanded or lobed . . . .*C. peregrina*
3. Walls of mostly more than 5 cell layers, slightly roughened, drying to
brown; thallus commonly much lobed and expanded . . . .*C. sinuosa*

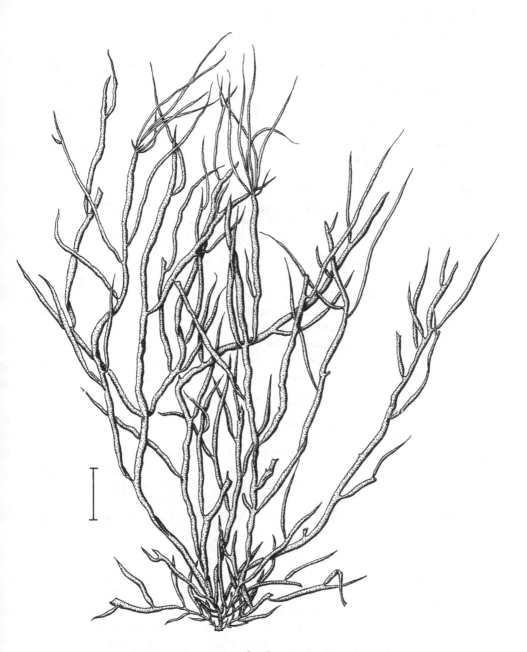

Fig. 165. *Rosenvingea floridana*. Rule: 2 cm. (NLN)

Colpomenia bullosa (Saund.) Yamada

Scytosiphon bullosus Saunders 1898: 163. *Colpomenia bullosa* (Saund.) Yamada 1948: 6; Hollenberg & Abbott 1966: 22. *C. sinuosa* f. *deformans* Setchell & Gardner 1903: 242.

Thalli olive tan to medium brown, to 10 cm tall, with 1 to several finger-like hollow divisions from a common base, usually only 1 or 2 of these maturing; detailed structural and reproductive features not well known.

Occasional on midtidal rocks, N. Japan and Alaska to C. Calif. Type locality: Pacific Grove, Calif.

This troublesome species is considered by Wynne (1972) to be con-specific with *C. peregrina*.

Colpomenia peregrina (Sauv.) Ham.

Colpomenia sinuosa var. *peregrina* Sauvageau 1927b: 321. *C. peregrina* (Sauv.) Hamel 1931–39 (1937): 201; Blackler 1964: 50. *C. sinuosa sensu* Saunders 1898: 164; *sensu* Setchell & Gardner 1925: 539; *sensu* Smith 1944: 128 (as applied to C. California specimens).

Thalli globular, with little or no division, thin and smooth, often narrowly attached, drying to greenish; with 1 or 2 rows of cortical cells and 2 or 3 rows of subcortical cells; colorless hairs superficial, in clusters; plurangia forming extensive continuous areas on lower parts of vesicle, with 2 rows of loculi and sometimes with bifurcate apices, 18–22 μm long, 5–8 μm diam.; colorless paraphyses abundant.

Common on rocks or other algae, mostly lower intertidal, Alaska to La Jolla, Calif. Common in N. Atlantic. Type locality: Mediterranean.

Colpomenia sinuosa (Roth) Derb. & Sol.

Ulva sinuosa Roth 1806: 327. *Colpomenia sinuosa* (Roth) Derbès & Solier 1851: 11; Blackler 1964: 50.

Thalli globular, becoming expanded and commonly deeply convoluted, 8–10(15) cm broad, relatively thick-walled, mostly over 5 cell layers thick, with slightly roughened surface, drying to brownish; plurangia biseriate, 5–8 μm diam., 18–22 μm long.

On rocks or other algae, midtidal, Anacapa I., Calif., to Mexico and Chile. Frequent in N. Atlantic. Type locality: Spain.

Colpomenia tuberculata Saund.

Saunders 1898: 164. *Colpomenia sinuosa* f. *tuberculata* (Saund.) Setchell & Gardner 1903: 242; 1925: 541.

Thalli fleshy, crisped, and somewhat coriaceous, medium to light brown, usually slightly flattened, 5–10+ cm diam., broadly attached, often with folded or convoluted surface; thallus at maturity covered with blunt projections or pointed tubercles protruding 4–10 mm; outer layers of 3–5 cell rows; inner cortex of 5–8 rows of large, thin-walled cells. Plurangia 3–4 μm diam., 20–25 μm long.

Mostly on rocks, San Pedro (type locality), Calif., to La Paz, Baja Calif.,

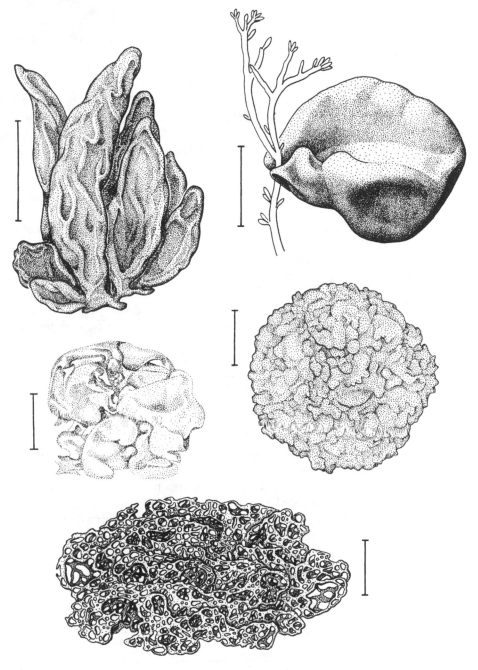

Fig. 166 (above left). *Colpomenia bullosa.* Rule: 2 cm. (DBP) **Fig. 167** (above right). *C. peregrina.* Rule: 2 cm. (JRJ/S) **Fig. 168** (middle left). *C. sinuosa.* Rule: 2 cm. (SM, after Dawson) **Fig. 169** (middle right). *C. tuberculata.* Rule: 2 cm. (DBP, after Saunders) **Fig. 170** (below). *Hydroclathrus clathratus.* Rule: 2 cm. (NLN)

and Ecuador; more abundant in southern part of range; very abundant in places in Gulf of Calif.

Hydroclathrus Bory 1825

Thalli globose, hollow, increasingly fenestrate with aging, tending to rupture and form irregular, fenestrate sheet, or by repeated convolution assuming a compact, massive form. Colorless hairs grouped in shallow depressions. Plurangia formed over entire surface of younger plants.

Hydroclathrus clathratus (C. Ag.) Howe

Encoelium clathratum C. Agardh 1822a: 412. *Hydroclathrus clathratus* (C. Ag.) Howe 1920: 590; Setchell & Gardner 1925: 543.

Thalli irregularly globular when young, hollow, sessile by broad attachment, soon developing numerous perforations, these to 8 mm diam., later by successive convolutions and by coalescence forming compact, massive thallus to 1 m diam.; margins of perforations involute; thallus wall of small, pigmented cells; inner cells in several layers, large, colorless; hairs in small, depressed tufts in center of plurangial sori; reported to bear plurangia only in juvenile stages before perforations are formed; reproduction not observed in Calif. material.

Frequent to locally abundant, May through November, on rocks, lower intertidal to subtidal (to 6 m), Portuguese Bend (near San Pedro) and Santa Catalina I., Calif., to Baja Calif. and Ecuador. Widely distributed in tropical and subtropical seas. Type locality: S. Pacific.

Order DICTYOTALES

Thalli erect, saxicolous, with 1 to several erect axes arising from thickened, felted base. Erect axes complanate, parenchymatous, with apical meristem and branching in 1 plane; chloroplasts discoid, numerous, mostly without pyrenoid. Sporangial and gametangial plants usually morphologically similar. Dioecious; reproductive structures fully or partially exserted, scattered or in sori; aplanospores large, nonmotile, 4 or 8 borne in unilocular sporangia. Gametangia in sori, the gametes oogamous, the eggs large, 1 per oogonium. Antheridia plurilocular; sperms with single flagellum. Paraphyses present or absent. Plants mostly tropical or subtropical.

Family DICTYOTACEAE

With characters of the order.

1. Each branch with terminal row of apical cells...................... 2
1. Each branch with single, lens-shaped, apical cell 5
   2. Thalli mostly less than 30 mm tall, 1 or 2 cells thick.............
    .... *Syringoderma* (p. 210)
   2. Thalli mostly 50+ mm tall, 5+ cells thick 3

3. Major portions of thallus with distinct midrib........*Dictyopteris* (p. 211)
3. Major portions of thallus without midrib, or apparent midrib present
only below 4
    4. Mature sori mostly in distinct concentric zones, partially exserted...
    .... *Taonia* (p. 213)
    4. Mature sori not forming concentric lines, entirely superficial......
    .... *Zonaria* (p. 215)
5. Thalli dark brown, coriaceous; branches with rounded apices and 4 or 5
cells thick at margins ....*Pachydictyon* (p. 209)
5. Thalli light or medium brown, delicate; branches with truncate or rounded apices and mostly 3 cells thick throughout...........*Dictyota* (below)

Dictyota Lamouroux 1809

Thalli erect, complanate, without midrib, typically dichotomo-flabel-lately branched, arising from branched stolons. Branches each with single large apical cell. Medulla of a single layer of large colorless cells. Cortex usually of a single layer of much smaller pigmented cells in longitudinal rows; hairs common on sterile thalli. Dioecious; oogamous; reproductive structures exserted. Sporangia unilocular, scattered over both surfaces of blades, each producing 4 aplanospores; oogonia in sori, each oogonium producing a single relatively large egg. Antheridia plurilocular, in sori with sterile margin.

Dictyota binghamiae J. Ag.

J. Agardh 1894: 72; Dawson 1950d: 268; Hollenberg & Abbott 1966: 19 (incl. synon-ymy).

Thalli light to medium brown, darker near base, 15–25(35) cm tall, with slight to pronounced tendency to pinnate branching, even in young plants, owing to unequal growth of branches; branches 1–1.5(2) cm broad, with broadly rounded apices and usually with numerous minute marginal teeth; medulla monostromatic throughout; cortex monostromatic except near holdfast, where composed of 3 or 4(6) cell layers; antheridial sori oval, on both surfaces; oogonial sori scattered on both surfaces, round, mostly 140–440 μm diam.; oogonia 80–95 μm diam.

Occasional to locally common, mostly subtidal (6–9(30) m), on rocks, Queen Charlotte Is., Br. Columbia, to C. Baja Calif.; in Calif., known chiefly from Monterey region, especially Carmel Submarine Canyon. Type local-ity: Santa Barbara, Calif.

Dictyota flabellata (Coll.) S. & G.

Dilophus flabellatus Collins, P.B.-A., 1895–1919 [1901]: no. 834. *Dictyota flabellata* (Coll.) Setchell & Gardner 1924b: 12 (in part); 1925: 652 (in part); Dawson 1950b: 89 (incl. synonymy).

Thalli medium brown, 10–20(26) cm tall, with little or no tendency to pinnate branching; branches (2)6–10(15) mm broad, with rounded apices,

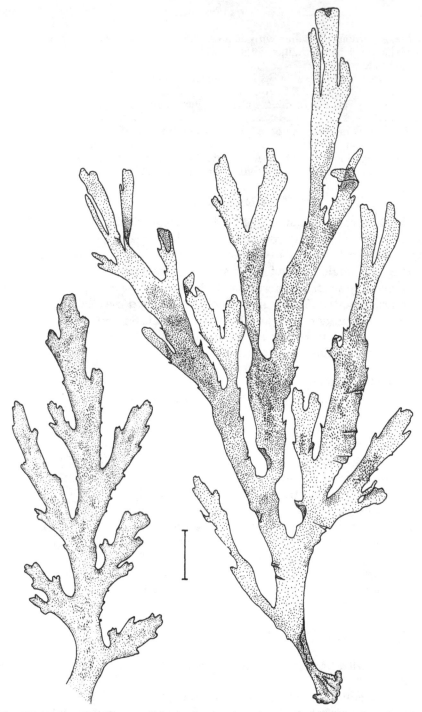

Fig. 171. *Dictyota binghamiae*: left, sterile specimen, showing marginal teeth; right, fertile specimen. Rule: 2 cm. (both SM)

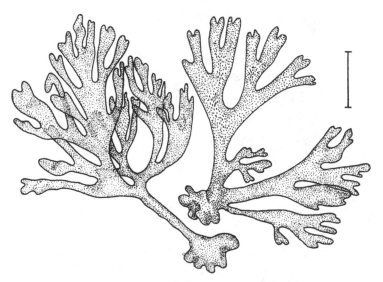

Fig. 172. *Dictyota flabellata.* Rule: 2 cm. (SM)

without marginal teeth, occasionally with numerous surface proliferations; medulla monostromatic throughout; cortex essentially monostromatic throughout; antheridia in scattered oval sori on both surfaces; oogonia in scattered sori, these mostly elongate, to 380 μm long; oogonia 35–38 μm diam.

Frequent, saxicolous, intertidal or subtidal (to 22 m), S. Calif. to Gulf of Calif. and Panama; in Calif., known from vicinity of Ventura southward, formerly common. Type locality: La Jolla, Calif.

Pachydictyon J. Agardh 1894

Thallus much divided, mostly flabellate, relatively large and coarse. Branches strict, with smooth margins and rounded apices and axils; mostly 3 cells thick, but the margins 4+ cells thick; branches terminating in single large apical cells. Reproductive structures borne in sori; sporangia unilocular, with 4 aplanospores; oogonia and antheridia as in *Dictyota*.

Pachydictyon coriaceum (Holmes) Okam.

Glossophora coriaceum Holmes 1896: 251. *Pachydictyon coriaceum* (Holmes) Okamura 1899: 39; Dawson 1950d: 268 (incl. synonymy).

Thalli dark brown, 20–30(44) cm tall, arising from thickened mass of dark brown hairs, these closely appressed and extending some distance up main axes; branches mostly 9–13 mm broad, 400–500 μm thick, somewhat coriaceous.

Frequent on rocks, lower intertidal to subtidal (to 13 m), Cape Arago,

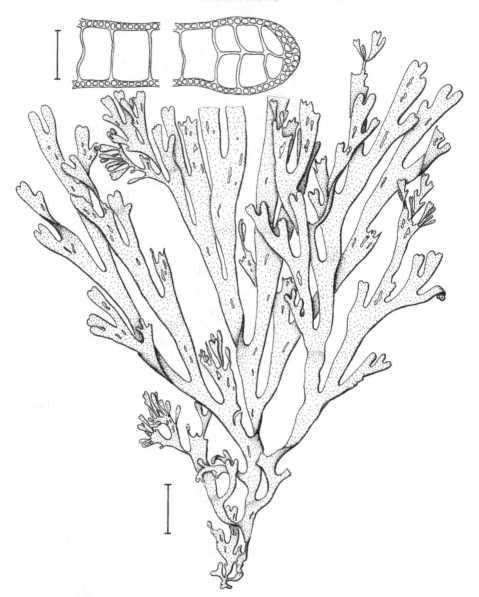

Fig. 173. *Pachydictyon coriaceum.* Rules: habit, 2 cm; cross section of blade, 200 μm. (both NLN)

Ore., and Diablo Canyon (San Luis Obispo Co.) to Channel Is. and San Diego, Calif., and to Baja Calif. and Panama. Type locality: Japan.

Syringoderma Levring 1940

Thalli erect, with several to numerous multifid blades arising from com-

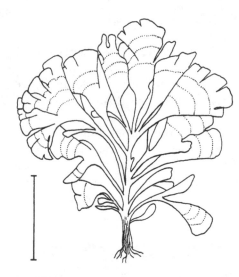

Fig. 174. *Syringoderma abyssicola*. Rule: 1 cm. (SM, after Setchell & Gardner)

pact rhizoidal base. Blade divisions plane, flabellate, with marginal row of initials, these forming laterally conjoined cell rows. Blade divisions monostromatic within 1–3 cells of margin, becoming distromatic by division of cells and ultimately covered by cortical layer of rhizoidal filaments. Chloroplasts numerous, spherical. Sporangia sparse, superficial, scattered, with many(?) aplanospores. Gametangia unknown.

Syringoderma abyssicola (S. & G.) Levr.

Chlanidophora abyssicola Setchell & Gardner 1924b: 11. *Syringoderma abyssicola* (S. & G.) Levring 1940: 8.

Thalli erect, 2–3 cm tall, with several to many broadly flabellate divisions; cells of monostromatic margins cylindrical, mostly rectangular in surface view, 30–50 μm long, 10–12 μm diam., those in distromatic parts about half as long; sporangia 32–38 μm long, 28–32 μm diam., scattered among short clavate paraphyses 4–7 cells long.

Rare, mostly subtidal (9–28 m) on rocks and shells at San Juan I. (type locality), Wash., to low intertidal near Trinidad (Humboldt Co.), Calif.

Dictyopteris Lamouroux 1809

Thalli densely and irregularly dichotomous, with percurrent midrib, tomentose below. Growth in apical row of cells. Sporangial sori numerous along both sides of midrib. Oogonial and antheridial sori scattered. Plants exhibiting a bluish iridescence under water.

Dictyopteris johnstonei Gardn.

Gardner 1940: 270; Silva 1957b: 42.

Thalli similar to those of *D. undulata* but smaller, mostly under 7 cm tall; branches much narrower, mostly 1–2 mm broad, with plane, mono-stromatic margins, grading into inconspicuous midrib; branching dichoto-mo-flabellate, with moderately wide axils; sporangia sparse, scattered on 1 surface.

Infrequent to rare, on rocks, lower tidepools to subtidal (45 m), Lone Cove (Santa Cruz I.; type locality), Calif., to Is. San Benito, Baja Calif., and possibly into Gulf of Calif.

Silva expressed doubt about the validity of the species, but large num-bers of specimens are not available for comparison. Winter specimens of *D. undulata* from the upper Gulf of Calif. appear to be shorter and with more narrow branches; thus temperature may have a strong influence on the appearance of the thalli, hence the validity of the species.

Dictyopteris undulata Holmes

Holmes 1896: 251; Abbott 1972b: 260. *Dictyopteris zonarioides* Farlow 1899: 73. *Neurocarpus zonarioides* (Farl.) Howe 1914: 69; Setchell & Gardner 1925: 656.

Thalli irregularly dichotomous, yellowish-brown to olive, drying to nearly black, 8–12(24) cm tall; short terminal branches with prominent midrib, the apices broad-obtuse or slightly retuse. Sporangia common.

Frequent on rocks, lower intertidal pools to subtidal (36 m), Santa

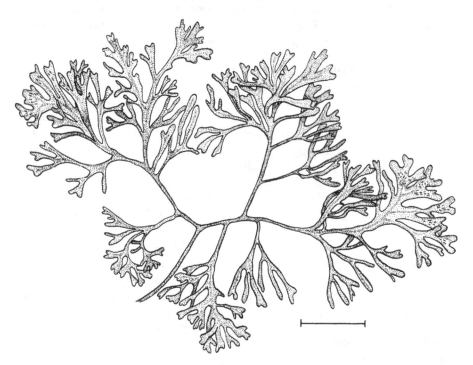

Fig. 175. *Dictyopteris johnstonei.* Rule: 2 cm. (SM)

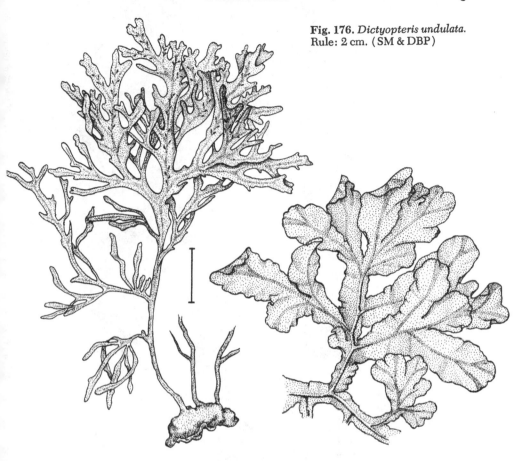

Fig. 176. *Dictyopteris undulata.* Rule: 2 cm. (SM & DBP)

Monica and Santa Catalina I., Calif., to San Jose del Cabo and into upper Gulf of Calif., Mexico. Type locality: Japan.

Taonia J. Agardh 1848

Thallus bladelike, with plane margins, without midrib, and with 1 to several flabellate divisions, often deeply dissected into narrow laciniae. Medulla with ±4 layers of large, colorless cells irregularly arranged; surface a single layer of pigmented cells, mostly in longitudinal rows. Growth in apical row of cells. Sori of sporangia and gametangia scattered or in somewhat concentric zones on both surfaces of blades, partially embedded. Sexual thalli rare, shorter than sporangial thalli.

Taonia lennebackeriae J. Ag.

J. Agardh 1894: 30; Setchell & Gardner 1925: 657; Mathieson 1966: 65.

Thalli erect, 10–30(90) cm tall; blades 10–60 mm broad, 130–180 μm

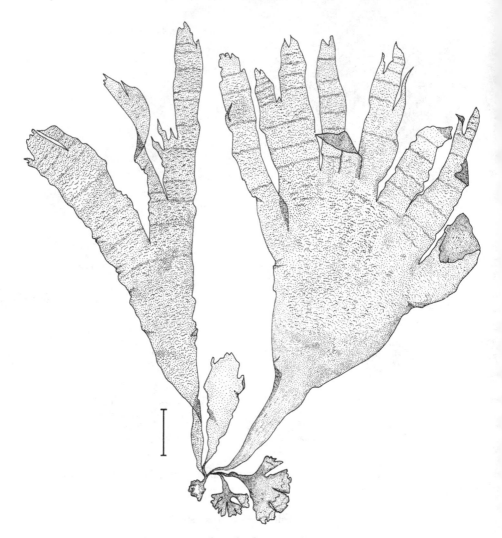

Fig. 177. *Taonia lennebackeriae.* Rule: 2 cm. (CS)

thick, olive to dark brown; surface cells 20–80 μm diam., 15–50 μm long;
sporangia scattered or in small groups, in vaguely concentric lines, not
in definite sori; antheridial sori infrequent but conspicuous, regularly
bounded by deeply pigmented cells, occurring only on plants that also
bear sporangia; oogonia unknown. Adventitious thalli frequent, formed on
margins and basal rhizoids.

Frequent on rocks, often partially embedded in sand, lower intertidal
to upper subtidal, Channel Is., Lechuza Pt. (Los Angeles Co.), Calif., to
Bahía Asunción, Baja Calif. Type locality: "California" (probably near
Santa Barbara).

Zonaria C. Agardh 1817

Thalli erect to somewhat decumbent, with prominently felted base. Growth in apical row of cells. Blades flabellately divided, without midrib, becoming stipitate below with eroding of marginal tissue. Medulla with several cell layers, the cells arranged bricklike in precise rows in 3 dimensions; surface layer with pigmented cells in mostly longitudinal rows. Sporangia and gametangia on separate thalli, in scattered sori on both surfaces of blades; sporangia with paraphyses; aplanospores 8 per sporangium.

Zonaria farlowii S. & G.

Setchell & Gardner 1924b: 11; 1925: 660; Haupt 1932: 239.

Thalli profusely flabellate, 8–15(30) cm tall, the terminal lobes often split into many narrow divisions, the lobe apices usually lighter-colored than main portions of blades; sori scattered; sporangial sori prominent, with multicellular paraphyses; gametangial sori on different plants, both kinds without paraphyses.

Common on rocks, lower intertidal to subtidal (20 m), Santa Barbara, Calif., to I. Magdalena, Baja Calif. Type locality: "So. Calif."

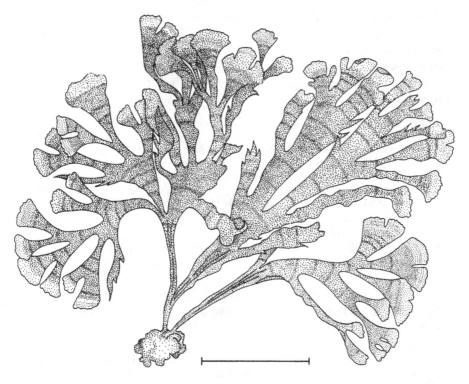

Fig. 178. *Zonaria farlowii*. Rule: 2 cm. (CS)

Oogonia require a lunar month to develop and are produced in periodic crops. Sporangia require a longer period to develop and are not mature at the same time on a given thallus. Unlike other Dictyotales that have been investigated, *Zonaria* possesses pyrenoids in the discoid chloroplasts.

Order SPHACELARIALES

Thalli filamentous, mostly tufted, relatively stiff. Each axis and branch with a conspicuous apical cell. Branches terete, uniseriate near apices, becoming multiseriate below by formation of longitudinal walls; branches of some genera with outer cortical layer. Attachment by rhizoidal filaments or crustose base. Chloroplasts discoid, numerous, pyrenoids reported for zoospores but none in vegetative cells. Unilocular sporangia and plurilocular gametangia rare, on separate but similar plants. Commonly with vegetative propagation by characteristic deciduous branchlets (propagula), this used for purposes of taxonomy.

Family SPHACELARIACEAE

With characters of the order.

Sphacelaria Lyngbye 1819

Thalli erect, tufted, attached by mass of rhizoidal filaments or sometimes by somewhat discoid base. Erect branches cylindrical, variously branched, mostly without cortication, with or without phaeophycean hairs. Branches arising near apex or uniseriate portion of axes. Sporangia and gametangia arising from multiseriate portion of axes, on short pedicels. Usually with propagula of distinctive form.

1. Erect branches with primary transverse and longitudinal septa, also with
   secondary transverse septa ....*S. racemosa*
1. Erect branches usually without secondary transverse septa............ 2
   2. Propagula short, without projecting branches.............*S. californica*
   2. Propagula slender, with elongate branches 3
3. Arms of propagula bifurcate ....*S. didichotoma*
3. Arms of propagula simple ....*S. furcigera*

Sphacelaria californica (Sauv.) S. & G.

Sphacelaria plumula var. *californica* Sauvageau 1901: 108. *S. californica* (Sauv.) Setchell & Gardner 1925: 395. *S. tribuloides sensu* Saunders 1898: 158. *S. hancockii* Dawson 1944a: 225.

Thalli brown, stiff, tufted, to 3+ cm tall, the lower third usually devoid of branches, the upper branches frequent but irregular, tapering from base to apex; with small discoid base, this producing contorted, partly colorless rhizoids; secondary transverse walls occasionally present; phaeophycean hairs usually absent, but abundant during production of propagula; propagula obovoid to club-shaped, 140–160 μm long, 80–120 μm diam., with

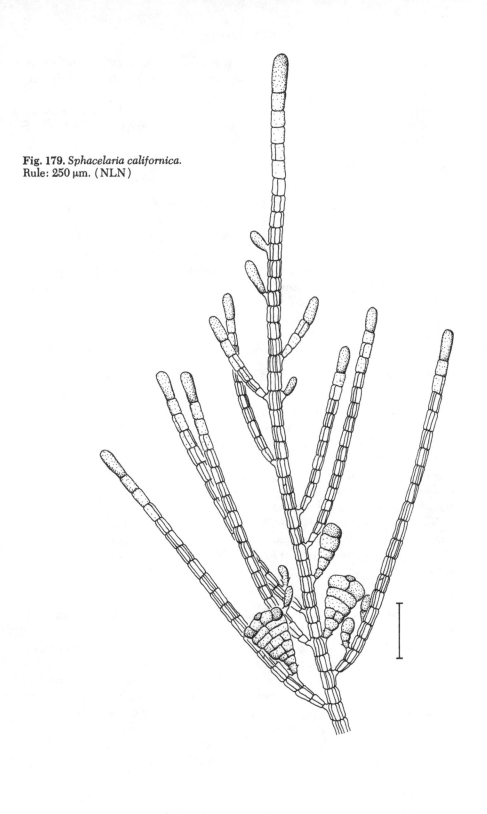

Fig. 179. *Sphacelaria californica.*
Rule: 250 μm. (NLN)

branches reduced to single large cell projecting slightly at each of 3 corners; sporangia short, ellipsoidal to obovate, 75–150 μm long, 50–70 μm diam.; gametangia unknown.

Frequent on rocks, midtidal to subtidal (20 m), San Pedro and Santa Catalina l. to San Diego (type locality), Calif., and Baja Calif. into Gulf of Calif.

Sphacelaria didichotoma Saund.

Saunders 1898: 158; Setchell & Gardner 1925: 397.

Thalli to 4 mm tall, densely tufted, with creeping, partly endophytic base; segments between septa 25–35 μm diam., slightly longer than diam., with 2 or 3 longitudinal walls; hairs unknown; propagula slender, bifurcate, each branch again bifurcate; main axis of propagulum 200–300 μm long, the branches 100–200 μm long; sporangia and gametangia unknown.

Rare, usually epiphytic, low intertidal to subtidal (to 6 m), Monterey to San Diego, Calif. Type locality: Carmel Bay, Calif.

Sphacelaria furcigera Kütz.

Kützing 1855: 27; Setchell & Gardner 1925: 396. S. subfusca S. & G. 1924b: 1; 1925: 395.

Thalli tufted, from less than 5 mm to 1 cm tall; base of closely intertwined, creeping filaments, often penetrating host; branching distant, sparse; hairs lateral and subterminal, sometimes usurping terminal position; segments between septa 16–30(50) μm diam., 2–3 times as long; propagulum slender, bifurcate, with rudiment of third branch between the 2 arms, containing 6–9(13) segments, ±20 μm diam. above, slightly attenuate below; branches of propagula with 5–8(20) cells, not attenuate apically; sporangia spherical, 50–70 μm diam., on short, 1-celled pedicels; gametangia with small loculi in male thalli and large loculi in female thalli.

Mostly epiphytic: in E. Pacific known from Alaska, Calif., and Chile; rare in Calif., low intertidal and subtidal, Carmel Bay, San Pedro, Redondo Beach, and Laguna Beach. Widely distributed in temperate and tropical seas. Type locality: Persian Gulf.

Sphacelaria racemosa Grev.

Greville 1824: pl. 96; Setchell & Gardner 1925: 393. S. arctica Kjellman 1889a: 51. S. racemosa var. arctica sensu Saunders 1901a: 419.

Thallus dark brown, coarse, tufted, 1–7 cm tall, opposite to irregularly branched above; base very small, prostrate; older primary segments between septa 55–70(130) μm diam., with secondary transverse as well as longitudinal walls; propagula unknown; sporangia 40–50 μm long, 40 μm diam.; gametangia not reported for Calif., but ovate-cylindrical to cylindrical and racemose elsewhere.

Mostly saxicolous, lower intertidal to upper subtidal, Alaska to Cape Mendocino, Calif.; rare in Calif. Type locality: Scotland.

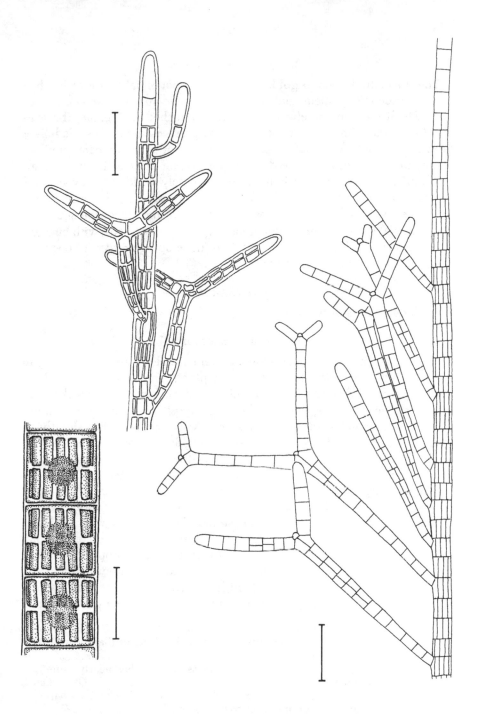

Fig. 180 (below right). *Sphacelaria didichotoma*. Rule: 100 μm. (JRJ/S)   **Fig. 181** (above left). *S. furcigera*. Rule: 100 μm. (SM, after Setchell & Gardner)   **Fig. 182** (below left). *S. racemosa*. Rule: 40 μm. (DBP, after Greville)

Order DESMARESTIALES

Thalli usually brown to golden, nonfleshy. Sporangial phase of life history macroscopic; gametangial phase microscopic and only known in culture. Thalli erect, pinnately or irregularly branched in 1 plane, the axes and branches cylindrical or markedly compressed, sometimes bladelike; growth in length and width principally trichothallic, in cambium-like meristem at base of hairs, the divisions cutting off more cells toward thallus than toward free end of hair. Hairs ultimately deciduous; toward end of growing season the hairs absent and plants markedly different in appearance. With unangia only, these isolated or in small inconspicuous sori. Gametangial thalli minute, filamentous, dioecious; oogamous, with biflagellate sperms; in culture, young sporophyte developing at apex of oogonial mass (as in order Laminariales).

Family DESMARESTIACEAE

With characters of the order.

Desmarestia Lamouroux 1813

Sporangial thalli perennial, elongate, remarkable in the field owing to large amounts of acids produced when plants are collected, bleaching nearby plants and producing an acrid odor. Holdfast stout, disklike, producing single erect axis, this soon forming opposite or alternate branchlets according to species, rarely unbranched. Axes and branches slightly to markedly compressed; some species remaining finely branched throughout, others becoming foliose. In Calif., a seasonal occurrence of hairs at margins of branches. Thalli containing unangia rarely collected in Calif.

Despite large size of some species (to 3 m long), internal structural features and cell modifications not as elaborate as in order Laminariales. Center of thallus with several axial strands having thick walls and surrounded by thick cortex of colorless cells, interspersed by colorless rhizoids. Surface layer of cells containing many lenticular chloroplasts, without pyrenoids. Tufts of hairs in which the intercalary meristem is found are terminal on branches. In older thalli, some transverse and longitudinal walls frequently pitted; cytoplasmic threads traversing the pits probable.

1. Axes and branches less than 2 mm wide throughout 2
1. Axes 4–15 (20) mm wide, the successive branches smaller than bearing branch or the thallus not branched 3
    2. Lower stipitate portion and axis compressed; branches nearly same width throughout length ....*D. kurilensis*
    2. Lower stipitate portion and axis cylindrical, or axis slightly compressed; branches tapering from base to apex...............*D. viridis*
3. Main axes linear, mostly alternately branched................*D. latifrons*
3. Main axes not linear, oppositely branched 4

Fig. 183. *Desmarestia kurilensis*. Rule: 2 cm. (SM)

4. Axes and main branches narrowing from base to apex........*D. ligulata*
4. Axes and branches (rare) foliose...............*D. ligulata* var. *firma*

Desmarestia kurilensis Yamada

Yamada 1935: 14.

Thallus 30–50 cm tall, arising from small, disk-shaped base, with per-current principal axis; bipinnate; principal axis about 2 mm wide near base; branches opposite, crowded, 0.5–1 mm wide, compressed through-out, wider through middle than at either end; hairs tapering, with hooked terminal cells.

Infrequent, saxicolous, subtidal (to 45 m), Br. Columbia and Carmel Submarine Canyon, Calif. Type locality: Kurile Is.

Much like *D. viridis*, but compressed throughout and having branches that are narrow throughout the plant. Calif. thalli have not been observed with hairs having a hooked terminal cell.

Desmarestia latifrons Kütz.

Kützing 1859: 40; Smith 1944: 120.

Thalli to 2 m tall, arising from disk-shaped base, dark brown to black; axis linear, compressed to flattened, 1.5–3 mm wide; branching of axis profuse, the long branches frequently formed near base, the shorter ones above, mostly alternate, with intervals of 10–30 mm between successive branches; axis and major branches with inconspicuous percurrent midrib; filaments along axis and branches 2–4 mm long, not found after July in C. Calif.; bare branches coarse and stringy.

Locally frequent, saxicolous, low intertidal to subtidal (15 m), in areas of strong surf, Coos Bay, Ore., to Government Pt. (Santa Barbara Co.), Calif. Type locality: Fort Ross (Sonoma Co.), Calif.

Desmarestia ligulata (Lightf.) Lamour.

Fucus ligulatus Lightfoot 1777: 946. *Desmarestia ligulata* (Lightf.) Lamouroux 1813: 45; A. Chapman 1972: 1 (incl. synonymy). *Fucus herbaceus* Turner 1808: 77. *D. her-bacea* (Turn.) Lamour. 1813: 45; Smith 1944: 121. *D. munda* Setchell & Gardner 1924b: 7. *D. jordanii* Gardner 1940: 269. *D. linearis* Smith 1944: 120. *D. mexicana* Dawson 1944a: 236.

Thalli annual, some perhaps perennial, to 8 m tall, the holdfast lobed or smooth, conical or flattened, 2–4 cm diam.; stipe flattened; axes usually broader at base and tapering upward or broad throughout, (0.5)1–2.5 cm broad; midrib usually prominent in mature plants, with opposite veins leading to branches; second-, third-, or rarely fourth-order branches appearing stipitate from these veins, the "stipes" becoming spines with loss of higher order of branching; trichothallic filaments forming dense margi-nal fringe in some forms, quickly formed and early deciduous in others.

Any set of characters chosen for taxonomic purposes shows an overlap

Fig. 184. *Desmarestia latifrons.*
Rule: 10 cm. (JRJ/S)

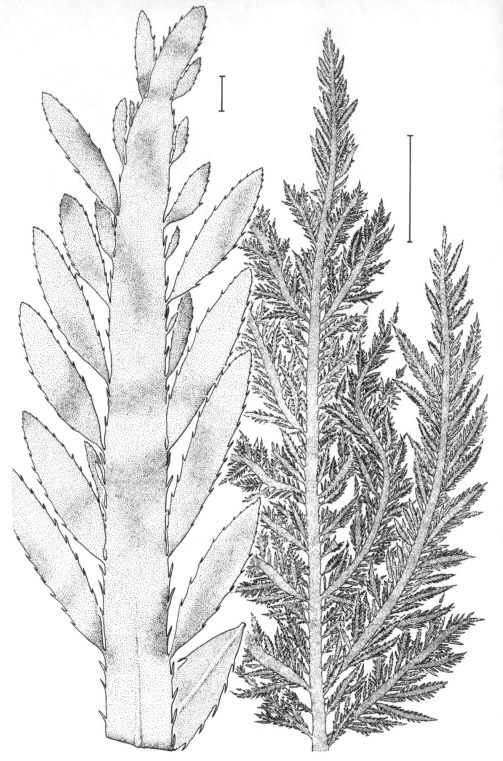

Fig. 185. *Desmarestia ligulata* var. *ligulata*. Rules: both 5 cm. (JRJ/S)

in the young thalli of *D. herbacea* and *D. munda*, whereas mature thalli show quite distinct habits. When examined in large series and compared with specimens from the N. Atlantic, these taxa do not remain distinct. In our own flora, the taxon previously known as *D. linearis* (which resembles the earlier *D. jordanii*) is known only from subtidal collections; in the same way, the supposedly distinct *D. mexicana* is separated geographically from other populations. Thus, although it is possible to select out certain specimens as distinct, such distinctions blur with increased numbers.

Desmarestia ligulata var. ligulata

Margins of young plants densely fringed with hairs, these deciduous with age; thalli compressed, less than 1 cm to 2–3 cm broad at base when young, to 1 m wide in some forms when mature, with several orders of branching.

Locally abundant, saxicolous or on wood, low intertidal to subtidal (15 m), Alaska to S. America; in Calif., from Humboldt Co. to La Jolla. Widely distributed in N. Hemisphere. Type locality: Scotland.

Desmarestia ligulata var. firma (C. Ag.) J. Ag.

Sporochnus herbaceus var. *firma* C. Agardh 1824: 261. *Desmarestia ligulata* var. *firma* (C. Ag.) J. Agardh 1848: 169; A. Chapman 1972: 2 (incl. synonymy). *D. foliacea* Pease 1920: 322. *D. tabacoides sensu* Hollenberg & Abbott 1966: 20.

Thalli leaflike, with 1 or more blades arising from short stipe and expanding into broad oval to subcordate blades, to 34 cm tall, 21 cm broad, mostly broad-oval; blades occasionally irregularly lobed or with the lateral margin torn; midrib clearly marked throughout length of blade, with 7–20 pairs of opposite veins.

Infrequent, saxicolous, subtidal (3 to 40 m), Br. Columbia, Wash., and Monterey Peninsula, Carmel Submarine Canyon, Granite Canyon (Monterey Co.), and La Jolla, Calif. Type locality: Atlantic Coast of France.

Desmarestia viridis (Müll.) Lamour.

Fucus viridis Müller 1782: 5. *Desmarestia viridis* (Müll.) Lamouroux 1813: 45; Smith 1944: 119; A. Chapman 1972: 1 (incl. synonymy applicable to Pacific coast). *D. pacifica* Setchell & Gardner 1924b: 6. *D. filamentosa* Dawson 1944a: 236.

Thalli 15–45(100) cm tall, with cylindrical stipe; extremely fragile; axis percurrent, subcylindrical, 0.75–1.5 mm wide at base, oppositely branched; primary branches lax, pinnately branched, occasionally bipinnate; branches clothed with short branchlets, very rarely found with hairs.

Occasional to rare from late spring into late summer, saxicolous, subtidal (to 45 m), Alaska to Baja Calif. and Gulf of Calif.; in Calif., at Mendocino Co., Monterey Peninsula, and Santa Catalina I. Type locality: N. Atlantic.

A close examination of the Monterey Peninsula specimens indicates that most given this name are probably *D. kurilensis* Yamada.

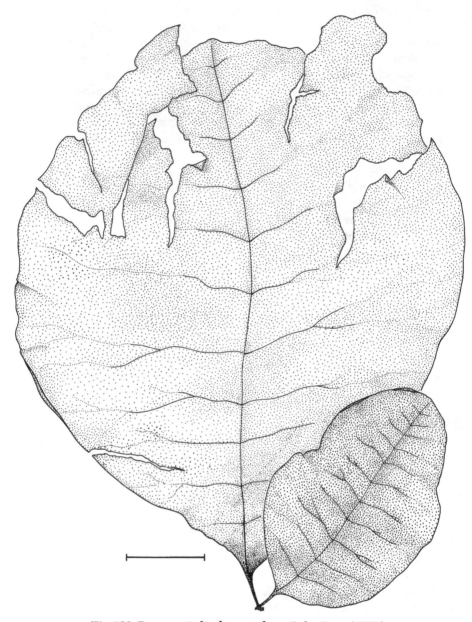

Fig. 186. *Desmarestia ligulata* var. *firma*. Rule: 5 cm. (DBP)

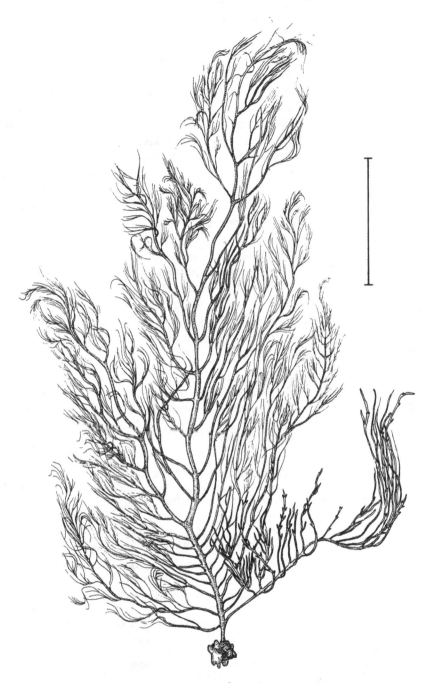

Fig. 187. *Desmarestia viridis*. Rule: 5 cm. (SM)

Order Laminariales

(Contributed by Nancy L. Nicholson)

Life history of two morphological phases, nearly uniform in species of this order. Sporangial thalli macroscopic; gametangial thalli microscopic. Basic macroscopic morphological pattern a blade, stipe, and holdfast, usually showing this distinction in juvenile stages, the tissues differentiating internally into epidermis, cortex, and medulla. Epidermal cells with numerous discoid chloroplasts, without pyrenoids. Cortex colorless, of many cell layers; medullary layers modified by elongation into trumpet-shaped cells whose crosswalls contain simple sievelike plates, the medullary tissue serving as route for translocation of photosynthate from blades through stipe to holdfast. Cortical mucilage ducts in blade or stipe present or absent. Growth in several intercalary meristems. Sori usually appear as well-defined, dark, regular or irregular patches, sometimes borne on special blades (sporophylls) in restricted portions of thallus, or scattered; sorus of unangia and closely packed paraphyses. Gametangial thalli (not known in nature) filamentous, dioecious; gametangia unilocular; sperms pyriform and laterally biflagellate; eggs nonmotile, fertilized and germinating at tip of oogonium.

1. Transition between stipe and blade neither splitting nor giving rise to outgrowths (but certain species of *Laminaria* characterized by dissected blades that may superficially appear to have splits originating at blade bases) .... Laminariaceae (below)
1. Transition between stipe and blade cloven into several parts, the cleavage extending well into growing region, or the transition area bearing appendages 2
   2. Cleavage arising at or near transition between stipe and blade......
   .... Lessoniaceae (p. 245)
   2. Outgrowths or appendages arising at or near transition area........
   .... Alariaceae (p. 237)

Family Laminariaceae

Fronds simple, the blades sometimes fraying or splitting with age. Stipe without true dichotomous branching, occasionally with adventitious branching from blade. Holdfasts simple or branched.

1. Stipe absent or rudimentary; attachment by haptera arising directly from blade .... *Hedophyllum* (p. 233)
1. Stipe present 2
   2. Midrib or ribs lacking, the blades simple or digitately divided into straplike sections, smooth or bullate...............*Laminaria* (p. 229)
   2. Midrib or longitudinal ribs present............................. 3
3. Blade with five longitudinal ribs......................*Costaria* (p. 237)
3. Blade with midrib 4

4. Stipe usually with lateral projections (in Calif. species); blade crisp, with irregular perforations on lateral areas...........*Agarum* (p. 234)
4. Stipe without lateral projections; blade flabby, lacking perforations on lateral areas ....*Pleurophycus* (p. 234)

Laminaria Lamouroux 1813

Spore-bearing phase annual or perennial, usually with single stipe attached either by mass of branched haptera or by discoid holdfast; more rarely with several stipes and blades borne on prostrate branched rhizome. Stipe cylindrical or compressed, always terminating in single blade, this sometimes further divided. Perennial species either with persistent blade or shedding blade each autumn and regenerating new blade. Blades entire or incompletely and palmately cleft into several divisions, usually smooth but sometimes ruffled or bullate, without longitudinal ribs or midrib. Sporangia produced in irregularly shaped, relatively inconspicuous sori.

1. Holdfast discoid ....*L. ephemera*
1. Holdfast of distinct, branched haptera 2
   2. Numerous stipes arising from creeping holdfast............*L. sinclairii*
   2. Single stipe and blade arising from each holdfast.................. 3
3. Blades divided and smooth in mature plants...............*L. dentigera*
3. Blades entire and bullate, or smooth............................. 4
   4. Blade covered entirely by bullae........................*L. farlowii*
   4. Blade smooth; mucilage ducts in blade and stipe, these rarely absent entirely ....*L. dentigera* (juvenile)

Laminaria dentigera Kjellm.

Kjellman 1889a: 45; Setchell & Gardner 1925: 604; Druehl 1968: 546. *Hafgygia andersonii* Areschoug 1883: 3. *Laminaria andersonii sensu* Smith 1944: 137. *L. cordata* Dawson 1950a: 153. *L. setchellii* Silva 1957b: 42.

Sporangial plants perennial from stipe, to 1.5(3) m long, with single stipe produced from each holdfast; holdfast of stiff, branched haptera compacted into conical mass 6–8 cm tall in older plants; stipe to 3 cm diam. near base, tapering and becoming slightly complanate to markedly flattened near blade; older stipes with concentric rings in outer cortex and mucilage ducts internal to innermost ring as seen from base of plant; mucilage ducts in upper stipe nearer surface and concentric rings lacking, the ducts entirely lacking in some specimens; blade ovate and entire in juvenile plants; blade smooth, 14–24 cm wide, in mature plants, appearing palmate by virtue of deeply incised linear divisions more or less equal in width, rarely to 1 m long; mucilage ducts present (except as noted above); sori appearing in late fall as irregularly linear dark patches; regeneration of new blades often beginning before old blades are shed.

Common in groups, saxicolous, lower intertidal to upper subtidal of

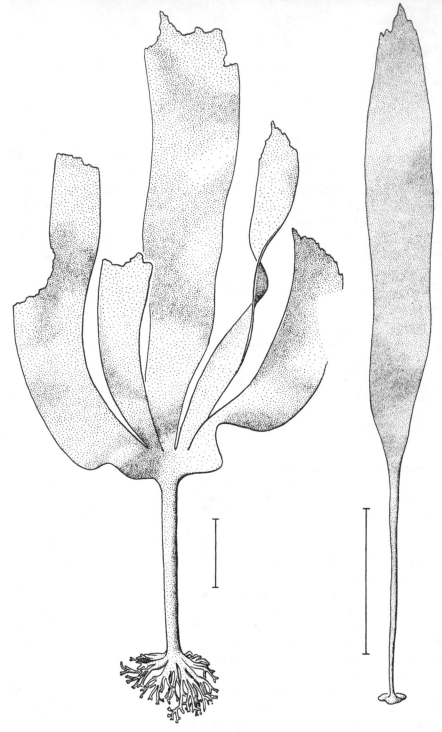

Fig. 188 (left). *Laminaria dentigera.* Rule: 5 cm. (JRJ/S)   **Fig. 189** (right). *L. ephemera.* Rule: 5 cm. (JRJ/S)

moderately exposed areas, Bering Strait to Ensenada, Baja Calif. Type locality: Bering I. (Komandorski Is.), U.S.S.R.

Laminaria ephemera Setch.

Setchell 1901: 121; 1908a: 92; Setchell & Gardner 1925: 603; Smith 1944: 136 (incl. synonymy).

Sporangial thalli annual, to 2 m tall, with single stipe and blade; holdfast discoid, without haptera; stipe cylindrical, 6–10 cm long, 2–4 mm diam.; blade narrowly cuneate to broadly rounded at base, above this narrowly to broadly linear, entire or with 1–3 longitudinal clefts extending nearly to base, the blade surface smooth; sori in longitudinal strips covering both surfaces of blade, medium brown in most plants; fertile late in the spring or very early summer, then disappearing.

Infrequent on rocks in lower intertidal tidepools, Volga I., Alaska, and Br. Columbia to Little Sur River (Monterey Co.), Calif. Type locality: Carmel Bay, Calif. (but rare there).

Laminaria farlowii Setch.

Setchell 1893: 355; Setchell & Gardner 1925: 599; Smith 1944: 136.

Sporangial thalli perennial, fertile in late summer, to 5 m tall, dark chocolate brown; holdfast of strong, compact, branching haptera; single stipe short, terete, flattening suddenly to single blade, the stipe 4–7 cm long, 4–6 mm diam., without mucilage ducts; blade thick, coriaceous, abundantly bullate, with relatively deep depressions, these more or less in longitudinal rows over both sides of blade; mucilage ducts in blade scanty; original blade persistent, not shed annually.

Isolated populations, saxicolous, low intertidal in N. Calif., subtidal (to 50 m) in Channel Is.; Br. Columbia to Bahía del Rosario, Baja Calif. Type locality: Santa Cruz, Calif.

Laminaria sinclairii (Harv.) Farl., Anders. & Eaton

Lessonia sinclairii Harvey 1846: 460. *Laminaria sinclairii* (Harv.) Farlow, Anderson & Eaton 1877–89 (1878): no. 118.

Sporangial thalli perennial, to 1 m tall, 3 cm broad, rich dark brown; with prostrate, branched rhizome bearing many erect stipes, each terminating in single blade; stipes cylindrical, to 10 cm long, 5 mm diam., with abundant mucilage ducts; blades linear, entire, with smooth surface.

Locally abundant, making dense patches on rocks (sometimes partially buried in sand), low intertidal, Vancouver I., Br. Columbia, to Ventura Co., Calif. Type locality: San Francisco, Calif. Absent on Monterey Peninsula.

Because of the distinctive rhizomes, this is the most readily recognized Calif. species of *Laminaria*.

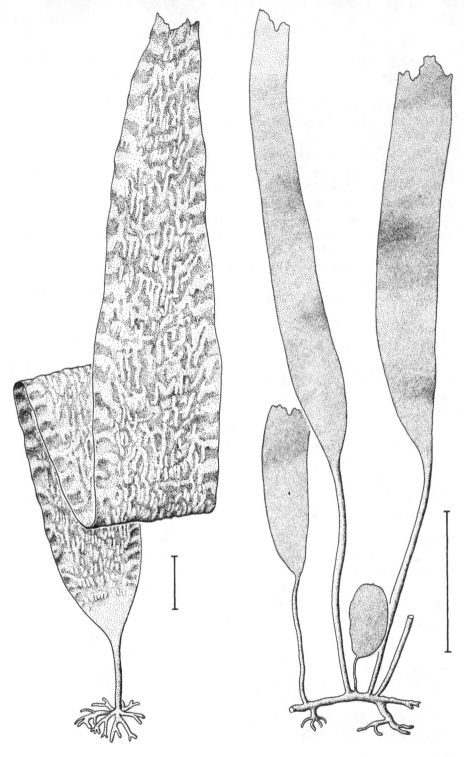

Fig. 190 (left). *Laminaria farlowii*. Rule: 5 cm. (JRJ/S)   Fig. 191 (right). *L. sinclairii*.
Rule: 5 cm. (JRJ/S)

Hedophyllum Setchell 1901

Thallus a simple unbranched blade, without visible stipe and with haptera from lower blade margins. Sporangial thalli with holdfast of branched haptera. Stipe distinct at first, later disappearing in some specimens; blade short, broad and plane at first, often becoming bullate, not auriculate at base, later giving rise to branched haptera from lower margin. Haptera either scattered or whorled. Sori basal, irregular in outline.

Hedophyllum sessile (C. Ag.) Setch.

Laminaria sessilis C. Agardh 1824: 270. *Hedophyllum sessile* (C. Ag.) Setchell 1901: 121; Setchell & Gardner 1925: 617. *L. apoda* Harvey 1862: 167.

Stipe of young sporangial plant very short and much flattened, soon disappearing entirely; blade at first ovate and entire, soon splitting deeply, even to base, and becoming cucullate, in age becoming sessile, not greatly thickened at base, giving rise to haptera along sessile margin, 30–50(150) cm tall and 80 cm wide in quiet water, rarely more than 30 cm tall on ex-

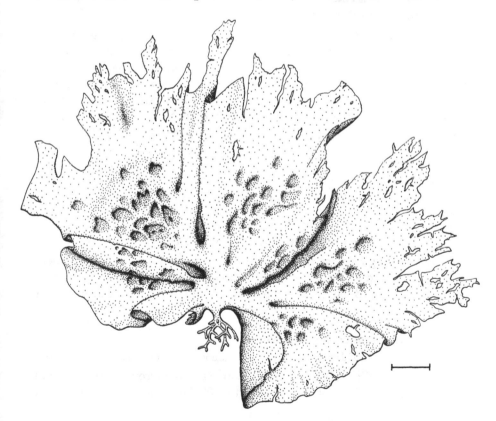

Fig. 192. *Hedophyllum sessile.* Rule: 5 cm. (NLN)

posed coasts; surface of blade either perfectly smooth or with irregular bullate swellings scattered over entire blade or only basal portion, with numerous mucilage ducts, these surrounded by small secretory cells; sori extensive, basal, irregular in outline.

Saxicolous, midtidal to subtidal, Attu I., Alaska, to Pt. Sur (Monterey Co.), Calif.; rare in Calif. Type locality: "Mari Australi" (probably Pacific Ocean).

Agarum Bory 1826

Sporangial thalli with holdfast composed of small, branched haptera. Unbranched stipe short, cylindrical or flattened, at times fimbriated, supporting single blade. Blade relatively thin, 0.5–1 mm thick in mature specimens, with narrow to broad percurrent midrib and perforated lateral wings (sometimes also bullate and undulate). Sori forming broad patches on both surfaces of blade.

Agarum fimbriatum Harv.

Harvey 1862: 166; Setchell & Gardner 1925: 616.

Thalli with haptera profuse and slender in specimens growing on wood in sheltered locations, larger and denser in old plants growing in exposed areas; stipe usually flattened, 2–6 cm long, 4–7 mm wide, with numerous, often branched fimbriae on margins, particularly near blade; blade thin and bullate, nearly circular to narrowly elliptic in outline, 20–80 cm long, 15–26 cm wide, the base rounded or slightly cordate, the margin crisp and fimbriate, the midrib fairly broad, complanate; perforations few and irregular in outline; sori borne on blades as irregular dark patches.

Frequent on rocks, wood, or other algae: midtidal or subtidal from Alaska through Puget Sd.; subtidal (to at least 115 m) in Calif. only south of Pt. Conception and around Channel Is. Type locality: Esquimalt, Br. Columbia.

Pleurophycus Setchell & Saunders 1900

Holdfast of sporangial plant branched. Stipe simple. Blade undivided, with single broad longitudinal midrib; lateral portions of blade broad and ruffled. Mucilage ducts absent throughout plant; sori produced on either side of midrib.

Pleurophycus gardneri Setch. & Saund.

Setchell & Saunders 1900: no. 346; Setchell & Gardner 1925: 606; Kjeldsen 1972: 416.

Sporangial thalli probably annual; holdfast of numerous whorls of rigid haptera, usually 10 cm or less broad; stipe 39–50 cm long, solid, terete at

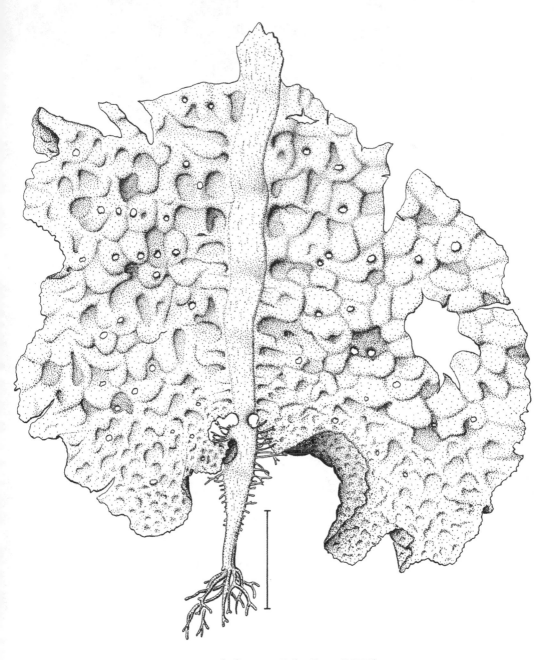

Fig. 193. *Agarum fimbriatum*. Rule: 5 cm. (NLN)

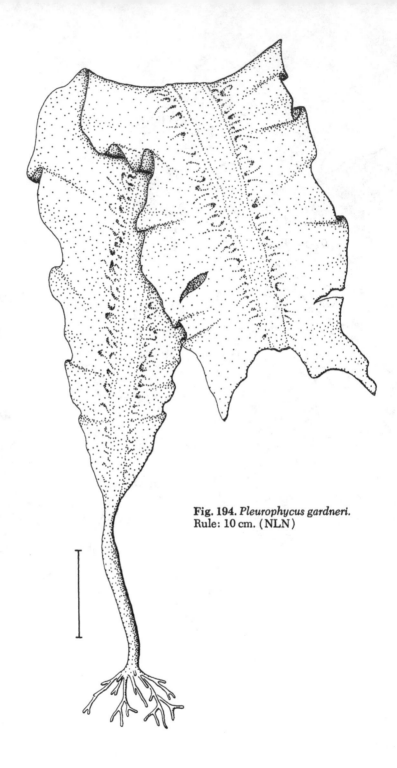

Fig. 194. *Pleurophycus gardneri.*
Rule: 10 cm. (NLN)

base, gradually flattening toward base of blade; blade flabby, elastic, delicately wrinkled near midrib, the margin entire in healthy plants, the blade 60–90 cm long, to 40 cm broad, dark olive green in life.

Frequent in northern part of range, saxicolous, low intertidal to upper subtidal, Montague I., Alaska, to Coos Bay, Ore.; rare in Calif., from Ft. Bragg and Salt Pt. (Sonoma Co.). Type locality (lectotype): Whidbey I., Wash.

Costaria Greville 1830

Sporangial thalli perennial, with single stipe attached by broad mass of haptera. Stipe unbranched, cylindrical, terminating in single, perennial, undivided, and broadly linear blade with 5 percurrent longitudinal ribs; alternate ribs projecting on opposite surfaces of blade; blade bullate in region between ribs. Sori in irregular patches, covering most of bullate area of blade.

Costaria costata (C. Ag.) Saunders

Laminaria costata C. Agardh 1817: xiii. Costaria costata (C. Ag.) Saunders 1895: 57; Setchell & Gardner 1925: 610; Smith 1944: 137.

Sporangial thalli usually 1.5–2 m tall when mature, occasionally to 3 m, dark chocolate brown; stipe cylindrical to somewhat compressed, to 50 cm long; blades to 35 cm broad when mature, either broadly linear (length 8–12 times breadth) or broadly ovate (length less than 4 times breadth); bullation of areas between ribs pronounced in mature blades; sori produced from midsummer to late autumn.

Common on rocks, lower intertidal to subtidal, N. Japan and Alaska (Shumagin Is.) to San Pedro, Calif.; subtidal in S. Calif. Type locality: "South" America.

Family ALARIACEAE

Sporangial thalli at maturity with branched or unbranched stipes. Growing region producing lateral blades, these borne either on terminal blade or along stipe. Terminal blade with a midrib or centrally thickened, or unmodified; terminal blade often overgrown by laterals, or lost in age. Sporangia restricted to sporophylls, these on stipe or blade. Gametangial thalli as in other Laminariales.

1. Stipe branched near base; lateral blades borne along most of stipe;
   pneumatocysts present .... Egregia (p. 242)
1. Stipe unbranched or forked distally; lateral blades restricted to small
   region of stipe; pneumatocysts absent 2
   2. Terminal blade soon replaced by forked continuation of stipe; blades
      corrugate .... Eisenia (p. 242)
   2. Terminal blade persistent; blades smooth or with midrib........... 3

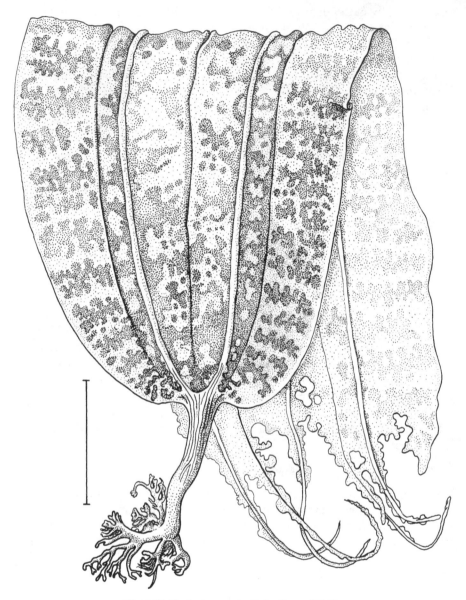

Fig. 195. *Costaria costata.* Rule: 2 cm. (SM)

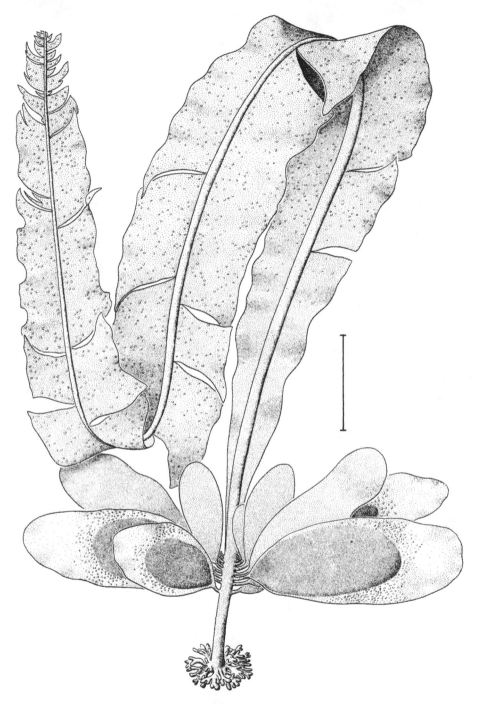

Fig. 196. *Alaria marginata*. Rule: 5 cm. (JRJ/S)

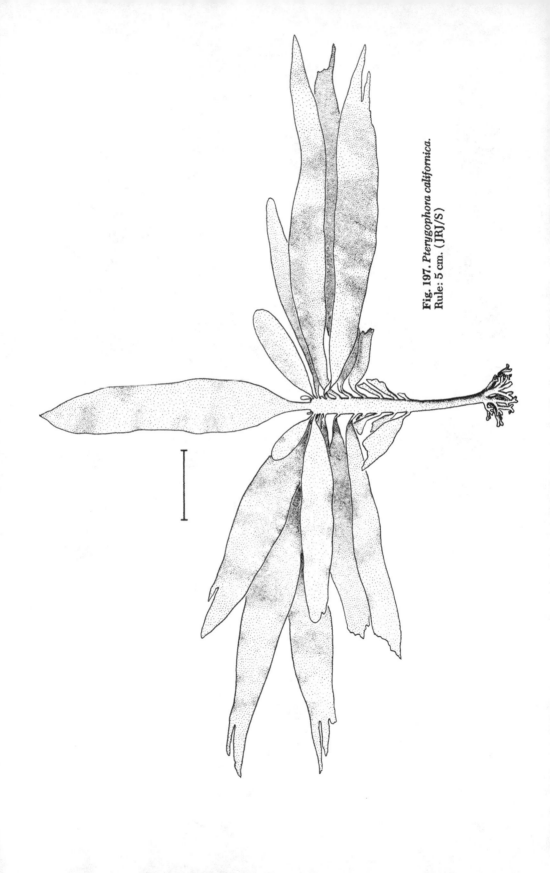

Fig. 197. *Pterygophora californica.*
Rule: 5 cm. (JRJ/S)

3. Terminal blade with prominent midrib; sporophylls differing from terminal blade in shape and size . . . . *Alaria* (below)
3. Terminal blade with barely distinguishable midrib; sporophylls similar to terminal blade in shape and size. . . . . *Pterygophora* (below)

Alaria Greville 1830

Sporangial plants perennial, with single unbranched stipe attached by mass of widely spreading, branched haptera. Stipe short, cylindrical or flattened, terminating in single blade. Blade broadly expanded, undivided, with conspicuous percurrent midrib; surface of blade smooth or bearing cryptostomata. Sporangia produced on linear to obovate flattened sporophylls, these borne on opposite sides of upper end of stipe. Sporophylls usually with each flattened face nearly covered with single large sorus.

Alaria marginata Post. & Rupr.

Postels & Ruprecht 1840: 11; Setchell & Gardner 1925: 640; Smith 1944: 147 (incl. synonymy). *Alaria curtipes* Saunders 1901b: 561.

Mature sporangial thalli usually 2.5–4(6) m tall, dark tan; stipe terete, 2–7 cm long; blades linear-lanceolate, the length 10–15 times breadth, the broadest (15–30 cm) about one-third distance from base to apex, tapering gradually or somewhat abruptly below broadest portion; all except lowermost portion of blade with cryptostomata; midrib conspicuous, varying considerably in breadth; sporophylls 20–40, sublinear to broadly elliptical, stipitate, with broadly rounded apices, 10–25 cm long, 2–6 cm broad.

Locally abundant on exposed rocks, Alaska to Carmel Highlands (Monterey Co.), Calif. Type locality: Unalaska, Alaska.

Forms approaching *A. nana* (not recognized in Calif. by Widdowson (1971)) may often be found in areas with exceptional exposure to surf.

Pterygophora Ruprecht 1852

Sporangial thalli perennial, the holdfast composed of stout, branched haptera. Stipe long, erect, unbranched, cylindrical in basal portion, flattened in upper portion, woody in texture and with concentric rings in cortex. Stipe terminating in single lanceolate blade, without midrib but with axial portion slightly thickened, surface of blade smooth. Sporophylls approximately same size and shape as blade but stipitate at base, borne beneath terminal blade on both margins of flattened portion of stipe, each margin with vertical row of 5–10 sporophylls. Sporophylls shed after fruiting and new series developed at higher level the following year. Sporangia in irregular patches on or nearly covering both surfaces of sporophyll, at times also developing on terminal blade.

Pterygophora californica Rupr.

Ruprecht 1852: 73; Setchell & Gardner 1925: 634; Smith 1944: 148.

Sporangial thalli to 2.3 m tall; stipes to 1.5 m long, to 4 cm broad in upper flattened portion; terminal blade to 80 cm long, 10 cm broad; fertility usually ensuing in October; terminal blade and sporophylls shed in winter.

Saxicolous, forming extensive beds at depths of 7–20 m, occasionally low intertidal, Vancouver I., Br. Columbia to Bahía del Rosario, Baja Calif. Type locality: Fort Ross (Sonoma Co.), Calif.

Eisenia Areschoug 1884

Sporangial thalli with holdfast of dichotomously branched haptera. Stipe elongating and persistent, bifurcate above, the 2 apparent branches actually the thickened lower margins of the original and subsequently eroded terminal blade. Small partial blade persisting at outer extremity of each false stipe throughout life of plant, this giving rise to numerous sporophylls along lower outer margin.

Eisenia arborea Aresch.

Areschoug 1876: 69; Setchell & Gardner 1925: 646; Hollenberg 1939a: 34; Hollenberg & Abbott 1966: 29.

Whole sporangial plant with conspicuous stipe topped by numerous blades and sporophylls. Holdfast stout, much branched, the branches intertwined but compact; stipe nearly terete at base, becoming much flattened above, very dark brown, almost black, 1–2 m long, rough and rigid, the upper portions containing many mucilage ducts; blades entire when young, later becoming toothed along margins and (with age) corrugate on surface; blades of mature plants golden olive, more or less longitudinally grooved, lanceolate with variable denticulations along margin, some of these to 4+ cm long, 1 cm broad; sporophylls borne on partial blades, 30–50 per blade; sori irregular in shape, covering most of surface of each sporophyll.

Saxicolous, forming dense subtidal "groves," low intertidal to subtidal (10 m), Vancouver I., Br. Columbia, Monterey region, Santa Catalina I. (type locality), Calif., and from S. Calif. to I. Magdalena, Baja Calif.

Egregia Areschoug 1876

Sporangial thalli perennial, with conspicuous conical holdfast of compacted, branched haptera; holdfasts of older plants much abraded. Stipe subcylindrical to markedly flattened, irregularly branched near base and with apex of each branch continued in long, ligulate, bladelike rachis. Lateral margins of stipe and rachis densely fringed with short to long papillae (in populations N. of Pt. Conception), irregularly covered or bald (Channel Is.), or completely bare (Ventura to Baja Calif.). Certain blades eventually developing spindle-shaped to subspherical pneumatocysts. Sporangia completely covering sporophylls, these without ridges,

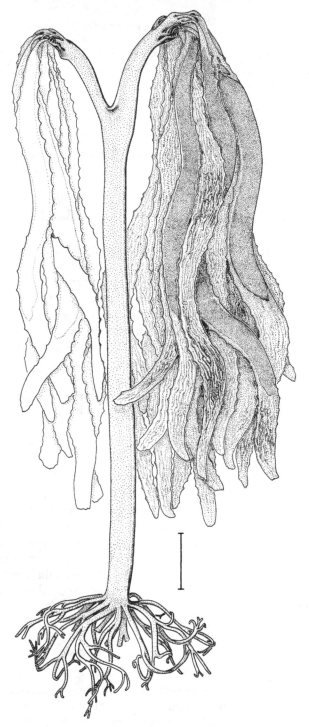

Fig. 198. *Eisenia arborea*. Rule: 10 cm. (DBP)

typically confined to grooves between ridges (where ridges occur). Sporophylls intermingled with sterile blades, distinguishable by deeper brown color or pattern of ridges.

Egregia menziesii (Turn.) Aresch.

Fucus menziesii Turner 1808: 57. *Egregia menziesii* (Turn.) Areschoug 1876: 67; Setchell & Gardner 1925: 647; Smith 1944: 149. *E. menziesii* ssp. *insularis* Silva 1957b: 45. *E. laevigata* Setchell 1925: 648; Smith 1944: 150. *E. laevigata* f. *borealis* Setch. 1925: 649. *E. laevigata* ssp. *borealis* (Setch.) Silva 1957b: 46. *E. planifolia* V. Chapman 1962: 36.

Mature sporangial thalli with several long, flattened branches densely covered distichously with broad to linear blades. Thalli 5–15 m long, the larger specimens subtidal, chocolate brown to olive green; holdfast to 25 cm diam.; stipe flattened, 1–3.5 cm broad; branches 4–25, arising principally within 1 m of base; stipe surface smooth, or patchily to densely covered with tubercles, these varying from short and blunt (1 mm or less) to long and slender (to 3 mm long); blades to 8 cm long, stipitate at base and broadly spatulate to filiform; blade surface sometimes with tubercles, particularly near base; pneumatocysts ellipsoid to subspherical, with or without papillae; sporophylls variable, from smooth, spatulate, or filiform to wrinkled pods and flattened terminal blades with sterile ridges; sporangia occurring in definite pattern or partly random pattern, fertile any time of year, but most abundantly so between April and November. Sporophylls shorter and more narrow than average vegetative blade and borne among them.

Common, saxicolous, in protected to moderately exposed areas, midtidal to subtidal (20 m), frequently forming continuous belts in intertidal, often mixed with *Macrocystis* in deeper water, Alaska to Pta. Eugenio, Baja Calif. Type locality: Nootka Sd., Vancouver I., Br. Columbia.

Variability in this species is approximately correlated with geographical distribution; northern populations (Alaska to Cape Mendocino) have tuberculate stipes and smooth sporophylls; populations from Los Angeles to Baja Calif. have smooth stipes and wrinkled sporophylls; populations from the middle coast (the Channel Is. and from the vicinity of Cape Mendocino to Ventura Co.) possess every possible combination of features observed in the geographic extremes and include a few unique vegetative and reproductive morphologies as well.

Family LESSONIACEAE

Sporangial thalli with branched stipes, these produced by split originating in meristematic region of blade and progressing toward apex of blade. Each branch terminating in single blade of determinate or indeterminate growth. Sori developing on sporophylls or on blades identical with vegetative blades. Gametangial thalli as in other Laminariales.

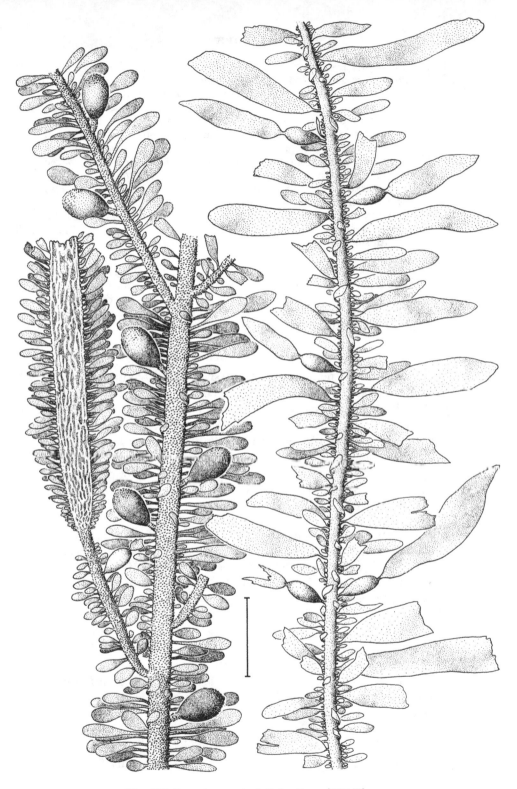

Fig. 199. *Egregia menziesii*. Rule: 5 cm. (JRJ/S)

1. Pneumatocysts present 2
1. Pneumatocysts absent (stipe may be hollow)..................... 4
   2. Numerous pneumatocysts per mature sporangial plant...........
    .... *Macrocystis* (p. 255)
   2. Single large pneumatocyst per mature sporangial plant............. 3
3. Blades arising from several clusters of short, dichotomous branches at
   top of pneumatocyst .... *Nereocystis* (p. 253)
3. Blades arising singly from sympodially branched "antlers," these borne
   at top of pneumatocyst .... *Pelagophycus* (p. 251)
   4. Stipe markedly flattened, more or less prostrate and with haptera
    along lateral margins; sori scattered............................. 5
   4. Stipe terete and erect or irregular in cross section; sori in flat or
    grooved sporophylls 6
5. Blade with smooth midrib, usually rounded at base.................
   .... *Dictyoneuropsis* (p. 248)
5. Blade lacking midrib, usually narrow at base....... *Dictyoneurum* (below)
   6. Stipe 1 per holdfast, terete and hollow; sori in grooves on blades....
    .... *Postelsia* (p. 251)
   6. Stipes numerous per holdfast, irregular in outline, solid; sori in sporo-
    phylls shorter than blades...................... *Lessoniopsis* (below)

Lessoniopsis Reinke 1903

Sporangial thalli perennial, with massive conical holdfast of compacted, repeatedly dichotomous haptera. Stipe conical, tough and dense in lower portion, branching repeatedly and dichotomously above, each branch continued in single, narrowly linear blade, new dichotomies produced by longitudinal splitting of blade, both activities resulting in densely divided terminal portions. Older blades with conspicuous flattened midrib. Sporangia produced on sporophylls borne laterally near base of primary dichotomies of stipe. Sporophylls without midrib, usually shorter and broader than sterile blades.

Lessoniopsis littoralis (Tild.) Reinke

Lessonia littoralis Tilden 1894–1909 (1900): no. 342. *Lessoniopsis littoralis* (Tild.) Reinke 1903: 25; Setchell & Gardner 1925: 632; Smith 1944: 145.

Sporangial thalli to 2 m tall, blackish-brown, densely bushy, with as many as 500 blades; holdfast to 15 cm broad at base; dichotomous portion of stipe 20–30 cm tall, the dichotomies 4–8 mm broad, the successive dichotomies 3–5 cm apart; stipe tissue dense, almost woody, or cartilaginous; blades occurring densely, narrowly linear, to 1 m long, 6–12 mm broad, the midribs flattened and 2–3 mm broad.

Locally common in areas exposed to full force of surf, saxicolous, Alaska to Malpaso Creek (Monterey Co.), Calif. Type locality: Cypress Pt. (Monterey Co.), Calif.

Dictyoneurum Ruprecht 1852

Sporangial thalli perennial, gregarious. Stipe markedly flattened, pros-

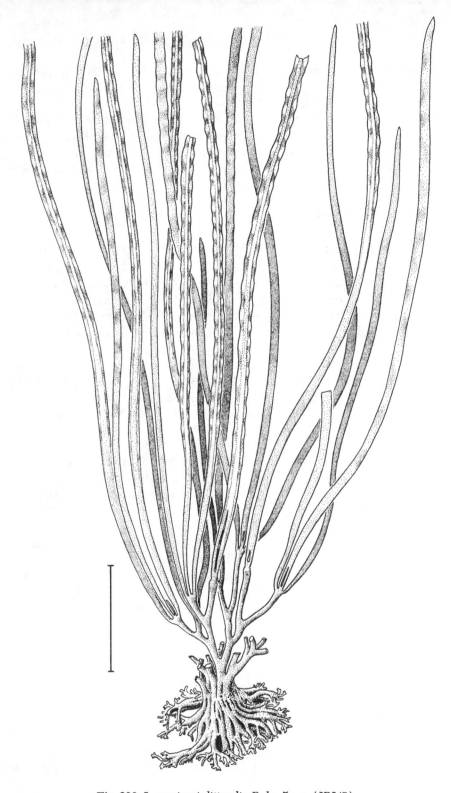

Fig. 200. *Lessoniopsis littoralis*. Rule: 5 cm. (JRJ/S)

trate, with forked haptera along lateral margins. Stipes rarely with more than 2 or 3 dichotomies, owing to progressive death and decay at lower end. Ultimate dichotomies of stipe upwardly bent, dividing dichotomously, each continued in a single blade, this splitting longitudinally from base to apex into 2 blades, 1 borne on each arm of new dichotomy. Blades linear, without midrib, both flattened surfaces covered with irregular reticulum of narrow ridges. Sporangia in irregularly shaped, inconspicuous sori, these produced on both flattened surfaces of blade.

Dictyoneurum californicum Rupr.

Ruprecht 1852: 80; Setchell & Gardner 1925: 622; Smith 1944: 139.

Sporangial thalli 0.5–2(4) m tall, light to dark yellowish-brown, in gregarious clumps, 25–100 per clump; blades 4–8 cm broad, linear, gradually tapering at both upper and lower ends, with or without marginal denticulations; reticulations on surface of blade more or less rectangular and in longitudinal series; upward splitting of blade into 2 blades frequent; sporangial areas first appearing during midsummer, darker in color than vegetative areas.

Locally frequent, saxicolous, low intertidal (where more or less directly exposed to surf) to subtidal (10 m), Vancouver I., Br. Columbia, to Pt. Conception and Channel Is., Calif.; common on Monterey Peninsula, less frequent elsewhere in Calif. Type locality: Fort Ross (Sonoma Co.), Calif.

Dictyoneuropsis Smith 1942

Sporangial thalli perennial, at first with unbranched prostrate stipe, the anterior end bending upward and terminating in single blade. Decumbent portion of stipe flattened and with branched haptera along lateral margins. Blades sublinear, with midrib; surface of blade, except for midrib, with coarse reticulum of narrow ridges. Blade and erect portion of stipe splitting longitudinally with age to form forked stipe, each branch terminating in a blade, the process repeated until stipe becomes 3 or more times dichotomous. Longitudinal division of blade through midrib followed by regeneration of new half blade along exposed edge of midrib. Sporangia in irregularly shaped sori between reticulations of blade. Sori at times covering midrib.

Dictyoneuropsis reticulata (Saund.) Smith

Costaria reticulata Saunders 1895: 58. *Dictyoneuropsis reticulata* (Saund.) Smith 1942: 651; 1944: 140.

Thalli to 95 cm tall, yellowish-brown; prostrate portion of stipe to 3 times dichotomous, the dichotomies 4–6 mm broad, 1 mm thick; haptera along lateral margins of dichotomies, finely divided, fibrous, 3–5 times dichotomous, 25–30 mm long; mature blades linear, slightly narrower at upper end, broadly rounded at base; reticulation of blade restricted to

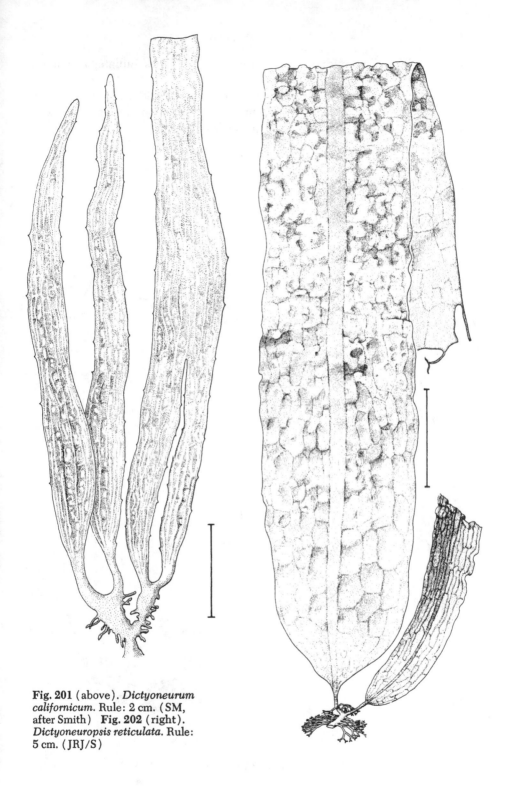

Fig. 201 (above). *Dictyoneurum californicum*. Rule: 2 cm. (SM, after Smith) **Fig. 202** (right). *Dictyoneuropsis reticulata*. Rule: 5 cm. (JRJ/S)

portions lateral to midrib, but entire blade at times bullate; maximum breadth of blade 15–25 cm.

Infrequent in sheltered areas, saxicolous, low intertidal (to 20 m), Vancouver I., Br. Columbia, to N. Channel Is., Calif.; common in vicinity of type locality, Monterey Bay near Pacific Grove, Calif.

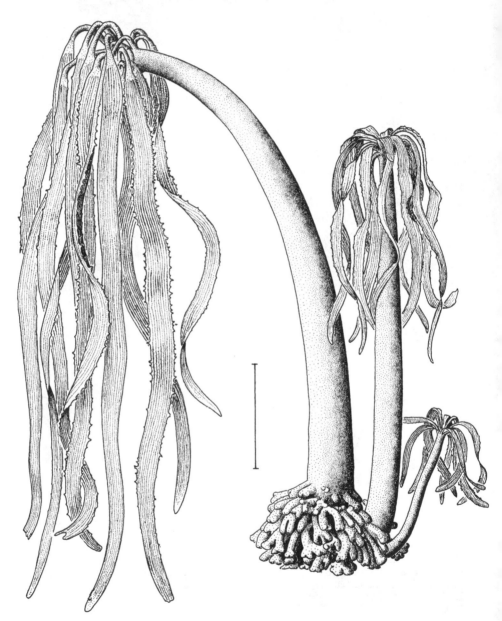

Fig. 203. *Postelsia palmaeformis.* Rule: 5 cm. (JRJ/S)

Postelsia Ruprecht 1852

Sporangial thalli with relatively small holdfast of stout, branched haptera. Stipe erect, cylindrical and hollow, tapering slightly from base to apex. Apex of stipe with many short, radially disposed, simple branches, each terminating in single blade. Branch and blade splitting longitudinally into 2 blades and branches, usually of equal size, the splitting beginning at junction of branch and blade. Blades sharply pointed, narrowly linear, the margins dentate. Both flattened surfaces of blade with deep, parallel, longitudinal grooves, the grooves of 1 surface alternating with those of other. Sporangia in linear sori, these lining grooves of blades.

Postelsia palmaeformis Rupr.

Ruprecht 1852: 75; Setchell & Gardner 1925: 625; Smith 1944: 142.

Sporangial thalli to 60 cm tall, usually growing in extensive stands; stipes erect and blades pendant when plants are exposed by recession of tide; mature plants golden brown, with 100 or more blades, these to 25 cm long; sporangia first produced in late spring; blades becoming eroded after fruiting; spores released during low tide and remaining in grooves of blades, dripping off the slender tips onto holdfast or nearby rock.

Locally abundant in areas exposed to surf, saxicolous, high to midtidal, Vancouver I., Br. Columbia, to Morro Bay (San Luis Obispo Co.), Calif. Type locality: Bodega Bay (Sonoma Co.), Calif.

Pelagophycus Areschoug 1881

Sporangial thalli with holdfast of widely spreading, frequently branched, slender haptera, the ultimate branchlets fine (often 1 mm diam.). Haptera produced in irregular whorls at junction of lower stipe and holdfast. Stipe solid and cylindrical throughout much of length, the upper portion inflated into stiff, elongate apophysis, this sharply constricted near distal end and terminating in ellipsoid to spherical pneumatocyst. Blades borne singly at ends of a system of sympodial branches ("antlers"), these arising from short, thick stalk at top of pneumatocyst. Mature blades to 1 m broad, markedly bullate, with short superficial spines. Sori in irregular patches on blades.

Pelagophycus porra (Lem.) Setch.

Laminaria porra Leman 1822: 189. Pelagophycus porra (Lem.) Setchell 1908b: 134; Setchell & Gardner 1925: 630. Nereocystis gigantea Areschoug 1876: 71. P. giganteus (Aresch.) Aresch. 1881: 49. P. intermedius Parker & Bleck 1965: 61.

Large, distinctive thalli with "antlers" formed from branches, these suggesting the common name "elk kelp," the plants further distinguished from Nereocystis by the large, corrugated blades adorned by spines. Sporangial thalli with haptera confined to short length of lowermost stipe,

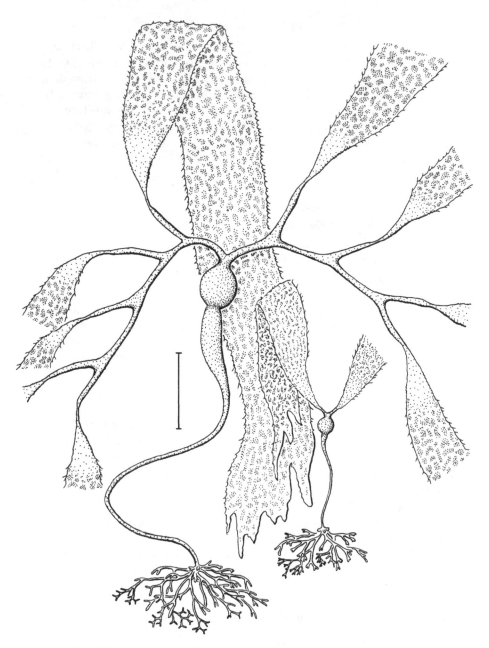

Fig. 204. *Pelagophycus porra.* Rules: large habit, 100 cm; inset habit, 20 cm. (NLN)

usually spreading 10–40 cm, often attached to rock and spreading into sand and shells; stipe variable in length, 7–27 m long, terminated by tubular apophysis and spherical to ellipsoid pneumatocyst, this 10–20 cm diam.; flattened branches arising from top of pneumatocyst and branching dichotomously after 1–4 cm in single plane, forming 2 sympodial members, each branch bearing single blade; blades on mature sporophytes 6–20 m long, to 1 m broad, slightly undulate, coarsely bullate, the convex surfaces of bullations on both sides of blades usually bearing spines, these curving distally and 1–3 mm long; margins of blades with short, spiny protuberances; sori in scattered, irregular patches on blades.

Populations scattered, subtidal on rock in gravel, Pt. Conception, Calif., to Is. San Benito, Baja Calif.; sporophytes in 30–90 m depth usually of uniform size. Drifted specimens recorded from Tomales Bay (Marin Co.), Pacific Grove, and near Pismo Beach, Calif., and from Pta. San José, Baja Calif. Type locality: near Santa Catalina I., Calif.

Nereocystis Postels & Ruprecht 1840

Sporangial thalli annual or persisting up to 18 months. Holdfast of frequently branched, tough haptera, these produced in irregular whorls at junction of lower stipe and holdfast. Stipe solid and cylindrical through most of length, the upper portion inflated into elongate rigid apophysis, this hollow, sharply constricted at upper end, terminating in ovoid to spherical pneumatocyst. Blades borne in 2–4 (rarely more) clusters at top of pneumatocyst, arising from short (0.5 cm) dichotomous branches and quickly becoming thin and ribbonlike. Sporophylls with irregularly rectangular sori, these progressively shed entire.

Nereocystis luetkeana (Mert.) Post. & Rupr.

Fucus luetkeanus Mertens 1829: 48. *Nereocystis luetkeana* (Mert.) Postels & Ruprecht 1840: 9; Setchell & Gardner 1925: 623.

Sporangial thalli golden to dark brown, usually produced in early spring and summer, occasionally persisting for more than 1 season, holdfast to 40 cm diam., a single stipe produced per holdfast; sporangial thallus to 40 m long, of which 36 m may be stipe; stipe 1 cm diam. throughout most of length, 2–7 cm diam. at apophysis; pneumatocysts 9–15 cm diam.; walls of apophysis and bulb to 2 cm thick in larger plants, the gas volume of apophysis and pneumatocyst to 3 liters in larger specimens; blades usually 30–64, to 4 m long, 15 cm broad, of variable width on single plant; sori produced as early as May and through December; fertile areas usually rectangular and produced in linear series on sporophylls, the youngest sori near blade bases; mature sori dark brown against paler olive drab of blade, dropping from blade and sinking prior to release of spores.

Frequent north of Carmel, Calif., forming extensive beds on rocks at

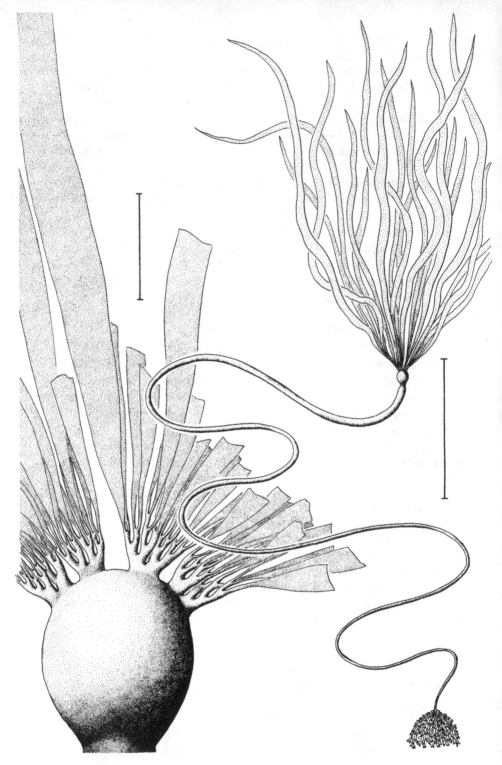

Fig. 205. *Nereocystis luetkeana*. Rules: habit, 100 cm; detail, 5 cm. (JRJ/S)

10–17 m; isolated plants in lower tidepools, these with stipes of normal diam., but only 1–2 m long, with blades of normal size and number, frequently producing sori; Alaska (Shumagin Is.) to San Luis Obispo Co., Calif.; drifted specimens to Kurile Is. and to San Diego. Type locality: Alaska.

Although generally shorter than *Macrocystis*, which is perennial, *Nereocystis* shows impressive growth rates, averaging 10 cm per day, whereas *M. pyrifera* in early spring may average 27 cm daily. Despite being an annual, *Nereocystis* usually carries a host of epiphytes on the upper stipe and apophysis, the commonest being *Porphyra nereocystis*, *Antithamnionella pacifica*, and *Enteromorpha linza*.

Macrocystis C. Agardh 1820

Sporangial thalli with basal portion perennial; holdfast conical and consisting of long-branched haptera, or prostrate, branched, subligulate, and creeping, with short, forked haptera along lateral margins. Stipe erect, cylindrical, dichotomously branched 2–4 times near base. Blades at regular intervals along stipe, unilaterally arranged immediately behind growing tips, soon becoming spirally arranged. Mature lateral blades undivided, more or less bullate, the margins denticulate, the stipe short, the basal pneumatocyst subglobose to fusiform. Sporangia (in Calif.) restricted to sporophylls borne on stipe near holdfast, with or without pneumatocysts. Sporophylls at first entire, later splitting from base to apex into 2 equal parts, eventually dividing 4 or 5 times, the divisions narrowly linear, with sporangia covering most of surface. In Australia and S. Africa, sori usually on undifferentiated blades.

Macrocystis integrifolia Bory
Bory 1826b: 9; Smith 1944: 143.

Sporangial thalli to 6 m long; holdfast flattened, subligulate, irregularly subdichotomous, creeping, with numerous branched haptera along lateral margins; stipe dichotomously branched 1–4 times, the terminal region of each branch narrowly falcate, becoming more narrow as growth slows; mature lateral blades lanceolate, bullate; pneumatocysts narrowly to broadly pyriform; lateral blades to 40 cm long, 8 cm broad; sporophylls usually without pneumatocyst, ultimately dividing dichotomously 4 or 5 times, the ultimate divisions to 30 cm long, 4 cm broad.

Infrequent in tidal channels and on gently sloping rocky ledges, lower intertidal to shallow subtidal, Br. Columbia to C. Calif. Type locality: western coast of S. America.

The growing tips change their general outline from falcate (with numerous young blades in the process of differentiation) to narrowly falcate (with few young blades), as illustrated in Smith (1944: pl. 26, fig. 2).

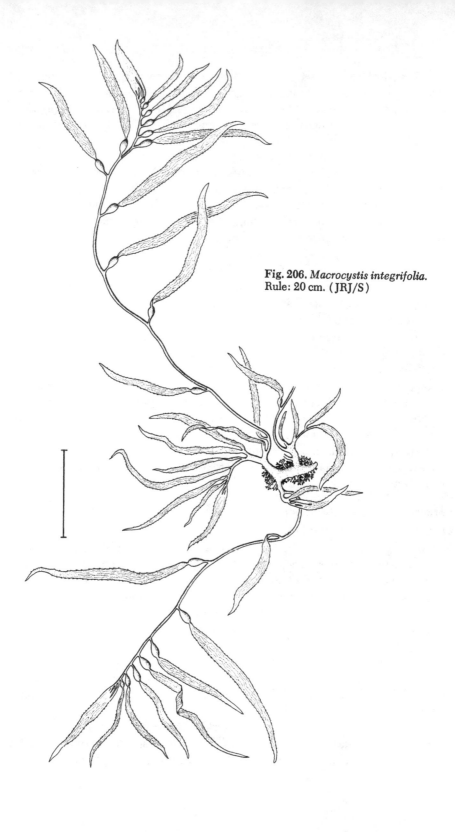

Fig. 206. *Macrocystis integrifolia.*
Rule: 20 cm. (JRJ/S)

Macrocystis pyrifera (L.) C. Ag.

Fucus pyriferus Linnaeus 1771: 311. *Macrocystis pyrifera* (L.) C. Agardh 1820: 47; Setchell & Gardner 1925: 627; Smith 1944: 144.

Sporangial thalli to 45.7 m long; holdfast of old plants conical, to 1 m tall, new holdfast produced in irregular whorls at junction of lower stipes and holdfast, growing throughout life of plant; stipe usually 4 or 5 times dichotomously divided near base; terminal blade of each branch broadly falcate when growing rapidly, becoming narrower with age, length 3–5 times breadth of 7–15(20) cm; mature lateral blades lanceolate, strongly bullate in most individuals, the blade to 80 cm long, 40 cm broad; sporophylls borne near base of branches, usually with pneumatocyst, the ultimate segments to 30 cm long, 4 cm broad.

Frequent in extensive stands (kelp "forests") on rocky substrata or occasionally anchoring in coarse sand, subtidal (6–20(80) m), forming most of the kelp forests in N. American part of its range; distribution nearly continuous (if undisturbed), except in Baja Calif., where confined to localities of cool, upwelling water; Alaska to Bahía Magdalena, Baja Calif., and S. America. Also S. Africa and S. Australia. Type locality: S. Atlantic.

A form of *Macrocystis* from Calif. resembling *M. angustifolia* from Australia has been described by Neushul (1971) from S. Calif. The presence of a rhizomelike holdfast, usually flattened in cross section, distinguishes *M. angustifolia* from *M. pyrifera*.

Upwards of 140,000 tons wet weight of *M. pyrifera* are harvested each year from the state-owned kelp beds off the S. Calif. coast for the extraction of alginates, colloids widely used in industry and in the preparation of certain foods. Since current demand exceeds supply, many new management methods have been developed relative to the ecology of the species. Important new basic as well as practical information has been gathered by North (1971) and others, contributing to an understanding of the *Macrocystis* community, which contains numerous species of sport fish as well as commercial quantities of sea urchins.

Order FUCALES

(Contributed by Nancy L. Nicholson)

Thalli macroscopic, usually perennial, flattened to cylindrical. Branching dichotomous throughout or radial about main axis. Growth of branches initiated by single apical cell, persistent throughout life of thallus of most species: secondary meristem intercalary, chiefly concerned with increase in girth. Outer layers of cells with several discoid chloroplasts per cell; pyrenoids apparently lacking except in *Fucus*. Colorless vesicles containing fucosan, a tannin-like compound very common in outer cell layers, meristematic cells, and reproductive cells, accounting for dark-brown color of most thalli, darker than that of those containing only fucoxanthin, the

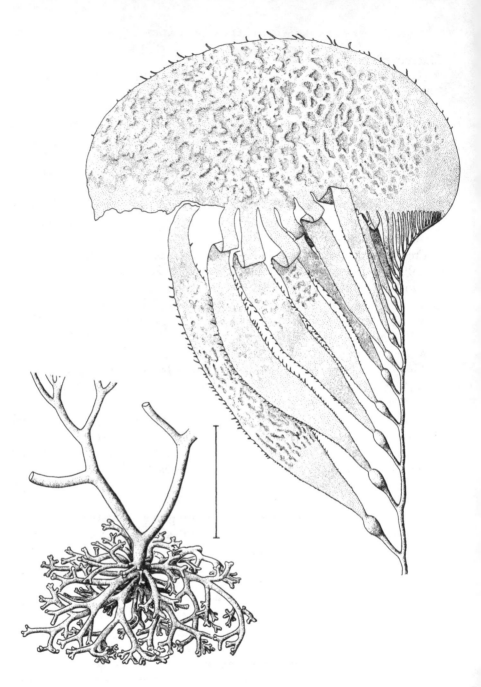

Fig. 207. *Macrocystis pyrifera.* Rule: 5 cm. (JRJ/S)

pigment in all brown algae. Reproductive organs developed within conceptacles, these embedded at apices of vegetative branches or borne on special branches; branch area containing fertile conceptacles constituting receptacle, this distintegrating after reproduction. Gametangia unilocular. Oogonia with 1–8 functional, uninucleate, nonflagellate eggs. Antheridia producing 64–128 small, biflagellate sperms; liberation of eggs and sperms followed by fertilization, the resulting zygotes developing directly into new diploid thalli. Plants monoecious or dioecious.

1. Branching of thallus dichotomous and in 1 plane.......Fucaceae (below)
1. Branching of thallus radial to central axis, at times appearing alternate and in 1 plane 2
  2. Branching radial only; differentiation into basal and apical portions negligible .... Sargassaceae (p. 272)
  2. Branching radial or alternate; differentiation into basal and apical portions marked (the apical perhaps not present all seasons of year) .... Cystoseiraceae (p. 267)

Family Fucaceae

Thalli lacking main central axis, dichotomously branched above holdfast, the branches lying in single plane. Eggs 1–8 per oogonium. Antheridia producing 64 or 128 sperms.

1. Thallus with two conspicuous rows of cryptostomata (appearing as white dots) on each side of midrib; oogonium with 1 functional egg........ .... Hesperophycus (p. 265)
1. Thallus otherwise 2
  2. Thallus with midrib for at least part of upper portion; oogonium with 8 functional eggs .... Fucus (below)
  2. Thallus without midrib 3
3. Oogonium with 1 functional egg .... Pelvetiopsis (p. 264)
3. Oogonium with 2 functional eggs .... Pelvetia (p. 261)

Fucus Linnaeus 1753

Thalli perennial, with disk-shaped or irregularly shaped holdfast. Erect portion of thallus dichotomously branched, flattened, with more or less distinct percurrent midrib; gas-filled pneumatocysts of definite form sometimes present, in pairs lateral to midrib; erect portion also bearing cryptostomata and caecostomata. Base of thallus becoming stipelike through abrasion of tissues lateral to midrib. Gametangia developed in conceptacles embedded at apices of final dichotomies. Plants monoecious or dioecious.

Fucus distichus L.

Linnaeus 1767: 716; Turner 1808: 7.

The type subspecies, not known from the Pacific, differs from ssp. *edentatus* in its relatively smaller size, delicate structure, and narrow blades.

Fig. 208 (above). *Fucus distichus* ssp. *edentatus*. Rule: 5 cm. (JRJ/S)
Fig. 209 (right). *F. distichus* ssp. *edentatus* f. *abbreviatus*. Rule: 5 cm. (DBP)

Fucus distichus ssp. edentatus (de la Pyl.) Pow.

Fucus edentatus de la Pylaie 1829: 84. *F. distichus* ssp. *edentatus* (de la Pyl.) Powell 1957a: 424 (incl. synonymy); Hollenberg & Abbott 1966: 31 (incl. synonymy).

Thalli 10–25 cm tall, rarely taller, regularly dichotomous, the divisions 1.5–2.5 cm broad, the thalli with prominent midrib in older portions, olive brown to dark brown; receptacles broadening at apices in most specimens, often constituting one-fourth to one-third length of mature thallus, swollen at reproductive maturity; plants monoecious.

Abundant in upper intertidal and midtidal, on rocks, sometimes growing mixed with *Hesperophycus* (which it resembles), N. Wash. to Pt. Conception, Calif. Also N. Atlantic. Type locality: Newfoundland.

Fucus distichus ssp. edentatus f. abbreviatus (Gardn.) Hollenb. & Abb.

Fucus furcatus f. *abbreviatus* Gardner 1922: 19. *F. distichus* ssp. *edentatus* f. *abbreviatus* (Gardn.) Hollenberg & Abbott 1966: 31 (incl. synonymy).

Thalli 10–15(25) cm tall, regularly dichotomous, with divisions 0.3–1 cm broad, with prominent midrib below, olive brown to dark brown; receptacles mostly tapering to acute apices, constituting half or more of length of mature thallus; caecostomata abundant; conceptacles numerous, prominent.

Locally common, discontinuously distributed, saxicolous, upper intertidal, Waldron I. (San Juan Co.; type locality), Wash., to Carmel River (Monterey Co.), Calif.

Pelvetia Decaisne & Thuret 1845

Thalli perennial. Holdfast conical, tough and resilient, producing 1 to several erect, dichotomously branched axes. Axes subcylindrical to compressed and channeled on 1 side, without midrib, with or without gas-filled vesicles. Cryptostomata inconspicuous. Gametangia unilocular and developed within conceptacles at apices of final dichotomies. Oogonium producing 2 (rarely 3–5) large, nonflagellate eggs, the division either longitudinal or transverse. Antheridia producing 128 small, biflagellate sperms. Plants monoecious.

Pelvetia fastigiata (J. Ag.) DeToni

Fucus fastigiatus J. Agardh 1841: 3; Harvey 1852: 68. *Pelvetia fastigiata* (J. Ag.) DeToni 1895: 215; Gardner 1910: 126; Smith 1944: 153.

Thalli to 90 cm long, dark greenish-olive to yellowish-brown; several erect, subcylindrical to compressed (particularly in upper regions) branches arising from each holdfast; branching regularly dichotomous throughout, but the 2 arms of the dichotomy generally unequal in length; successive dichotomies 3–5 cm apart, the dichotomies 2–5 mm broad, without gas-filled vesicles.

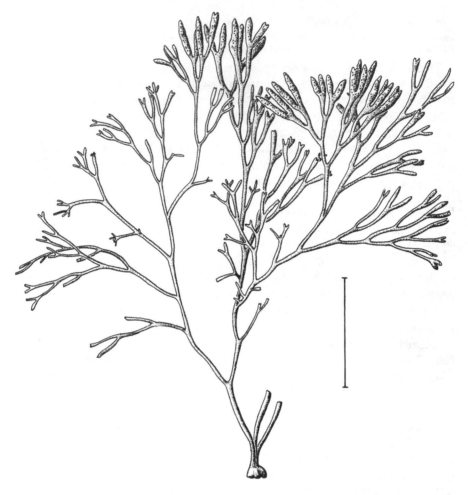

Fig. 210. *Pelvetia fastigiata.* Rule: 5 cm. (JRJ/S)

Locally abundant, forming beds on rocks somewhat protected from open surf, midtidal, Horswell Channel, Br. Columbia, to Pta. Baja, Baja Calif. Type locality: Calif. (probably Monterey).

Pelvetia fastigiata f. gracilis S. & G.

Setchell & Gardner 1917a: 386; Smith 1944: 154.

Thalli slender as compared to those of f. *fastigiata*, and more profusely branched, with receptacles 0.5–1 cm long; resembling *Pelvetiopsis arborescens.*

Locally abundant, on rocks, Pebble Beach (type locality) and Channel Is., Calif., where it is abundant and replaces *P. fastigiata* f. *fastigiata* in narrow band below *Hesperophycus harveyanus.*

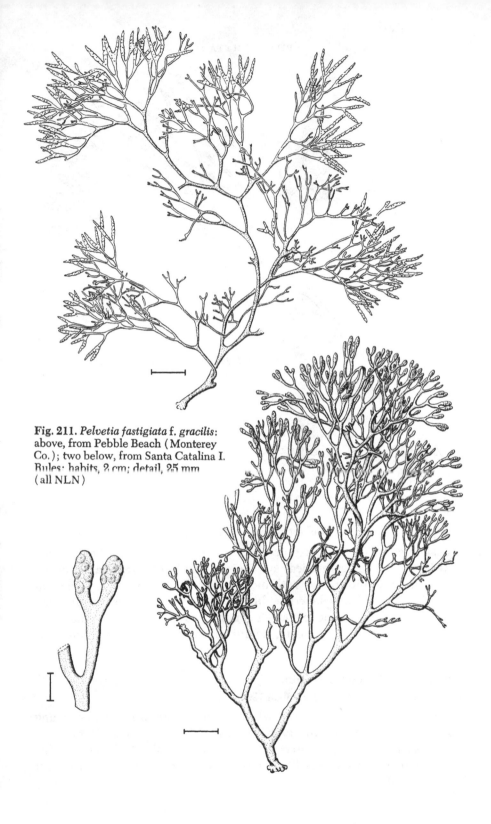

Fig. 211. *Pelvetia fastigiata* f. *gracilis*: above, from Pebble Beach (Monterey Co.); two below, from Santa Catalina I. Rules: habits, 2 cm; detail, 25 mm (all NLN)

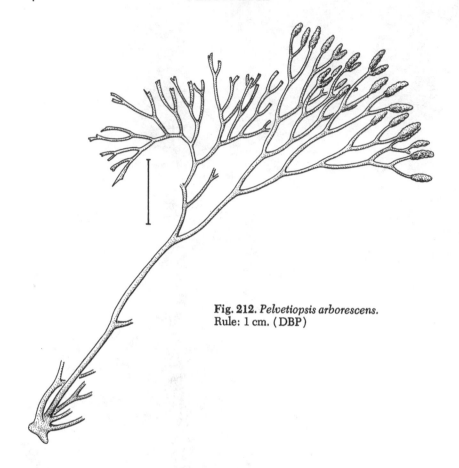

Fig. 212. *Pelvetiopsis arborescens.*
Rule: 1 cm. (DBP)

Pelvetiopsis Gardner 1910

Thalli perennial. Holdfast producing 1 to many erect, dichotomous branches. Branches cylindrical at base, flattened or cylindrical in upper portion, lacking midrib and with inconspicuous cryptostomata. Gametangia unilocular and developed within conceptacles at apices of ultimate dichotomies. Thalli monoecious. Oogonium producing 1 large, uninucleate egg and 1 small, 7-nucleate, nonfunctional cell. Antheridia with 128 small, biflagellate sperms.

Pelvetiopsis arborescens Gardn.

Gardner 1940: 270; Hollenberg & Abbott 1966: 33.

Thalli 10–15 cm tall; branches cylindrical up to half of length, becoming somewhat flattened upward, rarely to 3 mm wide; receptacles blunt.

Rare, saxicolous, upper intertidal of exposed headlands, Monterey Co. (Pacific Grove and Cypress Pt. to Little Sur River), Calif. Type locality: Pt. Lobos, Calif.

Pelvetiopsis limitata Gardn.

Gardner 1910: 127; 1913: 321; Smith 1944: 155.

Thalli 4–8 cm tall, light tan; erect flat branches usually arcuate, 3–5 times dichotomous, the dichotomies 2–4 mm broad, 5–15 mm apart; receptacles 5–10 mm long, inflated, simple or bifurcate, with acute apices.

Infrequent, saxicolous, usually atop rocks, more rarely on sides; upper intertidal, exposed to surf; Vancouver I., Br. Columbia, to Cambria (San Luis Obispo Co.), Calif. Type locality: San Francisco, Calif.

The form *lata* (Gardner 1910: 127; 1913: 321), collected at Asilomar Pt., Calif., and illustrated by Smith (1944: pl. 33, fig. 1), is longer (to 15 cm) than f. *limitata*, has broader branches (to 3 mm broad), and larger receptacles; the branches, moreover, are not arcuately curved, as in typical *P. limitata*.

In general, *P. arborescens* resembles a dwarf *Pelvetia*, *P. limitata* a dwarf *Fucus*.

Hesperophycus Setchell & Gardner 1910

Thalli perennial. Holdfast disk-shaped. Erect thallus dichotomously

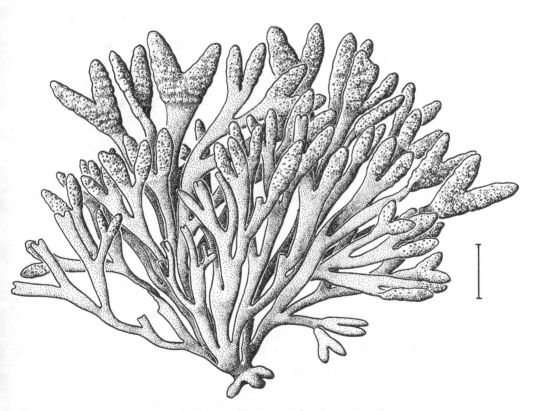

Fig. 213. *Pelvetiopsis limitata*. Rule: 1 cm. (JRJ/S)

branched, markedly flattened. Branches with distinct percurrent midrib, the cryptostomata lying in 2 parallel rows on either side of midrib. Base of thallus becoming stipelike through abrasion of tissue lateral to midrib. Gametangia unilocular, developed in conceptacles embedded at apices of final dichotomies. Plants monoecious. Oogonium producing 1 large, non-flagellate egg and 1 small nonfunctional, 7-nucleate cell. Antheridium producing 128 biflagellate sperms.

Hesperophycus harveyanus (Decne.) S. & G.

Fucus harveyanus Decaisne 1864: 9. *Hesperophycus harveyanus* (Decne.) Setchell & Gardner 1910: 127; Gardn. 1913: 317; Smith 1944: 152.

Thalli 10–40(60) cm tall, at times densely clumped, greenish-olive to yellowish-brown; erect axes regularly dichotomous; thallus between di-

Fig. 214. *Hesperophycus harveyanus.*
Rule: 2 cm. (LH)

chotomies 3–12 mm broad, linear, sometimes undulate; sterile apices of ultimate segments rounded; cryptostomata with conspicuous, protruding white hairs; receptacles swollen, simple or bifurcate, to 1.5 cm broad, 0.5–3 cm long.

Locally abundant to infrequent in upper intertidal, usually at higher level but sometimes mixed with *Fucus* on rocks in C. Calif., replacing *Fucus* south of Pt. Conception; Santa Cruz, Calif., to Is. San Benito, Baja Calif.; at present more common in Channel Is. than on facing mainland. Type locality: Monterey, Calif.

Family CYSTOSEIRACEAE

Thalli frequently bilateral, becoming radially or alternately branched, clearly differentiated into apical and basal regions, each with different external morphology. Apices annual in Calif., shed in winter. Basal portion perennial, leaflike, usually dissected, attached closely to angular stemlike stipe, leaving triangular scar when dropped away. Pneumatocysts intercalary on upper branches, solitary or in chains. Conceptacles internally developing conspicuous row of cells (tongue filament) before formation of oogonia and antheridia. Oogonium with 1 functional egg nucleus, 7 being expelled before oogonial discharge. Antheridia producing 64 sperms.

With two genera in the California flora, *Cystoseira* (below) and *Halidrys* (p. 272).

Cystoseira C. Agardh 1820

Thalli perennial, differentiated into holdfast, stipe, and branches. Lower thallus persisting (in protected areas) and renewing upper thallus, this of distinctly different morphology. Stipe erect, woody, angled where lateral branches have dropped away, triangular in cross section. Branches radially arranged; lower branches flattened throughout and with slight midrib; upper branches with occasional flattened regions but mainly cylindrical, bearing small pneumatocysts along length, either singly or in catenate series. Gametangia unilocular, developed within conceptacles on receptacles at apices of ultimate branchlets. Plants monoecious or dioecious. Oogonium containing 1 uninucleate egg and 7 residual nuclei, these expelled from cytoplasm before oogonial discharge. Oogonia in *C. osmundacea* produced successively. Antheridium containing 64 biflagellate sperms.

1. Pneumatocysts nearly all borne singly or at short intervals on intricate, slender, divaricate branches; lower stipe tripinnate............*C. neglecta*
1. Pneumatocysts in catenate series on stouter alternate branches; lower stipe 1 or 2 (rarely 3) times pinnate.............................. 2
  2. Upper portion of thallus 50+ cm tall when mature; basal portion of thallus once pinnate (intermediate forms with bipinnate lower thalli) .... *C. osmundacea*
  2. Upper portion of thallus less than 50 cm tall, the plants bushy in appearance; basal portion of thallus usually bipinnate........*C. setchellii*

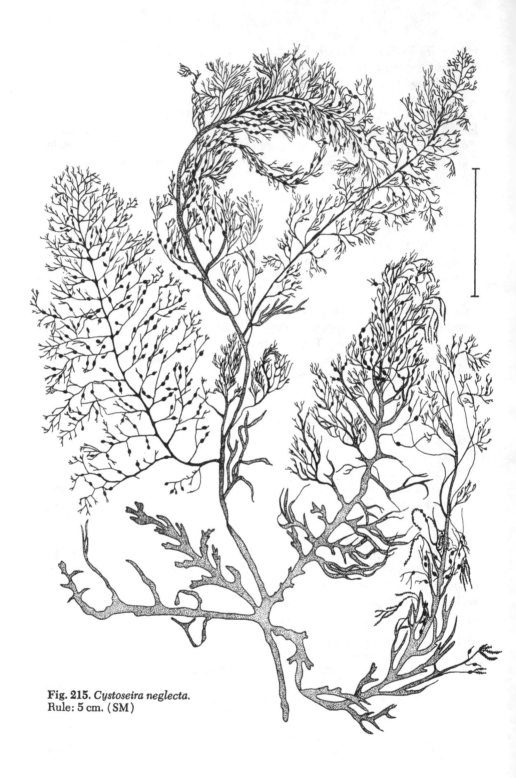

Fig. 215. *Cystoseira neglecta.*
Rule: 5 cm. (SM)

Cystoseira neglecta S. & G.

Setchell & Gardner 1917b: 388.

Thalli to 1(1.5) m tall, medium brown in life; lower thallus (2)3 or 4 times pinnately divided, the ultimate divisions very slender or with acute, trailing apices; upper thallus intricately branched, divaricate; pneumatocysts developing on separate filiform branches, solitary, or in series but 1–3 mm apart, spherical, smooth, 2–3 mm diam.; plant dioecious; receptacles sparingly branched, terminal on upper branches, nearly filiform, with conceptacles strung along length and producing beaded appearance.

Infrequent, saxicolous, shallow subtidal, Santa Catalina and San Clemente Is., Calif. Type locality: Avalon (Santa Catalina I.), Calif.

Cystoseira osmundacea (Turn.) C. Ag.

Fucus osmundaceus Turner 1809: 91. *Cystoseira osmundacea* (Turn.) C. Agardh 1820: 69; Gardner 1913: 333; Smith 1944: 156 (incl. synonymy).

Thalli to 8 m tall, blackish-brown below, light tan above; primary branches flattened in lower portion, about 1–1.5 cm broad, usually once pinnate (north of Pt. Conception) or bipinnate (south of Pt. Conception); ultimate branches of upper portion with tapering, catenate series of 5–12 subspherical to ellipsoid pneumatocysts; receptacles branched, flattened or terete, developing from branchlets distal to pneumatocysts; plant dioecious.

Occasional on rocks in lower intertidal, common subtidally (to 10 m), frequently mingled with *Macrocystis* beds along Monterey Peninsula; Seaside, Ore., to Ensenada, Baja Calif. Type locality: Trinidad (Humboldt Co.), Calif.

C. osmundacea intergrades gradually with the other two species in Calif., and intermediates should be expected. Nonetheless, it seems worthwhile to retain the species *C. neglecta* and *C. setchellii* on the grounds that they represent morphological types characteristic of certain environments.

Cystoseira setchellii Gardn.

Gardner 1913: 329.

Thallus appearing bushy, rarely to 1 m tall, medium brown; lower stipe usually bipinnate, the apices of pinnae acute or bluntly rounded; upper stipe to 50 cm long; pneumatocysts in 2–5 series, occasionally solitary, concentrated toward main axis of terminal portion of stipe, spherical to ellipsoid; receptacles terminal on branchlets, slightly tumid, sparingly to densely branched; plant dioecious.

Infrequent, on rocks near or covered by coarse sand, subtidal (to about 10 m) in areas of relatively calm water, Shell Beach (San Luis Obispo Co.) and Redondo Beach to San Diego, Calif.; abundant at Little Harbor and Pin Rock (Santa Catalina I.), Calif. Type locality: San Pedro, Calif.

No plants from the San Pedro region were found either subtidally or cast

Fig. 216. *Cystoseira osmundacea.*
Rule: 5 cm. (JRJ/S)

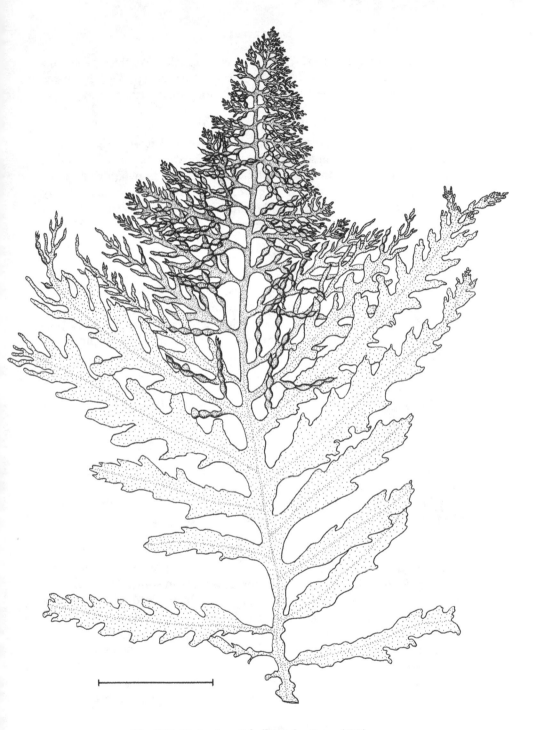

Fig. 217. *Cystoseira setchellii*. Rule: 5 cm. (CS)

ashore during the years 1969–72, and it is probable that this species no longer grows at the type locality. Gardner's discussion (p. 330) indicates that the material he studied was frequently found cast ashore.

Halidrys Lyngbye 1819

Thallus perennial, consisting of solid holdfast and lower stipe of flattened alternate fronds, these giving rise to pinnately branched, slender, terete branches. Pneumatocysts flattened, coalesced into tapering pods, usually with a narrow margin. Receptacles branched, developing distally to pods. Eggs 1 per oogonium. Number of sperm produced by antheridium unknown. Plants monoecious or dioecious.

Halidrys dioica Gardn.

Gardner 1913: 323.

Thalli to 2 m tall, medium brown to pale olive drab; lower stipe circular in cross section, 4–6 mm diam., solid, flexible, the loss of lateral branches producing a zigzag appearance; lower stipe perennial, with flattened pinnae, giving rise to terete bilateral upper branches; upper stipe pinnately branched, bearing tapering pneumatocysts, these with 5–12 chambers; receptacles slightly tumid, often intricately branched. Plant dioecious.

Locally abundant, lower intertidal on exposed rocks or subtidal (to 5 m), Redondo Beach, Calif., to Bahía Asunción, Baja Calif. Type locality: San Pedro, Calif.

A more openly branched and regularly pinnate form (Gardner 1913: pl. 45) is typical of populations on Santa Catalina and San Clemente Is. Plants to 2 m tall and densely branched have been collected from Santa Cruz I. and resemble the specimens illustrated (Gardner 1913: pls. 42–44); plants from Baja Calif. are less than 70 cm tall, with reduced upper stipes and markedly flattened and broadened pneumatocysts.

Family SARGASSACEAE

Thalli with radial branching, the branching more or less continuous from base to apex, modified by sterile branches, leaflike in some. "Leaves" with midrib and cryptostomata, and with smooth or denticulate margins. Pneumatocysts solitary, terminal on branchlets. Receptacles terminating fertile branchlets; oogonia and antheridia mixed in same conceptacles, or distinct in same receptacle, on same or different thalli. Conceptacles internally differentiating a conspicuous cell or row of cells from floor (tongue), their development of generic discrimination. Oogonia with 1 functional egg nucleus, 7 nuclei disintegrating. Antheridia producing 64 sperms.

Sargassum C. Agardh 1820

Thalli annual or perennial, attached by irregular, solid holdfast or by rhizoidal outgrowths from main axis. Main axis sometimes slightly differ-

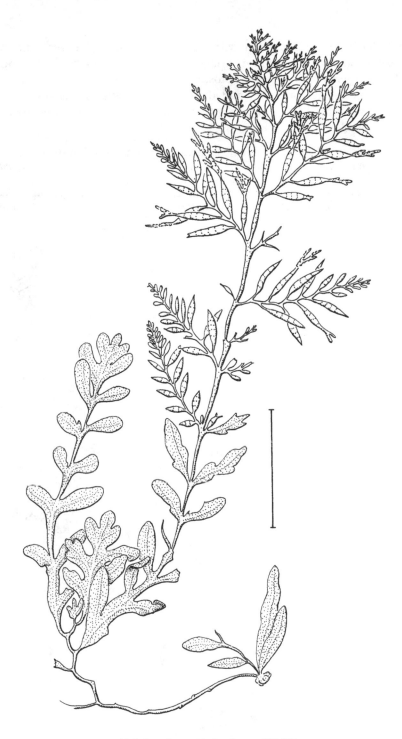

Fig. 218. *Halidrys dioica*. Rule: 5 cm. (NLN)

entiated into basal portion with flattened, elongate "leaves" and upper portion bearing shorter appendages, these sometimes radially arranged. "Leaves" with midrib and cryptostomata. Pneumatocysts common. Receptacles on special branches and developing in axils of "leaves." Oogonia and antheridia as in family. Both monoecious and dioecious species in Calif. flora.

1. Blades on all orders of branches divided into several filiform flattened
   sections ....S. *palmeri*
1. Blades simple ....,.................................. 2
    2. Branchlets in clusters along main axis...............S. *agardhianum*
    2. Branchlets remote from one another....................S. *muticum*

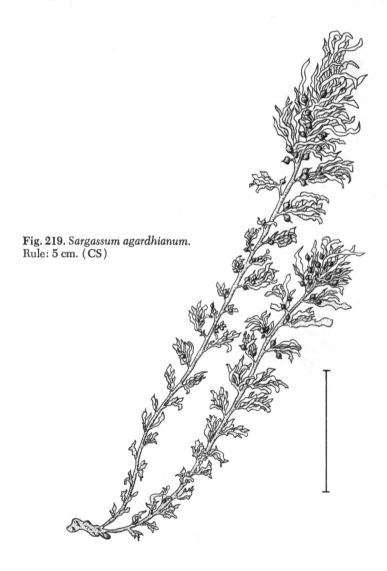

Fig. 219. *Sargassum agardhianum*.
Rule: 5 cm. (CS)

Sargassum agardhianum J. Ag.

J. Agardh 1889: 93; Setchell & Gardner 1925: 716.

Thalli to 40(100) cm tall, attached by irregular, lumpy holdfast, the basal portions perennial, much reduced in winter; young plants simple, branching later from basal area; branches few to many, 1 mm diam., terete, the main stipe shorter than laterals; branches with clusters of branchlets, pneumatocysts and/or receptacles in axils; leaves alternate to radial, linear-lanceolate with toothed margins; cryptostomata conspicuous. Pneumatocysts spherical or slightly ellipsoid, smooth or toothed, apiculate, 1.5–2.5 mm diam., on pedicels 1–1.5 mm long. Receptacles irregularly forked, with numerous conspicuous conceptacles; antheridia and oogonia usually in separate conceptacles, sometimes mixed in 1 receptacle.

Locally frequent, saxicolous, often forming dense stands, lower intertidal to upper subtidal, Pt. Dume (Los Angeles Co.) and Santa Catalina I., Calif., to Pta. Eugenio, Baja Calif. Type locality: San Diego, Calif.

Some forms have been reported to be almost entirely devoid of pneumatocysts.

Sargassum muticum (Yendo) Fensh.

Sargassum kjellmanianum f. *muticum* Yendo 1907: 104. *S. muticum* (Yendo) Fensholt 1955: 306.

Thalli perennial or annual (in Calif.) to 2(10) m tall, arising from felty, fibrous holdfast; main branches arising from basal 5 cm of plant, repeatedly and alternately branched to form intricate, bushy thallus; leaves linear-lanceolate in basal portion of stipe, to 10 cm long, the margins toothed; leaves of upper stipe narrow, often only 4 mm long, the margins entire or toothed; pneumatocysts borne in clusters or single in leaf axils; receptacles arising from leaf axils, occasionally forked, cylindrical, 1–2 mm diam.; some plants fertile all year, but especially from March through June at Santa Catalina I. where developing embryos may be found on the mucosoid receptacles.

Locally abundant, saxicolous, forming dense stands in quiet water, lower intertidal to subtidal (3 m to 5 m), Nanaimo, Br. Columbia, to San Diego, Calif. Type locality: Japan.

This is a Japanese species introduced to the West Coast of N. America about 1945, apparently on shells of young oysters. It has spread steadily southward from Puget Sd. and was reported from Humboldt Co., Calif., by E. Y. Dawson in 1965. First noticed in S. Calif. by graduate students at Santa Catalina I. in late winter 1970, it spread rapidly during the following year to Orange and San Diego Cos. and is still apparently expanding its territory southward. At this writing, *S. muticum* is known from San Francisco Bay, but is not yet known on the outer coast of C. California. It has also been reported in S. England.

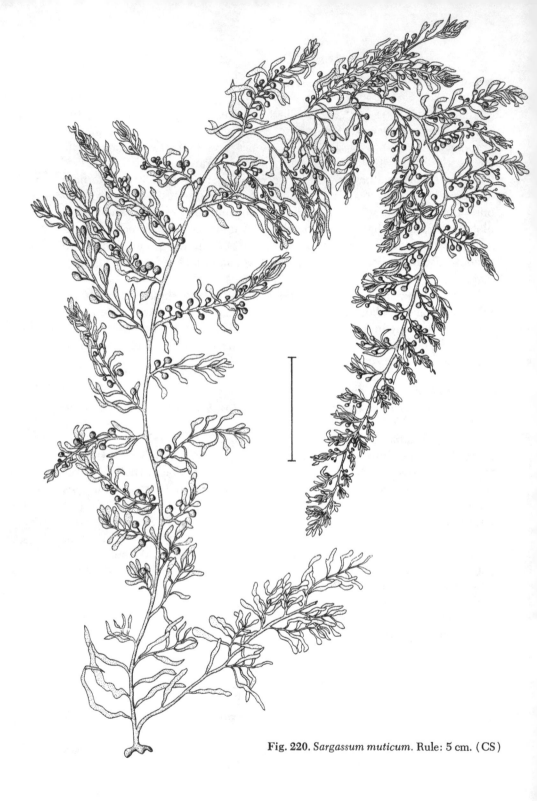

Fig. 220. *Sargassum muticum.* Rule: 5 cm. (CS)

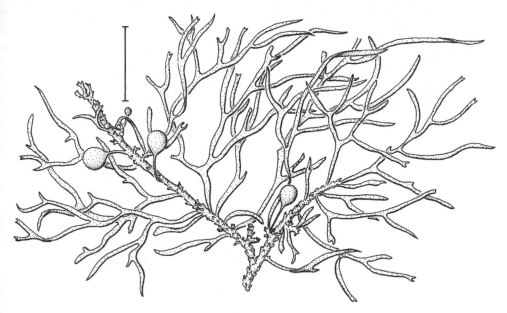

Fig. 221. *Sargassum palmeri*. Rule: 1 cm. (NLN)

Sargassum palmeri Grun.

Grunow 1915: 338; Setchell & Gardner 1925: 712. *Sargassum piluliferum sensu* Farlow, Anderson & Eaton, 1877–89 (1878): 102. *S. dissectifolium* S. & G. 1917c: 136.

Thalli perennial, 45–70 cm tall, arising from solid, rugose, more or less discoid holdfast; stipe terete, verrucose, to 1.2 m long, bearing 2–5 terete or slightly angled, alternate branches at apex, these disintegrating following liberation of gametes; primary branches producing sterile "leaves," these alternately dissected into 15–25 slightly flattened divisions 2–3 cm long; older branches, and at times ramuli, more or less spiny; numerous lateral terete secondary branches arising in axils of "leaves," longer below, shorter above; "leaves" with indistinct midrib and scattered cryptostomata, the pneumatocysts and receptacles produced in blade axils; pneumatocysts usually solitary, smooth, elliptic-spherical, 3.5–6 mm diam., on ends of pedicels as long as or longer than diam.; receptacles densely racemose, substipitate, often associated with pneumatocyst or reduced "leaf"; fertile fronds often appearing nude as "leaves" wear away; conceptacles conspicuous; plant dioecious; best developed in winter.

Occasional on rocks, low intertidal to subtidal, Santa Catalina I., Calif., to Is. San Benito, Baja Calif. Type locality: I. Guadalupe, Baja Calif.

Division Rhodophyta

The Red Algae

PREDOMINANTLY marine algae, high intertidal to deep subtidal; mostly multicellular, filamentous or pseudoparenchymatous, branched or unbranched, cylindrical to compressed or foliaceous; many species microscopic, unicellular or short-filamentous, very few species large. Cells mostly uninucleate, with 1 or more platelike, band-shaped, stellate or discoid chloroplasts in which a red phycobilin pigment, phycoerythrin, is usually predominant; in a few species phycocyanin predominant, giving a steel-gray color; carbohydrate food reserve known as floridean starch. Growth chiefly apical. Thalli rose-red or violet to olive or nearly black. Mostly perennial, usually with separate sporangial and gametangial stages in life history, the stages similar or dissimilar in appearance. Plants mostly dioecious. Spores and male gametes (spermatia) nonmotile. Reproductive structures and life histories in one class (the Bangiophyceae) difficult to interpret; in the other (the Florideophyceae), the reproductive structures more nearly uniform and life histories usually involving a gametangial plant followed by a sporangial plant with an interpolated carposporophyte.

Two classes in the California flora, the Bangiophyceae (below) and the Florideophyceae (p. 306).

Class BANGIOPHYCEAE

Thalli mostly filamentous or membranous. Cells uninucleate, mostly with single stellate chloroplast containing 1 pyrenoid. Cell division intercalary; pit-connections rare; no cells with flagella. Asexual reproduction by modified vegetative cells or by sporelike cells cut off singly from vegetative cells. Sexual reproduction claimed for only a few genera.

Two orders in the California flora, the Goniotrichales (below) and the Bangiales (p. 283).

Order GONIOTRICHALES

Thalli initially filamentous, often becoming multiseriate. Cells separated from one another in a gelatinous, sheathlike matrix. Cells spherical to cylindrical, uninucleate, mostly with single stellate chloroplast in each cell. Nonmotile monospores formed from individual vegetative cells. Sexual reproduction unknown.

Family GONIOTRICHACEAE

Thalli filamentous, branched. Cells seriate, separated from one another by gelatinous material. Cells ovoid or cylindrical, uninucleate, with single stellate chloroplast (except *Goniotrichopsis*), with or without pyrenoid. Asexual reproduction by naked monospores, and by direct metamorphosis of vegetative cells into spores, not preceded by cell division.

1. Cells with numerous discoid chloroplasts.........*Goniotrichopsis* (below)
1. Cells with single stellate chloroplast 2
   2. Some portions of filaments with cells loosely and irregularly arranged
   .... *Goniotrichum* (below)
   2. All portions of filaments uniseriate ....*Asterocytis* (p. 283)

Goniotrichum Kützing 1843

Thalli mostly epiphytic, mostly microscopic, erect, irregularly branched, with cells in 1 or more series. Chloroplast single, stellate.

Goniotrichum alsidii (Zan.) Howe

Bangia alsidii Zanardini 1839: 136. *Goniotrichum alsidii* (Zan.) Howe 1914: 75 (incl. synonymy).

Filaments to 5 mm long, 12–35 µm diam., including sheath; cells normally rose red, mostly uniseriate.

Common, epiphytic on a large variety of algae, lower intertidal to subtidal (to 12+ m), N. Wash. to Chile. Widely distributed. Type locality: Italy.

Goniotrichum cornu-cervi (Reinsch) Hauck

Stylonema cornu-cervi Reinsch 1875: 40. *Goniotrichum cornu-cervi* (Reinsch) Hauck 1885: 519; Kylin 1925: 6.

Thalli irregularly dichotomous, 0.2–1.5 mm tall, 20–120 µm diam., enlarging abruptly above base, with 2 to many rows of cells irregularly arranged in common sheathlike matrix, the cells mostly 7–14 µm diam, branches attenuate toward apices.

Occasional, epiphytic, subtidal (to 10 m), Puget Sd., Wash., to Peru. Widely distributed. Type locality: Yugoslavia.

Goniotrichopsis Smith 1943a

Thalli microscopic, erect, dichotomously branched. Filaments at first uniseriate, soon becoming multiseriate except for branch apices. Cells at first cylindrical, later spherical, each containing numerous discoid chloroplasts, without pyrenoids. Reproduction by monospores.

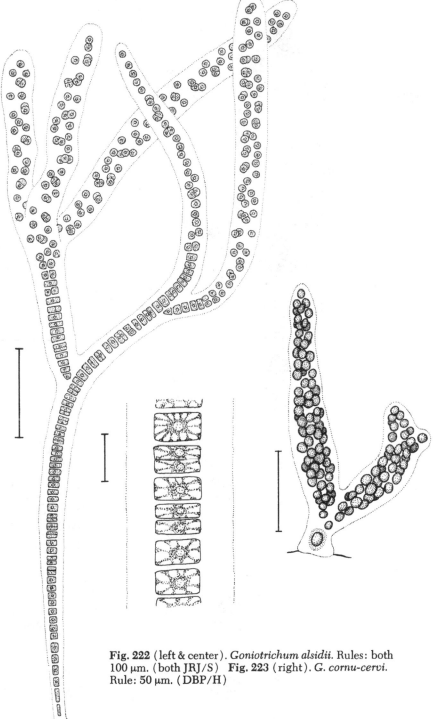

Fig. 222 (left & center). *Goniotrichum alsidii.* Rules: both 100 μm. (both JRJ/S)   **Fig. 223** (right). *G. cornu-cervi.* Rule: 50 μm. (DBP/H)

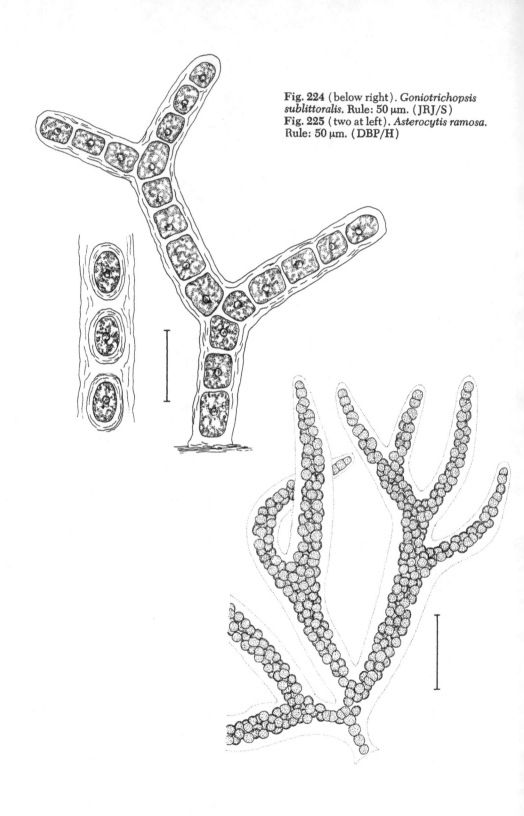

Fig. 224 (below right). *Goniotrichopsis sublittoralis*. Rule: 50 µm. (JRJ/S)
Fig. 225 (two at left). *Asterocytis ramosa*. Rule: 50 µm. (DBP/H)

Goniotrichopsis sublittoralis Smith

Smith 1943: 211.

Filaments bright rose red, to 850 µm long; branches to 40 µm diam., 3 or 4 times dichotomous; cells 8–12 µm diam.

Rare; mostly epiphytic on Florideae, especially *Callophyllis*, sometimes saxicolous, subtidal (to 15 m), San Juan I., Wash., and Monterey area, Calif. Type locality: Carmel Bay, Calif.

Asterocytis (Hansgirg) Schmitz 1896

Thalli filamentous, uniseriate or with frequent "false" branches, usually epiphytic. Cells enclosed in gelatinous walls, each cell with well-defined nucleus. Chloroplast single, blue-green, stellate, with central pyrenoid. Reproduction by monospores or akinetes. Sexual reproduction unknown.

Asterocytis ramosa (Thwaites) Schmitz

Hormospora ramosa Thwaites 1849: pl. 213. *Asterocytis ramosa* (Thwaites) Schmitz 1896–97 (1896): 314; Lewin & Robertson 1971: 236 (as *A. ornata*).

Filaments tufted, branched, to 300+ µm long, to 25 µm diam.; cells mostly 16–18 µm diam., axially compressed to oval, enclosed in thick gelatinous walls.

Reported in Calif. only from a shore collection at La Jolla, Calif. Widely distributed in temperate seas. Type locality: British Isles.

Order BANGIALES

Thalli filamentous, discoid or bladelike. Walls gelatinous, but cells not widely separated. No cells with flagella formed. Asexual reproduction by monospores. Other reproduction, claimed to be sexual, reported for certain genera, but difficult to interpret.

The use of quotes around "spermatia" and "carpospores" expresses the doubt now held concerning the sexual nature of these structures if interpreted along traditional lines. Culture studies demonstrate a variety of events and products not expected from cells participating in sexual reproduction as observed in other algal groups.

Two families in the California flora, the Erythropeltidaceae (below) and the Bangiaceae (p. 293).

Family ERYTHROPELTIDACEAE

Thalli uniseriate to bladelike. Asexual reproduction mostly by monospores formed from unequal division of vegetative cells. "Spermatangia," in certain genera, cut off singly from vegetative cells. "Carpospores" unknown or questionable.

1. Thalli without erect branches, minute, discoid, epiphytic . . . .
. . . . *Erythrocladia* (p. 284)
1. Thalli chiefly erect, with or without discoid base 2

Erythrocladia Rosenvinge 1909

Thalli microscopic, mostly epiphytic, monostromatic disks composed of radially branched filaments. Chloroplast single, platelike, parietal, containing 1 pyrenoid. Asexual reproduction by monospores. Sexual reproduction unknown.

Erythrocladia irregularis Rosenv.

Rosenvinge 1909: 72; Smith 1944: 166; Dawson 1953a: 5.

Thalli discoid, monostromatic, 50–300 µm diam., composed of radiating branching filaments, these closely adjoined at center of disk but free at irregular margins, the cells there not bifurcate.

A frequent epiphyte, low intertidal and subtidal, Br. Columbia to Gulf of Calif. Widely distributed in temperate waters. Type locality: Denmark.

Erythrocladia subintegra Rosenv.

Rosenvinge 1909: 73; Smith 1944: 166; Dawson 1953a: 5; Nichols & Lissant 1967: 6.

Thalli discoid, monostromatic, to 125 µm diam.; cells laterally adjoined and in radiating rows; central cells isodiametric, to 15 µm diam.; margin of disk entire, the marginal cells commonly bifurcate.

Common epiphyte, low intertidal to subtidal, Wash. to Baja Calif. Widely distributed, including tropics. Type locality: Denmark.

According to Heerebout (1968), *E. subintegra* may be conspecific with *E. irregularis*.

Erythrotrichia Areschoug 1850

Thalli erect, filamentous throughout or only toward base, attached by basal cell, basal disk, or short, rhizoidal filaments. Erect parts usually unbranched, with apex cylindrical or flattened. Cell division intercalary. Chloroplast usually single, usually stellate, containing 1 central pyrenoid. Reproduction by monospores, these cut off singly from vegetative cells by curving wall. "Spermatangia" reportedly cut off from outer end of vegetative cells. "Carpogonia" described as slightly modified vegetative cells with short trichogynes. Zygote reported to form single "carpospore" or to divide into several "carpospores."

1. Thallus with several to many erect fronds from basal disk or cushion..... 2
1. Thallus without basal disk or cushion 3

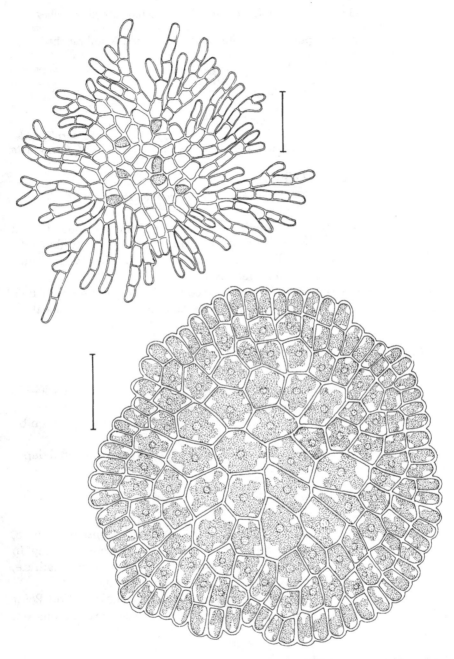

Fig. 226 (above left). *Erythrocladia irregularis.* Rule: 20 μm. (JRJ/S)   **Fig. 227** (below right). *E. subintegra.* Rule: 100 μm. (JRJ/S)

2. Basal attachment a thin disk; erect fronds with a few lateral branches
   near bases .... *E. welwitschii*
2. Basal attachment pulvinate, multistratose; erect fronds unbranched,
   multiseriate .... *E. pulvinata*
3. Erect fronds 80–120 mm tall .... *E. parksii*
3. Erect fronds mostly less than 15 mm tall 4
   4. Thallus uniseriate throughout .... *E. carnea*
   4. Upper thallus multiseriate 5
5. Upper thallus flattened, with local expansions 6–8+ cells wide ....
   .... *E. porphyroides*
5. Upper thallus cylindrical, without local expansions .... *E. tetraseriata*

Erythrotrichia carnea (Dillw.) J. Ag.

Conferva carnea Dillwyn 1807: pl. 84. Erythrotrichia carnea (Dillw.) J. Agardh 1883: 15; Smith 1944: 164.

Thallus bright pink, 1–8 mm tall, attached by lobed extensions of basal cell, uniseriate throughout; cells mostly 6–10 μm diam. at base of filaments, to 25 μm diam. above, the cells there 2–3 times as long as diam. Chloroplast single, stellate. Reproduction by monospores.

Common epiphyte, all tidal levels, Vancouver I., Br. Columbia, to Baja Calif. and Costa Rica. Widely distributed, including tropics. Type locality: Wales.

Erythrotrichia parksii Gardn.

Gardner 1927a: 239.

Thallus densely caespitose, pinkish-purple in mass, erect filaments 80–120 mm long, 15–18 μm diam. throughout; attached by short, rhizoidal processes from basal cells; chloroplasts reported to be band-shaped; reproduction by monospores.

Known only from type locality, Eureka, Calif., on other algae; tidal depth unknown.

Erythrotrichia porphyroides Gardn.

Gardner 1927a: 237.

Thallus minute, to 2 mm tall, with a few rhizoidal extensions at base, chiefly uniseriate, and with numerous meristematic regions resulting in local monostromatic expansions, these 6–8(16) cells wide; cells quadrate, the chloroplasts described as band-shaped; reproduction unknown.

Epiphytic, low intertidal, known only on *Laminaria sinclairii*, Fort Point (San Francisco; type locality), and on *Halidrys dioica*, near Corona del Mar (Orange Co.), Calif.

Erythrotrichia pulvinata Gardn.

Gardner 1927a: 238; Smith 1944: 184.

Thallus of numerous erect fronds arising from pulvinate base; fronds

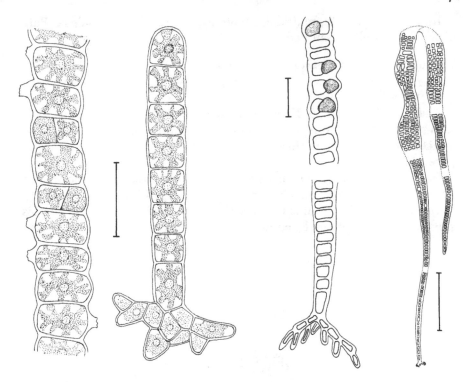

Fig. 228 (two at left). *Erythrotrichia carnea*. Rule: 20 µm. (JRJ/S) **Fig. 229** (middle right). *E. parksii*. Rule: 20 µm. (SM/H) **Fig. 230** (right). *E. porphyroides*. Rule: 100 µm. (SM/H)

uniseriate below, monostromatic and 6–8 cells broad above; chloroplast single, stellate, containing 1 pyrenoid; reproduction by monospores; sexual reproduction unknown.

Occasional epiphyte on tips of utricles of *Codium fragile*, low intertidal, Hope I., Br. Columbia, to Baja Calif. Type locality: Pebble Beach (Monterey Co.), Calif.

Heerebout (1968) suggested that *E. pulvinata* should probably be included under *E. boryana* (Mont.) Berthold. We prefer to recognize it as distinct for the present. The thick, pulvinate base is probably perennial, and the species seems to be host-specific.

Erythrotrichia tetraseriata Gardn.

Gardner 1927a: 240; Dawson 1953a: 12. *Erythrotrichia californica* Kylin 1941: 3; Smith 1944: 165.

Thallus 5–15 mm tall, attached by basal cell or by short rhizoidal extensions from it, uniseriate below, cylindrical above, there with cells more or less in transverse tiers, 4–8+ cells per tier; mature fertile parts 38–44 µm diam., the cells mostly quadrate; chloroplasts band-shaped, mostly entire.

Infrequent epiphyte, midtidal, Monterey, Calif., to I. Magdalena, Baja Calif. Type locality: San Pedro, Calif.

Erythrotrichia welwitschii (Rupr.) Batt.

Cruoria welwitschii Ruprecht 1851: 332. *Erythrotrichia welwitschii* (Rupr.) Batters 1902: 55; Smith 1944: 163.

Thallus to 3 mm tall, 9–10 μm diam. below, to 30 μm diam. above, reddish-pink, uniseriate from base to apex, with a few branches near base, this attached by a few short rhizoids. Cells with 1 stellate chloroplast, with pyrenoid.

Infrequent, epiphytic or saxicolous, South Bay, Ore., to Carmel Beach (Monterey Co.), Calif. Also Europe. Type locality: Lisbon, Portugal.

Heerebout (1968) doubts the records of *E. welwitschii* and thinks this species as well as others listed here are conspecific with *E. carnea*.

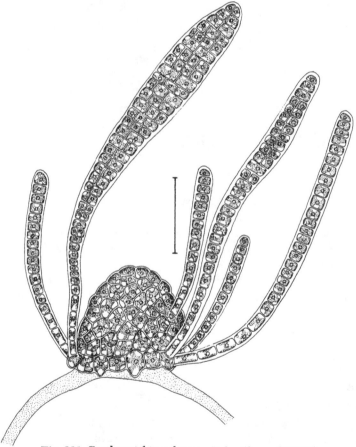

Fig. 231. *Erythrotrichia pulvinata.* Rule: 50 μm. (JRJ/S)

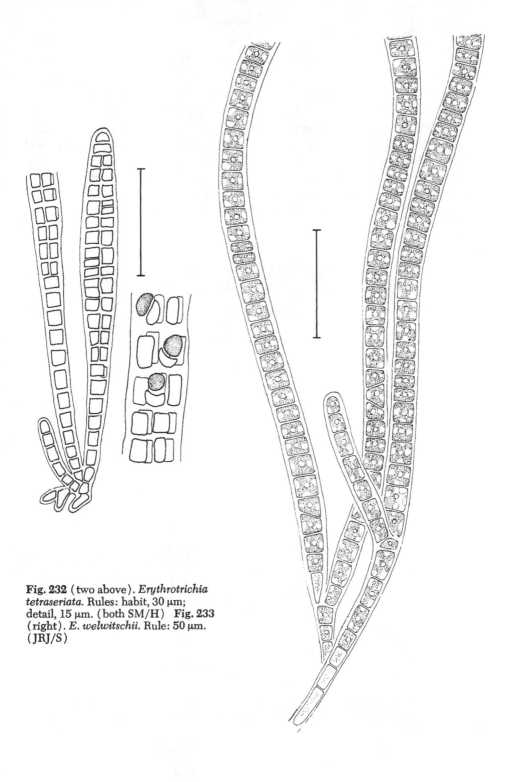

Fig. 232 (two above). *Erythrotrichia*
tetraseriata. Rules: habit, 30 µm;
detail, 15 µm. (both SM/H) Fig. 233
(right). *E. welwitschii*. Rule: 50 µm.
(JRJ/S)

Smithora Hollenberg 1959

Thalli host-specific epiphytes, with several to many blades arising from basal cushion. Blades monostromatic, with short, narrow attachment lacking rhizoidal filaments. Chloroplast single, stellate, with single pyrenoid. Asexual reproduction by monospores released from small, mostly marginal, wartlike sori, possibly also by characteristic deciduous, terminal to lateral sori, these periodically shed as gelatinous masses of cells and giving rise to new basal cushions. "Spermatangia" relatively large, cut off outwardly as single cells from pigmented cells of localized distromatic portions of blades; identity and function of these and corresponding female structures, if any, uncertain.

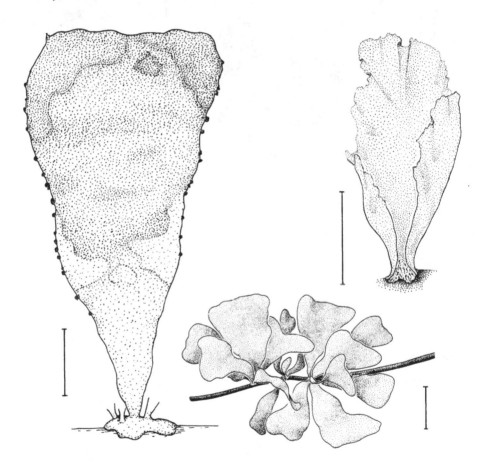

Fig. 234 (left & below). *Smithora naiadum*. Rules: left, 1 cm; below, 2 cm. (DBP/H & JRJ/S) **Fig. 235** (above right). *Porphyropsis coccinea* var. *dawsonii*. Rule: 500 μm. (DBP/H)

Smithora naiadum (Anders.) Hollenb.

Porphyra naiadum Anderson 1892: 148; Hus 1902: 212; Smith 1944: 169. *Smithora naiadum* (Anders.) Hollenberg 1959: 3. *S. naiadum* var. *australis* Dawson 1953a: 15.

Thallus purplish-red to deep purple, with 5–30 blades 10–50 mm tall arising from multicellular, pulvinate, perennial bases; young blades obovate, the older blades broadly cuneate, with slender, stipelike attachment; blades monostromatic, 25–30 μm thick in vegetative parts; cells isodiametric, irregularly angular, 15–20 μm diam. in surface view; reproduction as described for genus.

Locally abundant in season, epiphytic exclusively on *Phyllospadix* and *Zostera*, low intertidal, N. Br. Columbia to I. Magdalena, Baja Calif. Type locality: probably Farallon Is., Calif.

The occurrence of the several types of reproductive structures varies greatly in different parts of the range. Two forms are recognized by Hus. Forma *major*, with blades to 10 cm long, occurs on *Zostera* mostly in more sheltered water; common from N. Br. Columbia to C. Calif., it seems not to produce the marginal, wartlike sori often present on blades of forma *minor*, which occurs mostly on *Phyllospadix* in more exposed areas. Forma *minor* is much smaller, especially in S. Calif. and Mexico, and usually lacks the deciduous sori and wartlike sori.

Porphyropsis Rosenvinge 1909

Thalli minute, becoming vesicular from discoid or pulvinate bases, ultimately rupturing to form delicate, monostromatic blades. Reproduction by monospores. Sexual reproduction unknown.

Porphyropsis coccinea (Aresch.) Rosenv.

Porphyra coccinea Areschoug 1850: 407. *Porphyropsis coccinea* (Aresch.) Rosenvinge 1909–31 (1909): 69.

The type variety does not occur in California.

Porphyropsis coccinea var. **dawsonii** Hollenb. & Abb.

Hollenberg & Abbott 1968: 1239.

Thalli clustered, blades rose pink, to 2.5 mm tall, 12–15 μm thick, with ragged margins and involute-cuculate bases, broadly attached, with a few very short rhizoidal processes extending from lower part of blades to basal disk or cushion; growth zone clearly evident above zone of rhizoids; cells of blade 3–5 μm diam., with single parietal chloroplast; monospores formed singly by curving wall.

Infrequent, known from stipe of *Laminaria*, low intertidal, Patrick's Pt. (Humboldt Co.; type locality), subtidal from Carmel Bay, on *Callophyllis*, and from Cayucos (San Luis Obispo Co.), Calif., on *Ptilota*.

Similar plants reported from Wash. and Japan may represent this variety.

Membranella Hollenberg & Abbott 1968

Thallus small, membranous, with several monostromatic blades arising from pulvinate base, without rhizoidal processes from lower cells. Chloroplast single, stellate, with 1 central pyrenoid. Reproduction by monospores (?), these formed by very unequal division of vegetative cells in marginal areas, the larger half becoming the spore; monospores also formed in small, marginal, protuberant sori.

Membranella nitens Hollenb. & Abb.

Hollenberg & Abbott 1968: 1240.

Thallus purplish-red to violet, with orbicular blades 8–13(25) mm broad, or ovate and to 25 mm long; blades 20–25 µm thick, with cells 4.5–5.8 µm

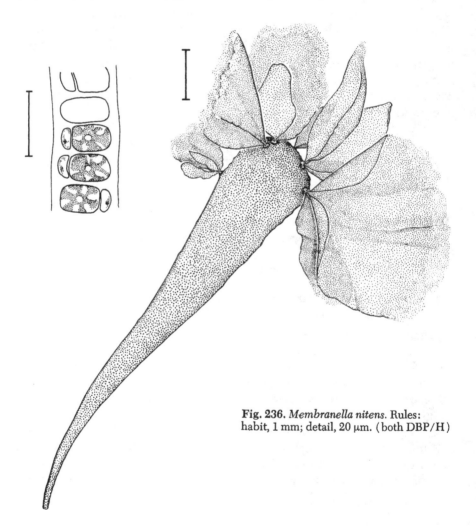

Fig. 236. *Membranella nitens*. Rules: habit, 1 mm; detail, 20 µm. (both DBP/H)

diam., 15–18 μm long in anticlinal section; margins of blades plane or slightly ruffled; fertile areas to 3 mm broad, along nearly entire margin; monospores(?), when released, leaving remains of cell walls as colorless margin of blade.

On tips of blades of *Egregia menziesii*, low intertidal, Crescent Beach (Olympic Peninsula), Wash. (uncertain record), and at various localities on Monterey Peninsula, Calif. Type locality: Moss Beach (Monterey Co.), Calif.

Family BANGIACEAE

Thalli cylindrical or bladelike. Growth intercalary to diffuse. Attachment by basal disk, this formed by numerous rhizoidal outgrowths from lower cells of erect parts. Cells uninucleate, mostly with single, stellate chloroplast with central pyrenoid. Obvious cytoplasmic connections usually lacking. Asexual reproduction in some cases by monospores cut off singly from vegetative cells. Sexual reproduction claimed for certain genera in accordance with several hypotheses, typically involving minute, nonmotile, pale "spermatia" (these thought to function as male gametes) and much larger "carpospores," likewise nonmotile but deeply pigmented, these formed in *Bangia* and *Porphyra* by repeated division of cell presumed to have been fertilized in some unknown manner by a "spermatium."

Spores, presumably "carpospores," in *Bangia* and *Porphyra* giving rise to rose red, branched, microscopic stage (known as *"Conchocelis rosea"* Batters, 1892), this consisting of filaments of uninucleate cells, each with single chloroplast; typically penetrating mollusk shells or other calcareous substrata; reproducing by monospores (conchospores), these borne singly in cells of enlarged fertile branches, either giving rise to macroscopic plant or repeating *"Conchocelis"* stage.

1. Thallus cylindrical .... *Bangia* (below)
1. Thallus bladelike 2
  2. "Carposporangia" in packets of 4–32 .... *Porphyra* (p. 294)
  2. "Carposporangia" formed singly, not in packets..... *Porphyrella* (p. 305)

Bangia Lyngbye 1819

Macroscopic thalli and microscopic *"Conchocelis"* stages present in life history. Larger stage erect, cylindrical, unbranched, uniseriate when young, later becoming multiseriate above, very gelatinous when moist. Cells of uniseriate part cylindrical, mostly shorter than broad, those of multiseriate part cubical to polyhedral, more or less in transverse tiers. Basal disk formed by tips of intramatrical rhizoidal extensions of lower cells. Asexual reproduction common, by direct division of vegetative cells into 2–4 or more sporelike cells. "Spermatangia" formed by repeated division of vegetative cells in 3 planes. "Carpospores" similarly formed by successive division of vegetative cells.

Bangia fusco-purpurea (Dillw.) Lyngb.

Conferva fusco-purpurea Dillwyn 1802–9 (1807): 54. *Bangia fusco-purpurea* (Dillw.) Lyngbye 1819: 83; Dawson 1953a: 13. *B. vermicularis* Harvey 1853: 55; Smith 1944: 167. *B. maxima* Gardner 1927a: 235.

Thalli aggregated, filamentous below, dull purplish to brownish, lubricous, mostly 10–30 mm tall (the fall and winter thalli to 100 mm tall), 40–60 µm diam. below, to 150 µm diam. above; thalli dioecious.

Frequent on large rocks, rarely epiphytic, upper intertidal, N. Br. Columbia to Baja Calif. and Chile. Widely distributed. Type locality: Wales.

Porphyra C. Agardh 1824

With thalli of differing morphologies during life history, the larger stages leafy and erect, the smaller stages microscopic and filamentous. "Gametangial" thalli with monostromatic or distromatic blades, arising singly from discoid attachment. Blades sessile or with brief stipe, normally entire, sometimes with much-ruffled margins. Cells embedded in colorless, firm, but gelatinous matrix. Chloroplasts 1 or 2, stellate, each with central pyrenoid. Plants monoecious or dioecious. "Spermatangia" in packets of 16–128, mostly in marginal areas. "Carpospores" in packets of 4–64, usually in larger marginal areas. Filamentous *"Conchocelis"* stage (see below, p. 304) probably perennial, commonly growing in matrix of mollusk shells.

1. Blades monostromatic 2
1. Blades distromatic 7
    2. Plant dioecious .... *P. lanceolata*
    2. Plant monoecious 3
3. Plant growing on stipe of *Nereocystis* .... *P. nereocystis*
3. Plant not growing on *Nereocystis* 4
    4. Cells each with 2 chloroplasts 5
    4. Cells each with 1 chloroplast 6
5. Blades lanceolate to ovate, growing on *Phyllospadix* .... *P. pulchra*
5. Blades mostly orbicular, growing mostly on *Gigartina* or *Pelvetia*...*P. smithii*
    6. Blades deeply lobed and amply ruffled, greenish to steel-gray tinged with purple .... *P. perforata*
    6. Blades plane, rusty pink .... *P. thuretii*
7. Plant monoecious, with "spermatangia" in minute area within marginal "carposporangial" areas ....*P. miniata*
7. Plant dioecious, with "spermatangia" forming yellowish marginal band.... 8
    8. Blades mostly less than 12 cm long, pale yellowish-gray, with pinkish "carposporangial" margin; high intertidal ....*P. schizophylla*
    8. Blades to 100 cm long, brick-red; subtidal .... *P. occidentalis*

Porphyra lanceolata (Setch. & Hus) Smith

Porphyra perforata f. *lanceolata* Setchell & Hus 1900: 65; Hus 1902: 208. *P. lanceolata* (Setch. & Hus) Smith 1943b: 213; 1944: 170.

Thalli in thick groups overlying each other; blades olive green to steel-gray with purplish tinge, monostromatic, to 50+ cm long, 100–200 µm thick; chloroplasts 2 per cell; plant dioecious; "male" plants to 6 cm broad, linear,

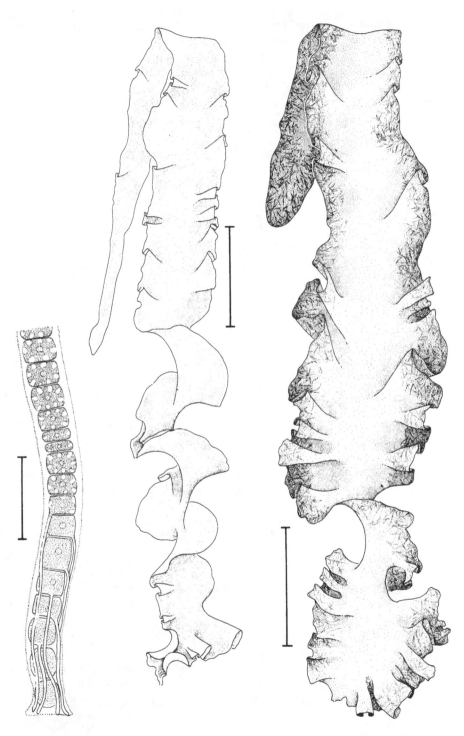

Fig. 237 (below left). *Bangia fusco-purpurea*. Rule: 50 μm. (JRJ/S) **Fig. 238** (two at right). *Porphyra lanceolata*. Rules: both 5 cm. (both JRJ/S)

with attenuated apex and continuous, cream-colored fertile margin 1–3 mm broad; "female" plants mostly broader, to 12 cm broad, lanceolate, with purplish, ruffled fertile margins 10–25 mm wide, frequently with lacerate apex; "spermatangia" formed in packets of 128; "carpospores" in packets of 32, developing in groups intermingled with groups of vegetative cells and commonly forming distinctive patterns; thalli annual, fertile in early winter when still 150–200 mm long, continuing growth and remaining fertile until fall.

Common on rocks, upper intertidal, Wash. to San Luis Obispo Co., Calif. Type locality: Carmel Bay, Calif.

Porphyra miniata (C. Ag.) C. Ag.

Ulva purpurea var. *miniata* C. Agardh 1817: 42. *Porphyra miniata* (C. Ag.) C. Agardh 1824: 191; Hollenberg 1972: 43 (incl. synonymy).

Thalli occurring in small groups or isolated; blades reddish-purple to pink, elliptical to ovate-lanceolate, mostly 15–80(600) cm long, 8–20(30) cm broad, 30–80 μm thick, distromatic, commonly deeply ruffled, especially below, the base deeply umbilicate to cuneate, with small, discoid holdfast; cells with 1 chloroplast; cells squarish in section to slightly elongate periclinally; plants mostly monoecious, with "spermatangia" in microscopic groups in marginal "carposporangial" zone, but occasional plants with extensive marginal "spermatangial" areas or divided into "spermatangial" and "carposporangial" halves; "spermatangia" mostly in packets of 8 in 2 anticlinal tiers, occasionally in packets of 16 in 4 tiers; "carpospores" mostly in packets of 4 in single tier.

Infrequent, saxicolous or epiphytic, lower intertidal to upper subtidal, Alaska to Humboldt Co. and Monterey, Calif. Also frequent in N. Atlantic. Type locality: Greenland.

Plants in sheltered water may grow to 6 m long. No constant features seem to justify recognition of more than a single very variable species.

Porphyra nereocystis Anders.

Anderson 1892: 149; Hus 1902: 210; Smith 1944: 171.

Thalli deep pink to dull purplish-red, in groups, 25–90(300) cm long; blades monostromatic, 25–60 μm thick in vegetative parts, subsessile, with discoid attachment, at first plane and broadly linear to broadly ovate, later with inrolled margins basally and ill-defined in shape owing to marginal laciniation; blades monoecious; "spermatangia" in packets of 128, forming sharply defined spots and streaks within marginal "carposporangial" areas; "carpospores" in packets of 32.

On stipes of *Nereocystis*, usually 3–6 m below pneumatocyst; occasional on other Laminariales; frequent to abundant, November to June, Alaska to S. Calif., usually found cast ashore. Type locality: probably Santa Cruz, Calif.

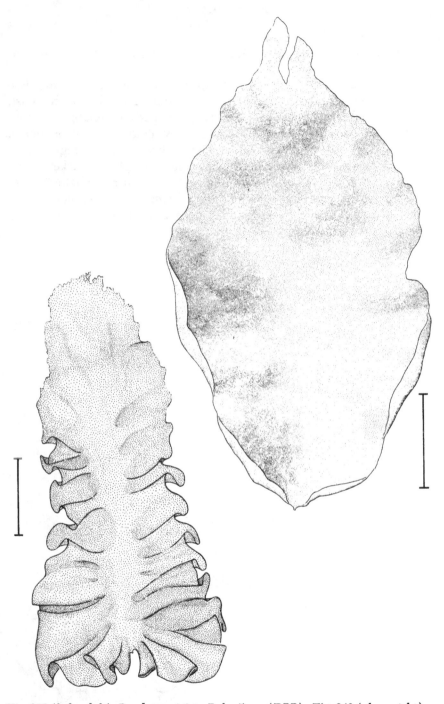

Fig. 239 (below left). *Porphyra miniata*. Rule: 5 cm. (DBP) **Fig. 240** (above right).
P. nereocystis. Rule: 5 cm. (JRJ/S)

Porphyra occidentalis Setch. & Hus

Setchell & Hus 1900: 69; Hus 1902: 228; Smith 1944: 174; Hollenberg & Abbott 1966: 37 (incl. synonymy).

Thalli isolated, 25–40(110) cm tall, 90–130 μm thick, distromatic, lanceolate, sessile, with broadly rounded base and small, discoid holdfast; blades rounded or acute at apex, the margins slightly undulate; cells subcubical, 12–15 μm diam.; chloroplast single, stellate; cell wall laminate, gelatinous; "spermatangial" plants, previously known in this region as *P. occidentalis,*cerise, 50–100 μm thick, with "spermatangia" in packets of 64, forming continuous yellowish margin along upper parts of blade; "carposporangial" plants previously known as *P. variegata* in this region brick red to purplish-red, with "carpospores" in packets of 16 or 32, intermingled with vegetative cells; apparently a spring annual.

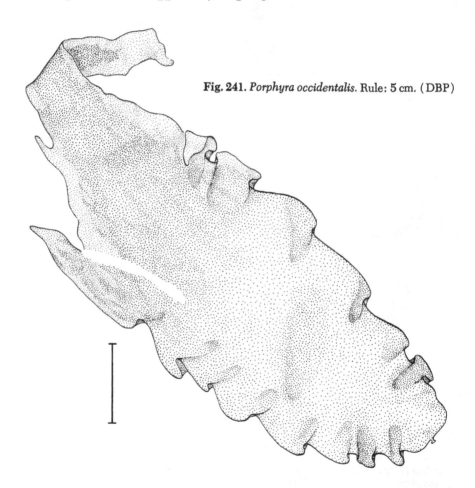

Fig. 241. *Porphyra occidentalis.* Rule: 5 cm. (DBP)

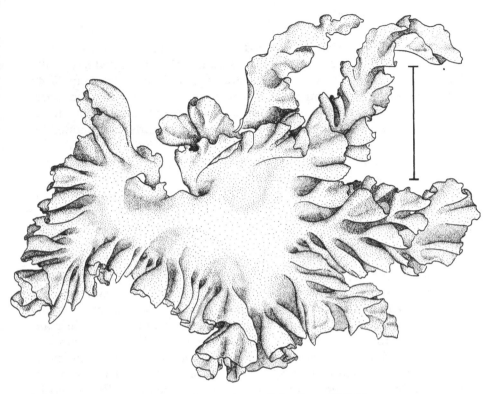

Fig. 242. *Porphyra perforata.* Rule: 5 cm. (JRJ/S)

Occasionally abundant as isolated thalli on rocks or epiphytic, subtidal (to 30 m), Ore. to Little Sur River (Monterey Co.), Calif. Type locality: Carmel Bay, Calif.

Porphyra perforata J. Ag.

J. Agardh 1883: 69; Hus 1902: 202; Smith 1944: 172 (incl. synonymy). *Porphyra perforata* f. *segregata* Setchell & Hus 1900: 64.

Thalli in dense groups, 15–30(110) cm tall, gray-green to brownish-purple, sessile, with small discoid attachment; blades ultimately often as broad as long, deeply ruffled and often deeply lobed or divided, monostromatic, 45–140 μm thick, with anticlinally ellipsoidal cells; chloroplast single, stellate; plant monoecious, with "spermatangia" and "carposporangia" in marginal patches, and with vegetative cells among latter; "spermatangia" in packets of 128, "carpospores" in packets of 32.

Common to abundant, usually on rocks, upper intertidal, Alaska to Baja Calif. Type locality: "California."

Much smaller in southern part of range.

Porphyra pulchra Hollenb.

Hollenberg 1943b: 213; Smith 1944: 170.

Thalli (8)20–30 cm tall, mostly 8–11 cm broad, pinkish-violet and often somewhat mottled in appearance; blades sessile or briefly stipitate, typically lanceolate or oblong-lanceolate, monostromatic, 50–90 μm thick; cells 25–40 μm long anticlinally; chloroplasts 2, stellate; plant monoecious, the "spermatangia" in packets of 64 or 128, grouped in oblong to linear sori extending lengthwise along margins and inward at apex of blades, surrounded by "carposporangial" and vegetative areas; "carpospores" in packets of 16 or 32.

Occasional or at times locally abundant, growing on *Phyllospadix*, lower to midtidal; known only from Bodega Bay (Sonoma Co.), Santa Cruz, and Monterey Peninsula, Calif. Type locality: Moss Beach (Monterey Co.), Calif.

Porphyra schizophylla Hollenb.

Hollenberg 1943d: 213; Smith 1944: 173.

Thalli distromatic to base, 200–250 μm thick near base; gelatinous matrix of blades with conspicuous line of separation between the 2 cell layers and with eccentric stratifications prominent on inner side of each cell; cells 20–25 μm diam.; chloroplast single, pale, in broader outer third of cell; plant monoecious, the "spermatangial" plants to 17 cm tall, 20–30(40) mm wide, light yellowish to gray, narrowly lanceolate, not much ruffled, sometimes spirally twisted; "spermatangia" in packets of 16(32) and forming continuous whitish to yellowish marginal band in which all cells are fertile; "carposporangial" plants mostly shorter and broader, often much ruffled, with pinkish to purplish-brown fertile marginal band in which frequent vegetative cells occur; "carpospores" in packets of 8.

Locally frequent, on high intertidal rocks swept by heavy surf, Wash. to Monterey Peninsula, Calif. Type locality: Pescadero Pt. (Monterey Peninsula), Calif.

Porphyra smithii Hollenb. & Abb.

Hollenberg & Abbott 1968: 1241. *Porphyra pulchra sensu* Smith 1944: 170 (in part).

Thalli mostly orbicular, 9–13(20) cm broad, rarely to 20 cm long, usually with very ample base obscuring point of attachment; margins smooth or undulate, mostly deeply ruffled; blades monostromatic, 125–135 μm thick; cells in vegetative regions 40–60 μm long, 20–25 μm diam., with long axis perpendicular to plane of blade, closely apposed; chloroplasts 2, widely separated, stellate, 1 in each end of cell; plant monoecious, the "spermatangia" in packets of 64 or 128; "carpospores" in packets of 32 or 64, forming dark purplish-red marginal band to 20 mm broad, this interrupted by relatively small, yellowish "spermatangial" patches, these gradually becoming deep marginal indentations as "spermatia" are liberated.

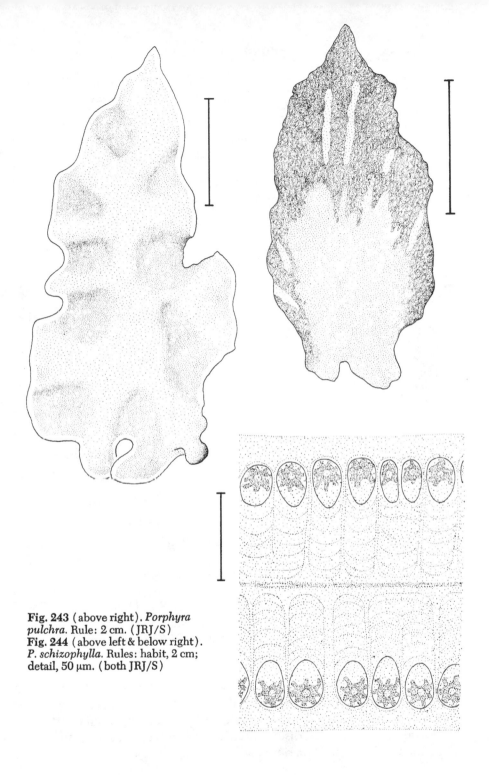

Fig. 243 (above right). *Porphyra pulchra*. Rule: 2 cm. (JRJ/S)
Fig. 244 (above left & below right). *P. schizophylla*. Rules: habit, 2 cm; detail, 50 μm. (both JRJ/S)

Fig. 245. *Porphyra smithii.* Rule: 2 cm. (DBP)

Occasional to locally frequent epiphyte, Roller Bay (Hope I.), Br. Columbia, Brown I., Wash., and Monterey Peninsula south to Malpaso Creek (Monterey Co.), Calif. Type locality: Mission Pt. (Monterey Co.), Calif.

Porphyra thuretii Daws.

Dawson 1944a: 253; Smith 1944: 171. *Porphyra leucosticta sensu* Hus 1902: 199.

Thalli dull rose to light brownish-red; blades oblong to lanceolate, delicate, monostromatic, 20–45(75) cm tall, 10–22 cm broad, 25–50 µm thick, cordate at base, with small discoid attachment; chloroplast single, stellate; plant monoecious, the "spermatangia" forming longitudinal streaks or

Fig. 246. *Porphyra thuretii.* Rule: 5 cm. (DBP)

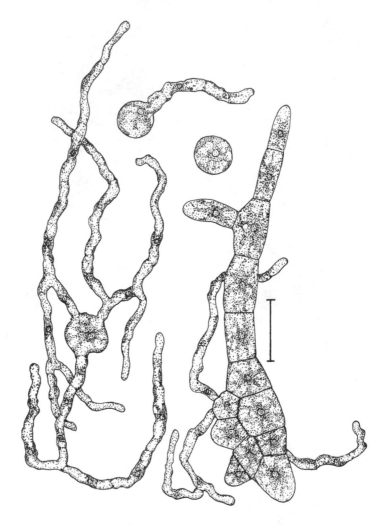

Fig. 247. *"Conchocelis rosea."* Rule: 25 μm. (DBP/H)

patches in marginal "carposporangial" areas; "spermatangia" in packets of 64; "carpospores" in packets of 8.

Infrequent to rare, on rocks or other algae, intertidal to subtidal, Ore. to C. Calif. and Baja Calif. and Costa Rica. Type locality: Pacific Grove, Calif.

Filamentous Stage of *Porphyra*

"Conchocelis rosea" Batt.

Batters 1892: 25.

Microscopic plants with uninucleate cells, containing single chloroplast; filaments much branched, typically penetrating mollusk shells, reproducing by monospores, known as conchospores, borne singly in cells of enlarged fertile branches.

Porphyrella Smith & Hollenberg 1943

Thalli epiphytic or saxicolous, with 1 to several blades growing on discoid to conical holdfasts, these composed of rhizoidal processes from lower cells. Blades monostromatic, thin and delicate. Chloroplast stellate, 1 per cell. Plants monoecious; "spermatangia" in packets grouped in small marginal patches; "carpospores"(?) released singly, during gelatinization of distal margins of blades.

Porphyrella californica Hollenb.

Hollenberg 1945a: 450.

Thalli 1–2 cm tall, usually in clusters, medium to light reddish-brown, iridescent in life when submerged; blades oval to orbicular, elongate with cuneate base when young, commonly cordate at maturity, attached by very short stipe; blades to 25 mm broad, 15–20 µm thick; cells close-packed, 5–10 µm diam. in surface view, slightly longer than diam. in anticlinal section; "spermatangia" in packets of 32 or 64, in small, irregular marginal patches; "carpospores"(?) marginal, liberated as single cells.

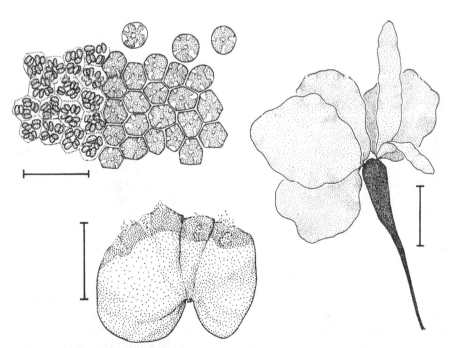

Fig. 248 (two at left). *Porphyrella californica*. Rules: detail, 20 µm; habit, 1 cm. (both DBP) Fig. 249 (right). *P. gardneri*. Rule: 1 cm. (JRJ/S)

Occasional, saxicolous or attached to limpets, colonial hydrozoans, or byssal fibers of mussels, intertidal, Santa Cruz I. (type locality) and Laguna Beach, Calif.

Porphyrella gardneri Smith & Hollenb.

Smith & Hollenberg 1943: 215; Smith 1944: 175.

Thalli mostly 2–3.5 cm tall, purplish-red, not noticeably iridescent, usually with several blades from single holdfast; blades cylindrically stipitate, oblong to lanceolate when young, cuneate to orbicular at maturity, sometimes cordate at base; cells of blade subcubical, 10–15 μm diam.; "spermatangia" in packets of about 64, in small irregular marginal patches; "carpospores"(?) released singly at margin during gelatinization of walls.

Common epiphyte on tips of blade divisions of *Laminaria dentigera*, less frequently on tips of leaflets of *Egregia*, in semi-exposed areas, Adak I., Alaska, to San Luis Obispo Co., Calif., and Pta. Banda, Baja Calif. Type locality: Pt. Joe (Monterey Peninsula), Calif.

Class FLORIDEOPHYCEAE

Thalli multicellular. Cells arranged in branched or unbranched filaments, or with filaments aggregated, forming terete, leaflike, or pseudoparenchymatous plants, some polysiphonous. Plants calcified or uncalcified. Growth of filaments by division of apical cells; intercalary cell division absent except in Delesseriaceae (Ceramiales) and Corallinaceae (Cryptonemiales). Pit-connections between all cells of kindred origin. Cells uninucleate or multinucleate. Chloroplasts simple, band-shaped, reticulate, or stellate, with or without pyrenoid; with phycobilin pigments (phycoerythrin and phycocyanin) of red, brown, or rarely bluish coloration. Sporangia of many types, determined by number of spores produced: sporangia having 4 spores (tetrasporangia) arranged tetrahedrally, cruciately, or zonately, usually involving meiotic division; sporangia having more than 4 spores sometimes with meiotic division (polysporangia), sometimes without (parasporangia); sporangia having 2 spores (bisporangia) with or without meiotic division; sporangia having 1 spore (monosporangia) entirely mitotic. Sexual reproduction oogamous. Male gamete (spermatium) spherical, colorless, nonmotile, formed 1 per spermatangium. Spermatangia formed singly, or in groups, or on special branches of thallus. Female gamete (carpogonium) with elongate receptive protuberance (trichogyne) firmly attached to female gametangial plant, with fertilization *in situ*. Fertilized carpogonium giving rise to tissue mass (carposporophyte, gonimoblast, or cystocarp) that may or may not involve fusion with other cells (auxiliary cell), this giving rise to 1 or more clusters of sporangia (carposporangia) either in terminal or intercalary position. Development of carposporophyte usually involving sterile filaments or axes, these surrounding carposporophyte as involucre or as compact cellular enclosure (pericarp).

Order NEMALIALES (NEMALIONALES)

Thalli uniaxial or multiaxial, with single or many apical cells, erect, and unbranched or little or richly branched. Some thalli filamentous, microscopic; some macroscopic, compressed, with branched united filaments; some cylindrical, gelatinous; most uncalcified (with exceptions); none bladelike. For most macroscopic species, medulla of elongate, tightly interwoven, longitudinal filaments, these forming cortical assimilatory filaments radial to axis. Structure of cortical filaments clearly of branched filaments or of branched filaments laterally appressed. Reproductive structures in some genera very simple; monosporangia, common in some species, are only known reproductive structure. Tetrasporangia in isomorphic tetrasporophyte in some species; in microscopic, filamentous tetrasporophyte in other species. No sporangial phase known for some species. Spermatangia usually in clusters. Carpogonial branch in terminal or lateral position on cortical filaments, consisting of carpogonium only or of 2–6 cells, the number usually inconstant in given species. Zygote after fertilization producing spores directly, or dividing transversely or longitudinally and by repeated divisions producing gonimoblast, all or most cells becoming carposporangia (or rarely carpotetrasporangia). Carposporangia on germination producing microscopic, filamentous stage, this bearing tetrasporangia in some species; fate unknown in other species; or carposporangia giving rise to tetrasporophytes resembling gametophytes. Fusion of cells basal to carpogonium common in most genera. Sterile filaments protective in function frequently produced by cells adjacent to carpogonial branch.

A heterogeneous assemblage of algae now being studied from various viewpoints; the systematic arrangement used here is therefore tentative.

3. Thalli with delicate axes; gametophytes usually followed by microscopic
   tetrasporophytes .... BONNEMAISONIACEAE (p. 336)
3. Thalli with coarse axes; gametophytes usually followed by macroscopic
   tetrasporophytes .... GELIDIACEAE (p. 340)
   4. Thalli gelatinous, wormlike cylinders, little or not branched; carpo-
      gonial branch terminal on vegetative lateral .... NEMALIACEAE (p. 324)
   4. Thalli not wormlike or cylindrical, repeatedly branched; carpogonial
      branch lateral on vegetative lateral 5
5. Surface of thalli with tightly appressed, regularly arranged cells; vegetative filaments difficult to separate............. CHAETANGIACEAE (p. 331)
5. Surface of thalli with loosely arranged cells; vegetative filaments easily
   separated .... HELMINTHOCLADIACEAE (p. 325)

Family ACROCHAETIACEAE

Thalli microscopic, filamentous, mostly epiphytic or epizoic; erect, branched thalli wholly epiphytic; in partly endophytic or endozoic thalli, both the external and the internal portions sometimes richly branched; thalli sometimes wholly endophytic or endozoic. Monosporangia in some species laterally or terminally placed, stalked or unstalked. Tetrasporangia, if present, cruciately or irregularly divided, usually unstalked. Spermatangia of some species in small colorless clusters. Carpogonial branch simple, unicellular, after fertilization not dividing, or dividing transversely or longitudinally, forming cystocarp, this with few terminal carposporangia and no sterile filaments.

The systematics of the Acrochaetiaceae are confused and have been attacked from various viewpoints since the discovery of *Acrochaetium*-like stages in the life histories of a great many macroscopic algae. Names to be applied are frequently not clear; moreover, under controlled conditions some species show nearly every range of variation in the characters used for taxonomic purposes, thus documenting the instability previously noted in field material. We choose to follow a conservative taxonomic treatment until definitive studies are made.

1. Thalli principally creeping filaments; spermatangia few, on very long
   stalks .... *Kylinia* (p. 323)
1. Thalli principally erect filaments; spermatangia in short clusters 2
   2. Cells with 1 to several chloroplasts, without pyrenoids ....
      .... *Rhodochorton* (p. 321)
   2. Cells with 1 large chloroplast, with or without pyrenoids ....
      .... *Acrochaetium* (below)

Acrochaetium Nägeli 1862

Thalli mostly microscopic, filamentous, to 5 mm tall, epiphytic, epizoic, endophytic, or endozoic or partly so. Chloroplasts usually present in endophytic parts except for rhizoids, conspicuous in epiphytic parts as 1 large laminate or stellate structure, these forms sometimes combined in same thallus; pyrenoids 1 or 2 or none. Thallus germinating from persistent or nonpersistent basal spore, forming creeping filament or basal disk. Monosporangia common, occurring as sole reproductive structure, or occurring on tetrasporophytes or gametophytes. Tetrasporangia in some species, but rare. Spermatangia in terminal or lateral clusters, tufted or in panicles. Carpogonium simple, usually sessile or intercalary, after fertilization dividing transversely and producing pyriform carposporangia directly, or 1 or 2 further divisions forming carposporangia. No sterile filaments around cystocarp.

On purely morphological grounds, and on the basis of material from a wide variety of habitats and hosts, certain conservative changes are made here in the status of Pacific West Coast species. In the following account only the basionyms of various species or names of recent wide usage are employed.

1. Thalli mostly epiphytic; if bases endophytic, not massively so 2
1. Thalli mostly endophytic or decumbent; erect filaments barely free of host 10
   2. Erect filaments relatively short, sparingly branched 3
   2. Erect filaments regularly branched 5
3. Most cells nearly square . . . . A. amphiroae
3. Most cells oval-spherical 4
   4. Erect filaments with terminal hairs . . . . A. microscopicum
   4. Erect filaments without terminal hairs . . . . A. arcuatum
5. Thalli with tetrasporangia and monosporangia 6
5. Thalli with monosporangia only 8
   6. Tetrasporangia terminal on main filaments . . . . A. coccineum
   6. Tetrasporangia on lateral branches 7
7. Terminal cells ending in hairs . . . . A. thuretii
7. Terminal cells without hairs . . . . A. pectinatum
   8. Base a penetrating endophytic filament . . . . A. barbadense
   8. Base usually an epiphytic horizontal filament 9
9. Monosporangia borne singly or more commonly in groups on second-order branches . . . . A. pacificum
9. Monosporangia borne singly on third-order branches . . . . A. daviesii
   10. Thalli creeping on surface of host, at times penetrating intercellular spaces 11
   10. Thalli creeping within host tissues 12
11. Erect portions without branches; monosporangia lateral on main axes. . . . . A. tenuissimum
11. Erect portions branched; monosporangia terminal on erect filaments. . . . . A. desmarestiae
   12. Thalli mostly within cell walls of host . . . . A. porphyrae
   12. Thalli mostly between host cells. 13
13. Thalli with massive, colorless rhizoidal system; monosporangia in lateral clusters . . . . A. rhizoideum
13. Thalli without rhizoids 14
   14. Endophytic portion photosynthetic; monosporangia terminal on branched filaments . . . . A. subimmersum
   14. Endophytic portion nonphotosynthetic; monosporangia terminal or lateral on unbranched filaments . . . . A. vagum

Acrochaetium amphiroae (Drew) Papenf.

Rhodochorton amphiroae Drew 1928: 179. *Acrochaetium amphiroae* (Drew) Papenfuss 1945: 312. (Not *R. amphiroae* of Smith 1944: 183, which is *Ptilothamnionopsis lejolisea* (Farlow) Dixon.)

Thalli with creeping and erect filaments forming tufts 0.5–1.25(2) mm tall, the basal portions of short, colorless endophytic filaments; erect axes irregularly branched, usually unilateral, the apices of the branchlets curving toward axis; cells cylindrical to squarish, 10–15 μm diam., (18)20–25 μm long, tapering downward in basal portions of axes and near branchlets, otherwise scarcely tapering in any given thallus; chloroplast single, parietal, usually with 1 pyrenoid, but sometimes none; monosporangia terminal in lateral clusters borne adaxially at base of secondary branches.

Occasional on various species of articulated coralline algae (shorter thalli on *Corallina officinalis*, longer thalli on *Serraticardia* or *Calliarthron*), low intertidal, N. Br. Columbia to I. Cedros, Baja Calif. Also known from Penikese I., Mass. Type locality: San Pedro, Calif.

Acrochaetium arcuatum (Drew) Tseng

Rhodochorton arcuatum Drew 1928: 165. *Acrochaetium arcuatum* (Drew) Tseng 1945: 158. *R. densum* Drew 1928: 168; Nakamura 1944: 101.

Thalli less than 1 mm tall, arising from conspicuous basal cells 12–15 μm diam., these producing directly, or first dividing and then producing, 1 or more short and usually unbranched erect filaments, the cells of these 8–10 μm wide, 1.5–2 times as long; chloroplast parietal, rarely stellate, with single pyrenoid; monosporangia secund on main axes, usually crowded, 10–15 μm wide; basal cell also producing creeping filaments, these sometimes partly endophytic; secondarily produced erect filaments sometimes formed from creeping filaments.

Frequent, midtidal, epiphytic on a variety of algae, Wash. to Pebble Beach (Monterey Co.), Calif. Type locality: Moss Beach (San Mateo Co.), Calif.

Acrochaetium barbadense (Vick.) Børg.

Chantransia barbadense Vickers 1905: 60. *Acrochaetium barbadense* (Vick.) Børgesen 1915–20 (1915): 43; Abbott 1962b: 83 (incl. synonymy). *C. macounii* Collins 1913: 113. *A. dictyotae* Coll. 1906b: 193.

Thalli epiphytic; bases partly to wholly endophytic, the endophytic portions formed from single or divided basal spore, forming short, uniseriate, creeping filaments radiating from spore, or such filaments laterally appressed, forming disk, or endophytic filaments more or less penetrating host, either producing occasional erect photosynthetic filaments or not further branched; erect filaments 1-2 mm tall, unbranched, irregularly branched, or the branching opposite, alternate, secund, or all 3 kinds on same filament, the apices sometimes colorless and hairlike; cells 7–12 μm diam., 2–4 times as long, the upper filaments with narrower cells; chloro-

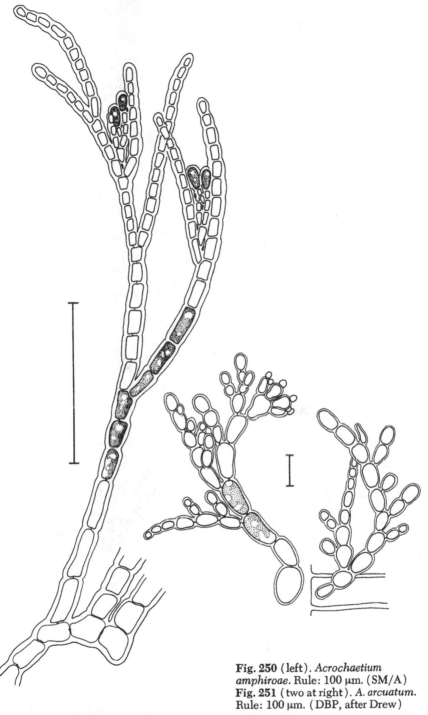

Fig. 250 (left). *Acrochaetium amphiroae.* Rule: 100 μm. (SM/A)
Fig. 251 (two at right). *A. arcuatum.* Rule: 100 μm. (DBP, after Drew)

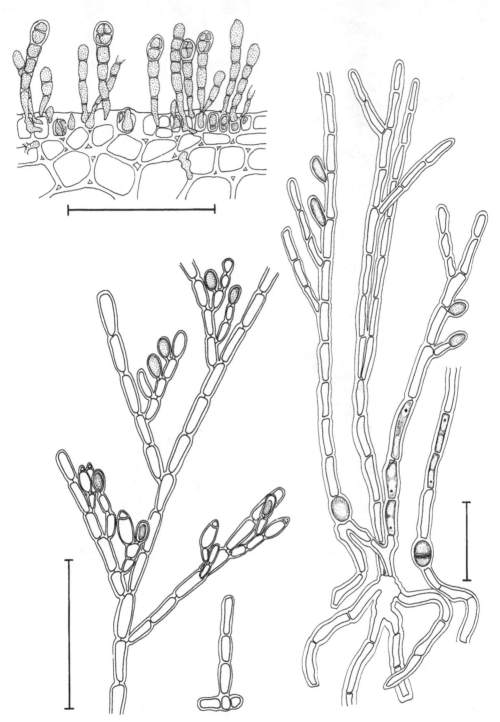

Fig. 252 (right). *Acrochaetium barbadense.* Rule: 100 μm. (SM/A) **Fig. 253** (above left). *A. coccineum.* Rule: 100 μm. (DBP, after Drew) **Fig. 254** (below left). *A. daviesii.* Rule: 100 μm. (DBP/A)

plast parietal with 1 or 2 pyrenoids; monosporangia opposite, alternate, or secund, terminal or lateral, pedicellate or not; carpogonia lateral and sessile; spermatangia in terminal clusters.

Frequent, on a variety of hosts showing multiaxial or parenchymatous structure, lower intertidal, Br. Columbia to S. Calif., including Channel Is., and into tropics. Type locality: Barbados.

Sexual plants are found only very occasionally, but then in abundance.

Acrochaetium coccineum (Drew) Papenf.

Rhodochorton coccineum Drew 1928: 192. *Acrochaetium coccineum* (Drew) Papenfuss 1945: 313. *R. obscurum* Drew 1928: 193.

Thalli conspicuous, the epiphytic portions forming bright red, felty stripes on stipes and blades of *Laminaria* or on blades of other brown algae, the erect portions 1–1.5 mm tall; tufts 1–3 cm wide, the erect filaments irregularly branched or unbranched, the cells 4–8 μm wide, 1.5–2.5 times as long, tapering upward, with 1 parietal chloroplast and no pyrenoids in each cell; endophytic portions superficially or extensively invading host; basal portions forming tight, irregular disks on surface of host, the disks 1 to several cells thick in section, mostly producing irregularly branched or unbranched filaments, these penetrating between and within cells of host; most penetrating cells soon losing chloroplasts and becoming more elongate and irregular; tetrasporangia terminal, 12–15 μm diam.; monosporangia terminal on lateral branchlets, 8–12 μm wide, oval.

Frequent on *Laminaria sinclairii, L. dentigera, Desmarestia latifrons*, and *Egregia menziesii*, Duxbury Reef (Marin Co.) to Little Sur (Monterey Co.), Calif. Type locality: San Francisco, Calif.

As conspicuous but not as common as *A. pacificum*, which occurs on many of the same hosts.

Acrochaetium daviesii (Dillw.) Näg.

Conferva daviesii Dillwyn 1802–9 (1809): 73. *Acrochaetium daviesii* (Dillw.) Nägeli 1861: 405; Hollenberg & Abbott 1966: 41. *Rhodochorton daviesii* (Dillw.) Drew 1928: 172 (incl. synonymy); Smith 1944: 184.

Thalli 4–6(8) mm tall, in bright pink tufts; base usually epiphytic, of creeping filaments tightly matted together, occasionally penetrating and becoming slightly endophytic; erect filaments entangled, without branches below, the branching above irregular, sometimes alternate or secund, the branches sometimes rebranching in same way but the secondary branchlets usually adaxial on first or second cell of the lateral, and short, uniseriate, or in loose clusters; cells cylindrical, 8–12 μm diam., 2(3) times as long; chloroplast single, parietal, with single pyrenoid; monosporangia common, clustered, usually within axil of branch; tetrasporangia infrequent, borne in same location as monosporangia.

Common on a variety of algae or animals, low intertidal to subtidal,

Wash. to La Jolla, Calif. Widely distributed in temperate waters. Type locality: Britain.

Acrochaetium desmarestiae Kyl.

Kylin 1925: 10; Hollenberg & Abbott 1966: 43. *Rhodochorton desmarestiae* (Kyl.) Drew 1928: 168. *Acrochaetium erythrophyllum* Jao 1937: 112. *A. scinaiae* Dawson 1949b: 3.

Thalli forming bright red patches on host, these 2–3 cm diam.; basal layer usually penetrating host between cells, otherwise creeping over surface of host, the creeping cells 3–8 μm wide; plants forming short, erect, rarely branched filaments less than 0.5 mm tall, the cells of these 6–8 μm

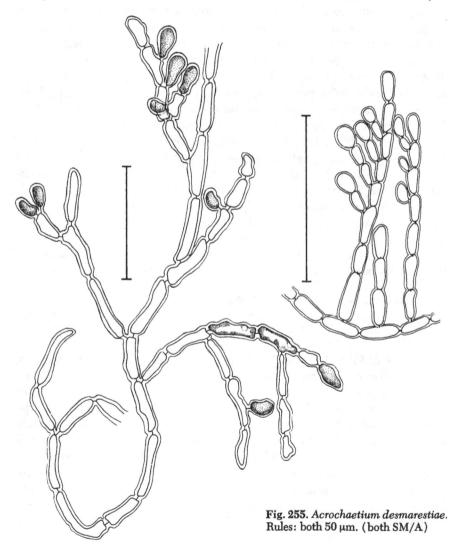

Fig. 255. *Acrochaetium desmarestiae.*
Rules: both 50 μm. (both SM/A)

diam.; chloroplast single, platelike, parietal, without pyrenoids; monosporangia 8.5–10 μm long, 17.5–22 μm diam., sessile or pedicellate, terminal on creeping or erect filaments.

Locally frequent on *Desmarestia ligulata* (especially), on other species of *Desmarestia*, or on any of a variety of bladelike algae, abundant late in season on old plants of hosts, low intertidal, Hope I., Br. Columbia, Canoe I. and Turn I. (type locality), Wash., through Ore. to Channel Is., Calif., and Gulf of Calif.

Acrochaetium microscopicum (Kütz.) Näg.

Callithamnion microscopicum Kützing 1849: 640. *Acrochaetium microscopicum* (Kütz.) Nägeli 1861: 407; Hollenberg & Abbott 1966: 42. *Rhodochorton hirsutum* Drew 1928: 166.

Thalli epiphytic, with spreading, disklike bases wholly superficial to hosts, or producing short, irregular, creeping filaments from single basal spore; erect filaments usually less than 1 mm tall, but to 1.5 mm, little branched, opposite, alternate, or irregular, frequently terminating in colorless hairs; cells of filaments somewhat moniliform, 8–10 μm diam.; chloroplast single, stellate, usually with 1 pyrenoid; monosporangia most common reproductive structures; sporangia lateral, or terminal, sessile or pedicellate; spermatangia rare, in dense terminal clusters on lateral branchlets; carpogonia pedicellate or sessile; development of cystocarp unknown.

Frequent, low intertidal, on *Ceramium* or *Polysiphonia* spp., Alaska to Vancouver I., Br. Columbia; and on *Botryocladia*, *Erythrophyllum*, or *Gymnogongrus* in region of Monterey Peninsula, Calif. Widely distributed. Type locality: England.

Acrochaetium pacificum Kyl.

Kylin 1925: 11. *Rhodochorton pacificum* (Kyl.) Drew 1928: 169. *R. elegans* Drew 1928: 181. *R. magnificum* Drew 1928: 180. *R. plumosum* Drew 1928: 173. *Acrochaetium plumosum* (Drew) Smith 1944: 180; Hollenberg & Abbott 1966: 41. *R. variabile* Drew 1928: 174.

Thalli epiphytic; base occasionally partially endophytic; erect filaments 1–3 mm tall, branching irregular, alternate, opposite, or secund; cells 4–8 μm wide, 1–4 times as long, the branches bearing monosporangia alternately or in secund series directly on main axes, on branchlets, or pedicellate or tufted in same places; axes simple or branched in more than 1 plane; plant never penetrating host with rhizoids, or occasionally entire basal filaments intercellular in host; at other times portions of photosynthetic filaments intracellular in host.

Common on variety of algae, especially Laminariales, or occasionally animals, low intertidal to subtidal, Br. Columbia to Baja Calif. Type locality: Brown I., Wash., on *Sertularia* at 10–20 m depth.

The commonest *Acrochaetium* on large brown algae, Monterey Peninsula to Channel Is., Calif.

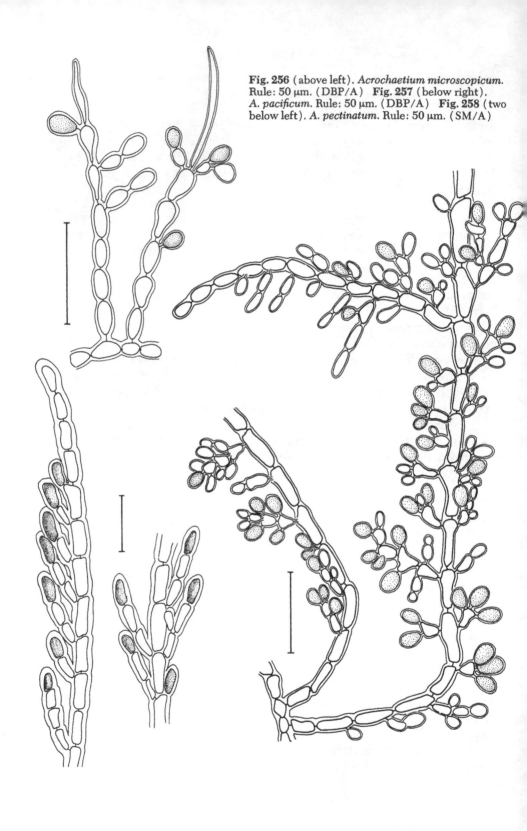

Fig. 256 (above left). *Acrochaetium microscopicum.* Rule: 50 µm. (DBP/A) Fig. 257 (below right). *A. pacificum.* Rule: 50 µm. (DBP/A) Fig. 258 (two below left). *A. pectinatum.* Rule: 50 µm. (SM/A)

Many specimens of all of the older-named species show that characters supposedly separating them overlap, not only in the same group of thalli but often in the same plant; in the accompanying illustration, monosporangial clusters representative of both *Rhodochorton plumosum* and *R. variabile* occur on the same thallus.

Acrochaetium pectinatum (Kyl.) Ham.

Chantransia pectinatum Kylin 1906: 120. *Acrochaetium pectinatum* (Kyl.) Hamel 1927: 103; West 1968: 89. *Rhodochorton gymnogongri* Drew 1928: 175. *R. simplex* Drew 1928: 165.

Thalli 1.5–2 mm tall; base of tetrasporophyte multicellular creeping filaments, these forming small disk; base of gametophyte from single persistent spore; branching secund, alternate, or opposite, remote or crowded; cells elongate, 4.5–8.5(12) μm diam., 12–50 μm long; chloroplasts lobed, spiral in young growing apices and 1 or 2 platelike and parietal in older cells, without pyrenoids; tetrasporangia or monosporangia terminal or lateral on same or different thalli, solitary or clustered on lateral branches; carpogonia sessile, borne low on thallus; spermatangia terminal on branches, clustered.

Infrequent, epiphytic or epizoic on a variety of algae or animals, subtidal, Wash. to Channel Is. and Santa Monica, Calif.; common in Channel Is. Type locality: Sweden.

The morphological responses shown by this species to physiological changes under controlled laboratory conditions have been well documented (West, 1968).

Boillot & Magne (1973) suggest that the Pacific material identified as *A. pectinatum* is not identical with *A. pectinatum* from Europe.

Acrochaetium porphyrae (Drew) Smith

Rhodochorton porphyrae Drew 1928: 188. *Acrochaetium porphyrae* (Drew) Smith 1944: 179.

Thalli making bright red patches 1–5 cm across on *Porphyra* and other membranous algae, mostly endophytic, germinating from single spore and producing endophytic filaments throughout host, especially within cell walls; endophytic filaments producing erect filaments of 3 or 4(8) cells, these usually unbranched, bearing sporangia external to host; sporangia borne singly or in small, irregularly placed clusters; some of the deeply penetrating cells with irregular chloroplasts; superficial cells with single, parietal chloroplast, each with 1 pyrenoid; cells of filaments traversing medulla of host usually more slender (3–5 μm) than those on surface of host (8–12 μm), both kinds irregularly shaped, usually thicker through the center than at either end; cells making irregular reticulum in polar view.

Occasional on various species of *Porphyra* or other membranous red algae, low intertidal to subtidal (10 m), Hope I., Br. Columbia, to Bahía San Quintín, Baja Calif. Type locality: San Francisco, Calif.

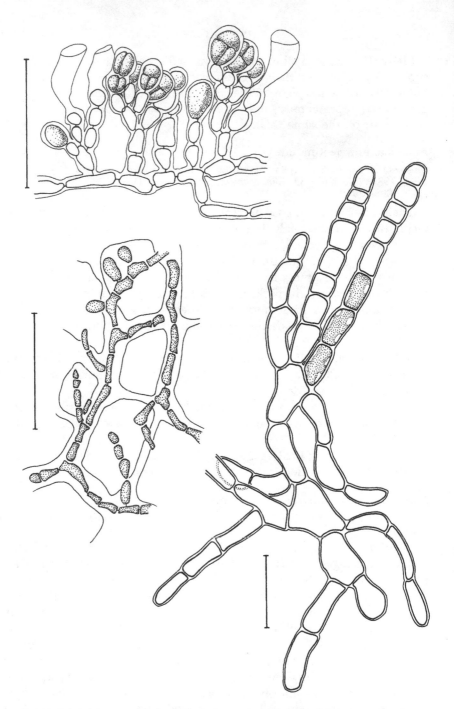

Fig. 259 (below left). *Acrochaetium porphyrae*. Rule: 50 μm. (SM/A) **Fig. 260** (below right). *A. rhizoideum*. Rule: 50 μm. (DBP/A) **Fig. 261** (above left). *A. subimmersum*. Rule: 50 μm. (CS/A)

Acrochaetium rhizoideum (Drew) Jao

Rhodochorton rhizoideum Drew 1928: 182. *Acrochaetium rhizoideum* (Drew) Jao 1937: 102 (incl. var. *rhizoideum* and var. *patens* (Drew) Smith 1944: 180).

Thalli epiphytic or endophytic in *Codium*, forming conspicuous tufts 3–4 mm tall; basal endophytic system extensively developed, entangled, penetrating deeply between utricles of host, usually colorless; cells irregular in shape, 25–33 μm diam.; erect branches produced from endophytic filaments, nearly uniform in width from base, branched strongly as they emerge from host; cells 16–19 μm wide, 2–4 times as long; chloroplast single, band-shaped or laminate, parietal, with or without pyrenoids; monosporangia terminal or lateral, rarely sessile.

Frequent on *Codium fragile*, occasional on other algae, low intertidal, Wash. to I. Cedros, Baja Calif. Type locality: Santa Catalina I., Calif.

Acrochaetium subimmersum (S. & G.) Papenf.

Rhodochorton subimmersum Setchell & Gardner 1903: 347. *Acrochaetium subimmersum* (S. & G.) Papenfuss 1945: 318.

Thalli forming small, deep red patches, these less than 1 cm broad, but frequently coalescing; greater part of thallus endophytic within cortical layers of host, occasionally penetrating to medulla and crossing to cortical layer of other side; most cells irregular in shape, thicker through middle than at either end, of at least 2 size ranges, 3–5 μm diam. and 5–8 μm diam., these sizes intermingled in outer cortical cells of host; free filaments usually unbranched if not fertile, the cells 4–6 μm diam. and long, nearly uniform in diam. from base to apex; chloroplast single, parietal, without pyrenoid; fertile filaments unilaterally branched near apex, the tetrasporangia borne terminally in these branching clusters.

Frequent endophyte in bladelike red algae, particularly *Grateloupia*, *Halymenia*, *Prionitis*, and *Schizymenia*, probably throughout the range of these hosts, low intertidal, Br. Columbia to Channel Is., Calif. Also known from Japan. Type locality: Whidbey I., Wash.

Acrochaetium tenuissimum (Coll.) Papenf.

Chantransia virgatula f. *tenuissima* Collins, P.B.-A., 1895–1919 [1900]: no. 741. *Acrochaetium tenuissimum* (Coll.) Papenfuss 1945: 319. *Rhodochorton tenuissimum* (Coll.) Drew 1928: 170.

Thalli microscopic, the bases of creeping filaments, producing erect filaments with few branches and no branchlets; cells cylindrical, 6.5–8 μm diam., decreasing slightly from base to apex of filaments, the length 4–6 times diam.; chloroplast single, parietal in cells of filaments, apical in sporangia; sporangia sessile, scattered along main filaments and branches, elliptical, 12–15 μm wide, 21–23.5 μm long. (After Drew.)

Known from a single collection on *Phyllospadix* near entrance to harbor at San Pedro, Calif.

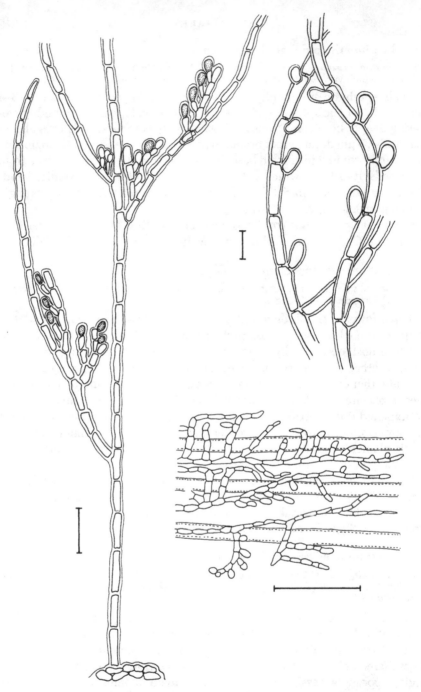

Fig. 262 (above right). *Acrochaetium tenuissimum*. Rule: 50 μm. (SM, after Drew)
Fig. 263 (left). *A. thuretii*. Rule: 50 μm. (SM/A)   **Fig. 264** (below right). *A. vagum*.
Rule: 50 μm. (DBP, after Drew)

Acrochaetium thuretii (Born.) Coll. & Herv.

Chantransia efflorescens var. *thuretii* Bornet 1904: xvi. *Acrochaetium thuretii* (Born.) Collins & Hervey 1917: 97. *C. thuretii* β *agama* Rosenvinge 1909–31 (1909): 102. *Rhodochorton thuretii* var. *agama* (Rosenv.) Drew 1928: 171.

Thalli wholly epiphytic; filaments to 4–5 mm long; base composed of creeping filaments entirely superficial to host; erect filaments branched; branches alternate and secund, usually ending in hairlike prolongations; cells cylindrical, 10–11 μm diam., 3–5 times as long; chloroplast single, parietal with 1 pyrenoid; monosporangia usually in pairs, 1 terminal and 1 lateral on single unicellular branchlet; branchlets in short, secund series of 2 or 3 on adaxial side of base of branches, very rarely rebranched; monosporangia 10–11 μm diam., 17–22 μm long; tetrasporangia subglobular, 21–22 μm diam., 25–26 μm long; sexual reproduction unknown for Calif. material. (After Drew.)

Infrequent, low intertidal and subtidal, epiphytic on *Zostera* or *Egregia*, Santa Cruz I. and San Pedro, Calif. Also N. Atlantic, S. America, and Australia. Type locality: Atlantic coast of France.

Acrochaetium vagum (Drew) Jao

Rhodochorton vagum Drew 1928: 188. *Acrochaetium vagum* (Drew) Jao 1937: 111. *R. implicatum* Drew 1928: 190. *R. penetrale* Drew 1928: 187. *A. ascidiophilum* Dawson 1953a: 21.

Thalli wholly or partly endophytic or endozoic, with creeping, irregularly branched filaments, these sometimes forming a pad; filaments either superficially embedded only or deeply embedded only; erect filaments occasionally sparingly branched, branchlets with terminal or lateral sessile or pedicellate sporangia; chloroplast single, mostly parietal, rarely stellate, the pyrenoids 1 or none; cells 4–8 μm diam., varying in width on same thallus.

Rare, but abundant when found, low intertidal (on *Pterosiphonia*, *Grateloupia*, *Halosaccion*, or tunicates) to subtidal (on *Sertularia*), Wash. to Channel Is., Calif., and Baja Calif.; in C. Calif., usually found in old, eroded thalli of *Halosaccion*. Type locality: Cape Flattery, Wash.

Rhodochorton Nägeli 1862

Thalli diminutive, 1–3(4) cm tall, filamentous, tufted, with prostrate creeping portions bearing erect filaments, these sparingly branched. Chloroplasts discoid, few to many per cell, small, without pyrenoids. Tetrasporangia cruciately divided, or plants rarely bearing bisporangia in field collections. Gametophytes in some species, these known only in culture, less than 200 μm long, dioecious. Spermatangia clustered terminally on short, erect axes. Carpogonia sessile on prostrate filaments, or on short, secondary, erect branches, single or clustered and terminal. Intercalary cells of prostrate filaments occasionally differentiated into carpogonia. Zygote after fertilization dividing transversely; upper cell producing primary goni-

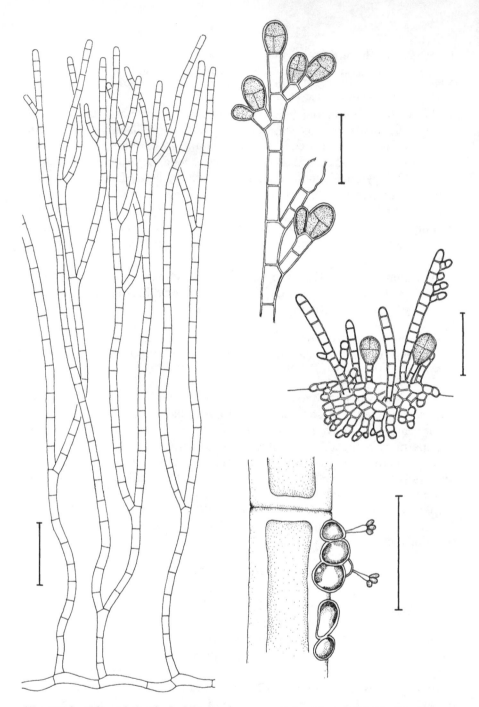

Fig. 265 (middle right). *Rhodochorton concrescens.* Rule: 50 µm. (JRJ/S) **Fig. 266** (left & above right). *R. purpureum.* Rules: habit, 100 µm; detail, 50 µm. (both JRJ/S) **Fig. 267** (below right). *Kylinia rosulata.* Rule: 25 µm. (SM/A)

moblast initial; lower cell repeatedly producing secondary gonimoblast initials, these dividing without branching and forming terminal tetrasporangia; at this stage this portion of plant sometimes becoming independent.

Rhodochorton concrescens Drew

Drew 1928: 167; Smith 1944: 183.

Thalli small, discoid, the prostrate portions several cells thick; erect filaments few, usually unbranched, if branched the branches of short cells 6–10.5 µm diam., these nearly uniform in diam. from base to apex of branch; tetrasporangia terminal on erect axes or terminal on short, subterminal laterals, cruciately divided, (16)22 µm diam., 20–28 µm long; sexual plants unknown, not produced in culture.

Infrequent, epizoic on various invertebrates, low intertidal to subtidal (10–20 m), Sonoma Co., Monterey Peninsula, and Santa Rosa I., Calif. Known also from New Zealand and Scotland. Type locality: Carmel Bay, Calif.

Rhodochorton purpureum (Lightf.) Rosenv.

Byssus purpurea Lightfoot 1777: 1000. *Rhodochorton purpureum* (Lightf.) Rosenvinge 1900: 75; Papenfuss 1945: 327 (incl. synonymy). *R. tenue* Kylin 1925: 44; West 1969: 12.

Thalli tufted, densely intertwined basally, with erect filaments 10–15 mm tall; filaments sparingly branched, the secondary branchlets nearly the size of primary axes; cells (10)15–20 µm wide, 1.5–3 times as long; tetrasporangia in compact clusters in upper portions of plants; sexual reproduction as described for genus.

Common, especially in shaded areas such as shallow caves, high intertidal, Humboldt Co. to San Diego Co., Calif.; especially common on shale in Santa Cruz Co. Widely distributed in N. Hemisphere. Type locality: Scotland.

Kylinia Rosenvinge 1909

Thalli minute, epiphytic, creeping, little branched, growing from single attaching cells. Each cell with 1 parietal chloroplast, occasionally with a pyrenoid-like body. Monosporangia spherical, terminal or lateral. Spermatangia frequent, 1 to several on long, hyaline stalks (androphores). Carpogonia rare, sessile, 1-celled. Zygote after fertilization not dividing but producing carposporangia directly.

Kylinia rosulata Rosenv.

Rosenvinge 1909–31 (1909): 141; J. Feldmann 1958: 493.

Thalli microscopic creeping filaments, with spherical to ovoid cells 5–8 µm diam. and with short prostrate branches formed in 1 plane, closely appressed to substratum; monosporangia spherical, 6–8 µm diam., terminal or lateral on short branches, occurring singly or in pairs; androphores of spermatangia to 10 μm long; carposporangia 10–15 μm diam.

Locally abundant, on filaments of *Sporochnus pedunculatus*, subtidal at 6 m off Little Harbor (Santa Catalina I.), Calif., and on same host at 10 m off Bahía Viscaino, Baja Calif. Type locality: Denmark, at 16 m, also on *S. pedunculatus*.

Boillot and Magne (1973), using plants identified with this species, were able to grow carpospores that germinated to form tetrasporangial thalli they identified with those of *Acrochaetium pectinatum*.

Family NEMALIACEAE

Thalli multiaxial, usually gelatinous. Medulla consisting of firm core of elongate, parallel, periclinal filaments, these branching radially to form cortex of photosynthetic filaments. Cortex branched dichotomously in lower portions, the branching irregular and clustered terminally. Chloroplast 1 per cell, lobed, axile or parietal, usually with single pyrenoid. Tetrasporangial phase (in *Nemalion*) an alternate microscopic phase resembling *Acrochaetium*, with tetrasporangia cruciately divided. Gametangial phase monoecious, protandrous, or dioecious. Spermatangia in dense clusters terminal on cortical branches. Carpogonial branch of 2 or 3 cells, these usually terminating modified vegetative filament. First division of zygote after fertilization transverse, the upper cell producing by repeated divisions a small gonimoblast, all or most cells becoming carposporangia. Few or no sterile filaments produced basal to carpogonium and hypogynous cell.

Nemalion Targioni-Tozetti 1818

Thalli cylindrical. Gametangial phase usually occurring annually, unbranched, little branched, or much branched, if branched the branching irregular, usually widely spaced, or mostly at tops of plants. Multiaxial; medulla of elongate, colorless filaments; cortex of branched pigmented filaments; chloroplast stellate, 1 per cell. Gametangial plants monoecious, protandrous, or dioecious; carpogonial branches of 3–4 cells with 0–2 subtending sterile cells. Sporangial plants microscopic, with erect filaments produced from creeping portion, these bearing cruciately divided tetrasporangia.

Nemalion helminthoides (Vell.) Batt.

Fucus helminthoides Velley 1792: 255. *Nemalion helminthoides* (Vell.) Batters 1902: 59; Hollenberg & Abbott 1966: 45.

Thalli usually growing in groups, 20–45(135) cm tall, golden brown; plants frequently overgrown with epiphytes; thalli soft, cylindrical, 1–4(6) mm diam., usually unbranched, occasionally branched in upper portions; acrochaetioid tetrasporangial phase unknown in nature.

Seasonally abundant, discontinuously distributed from Sitka, Alaska, to I. Cedros, Baja Calif.; in Calif., from Humboldt Co. to San Diego on high rocks. Widely distributed in temperate seas. Type locality: England.

In California, generally occurring in somewhat sheltered areas, commonly found with, but not as common as, *Cumagloia andersonii*.

Family HELMINTHOCLADIACEAE

Thalli soft and gelatinous or not gelatinous; some species calcified. Vegetative structure multiaxial with radiating cortical filaments, containing cells with single, large, parietal chloroplast, usually without pyrenoid. Microscopic phase bearing tetrasporangia known for some species, but not for Calif. species. Tetrasporangia unknown in some genera. Spermatangia in terminal or subterminal clusters on outermost cortical cells, not as dense as in *Nemalion*. Carpogonia differing from those of Nemaliaceae in ontogeny and final position of carpogonial branch, this being lateral to a vegetative filament and not itself a modified vegetative filament. Development of gonimoblast like that of *Nemalion* but usually only terminal cells becoming carposporangia; relatively large and variously derived group of sterile filaments surrounding gonimoblast.

1. Thalli calcified, chalklike, not gelatinous; regularly to irregularly dichotomously branched .... *Liagora* (p. 328)
1. Thalli not calcified, soft and gelatinous; radially to dichotomously branched 2
  2. Axes frequently compressed and saccate; plants in high intertidal....
  .... *Cumagloia* (p. 329)
  2. Axes cylindrical; plants in low intertidal 3
3. Plants epiphytic on the surfgrass *Phyllospadix*, less than 8 cm tall; cystocarps with dense involucre of sterile filaments........ *Helminthora* (below)
3. Plants saxicolous, more than 8 cm tall; cystocarps with weakly developed involucre of sterile filaments or involucre lacking ....
.... *Helminthocladia* (p. 327)

Helminthora J. Agardh 1851

Thalli of small to moderate size, erect, either dichotomously or radially branched. Medulla of periclinally directed filaments, these easily separated from one another. Cortex of few to many subdichotomously divided filaments. Chloroplast single, without pyrenoid. Tetrasporangia on microscopic acrochaetioid phase known only in culture. Terminal cells of cortex forming spermatangia in tufted clusters, or clusters paniculate. Carpogonial branch lateral to a vegetative filament, borne low on branch, of 4 or 5 cells. Carpogonium distinctively dumbbell-shaped, the zygote after fertilization dividing transversely, the upper cell forming gonimoblast, the terminal cells and a few subterminals becoming carposporangia. Elaborate sterile filaments formed by cells adjacent to carpogonial branch, most of these pro-

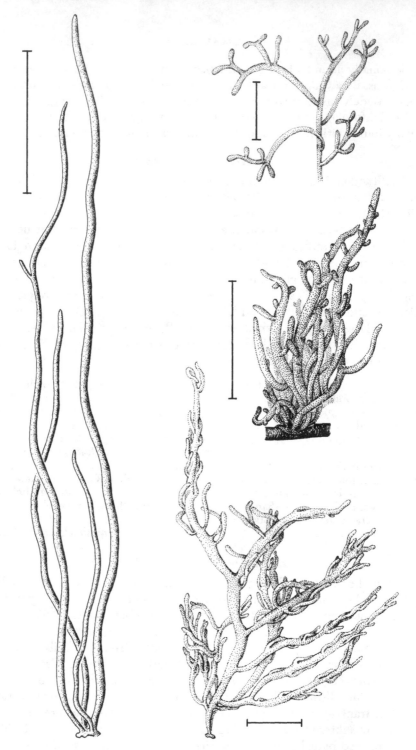

Fig. 268 (left). *Nemalion helminthoides.* Rule: 5 cm. (JRJ/S) **Fig. 269** (two above right). *Helminthora stricta.* Rules: both 5 mm. (both DBP/A) **Fig. 270** (below right). *Helminthocladia australis.* Rule: 10 mm. (DBP)

duced toward thallus surface and surrounding developing gonimoblast, a smaller number produced downward from the initial sterile filaments that form ring around carpogonial branch.

Helminthora stricta Gardn.

Gardner 1926: 205; Abbott 1965: 90. *Helminthora saundersii* Gardn. 1926: 205.

Thalli lubricous, gregarious, attached by small disk 0.25 mm diam., pinkish tan, showing 2 types of branching (which may indicate sexual nature); cystocarpic thalli regularly radially branched, to 5 cm tall; monoecious thalli loosely and irregularly branched, to 3 cm tall; both types of thalli appearing pinnate when pressed; spermatangia in terminal clusters; carpogonial branch of 4(5) cells; very conspicuous sterile filaments produced around gonimoblast; in some specimens gonimoblast remaining after discharge of terminal and subterminal carposporangia.

Infrequent, on *Phyllospadix*, low intertidal, Monterey, Ventura, Santa Barbara, and Channel Is. to La Jolla (type locality), Calif. *H. saundersii*, known from the type specimen from Monterey, forms the northern distribution record.

Helminthocladia J. Agardh 1851

Thalli little to much branched, but only to 3 or 4 orders of branches. Medulla forming a core of slender, tightly woven longitudinal filaments. Cortex of dichotomously divided filaments, the outermost of these obovate to clavate in shape, this characteristic of genus. Chloroplast 1 per cell, parietal or irregularly lobed, with or without pyrenoid. Tetrasporangia rare, in some species occurring in position normally occupied by carposporangia. Spermatangia in terminal tufts, some occurring in short panicles. Carpogonial branch of 3 or 4 cells, lateral on low or medianly placed vegetative filament. Zygote after fertilization dividing obliquely (occasionally appearing to divide longitudinally) and forming relatively small gonimoblast, this with terminal carposporangia. Sterile filaments varying from few to many.

Helminthocladia australis Harv.

Harvey 1863: 39; Womersley 1965: 470. *Helminthocladia australis* f. *californica* J. Agardh 1899: 96. *H. californica* (J. Ag.) Kylin 1941: 6; Abbott 1965: 95. *H. gracilis* Gardner 1926: 206.

Thalli pyramidal, 15–20(36) cm tall, with cylindrical branches 2–4(6) mm diam., the branching of 3 or 4 orders, the density of branching varying with specimens and possibly related to habitat; older thalli sometimes hollow owing to loss of medullary filaments, otherwise the medulla consisting of slender, tightly woven filaments, these very tough and slippery; cortical filaments 4 or 5 times dichotomously branched, the terminal branches frequently clustered irregularly; terminal cells of filaments in Calif. material obovate, 50 μm diam. at top, tapering to 10(12) μm; spermatangia in short, paniculate clusters; carpogonial branch of 3 cells, the carpogonium dividing longitudinally after fertilization; small gonimoblast formed with terminal carposporangia; basal and adjacent cells of carpogonial branch frequently fusing; sterile filaments varying in number, in Calif. material not conspicuous because of resemblance to surrounding vegetative filaments.

Infrequent, saxicolous, low intertidal, Santa Barbara, Calif., to Pta. Pequeña, Baja Calif., into Gulf of Calif. Also known from S. Africa, Japan, New Zealand, and Australia. Type locality: W. Australia.

Liagora Lamouroux 1812

Thalli calcified, erect, cylindrical to slightly flattened, repeatedly branched. Branches usually tapering to apex, either unbranched at apex or bifurcate. Medulla of slender to moderately thickened elongate cells, these interwoven, frequently calcified. Cortex lightly to heavily calcified, of dichotomously branched ovate to spherical cells, the terminal cells usually smaller than the subterminal. Chloroplast single, large, parietal, usually without pyrenoid. Plants usually dioecious. Spermatangia in clusters terminating cortical cells. Carpogonial branches of 3–5 cells, borne laterally. Zygote dividing transversely after fertilization. Gonimoblast relatively large, usually surrounded by branched, sterile filaments, these in some species stopping below carposporangia, in other species exceeding

Fig. 271. *Liagora californica*. Rule: 2 cm. (SM)

carposporangia. Carposporangia usually of undivided terminal cells, but some species divided cruciately (carpotetraspores). A microscopic, *Acrochaetium*-like, filamentous stage with tetrasporangia known for 2 species.

Liagora californica Zeh

Zeh 1912: 271; Dawson 1953a: 42.

Thalli bushy, compressed below, terete above, repeatedly dichotomously branched, 6–12(16) cm tall, with bright red apices, grayish-white below where heavily calcified; medulla calcified throughout thallus; cortical filaments branched 3 or 4 times; cells ovate to rectangular; tetrasporangia unknown; plant dioecious; spermatangia clustered on modified terminal cortical cells; carpogonial branch of 4 or 5 cells; gonimoblast surrounded by large number of sterile filaments.

Infrequent, saxicolous, lower intertidal to upper subtidal, Santa Catalina I. and Pt. Loma (San Diego Co.), Calif., and I. Guadalupe, Baja Calif. Type locality: Avalon (Santa Catalina I.), Calif.

Cumagloia Setchell & Gardner 1917

Thalli gregarious, with 1 or more fronds pendant from small, discoid holdfasts. Fronds cylindrical to strongly flattened, unbranched or sparingly branched, densely covered with fine, spinelike branchlets with furcate apices. Internally multiaxial, the longitudinally directed filaments slender and bearing cortical filaments dichotomously or trichotomously branched; outermost cells frequently larger than subterminal cells. Chloroplast single, laminate, parietal, without pyrenoid. Tetrasporangia unknown. Plants monoecious or dioecious. Spermatangia in irregular terminal clusters on cortical cells. Carpogonial branch 3-celled, borne laterally, low on cortical filaments. Zygote after fertilization dividing transversely, or not dividing, then forming horizontal filaments, which in turn bear short vertical filaments, these cutting off single carposporangia in sympodial fashion. Carpospores on germination producing acrochaetioid growth, their fate unknown. Poorly developed group of sterile filaments produced immediately adjacent to basal portion of carpogonial branch and not developing further.

Cumagloia andersonii (Farl.) S. & G.

Nemalion andersonii Farlow 1877: 240. *Cumagloia andersonii* (Farl.) Setchell & Gardner 1917d: 399; Smith 1944: 188.

Thalli (in Calif. material) annual, 15–30(90) cm tall, olive brown to deep purplish-red, gelatinous but tough; young axes cylindrical, 1–8 mm diam., older axes flattened, 1–3 cm wide, reaching maximum development in August in C. Calif.

Locally abundant in season, saxicolous, upper intertidal, Hope I., Br. Columbia, to Cabo Colnett, Baja Calif.; commonly found year after year on same rocks. Type locality: Santa Cruz, Calif.

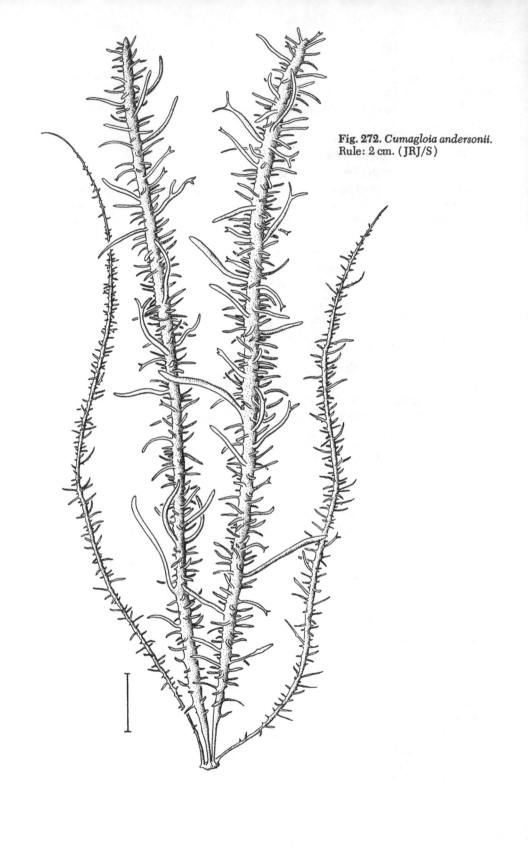

Fig. 272. *Cumagloia andersonii.*
Rule: 2 cm. (JRJ/S)

Family CHAETANGIACEAE

Thalli erect, dichotomously branched, multiaxial with central axis of stout, periclinally directed filaments. Cortex filamentous, of several simple or divided cell rows, some showing limited growth and becoming outermost layers, others extending through outermost layer and showing unlimited growth. Outermost layer of large, colorless, inflated cells ("utricles"), these appearing parenchymatous in surface view, calcified in some genera. Cells usually with 1 stellate chloroplast. Reproduction in some species by monospores. Tetrasporangia in some species borne on microscopic filamentous phase; in other species spore-bearing plants similar to gametangial plants; in some genera tetrasporangia unknown. Spermatangia in sori or scattered over thallus surface. Carpogonial branch 3-celled, distinguished by ring of nutritive cells surrounding hypogynous cell. Gonimoblasts embedded in modified conceptacles with terminal cells or rows of cells forming carposporangia. Sterile filaments surrounding spore masses, functioning as pericarp; without conspicuous ostiole.

1. Outermost cortical layer of inflated colorless cells overtopped with short
   filaments of moniliform cells ....*Pseudogloiophloea* (p. 334)
1. Outermost cortical layer of angular, inflated cells, these not overtopped
   with filaments of moniliform cells 2
   2. Gonimoblast radiating from 1 central point ....*Scinaia* (below)
   2. Gonimoblast radiating from central and several lateral points......
   .... *Pseudoscinaia* (p. 333)

Scinaia Bivona 1822

Thalli cylindrical, constricted or not constricted, multiaxial. Medulla consisting of relatively few longitudinal filaments; cortex differentiated into inner layer of loosely entwined, slender filaments, and an outer layer of variously shaped photosynthetic cells 2 or 3 cells thick. Outermost layer in mature portions of axes consisting of colorless inflated cells, these squarish in cross section in some species, palisade-like in others and forming an epidermis. Monosporangia present in some species. Tetrasporangia borne on microscopic, filamentous phase known in 2 species, but unknown in material from Calif. Spermatangia in continuous surface sori. Gonimoblasts embedded in outer cortical cells, surrounded by sterile filaments.

1. Thallus not constricted; branches fanlike ....S. *johnstoniae*
1. Thallus regularly and clearly constricted 2
   2. Branches cylindrical, 3–5 mm diam.; plants short ....S. *articulata*
   2. Branches flattened, 5–10 mm diam.; plants tall and large.....S. *latifrons*

Scinaia articulata Setch.

Setchell 1914b: 109; Dawson 1949b: 23.

Thalli 10 cm tall, regularly dichotomous, constricted; segments cylindrical, 3–5 mm wide; branches slightly attenuated below; axis conspicuous throughout; utricles slightly rectangular in section; spermatangia in sori, these forming caps over apices of mature branches; cystocarps sparse, scattered, broadly pyriform, flattened-globular, abruptly narrowed to short, broad openings; gonimoblast filaments numerous. (After Setchell.)

Rare, saxicolous, subtidal, Santa Barbara I. and Santa Barbara (type locality), Calif.

Scinaia johnstoniae Setch.

Setchell 1914b: 97; Dawson 1953a: 44.

Thalli 6–13 cm tall, arising from small, discoid holdfast of 1–2 mm diam.;

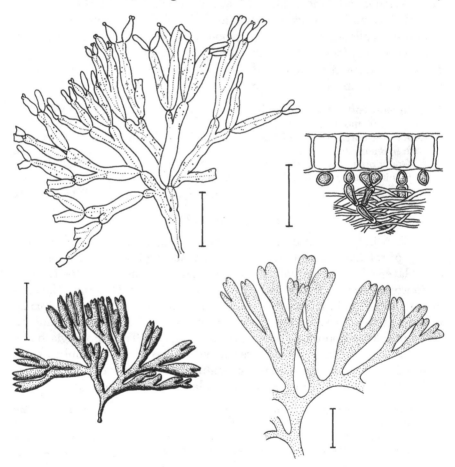

Fig. 273 (two at top). *Scinaia articulata*. Rules: habit, 2 cm; cross section of cortex, 50 μm. (SM & CS; both after Setchell)   Fig. 274 (below left). *S. johnstoniae*. Rule: 2 cm. (CS)   Fig. 275 (below right). *S. latifrons*. Rule: 2 cm. (CS)

axes dichotomously branched, without constrictions, the branches cylindrical, attenuated downward from 3.5–5 mm to 1–2 mm diam., penultimate branches fan-shaped, the apices acute; utricles squarish to slightly flattened in section; monosporangia produced on cystocarpic plants, borne in groups of 3 or 4 on special cortical filaments extending between utricles; cystocarps scattered throughout thallus, embedded in medulla. (After Setchell & Dawson.)

Saxicolous, upper subtidal, San Pedro (type locality), Balboa (Orange Co.), and La Jolla, Calif., and Baja Calif. to Gulf of Calif.; rare in Calif., more common southward.

Scinaia latifrons Howe

Howe 1911: 500; Setchell 1914b: 102; Dawson 1953a: 46.

Thalli 10–20 cm tall, arising from small, fleshy holdfasts of 2–3 mm diam., short-stipitate; axes dichotomously to irregularly dichotomously branched several times; branches constricted, flattened, 5–10 mm wide, usually broader through center of a segment than at either end; utricles flattened, square to rectangular in section; spermatangia in dense sori, occupying margins from base to apex; cystocarps conspicuous, aggregated along margins. (After Dawson.)

Saxicolous, subtidal, Channel Is., Calif., and Gulf of Calif.; rare in Calif., more common to south. Type locality: La Paz, Mexico.

Pseudoscinaia Setchell 1914b

Thalli arising from disk, cylindrical, repeatedly dichotomously branched, with blunt apices and branches of same diam. throughout; medulla of numerous stout, broad, parallel filaments; slender photosynthetic filaments protruding between utricles; spermatangia isolated or in small clusters scattered over outer layers; gonimoblasts scattered, originating within cortex, more or less globular to pyriform, with more or less elongated carpostome; gonimoblast filaments projecting into and also lining walls of sporiferous cavity, variously grouped and successively cutting off more or less elongate spores. (After Setchell.)

Pseudoscinaia snyderiae Setch.

Setchell 1914b: 120.

Plant dark red, 12–20 cm tall, 9–13 times regularly dichotomously branched, the branches not narrowed downward but nearly uniform in diam. throughout; cortex broad, the cells compact; utricles colorless, irregular in shape with frequent, scattered, slender, protruding pigmented cells.

Rare, cast ashore at San Pedro, La Jolla, and Pacific Beach (San Diego Co.; type locality), Calif.

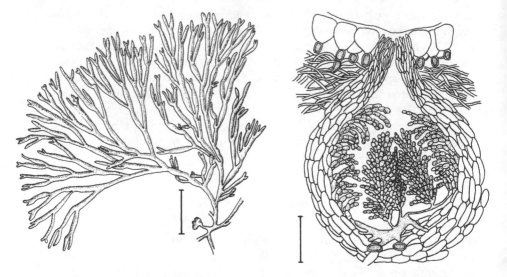

Fig. 276. *Pseudoscinaia snyderiae.* Rules: habit, 2 cm; longitudinal section of female conceptacle, 100 μm. (both CS, after Setchell)

Distinguished from *Scinaia* chiefly in having the gonimoblast filaments radiating from several focal points; in *Scinaia* there is a single central focus for gonimoblast initiation. However, detailed studies of these processes have never been made, and *Pseudoscinaia* has never been collected since first described.

Pseudogloiophloea Levring 1956

Thalli erect, solitary, repeatedly dichotomously branched. Branches cylindrical, tapering toward apex, or not tapering. Medulla thin, of few filaments. Cortex occupying most of axis, the outermost cells inflated and forming colorless utricles, these interspersed and overtopped with photosynthetic filaments continuous with inner cortical filaments, the successive layers of colored and colorless cells occurring on mature thallus. Surface of young thallus appearing parenchymatous in surface view. Monosporangia formed on gametophyte and on tetrasporophyte. Plants monoecious or dioecious. Spermatangia terminal on cortical cells protruding beyond utricles, forming nearly continuously over thallus surface. Carpogonial branch 3-celled, with ring of hypogynous cells. Gonimoblasts embedded, only the terminal cells becoming carposporangia. Young gametangial and sporangial plants microscopic, filamentous, each at times bearing monosporangia. Germinating monospores sometimes reproducing filamentous phase directly, under other conditions producing gametangial phase directly, or

producing tetrasporangia, these germinating to form microscopic gametangial phase, followed by adult plant.

Pseudogloiophloea confusa (Setch.) Levr.

Gloiophloea confusa Setchell 1914b: 118; Smith 1944: 190. *Pseudogloiophloea confusa*
(Setch.) Levring 1956: 8; Ramus 1969: 1.

Thalli dark red, 3–15 cm tall; subtidal plants taller .than intertidal specimens, those of intertidal exposed to wave shock shorter than all others and
tending to dark brownish-red; axes regularly dichotomously branched,
often with proliferous clusters in basal portions of branches; apices of
branches tapering, otherwise branches (1)2–5 mm diam., of nearly uniform diam. throughout; cortex thick, loosely arranged.

Infrequent, but widely distributed, saxicolous, mid- to subtidal (10 m),
Hope I., Br. Columbia, to Pta. San Quintín, Baja Calif.; more collections
known from Monterey southward than to north. Type locality: Monterey,
Calif.

Except for being shorter, these plants are externally very similar to
Pseudoscinaia snyderiae. Ramus (1969) found monoecious plants from
Jalama (Santa Barbara Co.), Calif., to be slightly different in anatomy and
response to culture conditions from dioecious plants collected at Moss
Beach (San Mateo Co.), Calif.; these differences have not been resolved
as taxonomic distinctions.

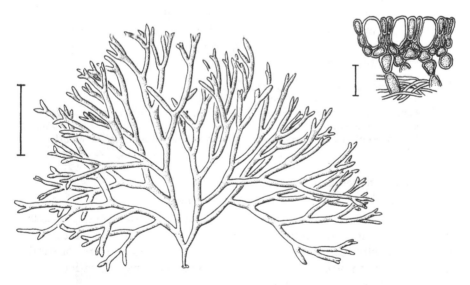

Fig. 277. *Pseudogloiophloea confusa*. Rules: habit, 2 cm; cross section of cortex, 100 μm.
(JRJ/S; CS, after Setchell)

Family BONNEMAISONIACEAE

Gametangial thalli erect, uniaxial, either cylindrical and radially branched or flattened and bilaterally branched. Cortex appearing parenchymatous, externally with mixture of large and small cells, occasionally with secretory (vesicular) cells, these with highly refractive contents. Chloroplasts many per cell, discoid, without pyrenoid. Plants usually dioecious. Spermatangia occupying nearly entire branch or branchlet. Carpogonial branch 3-celled, surrounded by tuft of nutritive filaments; after fertilization numerous fusions of these structures occur during development of gonimoblast, only the terminal cells becoming carposporangia. Tetrasporangia formed on independent, microscopic, filamentous phase.

Two genera in the California flora, *Bonnemaisonia* (below) and *Asparagopsis* (p. 340).

Bonnemaisonia C. Agardh 1822

Thallus flattened, bilaterally branched. Apical cell of branch dividing obliquely and producing axial cell rows, each cutting off 2 opposite pericentral cells, these becoming basal cells of determinate (short) and indeterminate (long) secondary branches, 1 flank of branch thus bearing alternately short and long secondary branches, opposite flank bearing long and short branches. Short branches frequently reduced to spinelike structures. Sporangial stage diminutive, filamentous, described (on p. 338) as *"Trailliella intricata."* Spermatangia on modified, clavate, indeterminate branch. Gonimoblasts also on modified indeterminate branches, urnshaped; carposporangia relatively few, large, these surrounded by pericarp, frequently with excentrically placed ostiole.

1. Branches without tendril-like hooks . . . . *B. geniculata*
1. Some branches with tendril-like hooks 2
    2. Plants saxicolous; axes delicate, mostly less than 1 mm diam.; branching from base to apex . . . . *B. nootkana*
    2. Plants epiphytic; axes more than 1 mm diam.; branching frequently interrupted . . . . *B. hamifera*

Bonnemaisonia geniculata Gardn.

Gardner 1927a: 336; Smith 1944: 192; Hudson & Wynne 1969: 207.

Thalli 10–12 cm tall, pinkish-red; central axis percurrent, markedly flattened, 2–6 mm wide; branching subopposite, distichous; plant dioecious; male plants with spermatangia occupying most of branch except for sterile apex and lowest portion; gonimoblasts elongate, urn-shaped, with delicate pericarp and few carposporangia.

Rare, epiphytic on *Erythrophyllum delesserioides*, Mendocino Co. (Mendocino and Ft. Bragg) and Monterey Co. (Pescadero Pt. and Pt. Sur, type locality), Calif.

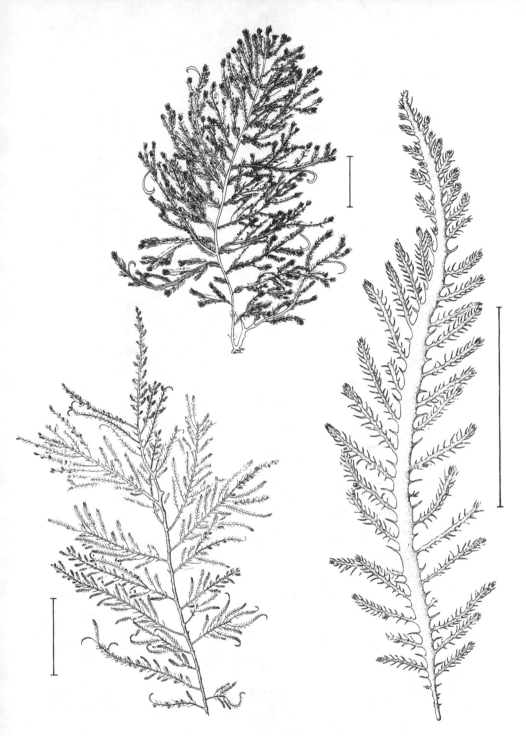

Fig. 278 (below right). *Bonnemaisonia geniculata*. Rule: 2 cm. (JRJ/S) **Fig. 279** (above).
B. hamifera. Rule: 2 cm. (SM, after Okamura) **Fig. 280** (below left). *B. nootkana*. Rule:
2 cm. (JRJ/S)

Bonnemaisonia hamifera Har.

Hariot 1891: 223; Dawson 1953a: 55; Chihara 1961: 125.

Thalli 7–15 cm tall, pinkish-tan; branching opposite, distichous but mostly obscure because of density and frequent entangling; branchlets often with conspicuous tendril-like, hooked ends; main axes conspicuous, 1–2 mm wide, frequently lacking branches in lower portions; upper portions plumose; tetrasporangia borne on alternating microscopic stage (*"Trailliella intricata"*); spermatangia occupying entire somewhat clavate branch; gonimoblasts urceolate; sexual plants not observed in Calif. specimens.

Infrequent, epiphytic, low intertidal to subtidal (20 m), Goleta (Santa Barbara Co.) Channel Is. (Santa Rosa, San Nicolas, and Santa Cruz), Calif., to Pta. San Quintín, Baja Calif. Known also from N. Atlantic and W. Pacific. Type locality: Yokosuka, Japan.

Bonnemaisonia nootkana (Esp.) Silva

Fucus nootkanus Esper 1797–1808 (1802): 30. *Bonnemaisonia nootkana* (Esp.) Silva 1953: 225. *B. californica* Buffham 1896: 181; Smith 1944: 192.

Thalli delicate, finely branched, 15–20(28) cm tall, orange-red to dark red when fresh, drying to dark rose or red; main axes rarely more than 1 mm diam., with 3 or 4(5) orders of branching; branching opposite, distichous, with 1 branch of each opposite pair simple, the other branched; occasional branch ends inflated and forming a hook; some thalli partly ball-like owing to hooks attaching to adjacent branches and thalli; tetrasporangia borne on microscopic stage (*"Trailliella intricata"*); spermatangia borne on and entirely occupying an indeterminate branch; gonimoblasts rounded, not as elongate as those of *B. geniculata*.

Locally abundant, saxicolous, subtidal (5–40 m), Vancouver I., Br. Columbia (type locality), to Pta. San Quintín, Baja Calif.

Tetrasporophyte Stage of *Bonnemaisonia*

"Trailliella intricata" Batt.

Batters 1896: 10; Dixon 1959: 342.

Thalli forming soft, delicate balls 1–2 cm diam., filamentous, uniseriate, irregularly branched; cells 25–35 μm diam., 1.5–2.5 times as long, constricted between successive cells with globular, highly refractive vesicular cells wedged between lateral corners of constrictions, these alternate or irregular; tetrasporangia occasional, cruciately divided. Calif. material usually sterile.

This entity has been definitely demonstrated to be the sporophytic phase of *Bonnemaisonia hamifera* and *B. nootkana*; the sporophytic phase of *B. geniculata* has never been cultured.

Rare, epiphytic, Vancouver I., Br. Columbia; subtidal (10 m) on *Dictyopteris*, Santa Catalina I., Calif.; low intertidal to subtidal on other algae,

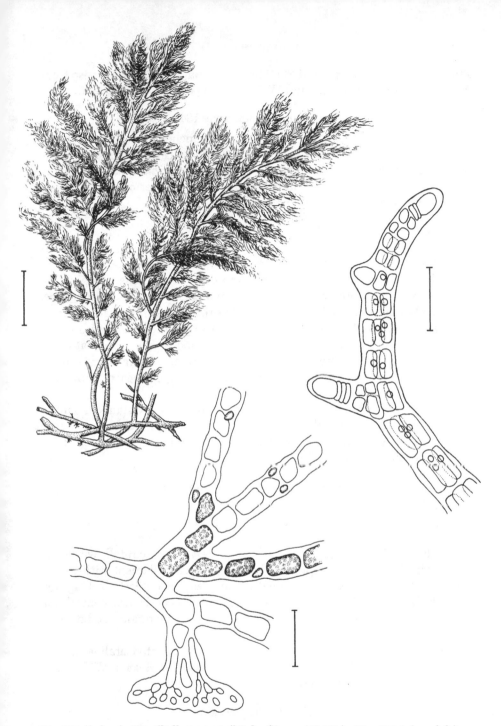

Fig. 281 (below). *"Trailliella intricata."* Rule: 50 µm. (SM/A)   **Fig. 282** (above left). *Asparagopsis taxiformis.* Rule: 2 cm. (SM)   **Fig. 283** (right). *"Falkenbergia hillebrandii."* Rule: 50 µm. (SM/A)

Pta. Banda, I. Guadalupe, and Pta. Malarrimo, Baja Calif. Known also from
N. Atlantic, Mediterranean, and W. Pacific. Type locality: England.

Asparagopsis Montagne 1840

Thalli with axes cylindrical, erect, radially branched, arising from creeping stoloniferous portion, this anchored by rhizoids. Tetrasporangia on free-
living, microscopic, filamentous stage (*"Falkenbergia hillebrandii"*). Spermatangia in short, catkinlike branchlets borne on indeterminate branches.
Gonimoblasts terminal on similar short branchlets.

Asparagopsis taxiformis (Del.) Trev.

Fucus taxiformis Delile 1813: 295. *Asparagopsis taxiformis* (Del.) Trevisan 1845: 1;
Dixon 1964b: 902. *A. sanfordiana sensu* Okamura 1907–42 (1909): 135.

Thalli 10–20(43) cm tall, with several to many erect, generally plumose
fronds arising from creeping, cylindrical, entangled stoloniferous portions,
these anchored by rhizoids; erect axes cylindrical, 1.5–2 mm diam., bare of
branches in lower third, densely branched in pyramidal shape upwards;
subtidal specimens soft and silky; low-intertidal specimens somewhat woolly and felted; tetrasporangial stage described as *"Falkenbergia hillebrandii"*; spermatangia in clusters borne close to axis; cystocarps appearing
pedicellate, clavate in shape.

Rare in Calif., saxicolous: subtidal (5–10 m), Santa Catalina I.; intertidal to subtidal, I. Guadalupe to Pta. Malarrimo, Baja Calif.; common in
Gulf of Calif. Widespread throughout warm Pacific, warmer parts of Atlantic, and Mediterranean. Type locality: Egypt.

Tetrasporophyte Stage of *Asparagopsis*

"Falkenbergia hillebrandii" (Ard.) Falk.

Polysiphonia hillebrandii Ardissone 1883: 376. *Falkenbergia hillebrandii* (Ard.) Falkenberg 1901: 689.

Thalli creeping, microscopic, filamentous, less than 1 cm tall, with disklike holdfasts; prominent apical cell cutting off axial filament with 3 pericentral cells, the filament occasionally branched; trichoblasts lacking; minute, highly refractive glandlike cells lying between slender central filament and pericentrals; tetrasporangia cruciately divided, modified directly
from a pericentral cell.

Rare in Calif., subtidal (10 m) on *Codium fragile*, Santa Catalina I.; more
common into Baja Calif. and warmer parts of Pacific Mexico. Widely distributed in warmer seas. Type locality: Italy.

Family GELIDIACEAE

(Contributed by Joan G. Stewart)

Thalli with terete to flattened erect axes, these variously arranged; greenish, bright red to brownish-red, or pink to black in color. Prostrate axes

cylindrical and paler in color. Single principal apical cell present, often distinct. Cortex with several layers of pigmented cells, the cells with 1 large parietal chloroplast, without pyrenoid; medulla of larger, nonpigmented cells. Tetrasporangial and gametangial plants indistinguishable when not fertile. Tetrasporangia cruciately or tetrahedrally divided, located in mostly terminal sori. Spermatangial sori on ultimate branchlets. Cystocarps developed apically (but mature structures may be more proximal), protruding above surface of branch. Carpospores released through 1 pore (rarely more), released on 1 or both surfaces of thallus.

1. Thalli with little or no pigment, epiphytic or parasitic on *Gelidium* and
   Pterocladia spp. . . . . *Gelidiocolax* (below)
1. Thalli pigmented, free-living 2
   2. Cystocarps bilocular, protruding from and opening on both surfaces
      of branch . . . . *Gelidium* (p. 342)
   2. Cystocarps unilocular, protruding from and opening on 1 surface of
      branch . . . . *Pterocladia* (p. 349)

Gelidiocolax Gardner 1927a

Thalli pale pink or white, forming spherical, compact tubercles on *Gelidium* and *Pterocladia*. Filaments branched, both embedded in and extending above host surface. Tetrasporangia, spermatangia, or carpogonial branches present in upper part of thallus.

Fan and Papenfuss (1959) have suggested assignment of the genus to the Choreocolacaceae (Cryptonemiales).

Gelidiocolax mammillata Fan & Papenf.

Fan & Papenfuss 1959: 34.

Thallus a wartlike tubercle, to 1 mm diam., protruding to 0.5 mm above host surface; tetrasporangia, spermatangia, and cystocarps produced primarily in projections.

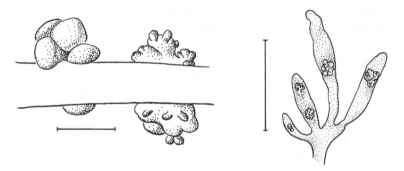

Fig. 284 (left). *Gelidiocolax mammillata.* Rule: 3 mm. (SM/A) Fig. 285 (right). *G. microsphaerica.* Rule: 3 mm. (SM)

Rare, subtidal off Santa Barbara I., Calif., on *Pterocladia capillacea.* Type locality: Hawaii.

Gelidiocolax microsphaerica Gardn.

Gardner 1927a: 341; Fan & Papenfuss 1959: 33.

Thallus a smooth, more or less hemispherical mound, 0.18–0.23 mm diam. Occasional, epiphytic: subtidal (7 m) on *Gelidium robustum,* Pacific Grove, Calif.; intertidal on *G. coulteri* and subtidal on *G. nudifrons* or *G. purpurascens,* S. Calif.; also I. San Martín, Mexico, and Senegal, W. Africa. Type locality: Balboa Beach (Orange Co.), Calif., on *G. pulchrum* (= *G. purpurascens*).

Gelidium Lamouroux 1813

Thalli cartilaginous, to 30+ cm tall. Axes erect, terete to compressed, variously branched, sometimes strikingly distichous, red to deep purple or black. Plants saxicolous, attached to substratum by branched prostrate axes; proportions of prostrate and erect axes varying; plants sometimes occurring in mats of algal turf with extensive basal parts or in more discrete clumps with 1 or more axes arising from limited basal system. Cortex with several rows of pigmented cells, the smaller toward the outside, mostly 2–12 μm diam., irregularly arranged. Medullary cells in cross section generally rounded, 20–27 μm diam., colorless, compacted or loosely appressed, with or without evident starch granules. Rhizoidal filaments thick-walled, 2–5 μm diam., in medulla and/or cortex, varying in number and position within species and even within single plant. Tetrasporangia in sori at apices of branches, or extending over entire flattened lateral branchlet, or also extending into supporting branch beneath; tetrasporangial branchlets with or without distinct sterile margins; tetrasporangial plants often recognizable by dark, granular appearance of fertile branchlets, this resulting from size and intense pigmentation of spores. Spermatangial sori sometimes apparent as relatively unpigmented areas on apices of branchlets, usually conspicuous by presence of sterile darker margin. Carpogonial filament unicellular, fusing with adjacent cells after fertilization. Mature cystocarps protruding equally on both surfaces of branch, usually with single pore on each surface, rarely with 2 or 3; vegetative growth continuing apically beyond developing cystocarp, making its position more proximal.

The following are known only from the type collections from Calif., and their relationships cannot be ascertained; for the present they must be thought of as doubtful species: *Gelidium contortum* Loomis (1960), San Francisco; *G. umbricolum* Dawson & Neushul (1966), Anacapa I.; and *G. venturianum* Dawson (1958), Ventura Co.

1. Thalli taller than 10 cm, mostly low intertidal and subtidal 2
1. Thalli shorter than 8 cm; mostly mid- to high-intertidal 5

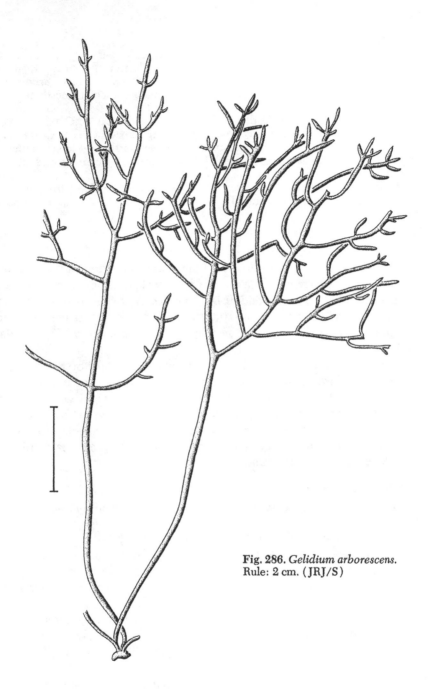

Fig. 286. *Gelidium arborescens.*
Rule: 2 cm. (JRJ/S)

Gelidium arborescens Gardn.

Gardner 1927b: 276; Smith 1944: 197.

Thalli 8–15(25) cm tall, the branching sparse and irregular throughout, infrequently more regularly alternate and geniculate; axes cylindrical to compressed, 0.5–1.5 mm broad, 0.4–1.5 mm thick; fertile plants unknown.

Common, saxicolous, low intertidal to subtidal (3 m) at Pebble Beach (Monterey Co.), Calif., and in nearby drift. Type locality: Carmel Bay, Calif.

These plants are separable from *G. purpurascens* only by the sparse irregular branching. Pebble Beach is the site of plants showing branching more typical of *G. purpurascens*; but since several algal genera exhibit markedly different forms in this bay, and since the physical parameters of the area are quite different from more exposed areas, ecological factors ought to be considered before the status of *G. arborescens* is further assessed. See also discussion under *G. nudifrons*.

Gelidium coulteri Harv.

Harvey 1853: 117; Smith 1944: 196; Dawson 1953a: 70. *Gelidium undulatum* Loomis 1960: 4.

Thalli in thick clumps to 10 cm tall, occasionally taller in less exposed sites; erect axes with few major branches but with numerous distichous branchlets, these mostly short and equal in length from base to apex of axis; prostrate axes fewer and less extensive than erect parts; main axes 0.3–1.1 mm broad, 0.2–0.5 mm thick, cylindrical basally, more compressed or flat above; fertile apices often broad and of various irregular shapes.

Abundant intertidally, in dense mats or clumps on rocks, or in tufts on mussel shells, Wash. to Pta. Pequeña, Baja Calif. Type locality: Monterey, Calif.

The diagnosis of the "typical" branching pattern for this species is based on selected specimens and is illustrated here. The species varies most in compression of axes, number and length of branches, and size of plants.

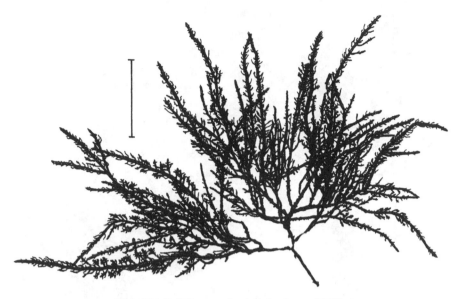

Fig. 287. *Gelidium coulteri*. Rule: 2 cm. (JGS)

Gelidium nudifrons Gardn.

Gardner 1927b: 274; Dawson 1953a: 65.

Plants to 35 cm tall, clumped, sparsely and irregularly branched; axes compressed to flattened throughout, with all parts of similar dimensions, 0.4–0.9 mm broad, 0.3–0.6 mm thick.

Occasional, on rocks, subtidal (to 30 m), S. Calif. and Channel Is. to Pta. Santa Rosalía, Baja Calif. Type locality: Ensenada, Baja Calif.

G. nudifrons and *G. arborescens* as diagnosed here do not overlap in geographic distribution; at present they are also separable by depth of subtidal occurrence and by the dimensions of branches. *G. arborescens* is more robust and stiff than *G. nudifrons*; but the latter, when grazed back to basal portions, can be deceptively like plants in the Pebble Beach population.

Gelidium purpurascens Gardn.

Gardner 1927b: 275; Smith 1944: 197; Dawson 1953a: 72. *Gelidium densum* Gardn. 1927b: 278. *G. distichum* Loomis 1949: 2. *G. gardneri* Loom. 1960: 5. *G. papenfussii* Loom. 1949: 1; Daws. 1953a: 69. *G. polystichum* Gardn. 1927b: 276; Daws. 1953a: 66. *G. pulchrum* Gardn. 1927b: 279; Daws. 1953a: 74. *G. ramuliferum* Gardn. 1927b: 279. *G. setchellii* Gardn. 1927b: 275; Hollenberg & Abbott 1966: 47.

Plants 8–15(30) cm tall, mostly with rather regular distichous branching, sometimes in part markedly polystichous, alternate or opposite, with or without geniculate appearance; axes basally nearly cylindrical or compressed, usually more compressed above; ultimate sterile branchlets occasionally terete and with pointed apices; branching highly variable; lower

Fig. 288. *Gelidium nudifrons.*
Rule: 2 cm. (JGS)

parts unbranched, or with branching near base; distal branching sparse to dense; main axes 0.6–1.6(2) mm broad, 0.25–0.8(1.2) mm thick near base; ultimate branches 0.27–0.95 mm broad, 0.13–0.6 mm thick.

Common, on rocks, intertidal to upper subtidal, Wash. to I. Cedros, Baja Calif. Type locality: Moss Beach (San Mateo Co.), Calif.

Several previously described Calif. species are included here as morphological variants. Extensive sampling indicates the presence of continuous variation in this assemblage, with no entities separable from the group. See also the discussion under *G. arborescens*.

Gelidium pusillum (Stackh.) Le Jol.

Fucus pusillus Stackhouse 1801: 17. *Gelidium pusillum* (Stackh.) Le Jolis 1863: 139; Feldmann & Hamel 1936: 112; Smith 1944: 195; Dawson 1953a: 62.

Thalli 1–2(5) cm tall; prostrate axes terete, producing terete erect axes and/or compressed erect branches; erect parts sometimes becoming secondarily attached to substratum, at which points more prostrate parts may develop; erect branching generally distichous, predominantly irregular; flattened branches 0.4–1.5 mm broad, 0.1–0.3 mm thick; more terete branches 0.1–0.3 mm diam; tetrasporangial and cystocarpic ultimate branchlets often conspicuously flatter, broader, and more irregular in outline than sterile parts.

Often abundant, in algal turf or separate clumps, throughout intertidal and into subtidal, Br. Columbia to Ecuador. Common and widely distributed in most warm seas. Type locality: England.

Several specific names (*latifolium, caloglossoides, parva, sinicola*) and varietal forms of the species have been applied to the small, tangled gelidiaceous algae described here as *G. pusillum*. This name has been previously used for collections from the Pacific Coast of the United States and Mexico, and it is adopted here for the bilocular thalli as most expedient until a critical evaluation can be made of material from other locations and many other habitats. For plants vegetatively similar but bearing unilocular cystocarps, see *Pterocladia*.

Gelidium robustum (Gardn.) Hollenb. & Abb.

Gelidium cartilagineum var. *robustum* Gardner 1927b: 280; Smith 1944: 196; Dawson 1953a: 71. *G. robustum* (Gardn.) Hollenberg & Abbott 1965: 1179.

Thalli to 40(100) cm tall, the branches mostly compressed, the basal axes cylindrical; branching most frequently distichous, often strikingly geniculate, occasionally showing a more polystichous arrangement whereby some lateral branchlets produce short, generally unbranched branchlets from their flattened surfaces; young, short axes often once-pinnate; older, tall fronds having distal parts much branched, the lower part of axes unbranched; axes 0.9–2.2 mm broad, 0.4–1.7 mm thick.

Infrequent, saxicolous, from Br. Columbia to N. Calif., common through

Fig. 289 (below right). *Gelidium purpurascens.* Rule: 2 cm. (JRJ/S) **Fig. 290** (two above right). *G. pusillum.* Rule: 2 cm. (both JGS) **Fig. 291** (left). *G. robustum.* Rule: 2 cm. (JRJ/S)

remainder of Calif. to I. Magdalena, Baja Calif.; in Calif., low intertidal to subtidal along entire coast. Type locality: Ensenada, Baja Calif.

This is the coarsest and largest of Calif. species of *Gelidium*, usually easy to identify. Yet even in this well-known and relatively well-delimited species it is sometimes difficult to decide whether a plant should be assigned here or should be considered a broad-geniculate *G. purpurascens*.

Pterocladia J. Agardh 1851

Thalli similar to those of *Gelidium* species, uniaxial with prominent apical cell, the axial row producing lateral branches, these rebranching and forming pigmented cortex of several cell layers. Lower cells of laterals producing colorless rhizoids, these filling medulla; rhizoids often packed with hyaline storage material. Thalli usually less rigid, more compressed than in *Gelidium*, with distichous branches tending to be constricted at their bases. Development of tetrasporangia and spermatangia as in *Gelidium*. Cystocarps with 1 locule, opening on 1 side of thallus.

The three presently known species of *Pterocladia* from Calif. are quite distinct among themselves, but each is quite similar to *Gelidium* species of corresponding morphology.

1. Thalli sparsely branched, the parts slender, of relatively uniform dimensions throughout .... *P. media*
1. Thalli frequently branched, the parts relatively stout, the branches varying in length and width throughout 2
  2. Main axes percurrent, the higher orders of branches often symmetrical, with progressively shorter branchlets; thalli to 30 cm tall. . . .*P. capillacea*
  2. Main axes and other orders of branches (when present) all of similar size and with irregular arrangement; thalli commonly less than 3 cm tall .... *P. caloglossoides*

Pterocladia caloglossoides (Howe) Daws.

Gelidium caloglossoides Howe 1914: 96. *Pterocladia caloglossoides* (Howe) Dawson 1953a: 76.

Thalli mostly 1–2 cm tall, the erect portion arising from creeping, rhizomelike portion, this with branches less than 200 μm broad, fastened at frequent intervals by peglike or disklike attachments; erect portion with 1 to several free, determinate or indeterminate branches, these flattened except at base, 200–500 μm broad, unbranched, or once or twice pinnate, or branchlets irregularly arranged and of irregular length; apices usually acute, showing clear apical cell; cells usually in distinct upward rows; numerous rhizoidal filaments densely grouped around axial cell; tetrasporangia frequently grouped in V-shaped stripes; spermatangia unknown; gonimoblasts usually apical but sometimes lateral.

Infrequent, saxicolous, intertidal, C. Calif. to Baja Calif. and Gulf of Calif. Type locality: San Felipe, Gulf of Calif.

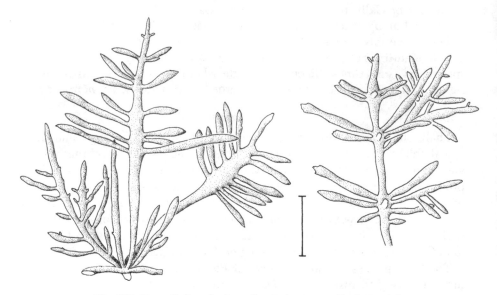

Fig. 292. *Pterocladia caloglossoides.* Rule: 2 mm. (both JRJ/S)

The unilocular cystocarps on Calif. thalli confirm placement in *Ptero-cladia*. The abundant, small, turflike, gelidiaceous algae found intertidally most commonly represent young *P. capillacea* (in S. Calif.) or *Gelidium pusillum*, or perhaps small *G. coulteri*. Broader specimens of *P. caloglossoides* may be found in the same assemblage, but at present they cannot easily be separated from the other three species. Narrower thalli are more distinctive.

Pterocladia capillacea (Gmel.) Born. & Thur.

Fucus capillaceus Gmelin 1768: 146. *Pterocladia capillacea* (Gmel.) Bornet & Thuret 1876: 57; Stewart 1968: 76. *Gelidium pyramidale* Gardner 1927b: 273. *P. pyramidale* (Gardn.) Dawson 1953a: 79.

Thalli to 30 cm tall; branching distichous, often with very regular and pinnate appearance, sparse to dense throughout, or different in proximal and distal parts, with 1–5 orders of branching, this yielding plants of various shapes; axes compressed to flat basally and flattened above, to 2 mm broad, 0.9 mm thick below; ultimate branchlets to 1.3 mm broad, 0.3–0.7 mm thick, usually with constriction at junction of laterals and main axes.

Common, saxicolous, throughout intertidal and to 20 m subtidally, Santa Barbara, Calif., to I. Magdalena, Baja Calif., and Is. Revillagigedo. Common also in Atlantic and Pacific. Type locality: Mediterranean.

Very small *P. capillacea* plants can be distinguished from *Gelidium pusillum* plants of comparable size by their flatter branches and/or more regular branching.

Pterocladia media Daws.

Dawson 1958: 68. *Gelidium crinale* f. *luxurians* Collins, P.B.-A., 1895–1919 [1906]: no. 1138; Gardner 1927b: 277; Dawson 1953a: 65.

Thalli 3–5(8) cm tall, tufted; simple or branched erect axes arising from system of stoloniferous prostrate parts attached to substrate by short, peg-like rhizoids; branching, when present, irregular and sparse, generally

Fig. 293 (right). *Pterocladia capillacea.* Rule: 1 cm. (JGS) Fig. 294 (two at left). *P. media.* Rule: 1 cm. (both JGS)

distichous, often secund or more regularly alternate or pinnate terminally; lateral branchlets with or without slight constrictions at point of junction with proximal axis; all axes and branches very thin, narrow, mostly of uniform dimensions from base to apex, 150–300 μm broad, 100–200 μm thick, the rhizoidal filaments few to numerous, more often occurring in inner medulla, but when relatively abundant, present throughout medulla; cortex with several rows of pigmented cells, all cells of similar size or with cells of inner rows to twice as large as cells in outer layer; medulla of irregularly shaped cells, fewer than cortical cells; tetrasporangia in some cases arranged in V-shaped rows apically on branches, more often not so arranged and within sorus; cystocarps developing and protruding on 1 surface of apical region of branch, mostly with single ostiole.

Occasional in sandy patches on rock, intertidal, San Francisco, Calif., to Baja Calif. and Gulf of Calif. Type locality: La Jolla, Calif.

Plants collected in N. Calif. labeled *Gelidium sinicola* may belong here, or may represent more terete *G. pusillum*; final attribution of *G. sinicola* is not possible in the absence of cystocarpic material.

The plants are soft and lax rather than stiff or cartilaginous, and are more slender and more irregularly branched than those of *G. pusillum*, which can be similar in height and habitat, but are generally found in thicker clumps or dense mats. *P. capillacea* is more strikingly distichous and pinnate than *P. media*, with the percurrent main axes of larger dimensions than the ultimate branchlets; usually, all parts are broader, thus appearing more flattened than those of *P. media*.

Order CRYPTONEMIALES

Thalli uniaxial with single, dome-shaped apical cell, or multiaxial with many apical cells; erect or repent; unbranched, richly branched, bladelike, or crustose; calcified or uncalcified. Medulla and cortex derived from primary filaments, sometimes appearing parenchymatous in section. Cells usually uninucleate with 1 to few chloroplasts, without pyrenoids. Reproductive structures scattered, or in nemathecia, or in conceptacles. Tetrasporangia, when scattered, embedded in cortex, otherwise in nemathecia or conceptacles, cruciately, irregularly, or zonately divided. Spermatangia in superficial patches, or in sori, or in internal chains or in branched clusters within conceptacles. Carpogonial branches separated from auxiliary cell branch (nonprocarpic) as specially formed branches of accessory origin, or close to auxiliary cell (procarpic). Gonimoblasts small and wholly embedded or in conceptacles, or large and protruding; with or without surrounding sterile tissue.

The families of the Cryptonemiales are distinguished from one another primarily by the morphology of the female reproductive structures and by postfertilization events. Since these are not readily analyzed, and the female plants not always available, other characters, chiefly vegetative, have been emphasized in the following key. It will become clear that a key based on vegetative structures, e.g., uniaxial vs. multiaxial growth, separates members of a family that would otherwise be united on similarity of reproductive structures, and accounts for the occurrence of a family more than once in the key.

1. Thalli uniaxial, erect, compressed or terete, not bladelike.............. 2
1. Thalli multiaxial, erect and compressed-cylindroid, or flat crusts, or bladelike 5
   2. Branches with minute spines; cystocarps very conspicuous.........
   .... ENDOCLADIACEAE (p. 421)
   2. Branches without spines; cystocarps relatively inconspicuous......... 3
3. Auxiliary cell branches and carpogonial branches borne on same supporting cell .... GLOIOSIPHONIACEAE (p. 419)
3. Auxiliary cell branches and carpogonial branches borne separately....... 4
   4. Gonimoblast arising from auxiliary cell in auxiliary cell branch.....
   .... DUMONTIACEAE (in part; p. 354)
   4. Gonimoblast arising from cell in carpogonial branch ....
   .... WEEKSIACEAE (in part; p. 363)
5. Thalli crustose, closely adhering to rock; rarely epiphytic.............. 6
5. Thalli erect, branched, or bladelike 8
   6. Crusts stony, calcified, brittle when dry, lavender to deep rose; reproductive structures only in conceptacles ....
   .... CORALLINACEAE (in part; p. 379)
   6. Crusts tough, leathery, or soft, dark red to dark purple or blackish.... 7
7. Reproductive structures in elevated sori....... PEYSSONNELIACEAE (p. 368)
7. Reproductive structures in pitlike conceptacles...................
   .... HILDENBRANDIACEAE (p. 377)
   8. Thalli erect but from crustose base, and with calcified segments joined by uncalcified joints ...........CORALLINACEAE (in part; p. 379)
   8. Thalli uncalcified, unsegmented; erect axes terete, compressed, or expanded into blades 9
9. Plants "parasitic" 10
9. Plants not parasitic 11
   10. Thalli nearly globular, with irregularly cylindrical branches; outer cells elongate in section................CHOREOCOLACACEAE (p. 469)
   10. Thalli globular, without branches; all cells nearly isodiametric.....
   .... KALLYMENIACEAE (in part; p. 451)
11. Plants erect, wiry or terete to compressed, repeatedly branched......
   .... CRYPTONEMIACEAE (in part; p. 424)
11. Plants erect, mostly or completely bladelike, sometimes branched..... 12
   12. Blades mostly simple, occasionally lobed or with proliferations; medulla thin and filamentous....... CRYPTONEMIACEAE (in part; p. 424)
   12. Blades usually divided, usually without proliferations 13
13. Blades perfoliate .... WEEKSIACEAE (in part; p. 363)
13. Blades not perfoliate, usually lobed or divided......................14
   14. Blades strap-shaped, or repeatedly branched and finely divided, mostly 8–10 cm tall............. KALLYMENIACEAE (in part; p. 451)
   14. Blades lobed, mostly 15+ cm tall 15

Family DUMONTIACEAE

Thalli mostly uniaxial (multiaxial in *Dilsea*), frequently soft, slippery, or mucosoid. Apical cell dome-shaped. Each axial cell producing 4 repeatedly branched vegetative laterals, the outermost cell rows of these apposed and making up cortex. Basal cells of laterals producing rhizoidal filaments from abaxial surfaces, fertile filaments from adaxial surfaces. Tetrasporangia on tetrasporophytes similar to gametophytes, cruciately, irregularly, or zonately divided; not known in some genera, borne on alternating crustose phase in other genera. Carpogonial branch usually of 8–18 cells, bent in hooklike fashion. Auxiliary cell borne in equally elaborate branch, either terminal or intercalary. Carpogonium after fertilization fusing directly or indirectly with cell (nutritive cell) lying near it in its branch; 1 or more connecting filaments produced by this fusion growing to and connecting with auxiliary cell, sometimes sending out further connecting filaments to other auxiliary cells; small mass of carposporangia produced from each such auxiliary cell. All cells of gonimoblast becoming carposporangia; gonimoblast without surrounding sterile tissue.

1. Thalli multiaxial, bladelike; cortex in cross section 2+ times as thick as
   medulla .... *Dilsea* (p. 361)
1. Thalli uniaxial 2
   2. Plants markedly gelatinous to mucosoid, the branching irregular in
      upper portions of thallus .... *Dudresnaya* (below)
   2. Plants not gelatinous, repeatedly branched from base to apex........ 3
3. Thalli usually compressed, the branching distichous and subalternate;
   reproductive structures inconspicuous ....*Farlowia* (p. 355)
3. Thalli cylindrical or subcylindrical; reproductive structures conspicuous.. 4
   4. Thalli with percurrent axis, the branching distichous......*Pikea* (p. 359)
   4. Thalli without percurrent axis, the branching radial, the ultimate
      order short and spinelike ....*Cryptosiphonia* (p. 361)

Dudresnaya Crouan & Crouan 1835

Thalli uniaxial, markedly gelatinous and mucosoid, irregularly radially branched, compressed. Internally repeatedly branched laterals formed from each cell of axis, lying within gelatinous matrix. Cells elongate and cylindrical nearest axis, becoming shorter and finally somewhat spherical at outside. Tetrasporophytes rare; tetrasporangia zonately divided. Spermatangia in superficial clusters on cortical cells. Carpogonial branches of 5–9(19) cells. Auxiliary cell branch of 9–14 cells, the auxiliary cell an

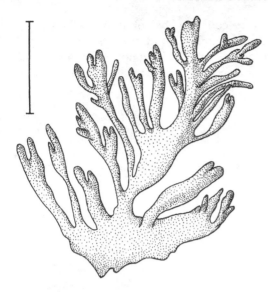

Fig. 295. *Dudresnaya colombiana.*
Rule: 1 cm. (SM)

intercalary cell. Carpogonium after fertilization fusing with 1 or more nutritive cells behind it, sending out 1–3 connecting filaments to 1 or more successive auxiliary cells. Cystocarps small, all cells becoming carposporangia.

Dudresnaya colombiana Tayl.

Taylor 1945: 162; Mower & Widdowson 1969: 76.

Thalli 5+ cm tall, compressed, to 6 mm diam., the branching radial and irregular; branches tapering to acute apices, very softly gelatinous, pale rose pink; axial filaments 12–16(20) μm diam., surrounded by numerous delicate rhizoidal filaments; female reproductive structures imperfectly known; tetrasporangia and spermatangia unknown.

Rare, saxicolous, subtidal (7 m), Little Harbor (Santa Catalina I.), Calif. Type locality: Colombia.

Farlowia J. Agardh 1876

Thalli erect, solitary or in clusters, arising from small, discoid bases. Erect portion freely branched, the branching distichous, alternate to subopposite. Branches compressed, with or without percurrent axes; on drying, some branches appearing to have midrib and broken lateral veins owing to formation of many rhizoidal filaments around axial strands. Cortex of compacted filaments in rows of 6–8 cells. Medulla of periclinally directed rhizoidal filaments grouped around uniaxial strand. Tetrasporangia rare, irregularly cruciately divided, forming irregular sori, scarcely modifying surrounding cortical cells. Spermatangia in superficial patches.

Carpogonial branches and auxiliary cell branches of 10–18 cells. Carpogonium fusing with nutritive cell through a short process, producing 1
filament that connects with 1 or more successively remote intercalary auxiliary cells. Gonimoblasts small, formed in groups, raising thallus surface
in irregular nemathecioid areas.

1. Lower axes mostly 4+ mm wide........................*F. compressa*
1. Lower axes mostly less than 3 mm wide 2
    2. Main and second-order branches with filiform proliferations; axes usu-
      ally terete ....*F. conferta*
    2. Main and second-order branches without proliferations; axes usually
      compressed throughout ....*F. mollis*

Fig. 296. *Farlowia compressa.*
Rule: 2 cm. (JRJ/S)

Farlowia compressa J. Ag.

J. Agardh 1876: 262; Smith 1944: 204; Abbott 1962a: 29.

Thalli 20–40(60) cm tall, coarse, rust red to deep red, drying to black; branching predominantly distichous, the ultimate branches expanded in upper portions of plants, 0.5–1.5 cm diam.; main axis 3–12 mm wide, varying throughout plant; lower axes thicker through midline than upper axes, appearing veinlike owing to increase in numbers of rhizoidal filaments arising from basal cells of laterals; broken, obliquely placed veins also at times conspicuous; reproductive structures rare, when present as for the genus.

Infrequent, saxicolous, low intertidal to subtidal (3 m), Hope I., Br. Columbia, to San Luis Obispo Co., Calif. Type locality: Monterey Bay, Calif.

Farlowia conferta (Setch.) Abb.

Leptocladia conferta Setchell 1912: 252; Smith 1944: 206. *Farlowia conferta* (Setch.) Abbott 1968: 186. *Pikea nootkana sensu* Doty 1947b: 164.

Thalli 15–20(30) cm tall, dark red, drying to black, with several axes arising from fleshy, peglike holdfasts. Axes 1–3 mm wide, dichotomously or subdichotomously branched, at times percurrent, terete to somewhat compressed but with raised central portion occasionally resembling midrib, with frequent fasciculate proliferations on main axes and on second-order branches; medulla well defined, of densely interwoven, periclinally directed filaments, with single axial strand; cortex of 6–8 dichotomously branched cell layers; tetrasporangia irregularly cruciately divided, in unmodified sori near apices of branches; fertile female areas throughout terminal branchlets, resembling nemathecia; development of gonimoblast as for genus.

Locally common, saxicolous in sand-swept areas, low intertidal, S. Ore. to Cayucos (San Luis Obispo Co.), Calif. Type locality: Dillon Beach (Marin Co.), Calif.

Farlowia mollis (Harv. & Bail.) Farl. & Setch.

Gigartina mollis Harvey & Bailey 1851: 372. *Farlowia mollis* (Harv. & Bail.) Farlow & Setchell, P.B.-A., 1895–1919 [1901]: no. 898; also P.B.-A., 1895–1919 [1901]: no. 1150; Smith 1944: 204; Abbott 1962a: 29. *F. crassa* J. Agardh 1876: 262.

Thalli 10–20 cm tall, bright red to blackish-red; main axis flattened, 1–3 mm wide, usually compressed; branches of higher orders frequently subcylindrical, soft to touch and slippery, predominately distichous, alternate; ultimate branches on rapidly growing thalli to 1 mm wide, becoming eroded on older thalli, then appearing paddle-shaped with smooth margins; obscure midrib and veins common to other 2 species usually not seen in this species; reproductive structures as for genus.

Locally frequent although usually as isolated thalli, saxicolous in sand-

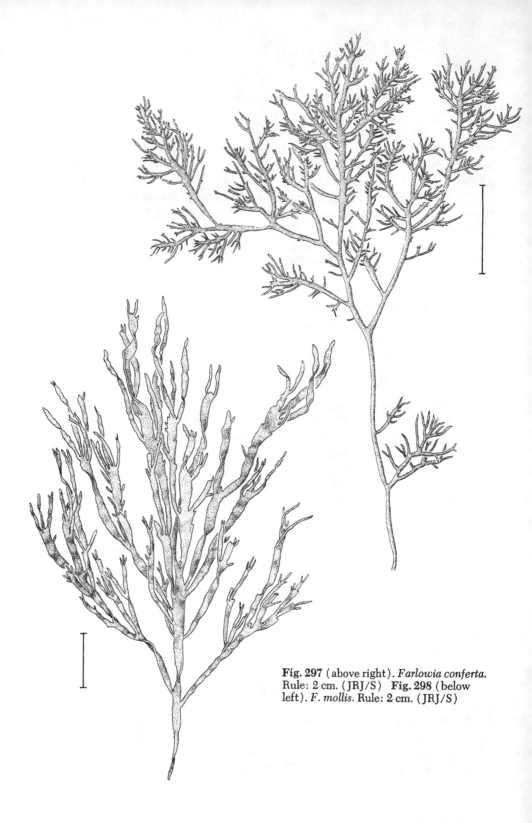

Fig. 297 (above right). *Farlowia conferta.* Rule: 2 cm. (JRJ/S) **Fig. 298** (below left). *F. mollis.* Rule: 2 cm. (JRJ/S)

swept areas, midtidal to subtidal (20 m), Alaska to San Diego, Calif. Also reported from Kurile Is. Type locality: Puget Sd., Wash.

Pikea Harvey 1853

Thalli with 2 morphological phases: larger, gametangial plants erect, isolated or tufted and arising from small, conical holdfasts; sporangial plants small, crustose, bearing cruciately divided tetrasporangia. Gametangial plants with or without distinct stipe, with or without percurrent axis; major branches somewhat compressed, pinnately branched. Medulla with 1 prominent axial filament, surrounded by rhizoidal filaments, these in mature portions appearing pseudoparenchymatous. Cortex filamentous, of 6–8 dichotomously branched cell rows, forming rigid exterior. Spermatangia superficially produced, in whitish patches at apices of young branches. Development of carpogonial and auxiliary-cell branches as in *Farlowia*, but differing in formation of gonimoblast in having the cells adjacent to auxiliary cell participating in forming several gonimolobes. Nearly all cells of gonimoblast become carposporangia.

Pikea californica Harv.

Harvey 1853: 246; Smith 1944: 202; Abbott 1968: 182 (incl. synonymy); Scott & Dixon 1971: 295.

Gametangial thalli 5–10(24) cm tall, bushy; several fronds arising from 1 holdfast, deep red, drying to blackish. Branching pinnate; major branches 1–1.5 mm broad, the final orders acutely pointed and spinelike. Tetrasporangia cruciately divided, in crustose phase; development of other reproductive structures as for genus; mature gonimoblasts crowding apices of branches, each apex subterminally swollen and spindle-shaped; gonimoblasts in nemathecioid rows.

Locally frequent, saxicolous, rarely epiphytic, low intertidal to subtidal (20 m), S. Br. Columbia to Pta. Baja, Baja Calif. Also reported from Japan. Type locality: San Francisco, Calif.

Pikea robusta Abb.

Abbott 1968: 184. *Pikea pinnata sensu* Smith 1944: 202.

Thalli 20–30(40) cm tall, usually only 1 frond per holdfast, rust red, retaining color on drying; several major branches arising from short stipe, these 1.5–4 mm wide, percurrent, strongly flattened, pinnately branched to 3 or 4 orders, the branches mostly opposite but also irregular; second-order branches as wide as first-order; ultimate branchlets dense, fringing axes closely; tetrasporangia unknown; development of other reproductive structures as for the genus; gonimoblast position on branchlets subapical and differing from that of *P. californica* in forming wartlike areas and eroding apices of branches.

Infrequent, saxicolous, subtidal (to 20 m), Wash. to Santa Barbara Co., Calif. Type locality: Carmel, Calif.

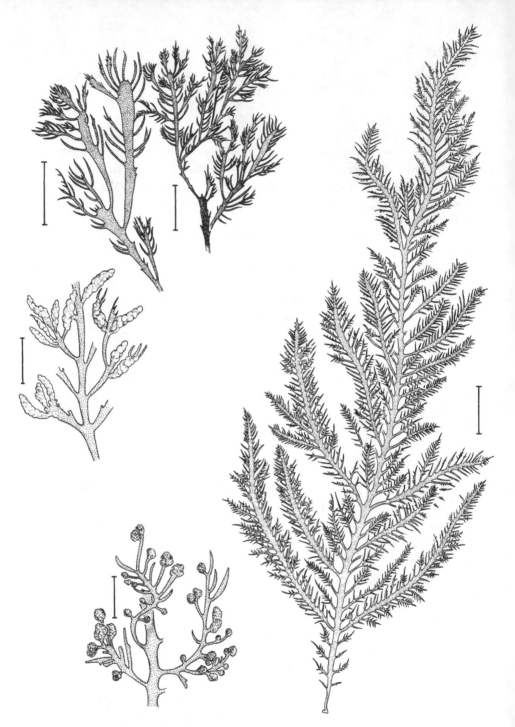

Fig. 299 (three above left). *Pikea californica*. Rules: habits, both 1 cm; cystocarpic nemathecia, 2 cm. (all DBP) **Fig. 300** (right & below left). *P. robusta*. Rules: habit, 1 cm; cystocarpic nemathecia, 2 cm. (JRJ/S & DBP)

Cryptosiphonia J. Agardh 1876

Thalli erect, mostly in tufts, several fronds arising from discoid hold-fasts. Fronds cylindrical, radially branched. Ultimate branchlets short and pointed, uniaxial, lower portions of axis surrounded by rhizoidal filaments. Each axial cell bearing 2 laterals at right angles to each other, laterals alternating position in successive axial cells. Cortex pseudoparenchymatous, firmly compacted. Tetrasporangia irregularly cruciately divided, embedded in outer cortex. Plants dioecious. Spermatangia superficial, covering entire thallus except basal portions. Carpogonial branches and auxiliary-cell branches 7–12 cells long. Development of cystocarp as in *Farlowia*; mature cystocarps in swollen, fusiform branchlets, globose to reniform, embedded in thallus.

Cryptosiphonia woodii (J. Ag.) J. Ag.

Pikea woodii J. Agardh 1872: 15. *Cryptosiphonia woodii* (J. Ag.) J. Ag. 1876: 251; Smith 1944: 200.

Thalli 10–25(35) cm tall, slender, axes 1–2 mm diam., olive brown to deep blackish-purple; usually profusely branched; female gametangial plants more densely branched and shorter than tetrasporangial plants; fertile branches golden.

Locally abundant but inconspicuous element in lower *Gigartina papillata–Rhodoglossum affine* association, saxicolous, midtidal, Alaska to San Pedro, Calif. Type locality: Vancouver I., Br. Columbia.

Dilsea Stackhouse 1809

Thalli erect, mostly clustered; blades simple, entire, or dissected nearly to base, thick and leathery (almost never adhering to paper after drying), multiaxial, with or without short stipe. Medulla dense, of thick, longitudinally and radially intertwined filaments occupying small portion of blade cross section. Cortex thick, with 2 clear layers, the outer of 6–8 cells containing plastids, the inner of 4–6 cells containing granular substances. Tetrasporangia cruciately divided, modified from vegetative cells of cortex, in chains forming horizontal band in cortex. Spermatangia in short, superficial chains composing a sorus. Carpogonial branch of 10–22 cells, the nutritive cells especially conspicuous for their large size and glistening contents. Auxiliary-cell branches of 10–18 cells, the auxiliary cell intercalary. Development of cystocarp as in *Farlowia*; cystocarp globose, the carposporangia lying beneath carpostome.

Dilsea californica (J. Ag.) Kuntze

Sarcophyllis californica J. Agardh 1876: 263. *Dilsea californica* (J. Ag.) Kuntze 1891: 892; Doty 1947b: 165; Abbott 1968: 186. *S. californica* f. *pygmaea* Setchell, P.B.-A., 1895–1919 [1897]: no. 396.

Thalli 9–10(45) cm tall, arising from fleshy holdfasts, with or without

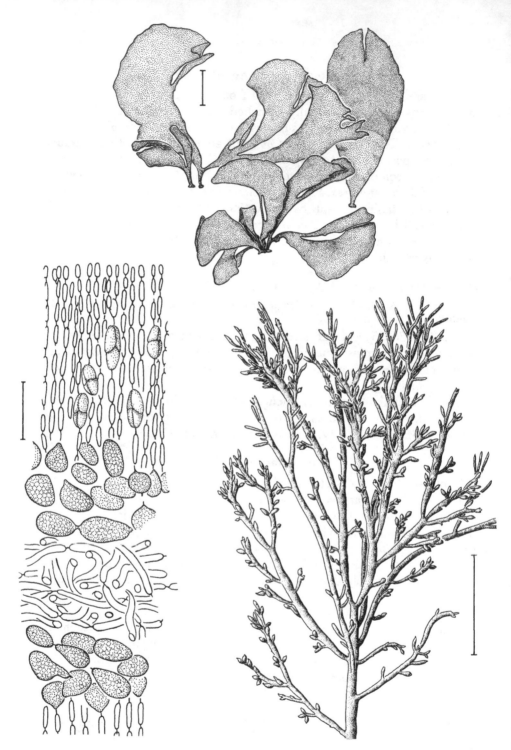

Fig. 301 (below right). *Cryptosiphonia woodii*. Rule: 2 cm. (JRJ/S) **Fig. 302** (below left & above). *Dilsea californica*. Rules: habit, 2 cm; section of thallus, 50 μm. (DBP & DBP/A)

short stipes; blades clustered, simple, entire, or deeply dissected into sickle-shaped divisions, about 1 mm thick; brownish-purple to deep purple-red, drying to dark brown or black; tetrasporangia formed in intercalary chains, these to 150 μm long, occupying much of cortical layer; spermatangia of this species not observed; development of gonimoblast as in *Farlowia*, except for direct fusion of carpogonium with nutritive cell.

Locally frequent, saxicolous, low intertidal to subtidal (20 m), Alaska to Bushnell's Beach (San Luis Obispo Co.), Calif.; most common between S. Ore. and N. Marin Co., Calif. Type locality: Ore.

Family WEEKSIACEAE

Thalli initially uniaxial, sometimes becoming multiaxial and bladelike. Auxiliary-cell branches remote from carpogonial branch, usually several on each supporting cell, resembling those of Dumontiaceae, sometimes functional. Carpogonial branches with extremely large nutritive cells, 1 of these after fertilization of carpogonium producing gonimoblast, all cells of which become carposporangia.

1. Uniaxial throughout life ....*Leptocladia* (below)
1. Uniaxial when young only, multiaxial at maturity 2
  2. Blades perfoliate ....*Constantinea* (p. 365)
  2. Blades marginally attached, expanded, usually dissected or lobed...
.... *Weeksia* (p. 365)

Leptocladia J. Agardh 1892

Thalli uniaxial, with dome-shaped apical cell; cells of axis each producing 4 laterals transversely at right angles to each other, those in one plane more strongly developed than other. Rhizoidal filaments arising from basal cells of laterals. Ultimate branches of laterals forming cortex of 5–8 cell rows. Tetrasporangia irregularly cruciately divided, in small nemathecia. Spermatangia in small, superficial patches. Carpogonial branches of 12–18 cells. Auxiliary-cell branches of 8–18 cells, apparently without function. Development of gonimoblast similar to that in *Weeksia*, all cells becoming carposporangia.

Leptocladia binghamiae J. Ag.

J. Agardh 1892: 96; Dawson 1953a: 89; Abbott 1968: 194.

Thalli 15–30(60) cm tall, compressed, deep red, drying to blackish-red; main axes (1.5)4–5 mm wide, repeatedly branched dichotomously, sub-dichotomously, and irregularly to 6–8 orders; margins of branches with short, pointed teeth; final orders of branching short and toothlike or slightly expanded; tetrasporangial plants more densely branched and taller than cystocarpic plants; tetrasporangia irregularly cruciately divided, sur-rounded by paraphyses in irregularly shaped nemathecia near apices of

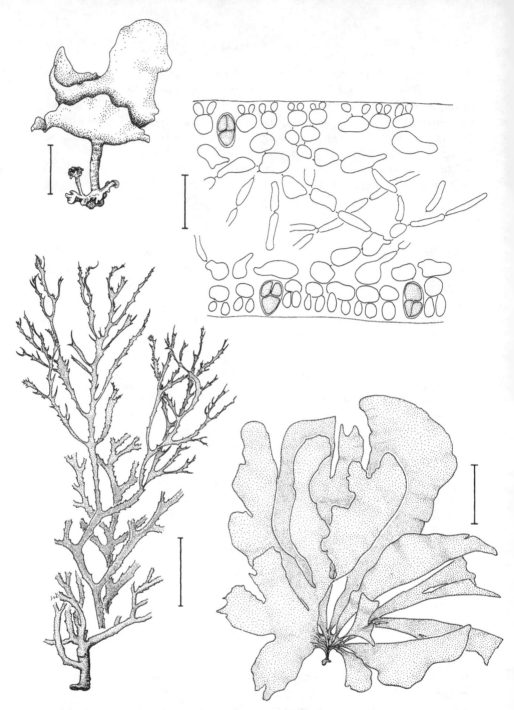

Fig. 303 (below left). *Leptocladia binghamiae*. Rule: 2 cm. (DBP) Fig. 304 (above left). *Constantinea simplex*. Rule: 2 cm. (LH) Fig. 305 (above right & below right). *Weeksia digitata*. Rules: habit, 4 cm; cross section of fertile area, 50 μm. (DBP & DBP/A)

thallus; gonimoblasts developing as in *Weeksia*, with gonimoblast produced on carpogonial branch; cystocarps distorting cortex, with carpostome.

Frequent, saxicolous, subtidal (to 30 m), Cambria (San Luis Obispo Co.), Calif., through S. Calif., including Channel Is., to Baja Calif. and Galápagos Is. Type locality: Santa Barbara, Calif.

Constantinea Postels & Ruprecht 1840

Thalli stipitate, the cylindrical stipes branched or unbranched, with peltate or perfoliate blades; plants perennial; lower portion of stipe with scars of old blades. Blades at first entire and membranous, coriaceous when mature, becoming split and eroded with age, deep red. Tetrasporangia in nemathecia with paraphyses, irregularly zonately divided; nemathecia forming smooth, confluent, sometimes completely continuous zone up to half diam. of blade. Spermatangia unknown. Carpogonial and auxiliary-cell branches of 9–12 cells. Nutritive cells among largest in Cryptonemiales, arranged in several planes. Gonimoblasts produced by nutritive cell in carpogonial branch and by auxiliary cell in nearby auxiliary-cell branch. Gonimoblasts forming fertile band on blade, this not as wide as tetrasporangial nemathecial band.

Constantinea simplex Setch.

Setchell 1901: 127; 1906: 171; Smith 1944: 207; Abbott 1968: 192.

Thalli 2–8 cm tall, the stipes usually unbranched, 0.6–1.2 cm diam.; blades 6–12 cm wide; other characters as for genus.

Locally abundant, saxicolous, usually in groups, low intertidal to subtidal (15 m), N. Br. Columbia to Pt. Conception, Calif. Type locality: Dillon Beach (Marin Co.), Calif.

Weeksia Setchell 1901

Blades large, membranous, entire, dissected, or lobed, with smooth or puckered surface, or puckered basally only; at first uniaxial, then multiaxial. Medulla of stout, periclinally directed filaments, with occasional anticlinal filaments and giant cells in longitudinal rows representing primary axial filaments; rows contributing to veinlike thickenings frequent at base of blades. Cortex of 3–5(7) layers of rounded cells, the innermost layers becoming enlarged on connecting with medullary filaments. Tetrasporangia small, inconspicuous, cruciately to irregularly cruciately divided. Spermatangia in superficial patches. Carpogonial and auxiliary-cell branches of 8–15 cells, 4 or 5 of those in carpogonial branch very conspicuous nutritive cells. Gonimoblast formed from nutritive cell, all cells becoming carposporangia, lying beneath a carpostome.

1. Blades shaped like cabbage leaves, usually semicircular.......*W. reticulata*
1. Blades lobed or dissected, fan-shaped 2
   2. Blades thin (150–200 μm), with no basal veins............*W. howellii*
   2. Blades thick (300–350 μm), with basal veins.............*W. digitata*

Weeksia digitata Abb.

Abbott 1968: 191.

Blades to 60 cm tall, dissected or cleft, the divisions 3–15 cm wide, reddish-purple, the margins entire, the edges ridged; lower portions usually veined; holdfast fleshy, with or without short stipe; sections near upper margin 300–350 μm thick; medulla thick, occupying about half of cross section, of stout periclinal and anticlinal filaments; cortex of 4 or 5 cell layers; tetrasporangia cruciately or irregularly cruciately divided; spermatangia unknown; gonimoblasts compact, about 120 μm diam.

Rare (except at type locality), saxicolous, subtidal (to 20 m), Monterey to San Diego, Calif.; frequently cast ashore near Carmel (type locality) in summer.

Weeksia howellii S. & G.

Setchell & Gardner 1937a: 77; Abbott 1968: 191.

Blades elongate to suborbiculate, 20–30 cm tall, 150–200 μm thick, reddish-brown; medulla of mostly periclinally directed filaments; cortex of 3 or 4 cell layers, compact; tetrasporangia cruciately divided; spermatangia unknown; gonimoblasts compact, 60–70 μm diam.

Subtidal (30 m) off Anacapa I., Calif. Type locality and only other collection: dredged off I. Natividad, Mexico.

Weeksia reticulata Setch.

Setchell 1901: 128; Smith 1944: 206; Abbott 1968: 190.

Thalli 10–20 cm tall, growing in cabbagelike clusters, usually with fleshy peglike holdfasts, little or no stipe, bluish-purple; blades reniform to rounded, with basally prominent veins, or the veins running longitudinally throughout blades, nearly to margins; blades remaining simple and expanded, or occasionally cleft and appearing branched; medulla of stout periclinally and anticlinally directed filaments with dense contents; cortex of 4–6 layers of pigmented cells; plants monoecious; spermatangia in small, discontinuous, superficial patches; tetrasporangia inconspicuous, within outer cortical layers, cruciately to irregularly cruciately divided; carpogonial branches and auxiliary-cell branches borne on innermost cortical cells, 8–15 cells each, frequently with short, sterile laterals on lower cells; upper 5–7 cells of carpogonial branches highly modified by dumbbell-shaped nutritive cells; carpogonium after fertilization fusing directly with 1 of these cells to form gonimoblast, all cells of which become carposporangia, these lying beneath carpostome.

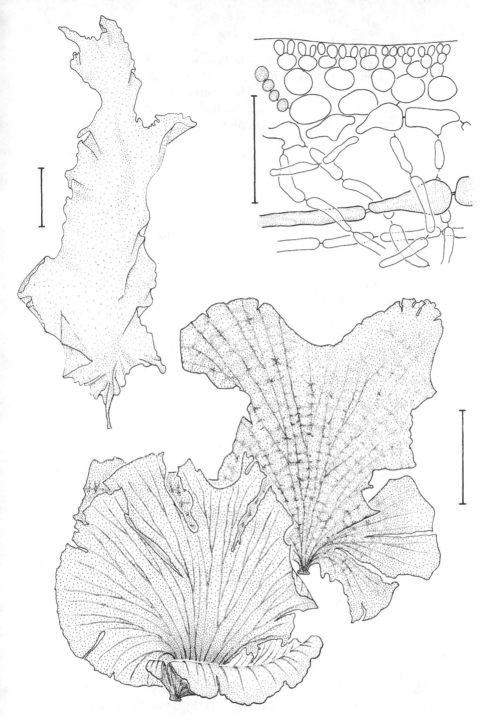

Fig. 306 (above left). *Weeksia howellii*. Rule: 4 cm. (SM) **Fig. 307** (above right & below). *W. reticulata*. Rules: habit, 4 cm; cross section of fertile area, 50 μm. (DBP & DBP/A)

Frequent subtidally (17–37 m) near Carmel, Calif.; infrequent sub-tidally in San Luis Obispo Co. and Channel Is., Calif., and Pta. Santo Tomas, Baja Calif. Type locality: Pacific Grove, Calif.

With shorter thallus and thicker blades than other species of *Weeksia*. The bluish cast of freshly collected specimens differs from the reddish cast of *W. digitata*, which grows in the same habitat and is far commoner in the Carmel area; the latter species has but few veins and a smoother, tougher thallus.

Family PEYSSONNELIACEAE

Thalli crustose, composed chiefly of erect filaments arising from distinct or indistinct basal layer, occasionally calcified, with or without rhizoids. Fertile structures found in nemathecia or sori, usually with paraphyses. Tetrasporangia cruciately divided, variously borne. Spermatangia in nem-athecia or in surface patches, mostly unknown. Female reproductive struc-tures known only for some species of *Peyssonnelia*.

A vaguely defined family of uncertain limits, in which some of the genera are only tentatively placed. Most noncorallinaceous crusts for which only tetrasporangia are known are placed in this family. Some (e.g. *Cruoriopsis* spp.) are suspected of being the sporangial phase of genera with a mark-edly different appearance in the gametangial phase.

1. Thallus with rhizoids on lower surface.............*Peyssonnelia* (below)
1. Thallus without rhizoids.. 2
   2. Tetrasporangia among curved paraphyses 3
   2. Tetrasporangia not among curved paraphyses 4
3. Crusts firm, relatively thick and leathery; cell rows directed toward as well as away from substratum....................*Coriophyllum* (p. 373)
3. Crusts thin, not leathery; all cell rows directed away from substratum
    .... *Rhodophysema* (p. 372)
   4. Plants epiphytic; tetrasporangia on 1-celled pedicels arising from monostromatic hypothallial layer ....*Pulvinia* (p. 374)
   4. Plants saxicolous; tetrasporangia lateral or terminal on multicellular erect filaments 5
5. Tetrasporangia strictly terminal, fully exposed on firmly adjoined erect filaments .... *Rhododiscus* (p. 374)
5. Tetrasporangia terminal or lateral, not arising above adjacent, loosely adjoined vegetative filaments ....*Cruoriopsis* (p. 376)

Peyssonnelia Decaisne 1841

Thalli crustose, uncalcified to strongly calcified, mostly closely adherent by numerous, typically unicellular rhizoids. Hypothallus monostromatic, deeply pigmented. Perithallus of firmly adjoined, assurgent to erect, usu-ally forked cell rows. Tetrasporangia cruciately divided, in nemathecial sori scattered among multicellular paraphyses. Spermatangia in cylindrical packets, these terminating erect filaments and grouped in superficial sori, each packet formed by cells cut off laterally from several uppermost cells. Carpogonial and auxiliary-cell branches usually numerous, in nemathecia, arising laterally at base of paraphyses, mostly of 4 cells each. Auxiliary cell intercalary, usually next to basal cell of branch. Carposporangia few, large, arising singly or in small groups from connecting filaments or from auxiliary cell and developing toward upper surface of crust.

1. Nemathecia extensive, not noticeably elevated................*P. hairii*
1. Nemathecia relatively small, mostly distinctly elevated............... 2
   2. Paraphyses often once forked ....*P. rubra*
   2. Paraphyses not forked 3
3. Intertidal; mostly less than 2 cm broad, or broader by coalescence; nemathecia circular, mostly less than 1 mm broad.........*P. meridionalis*
3. Subtidal; to 6 cm broad; nemathecia to 1–2 cm broad, often larger by confluence of smaller areas ....*P. profunda*

Peyssonnelia hairii Hollenb. & Abb.

Hollenberg & Abbott 1968: 1244.

Crusts dark red, 5–20+ cm broad, mostly 0.6–1 mm thick, with thin margins and numerous unicellular rhizoids; cells of hypothallus in anticlinal radial section 30–40 μm long, 15–22 μm high, in radiating, forking rows, lightly calcified; perithallial cells 10–12 μm wide, about twice as long in assurgent lower part, 5–7 μm wide and 1–2.5 times as long in upper erect part, which is commonly horizontally stratified; chloroplast single, without pyrenoid; tetrasporangial sori extensive and more or less continuous over much of mature thallus, not noticeably elevated; tetrasporangia 30–40 μm long, 18–20 μm diam.; paraphyses unbranched, 80–100 μm long, of 6–8 cells 4–5 μm diam., 3–4 times longer below, shorter and wider above; spermatangia and carposporangia not definitely known.

Saxicolous, lower intertidal; known only from Monterey Co., Calif. Type locality: vicinity of Little Sur River, Calif.

Peyssonnelia meridionalis Hollenb. & Abb.

Hollenberg & Abbott 1968: 1244.

Crusts brownish-red to purplish, circular, mostly less than 2 cm diam., 140–200 μm thick, surfaces often with radial lines, frequently by coalescence forming crusts to 15 cm diam.; cells of hypothallus lightly calcified, mostly isodiametric, 15–19 μm wide, in radiating, forking rows; rhizoids unicellular, numerous; perithallus with little or no calcification, composed of 8–12+ cell layers, the cells about 10 μm wide in upper part; tetrasporangial and carposporangial nemathecia prominently elevated, mostly less than 1 mm broad, 75–125 μm high; spermatangial sori not elevated; para-

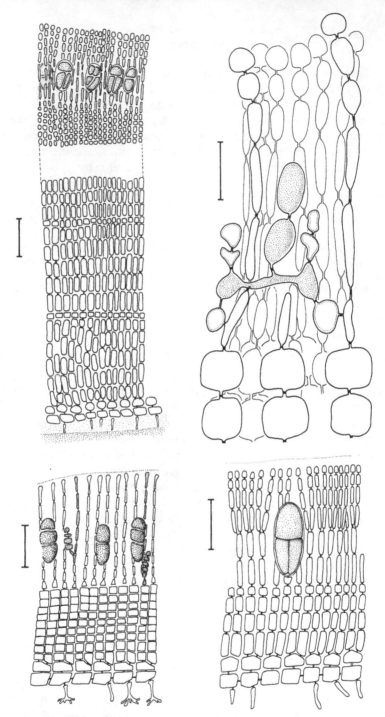

Fig. 308 (above left). *Peyssonnelia hairii*: longitudinal section. Rule: 50 μm. (DBP)
Fig. 309 (above right). *P. meridionalis*: cystocarpic area. Rule: 50 μm. (DBP/A-H)
Fig. 310 (below left). *P. rubra* var. *orientalis*. Rule: 50 μm. (DBP/H)   **Fig. 311** (below right). *P. profunda*. Rule: 50 μm. (DBP/H)

physes unbranched, about 60 μm long, of 6–8 cells; connecting filaments mostly joining several auxiliary cells in sequence by means of slender lateral branches; carposporangia in groups of 3 or 4, arising from connecting filaments.

Common on intertidal rocks and shells, especially on *Tegula brunnea*, Ore. to Monterey Peninsula and San Luis Obispo Co., and probably to S. Calif. Type locality: Mussel Pt. (Pacific Grove), Calif.

Peyssonnelia profunda Hollenb. & Abb.

Hollenberg & Abbott 1965: 1179; 1966: 54.

Crusts dark red, soft, with distinct radial lines, to 6 cm broad and 0.5–0.8 mm thick, only slightly calcified, firmly attached to substratum; cells of hypothallus mostly 12–15 μm wide, 1.5–2 times as long, in regularly radiating, forking rows; perithallus with assurgent to erect rows of cells, the cells mostly 10–12 μm wide, 1–1.5 times as long above, somewhat longer below; tetrasporangial and carpogonial nemathecia prominently elevated, 1–2 cm broad, often larger by confluence of smaller areas; tetrasporangia mostly 80–90 μm long, 40–60 μm diam.; paraphyses simple, 100–112 μm long, of 6–9 cells, those below ±5 μm wide and 3 times as long, those above shorter, 12–14 μm wide; spermatangial sori only slightly elevated.

Frequently dredged from shale at 14–18 m off Del Monte Beach (Monterey Bay; type locality), Calif.; also subtidal (10–15 m), San Luis Obispo Co. and La Jolla, Calif.

Peyssonnelia rubra (Grev.) J. Ag.

Zonaria rubra Greville 1826: 335. *Peyssonnelia rubra* (Grev.) J. Agardh 1852: 502.

The type variety, with larger, more leathery plants than var. *orientalis*, does not occur in California.

Peyssonnelia rubra var. orientalis Web. v. Bosse

Weber van Bosse 1913–28 (1921): 272; Dawson 1953a: 104; 1957: 3. *Peyssonnelia rubra* (Greville) J. Agardh *sensu* Setchell & Gardner 1930: 175; Taylor 1945: 168.

Crusts rose red to deep red, loosely attached at margins by numerous unicellular rhizoids, usually with distinct radial and faint concentric lines, 100–300 μm thick; hypothallial cells lightly calcified, 25–40 μm long, 12–16 μm wide, slightly taller in anticlinal section, in radiating, forking rows, with bases downwardly protuberant as individual cells; perithallus of unbranched, usually strictly erect rows of 6–10 cells, these isodiametric below and shorter above; tetrasporangial nemathecia distinctly elevated, slightly gelatinous, 80–100 μm thick, with slender, slightly clavate, simple or once-forked paraphyses; tetrasporangia 70–110 μm long, 25–55 μm diam.; sexual reproduction not observed.

Infrequent, saxicolous, intertidal to subtidal (18 m), S. Calif. and Channel Is. to Is. Revillagigedo, Mexico, and Galápagos Is. Type locality: East Indies, not specifically designated.

Rhodophysema Batters 1900

Thalli crustose, noncalcareous, with closely adjoined, unbranched erect filaments arising from monostromatic basal layer, this without rhizoids. Tetrasporangia cruciately divided, in elevated sori intermingled with curved, multicellular paraphyses, both terminal on erect filaments. Spermatangia on erect filaments, in continuous sori. Carpogonia and carposporangia unknown.

Rhodophysema elegans (Crouan & Crouan) Batt.

Rhododermis elegans Crouan & Crouan 1867: 148. *Rhodophysema elegans* (Crouan & Crouan) Batters 1890: 221.

The type variety, with monostromatic perithallus and straight paraphyses, does not occur in California.

Rhodophysema elegans var. polystromatica (Batt.) Dix.

Rhododermis elegans var. *polystromatica* Batters 1890: 310. *Rhodophysema elegans* var. *polystromatica* (Batt.) Dixon 1964a: 71. *Rhododermis elegans* Hollenberg 1948: 157.

Thalli bright rose red to dark brownish-red; plants epiphytic or saxicolous, to 5 mm broad (if epiphytic often much more extensive), 60–100 μm thick in sporangial areas; cells of hypothallus about 8 μm wide and high, 12–16 μm long; erect filaments polystromatic, firmly adjoined, mostly of 6–9 cells, these 5–9 μm wide and slightly shorter than wide; tetrasporangia 24–40 μm long, 13–15 μm diam.; paraphyses mostly 45–50 μm long, slightly to strongly curved, of 4–9 cells.

Occasional on rocks, low intertidal, common on stipes of *Cystoseira osmundacea*, sometimes on other algae, San Juan I., Wash., to Corona del Mar (Orange Co.), Calif. Frequent in N. Atlantic Ocean. Type locality: England.

Rhodophysema minus Hollenb. & Abb.

Hollenberg & Abbott 1965: 1181; 1966: 56.

Thalli bright red to dull red, thin, 8–10+ mm broad, monostromatic except for fertile areas; cells of basal layer 6–7 μm wide, and 1.2–2 times as long; tetrasporangial sori single and central or several and scattered; tetrasporangia 15–20 μm long, 14–16 μm diam., on short pedicels; paraphyses curved, 4 or 5 cells long, with outer cells 5.5–6.5 μm wide, mostly 1.5–2 times as long.

Occasional on subtidal rocks, shells, etc., or intertidal, rarely epiphytic;

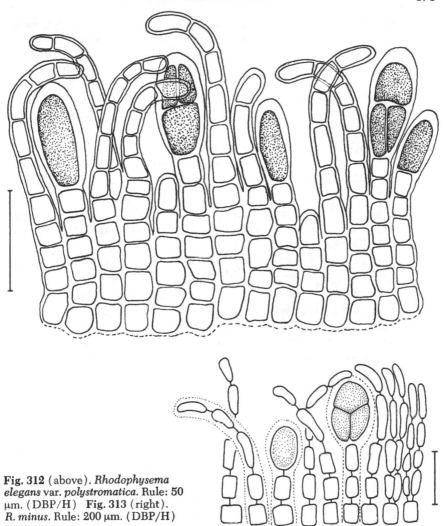

Fig. 312 (above). *Rhodophysema elegans* var. *polystromatica*. Rule: 50 μm. (DBP/H) **Fig. 313** (right). *R. minus*. Rule: 200 μm. (DBP/H)

known from Smith I., Wash., and Monterey Bay, Carmel Bay, and Corona del Mar (Orange Co.), Calif. Type locality: Monterey Harbor, Calif.

Coriophyllum Setchell & Gardner 1917

Crusts uncalcified, firmly cartilaginous or leathery, firmly adhering to substratum without rhizoids; cell rows erect, downwardly directed, firmly adjoined, arising from middle layer of closely interwoven, anastomosing filaments; tetrasporangia cruciately to tetrahedrally divided, numerous, in irregularly shaped nemathecia among curved, clavate, multicellular paraphyses; sexual reproduction unknown.

Setchell and Gardner (in Gardner 1927: 341) proposed the genus name *Asymmetria* for their genus *Coriophyllum* in the belief that this conflicted with *Coriophyllus*, an earlier-named flowering plant. But Kylin (1956) considered these names sufficiently different, and returned to *Coriophyllum*.

Coriophyllum expansum S. & G.

Setchell & Gardner 1917e: 396. *Asymmetria expansa* (S. & G.) S. & G. 1927: 341; Smith 1944: 214.

Crusts 5–8(40) cm broad, 0.5–1 mm thick, dull reddish-brown, with very irregular surface and margins; erect filaments 4–6 μm diam.; tetrasporangia 60–70 μm long, 22–28 μm diam.; paraphyses of 6–12 cells, these ±3 μm diam.

On vertical rock faces, commonly enclosing small rock particles, upper intertidal to subtidal; known only from Monterey Peninsula. Type locality: Cypress Pt. (Monterey Co.), Calif.

Pulvinia Hollenberg 1970

Thalli epiphytic, minute, cushionlike, noncalcareous, with short, erect filaments arising from cells of monostromatic basal layer, this without rhizoids. Tetrasporangia obliquely to irregularly cruciately divided, among short paraphyses, forming continuous sorus over most of thallus. Spermatangia unknown; cystocarps(?) ostiolate, distinctly protuberant.

Pulvinia epiphytica Hollenb.

Hollenberg 1970: 63.

Thalli deep pink, pulvinate, to 3+ mm broad, composed of radiating rows of cells 8–10 μm wide, 1–2 times as long, bearing short, erect filaments, sporangia, and paraphyses; tetrasporangia 12–16 μm diam., 35–38 μm long, on short, 1-celled pedicels; paraphyses 2 or 3 cells long; cystocarps (presumably of this species) ostiolate, protuberant, 250–300 μm diam.; carpospores 60–80 μm diam., in terminal chains of 2.

On other algae; known only from Redondo Beach (type locality) and Del Mar (San Diego Co.), Calif.

Rhododiscus Crouan & Crouan 1859

Thalli noncalcareous, crustose, firmly attached to rocks, without rhizoids. Thallus of unbranched, firmly cohering, erect filaments, these arising from ill-defined, monostromatic hypothallus. Tetrasporangia cruciately divided, terminal on erect filaments, fully exposed and without paraphyses, grouped in relatively extensive surface sori. Sexual reproduction unknown.

Rhododiscus carmelita Hollenb. & Abb.

Hollenberg & Abbott 1968: 1243.

Thalli dark rose red; erect filaments 4.5–6 μm diam., slightly assurgent;

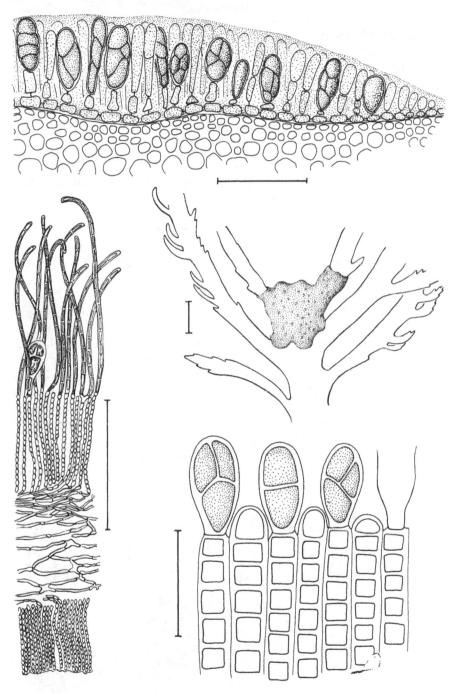

Fig. 314 (below left). *Coriophyllum expansum*. Rule: 500 µm. (SM) **Fig. 315** (above & middle right). *Pulvinia epiphytica*. Rules: longitudinal section through sporangial sorus, showing irregular arrangement of spores, 50 µm; habit, 500 µm. (DBP/H & DBP) **Fig. 316** (below right). *Rhododiscus carmelita*. Rule: 20 µm. (DBP/H)

cells of erect filaments mostly shorter than diam. in upper parts; tetrasporangia 8–10 µm diam., 16–20 µm long, pyriform to broadly clavate, obliquely cruciate, forming extensive irregular sori, without paraphyses.

Known only from shallow upper intertidal tidepool near mouth of Carmel River (Monterey Co.; type locality), Calif.

Cruoriopsis Dufour 1864

Thalli thin, noncalcareous crusts. Hypothallus of 1 or 2 cell layers, without rhizoids. Erect filaments mostly unbranched, loosely adjoined with gelatinous matrix. Tetrasporangia cruciately divided, terminal or lateral on erect filaments. Sexual reproduction unknown or uncertain; carpospores reported to occur in chains.

The status of this genus is uncertain, according to Denizot (1968). Culture studies by Edelstein (1970) show that a crustose stage in the life history of *Gloiosiphonia capillaris* is very similar to crusts commonly iden: tified as *Cruoriopsis* species.

Cruoriopsis aestuarii Hollenb.

Hollenberg 1970: 63.

Thalli to 6+ mm broad, 100 µm thick in fruiting portions, 1 cell thick at growing margins, closely attached to substratum; hypothallus monostromatic, bearing erect filaments; erect filaments mostly simple, gelatinous, mostly of 8–10 cells, these 8–10 µm diam. below and slightly narrower above, 0.6–1.5 times as long as diam.; tetrasporangia 50–55 µm long, 25–30 µm diam., on short, 1-celled pedicels, these laterally attached on 2d or 3d cell from base of erect filaments.

Infrequent, on rocks or shells, low intertidal to subtidal, near San Juan I., Wash., and upper Newport Harbor (Orange Co.; type locality), Calif.

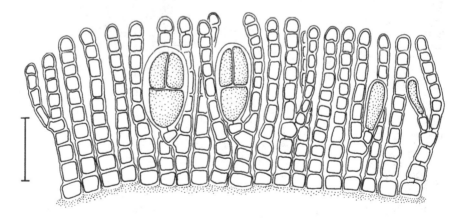

Fig. 317. *Cruoriopsis aestuarii.* Rule: 50 µm. (DBP/H)

Family Hildenbrandiaceae

Thalli uncalcified, crustose, closely adherent to substratum, without rhizoids, composed of closely adjoined erect filaments arising from ill-defined basal layers. Perithallus sometimes horizontally stratified. Tetrasporangia in pitlike conceptacles, zonately or cruciately divided, with or without paraphyses. Sexual reproduction unknown.

Hildenbrandia Nardo 1834

With characters of the family.

1. Tetrasporangial divisions irregular or obliquely cruciate......*H. prototypus*
1. Tetrasporangial divisions zonate 2
    2. Crusts extensive, mostly over 450 µm thick; conceptacles tubular, mostly over 200 µm deep; paraphyses absent..........*H. occidentalis*
    2. Crusts less extensive, usually less than 300 µm thick; conceptacles mostly less than 100 µm deep; paraphyses numerous........*H. dawsonii*

Hildenbrandia dawsonii (Ardré) Hollenb.

Hildenbrandia canariensis var. *dawsonii* Ardré 1959: 234. *H. dawsonii* (Ardré) Hollenberg 1971b: 286. *H. prototypus* var. *kerguelensis sensu* Dawson 1953a: 96.

Crusts dull red to brownish-red, to 300 µm thick, but mostly thinner; cells of erect filaments mostly quadrate, 3–5 µm diam.; conceptacles very shallow to globular, to 175 µm wide, 70–90 µm deep; tetrasporangia transversely zonately divided, 5.5–6.5 µm diam., 20–28 µm long, among numerous very slender, hyaline, unbranched paraphyses.

Frequent on rocks, midtidal to upper intertidal, San Juan I., Wash., to I. Cedros, Baja Calif. Type locality: Bahía Vizcaíno, Baja Calif.

Hildenbrandia occidentalis Setch.

Setchell 1917g: 393; Smith 1944: 215.

Crusts dark purplish-red, nearly black when dry, indefinitely expanded to 100 cm broad, typically 1–2 mm thick, when well developed of cartilaginous consistency; cells of erect filaments 3–4.5 µm wide, quadrate to twice as long anticlinally; conceptacles numerous, scattered, 100–150 µm diam., 200–300(800) µm deep, flask-shaped to mostly cylindrical; tetrasporangia transversely zonately divided, 9–10 µm diam., 25–32 µm long, mostly perpendicular to wall of conceptacle; paraphyses absent.

Common on upper intertidal rocks, N. Br. Columbia to I. San Gerónimo, Baja Calif., and Galápagos Is.; very common in C. Calif. Type locality: San Francisco, Calif.

Hildenbrandia prototypus Nardo

Nardo 1834: 676; Rosenvinge 1909–31 (1917): 202. *Hildenbrandia rosea* Kützing 1843: 384.

Crusts pale rose red to bright red, 3–5+ cm broad, 50–300(450) µm

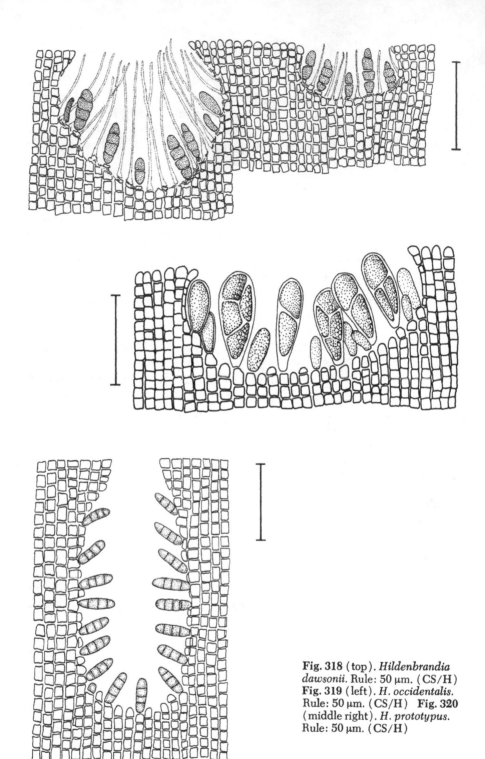

Fig. 318 (top). *Hildenbrandia dawsonii.* Rule: 50 μm. (CS/H)
Fig. 319 (left). *H. occidentalis.* Rule: 50 μm. (CS/H) **Fig. 320** (middle right). *H. prototypus.* Rule: 50 μm. (CS/H)

thick; cells of erect filaments compact, quadrate to slightly anticlinally elongate, mostly 3–4 μm diam.; conceptacles shallow surface depressions to subspherical, 35–110(200) μm diam., 35–80 μm deep; tetrasporangia obliquely cruciately to irregularly divided, 15–28 μm long, 10–12 μm diam.; paraphyses absent or mostly inconspicuous.

Common on rocks, lower intertidal to upper subtidal, Alaska to Oaxaca, Mexico. Nearly worldwide in distribution. Type locality: Italy.

Family CORALLINACEAE

(Contributed by H. William Johansen)

Thalli mostly calcified, with 2 main vegetative habits: either wholly crustose and entirely or partly adherent, nodular and unattached, or mostly erect, with flexible, articulate fronds arising from crustose or rhizomatous bases. Saxicolous, epiphytic, endophytic, or epizoic on shells. Crustose thalli or crustose parts of erect thalli differentiated into three tissues: hypothallium of filaments oriented approximately parallel to substratum; perithallium of filaments oriented approximately perpendicular to substratum; and epithallium consisting of 1 or more strata of cover cells. Perithallium in some species with large, hyaline glandlike structures (megacells or heterocysts), arranged in a variety of ways. Crustose portions sometimes smooth, more often rounded, knobby, folded, or irregularly papillate (with excrescences); some branched. Erect thalli articulated, consisting of regularly alternating, calcified intergenicula (segments) and uncalcified genicula; each frond made up of medulla, cortex, and epithallium; fronds branched, rarely unbranched. Cortical and perithallial cells with 1 to many irregular chloroplasts. Reproductive cells borne in conceptacles, these originating below epithallium in crustose species, the roofs of tetrasporangial conceptacles formed from filaments among tetrasporangia, the roofs of male and female conceptacles from surrounding epithallial tissue. In articulate species, conceptacles originating in medullary tissue axial or marginal in position; those originating in cortical tissue lateral in position. Conceptacles uniporate or in some nonarticulated genera multiporate, in these tetrasporangia developing thickened apices and plugging pores until discharge. Tetrasporangia zonately divided, sometimes replaced by bisporangia. Male conceptacles with spermatangia densely lining floor only, or floor and wall, in some genera these mixed with paraphyses; in some articulated species, overgrowing filaments forming roof producing attenuated beak on conceptacle. Female conceptacles containing numerous procarps, each consisting of 1–3 2-celled carpogonial filaments and 1 supporting cell. Carposporophyte forming from filaments growing from fusion cell, rarely developing directly from unfused supporting cells. Carposporangia large, terminal.

No genera or species of red algae are more difficult to identify without fertile material than the Corallinaceae. With fertile material and a little patience, it is possible to identify most of the West Coast species, owing to recent and current studies on these forms. The keys to genera and species presented here are designed to capitalize on the obvious vegetative and fertile structures that may be found in adequately collected specimens. The 24 genera recognized from California are treated in two groups: the crustose species (immediately following the key) and the articulated species (p. 400).

1. Geniculated fronds absent 2
1. Geniculated fronds present 14
  2. Vegetative parts endophytic in articulated Corallinaceae, only the
    conceptacles emergent ....*Choreonema* (p. 397)
  2. Plants not endophytic 3
3. Tetrasporangial conceptacle with several pores..................... 4
3. Tetrasporangial conceptacle with 1 pore 8
  4. Thalli usually epiphytic on articulated Corallinaceae, the edges free... 6
  4. Thalli saxicolous, epizoic, or epiphytic on noncorallinaceous plants by
    entire lower surface, sometimes nodular and unattached............ 5
5. Vegetative parts of thalli less than 200 µm and 10 cells thick; epiphytic
  on noncalcareous plants ....*Melobesia* (p. 387)
5. Vegetative parts of thalli more than 200 µm and 15 cells thick; usually
  saxicolous or epizoic, sometimes nodular and unattached............
  .... *Lithothamnium* (p. 381)
  6. Hypothallium thick, the cells in decumbently arched tiers.........
  .... *Mesophyllum* (p. 389)
  6. Hypothallium thin, the cells not in tiers 7
7. Epithallium polystromatic, pigmented; usually epiphytic on *Calliarthron*
  .... *Clathromorphum* (p. 391)
7. Epithallium monostromatic to sometimes polystromatic, not photosynthetic; usually epiphytic on *Corallina* or *Bossiella*...................
  .... *Neopolyporolithon* (p. 392)
  8. Hypothallial cells elongate, palisadelike ....*Tenarea* (p. 395)
  8. Hypothallial cells subcuboidal, not palisadelike 9
9. Thalli usually epiphytic; mostly less than 100 µm thick in vegetative
  parts 10
9. Thalli usually saxicolous or epizoic; mostly more than 300 µm thick in
  vegetative parts 11
  10. Megacells present ....*Fosliella* (p. 399)
  10. Megacells absent ....*Heteroderma* (p. 399)
11. Megacells present 12
11. Megacells absent 13
  12. Megacells scattered singly; Californian species lacking excrescences .... *Hydrolithon* (p. 399)
  12. Megacells usually grouped in vertical series; Californian species
    having excrescences ....*Neogoniolithon* (p. 400)
13. Secondary pit-connections present; Californian species pink and with
  excrescences more than 6 mm high..............*Lithophyllum* (p. 392)
13. Secondary pit-connections lacking; Californian species purple and with
  excrescences less than 2 mm high..........*Pseudolithophyllum* (p. 397)
  14. All or some conceptacles axial 15
  14. Conceptacles never axial 21

Crustose Species

Lithothamnium Philippi 1837

Thalli lacking genicula; saxicolous, epiphytic, epizoic, sometimes nodular or branching and free-living. Hypothallial cells not in tiers. Perithallial cells pigmented, in tiers, but these sometimes obscure. Epithallium nonpigmented, essentially monostromatic, of small cells distinctively angular in vertical section. Secondary pit-connections absent; intercellular fusions present. Male and female conceptacles uniporate; tetrasporangial conceptacles multiporate.

1. Thalli crustose, thin and smooth 2
1. Thalli crustose, or nodular, having excrescences or branches........... 4
  2. Tetrasporangial conceptacles protuberant, opening by 12–20 pores,
    roofs less than 100 μm diam. ....*L. volcanum*
  2. Tetrasporangial conceptacles low, opening by ±30 pores, roofs more
    than 150 μm diam. 3
3. Roofs of tetrasporangial conceptacles mostly 300+ μm diam. *L. californicum*
3. Roofs of tetrasporangial conceptacles mostly 150–300 μm diam. ......
  .... *L. microsporum*
  4. Thalli consisting of anastomosing branches, nodular lumps or crusts
    more than 5 mm thick when fully developed; excrescences coarse,
    sometimes branched 5
  4. Thalli crustose, mostly less than 2 mm thick; excrescences less than
    4 mm diam. at base, unbranched........................... 8

5. Thalli free or loosely attached, the extended branches radiating or spreading irregularly 6
5. Thalli crustose or nodular, lacking extended branches 7
   6. Branches slender, less than 3 mm diam., the arms overlapping and crossing because of angular branching pattern............*L. australe*
   6. Branches often more than 4 mm diam., the ultimate branches principally dividing in 1 plane, the arms of branches short, not overlapping ....*L. montereyicum*
7. Excrescences coarse, branched, 10–15 mm tall, 5–8 mm diam. .... *L. giganteum*
7. Excrescences rarely branched, less than 4 mm tall, less than 4 mm diam. .... *L. crassiusculum*
   8. Excrescences sparse and of variable size; hypothallium more than 300 μm thick ....*L. phymatodeum*
   8. Excrescences dense and relatively uniform in size; hypothallium less than 200 μm thick 9
9. Excrescences less than 2 mm tall, often tapering to a sharp apex...... .... *L. aculeiferum*
9. Excrescences 2–6 mm tall, round-topped ....*L. pacificum*

Lithothamnium aculeiferum Mason

Mason 1943a: 94; 1953: 326; Dawson 1960: 10; Masaki 1968: 10.

Thalli crustose, firmly adherent, sometimes completely covering small pebbles, 1–3 mm thick; excrescences small, closely spaced, uniform in size, shape, and distribution, 0.5–1.5 mm tall, 1–1.5 mm diam. at base, often tapering to acute apex; hypothallium 100–200 μm thick, poorly developed;

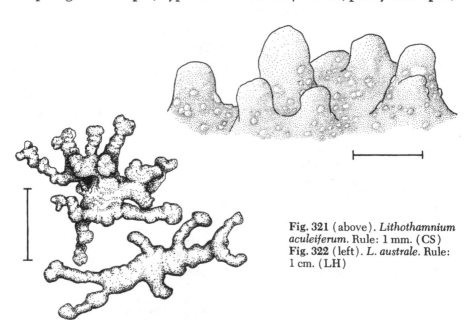

Fig. 321 (above). *Lithothamnium aculeiferum.* Rule: 1 mm. (CS)
Fig. 322 (left). *L. australe.* Rule: 1 cm. (LH)

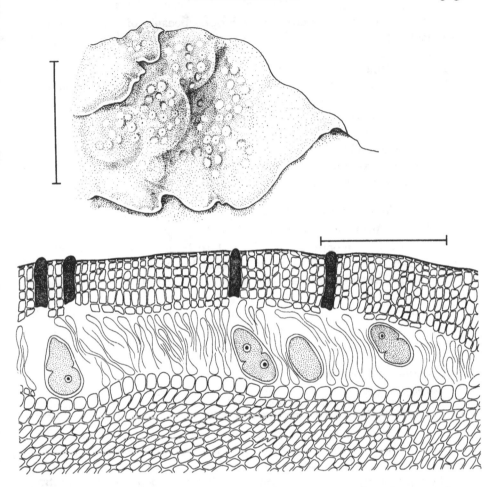

Fig. 323. *Lithothamnium californicum.* Rules: habit, 1 cm; section through sporangial sorus, showing bisporangia and multiporate conceptacle, 100 μm. (SM & CS/HWJ)

perithallium to 2.5 mm thick; tetrasporangial and bisporangial conceptacles mostly immersed, scattered, 140–220 μm diam., with 15–40 pores; carposporangial plants as for genus; male plants not recorded.

Occasional, saxicolous, intertidal to subtidal, Marin Co., Calif., to Scammon Lagoon, Baja Calif. Type locality: San Pedro, Calif.

Lithothamnium australe (Fosl.) Fosl.

Lithothamnium coralloides Crouan & Crouan f. *australis* Foslie 1895: 90. *Lithothamnium australe* (Fosl.) Foslie 1900b: 13; Dawson 1960: 11.

Thalli unattached, variable in form, composed of free, compact nodules having conspicuous excrescences, or irregularly organized clumps of mostly slender, subcylindrical branches 1–3 mm diam., the arms of branches spreading, often crossing each other because of angular pattern of branching; branch hypothallium consisting of tiered cells mostly 10–20 μm long; perithallium thin; tetrasporangial conceptacles scarce, protruding little, outside diam. 400–500 μm; sexual plants not recorded.

Locally common, unattached, subtidal (10–30 m), Santa Catalina and Anacapa Is., Calif.; N. Baja Calif., to Gulf of Calif. and Guerrero, Mexico; and into tropics to south. Type locality: La Paz, Gulf of Calif.

This species appears to be closely related to *Lithothamnium montereyicum*, an imperfectly known species from north of the center of distribution of *L. australe*. Dawson (1960) reports that under favorable conditions, this species may cover the ocean floor to the exclusion of other organisms.

Lithothamnium californicum Fosl.

Foslie 1900a: 3; Mason 1953: 324; Dawson 1960: 13.

Thalli crustose, smooth, firmly adherent, to 1+ mm thick; hypothallium thin, of 2–6 layers of cells; perithallium relatively thick; tetrasporangial conceptacles crowded, 300–400 μm diam., about 30 pores in roof; sexual plants not known.

Common, saxicolous, in tidepools and subtidally, S. Br. Columbia to I. Magdalena, Baja Calif.; in Calif., from Patrick's Pt. (Humboldt Co.), to La Jolla. Type locality: San Pedro, Calif.

Lithothamnium crassiusculum (Fosl.) Mason

Lithothamnium rugosum Foslie f. *crassiusculum* Fosl. 1901: 4. *L. pacificum* (Fosl.) Fosl. f. *crassiusculum* (Fosl.) Fosl. 1906: 10. *L. crassiusculum* (Fosl.) Mason 1943b: 93; 1953: 329. *L. fruticulosum sensu* Dawson 1960: 13.

Thalli crustose when young, later sometimes becoming nodular and unattached; crusts often superimposed or confluent, covered with excrescences, these coarse, warty, sometimes branched, frequently perforated by boring worms; branches 3–4 mm diam. below, 1.5–3 mm above; tetrasporangial conceptacles borne on branches, 300–800 μm outside diam., with 25–50 pores; male and female plants also known.

Occasional, saxicolous, intertidal to subtidal, N. Calif. to C. America. Type locality: White's Pt. (San Pedro), Calif.

Lithothamnium giganteum Mason

Mason 1943c: 93; 1953: 326; Dawson 1960: 15.

Thalli completely covering substratum, with age becoming unattached, nodular, subspherical, to 6+ cm diam., the crusts 6–8 mm thick; excrescences coarse, branched, with slightly swollen tips, 10–15 mm tall, 5–8 mm diam.; tetrasporangial conceptacles immersed, 450–500 μm diam., with 50–60 pores; female plants as for genus; male plants not recorded.

Apparently rare, on pebbles or shells, subtidal (6–10 m), La Jolla, Calif. (type locality), to Pta. Eugenio, Baja Calif.

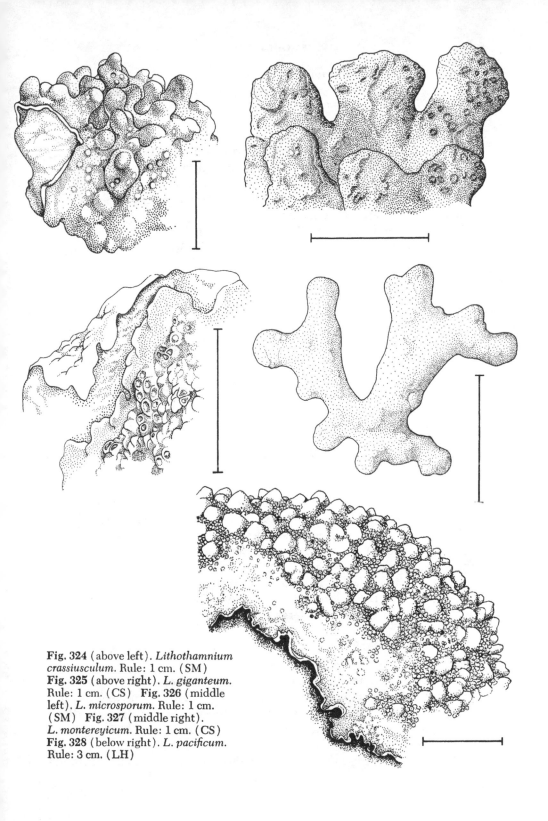

Fig. 324 (above left). *Lithothamnium crassiusculum*. Rule: 1 cm. (SM)
Fig. 325 (above right). *L. giganteum*. Rule: 1 cm. (CS)   **Fig. 326** (middle left). *L. microsporum*. Rule: 1 cm. (SM)   **Fig. 327** (middle right). *L. montereyicum*. Rule: 1 cm. (CS)
Fig. 328 (below right). *L. pacificum*. Rule: 3 cm. (LH)

Lithothamnium microsporum (Fosl.) Fosl.

Lithothamnium californicum f. *microspora* Foslie 1902a: 5. *L. microsporum* (Fosl.) Foslie 1929: 51. *L. lenormandii sensu* Dawson 1960: 20. (Not *Phymatolithon lenormandii* (Aresch.) Adey 1966: 325.)

Thalli saxicolous, to 200 µm thick, firmly adherent, surface macroscopically smooth, microscopically irregular, giving plants a dull appearance; on rough substrata the surface squamulose; conceptacles aggregated; tetrasporangial conceptacles sometimes raised (125)200–300 µm inside diam., to ±40 pores per conceptacle; carposporangial conceptacles about 300 µm diam.; male conceptacles unknown.

Infrequent, on rocks, intertidal to subtidal, Channel Rocks (Puget Sd.), Wash., to Central America. Type locality: Pacific Beach (San Diego Co.), Calif.

Lithothamnium montereyicum Fosl.

Foslie 1906: 14; Mason 1953: 325.

Thalli composed of unattached, sparsely branched, coralloid flattened branches, these oval in transection, the apices blunt, 2–3 cm long, about 1 cm diam. near base; tetrasporangial conceptacles 200–250 µm diam.; sexual plants not recorded.

Known only from specimens dredged in 22 m off Monterey, Calif. (type locality).

See also the discussion under *L. australe*, above.

Lithothamnium pacificum (Fosl.) Fosl.

Lithothamnium sonderi Hauck f. *pacificum* Foslie 1902a: 4. *L. pacificum* (Fosl.) Foslie 1906: 10; Mason 1953: 328 (incl. synonymy); Masaki 1968: 16.

Thalli crustose, deep rose red when fresh, 2+ mm thick; excrescences numerous, 2–6 mm long, 1–4 mm diam.; hypothallium weakly developed, to 80 µm thick; tetrasporangial conceptacles crowded, on excrescences and basal crust, 250–500 µm diam., with about 30 pores, becoming white with age; sexual plants not recorded.

Common, intertidal, on rocks or shells throughout Calif. and frequently associated with *Hydrolithon decipiens*, Br. Columbia to S. Calif. Type locality: Pebble Beach (Monterey Co.), Calif.

Lithothamnium phymatodeum Fosl.

Foslie 1902a: 3; Mason 1953: 327 (incl. synonymy).

Thalli crustose, reddish-purple, brittle, to 2 mm thick; excrescences irregular, 3+ mm tall, to 4 mm diam.; hypothallium and perithallium each 300–500 µm thick; tetrasporangial conceptacles 275–350 µm diam., protruding, forming conspicuous patches, each with 25–40 pores; sexual plants not recorded.

Frequent, on stones, shells, or other algae, lower intertidal to subtidal,

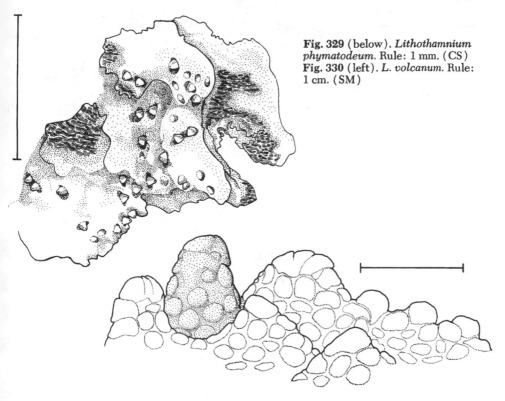

Fig. 329 (below). *Lithothamnium phymatodeum*. Rule: 1 mm. (CS)
Fig. 330 (left). *L. volcanum*. Rule: 1 cm. (SM)

Wash. to Pacific Grove, Calif., and probably to Pt. Conception, Calif. Type locality: Whidbey I., Wash.

Similar in appearance to *L. pacificum*, but distinguished by its thinner, more brittle thalli, irregularly shaped excrescences, and thicker hypothallium.

Lithothamnium volcanum Daws.

Dawson 1960: 23.

Thalli crustose, smooth, firmly adherent, about 250 μm thick; hypothallium weakly developed; tetrasporangial conceptacles protruding, 300–400 μm diam. or less, with 12–20 closely spaced pores; male and female plants as for genus.

Possibly locally abundant, on rocks or small stones, deep subtidal near Santa Cruz I., Calif., in 90 m at Cortez Bank (U.S.-Mexican boundary), and to Baja Calif. Type locality: I. Magdalena, Baja Calif.

Melobesia Lamouroux 1812

Thalli lacking genicula, epiphytic, crustose and thin, to 12 cell layers thick near conceptacles, thinner elsewhere. Hypothallium monostromatic;

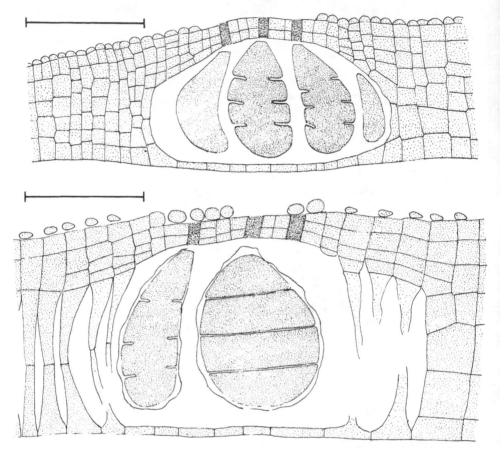

Fig. 331 (above). *Melobesia marginata.* Rule: 50 μm. (JRJ/S)   **Fig. 332** (below). *M. mediocris.* Rule: 50 μm. (JRJ/S)

perithallium to 10 cell layers thick; epithallium monostromatic. Secondary pit-connections lacking, intercellular fusions present. Male and female conceptacles uniporate; tetrasporangial conceptacles multiporate; roofs of male and female conceptacles formed by overgrowth of surrounding tissue.

Melobesia marginata Setch. & Fosl.

Setchell & Foslie 1902a: 10; Smith 1944: 219; Mason 1953: 321 (incl. synonymy); Dawson 1960: 6.

Thalli forming thin, irregular patches, 50–100 μm thick, on noncalcareous red algae; tetrasporangial conceptacles to 150 μm diam., but usually less than 120 μm, with 10–12 pores; male and female plants also known.

Common, epiphytic on *Gymnogongrus* or *Laurencia*, S. Br. Columbia to Peru. Type locality: Bodega Bay (Sonoma Co.), Calif.

Melobesia mediocris (Fosl.) Setch. & Mason

Lithophyllum zostericolum f. *mediocris* Foslie 1900a: 5. *Melobesia mediocris* (Fosl.) Setchell & Mason 1943: 95; Mason 1953: 320 (incl. synonymy); Dawson 1960: 7.

Thalli coalescent, forming thin, irregularly shaped patches; discrete thalli to 2 mm diam., to 200 µm thick; tetrasporangial conceptacles to 210 µm diam., with 9–14 pores; male and female plants also known.

Common on *Phyllospadix* throughout Calif., less common on *Zostera* in S. Calif.; N. Br. Columbia through Baja Calif. Type locality: Santa Cruz, Calif., on *Phyllospadix*.

Mesophyllum Lemoine 1928

Thalli lacking genicula, saxicolous, epiphytic, or epizoic; sometimes thick, more often thin (in Calif. species), leafy, and loosely attached to substratum. Hypothallium thick, cells in decumbently arched tiers. Epithallium essentially monostromatic, not photosynthetic, consisting of cells obscurely rectangular in vertical section. Secondary pit-connections absent, intercellular fusions present. Male and female conceptacles uniporate; tetrasporangial conceptacles multiporate. Roofs of male and female conceptacles formed by overgrowth of surrounding tissue.

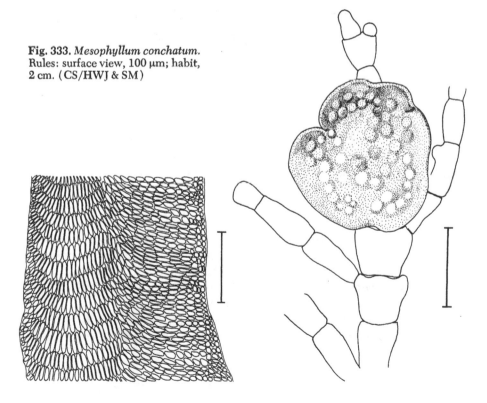

Fig. 333. *Mesophyllum conchatum.* Rules: surface view, 100 µm; habit, 2 cm. (CS/HWJ & SM)

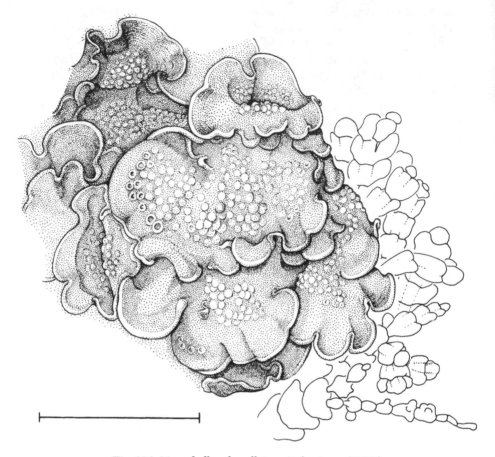

Fig. 334. *Mesophyllum lamellatum.* Rule: 1 cm. (DBP)

Mesophyllum conchatum (Setch. & Fosl.) Adey

Lithothamnium conchatum Setchell & Foslie 1902a: 5. *Mesophyllum conchatum* (Setch. & Fosl.) Adey 1970: 23. *Polyporolithon conchatum* (Setch. & Fosl.) Mason 1953: 317.

Thalli dark lavender, forming circular to semicircular crusts, to 3 cm diam., to 1 mm thick; growing on articulated corallines, attached to host by short central stalk on lower surface; upper surface plane or undulate, often with slight concentric ridges; hypothallium 200–800 μm thick; tetrasporangial conceptacles crowded, 0.5–1.3 mm diam., roofs slightly convex, with 30–70 pores; male and female plants also known.

Common on articulated corallines, especially *Calliarthron*, S. Br. Columbia to San Luis Obispo Co., Calif. Type locality: near Pacific Grove, Calif., on *Calliarthron.*

Mesophyllum lamellatum (Setch. & Fosl.) Adey

Lithothamnium lamellatum Setchell & Foslie 1903: 4; Mason 1953: 330; Dawson 1960: 18. *Mesophyllum lamellatum* (Setch. & Fosl.) Adey 1970: 25.

Thalli pale pink; crusts smooth, consisting of overlapping and shingle-like lobes, 2–3 cm diam., less than 500 μm thick; hypothallium constituting about three-fourths thickness of thallus; tetrasporangial conceptacles crowded, not protruding, 350–600 μm diam., with 40–60 pores; male and female plants also known.

On rocks or other algae, especially articulated corallines, Ore. to Baja Calif.; common intertidally in C. Calif., subtidally in C. and S. Calif. Type locality: Cypress Pt. (Monterey Co.), Calif.

Clathromorphum Foslie 1898

Thalli crustose, lacking genicula, saxicolous or epiphytic (as in Calif. species). Hypothallial and perithallial cells not in tiers; epithallium poly-

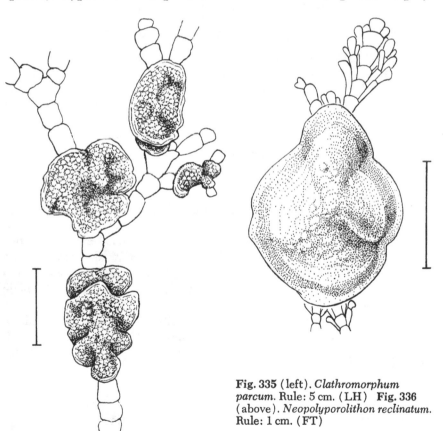

Fig. 335 (left). *Clathromorphum parcum.* Rule: 5 cm. (LH)   Fig. 336 (above). *Neopolyporolithon reclinatum.* Rule: 1 cm. (FT)

stromatic, to 5 cell layers thick, photosynthetic. Secondary pit-connections lacking, intercellular fusions present. Roofs of male conceptacles formed by horizontal separation of tissue in fertile area; roofs of female conceptacles formed by overgrowth of surrounding tissue; tetrasporangial conceptacles multiporate.

Clathromorphum parcum (Setch. & Fosl.) Adey

Lithothamnium parcum Setchell & Foslie 1907: 14. *Clathromorphum parcum* (Setch. & Fosl.) Adey 1970: 27. *Polyporolithon parcum* (Setch. & Fosl.) Mason 1953: 318.

Thalli discoid, lavender to purplish, to 2+ cm diam., to 3 mm thick, attached to hosts by central stalks; upper surface plane, slightly convex, or undulate, margins smooth; tetrasporangial conceptacles not crowded, 0.3–0.5 mm diam., the roofs with 15–25 pores; male and female plants also known.

Common on articulated corallines, especially *Calliarthron*, N. Wash. to San Luis Obispo Co., Calif. Type locality: Pacific Grove, Calif.

Neopolyporolithon Adey & Johansen 1972

Thalli lacking genicula, crustose, epiphytic on articulated corallines. Hypothallial and perithallial cells not in tiers; epithallium 1–3 cell layers thick, not photosynthetic; upper stratum of perithallium prominent. Secondary pit-connections absent, intercellular fusions present. Roofs of male conceptacles formed by horizontal separation of tissue in fertile area; roofs of female conceptacles formed by overgrowth of surrounding tissue; tetrasporangial conceptacles multiporate.

Neopolyporolithon reclinatum (Fosl.) Adey & Johans.

Lithothamnium conchatum f. *reclinatum* Foslie 1906: 6. *Neopolyporolithon reclinatum* (Fosl.) Adey & Johansen 1972: 160. *Polyporolithon reclinatum* (Fosl.) Mason 1953: 319 (incl. synonymy).

Thalli epiphytic, approximately discoid to irregularly encircling host; crusts attached near centers, the edges free, to 1.5 cm diam., 0.2–1 mm thick, frequently overlapping; tetrasporangial conceptacles 100–450 µm diam., the roofs flat or slightly convex, with 25–30 pores; male and female plants also known, sometimes monoecious.

Common on articulated corallines, especially small species of *Corallina* or *Bossiella*, in tidepools and in upper subtidal, C. Alaska to La Jolla, Calif. Also known from Japan. Type locality: Port Renfrew (Vancouver I.), Br. Columbia, on *Bossiella*.

Lithophyllum Philippi 1837

Thalli lacking genicula, crustose or becoming unattached, usually saxicolous or on shells. Hypothallium monostromatic or polystromatic, not palisadelike, megacells absent. Secondary pit-connections present; intercellular fusions absent or occasionally present. All conceptacles not thickened, not plugged, uniporate.

1. Thalli lacking excrescences, smooth, lichenoid ....*L. lichenare*
1. Thalli having excrescences, rough, not lichenoid 2
   2. Excrescences mostly less than 4 mm long; surface glazed......*L. imitans*
   2. Some excrescences more than 6 mm long; surface dull.............. 3
3. Thalli less than 2 mm thick; excrescences terminally enlarged and more
   or less flat-topped, sometimes with central depressions....*L. proboscideum*
3. Thalli frequently more than 3 mm thick; excrescences terminally rounded and narrower at top than at base.....................*L. grumosum*

Lithophyllum grumosum (Fosl.) Fosl.

Lithothamnium grumosum Foslie 1897: 16. *Lithophyllum grumosum* (Fosl.) Fosl. 1898: 10; Mason 1953: 339 (incl. synonymy); Dawson 1960: 39.

Thalli crustose, or becoming unattached and nodular with age, dark violet to purplish; crusts to 5+ mm thick; excrescences coarse to 1+ cm diam., to 15 mm long, sometimes branched, lobed, or anastomosed; hypothallium one-fourth to one-third thickness of crust; tetrasporangial conceptacles 225–350 µm diam.; male and female plants also known.

Uncommon, saxicolous or unattached, subtidal, Ore. to Baja Calif. Type locality: Carmel Bay, Calif.

Lithophyllum imitans Fosl.

Foslie 1909: 13; Mason 1953: 340 (incl. synonymy); Dawson 1960: 41; Dawson, Acleto & Foldvik 1964: 43.

Thalli crustose, pink when fresh, firmly adherent to substrate, the surfaces glazed, sometimes completely covering small stones and shells, to 2 mm thick; excrescences variable in size and shape, simple and round-topped, to 1 cm long but usually less, 2–6 mm diam.; hypothallium weakly developed, about one-eighth thickness of crust; tetrasporangial and bisporangial conceptacles scattered, protruding slightly, 180–330 µm inside diam.; sexual plants also collected.

Common, saxicolous, lower intertidal to subtidal, C. Calif. to Peru. Type locality: Pacific Beach (San Diego Co.), Calif.

Lithophyllum lichenare Mason

Mason 1953: 339; Dawson 1960: 42.

Thalli crustose, smooth, irregularly semicircular, forming undulate patches to 15 cm broad, to 2 mm thick, loosely attached, the margins usually free, rose pink when fresh; hypothallium more than one-half thickness of crust; tetrasporangial conceptacles 100–200 µm diam.; female plants also collected; male plants not known.

Locally abundant, saxicolous, especially in surfy areas of intertidal, also subtidal, Wash. to Baja Calif. Type locality: San Juan I., Wash.

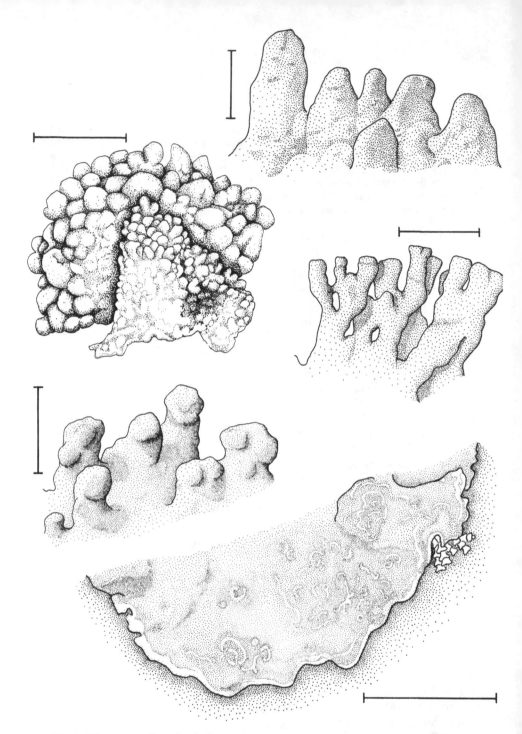

Fig. 337 (two at top). *Lithophyllum grumosum*. Rules: habit, 5 cm; detail of excrescences, 5 mm. (LH & CS) **Fig. 338** (below left). *L. imitans*. Rule: 5 mm. (CS) **Fig. 339** (below right). *L. lichenare*. Rule: 1 cm. (CS) **Fig. 340** (middle right). *L. proboscideum*. Rule: 5 mm. (CS)

Lithophyllum proboscideum (Fosl.) Fosl.

Lithothamnium proboscideum Foslie 1897: 14. *Lithophyllum proboscideum* (Fosl.) Fosl. 1900a: 18; Mason 1953: 342 (incl. synonymy); Dawson 1960: 47.

Thalli crustose, 1–2 mm thick, with crowded, subcylindrical, flat-topped excrescences, these simple, unbranched or anastomosed, to 3 cm long, 3 mm diam. at base, 4–5 mm at apices, a central depression sometimes present in apices; hypothallium well developed; tetrasporangial conceptacles scattered, on lower surfaces of excrescences, immersed, 200–300 μm diam.; sexual plants not known.

Uncommon, saxicolous, low intertidal to shallow subtidal, C. Calif. to Mexico; known from Bodega Bay (Sonoma Co.), Monterey Peninsula, and Cambria (San Luis Obispo Co.), Calif.; also found in an Indian midden on San Nicolas I., Calif. Type locality: off Monterey, Calif.

Tenarea Bory 1832

Thalli lacking genicula, crustose, saxicolous or (in Calif. species) epiphytic. Hypothallium of 1 tier of elongated, palisade-like cells, these oriented obliquely to vertically with respect to substratum. Perithallium clearly stratified, cells usually resembling hypothallial cells. Epithallium monostromatic. Megacells absent. Secondary pit-connections present; intercellular fusions absent or occasionally present. All conceptacles uniporate.

Tenarea ascripticia (Fosl.) Adey

Lithophyllum pustulatum f. *ascripticium* Foslie 1907a: 34; Nichols 1909: 354. *Tenarea ascripticia* (Fosl.) Adey 1970: 6. *Dermatolithon pustulatum* f. *ascripticium* (Fosl.) DeToni 1924: 665; Dawson 1960: 32 (incl. synonymy). *D. ascripticium* (Fosl.) Setchell & Mason 1943: 96; Mason 1953: 344; Ganesan 1962: 108. *Fosliella ascripticia* (Fosl.) Smith 1944: 224.

Thalli epiphytic, deep pink to violet, completely adherent, sometimes encircling host, to 200+ μm thick; hypothallial cells (15)20–80(115) μm long; perithallium weakly developed; tetrasporangial and bisporangial conceptacles protruding, 200–350(500) μm outside diam.; sexual plants also known.

Common, low intertidal to subtidal, on a variety of coarse algae, Wash. to C. America. Type locality: Monterey, Calif., on *Botryoglossum.*

Dawson (1960) reported two collections of *Dermatolithon canescens* (Foslie) Foslie [=*Tenarea canescens* (Foslie) Adey, 1970] from S. Calif., but further study is required before it can be accepted for the Calif. flora.

Tenarea dispar (Fosl.) Adey

Lithophyllum tumidulum f. *dispar* Foslie 1907b: 29; Nichols 1909: 357. *Tenarea dispar* (Fosl.) Adey 1970: 7. *Dermatolithon dispar* (Fosl.) Fosl. 1909: 58; Mason 1953: 343; Dawson 1960: 34; Masaki & Tokida 1960: 37; Masaki 1968: 52. *Fosliella dispar* (Fosl.) Smith 1944: 225.

Thalli epiphytic, of small, irregular crusts mostly less than 1 mm but rarely to 2 mm thick, often encircling host; hypothallial cells 15–100 μm

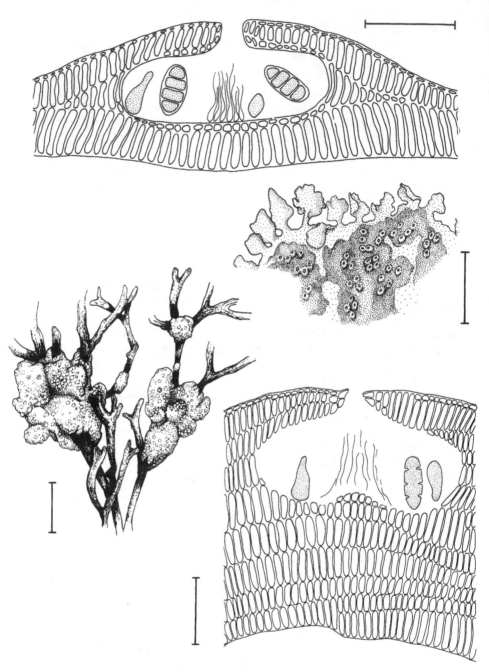

Fig. 341 (above & middle right). *Tenarea ascripticia*. Rules: section through uniporate tetrasporangial conceptacle, 100 µm; habit, 2 mm. (CS/HWJ & LH) **Fig. 342** (middle left & below). *T. dispar*. Rules: habit, 4 cm; section, 100 µm. (LH & CS/HWJ)

long; perithallium of 20(28) cell layers; tetrasporangial and bisporangial conceptacles immersed, to 350 µm inside diam.; dioecious and monoecious plants known.

Common, epiphytic on various red algae, including *Laurencia, Ahnfeltia, Gelidium,* and *Gymnogongrus,* as well as on articulated corallines, low intertidal, Whidbey I. (type locality), Wash., to Baja Calif.

Dawson (1960) reported the possible occurrence of *Dermatolithon corallinae* (Crouan & Crouan) Foslie, 1902b, at Santa Catalina I., but further study is required before it can be accepted for the Calif. flora.

Pseudolithophyllum Lemoine 1913

Thalli lacking genicula, coarse, crustose, sometimes branching. Hypothallium monostromatic or polystromatic. Perithallium well developed; megacells absent. Epithallium 1 or more cell layers thick, not photosynthetic. Secondary pit-connections absent; intercellular fusions present. All conceptacles uniporate.

Pseudolithophyllum neofarlowii (Setch. & Mason) Adey

Lithophyllum neofarlowii Setchell & Mason 1943: 95; Mason 1953: 341 (incl. synonymy); Dawson 1960: 45. *Pseudolithophyllum neofarlowii* (Setch. & Mason) Adey 1970: 13.

Thalli crustose, irregular, whitish (in high intertidal) to purple, growing over large areas of substrate, 0.3–1 mm thick, the surface rough, covered with subspherical excrescences to 2 mm diam., these often anastomosing; hypothallium weakly developed, only 2 or 3 cell layers thick; tetrasporangial conceptacles scattered, protruding moderately, about 250 µm outside diam.; sexual plants also collected.

Intertidal, saxicolous, Alaska to Baja Calif.; common in N. and C. Calif. Type locality: Monterey, Calif.

Choreonema Schmitz 1889

Thalli lacking genicula; endophytic in the intergenicula of articulated corallines, with only conceptacles external. Vegetative parts filamentous, with small, lateral cells (possibly modified cover cells) cut off at intervals. Secondary pit-connections absent. All conceptacles uniporate, protruding from intergenicula of host.

Choreonema thuretii (Born.) Schmitz

Melobesia thuretii Bornet 1878: 96. *Choreonema thuretii* (Born.) Schmitz 1889: 455; Suneson 1937: 53; Dawson 1960: 60.

Conceptacles ±100 µm diam.; other characters as for genus; bisporangial, tetrasporangial, male, and female plants known.

Infrequent, in warm water, low intertidal, endophytic in small articulated corallines such as *Jania* and *Haliptylon,* San Clemente I., Corona del

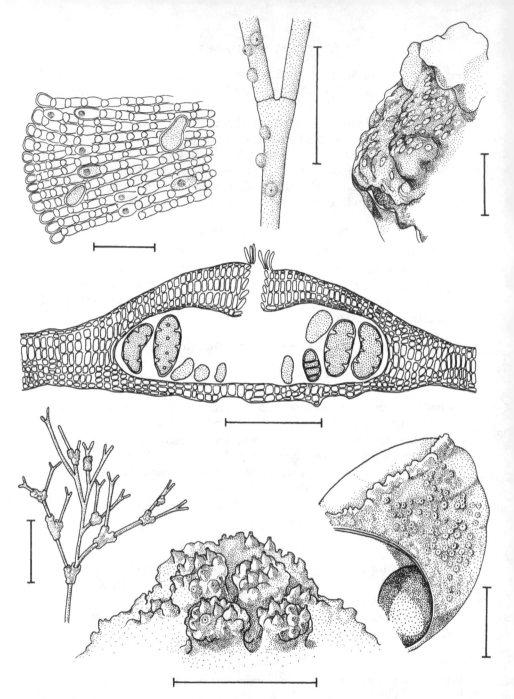

Fig. 343 (above right). *Pseudolithophyllum neofarlowii*. Rule: 5 mm. (SM) **Fig. 344** (above middle). *Choreonema thuretii*. Rule: 5 mm. (CS) **Fig. 345** (above left). *Fosliella paschalis*. Rule: 50 μm. (DBP/H) **Fig. 346** (center & below left). *Heteroderma nicholsii*. Rules: cross section, 100 μm; habit, 1 cm. (CS/HWJ & SM) **Fig. 347** (below right). *Hydrolithon decipiens*. Rule: 5 mm. (SM) **Fig. 348** (below middle). *Neogoniolithon setchellii*. Rule: 1 cm. (SM)

Mar (Orange Co.), Cardiff (San Diego Co.), and La Jolla, Calif., and south to Costa Rica. Type locality: Atlantic coast of France.

Fosliella Howe 1920

Thalli lacking genicula, epiphytic, crustose, thin. Vegetative parts a monostromatic hypothallium and a monostromatic epithallium. Megacells present. Secondary pit-connections lacking; intercellular fusions present. All conceptacles uniporate.

Fosliella paschalis (Lem.) S. & G.

Melobesia paschalis Lemoine 1924: 289. *Fosliella paschalis* (Lem.) Setchell & Gardner 1930: 176; Dawson 1960: 31; Hollenberg 1970: 65.

Thalli forming pinkish patches less than 25 μm thick on various algae; megacells producing stipitate, discoid outgrowths, which may become multicellular and may serve as disseminating units; bisporangial and tetrasporangial conceptacles to 160 μm diam.; sexual plants not known.

Infrequent, low intertidal, on *Rhodymenia pacifica* at Laguna Beach, Calif.; on *Eisenia* or *Laurencia* in Baja Calif., and Is. Revillagigedo, Mexico. Type locality: Easter I.

Heteroderma Foslie 1909

Thalli lacking genicula, crustose, very thin, of fewer than 5 cell layers except adjacent to conceptacles. Megacells and secondary pit-connections lacking; intercellular fusions present. All conceptacles uniporate.

Heteroderma nicholsii Setch. & Mason

Setchell & Mason 1943: 96; Mason 1953: 336 (incl. synonymy); Dawson 1960: 57; Dawson, Acleto & Foldvik 1964: 44. *Lithophyllum pustulatum* f. *australis* Foslie 1905a: 117; Nichols 1909: 356.

Thalli crustose, to 100(350) μm thick, pinkish; tetrasporangial conceptacles protruding, to 350 μm diam.; male and female plants known.

On various red and brown algae, N. Calif. to Peru; common subtidally in Calif. Type locality: La Jolla, Calif., on *Pachydictyon*.

Hydrolithon (Foslie) Foslie 1909

Thalli lacking genicula, crustose. Hypothallium monostromatic; cells approximately isodiametric. Perithallium thick; cells irregular in size and arrangement; isolated megacells in outer perithallium. Secondary pit-connections lacking; intercellular fusions present. All conceptacles uniporate.

Hydrolithon decipiens (Fosl.) Adey

Lithothamnium decipiens Foslie 1897: 20. *Hydrolithon decipiens* (Fosl.) Adey 1970: 11. *Lithophyllum decipiens* (Fosl.) Fosl. 1900b: 19; Mason 1953: 338 (incl. synonymy); Dawson 1960: 37; Masaki 1968: 33.

Thalli crustose, firmly adherent, to 500 μm thick, the surface macroscopically smooth; hypothallium to 5 cell layers thick; perithallium comprising most of crust; tetrasporangial conceptacles mostly immersed, to 200 μm diam.; male and female plants also known.

Common to frequent, on stones or shells, Br. Columbia to Galápagos Is. Type locality: San Pedro, Calif.

Neogoniolithon Setchell & Mason 1943

Thalli lacking genicula, crustose. Hypothallium polystromatic. Perithallium thick; megacells present in perithallium, singly or in vertical series. Secondary pit-connections absent; intercellular fusions present. All conceptacles uniporate.

Neogoniolithon setchellii (Fosl.) Adey

Lithothamnium setchellii Foslie 1897: 18. *Neogoniolithon setchellii* (Fosl.) Adey 1970: 9. *Hydrolithon setchellii* (Fosl.) Setchell & Mason 1943: 97; Mason 1953: 334 (incl. synonymy); Dawson 1960: 29.

Thalli conspicuous; crusts brittle, overlapping and becoming loosely cemented together; superimposed crusts to 2 mm thick in older parts, the individual crusts to 500 μm thick; excrescences numerous, 1–3 mm long, 1–3 mm diam.; tetrasporangial conceptacles 350–600 μm diam., protruding and rostrate; female plants also known; male plants unknown.

Locally abundant on rocks in low tidepools, Redondo Beach, Calif., to I. Guadalupe, Baja Calif. Type locality: San Pedro, Calif.

Articulated Species
Amphiroa Lamouroux 1812

Thalli of articulated fronds arising from crustose bases. Intergenicular medulla consisting of straight cells arrayed in numerous arching tiers of differing heights. Genicular medulla of straight cells arrayed in 1 to several arching tiers, also of differing heights; both intergenicula and genicula corticated. Branch apices with cover cells. Secondary pit-connections present; intercellular fusions lacking. Conceptacles lateral, uniporate, developing from cortical tissue. Roofs of tetrasporangial conceptacles sometimes formed by growth of tissue from within fertile area. Spermatangia developing only on floor of conceptacle.

Amphiroa zonata Yendo

Yendo 1902b: 10; Dawson 1953a: 146.

Thalli erect, to 6 cm tall, regularly to irregularly dichotomously branched; intergenicula cylindrical or compressed, with annular markings when young, to 1.2 mm diam., usually more than 1 cm long; intergenicular medulla usually with groups of 3 or 4 tiers of long cells alternating with single tier of short cells; genicula usually with 3 medullary tiers, the cortex discontinuous with intergenicular cortex; tetrasporangial and bisporangial conceptacles protruding slightly, 350–500 μm diam.; sexual plants not known.

Common on rocks in warm, sunny tidepools, Santa Catalina I., Laguna Beach, Corona del Mar (Orange Co.), and La Jolla, Calif., to S. America. Type locality: Japan.

Lithothrix Gray 1867

Thalli of crustose bases and articulated fronds. Intergenicular medulla of a single tier of short cells. Genicular medulla of a single tier of long cells, uncorticated but surrounded by calcified flange of cortical tissue, this produced by the surmounting intergeniculum. Branch apices with cover cells. Secondary pit-connections present; intercellular fusions lacking. Conceptacles lateral, uniporate, developing from cortical tissue.

Lithothrix aspergillum Gray

Gray 1867: 33; Manza 1940: 296; Smith 1944: 231 (incl. some synonymy); Segawa 1947: 87. *Amphiroa aspergillum* f. *nana* Setchell & Gardner 1903: 359. *Lithothrix aspergillum* f. *nana* (S. & G.) Yendo 1905: 14.

Thalli saxicolous, occasionally epizoic; fronds clustered, to 13 cm tall, dull purple; primary branching dichotomous, but irregularly distributed lateral branches sometimes abundant; lower intergenicula terete, ±0.5 mm long, 0.5–1 mm broad; upper intergenicula in main branches terete or often compressed, ±1 mm long, 1–1.5 mm broad, those in secondary branches smaller; conceptacles 350–450 μm diam.; most collected plants bisporangial; tetrasporangial, male, and female specimens known from San Diego Co.

Locally abundant on rocks or animals in sandy areas in lower intertidal, Br. Columbia to S. Baja Calif. Type locality: Vancouver I., Br. Columbia.

Yamadaea Segawa 1955

Thalli with extensive crustose bases and short articulated fronds. Fronds abbreviated, to 3 mm tall, each of 1(2) intergeniculum, this, if fertile, containing 1 conceptacle. Intergenicular medulla of arching tiers of straight cells of same length. Genicula of a single tier of long, thick-walled cells. Conceptacles axial in origin, the pore central. Spermatangial conceptacles beaked, the conceptacular canal more than 100 μm long. Fusion cell broad and thin, bearing carposporangial filaments, these mostly on periphery.

Yamadaea melobesioides Segawa

Segawa 1955: 241; Hollenberg & Abbott 1966: 63.

Basal crusts to 5+ cm broad, ±400 μm thick at margins; fronds dark violet, 2–3 mm tall, irregularly and sparsely distributed; other characters as for genus.

On rocks, subtidal (to 25 m), off Monterey, Calif; perhaps more widely

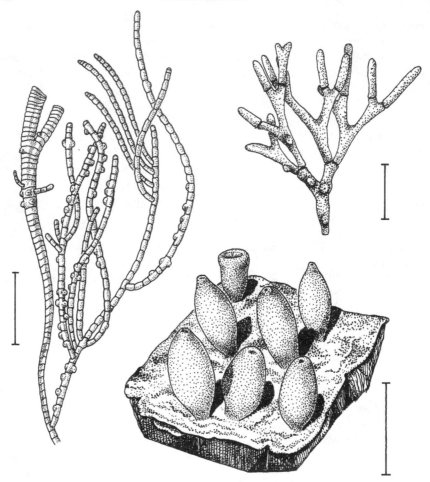

Fig. 349 (above right). *Amphiroa zonata*. Rule: 2 cm. (LH) Fig. 350 (left). *Lithothrix aspergillum*. Rule: 1 cm. (SM) Fig. 351 (below right). *Yamadaea melobesioides*. Rule: 1 mm. (DBP)

distributed, but this unknown owing to diminutive size of fronds. Type locality: Japan.

Corallina Linnaeus 1758

Thalli of crustose bases and articulated, pinnately branched fronds; usually saxicolous, sometimes epiphytic or epizoic. Intergenicular medulla with arching tiers of straight cells of same length. Genicula with single tier of long, thick-walled cells. Conceptacles axial in origin, 1 in each fertile intergeniculum, sometimes bulging from intergenicular surfaces, the pores central. Spermatangial conceptacles beaked, the conceptacular canal 100+ μm long. Fusion cell broad and thin, bearing carposporangial filaments, these mostly on periphery.

1. At least some intergenicula frequently with 3+ lateral branches........ 2
1. Axial intergenicula usually with 2 lateral branches................... 3
   2. Fronds stiff ....*C. polysticha*
   2. Fronds flaccid ....*C. vancouveriensis*
3. Intergenicula markedly flattened, with more or less conspicuous wings.... 4
3. Intergenicula not markedly flattened, without conspicuous wings........ 5
   4. Intergenicula ±1 mm long; plants intertidal or upper subtidal.....
   .... *C. frondescens*
   4. Intergenicula ±2 mm long; plants occurring in lower subtidal......
   .... *C. bathybentha*
5. Bulging conceptacles on intergenicular surfaces frequently present....
   .... *C. pinnatifolia*
5. Bulging conceptacles on intergenicular surfaces rarely present.......... 6
   6. Axial intergenicula more than 1 mm broad and 1.5 mm long; lateral
      branches progressively shorter toward tip of frond and not appressed
      to main axes ....*C. officinalis*
   6. Axial intergenicula less than 1 mm broad and 1 mm long; lateral
      branches short, all about same length, laterally appressed to main
      axes ....*C. vancouveriensis*

Corallina bathybentha Daws.

Dawson 1949b: 6.

Fronds to 6 cm tall, the branching regular, of 1 or 2 orders; axial intergenicula to 2+ mm long, flat, with pronounced wings; axial conceptacles sometimes appearing marginal when occurring on upper margins of intergenicular wings, to 0.5 mm diam.; tetrasporangial and spermatangial plants collected, characters of these as for genus.

Rare, saxicolous, subtidal (to 52 m), known from Mendocino (city) to Santa Barbara I. and Anacapa I. (type locality), Calif.

Corallina frondescens Post. & Rupr.

Postels & Ruprecht 1840: 20. *Arthrocardia frondescens* (Post. & Rupr.) Areschoug 1852: 549. *Amphiroa tuberculosa* f. *frondescens* (Post. & Rupr.) Setchell & Gardner 1903: 362. *Bossiella frondescens* (Post. & Rupr.) Dawson 1964: 540. *Joculator delicatulus* Doty 1947b: 167. *Corallina pinnatifolia* var. *digitata* Daws. 1953a: 125.

Fronds to 5 cm tall; axial intergenicula to 1 mm long, flattened, winged to the extent that laterals allow this; terminal intergenicula sometimes broadly expanded, irregularly incised; conceptacles axial, sometimes also bulging from intergenicular surfaces.

Saxicolous, upper subtidal, Aleutian Is. to Mexico; rare in Calif. Type locality: Unalaska I., Alaska.

Corallina officinalis L.

Linnaeus 1758: 805.

The type variety is not found in California; var. *chilensis* differs from var. *officinalis* in having larger intergenicula and in being more compoundly branched.

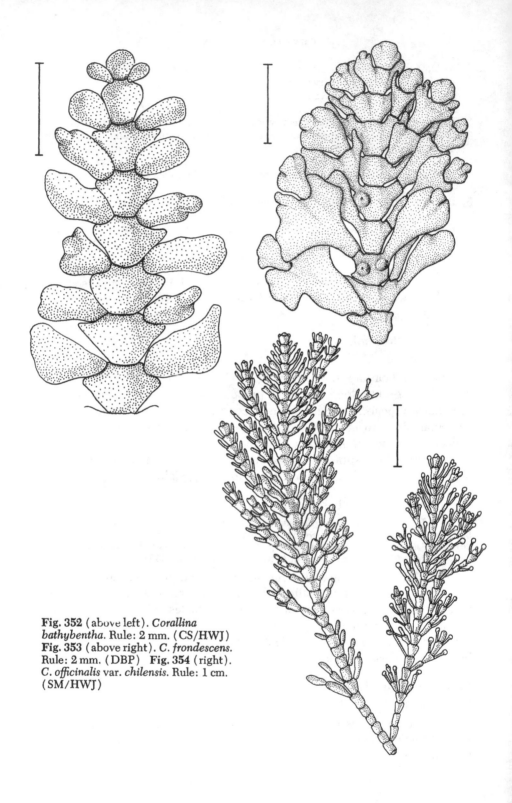

Fig. 352 (above left). *Corallina bathybentha*. Rule: 2 mm. (CS/HWJ)
Fig. 353 (above right). *C. frondescens*. Rule: 2 mm. (DBP) **Fig. 354** (right). *C. officinalis* var. *chilensis*. Rule: 1 cm. (SM/HWJ)

Corallina officinalis var. chilensis (Dec.) Kütz.

Corallina chilensis Decaisne 1847: 103; Smith 1944: 230. *C. officinalis* var. *chilensis* (Dec.) Kützing 1858: 32; Dawson 1953a: 132; Dawson, Acleto & Foldvik 1964: 45.

Fronds whitish, pinkish, purplish, to 15 cm tall, mostly bipinnate to tripinnate; branches progressively shorter near apex and tending to lie in 1 plane; axial intergenicula flat, usually not winged, 1–2 mm long, to 1.5 mm broad; intergenicula in lateral branches robust; conceptacles 0.5–0.6 mm diam.; tetrasporangial and male specimens collected.

Common, saxicolous, in low intertidal pools and subtidally, Alaska to Chile. Type locality: Chile.

A very variable species in habitat preference and morphology.

Corallina pinnatifolia (Manza) Daws.

Joculator pinnatifolius Manza 1937a: 47; 1940: 263. *Corallina pinnatifolia* (Manza) Dawson 1953a: 124.

Fronds to 7 cm tall, forming dense clumps; axes densely branched to 1 or 2 orders; axial intergenicula to 1.5 mm long, to 2 mm broad; non-axial intergenicula sometimes flattened and irregularly lobed to varying extents; conceptacles axial, sometimes also bulging from intergenicular surfaces; tetrasporangial and male plants known.

Common saxicolous, in tidepools, Carpinteria (Santa Barbara Co.), Calif., to Ecuador. Type locality: Orange Co., Calif.

Corallina polysticha Daws.

Dawson 1953a: 131.

Fronds to 4 cm tall, bunched, stiff and erect, 2–10 lateral branches arising from each branching intergeniculum in a polystichous or verticillate manner; axial intergenicula to 1 mm long; conceptacles axial, ±0.5 mm diam.

Uncommon, saxicolous, favoring intertidal warm-water areas, Santa Catalina I., Calif., to Bahía Vizcaíno, Baja Calif. Type locality: I. Guadalupe, Mexico.

Corallina vancouveriensis Yendo

Yendo 1902a: 719; Dawson 1953a: 126; Ganesan 1968a: 67. *Corallina aculeata* Yendo 1902a: 720. *C. vancouveriensis* var. *aculeata* (Yendo) Daws. 1953a: 128. *C. gracilis* f. *densa* Collins 1906a: 112; Smith 1944: 230. *C. densa* (Coll.) Doty 1947b: 167. *C. vancouveriensis* f. *densa* Yendo 1902a: 719. *C. gracilis* var. *lycopodioides* Taylor 1945: 200. *C. vancouveriensis* var. *lycopodioides* (Tayl.) Daws. 1953a: 129.

Fronds dark violet to purplish, to 14+ cm long, usually densely branched, congested in limp tufts pendant on rock faces or more or less erect in tidepools; axial intergenicula to 1 mm long and 1 mm broad, all or almost all axial intergenicula bearing 2–5 adaxially appressed branches,

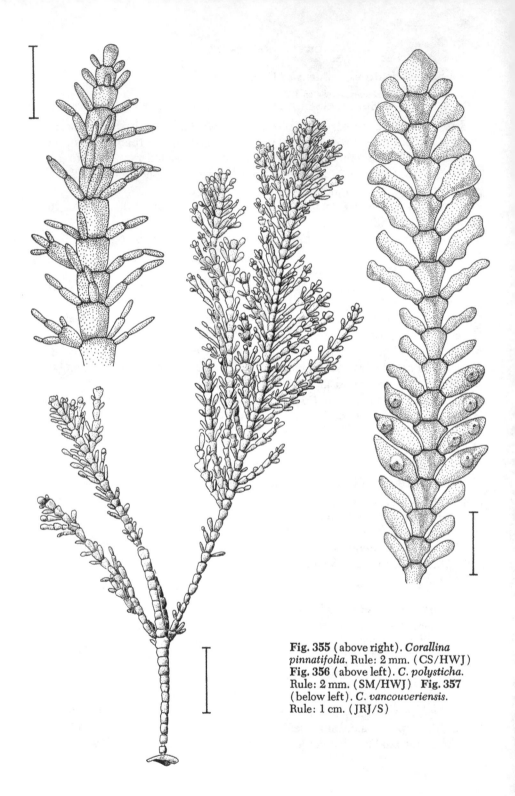

Fig. 355 (above right). *Corallina pinnatifolia*. Rule: 2 mm. (CS/HWJ) **Fig. 356** (above left). *C. polysticha*. Rule: 2 mm. (SM/HWJ) **Fig. 357** (below left). *C. vancouveriensis*. Rule: 1 cm. (JRJ/S)

these 2+ mm long, usually considerably narrower than main axes; ultimate or penultimate intergenicula sometimes broad, irregularly lobed or incised; conceptacles to 500 μm outside diam.

On rocks and in tidepools, Aleutian Is. to Galápagos Is.; extremely common on intertidal rocks in Calif. Type locality: Port Renfrew (Vancouver I.), Br. Columbia.

In Calif., there are several ecological variants approximately corresponding to varieties *vancouveriensis, densa, aculeata,* and *lycopodioides* as interpreted by Dawson (1953a).

This is the only articulate Calif. coralline able to withstand several hours of exposure.

Chiharaea Johansen 1966

Thalli saxicolous, with extensive crustose bases and articulated fronds. Fronds small, decumbent, compressed and dorsiventrally oriented, irregularly branched, of 14 or fewer intergenicula. Intergenicular medulla with arching tiers of straight cells of same height. Genicula with single tier of long, thick-walled cells. Conceptacles axial, 1–3 per intergeniculum; fertile intergenicula rarely branched; pore on dorsal surface in tetrasporangial plants, central in sexual plants. Spermatangial conceptacles beaked; conceptacular canal 100+ μm long. Fusion cell broad and thin, with carposporangial filaments arising from entire surface.

Chiharaea bodegensis Johans.

Johansen 1966: 59.

Fronds dark purplish, sparse or congested, each to 5 mm long, with up to 14 intergenicula; distal intergenicula broader and flatter than proximal ones, to 1.5 mm broad; tetrasporangial conceptacles deeply embedded; male and female plants also known.

Uncommon and inconspicuous, on rocks, low intertidal to subtidal, usually in areas of considerable surge, N. Calif. to Malpaso Creek (Monterey Co.), Calif. Type locality: Bodega Head (Sonoma Co.), Calif.

Arthrocardia Decaisne 1842

Thalli of crustose bases and articulated fronds. Fronds pinnately branched in vegetative parts, dichotomously branched in fertile parts. Intergenicular medulla with arching tiers of straight cells of same length. Genicula with a single tier of long, thick-walled cells. Conceptacles axial in origin; pore terminal or subterminal (in Calif. specimens). Spermatangial conceptacles beaked; conceptacular canal 100+ μm long. Fusion cell thin and broad, with carposporangial filaments arising from entire surface.

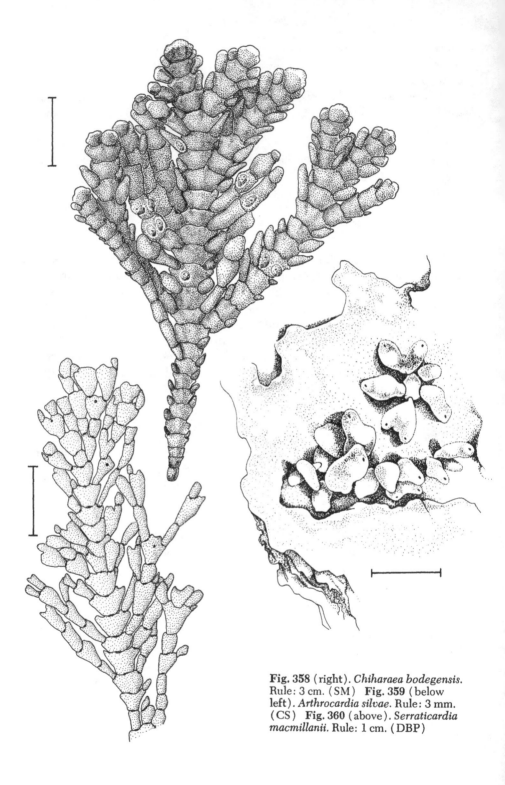

Fig. 358 (right). *Chiharaea bodegensis.*
Rule: 3 cm. (SM) **Fig. 359** (below
left). *Arthrocardia silvae.* Rule: 3 mm.
(CS) **Fig. 360** (above). *Serraticardia
macmillanii.* Rule: 1 cm. (DBP)

Arthrocardia silvae Johans.

Johansen 1971a: 241.

Thalli saxicolous; fronds purplish, to 2(3) cm tall, the axes percurrent, the lateral branches vegetative or fertile; axial intergenicula to 1(1.5) mm long, 0.5–1 mm broad, wings lacking or inconspicuous; fertile branches with 1–4 conceptacles, sometimes more in sexual plants; pores of tetrasporangial conceptacles subterminal on 1 surface of fertile intergenicula, this tendency less evident in carposporangial plants; male plants also known.

On rocks, Humboldt Co., Mendocino Co., Pt. Reyes, and from near Carmel River (Monterey Co.), Calif.; locally abundant on exposed intertidal rocks in N. Calif., where it superficially resembles *Bossiella plumosa* or *Corallina frondescens*. Type locality: Wheeler (Jackass Creek, Mendocino Co.), Calif.

Serraticardia (Yendo) Silva 1957

Thalli of crustose bases and geniculated, pinnately branched fronds; usually saxicolous. Lateral branches arising from every or almost every upper axial intergeniculum. Intergenicular medulla with arching tiers of straight cells of same length. Genicula with a single tier of long, thick-walled cells. Conceptacles axial on intergenicular surfaces. Spermatangial conceptacles beaked, the conceptacular canal 100+ µm long. Fusion cell broad and thin, with carposporangial filaments arising from periphery or surface.

Serraticardia macmillanii (Yendo) Silva

Cheilosporum macmillanii Yendo 1902a: 718. *Serraticardia macmillanii* (Yendo) Silva 1957b: 48 (incl. synonymy); Hollenberg & Abbott 1966: 64; Ganesan 1968b: 10; Johansen 1971a: 247.

Thalli saxicolous; fronds to 12 cm tall, pinnately to tripinnately branched; lateral branches arising from every or almost every upper axial intergeniculum (sometimes shedding some lateral branches, with result that branching appears intermittent); axial intergenicula hexagonal in outline, wingless, to 3 mm broad, (0.8)1–1.5(2) mm long; tetrasporangial conceptacles lateral, the pores sometimes excentric; sexual plants monoecious, the male conceptacles axial, the female conceptacles lateral.

On rocks in surge channels and in heavy surf, irregularly distributed but sometimes locally abundant, Alaska to Channel Is., Calif. Type locality: Port Renfrew (Vancouver I.), Br. Columbia.

Bossiella Silva 1957b

Thalli of crustose bases and articulated fronds. Fronds pinnately or dichotomously branched. Intergenicula flat, winged to varying degrees in

Calif. species. Intergenicular medulla with arching tiers of straight cells of same length. Genicula with a single tier of long, thick-walled cells. Conceptacles cortical in origin, usually originating in intergenicula somewhat below branch apices, the pores central or excentric; 2–8+ tetrasporangial or bisporangial conceptacles on surface of fertile intergeniculum, the sexual plants generally having more conceptacles per intergeniculum; more than 10 sporangia produced in tetrasporangial or bisporangial conceptacle at any one time. Spermatangial conceptacles beaked, the conceptacular canal 100+ µm long. Procarpic conceptacles containing 100+ supporting cells. Fusion cell thin and broad, with carposporangial filaments arising anywhere on surface.

1. Intergenicula in upper parts more than 2 mm long; frequently 4+ tetrasporangial or bisporangial conceptacles on each fertile intergeniculum; branching irregular, often a mixture of pinnate and dichotomous......
   .... B. californica
1. Intergenicula in upper parts less than 2 mm long; usually 2 (sometimes 4) tetrasporangial or bisporangial conceptacles on each fertile intergeniculum; branching usually clearly pinnate or dichotomous 2
   2. Branching primarily dichotomous .... B. orbigniana
   2. Branching primarily pinnate 3
3. Upper axial intergenicula mostly branched; some branches with only 1 intergeniculum .... B. plumosa
3. Many to fewer than half of upper axial intergenicula branched; lateral branches rarely with only 1 intergeniculum .... B. chiloensis

Bossiella californica (Dec.) Silva

Amphiroa californica Decaisne 1842: 124. *Bossiella californica* (Dec.) Silva 1957b: 46.

Plants forming fanlike tufts, erect or dorsiventrally arranged, the branching dichotomous, pinnate, or irregular (often a combination of these patterns); branches often showing midrib if wings are thin, or without midrib if wings are thick; wings close-set; intergenicula in upper parts more than 2.5 mm long. Tetrasporangial conceptacles more common than bisporangial conceptacles, 4+ of either on each fertile intergeniculum.

Bossiella californica ssp. californica

Silva 1957b: 46; Johansen 1971b: 388 (incl. synonymy).

Fronds to 12 cm tall; branching irregular, pinnate or dichotomous; stipe to 2 mm broad; intergenicula 2–5 mm long, to 6 mm broad, sometimes obscurely sagittate, wings blunt and thick; tetrasporangial and bisporangial conceptacles (2)4–6(14) on surface of each fertile intergeniculum, in 2 rows parallel to and on each side of axis, the pore central; male and female plants also known.

Locally common, low intertidal, known from Bodega Head (Sonoma Co.), Monterey Peninsula, and Channel Is., Calif., to Bahía Asunción, Baja Calif. Type locality: Monterey Peninsula, Calif.

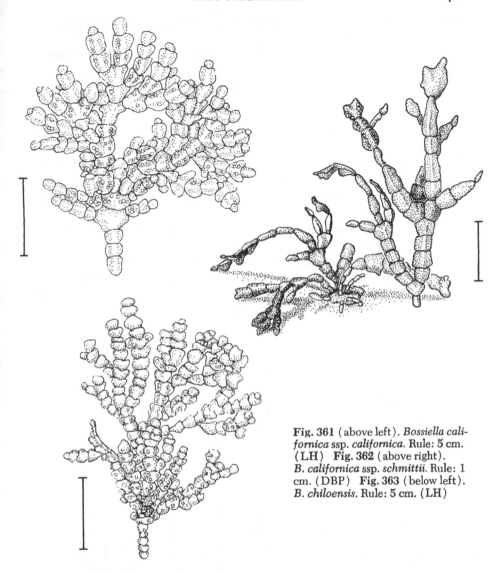

Fig. 361 (above left). *Bossiella californica* ssp. *californica*. Rule: 5 cm. (LH) **Fig. 362** (above right). *B. californica* ssp. *schmittii*. Rule: 1 cm. (DBP) **Fig. 363** (below left). *B. chiloensis*. Rule: 5 cm. (LH)

Bossiella californica ssp. schmittii (Manza) Johans.

Calliarthron schmittii Manza 1937b: 566. *Bossiella schmittii* (Manza) Johansen 1969: 61. *B. californica* ssp. *schmittii* (Manza) Johans. 1971b: 389.

Plants like subsp. *californica* with following differences: fronds usually growing approximately horizontally from ridges of rock; dorsal surfaces dark violet, ventral surfaces lighter shades; intergenicula 3–12 mm long, to 10(15) mm broad, the wings thin, flat, usually bending down, the midrib prominent on ventral surface; ventral surface frequently harboring bryozoans and tube worms; conceptacles usually restricted to dorsal surface, to 50+ on each fertile intergeniculum; tetrasporangial, bisporangial, male, and female plants known.

On rocks, characteristically subtidal, sometimes common in *Macrocystis pyrifera* beds, Br. Columbia to San Diego Co., Calif. Type locality: off Pt. Loma (San Diego Co.), Calif., in 40–46 m.

Bossiella chiloensis (Dec.) Johans.

Amphiroa chiloensis Decaisne 1842: 125. *Bossiella chiloensis* (Dec.) Johansen 1971b: 389 (incl. synonymy).

Plants to 20 cm tall, branching densely bi- and tripinnate, sometimes interrupted by unbranched intergenicula; intergenicula 1–2.5 mm long, to 4 mm broad, sagittate to broadly rounded; upper and lower wing margins tending to be parallel to and in contact with adjacent wings; 2 wings of intergeniculum sometimes unequal in size; tetrasporangial conceptacles (1)2–4(8) on each intergenicular surface, the pore central or excentric; tetrasporangial, bisporangial, male, and female plants known.

Rarely abundant, on upper subtidal rocks, Pacific Coast of N. and S. America; in Calif., known definitely at Bodega Head (Sonoma Co.), and Monterey, Calif. Type locality: I. Chiloe, Chile.

Bossiella orbigniana (Dec.) Silva

Amphiroa orbigniana Decaisne 1842: 124. *Bossiella orbigniana* (Dec.) Silva 1957b: 47.

Plants nearly always dichotomously branched (rarely pinnate); margins of branches incised, the degree owing to close or wide separation of wings of arrow-shaped intergenicula; intergenicula of upper parts less than 2.5 mm long; bisporangial conceptacles more common than tetrasporangial, 2+ on each fertile intergeniculum.

Bossiella orbigniana ssp. orbigniana

Silva 1957b: 47; Johansen 1971b: 392 (incl. synonymy).

Plants to 15(30) cm tall, the branching dichotomous, rarely with pinnate branching interspersed, unbranched intergenicula frequent; intergenicula sagittate with acute or subacute lobes, to 5 mm broad, (1)1.5–2(4.5) mm long; wings conspicuous or more or less reduced, thin or thick; outer margin of each wing curving convexly down to midrib and not distinguishable from lower margin, the midrib often projecting above and/or below wing margins; wings of successive intergenicula separated by gaps, and branches appearing serrate; conceptacles (1)2–4(8) on each intergenicular surface; plants mostly bisporangial, sometimes tetrasporangial, the pore occasionally excentric; male plants known, female plants unknown.

Infrequent to common, saxicolous, mostly subtidal (to 30+ m), Ore. to southern S. America; common off Monterey and S. Calif. Type locality: Patagonia, Argentina.

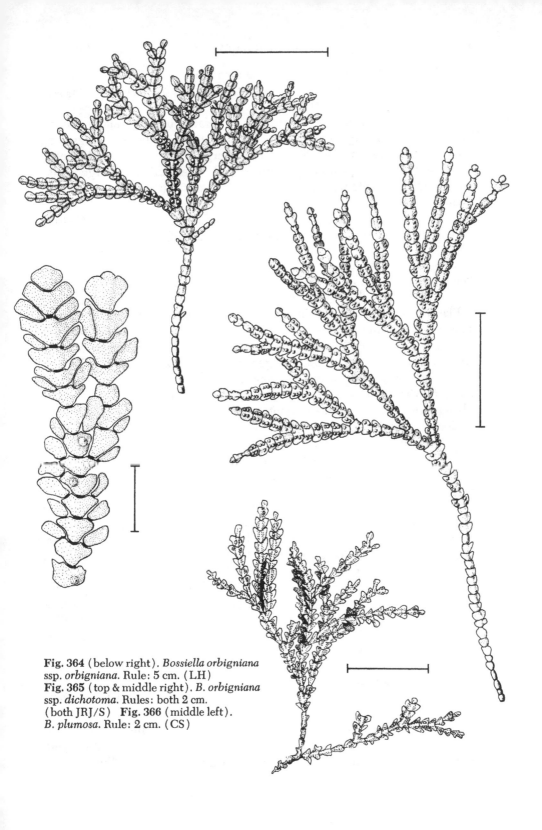

Fig. 364 (below right). *Bossiella orbigniana*
ssp. *orbigniana*. Rule: 5 cm. (LH)
Fig. 365 (top & middle right). *B. orbigniana*
ssp. *dichotoma*. Rules: both 2 cm.
(both JRJ/S) **Fig. 366** (middle left).
B. plumosa. Rule: 2 cm. (CS)

Bossiella orbigniana ssp. **dichotoma** (Manza) Johans.

Bossea dichotoma Manza 1937b: 562. *Bossiella orbigniana* ssp. *dichotoma* (Manza) Johansen 1971b: 394 (incl. synonymy).

Plants like subspecies *orbigniana,* but with following differences: fronds to 21+ cm tall; intergenicula 1–2.5 mm long, to 6 mm broad; wings conspicuous, the wing margins of adjacent intergenicula parallel and in contact, the midrib not projecting at upper and lower ends of intergenicula; most collected plants bisporangial, some tetrasporangial; male and female plants also known.

Locally frequent, saxicolous, low intertidal, Br. Columbia to I. Cedros, Baja Calif. Type locality: Moss Beach (San Mateo Co.), Calif.

Bossiella plumosa (Manza) Silva

Bossea plumosa Manza 1937a: 46; 1940: 303. *Bossiella plumosa* (Manza) Silva 1957b: 47; Ganesan 1968b: 16. *Bossea frondifera* Manza 1937b: 562.

Plants to 7 cm tall; branching pinnate from every or almost every upper axial intergeniculum, some pinnae consisting of a single intergeniculum; intergenicula generally 1 mm long or less, to 3 mm broad, prominently winged, the wings thin, sharp-edged, the upper part of wings lacking where branches arise; usually 2 conceptacles on face of fertile intergeniculum, the pore central; tetrasporangial plants common, bisporangial plants rare; male and female plants also known.

Common, saxicolous, lower intertidal, Alaska to Baja Calif. Type locality: Moss Beach (San Mateo Co.), Calif.

Easily confused with *Corallina frondescens,* but lacking the axial and pseudolateral conceptacles present in that species.

Calliarthron Manza 1937a

Thalli saxicolous, with crustose bases and geniculate fronds. Fronds dichotomously, pinnately, or irregularly branched. Intergenicula flat, terete, or subterete. Intergenicular medulla of undulate filaments, these forming arching tiers of cells of same length. Genicula with a single tier of long, thick-walled cells. Conceptacles marginal and, in lower parts of fronds, on intergenicular surfaces, 1 to several per intergeniculum, the pore central. Spermatangial conceptacles beaked, the conceptacular canal 100+ μm long. Fusion cell broad and thin, with carposporangial filaments arising from entire surface.

Calliarthron cheilosporioides Manza

Manza 1937a: 46; 1940: 266; Smith 1944: 237; Dawson 1953a: 113; Johansen 1969: 60.

Fronds to 30 cm tall; branching pinnate, sometimes dichotomous near base; lower intergenicula to 7 mm long; upper intergenicula flat, winged, to 6 mm broad, 2–3 mm long, the upper margins of wings angling upward

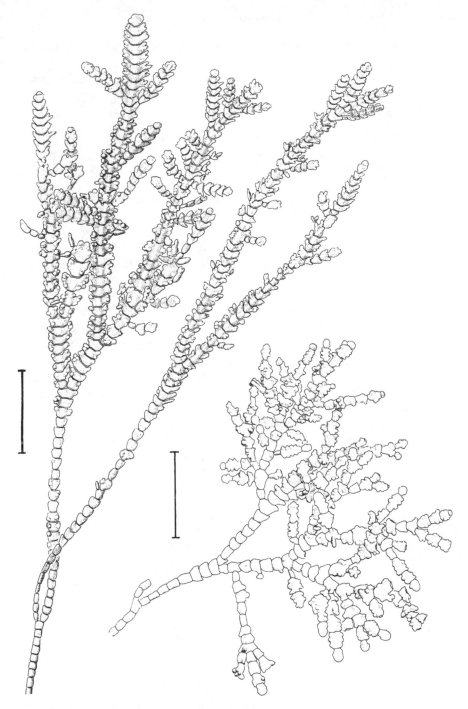

Fig. 367 (left). *Calliarthron cheilosporioides*. Rule: 2 cm. (JRJ/S) **Fig. 368** (right). *C. tuberculosum*. Rule: 2 cm. (JRJ/S)

at 45–55° to long axis of branch; tetrasporangial, male, and female plants known.

Frequent, saxicolous, low intertidal and upper subtidal in surfy areas, Pacific Grove, Calif. (type locality) to I. Cedros, Baja Calif.; common subtidally in kelp beds off S. Calif.

Calliarthron tuberculosum (Post. & Rupr.) Dawson

Corallina tuberculosa Postels & Ruprecht 1840: 20. *Calliarthron tuberculosum* (Post. & Rupr.) Dawson 1964: 540 (incl. synonymy); Johansen 1969: 1 (incl. synonymy). *Calliarthron setchelliae* Manza 1937b: 566; Smith 1944: 237.

Fronds to 20+ cm high; branching pinnate, dichotomous, or irregular; upper intergenicula flat or subterete; where upper margins of intergenicula are discernible, angle greater than 60° to axis of branch; intergenicula to 4 mm broad, 2–6 mm long; tetrasporangial, male, and female plants known.

Often extremely common in low tidepools or subtidally in kelp beds and elsewhere, Alaska to I. Cedros, Baja Calif. Type locality: Sitka, Alaska.

Haliptylon (Decaisne) Johansen 1971

Thalli with crustose or rhizomatous bases and articulated fronds. Fronds pinnately branched, sometimes dichotomous near branch tips. Nonaxial intergenicula terete or subterete. Intergenicular medulla with arching tiers of straight cells of same length. Genicula with a single tier of thick-walled cells. Conceptacles axial in origin, 1 in each fertile intergeniculum. Spermatangial conceptacles narrow, elongate, unbeaked; conceptacular canal less than 100 µm long. Fusion cell narrow, thick, biconvex, with carposporangial filaments arising strictly from margin. Intergenicula that bear tetrasporangial or carposporangial conceptacles producing 2+ surmounting branches.

Haliptylon gracile (Lamour.) Johans.

Corallina gracilis Lamouroux 1816: 288; Dawson 1953a: 129; Ganesan 1968a: 71. *Haliptylon gracile* (Lamour.) Johansen 1971a: 243.

Fronds to 10 cm tall, usually saxicolous, multipinnate; axial intergenicula 500–800 µm long, to 500 µm broad; intergenicula in branches smaller, as little as 100 µm diam. or less; tetrasporangial conceptacles to 400 µm diam.; sexual plants not known.

Locally common, saxicolous, low intertidal, warm-water areas, Channel Is. and S. Calif. mainland to Pta. Malarrimo, Baja Calif. Type locality: "Australasie."

Jania Lamouroux 1812

Thalli with crustose or rhizomatous bases and articulated fronds. Fronds dichotomously branched throughout, or occasionally bearing short pinnae or secondarily produced laterals. Intergenicula terete or subterete. Inter-

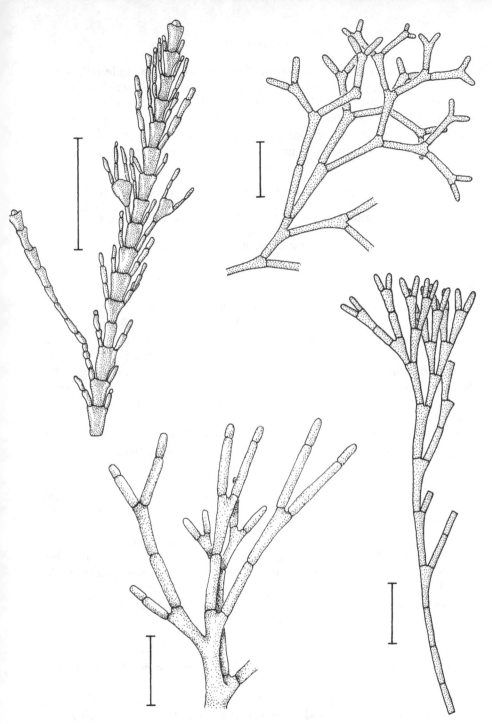

Fig. 369 (above left). *Haliptylon gracile*. Rule: 3 mm. (SM) **Fig. 370** (above right).
Jania adhaerens. Rule: 3 mm. (CS) **Fig. 371** (below left). *J. crassa*. Rule: 1 mm.
(CS/HWJ) **Fig. 372** (below right). *J. tenella*. Rule: 1 mm. (CS)

genicular medulla with arching tiers of straight cells of same length. Genicula with a single tier of thick-walled cells. Conceptacles axial in origin, 1 in each fertile intergeniculum. Spermatangial conceptacles narrow, elongate, unbeaked; conceptacular canal less than 100 μm long. Fusion cell thick, biconvex, with carposporangial filaments arising strictly from margin. Those intergenicula bearing tetrasporangial or carposporangial conceptacles producing 2+ surmounting branches.

1. Intergenicula mostly more than 200 μm diam.; plants usually more than 3 cm tall .... *J. crassa*
1. Intergenicula mostly less than 200 μm diam.; plants usually less than 2 cm tall 2
   2. Angles at dichotomies mostly more than 45° .... *J. adhaerens*
   2. Angles at dichotomies mostly less than 45° .... *J. tenella*

Jania adhaerens Lamour.

Lamouroux 1816: 270. *Corallina decussato-dichotoma* Yendo 1902b: 25. *Jania decussato-dichotoma* (Yendo) Yendo 1905: 37; Dawson 1953a: 117.

Thalli forming dense turfs, to 1+ cm thick, often intermingled with other algae; bases rhizomatous; intergenicula 100–200 μm broad, 2–5 times as long; branching dichotomous, the branch angles usually more than 45°; tetrasporangial conceptacles 200–300 μm outside diam.; spermatangial plants poorly known; cystocarpic conceptacles vasiform, with lateral branches on top margin.

Common on rocks in warm-water localities, Cardiff (San Diego Co.), Calif., to Gulf of Calif. and into tropics. Type locality: Mediterranean.

Jania crassa Lamour.

Lamouroux 1821: 23. *Jania micrarthrodia* var. *crassa* (Lamour.) Areschoug 1852: 555. *J. natalensis* Harvey 1847: 107; Dawson 1953a: 118; Dawson, Acleto & Foldvik 1964: 46. *Corallina natalensis* (Harv.) Kützing 1858: 38.

Fronds forming clumps to 7 cm tall, arising from crustose bases; intergenicula (150)200–300(400) μm diam., usually 3–5 times as long but length variable; branch angles less than 45°; tetrasporangial conceptacles 400–450 μm outside diam.; sexual plants not known for Calif.

Infrequent, on rocks in warm tidepools, Santa Barbara Co., Calif., to Peru. Type locality: New Zealand.

Jania tenella (Kütz.) Grun.

Corallina tenella Kützing 1858: 41. *Jania tenella* (Kütz.) Grunow 1873: 42; Dawson 1953a: 120.

Thalli usually epiphytic, sometimes saxicolous; fronds forming soft, dense tufts 1–2 cm tall, arising from small, crustose bases; branch angles less than 45°; intergenicula less than 150 μm diam., 3–5 times as long; tetrasporangial conceptacles ±250 μm outside diam.; spermatangial conceptacles ±150 μm outside diam., 300 μm long.

Uncommon, on rocks or other algae in warm tidepools, mid- to low intertidal, White Cove (Santa Catalina I.) and Cardiff (San Diego Co.), Calif., to C. America. Widespread in tropical regions. Type locality: Italy.

Family GLOIOSIPHONIACEAE

Plants structurally similar to those of Dumontiaceae, uniaxial, with 4 laterals arising from each cell of axial strand to form filamentous cortex, this at first loosely held in gelatinous matrix, later becoming less gelatinous and more compact. Plants differing from those of Dumontiaceae principally in uniting of carpogonial and auxiliary-cell branches, these borne on same supporting cell, usually the basal cell of a lateral. Tetrasporangia cruciately divided, borne on crustose alternating phase. Spermatangia in superficial patches. Carpogonial and auxiliary-cell branches of 3–5 and 4–7 cells, respectively; auxiliary cell terminal or intercalary. Immediate post-fertilization stages different from those in Dumontiaceae and varying in each genus in family. Gonimoblasts small, composed of a mass of carposporangia, with or without carpostome; surrounding sterile filaments lacking.

Two genera in the California flora, *Gloiosiphonia* (below) and *Schimmelmannia* (p. 421).

Gloiosiphonia Berkeley 1833

Thalli with large gametangial stages alternating with small, crustose, tetrasporangial phases. Gametangial plants gregarious, tufted, branched in whorls or radially. Branches slender, attenuate, very gelatinous to firm; younger branches internally with 4 loosely compacted laterals; older branches thickened by rhizoidal filaments formed from basal cells of laterals. Tetrasporangia cruciately divided, borne on crustose stage resembling *Cruoriopsis* (Peyssonelliaceae). Spermatangia in superficial patches. Carpogonial branches short, usually 3-celled; auxiliary-cell branch separated from carpogonial branch. Carpogonium, after fertilization, reorganizing into 2 or 3 connecting filaments, these fusing with 1 or more auxiliary cells intercalary in auxiliary-cell branch. Gonimoblasts small, all cells becoming carposporangia.

Gloiosiphonia capillaris (Huds.) Berk.

Fucus capillaris Hudson 1778: 591. *Gloiosiphonia capillaris* (Huds.) Berkeley 1833: 45; Edelstein 1972: 227. *Nemastoma californica* Farlow 1877: 243. *G. californica* (Farl.) J. Agardh 1884–85 (1885): 10; Smith 1944: 209. *Calosiphonia californica* (Farl.) J. Ag. 1899: 83.

Thalli light rose, 7–15(60) cm tall, repeatedly radially branched; central axis to 5 mm wide, gelatinous to firm, subterete to flattened; branchlets frequently attenuate, the apices filiform; plants monoecious or protandrous.

Locally common spring annual in C. Calif., infrequent elsewhere; sax-

Fig. 373 (above left). *Gloiosiphonia capillaris*. Rule: 2 cm. (SM)
Fig. 374 (below left). *G. verticillaris*. Rule: 2 cm. (LH)
Fig. 375 (right). *Schimmelmannia plumosa*. Rule: 2 cm. (JRJ/S)

icolous, low intertidal in deep tidepools, both sides of N. Pacific; in Calif., from Humboldt Co. to Santa Barbara Co. Also N. Atlantic. Type locality: England.

Gloiosiphonia verticillaris Farl.

Farlow 1889: 3; Smith 1944: 209.

Thalli deep rose, slippery, in gregarious tufts 20–30(60) cm tall; axes cylindrical, infrequently divided, to 3 mm wide; branches whorled; branchlets simple when young, frequently repeatedly divided to third and fourth order in upper parts of thallus, especially when cystocarpic; branches thickened by internal rhizoidal filaments when older; plants dioecious (protandrous?).

Locally abundant spring annual, saxicolous, low intertidal in areas scoured by sand, Alaska to San Luis Obispo Co., Calif. Type locality: Santa Cruz, Calif.

Far more common than *G. capillaris* in Calif.; the tetrasporangial stage, as yet unknown, is expected to be similar to that of *G. capillaris*.

Schimmelmannia Kützing 1847

Thalli uniaxial with percurrent axes. Branching distichous, dense; branches irregular, mostly at tops of thalli; margins serrate. Tetrasporangia cruciately divided, borne on crustose stage. Spermatangia in superficial patches. Carpogonial and auxiliary-cell branches borne on same supporting cell. Carpogonium after fertilization dividing once or twice, the terminal or median cell fusing as connecting cell with adjacent terminal auxiliary cell. Gonimoblast small, lying below carpostome, all cells becoming carposporangia.

Schimmelmannia plumosa (Setch.) Abb.

Baylesia plumosa Setchell 1912: 249; Okamura 1907–42 (1927): 167; Segawa 1936: 1987; Smith 1944: 201. *Schimmelmannia plumosa* (Setch.) Abbott 1961: 379.

Thalli 10–20(60) cm tall, dark wine red, branched irregularly, with percurrent axes bearing short, dense, fringing, pinnate branchlets; carpogonial branch of 4–6 cells; auxiliary-cell branch of 4 or 5 cells; reproduction as for genus.

Infrequent, saxicolous, low intertidal, Moss Beach (San Mateo Co.), to Santa Barbara Co., Calif. Also known from C. Japan. Type locality: Pacific Grove, Calif.

Family ENDOCLADIACEAE

Thalli uniaxial; axial filament percurrent, giving rise to lateral branches. Outer cortical cells small, compacted, appearing parenchymatous in surface or sectional view. Inner cortex with rhizoidal filaments, in older portions of thallus these forming bulk of tissue. Tetrasporangia in nemathecia or scattered throughout outer cortex, irregularly or regularly cruciately divided. Spermatangia in colorless sori, these in short chains or borne on terminal cortical cells. Carpogonial and auxiliary-cell branches borne on same accessory fertile branches. Carpogonial branches 2-celled, usually several borne together, after fertilization with or without connecting filaments. Gonimoblasts large, nearly all cells becoming carposporangia; ostiole lacking.

Two genera in the California flora, *Endocladia* and *Gloiopeltis* (both below).

Endocladia J. Agardh 1841

Thalli short, erect, profusely branched, bushy, with radial to subdichotomous branching. Branches cylindrical, all surfaces covered with minute, conical spines. Cells of central axial filament bearing 2 mutually opposed laterals, the pairs of laterals on successive cells alternating at right angles to each other. Tetrasporangia irregularly cruciately divided, with unbranched sterile cells in nemathecia. Spermatangia in short chains in small, irregularly shaped sori. More than 1 2-celled carpogonial branch, each borne on a cell connected to an auxiliary cell. Gonimoblast filaments developing inward; carposporangia in chains. Mature gonimoblast massive, projecting above surface.

Endocladia muricata (Post. & Rupr.) J. Ag.

Gigartina muricata Postels & Ruprecht 1840: 16. *Endocladia muricata* (Post. & Rupr.) J. Agardh 1847: 10; Smith 1944: 211.

Thalli densely bushy, 4–8 cm tall, dark red to blackish-brown; branches ±0.5 mm diam., cylindrical throughout, covered with spines to 0.5 mm long, giving thallus harsh texture; gonimoblasts subterminal on branchlets, near tops of main branches, usually swollen laterally, spherical, orange-gold when mature.

Locally abundant on tops or vertical faces of rocks, high to midtidal, Alaska to Pta. Santo Tomás, Baja Calif., including Channel Is., Calif. Type locality: Alaska.

In C. Calif., the most common alga of the upper intertidal; in S. Calif., in local areas sometimes forming low, tight turf with *Gelidium* spp. and blue-green algae.

Gloiopeltis J. Agardh 1842

Thalli erect to partially decumbent, usually in small tufts. Branches uniaxial, with axial filament and surrounding rhizoids frequently degenerating when plant is mature. Surface cells firmly compacted and appearing parenchymatous, the surface firm to gelatinous. Tetrasporangia scattered

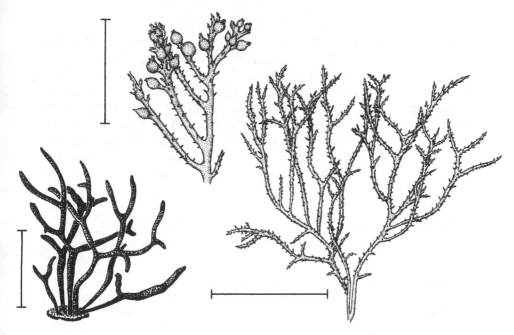

Fig. 376 (two at right). *Endocladia muricata.* Rules: sterile plant, 1 cm; apex of cystocarpic plant, 5 mm. (both JRJ/S)　**Fig. 377** (left). *Gloiopeltis furcata.* Rule: 1 cm. (DBP)

in outer cortex. Tetrasporangial plants slightly taller than cystocarpic plants. Spermatangia of converted terminal cortical cells, not in chains; carpogonial and auxiliary-cell branches borne in same branched filament; carpogonial branches 2-celled, frequently occurring in clusters (polycarpogonial). Auxiliary cell an enlarged cell borne in same fertile branch system, a connecting filament necessary for transfer of fertilized nucleus. Gonimoblasts prominently protruding, nearly all cells becoming carposporangia.

Gloiopeltis furcata (Post. & Rupr.) J. Ag.

Dumontia furcata Postels & Ruprecht 1840: 19. *Gloiopeltis furcata* (Post. & Rupr.) J. Agardh 1851: 235; Doty 1947b: 166; Hollenberg & Abbott 1966: 53 (incl. synonymy).

Thalli 2.5–5 cm tall, terete to slightly compressed, slippery, irregularly dichotomously branched, the apices simple or furcate, brownish-purple, drying to nearly black; reproduction as for genus.

Infrequent, high intertidal on rocks in areas exposed to surf, Aleutian Is. to Pta. Eugenio, Baja Calif.; in Calif., from Humboldt Co. to Malpaso Creek (Monterey Co.) and Santa Catalina Island. Type locality: North Pacific.

Family CRYPTONEMIACEAE

Thalli filamentous, multiaxial, erect, compressed or terete, foliaceous or much divided. Medulla of periclinal and sometimes anticlinal filaments. Cortex filamentous, outwardly compacted. Tetrasporangia usually isolated just beneath thallus surface or loosely aggregated in sori, cruciately divided. Spermatangia usually in superficial patches, these discoloring areas where they occur. Carpogonia borne in different clusters (ampullae) and nonprocarpic. Connecting filaments produced after fertilization of carpogonium and transferring nuclear material to auxiliary cell in separate ampullae. Gonimoblasts embedded and small, with or without small amount of sterile tissue around spore mass.

Halymenia C. Agardh 1817

Thalli erect, bladelike (in Calif. species), 1 or more blades arising from discoid holdfast. Blades entire, sometimes divided or lobed, occasionally with small, proliferous branches. Cortex of few to many rows of branched cells. Medulla of loosely interwoven, periclinally directed filaments, the 1 or 2 rows just beneath cortex usually stellate in shape; a few to many anticlinally directed filaments running perpendicularly from inner face of 1 cortex to that of opposite cortex. Tetrasporangia cruciately divided, scattered, embedded in outer cortex. Spermatangia in small, whitish sori, irregular in shape on thallus surface. Carpogonial branches and auxiliary-cell branches separated from each other, the auxiliary-cell cluster with moderate internal branching. Cystocarp relatively small, lying in outer medulla, with few sterile filaments surrounding it. All cells of gonimoblast becoming carposporangia; carpostome present; no fusion of basal cells to provide stalk (as occurs in *Prionitis* or *Grateloupia*).

1. Thalli in clusters, each blade lanceolate; medullary filaments many, arranged at random; low intertidal .... *H. schizymenioides*
1. Thalli single, each blade broadly lanceolate, falcate, or cordate; subtidal 2
  2. Thalli deep red when fresh, the surface slippery; medulla with a few widely spaced anticlinal filaments .... *H. coccinea*
  2. Thalli pink to deep rose; anticlinal medullary filaments closely spaced 3
3. Blades longer than broad, the margins ruffled, lacerated, or cleft; anticlinal medullary filaments stout, numerous .... *H. hollenbergii*
3. Blades nearly as broad as long, the margins entire 4
  4. Anticlinal medullary filaments slender, profuse and conspicuous....
  .... *H. californica*
  4. Anticlinal medullary filaments few and inconspicuous .... *H. templetonii*

Halymenia californica Smith & Hollenb.

Smith & Hollenberg 1943: 216; Smith 1944: 243; Abbott 1967a: 140 (incl. synonymy).

Thalli 20–70 cm tall, dark rose red; blades 4–18 cm wide, broadly lanceolate, sometimes falcate, occasionally branched from short, broad stipes; bases cuneate; blade 450–500 µm thick; outer cortex of 3 or 4 layers; inner cortex (or outer medulla) of 1 or 2 rows of stellate or irregular filaments; medulla traversed from cortex to cortex by many anticlinal filaments 2–4 µm wide, with few to a moderate number of periclinal filaments; tetrasporangia scattered over surface of thallus; spermatangial sori continuous; gonimoblasts hemispherical; carpostome present.

One of the most common subtidal (8–35 m), saxicolous, foliose algae, Hope I., Br. Columbia, south through Calif. to Gulf of Calif.; frequently cast ashore. Type locality: Pacific Grove, Calif.

Halymenia coccinea (Harv.) Abb.

Schizymenia coccinea Harvey 1862: 174. *Halymenia coccinea* (Harv.) Abbott 1967a: 141. *Aeodes gardneri* Kylin 1925: 17; Smith 1944: 241.

Thalli saxicolous, rarely epiphytic, usually solitary, membranous, lanceolate to broadly lanceolate, 15–40 cm wide, 30–45(110) cm long, bright cherry red when fresh, with soft, slippery texture, drying to brownish-red, cuneate to sometimes subcordate, with or without short. broad stipe, the margins ruffled, undulate, or plane; blade 300–500 µm thick, showing almost no periclinal filaments in medulla but with widely separated anticlinal filaments, these 2–4 µm diam., traversing from cortex to cortex; cor-

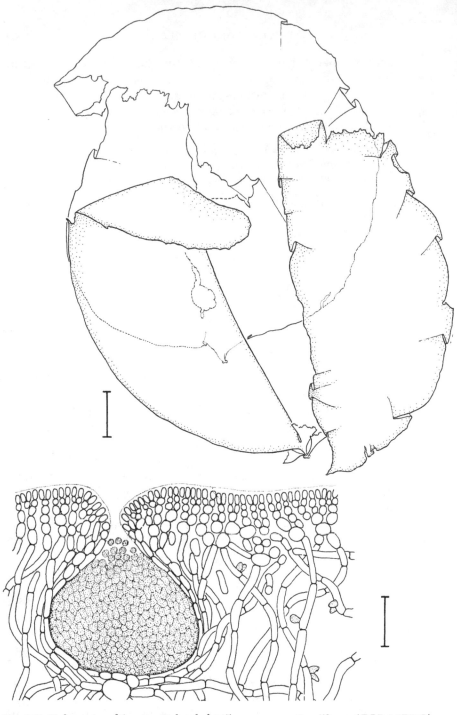

Fig. 378. *Halymenia californica.* Rules: habit, 5 cm; cross section, 50 μm. (DBP & JRJ/S)

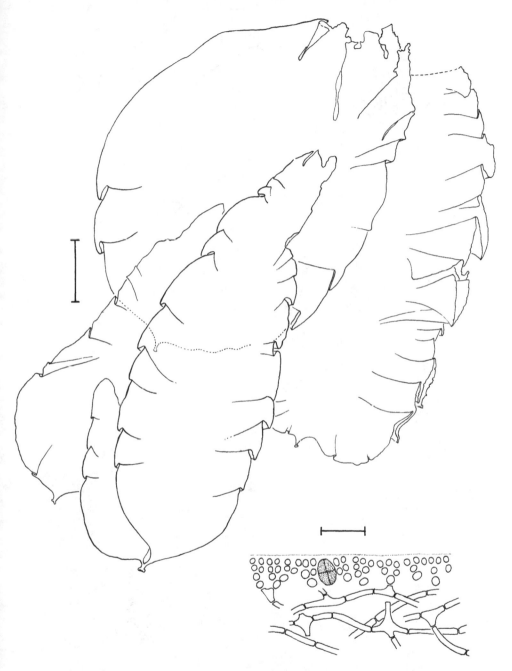

Fig. 379. *Halymenia coccinea.* Rules: habit, 5 cm; cross section, 50 μm. (DBP & JRJ/S)

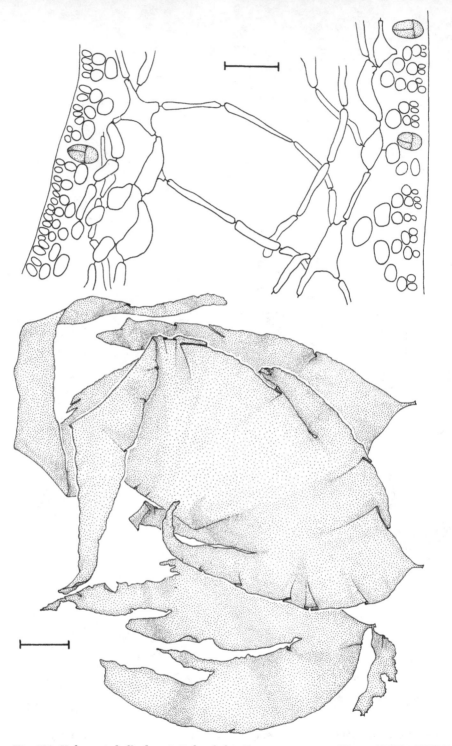

Fig. 380. *Halymenia hollenbergii.* Rules: habit, 5 cm; cross section, 50 μm. (DBP & DBP/A)

tex of 3 or 4 cell layers; tetrasporangia embedded in thallus; spermatangia superficial; gonimoblasts projecting beyond blade surface, surrounded by weakly developed sterile filaments.

Locally frequent, subtidal (8–35 m), S. Br. Columbia; San Juan I. (type locality) and other islands of Puget Sd., Wash.; and from Ore. through Calif. and Baja Calif. to Gulf of Calif.; frequently cast ashore throughout its range.

When fresh, *H. coccinea* is easily distinguishable from *H. californica* by its very slippery texture and its bright red color; *H. californica*, which also has a soft surface, looks like gelatin pudding and is a dark rose color.

Halymenia hollenbergii Abb.

Abbott 1967a: 143.

Thalli broadly lanceolate to subcordate, cleft from lateral margins into 1–6 lobes; margins smooth to crisped, ruffled, sometimes undulate; blades 60–70 cm tall, 16–18 cm broad, light rosy brown; blade 350–400 μm thick, with narrow cortex 4 or 5 cells deep; medulla twice as broad as cortical areas, the anticlinal filaments 4–10 μm diam., the periclinal filaments few and outer ones dumbbell-shaped; tetrasporangia in outer cortex; gonimoblasts and spermatangia unknown.

Infrequent, saxicolous, subtidal, Santa Barbara, Santa Cruz I., Redondo Beach, and Imperial Beach (San Diego Co.; type locality), Calif., to Pta. Abreojos, Baja Calif.

The light rosy color, deeply cleft margins, thicker anticlinal and periclinal filaments, and thinner cortex distinguish this species from *H. californica* and *H. coccinea*.

Halymenia schizymenioides Hollenb. & Abb.

Hollenberg & Abbott 1965: 1182; 1966: 70; Abbott 1967a: 141.

Thalli narrowly lanceolate to ovate-lanceolate, 20–25(46) cm tall, sometimes deeply cleft, wine red to rose red, the margins simple, rarely ruffled; blade 80–100 μm thick, both anticlinal and periclinal medullary filaments occupying more than half the section; anticlinal filaments arranged in crisscross fashion, only a few crossing from cortex to cortex; tetrasporangia scattered over surface; cystocarps with sparse, sterile filaments, with small carpostome.

Frequent, saxicolous, intertidal to upper subtidal, Mukkaw Bay, Wash., through Ore. and Calif. to Arroyo Honda (Santa Barbara Co.), Calif. Type locality: Mission Pt. (near Carmel, Monterey Co.), Calif.

Resembling some forms of *Schizymenia pacifica* (Kylin) Kylin, and extremely difficult to distinguish with sterile material or if the *Schizymenia* gland cells are not well developed in the cortex.

Halymenia templetonii (S. & G.) Abb.

Weeksia templetonii Setchell & Gardner 1937a: 76. *Halymenia templetonii* (S. & G.) Abbott 1967a: 143 (incl. synonymy).

Thalli saxicolous, 8–12(10–15) cm tall; blade orbicular in outline, mucilaginous, flaccid, attached by small disk, with faint, radiating, false veins at base but no differentiation of tissues forming them; blade 200–250(400) μm thick; medulla of network of filaments with relatively straight cells, these 5–7 μm diam.; cortex of 3 or 4 layers; tetrasporangia small and cruciately divided; gonimoblasts scattered and embedded; spermatangia unknown.

Rare, saxicolous, subtidal (to 35 m): in Calif., only off Anacapa I.; in Mexico, off I. Cedros, Baja Calif. (type locality), and I. del Espiritu Santo, Gulf of Calif.

Grateloupia C. Agardh 1822

Thalli erect, with slender, foliar main axes with branches, these occasionally branched again or proliferous, some thalli branching in 1 plane only, others in all directions, or with 1 or more blades growing from discoid holdfast. Blades usually stipitate, simple, or pinnately, palmately, dichotomously, or irregularly divided. Surface smooth, somewhat gelatinous, the lateral margins frequently with proliferous bladelets, or spiny outgrowths. Medulla of periclinally directed, colorless filaments and 1 or 2 layers of colorless, stellate cells. Innermost cortical layer also stellate; outermost cortex of 4–8 dichotomously divided cell rows. Rhizoidal filaments common in older thalli. Tetrasporangia scattered in outer cortex, sometimes loosely united in sori. Plants monoecious or dioecious. Spermatangia in superficial patches. Gonimoblasts with only outermost cells becoming carposporangia, or with several gonimolobes in which nearly all cells become carposporangia, these lying beneath carpostome. Fusion of basal cells of ampullae forming a stalk beneath spore mass.

1. Thallus of narrow or expanded simple blades, usually unbranched, frequently with distichous spiny or bladelike lateral proliferations . . . . *G. doryphora*
1. Thallus of foliar but not expanded main axes, branched, the proliferations slender, radially arranged 2
   2. Blade solitary, arising from 1 holdfast; main axis tubular to compressed, 1–3 mm diam.; upper portions of blade with dense tubular proliferations . . . . *G. filicina*
   2. Blades gregarious, arising from 1 holdfast; main axes cylindrical 3
3. Proliferations tubular to spinelike on same thallus, frequently proliferous in turn . . . . *G. setchellii*
3. Proliferations tubular, similar to axis, mostly of same length and width throughout thallus . . . . *G. prolongata*

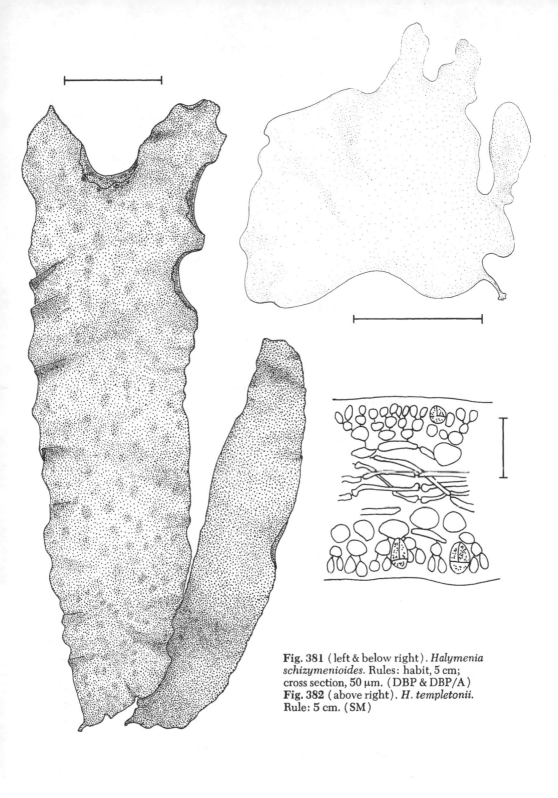

Fig. 381 (left & below right). *Halymenia schizymenioides*. Rules: habit, 5 cm; cross section, 50 µm. (DBP & DBP/A)
Fig. 382 (above right). *H. templetonii*. Rule: 5 cm. (SM)

Grateloupia doryphora (Mont.) Howe

Halymenia doryphora Montagne 1839: 21. *Grateloupia doryphora* (Mont.) Howe 1914: 169; André & Gayral 1961: 38 (incl. synonymy); Dawson, Acleto & Foldvik 1964: 49 (incl. synonymy). *G. cutleriae* f. *maxima* Gardner, P.B.-A. 1895–1919 [1911]: no. 124. *G. maxima* (Gardn.) Kylin 1941: 10. *G. abreviata* Kyl. 1941: 10. *G. multiphylla* Dawson 1954a: 251.

Thalli narrowly to broadly lanceolate, to 2 m tall, of soft gelatinous texture, wine red or olive, purple, to yellowish, several blades arising from common holdfast; if bladelike and expanded, upper half frequently incised; margin of blade sometimes plane, more frequently proliferous; proliferations spinelike (short, narrow, less than 1 mm wide) or bladelike (elongate, to 3 cm wide); if not expanded, but foliose, several narrow blades produced from broad flattened stipe, the margins with attenuate long spines.

Locally abundant in sheltered areas, saxicolous, low intertidal, Puget Sd., Wash., to Peru (type locality); in Calif., common in both harbor and exposed situations.

This is one of the most variable species of red algae in temperate and subtropical areas. Simple, bladelike forms resemble in color and texture some forms of *Schizymenia pacifica*, but in addition to lacking gland cells they are generally more lanceolate and more gelatinous than *Schizymenia*. Specimens resembling the narrow, proliferous forms of both *G. doryphora* and *Prionitis lyallii* are commonly found, and it is difficult to distinguish them except by critically examining sections through old and young portions of thalli. Usually, *Grateloupia* species show stellate cells in the outer medulla and have a narrow cortex, whereas in *Prionitis* the stellate cells are usually lacking and the cortex is very thick compared to that of *Grateloupia*.

Grateloupia filicina (Lamour.) C. Ag.

Delesseria filicina Lamouroux 1813: 38. *Grateloupia filicina* (Lamour.) C. Agardh 1822a: 223; Dawson 1954a: 252. *G. avalonae* Daws. 1949b: 4.

Thalli clumped, tufted, dark red or purplish to reddish-brown; individual plants 2–10(20) cm tall, arising from small holdfast; usually short-stipitate, with a percurrent, flattened axis giving off slender, thickly placed branched or unbranched laterals from base to apex, but these more abundant at tops of thalli; some populations weakly compressed, wholly distichously branched, others cylindrical, wholly radially branched, in still others these characters mixed; tetrasporangia scattered in cortex; spermatangia in superficial patches; gonimoblasts variously described as having only outer cells becoming carposporangia or with all filaments becoming carposporangia in gonimolobes of successive ages; with few or no surrounding sterile filaments.

Saxicolous, midtidal, numerous where found; rare in Calif. (Santa Catalina I.), more common in Mexico and in the tropics. Type locality: Adriatic.

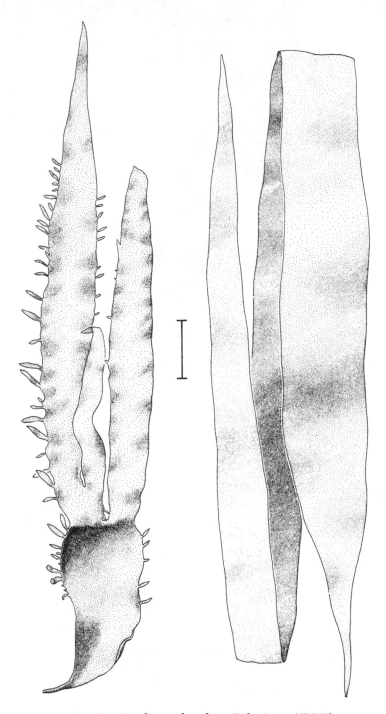

Fig. 383. *Grateloupia doryphora*. Rule: 3 cm. (JRJ/S)

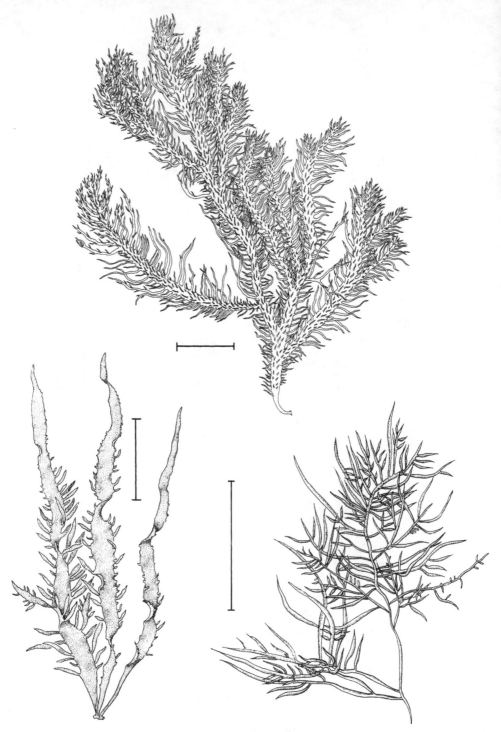

Fig. 384 (below right). *Grateloupia filicina*. Rule: 3 cm. (SM) **Fig. 385** (above). *G. prolongata*. Rule: 3 cm. (CS) **Fig. 386** (below left). *G. setchellii*. Rule: 3 cm. (JRJ/S)

Grateloupia prolongata J. Ag.

J. Agardh 1847: 10; Setchell & Gardner 1924a: 780; Dawson 1954a: 248.

Thalli dark red, coarse but firmly gelatinous, 8–20(40) cm tall; axes cylindrical to somewhat compressed, 2–3 cm diam., from short stipes, branched pinnately or radially, the proliferations slender, short, scattered throughout thallus; tetrasporangia and spermatangia as in other species; cystocarps embedded, raising thallus surface.

Saxicolous in tidepools, low intertidal: occasional in S. Calif., Goleta (Santa Barbara Co.), to La Jolla; common in Baja Calif. through Gulf of Calif. and south to Oaxaca, Mexico (type locality). Also known from Japan.

Grateloupia setchellii Kyl.

Kylin 1941: 10; Smith 1944: 240; Doty 1947b: 171.

Thalli soft, cylindrical-tubular, rose-colored, occasionally compressed, 5–15(45) cm tall, the stipes markedly intertwined; if tubular, thalli with a combination of narrow, tubular branchlets (similar to proliferations) and fine, spinelike proliferations, each of which may be proliferous in turn; if compressed, thalli bladelike, narrowly lanceolate, with proliferations appearing marginal on dried specimens.

Rare, saxicolous in sandy areas, low intertidal, Tofino, Br. Columbia; Ore.; in Calif., from Patrick's Pt. (Humboldt Co.), San Francisco, Santa Cruz Co., Monterey Peninsula, and San Luis Obispo Co. Type locality: Pt. Joe (Pyramid Pt.; Monterey Co.), Calif.

A poorly understood species, never collected in abundance beyond Monterey Co.; species limits not clear.

Cryptonemia J. Agardh 1842

Thalli erect, with 1 or more blades arising from discoid holdfasts, with or without stipes. Blade usually entire, sometimes irregularly laciniate or lobed; junction of stipe and blade usually somewhat thickened. Blade multiaxial. Medulla with loosely associated, periclinally directed filaments, a few highly refractive cells interspersed in these. Cortex thin, of 2–5(7) cell rows. Tetrasporangia scattered in cortex, cruciately divided. Spermatangia in large, irregular, superficial patches. Plants nonprocarpic. Auxiliary-cell cluster densely branched; connecting filaments necessary for transfer of diploid nucleus. Gonimoblasts scattered, small, with surrounding sterile filaments, most cells becoming carposporangia.

1. Blades oval, simple, the margins entire . . . . *C. ovalifolia*
1. Blades lobed, divided, or proliferous 2
   2. Blades simple when young, once or twice lobed when mature; thallus brownish-red . . . . *C. obovata*
   2. Blades deeply divided and frequently proliferous 3
   3. Divisions of blade irregular, mostly palmate, the margins of blades proliferous or ruffled . . . . *C. borealis*
   3. Divisions of blade mostly pinnate, the margins entire . . . . *C. angustata*

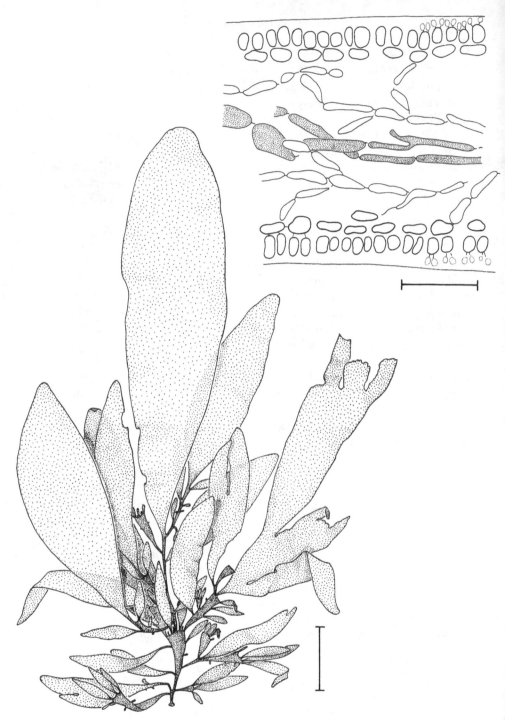

Fig. 387. *Cryptonemia angustata*. Rules: habit, 3 cm; cross section, 30 μm. (DBP & DBP/A)

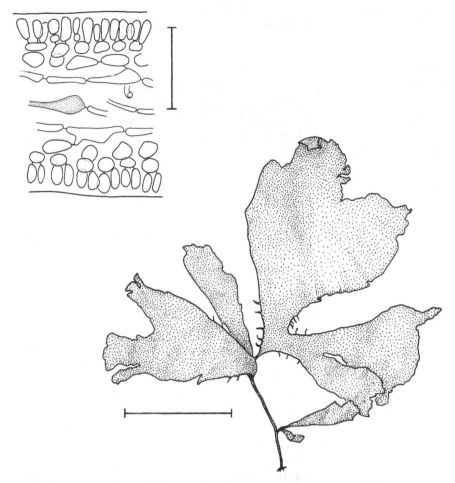

Fig. 388. *Cryptonemia borealis.* Rules: habit, 3 cm; cross section, 30 μm. (DBP & DBP/A)

Cryptonemia angustata (S. & G.) Daws.

Callymenia angustata Setchell & Gardner 1937a: 77. *Cryptonemia angustata* (S. & G.) Dawson 1954a: 285; Abbott 1967a: 147 (incl. synonymy).

Thalli rose to bluish-red, 5–30 cm tall; first and second orders of branches pinnately divided but frequently obscured by irregular proliferations; blades membranous, broadly lanceolate, each to 25 cm tall, 6 cm broad, sometimes divided, of various sizes on same thallus, stipitate; basal stipe thickened, cuneate, with small holdfast; blade 40–120 μm thick, the refractive cells few; cortex of 2–5 cell layers; tetrasporangia cruciate, scattered over thallus surface; gonimoblasts raising surface like blisters, discrete, with conspicuous carpostomes, internally surrounded by few thin-walled, sterile filaments.

Uncommon, saxicolous, subtidal (10–30 m), Monterey Bay and La Jolla, Calif., I. Magdalena, Baja Calif. (type locality), and Gulf of Calif. and Is. Revillagigedo to Peru.

Cryptonemia borealis Kyl.

Kylin 1925: 19; Hollenberg & Abbott 1966: 60; Abbott 1967a: 145.

Thalli 8–16 cm tall, membranous; intact specimens frequently palmate, but divided in various ways, each blade simple and obovate or irregularly cleft or split, the portion between stipe and blade thick; margins frequently ragged, occasionally ruffled or with proliferations; blades deep rose red to cerise, 60–100 μm thick, the refractive cells few to many; cortex of 3 or 4 layers; tetrasporangia scattered, sometimes in patches; gonimoblasts scattered, each with small carpostome and internally surrounded by a few sterile filaments.

Frequent, saxicolous and subtidal, in area of Puget Sd., Wash., less frequent elsewhere; low intertidal at Coos Head, Ore., and subtidal (10–30 m) in Carmel Submarine Canyon and La Jolla Submarine Canyon, Calif. Type locality: Canoe I., Wash.

Cryptonemia obovata J. Ag.

J. Agardh 1876: 681; Hollenberg & Abbott 1966: 68; Abbott 1967a: 145.

Thalli 20–30(45) cm tall, with or without stipes, 1 to several blades arising from peglike holdfasts; blade simple-ovate to obovate, or more commonly divided into several narrow or wide lobes, firm and crisp, reddish-brown, 120–450 μm thick; cortex of 3–6 cell rows; medulla varying in thickness, with numerous refractive cells; tetrasporangia cruciately divided, scattered through cortex singly or in small patches; gonimoblasts like small blisters, with sterile surrounding filaments.

Frequent to locally abundant, saxicolous, subtidal (10–50 m), Prince William Sd., Alaska, to I. de San Esteban, Gulf of Calif.; in Calif., from Bodega Head (Sonoma Co.) to San Diego, including Channel Is., more common from Santa Barbara southward. Type locality: San Francisco, Calif.

More easily identified than other species of *Cryptonemia* in this flora because of its reddish-brown color, large size, and thick blade.

Cryptonemia ovalifolia Kyl.

Kylin 1941: 11; Smith 1944: 241; Abbott 1967a: 145.

Thalli 4–5(8) cm tall, bluish-red, with inconspicuous stipes, mostly broadly ovate, rarely lobed, the margins entire or crisped; blade 60–120 μm thick, with few periclinal medullary filaments; refractive cells common; cortex of 2 or 3 cell layers; tetrasporangia and gonimoblasts as in other species.

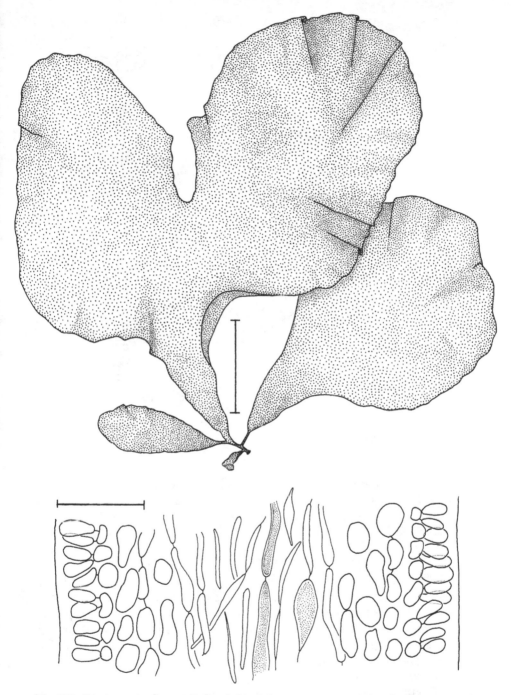

Fig. 389. *Cryptonemia obovata.* Rules: habit, 3 cm; cross section, 30 μm. (DBP & DBP/A)

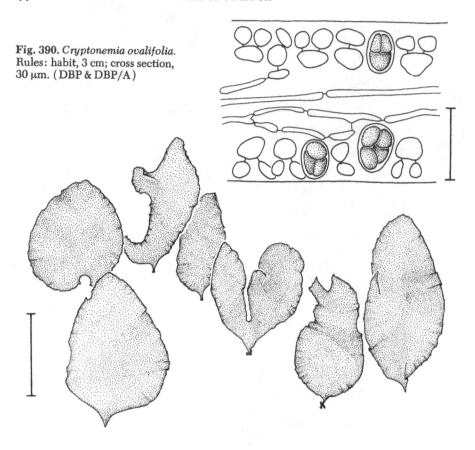

Fig. 390. *Cryptonemia ovalifolia.*
Rules: habit, 3 cm; cross section,
30 μm. (DBP & DBP/A)

Abundant where found, in dense, overlapping clusters in shaded over-hangs along channels, Cape Arago, Ore., and Farallon Is. and Monterey Peninsula to Little Sur River (Monterey Co.), Calif. Type locality: Pt. Pinos (Pacific Grove), Calif.

Dermocorynus Crouan & Crouan 1859

Thallus an expanded crust with anticlinal rows of closely adjoined cells of nearly uniform size and shape forming branched or unbranched fila-ments. When fertile, producing a few erect, flattened, cylindrical-papillate, or paddle-shaped branches, these consisting of filamentous medulla with elongate and stellate cells and cortex of 6–8 cell rows of branched fila-ments. Tetrasporangia cruciately divided, scattered over entire surface of branches. Spermatangia cut off from superficial cells of thallus, not in sori. Auxiliary-cell branches densely branched, separate from 2-celled car-pogonial branch. Gonimoblasts without carpostome, embedded in fertile branches.

Dermocorynus occidentalis Hollenb.

Hollenberg 1940: 868; Hollenberg & Abbott 1966: 72; Chiang 1970: 24.

Thallus dark red, inconspicuous, the crustose portion 50–100 μm thick, sometimes with modified hypothallus; erect branches 0.5–2 mm tall; tetrasporangia in groups; plants nonprocarpic; connecting filaments produced from carpogonium and hypogynous cell; on initiation of gonimoblast, fusion of auxiliary cell with cells in fertile cluster producing fusion cell basal to cystocarp; all other cells become carposporangia, produced in 2 or 3 gonimolobes, these surrounded by sterile filaments; carpostome and ostiole lacking.

Rare, on low intertidal rocks, and occasionally subtidal, N. Wash. and Pacific Grove, Redondo Beach, and Laguna Beach (type locality), Calif., to Pta. Banda, Baja Calif. Also Hawaii.

Carpopeltis Schmitz 1895

Thalli erect, arising from subcylindrical stipes, compressed to flattened in terminal portions, repeatedly branched. Branches congested near top, sometimes undulate to crinkly along margins. Medulla of periclinally directed filaments; young portions of branches with medulla of netlike filaments nearly uniform in diam., but these becoming of irregular width and irregular arrangement in older portions. Cortex of 5 or 6 rows of cells, the innermost oval to rectangular on bordering medulla. Tetrasporangia cru-

Fig. 391 (below). *Dermocorynus occidentalis*. Rule: 1 mm. (DBP)
Fig. 392 (right). *Carpopeltis bushiae*. Rule: 2 cm. (SM)

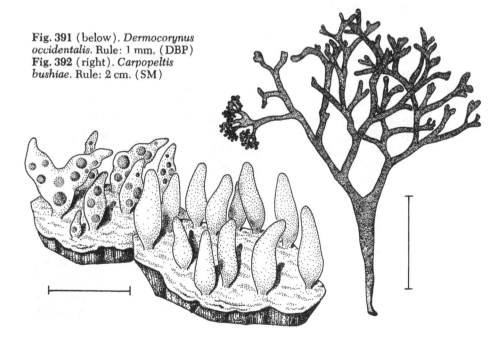

ciately divided, in nemathecia. Spermatangia in colorless, superficial sori. Gonimoblasts small, their development unknown.

Carpopeltis bushiae (Farl.) Kyl.

Polyopes bushiae Farlow 1899: 75; Dawson 1954a: 265. *Carpopeltis bushiae* (Farl.) Kylin 1956: 221.

Thalli deep red, 8–10 cm tall, arising from thick, discoid holdfasts; lower axes subcylindrical, the upper portions flattened, dichotomously divided, crowded, with short intervals between dichotomies; ultimate branches 3–4 mm wide, the lower 1–2 mm wide, the apices usually curled, or margins undulate with blunt to obtuse apices; medulla narrow, of mostly periclinally directed filaments; cortex of 5 or 6 rows of small, densely packed cells; tetrasporangia in somewhat sunken nemathecia on 1 or both surfaces of thallus, each sporangium surrounded by sterile filaments; spermatangia in superficial sori (as in *Prionitis*); gonimoblasts grouped on terminal portions of branches, each group with carpostome.

Frequent, saxicolous, low intertidal in shaded places to subtidal (4–20 m), Anacapa I., Santa Catalina I., and San Pedro (type locality), Calif., through S. Calif. to Is. San Benito, Baja Calif.

Development of the female reproductive structures leading to formation of the gonimoblast in *Carpopeltis* is unknown, and these structures have never been studied in either of the western N. American species attributed to this genus; thus their systematic position must remain uncertain until such studies are made.

Prionitis J. Agardh 1851

Thalli erect, 1 or more axes arising from discoid holdfasts. Axes terete below and compressed above, or terete or compressed throughout. Major branches approximately same breadth throughout, dichotomously or irregularly divided. Lateral margins of major secondary branches frequently with numerous peglike to foliar proliferations; branches lying in same plane, appearing pinnate. Surface of thallus smooth. Medulla of densely interwoven filaments. Cortex of small, tightly packed cells in deep rows. Transitional area between medulla and cortex sometimes with stellate cells. In general, medulla twice as thick as cortex (in *Carpopeltis bushiae*, thickness of medulla equal to that of cortex). Tetrasporangia cruciately divided, isolated in cortex or grouped in small sori. Spermatangia in extensive, whitish, superficial sori covering both surfaces of branches. Carpogonial branches 2-celled, arising in small cluster of sterile filaments; auxiliary cell intercalary, immersed in small, branched cluster of sterile filaments. Connecting filament necessary for transfer of diploid nucleus. Gonimoblasts in groups, modifying thallus externally, borne on internal stalks formed by fused basal cells of gonimoblast. Gonimoblast filaments developing toward surface, nearly all cells but basal ones becoming carposporangia, surrounded by sterile filaments.

1. Thalli slender, wiry; axes terete, less than 2 mm diam. . . . . *P. filiformis*
1. Thalli broader; axes flattened, strap-shaped, 2.5+ mm broad 2
  2. Branching regularly dichotomous; branches compressed 3
  2. Branching irregular; branches bladelike, flat in section 6
3. Dichotomies relatively short throughout, narrowly divergent 4
3. Dichotomies relatively long, spreading 5
  4. Branches tapering from base to apex . . . . *P. cornea*
  4. Branches not tapering, of irregular widths . . . . *P. angusta*
5. Thalli relatively stout, scarcely tapering from base to apex; proliferations few . . . . *P. australis*
5. Thalli narrow and thin, the branches frequently inflated terminally; proliferations many, especially on penultimate branches . . . . *P. linearis*
  6. Branches irregularly dichotomous; proliferations pinnate or irregular, rarely expanded bladelets . . . . *P. lanceolata*
  6. Branches not dichotomous, bladelike; proliferations inconspicuous or bladelike 7
7. Main axis producing a single, rarely divided blade; with small, undivided stipe . . . . *P. simplex*
7. Main axis producing terminal or lateral clusters of lanceolate blades, bladelike proliferations, or both; with conspicuous, divided stipe . . . . . .
. . . . *P. lyallii*

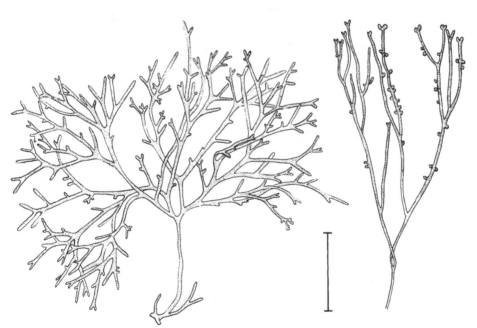

Fig. 393. *Prionitis angusta*: left, habit; right, detail of cystocarpic thallus. Rule: 3 cm. (both SM)

Prionitis angusta (Harv.) Okam.

Gymnogongrus ligulatus var. *angusta* Harvey 1859a: 332. *Prionitis angusta* (Harv.)
Okamura 1899: 4. *Carpopeltis angusta* (Harv.) Okam. 1907–42 (1912): 66. *Zanardinula cornea sensu* Dawson 1954a: 282 (in major part).

Thalli tufted, 10–20 cm tall, deep red, drying to reddish-tan, compressed
except at entangled, matlike bases; branches 1–2 mm broad, irregularly
dichotomously divided, sometimes appearing pinnate, frequently constricted at dichotomies, then flaring upward and narrowing again; upper
margins occasionally slightly undulate; proliferations rounded, peglike,
discrete, commonest on penultimate and ultimate branches; apices blunt
to slightly broadened or divaricate; tetrasporangia in sori at apices of terminal branches or in proliferations; cystocarps similarly located.

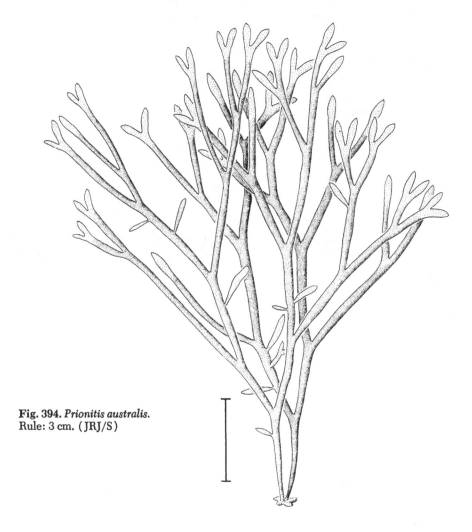

Fig. 394. *Prionitis australis.*
Rule: 3 cm. (JRJ/S)

Infrequent, saxicolous, low intertidal to subtidal (10 m), Santa Barbara and Channel Is., Calif., to N. Baja Calif. Type locality: Japan.

Prionitis australis (J. Ag.) J. Ag.

Phyllotylus australis J. Agardh 1847: 9. *Prionitis australis* (J. Ag.) J. Ag. 1851: 188; Kylin 1941: 12; Smith 1944: 245.

Thalli tufted, 15–25(40) cm tall, reddish-brown, compressed throughout; branches 2.8–4.2 mm broad, scarcely tapering from base to apex, the apices blunt to obtuse except where grazed; branches repeatedly dichotomously branched throughout, the lower intervals between dichotomies 1.5–2.5 times longer than succeeding upper intervals, the dichotomies of low-intertidal specimens more regular and intervals shorter than in subtidal specimens; proliferations lacking or relatively few, cuneate to cordate; tetrasporangia in main branches; spermatangia as for genus; gonimoblasts on apices of branches and in proliferations, with carpostomes.

Occasional to frequent, low intertidal on exposed rocks, or subtidal (to 8 m), Crescent City (Del Norte Co.) to San Luis Obispo Co., Calif. Type locality: "Pacific Ocean," possibly Monterey.

Prionitis cornea (Okam.) Daws.

Grateloupia cornea Okamura 1907–42 (1913): 63. *Prionitis cornea* (Okam.) Dawson 1958: 71. *Zanardinula cornea* (Okam.) Daws. 1954a: 282.

Thalli cartilaginous, 12–20 cm tall, dark red, arising from fleshy holdfasts, tufted, gregarious; main axes and branches 1–2 mm diam., approximately same diam. throughout, tapering slightly at apices, dichotomously branched; dichotomies remote, spreading, the ultimate dichotomies occasionally 2–3 times longer than those below, in some specimens short; proliferations constricted at lateral margin of branches, acutely pointed, pinnately placed; tetrasporangia in elongate sori on ultimate branches; other reproductive structures unknown in Calif.

Infrequent, saxicolous, low intertidal, Pacific Grove, Shell Beach (San Luis Obispo Co.), and Santa Catalina I., Calif., to N. Baja Calif. Type locality: Japan.

Prionitis filiformis Kyl.

Kylin 1941: 13; Smith 1944: 244 (incl. synonymy). *Zanardinula viscainensis* Dawson 1954a: 280.

Thalli tufted, 20–30 cm tall, orange-red to reddish-brown, the bases wiry and entangled; lower half of thallus terete, the upper portions weakly flattened, rarely more than 1 mm broad; thalli dichotomously branched 4 or 5 times, the apices acuminate, unbranched or furcate; proliferations numerous, inconspicuous, peglike to fusiform, 1–3 mm long, pinnately arranged in upper third of plants, some of the proliferations elongating as short secondary branches; tetrasporangia in main branches; spermatangia

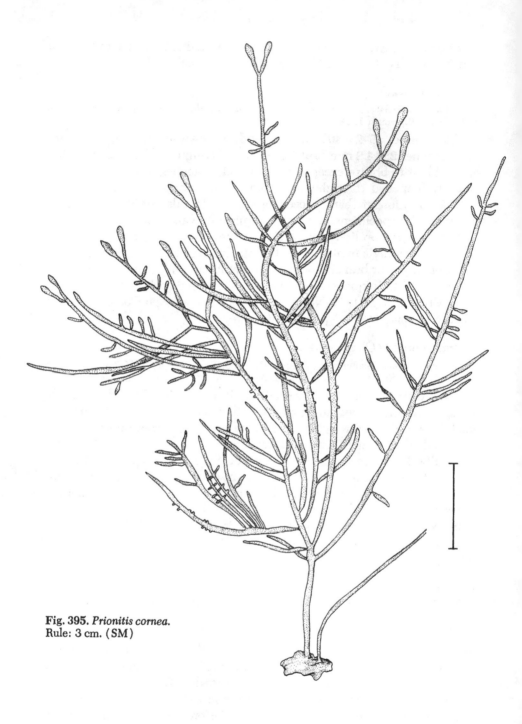

Fig. 395. *Prionitis cornea.*
Rule: 3 cm. (SM)

unknown; gonimoblasts on lateral proliferations only, with carpostome.

Infrequent, saxicolous, low intertidal, Pine I., Br. Columbia, and Cape Arago, Ore., to Pta. Abreojos, Baja Calif.; in Calif., from Crescent City (Del Norte Co.) to Carmel Bay. Also known from S. Africa. Type locality: Land's End (San Francisco), Calif.

Prionitis lanceolata (Harv.) Harv.

Gelidium lanceolatum Harvey 1833b: 164. *Prionitis lanceolata* (Harv.) Harv. 1853: 197; Smith 1944: 246.

The populations of this common and highly variable species might best be described separately. *Mid- and low-intertidal thalli*: stipes several to numerous, arising from discoid holdfasts, 20–35 cm tall; branches 2–4 mm wide, 5–7 times irregularly dichotomously divided, the ultimate dichotomy usually furcate; proliferations 3–20 mm long, pinnately placed, nearly equidistant, frequently longer toward base of plant, giving upper portions a pyramidal shape; plants reddish. *Subtidal thalli*: stipes 1 or 2, arising

Fig. 396 (left). *Prionitis filiformis.* Rule: 3 cm. (SM) Fig. 397 (right). *P. lanceolata.* Rule: 3 cm. (JRJ/S)

from holdfasts, 20–30(80) cm tall; axes and branches 2.5–5(7) mm wide, only 3 or 4 times dichotomous; proliferations lacking, or as wide as bearing axis and up to 4–5 times as long; occasional proliferations bearing further order of proliferations; plants resembling those of *P. australis* but more flaccid, with long intervals between branching, reddish-brown. *Populations of high intertidal and of high, deep, permanent tidepools subject to extensive insolation*: individual thalli 3–15 cm tall, congested, densely and irregularly branched, with dense proliferations; branches 1–2 mm wide; proliferations no longer than 1 cm; plants light red to brown. *Thalli growing in altered environment of an intertidal sewage outfall*: branching very irregular, the branchlets tufted above, congested, divided 3 or 4 times; proliferations and branchlets on all surfaces, swollen, 0.75–1 cm diam., 1.5–2.5 cm long; plants purplish to brownish-black. *All populations*: reproductive structures as for genus, found in mid- to low-intertidal populations most of year in C. Calif., in other populations mostly in summer.

Locally abundant, saxicolous, high intertidal to subtidal (30 m), Vancouver I., Br. Columbia, to Pta. Santa Rosalía, Baja Calif. Type locality: Monterey, Calif.

N. Wash. specimens tend to be shorter, with slender branches and proportionately long proliferations; S. Calif. specimens tend to be longer and to have the proliferations more irregular in occurrence than specimens from C. Calif.

Prionitis linearis Kyl.

Kylin 1941: 12; Smith 1944: 245.

Thalli clustered, 15–25 cm tall, dark red, 4 or 5 times dichotomously branched, the dichotomies distant; branches 1–2 mm broad; lower branches wiry; penultimate and ultimate branches usually inflated, 2–3 mm wide, acuminate; proliferations dense along upper margins, mostly short and pointed; tetrasporangia in main branches.

Locally abundant, saxicolous, low intertidal in areas of coarse sand, S. Alaska to Baja Calif.; in Calif., at Cape Mendocino, Moss Beach (San Mateo Co.), Monterey Peninsula, Cojo Pt. (Santa Barbara Co.), and La Jolla (type locality).

Prionitis lyallii Harv.

Harvey 1862: 173; Smith 1944: 247. *Prionitis andersonii* J. Agardh 1876: 159; Smith 1944: 246.

Thalli 20–35(75) cm tall, brownish to bright brick-red; stipes 1–10 mm diam., 2–70 mm long, little divided or dichotomously divided 3 or 4 times; stipes giving rise terminally to several generally lanceolate blades, these 1–5 cm wide, 3–45(60) cm long; occasionally stipes gradually grading into a broadened, nonfoliar portion throughout length of plant; primary

Fig. 398. *Prionitis linearis.* Rule: 3 cm. (SM)

blades frequently bearing irregular, narrowly stipitate, secondary and tertiary blades; blades firm or soft to slippery, rarely with marginal proliferations; tetrasporangia in modified nemathecia in blades; gonimoblasts scattered throughout blades, 1–1.5 mm diam., conspicuous for genus.

Common, on low intertidal rocks covered with coarse sand, to subtidal (to 35 m), Esquimalt (Vancouver I.; type locality), Br. Columbia, to Pta. María, Baja Calif.

Several forms of this common species have been recognized by Harvey and also by Setchell and Gardner. These authors realized, however, that the forms grade into each other. At the present, although we have more

Fig. 399. *Prionitis lyallii.* Rule: 3 cm. (JRJ/S)

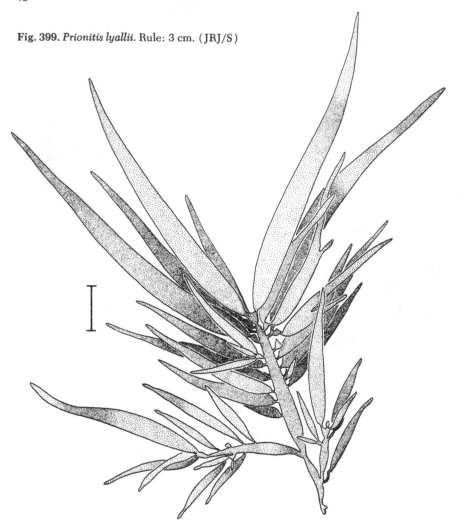

specimens from a wider variety of habitats and geographical ranges, these forms cannot be distinguished, as has been done with *P. lanceolata.* As with that species, it is almost impossible to choose a "typical" specimen.

Prionitis simplex Hollenb. & Abb.

Hollenberg & Abbott 1968: 1246.

Thalli dark rose red, of single, simple blades, 20–100 cm tall, 5–10 cm wide, narrowing gradually to blunt apices, the margins entire, plane, occasionally with a few elongate proliferations on lower lateral edges; stipe inconspicuous, usually unbranched; tetrasporangia scattered throughout cortex; cystocarps scattered, less than 1 mm diam.

Rare, saxicolous, subtidal (8 m), Lover's Pt. (Pacific Grove; type locality), and cast ashore in other localities on Monterey Peninsula, Calif.; also San Luis Obispo Co.

Resembling, because of long, broad lobes, *P. lyallii* (=*P. andersonii*), but far larger than specimens of that well-known and frequently collected species; also without the branched lower portions characteristic of *P. lyallii.*

Family KALLYMENIACEAE

Thalli multiaxial, usually erect, with simple, expanded, foliose blades, or the blades with few to many divisions. Cortex generally thin, of 2–4 cell

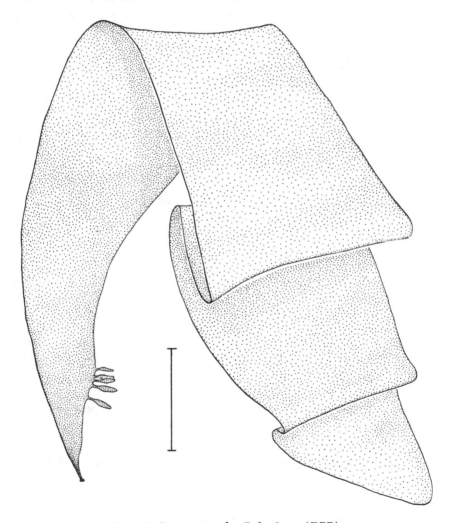

Fig. 400. *Prionitis simplex.* Rule: 3 cm. (DBP)

layers. Medulla 3–4 times thickness of cortex, with periclinally directed filaments and modified cells of stellate shape, or with large, pseudoparenchymatous cells interspersed with branched, pigmented filaments. Tetrasporangia cruciately divided, scattered in cortex. Spermatangia superficial, in patches. Carpogonial and auxiliary-cell branches nonprocarpic or procarpic; connecting filaments necessary for nuclear transfer if branches nonprocarpic. Carpogonial branch 3-celled, 1 to several borne on common supporting cell. Supporting cell functioning either as auxiliary cell of carpogonial branches or as supporting cell of another branched system, this bearing nonfunctional carpogonial branches or subsidiary cells representing these branches. Gonimoblast filaments growing toward center of blade; gonimoblasts globose, fairly massive, encircled by remains of nutritive tissue, frequently with 1 to several beaked ostioles.

1. Thalli parasitic on *Callophyllis*, pulvinate . . . . *Callocolax* (p. 467)
1. Thalli not parasitic 2
   2. Thallus dissected into many branches (except *C. firma*) . . . .
   . . . . *Callophyllis* (p. 459)
   2. Thallus bladelike 3
3. Principal blades with strong midribs, the blades frequently eroded. . . .
   . . . . *Erythrophyllum* (p. 457)
3. Blades without midribs 4
   4. Blades thin and soft; medulla of large cells surrounded by small filaments . . . . *Pugetia* (p. 455)
   4. Blades thick and firm 5
5. Medulla of elongate filaments, some of these stellate. . . . *Kallymenia* (below)
5. Medulla of large cells surrounded by small, pigmented filaments . . . . . .
   . . . . *Callophyllis firma* (p. 460)

Kallymenia J. Agardh 1842

Thalli bladelike, expanded, the blades usually lobed, the surfaces firm to somewhat slippery. Cortex thin, of 3 or 4 cell layers. Medullary filaments periclinally directed, not dense, associated with sparsely branched to much-branched, giant, stellate cells with highly refractive contents. Tetrasporangia cruciately divided. Spermatangia superficial, scattered over thallus. Carpogonial branches and auxiliary cells on separate supporting cells; cells of carpogonial branch after fertilization fusing to produce 1 or more connecting filaments, these connecting with auxiliary cell. Carposporangia in groups developed from large gonimoblast, interspersed with sterile filaments; most cells of gonimoblast becoming carposporangia.

1. Thalli deeply lobed, the lobes broadly spatulate and ruffled . . . . . . *K. norrisii*
1. Thalli usually without lobes, but if lobed each lobe cuneate and not ruffled 2
   2. Giant stellate cells few and obscure; sparsely branched . . . .
   . . . . *K. oblongifructa*
   2. Giant stellate cells more common; radially branched . . . . *K. pacifica*

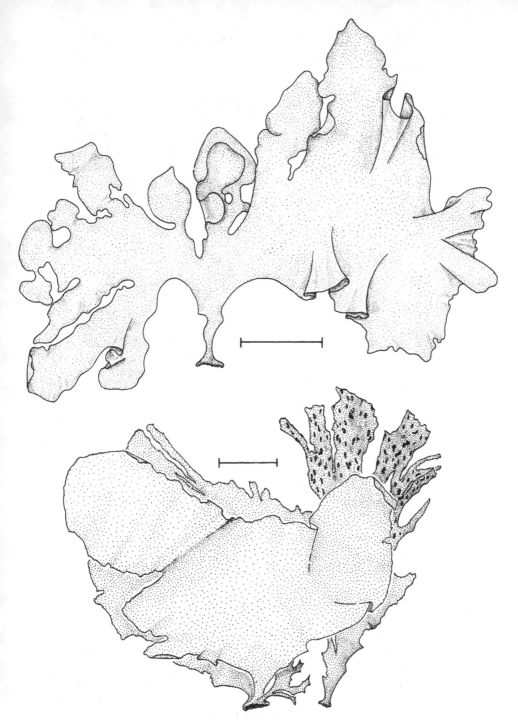

Fig. 401 (above). *Kallymenia norrisii*. Rule: 3 cm. (SM) **Fig. 402** (below). *K. oblongifructa*. Rule: 3 cm. (DBP)

Kallymenia norrisii Hollenb. & Abb.

Hollenberg & Abbott 1965: 1183; Abbott 1968: 197.

Thalli dark purple-red, having, when young and only 2–3 cm tall, spatulate to auriculate blades, attached by fleshy, disklike holdfast; blades, when mature and to 30 cm tall, broad with ruffled lobes, occasionally with *Opuntiella*-like proliferations from lower margins, sometimes with short stipe, mostly without stipe; 1 or more blades attached to fleshy marginal holdfast of 1–2 cm diam.; giant stellate cells common, to 4–5 mm across and visible with hand lens; tetrasporangia scattered throughout outer cortex, not modifying thallus where they occur; gonimoblasts 2–3 mm diam., scattered along outer, younger parts of blade.

Saxicolous, frequent, subtidal (15–20 m) in region of type locality, Monterey, Calif. Not known elsewhere.

Kallymenia oblongifructa (Setch.) Setch.

Iridaea oblongifructa Setchell 1901: 123. *Kallymenia oblongifructa* (Setch.) Setch. 1912: 234; Abbott 1968: 197.

Thalli brownish-red, 15–45(75) cm tall, 18–30 cm broad; blades arising singly or in small clusters from broad, fleshy discoid holdfasts; stipes inconspicuous or lacking; blades simple, elongate, and broadly lanceolate, or lobed with cuneate divisions; cortex of 3 or 4 cell layers; medulla with predominantly periclinally directed filaments, with anticlinal filaments near lateral margins; giant stellate cells few, mostly simple, unbranched and running parallel to periclinal filaments of medulla; tetrasporangia cruciately divided in outer cortex; spermatangia not known; cystocarps 3–4 mm diam., with 1 inconspicuous ostiole.

Common, saxicolous, subtidal (to 25 m) in region of Whidbey I., Wash. (type locality); rare from Seldovia, Alaska, to Crescent City (Del Norte Co.) and Kibesillah (Mendocino Co.), Calif.

Puget Sd. specimens have a very short stipe and a clear fleshy holdfast, both mostly lacking in *K. norrisii* and *K. pacifica*; the blades also tend to be simple, rather than divided or lobed as in the two southern species.

Kallymenia pacifica Kyl.

Kylin 1956: 233; Abbott 1968: 196 (incl. synonymy).

Thalli when young with circular, simple blades arising from very short, marginal stipes and holdfasts, bluish-red, becoming dissected by lobes when ±5 cm tall, and dissected to palmate in very old specimens, these 36(45) cm tall; thalli often remaining nearly circular if in calm subtidal habitat; thalli ±15 cm tall evidently most common, these with lobes broad-cuneate to 15 cm wide at outside margin of thallus, tapering toward center to fleshy, foliar portion, this attached to thickened holdfast, the margins crisp to crenulate; cortex 2–5 cell layers thick; medulla with relatively few periclinally and anticlinally directed filaments, with moderate

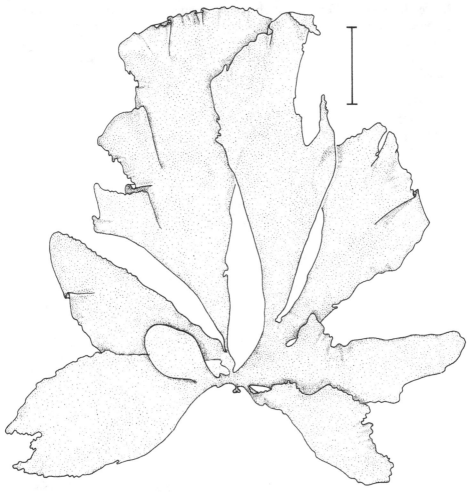

Fig. 403. *Kallymenia pacifica*. Rule: 3 cm. (SM)

number of giant stellate cells; tetrasporangia not known; spermatangia superficial, the cystocarps 3–4 mm diam., both occurring throughout thallus except basally.

Infrequent, saxicolous, subtidal (10–30 m), Anacapa I. to Imperial Beach (San Diego Co.), Calif., and to Pta. Santo Tomás, Baja Calif. Type locality: San Diego, Calif.

Pugetia Kylin 1925

Thalli bladelike, soft, filmy, membranous, usually circular but sometimes cleft. Cortex of 1 or 2 cell layers. Medulla wide, of large, parenchyma-like cells surrounded by branched, pigmented filaments. Tetrasporangia cruciately divided, in outer cortex. Spermatangia unknown. Carpogonial branch borne on different branch system from auxiliary cell, a connecting filament therefore necessary for nuclear transfer. Plants monocarpogonial; cystocarp without sterile filaments separating carposporangia. Most cells of gonimoblast become carposporangia, forming sequential spore clusters (gonimolobes).

Pugetia fragilissima Kyl.

Kylin 1925: 31; R. Norris 1957: 266; Hollenberg & Abbott 1966: 82.

Thalli membranous, deep rose red, to 15 cm diam., circular in outline to cleft, the margins entire, sometimes fimbriate; cystocarps to 1 mm diam., mostly smaller, densely crowding blade surface; tetrasporangia as for genus.

Locally abundant, saxicolous, subtidal (6–30 m), Br. Columbia to S. Calif. (Anacapa I.). Type locality: Canoe I., Wash.

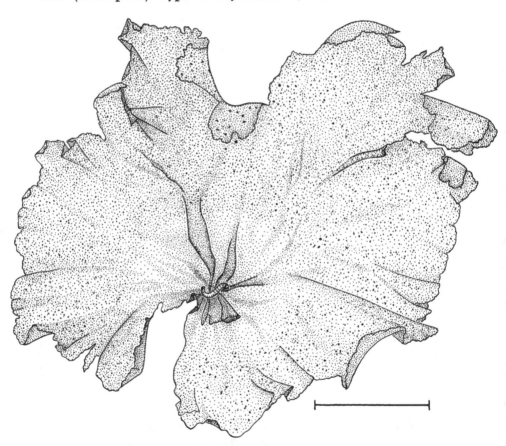

Fig. 404. *Pugetia fragilissima*. Rule: 5 cm. (DBP)

Erythrophyllum J. Agardh 1872

Thalli of 1 or more simple or divided lanceolate blades arising from discoid holdfasts. Blades stipitate, with conspicuous, percurrent midrib and with opposite conspicuous or inconspicuous forked veins diagonal to midrib, some more strongly developed than others; old blades lacerated along veins. Near end of growing season (fall and winter), blades eroded to within 1 cm on either side of midrib and major lateral veins. Cortex of 4–6 cell layers. Medulla of periclinally directed filaments without giant cells. Reproductive structures restricted to papillate outgrowths on blade surface. Tetrasporangia zonately divided, distributed throughout cortex. Spermatangia unknown. Plants monocarpogonial; auxiliary cell in cluster with several subsidiary cells, remote from carpogonial branch. After fertilization, connecting filament produced from fusion cell formed by union of subsidiary cells and supporting cell of carpogonial branch. Gonimoblasts globose; groups of carposporangia separated by sterile filaments.

Erythrophyllum delesserioides J. Ag.

J. Agardh 1872: 11; Smith 1944: 292; R. Norris 1957: 298.

Thalli annual, dark red, drying black, beginning growth (in C. Calif.) from late Jan. through early March; blades then simple, lanceolate, with percurrent midrib and acuminate apices; by June-July reaching maximum height, 30–50(100) cm, and maximum division of lateral veins, width then generally 4–8(10) cm; after July lateral margins lacerated nearly to midrib, some of lacerated portions further eroded to main divisions of lateral veins; papillae forming on remaining blade surfaces, further erosion occurring as papillae increase in number; in late winter only midrib and main lateral veins remaining, crowded with papillae.

Locally abundant, on rocks exposed to heavy surf, low intertidal, Alaska to Shell Beach (San Luis Obispo Co.), Calif. Type locality: Vancouver I., Br. Columbia.

Erythrophyllum splendens Doty

Doty 1947b: 184.

Thalli dark rose, 20–30(127) cm tall, at first forming simple, linear-lanceolate blades with nearly opposite lateral veins only basally; blades developing bladelets, these soon forming new veins distally, each pair of veins extending into new bladelets without lateral veins; plant at maturity completely pinnate in this manner; in winter, laminae eroded except within 2 cm of midribs of blades, these remnants with reproductive papillae.

Locally frequent, epiphytic on *Alaria, Laminaria*, or coarse algae, low intertidal to upper subtidal, Cape Arago, Ore. (type locality), to Duxbury Reef (Marin Co.), Calif.; frequently cast ashore.

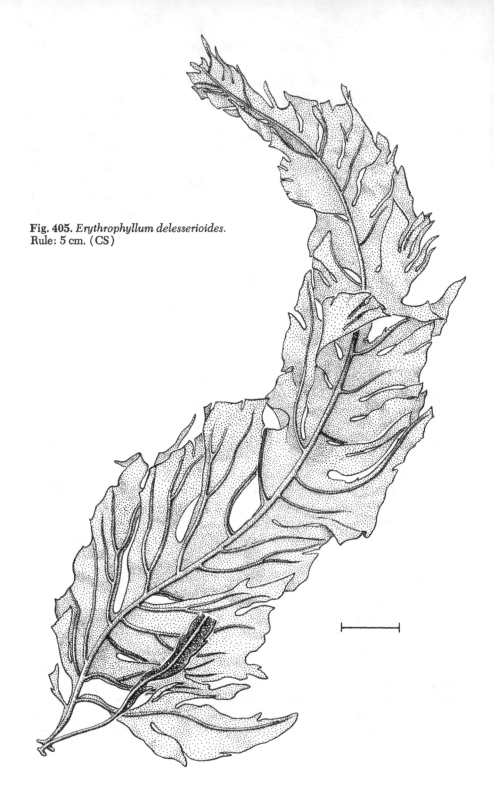

Fig. 405. *Erythrophyllum delesserioides.*
Rule: 5 cm. (CS)

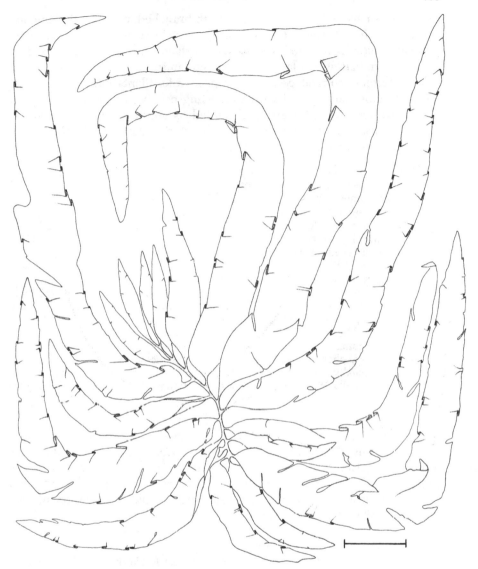

Fig. 406. *Erythrophyllum splendens*. Rule: 5 cm. (CS)

Callophyllis Kützing 1843

Thalli with circular to fan-shaped, much-divided blades arising from discoid holdfasts. Blades without midrib or veins, divided dichotomously, palmately, or pinnately, occasionally to many fine, narrow divisions. Margins of divisions smooth, crisped, dentate, or laciniate. Medulla of large, pseudoparenchymatous cells mingled with branched, pigmented filaments. Cortex 4 or 5 cells thick, the cells progressively smaller toward surface. Tetrasporangia embedded just below surface, cruciately divided, frequently germinating in place. Spermatangia in superficial patches, formed from outermost cortical cells. Plants procarpic. Plants either polycarpogonial or monocarpogonial with supporting cell becoming auxiliary cell. Cystocarp developing inward, surrounded by sterile tissue; carposporangial masses separated from each other by sterile filaments. Gonimoblasts unilaterally protuberant, each with 1 or more ostioles.

1. Thallus unbranched; blades expanded, simple or dissected .... *C. firma*
1. Thallus branched; blades not expanded 2
  2. Main branches 2–3 cm broad 3
  2. Main branches less than 2 cm broad 5
3. Ultimate divisions longer than penultimate divisions, usually paired, dichotomously divided .... *C. obtusifolia*
3. Ultimate divisions shorter than penultimate divisions 4
  4. Margins crisp to crenulate .... *C. crenulata*
  4. Margins plane, with pinnate bladelets in older thalli......... *C. pinnata*
5. Surface of thallus soft and slippery .... *C. heanophylla*
5. Surface of thallus firm and cartilaginous 6
  6. Plants crisp, closely tufted .... *C. linearis*
  6. Plants flabby, loosely tufted 7
7. Apices of branches broadly rounded, with few notches ...... *C. thompsonii*
7. Apices of branches not rounded, uneven, finely dissected 8
  8. Plants usually 4–6 cm tall; cystocarps to 1 mm diam ...... *C. flabellulata*
  8. Plants usually 15+ cm tall; cystocarps to 3 mm diam .... *C. violacea*

Monocarpogonial Species

There are two nearly parallel series (species "pairs") within the Calif. species of this genus, where vegetative appearance seems similar. The only distinguishing character that separates them is whether 1 carpogonial branch (monocarpogonial) or many branches (polycarpogonial) are borne on a single supporting cell. Microscopic examination of thin sections will reveal this condition.

Callophyllis firma (Kyl.) R. Norr.

Pugetia firma Kylin 1941: 15. *Callophyllis firma* (Kyl.) R. Norris 1957: 287; Abbott & Norris 1965: 73.

Thalli with blades broad and circular, simple, dissected, lobed, or peltate, 4–10(20) cm diam.; intertidal specimens deep rose, crisp, brittle, the large cells of medulla visible with unaided eye; subtidal specimens bluishred, slippery, tough, no medullary cells visible in surface view; margins at end of growing season irregularly eroded, often raised; gonimoblasts scattered over surface.

Frequent, saxicolous, low intertidal to subtidal (30 m), N. Br. Columbia

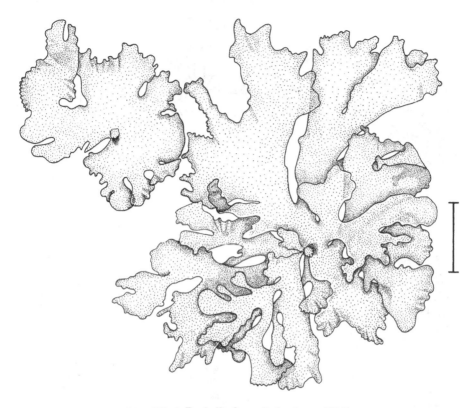

Fig. 407. *Callophyllis firma*. Rule: 3 cm. (SM)

to Pta. Santo Tomás, Baja Calif. Type locality: Pt. Pinos (Pacific Grove), Calif.

Callophyllis flabellulata Harv.

Harvey 1862: 171; Abbott & Norris 1965: 70; Hollenberg & Abbott 1966: 76 (incl. synonymy).

Thalli 4–10(18) cm tall (those of deep subtidal below 5 cm), orange red to dark red, some drying to black, in rounded lax tufts; thalli usually branching 5 or 6 times, the main axis to 5 times as broad as branches of higher orders, the ultimate branches finely dissected, 1–10+ mm wide; gonimoblasts umbonate, with several beaked ostioles, located on or within margin, or on faces of segments of branches, or in both places.

Common, saxicolous or epiphytic, subtidal (to 40 m), Br. Columbia to Baja California. Type locality: Vancouver I., Br. Columbia.

This is the most common species of *Callophyllis* in this range; the northern specimens tend to be more finely divided than S. Calif. specimens.

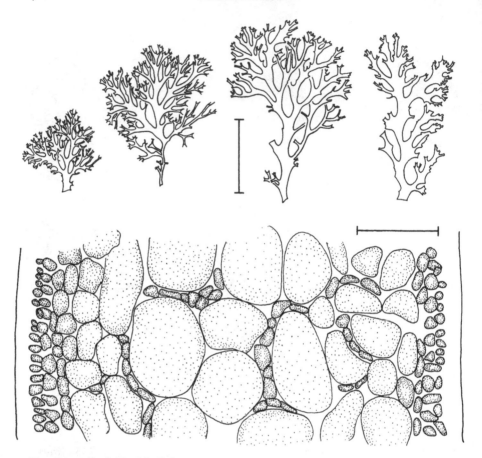

Fig. 408. *Callophyllis flabellulata*: four above, variation in branching pattern; below, cross section showing uniseriate filaments between larger medullary cells. Rules: above, 3 cm; below, 50 μm. (FT & LH/A)

Callophyllis linearis (Kyl.) Abb. & Norr.

Gracilaria linearis Kylin 1941: 22. *Callophyllis linearis* (Kyl.) Abbott & Norris 1965: 72.

Thalli 2–8 cm tall, purplish-brown, closely tufted, unequally dichoto-mously branched to 3 or 4 times, canaliculate, cartilaginous; apices acumi-nate, dentate, or bifurcate; tetrasporangia in irregular sori, occasionally spread beyond margins of sorus; spermatangia in small irregular patches in upper portion of plants; gonimoblasts few, to 1 mm diam., projecting but inconspicuous, borne on both surfaces of branches, some near margin.

Locally abundant on exposed headlands, saxicolous, low intertidal, Pt. Pinos (Pacific Grove; type locality), to 16 km south of Pt. Sur (Monterey Co.), Calif.; one subtidal (7 m) collection from off Pta. Banda, Baja Calif.

Fig. 409 (below right). *Callophyllis linearis*. Rule: 1 cm. (DBP)   **Fig. 410** (above). *C. obtusifolia*. Rule: 3 cm. (CS)   **Fig. 411** (below left). *C. thompsonii*. Rule: 3 cm. (DBP)

Callophyllis obtusifolia J. Ag.

J. Agardh 1851: 297; Abbott & Norris 1965: 75.

Thalli 15–50 cm tall, deep red to dark brown, usually pinnately but rarely palmately divided, proliferous in lower portions late in season, then also regenerative; blades long, linear, usually paired, the terminal dichotomies longer than penultimate dichotomies, nearly 1 cm wide throughout length of both dichotomies, gradually tapering to ±0.5 cm at blunt apices; tetrasporangia scattered over blades, sometimes germinating in place; gonimoblasts scattered, usually less than 1 mm diam.

Infrequent, mostly epiphytic, low intertidal, Bolinas (Marin Co.), Calif., to Pta. Entrada, Baja Calif. Type locality: presumably Monterey Peninsula, Calif.

Very similar to, but not as coarse as, *C. pinnata*.

Callophyllis thompsonii Setch.

Setchell 1923b: 399; Abbott & Norris 1965: 76; Hollenberg & Abbott 1966: 79.

Thalli 8–12(20) cm tall, dark purple-red; branches arising from small disk, 3–5 times dichotomously flabellate; ultimate lobes broadly rounded, 1.5–2 cm wide, with a few broad teeth or incisions; tetrasporangia scattered in upper parts of thallus; gonimoblasts strictly seriate within margin.

Infrequent, saxicolous, subtidal (to 20 m), Canoe I., Wash. (type locality), and Monterey, Calif.

Resembling *C. violacea* but with the ultimate divisions broader, rounder, and less incised.

Callophyllis violacea J. Ag.

J. Agardh 1885: 34; Abbott & Norris 1965: 74.

Thalli 5–27 cm tall (average 15), dark red to purplish-red, with 1 to several branches arising from discoid holdfast, fleshy to cartilaginous, sometimes furrowed; plants if epiphytic with only 1 or 2 fan-shaped branches, if saxicolous growing in tufts; main axis 1–4 cm wide at base, branching near base and expanding upward before branching again; branches usually alternate to 5 or 6 orders, the upper blades shorter and more crowded than lower, the apices toothed or much dissected; tetrasporangia scattered, the gonimoblasts to 3 mm diam., scattered, projecting prominently, especially when old.

Locally abundant, saxicolous, low intertidal to subtidal (20 m), Br. Columbia to Baja Calif. Type locality: Santa Barbara, Calif.

This is one of the three most commonly occurring species of *Callophyllis* on the Pacific Coast, the other two being *C. flabellulata* and *C. pinnata*. *C. violacea* is more common and more variable south of Pt. Conception than north of it. The commonest form in central Calif. is that repre-

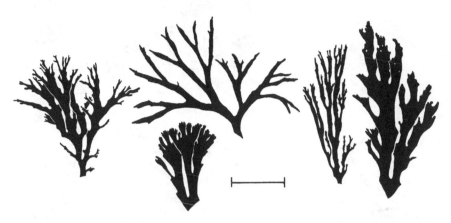

Fig. 412. *Callophyllis violacea*: variation in branching pattern. Rule: 3 cm. (FT)

sented by what has been known as *C. megalocarpa*; this, though recognizable as an isolate, cannot retain its integrity when compared with a suite of specimens.

Polycarpogonial Species

With several to many carpogonial branches borne on a single supporting cell.

Callophyllis crenulata Setch.

Setchell 1923b: 400; Smith 1944: 250; Abbott & Norris 1965: 77.

Thalli 3–14(20) cm tall, orange red to dark purplish-red, drying to dull rose; 1 or more branches arising from discoid holdfast, rarely stipitate; branches frequently canaliculate, the branching flabellate, rarely proliferous; penultimate blades broadest, to 2 cm wide; ultimate divisions 1(5) cm wide, with few irregular teeth, or with apices somewhat spatulate with crenulated margins; margins of all blades smooth, or undulate to crenulate, and at all times crisp and somewhat raised; tetrasporangia scattered; spermatangia in pale terminal patches; gonimoblasts scattered, 1–3 mm diam., flat, sometimes beaked, with 1–3 ostioles.

Locally abundant in northern part of its range, saxicolous, low intertidal to subtidal (30 m), Br. Columbia to Little Sur River (Monterey Co.), Calif. Type locality: Whidbey I., Wash.

Callophyllis heanophylla Setch.

Setchell 1923b: 401; R. Norris 1957: 281; Abbott & Norris 1965: 79.

Thalli to 7 cm tall, mostly shorter, short-stipitate, light rose, with dichotomously flabellate blades of 3 orders; blade soft and slippery, scarcely

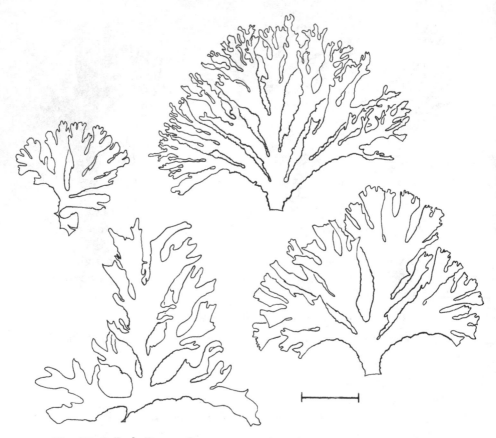

Fig. 413. *Callophyllis crenulata*: variation in branching pattern. Rule: 3 cm. (FT)

more than 2 cm wide, with apices obtuse to irregularly rounded, sometimes somewhat dentate; tetrasporangia at first restricted to angles of penultimate and ultimate dichotomies, spreading to broad faces of blades, in slightly modified submarginal sori; gonimoblasts 0.5–0.75 mm diam., distributed irregularly in upper blades, bulging on both sides of thallus, with 1–3 ostioles.

Infrequent, on worm tubes, arborescent bryozoans, or small rocks, subtidal (2–30 m), Wash. and Monterey to Santa Barbara and Channel Is., Calif. Type locality: Canoe I., Wash.

Callophyllis pinnata Setch. & Swezy

Setchell & Swezy 1923: 400; Smith 1944: 251; Abbott & Norris 1965: 80.

Thalli 12–30(45) cm tall, deep rose or red to almost black, coarse in texture; cystocarpic plants mostly palmately flabellately divided, with main divisions 2–3 cm wide, 10–30 cm long from main dichotomies, branching 4 or 5 times; blade apices straight to somewhat pointed, the margins smooth or dissected; tetrasporangial plants more loosely and irregularly branched, the main divisions usually not more than 2 cm wide, 15–45 cm long from main dichotomies, branching to 3 or 4 times; all older plants sometimes with proliferations on margins of main branches, these less than 1 cm wide, to 8 cm long, distichously placed along margins, rarely occurring on margins of higher-order branches; tetrasporangia scattered over blade, frequently germinating in place; gonimoblasts scattered over blade, less than 2 mm diam.

Locally abundant on all rocky headlands, epiphytic or epizoic, low intertidal to upper subtidal, Wash. to Baja Calif. Type locality: Duxbury Reef (Marin Co.), Calif.

Callocolax Batters 1895

Thalli "parasitic"; small tubercles with short stipe penetrating small portion of blades of hosts. Cells irregular, arranged in irregular network throughout tubercle, the outermost of 3 or 4 layers similar to and nearly continuous with cortex of host. Tetrasporangia cruciately divided, scattered through cortical layer. Spermatangia formed from superficial cortical cells. Plants procarpic; single 3-celled carpogonial branch borne on supporting cell functioning as auxiliary cell; procarp with only 1 subsidiary cell per carpogonial branch; direct fusion following fertilization. Gonimoblast globose, without ostiole; carposporangial masses separated by a few sterile filaments.

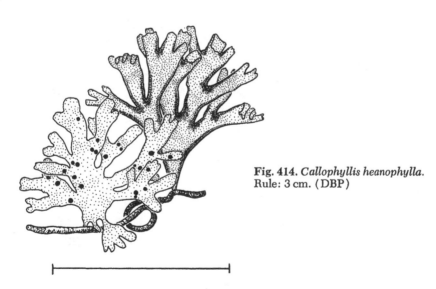

Fig. 414. *Callophyllis heanophylla.* Rule: 3 cm. (DBP)

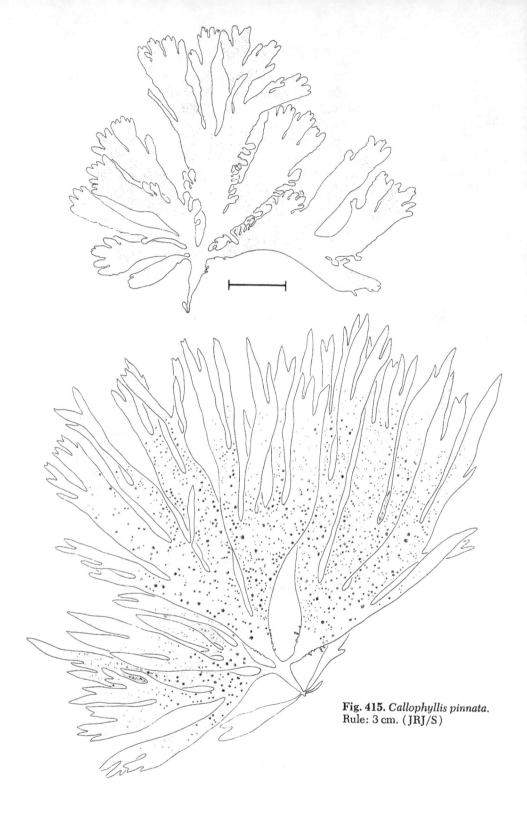

Fig. 415. *Callophyllis pinnata.*
Rule: 3 cm. (JRJ/S)

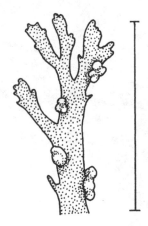

Fig. 416. *Callocolax fungiformis.*
Rule: 1 cm. (LH, after JRJ/S)

Callocolax fungiformis Kyl.

Kylin 1925: 35. *Callocolax neglectus sensu* Smith 1944: 253. *C. globulosis* Dawson 1945a: 94; Scagel 1957: 172.

Thalli "parasitic" on margins and flat surfaces of various species of *Callophyllis,* pale pink to cream-colored; tubercles 0.5–1 mm diam. on younger portions of host, to 3–4 mm diam. on older portions (particularly on flat surfaces rather than margins); stipe 0.5–1 mm long, depending on proportional size of tubercles; tubercle globose and simple or lobed several times, button-shaped to umbonate in lateral view; reproductive structures as for genus.

Infrequent, subtidal on various species of *Callophyllis* (*C. edentata, heanophylla,* or *pinnata,* but more commonly on *C. flabellulata*), Wash. to Pt. Loma (San Diego Co.), Calif.; in Calif., at Half Moon Bay (San Mateo Co.), Monterey Peninsula, Anacapa I., and Pt. Loma. Type locality: Turn I., Wash., on *C. edentata.*

Family CHOREOCOLACACEAE

Thalli minute, parasitic, with no plastids or remnants of them, thus frequently colorless or creamy, globose or with stubby, irregularly cylindrical branches. Tetrasporangia borne in outer cortex, cruciately divided. Spermatangia in chains, clustered to form continuous layer. Carpogonial and auxiliary-cell branches borne on common fertile filament, the supporting cell functioning as auxiliary cell; carpogonial branch of 2 or 4 cells. Gonimoblast growing toward thallus surface.

Choreocolax Reinsch 1875

Carpogonial branch 4-celled. Gonimoblast in conceptacle-like cavity; only terminal cells becoming carposporangia. Other characters as for the family.

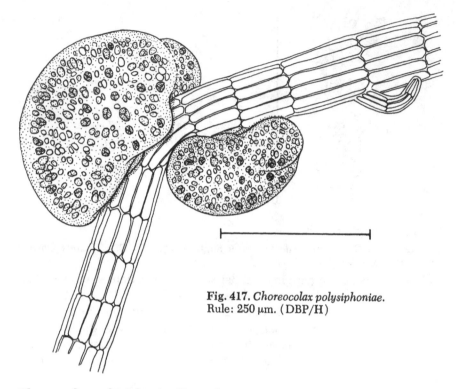

Fig. 417. *Choreocolax polysiphoniae.*
Rule: 250 μm. (DBP/H)

Choreocolax polysiphoniae Reinsch

Reinsch 1874–75 (1875): 61; Sturch 1926: 585; Smith 1944: 255.

Thalli small, globose to irregular, on margins and surfaces of hosts, attached by branched filaments penetrating host tissues; portions external to host of branched, laterally compacted filaments.

Infrequent, on *Polysiphonia, Pterosiphonia,* or *Pterochondria,* intertidal to subtidal (10 m), Sitka, Alaska, Duncan Bay, Br. Columbia, N. Wash., Monterey Peninsula and Laguna Beach, Calif., and Cabo Colnett, Baja Calif. Type locality: Atlantic coast of N. America.

Order GIGARTINALES

Thalli of various vegetative morphologies, mostly fleshy, always formed by compact aggregation of filaments into crusts, erect bushes, or blades. Cells generally small, close together, usually containing several small discoid chloroplasts without pyrenoids. Most species with separate gametangial and tetrasporangial plants, these similar to each other. Tetrasporangia zonately or cruciately divided, scattered and embedded in cortex or grouped in deep internal sori, occasionally in chains or in specialized branchlets (stichidia). Most families procarpic. In all species an ordinary cortical cell functioning as auxiliary cell, this usually producing gonimoblast, some or all cells becoming carposporangia. Gonimoblast with few to many sterile filaments, or with elaborate protective pericarp.

The Gigartinales are among the largest red algae. Because of their close vegetative resemblance to species in some other orders of red algae, reproductive material is critical for identification. The internal morphologies of crustose and multiaxial forms have analogs in the Cryptonemiales; those of the pseudoparenchymatous forms, in the Rhodymeniales. Despite these similarities, the development of reproductive structures is quite uniform in most families in this order, and details are used for familial and generic distinction. Species discrimination is usually dependent on external morphology.

1. Thalli mainly crustose, noncalcareous, mostly without erect branches 2
1. Thalli chiefly erect, arising from crustose base 5
  2. Tetrasporangia intercalary in erect filaments . . . .
    . . . . PETROCELIDACEAE (p. 476)
  2. Tetrasporangia terminal or lateral on erect filaments, or unknown 3
3. Tetrasporangia unknown . . . . PHYLLOPHORACEAE (*Besa*) (p. 515)
3. Tetrasporangia terminal or lateral on erect filaments 4
  4. Female plants with an evident auxiliary cell distinguishable before fertilization; gonimoblast arising from auxiliary cell, developing only toward thallus surface . . . . BLINKSIACEAE (below)
  4. Female plants without an evident auxiliary cell, as far as known; gonimoblast developing from connecting filaments, growing both toward and away from thallus surface . . . . CRUORIACEAE (p. 472)
5. Medulla evidently filamentous, at least in part 6
5. Medulla of large cells, without evident filaments 9
  6. Tetrasporangia unknown and probably lacking . . . .
    . . . . PHYLLOPHORACEAE (*Ahnfeltia*) (p. 503)
  6. Tetrasporangia commonly present 7
7. Tetrasporangia zonately divided, scattered. . . . . SOLIERIACEAE (p. 481)
7. Tetrasporangia cruciately divided 8
  8. Medulla a meshwork of fine filaments; tetrasporangia in dense groups in medulla or inner cortex; procarps present. . . . GIGARTINACEAE (p. 516)
  8. Medullary filaments not forming a meshwork; tetrasporangia scattered; procarps lacking . . . . NEMASTOMATACEAE (p. 478)
9. Tetrasporangia scattered in cortex . . . . GRACILARIACEAE (p. 494)
9. Tetrasporangia in nemathecia or in special branchlets (stichidia) 10
  10. Branching distichous . . . . PLOCAMIACEAE (p. 490)
  10. Branching not distichous 11
11. Tetrasporangial nemathecia on short, thornlike branchlets; tetrasporangia zonately divided . . . . HYPNEACEAE (p. 489)
11. Tetrasporangial nemathecia on ordinary branches; tetrasporangia cruciately divided, in chains or lacking. . . . . PHYLLOPHORACEAE (p. 502)

Family BLINKSIACEAE

Thalli crustose, noncalcareous. Hypothallus thin, of irregularly arranged, relatively large cells. Perithallus of erect, firmly adjoined filaments. Tetrasporangia zonately divided, terminal on erect filaments among paraphysis-like filaments. Carpogonial branches 4-celled, terminal on short erect filaments. Auxiliary cells intercalary in free upper portion of erect filaments. Gonimoblasts arising from evident auxiliary cell and developing outward; all cells forming carposporangia.

Blinksia Hollenberg & Abbott 1968

Firmly attached crusts without rhizoids; with the characters of the family.

Blinksia californica Hollenb. & Abb.

Hollenberg & Abbott 1968: 1248.

Crusts irregularly circular, light red to purplish, 2–5+ mm diam., 100–160 µm thick; cells of erect filaments 3–4 µm diam., 1–2 times as long as diam.; tetrasporangia 6–8 µm long, 4–5 µm diam.; paraphyses clavate, of 4–8(10) cells, the uppermost cells 3–3.5 µm diam., the lower cells more slender; spermatangia unknown; trichogyne straight, to 70 µm long; auxiliary cells numerous, to 7 µm diam.; carposporangia 15–20 in 1 obovoid group.

On rocks, lower intertidal, Moss Beach (Pacific Grove), Calif., the type and only known locality.

Family CRUORIACEAE

Thalli crustose, saxicolous, noncalcareous. Hypothallus 1 to several cells thick, composed of laterally adjoined filaments; rhizoids lacking. Perithallus of erect filaments, these firmly adjoined or loosely united by gelatinous matrix. Gland cells refractive, hyaline, scattered, or in rows in perithallus, occasionally lacking. Tetrasporangia zonately or cruciately divided, lateral or terminal on erect filaments. Sexual reproduction very imperfectly known, mostly given as for *Cruoria*. Carpogonial and auxiliary-cell filaments separated from one another, not grouped in nemathecia; auxiliary cells not readily distinguished. Gonimoblasts arising from connecting filaments and developing partly toward and partly away from surface of crust.

Two genera in the California flora, *Cruoria* (below) and *Haematocelis* (p. 474).

Cruoria Fries 1835

Crusts noncalcareous, saxicolous, firmly attached, without rhizoids. Basal layer monostromatic or polystromatic. Erect filaments simple to sparingly branched, closely adjoined. Tetrasporangia zonately divided, lateral on erect filaments. Spermatangia laterally tufted on upper parts of erect filaments. Carpogonial filaments 2-celled, lateral on erect filaments. Gonimoblasts developing from connecting filaments, forming groups of a few large carposporangia.

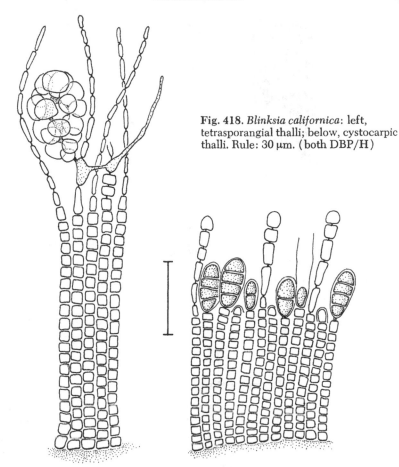

Fig. 418. *Blinksia californica*: left, tetrasporangial thalli; below, cystocarpic thalli. Rule: 30 μm. (both DBP/H)

Cruoria profunda Daws.

Dawson 1961a: 192.

Crusts dark red, to 3+ cm broad, to 0.5 mm thick; hypothallus of 2–4 layers of horizontally elongate cells, these 60–90 μm long, 7–9 μm diam.; perithallus of mostly unbranched, erect filaments, the upper cells 7–9 μm diam., the lowermost broader, angular or stellate, with occasional lateral anastomoses; gland cells in middle to lower perithallial layers, 50–200 μm long anticlinally, 10–20 μm diam., deep-staining; tetrasporangia slenderly fusiform, ±70 μm long, 9–11 μm diam., deeply embedded among erect filaments; sexual reproduction unknown.

Occasional to frequent on subtidal rocks, San Juan and Whidbey Is., Wash., and subtidally (to 48 m) on Cortez Bank (off U.S.-Mexican boundary; type locality).

Denizot (1968) questions the generic disposition of this alga.

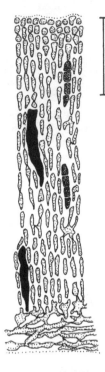

Fig. 419. *Cruoria profunda*. Rule: 100 μm. (LH, after Dawson)

Haematocelis J. Agardh 1851

Crusts noncalcareous, firmly attached to substratum without rhizoids. Hypothallus of 1 or 2 layers of radially disposed, branching filaments, these of horizontally elongate cells. Perithallus of assurgent to erect, branching, firmly united filaments, often stratified. Tetrasporangia terminal, zonately divided, embedded in uppermost layers. Sexual reproduction unknown.

The systematic position of this genus is uncertain, since sexual structures have not been observed.

Haematocelis rubens J. Ag.

J. Agardh 1851: 497; Dawson 1953a: 99; Denizot 1968: 241 (incl. synonymy).

Crusts dull red, to 60 mm broad, 180–360 μm thick, mostly firmly attached; cells of horizontal layer 5–7.5 μm diam., 15–27 μm long, those of erect filaments 5 μm diam., 5–10 μm long; tetrasporangial nemathecia superficial, 60–70 μm deep; tetrasporangia blunt-fusiform to oblong, 10–12 μm diam., 37–47 μm long, among densely packed, cylindrical paraphyses, these 3–5 μm diam., with 5–7 cells of variable length.

Infrequent on low intertidal rocks, shells, and other algae; Patrick's Pt. (Humboldt Co.) and La Jolla, Calif., to Baja Calif. Type locality: France.

Haematocelis zonalis Daws. & Neush.

Dawson & Neushul 1966: 176; Hollenberg 1948: 157 (as *Cruoria pacifica* Kjellman).

Crusts to 20+ mm broad, 200–350 µm thick, pale red to dark red; basal stratum of spreading filaments bearing ascending to erect, sparingly branched, and closely adjoined filaments with cells 5–10 µm diam., 1–3 times as long; horizontal growth bands evident in vertical section; tetrasporangia numerous, 12–20 µm diam., 40–50 µm long, terminal on erect filaments, later becoming shallowly to deeply embedded by continued growth of surrounding tissue.

Infrequent, on rocks and shells, intertidal to subtidal (30 m), Montano de Oro Beach State Park (San Luis Obispo Co.) to Laguna Beach, Calif. Type locality: off Anacapa I., Calif. (23–30 m).

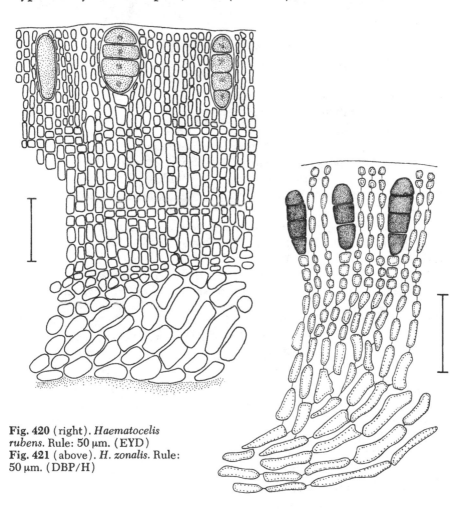

Fig. 420 (right). *Haematocelis rubens.* Rule: 50 µm. (EYD)
Fig. 421 (above). *H. zonalis.* Rule: 50 µm. (DBP/H)

Family PETROCELIDACEAE

Thalli crustose, saxicolous, noncalcareous, firmly attached to substratum without rhizoids. Hypothallus 1 to several cells thick, of prostrate, branched filaments. Perithallus of erect, simple to sparingly branched filaments loosely adjoined by a gelatinous matrix, typically with zones of anastomoses between cells of filaments. Tetrasporangia cruciately divided, intercalary, solitary or in series on erect filaments. Spermatangia on short branches of 1 or 2 cells at upper end of erect filaments. Carpogonial branches 2-celled, lateral, near apices of erect filaments. Auxiliary cells intercalary in erect filaments. Gonimoblasts developing from connecting filaments.

Sexual structures, described from European material, probably those of another taxon.

Petrocelis J. Agardh 1851

With characters of the family.

Petrocelis franciscana S. & G.

Setchell & Gardner 1917f: 391; Smith 1944: 217.

Crusts olive brown to reddish-black, to 0.25–1 m broad, typically 2–2.5 mm thick; hypothallial cells angular, mostly isodiametric; perithallial filaments of relatively uniform diam., 3.5–5 µm, 2.5 times as long, with 1 or more zones of anastomoses between cells of adjacent filaments; tetrasporangia 25–40 µm long, 20–28 µm diam., solitary, borne 15–25 cells below apices of filaments; sexual reproduction unknown.

Common, midtidal to upper intertidal zones on rocky headlands, Hope I., Br. Columbia, and Puget Sd. to Baja Calif. Type locality: San Francisco, Calif.

Culture studies (West, 1972) indicate that this species represents a sporangial phase in the life history of some species of *Gigartina*.

Petrocelis haematis Hollenb.

Hollenberg 1943a: 575. *Erythrodermis haematis* (Hollenb.) Denizot 1968: 223.

Crusts 8–12+ mm broad, 120–170 µm thick, deep red; erect filaments arising from single prostrate cell layer; cells of erect filaments ±10 µm diam., 1–1.5 times as long in fruiting portions; tetrasporangial nemathecia 1–2(4.5) mm diam., with gelatinous filaments readily separating under pressure; tetrasporangia 8–9 µm diam., 10–12 µm long, in catenate series of 4 or 5 in upper part of erect filaments, these terminating in a sterile cell; sexual reproduction unknown.

Frequent to rare on loose rocks, low intertidal, Pacific Grove and Corona del Mar (Orange Co.; type locality), Calif.

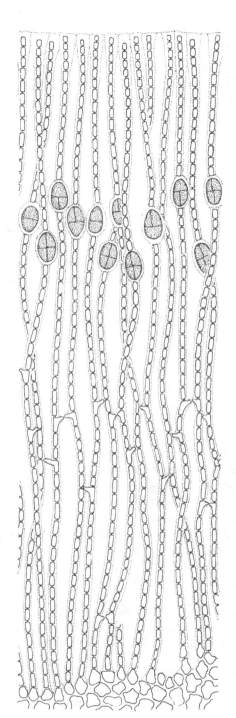

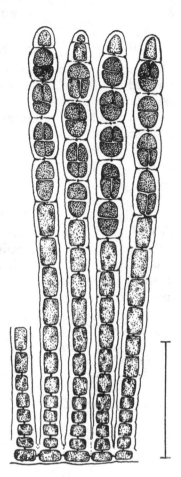

Fig. 422 (left). *Petrocelis franciscana.*
Rule: 50 μm. (JRJ/S) Fig. 423
(above). *P. haematis.* Rule: 50 μm.
(GJH)

Family NEMASTOMATACEAE

Thalli erect, cylindrical to bladed, simple or branched, multiaxial. Medulla predominantly filamentous. Cortex of anticlinal cell rows. Tetrasporangia typically cruciately divided, usually scattered in cortex. Spermatangia in superficial sori. Procarps absent. Auxiliary cell readily recognizable before fertilization. Gonimoblasts developing outward, nearly all cells becoming carposporangia.

Two genera in the California flora, *Schizymenia* (below) and *Predaea* (p. 481).

Schizymenia J. Agardh 1851

Thalli bladed, without veins. Medulla of prominent, tangled filaments. Large, hyaline, ovate gland cells usually abundant in outer cortex. Tetrasporangia mostly cruciately divided. Cystocarps globose, deeply embedded, with carpostome. Most gonimoblast cells becoming carposporangia; carposporangia in groups.

1. Blades with rough, wrinkled or puckered surfaces; tetrasporangia zonately divided .... *S. epiphytica*
1. Blades smooth; tetrasporangia cruciately divided 2
   2. Blades brownish-red, slippery; gland cells 10–20 µm diam. .....*S. pacifica*
   2. Blades bluish-rose, not slippery; gland cells ±4 µm diam. ...*S. dawsonii*

Schizymenia dawsonii Abb.

Abbott 1967b: 168.

Blades to 30 cm tall, 25 cm broad, 250–300 µm thick, bluish-rose, entire to deeply cleft 2 or 3 times, with smooth margins; stipes 2–3 mm long; cortex with 5 or 6 rows of closely packed cells; medulla of loosely arranged, mostly periclinal filaments, these 4–6 µm diam.; gland cells inconspicuous, 4 µm diam., 8–12 µm long; tetrasporangia cruciately divided, 36 µm diam., 28–36 µm long, scattered in outer cortex; spermatangia in wide, superficial patches; gonimoblasts 200–400 µm diam., barely emergent, surrounded by a few sterile filaments, with small carpostome.

Infrequent, saxicolous, subtidal (to 41 m), La Jolla, Calif., to Papalote Bay, Baja Calif. Type locality: Pta. Santo Tomás, Baja Calif.

Schizymenia epiphytica (Setch. & Laws.) Smith & Hollenb.

Peyssonneliopsis epiphytica Setchell & Lawson 1905: 63. *Schizymenia epiphytica* (Setch. & Laws.) Smith & Hollenberg 1943: 221; Smith 1944: 258; Abbott 1967b: 166 (incl. synonymy).

Blades bluish-red, darker when dry, to 40 cm tall, 150–300 µm thick, usually estipitate, 1 or sometimes several arising from small, fleshy holdfasts; blades entire, coarse and much wrinkled, reniform to orbicular, with

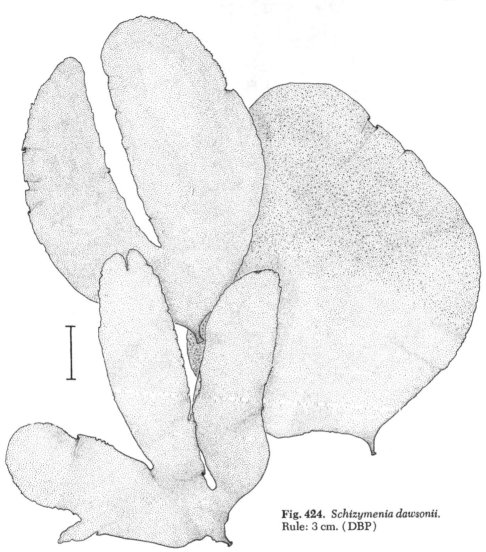

Fig. 424. *Schizymenia dawsonii.*
Rule: 3 cm. (DBP)

crisped rimlike marginal thickening in older plants; gland cells numerous, ellipsoidal to globose, 45–125 μm diam.; medullary filaments well separated; tetrasporangia 10–14 μm diam., 50–60 μm long, zonately divided, in nemathecioid groups, apparently arising from an auxiliary cell but details of sexual reproduction uncertain; free tetrasporophyte seemingly lacking.

Occasional, saxicolous, subtidal (to 30 m), Whidbey I., Wash., to Baja Calif.; frequently cast ashore near Pacific Grove, Calif. (type locality).

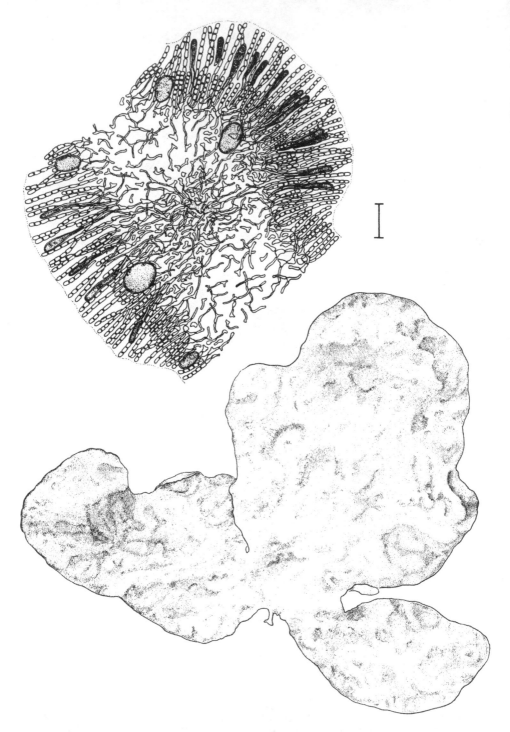

Fig. 425. *Schizymenia epiphytica.* Rules: habit, 3 cm; cross section, 50 μm. (above, GJH; below, JRJ/S)

Schizymenia pacifica (Kyl.) Kyl.

Turnerella pacifica Kylin 1925: 21. *Schizymenia pacifica* (Kyl.) Kyl. 1932: 10; Abbott 1967b: 162 (incl. synonymy).

Blades annual, usually saxicolous, growing in groups, 25–30(70) cm tall, 15–30 cm wide, 250–400 μm thick, brownish-red, slippery, the surfaces having appearance of tanned leather; blades ovate-lanceolate to cordate, often deeply split; stipe inconspicuous or lacking, with small, fleshy holdfast; gland cells obovate or pyriform, 10–20 μm diam., prominent in fresh material; tetrasporangia cruciately divided, 30 μm diam., 50 μm long, scattered in cortex; spermatangia in extensive surface areas; gonimoblasts 150–180 μm diam., with several gonimolobes maturing in sequence.

Common on rocks in exposed locations, occasionally epiphytic, lower intertidal to subtidal (18 m), Japan and Alaska to Baja Calif. and into Gulf of Calif.; very common in C. Calif. Type locality: Friday Harbor, Wash.

Predaea DeToni 1936

Thalli foliose to subpeltate, gelatinous to mucilaginous, with or without faint veins. Medulla of loose, slender, branched filaments. Cortex of anticlinal rows of pigmented cells. Cortical gland cells present or absent. Tetrasporangia unknown. Spermatangia in superficial patches. Plants nonprocarpic. Carpogonial branch 2-celled. Auxiliary cell developed by transformation of single inner cortical cell; nutritive cells borne adjacent to auxiliary cell. Gonimoblast arising from connecting filament near point of union with auxiliary cell, or from auxiliary cell, developing outward; mature gonimoblasts small, embedded in inner cortex.

Predaea masonii (S. & G.) DeToni

Clarionea masonii Setchell & Gardner 1930: 174. *Predaea masonii* (S. & G.) DeToni 1936: [5] (pages unnumbered); Dawson 1953b: 335; 1961a: 196.

Thalli very delicate, gelatinous to mucilaginous, saxicolous, to 12 cm tall, 9 cm broad, with small, adherent bases; stipes inconspicuous; blades pale red, to 8 mm thick below, 1–3 mm thick above, unbranched, irregularly ovate, with or without faint central vein and lateral veins; medullary filaments colorless, 2.5–4 μm diam., of very long cells; cortical filaments branched, the cells elliptical, ±4 μm diam., 2–4 times as long; carposporangial masses irregularly lobed, to 200 μm diam.; tetrasporangia and spermatangia unknown.

Rare, subtidal off Santa Catalina and Anacapa Is., Calif., I. Guadalupe, Baja Calif., I. Clarión (Is. Revillagigedo; type locality), Mexico, and Gulf of Calif.; probably also from Atlantic Mexico.

Family SOLIERIACEAE

Thalli cylindrical to foliaceous, erect, simple or variously branched, mul-

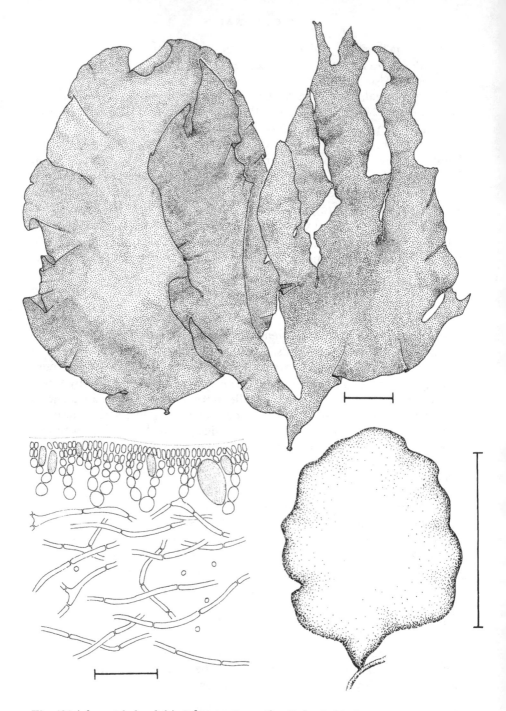

Fig. 426 (above & below left). *Schizymenia pacifica*. Rules: habit, 3 cm; cross section, 50 μm. (DBP & DBP/A) **Fig. 427** (below right). *Predaea masonii*. Rule: 3 cm. (LH, after Dawson; somewhat modified)

tiaxial; medulla obviously filamentous. Cortex of anticlinal cell rows, the cells progressively smaller toward surface. Tetrasporangia zonately divided, distributed in cortex. Spermatangia in surface sori. Plants nonprocarpial. Carpogonial branches of 3 or 4 cells. Gonimoblast filaments mostly developing inward. Carposporangia usually terminal. Gonimoblasts deeply embedded or protuberant, with or without investing sterile filaments, mostly with carpostome.

1. Thalli small, irregularly spherical, "parasitic" on *Neoagardhiella*. . . . . . .
. . . . *Gardneriella* (p. 485)
1. Thalli not "parasitic" 2
  2. Thallus with saccate branches on solid stipe. . . . .*Reticulobotrys* (p. 485)
  2. Thallus without saccate parts 3
3. Branches terete . . . .*Neoagardhiella* (below)
3. Branches complanate 4
  4. Branching dichotomous, the branches tapering, commonly with abruptly acute apices . . . .*Sarcodiotheca* (p. 487)
  4. Branching irregular, with rounded secondary blades arising from margin of primary blade . . . .*Opuntiella* (p. 485)

Neoagardhiella Wynne & Taylor 1973

Thalli erect, branched, terete, with 1 to many axes arising from discoid bases. Erect axes to 4 mm diam., branching at irregular intervals. Medulla with core of longitudinal filaments surrounded by large cells, grading into cortex of small, pigmented cells. Tetrasporangia zonately divided, distributed in outer cortex. Spermatangia in small superficial patches, ultimately confluent and continuous throughout most upper branches. Carpogonial branches 3-celled, arising in inner cortex, growing inwardly, with trichogyne bent sharply outward at its base. Cystocarps prominently protuberant at maturity, with ostioles. Gonimoblast with central mass of sterile cells. Carposporangia terminal.

Neoagardhiella baileyi (Kütz.) Wynne & Tayl.

Rhabdonia baileyi Kützing 1866: 26. *Neoagardhiella baileyi* (Kütz.) Wynne & Taylor 1973: 101 (incl. synonymy). *R. coulteri* Harvey 1853: 154. *Agardhiella coulteri* (Harv.) Setchell, P.B.-A., 1895–1919 [1897]: no. 333; Kylin 1941: 18; Smith 1944: 260. *A. tenera sensu* Dawson 1961a: 231; Hollenberg & Abbott 1966: 83.

Thalli 10–40 cm tall, axes several from small holdfasts, occasionally stoloniferous; branches mostly irregular, sparse or dense, radial or distichous, 1–4 mm diam., of varying length, mostly elongate-acuminate, with acute apices and narrowed bases, mostly without further branches, but occasionally bearing 1 or 2 further orders; tetrasporangia averaging 42 μm diam., 70 μm long, occasionally germinating *in situ*; spermatangia as for genus; cystocarps 1–2 mm diam., numerous, bulging the thallus.

Common and extremely variable, on rocks, intertidal to subtidal (to 30 m) mostly near sand, Br. Columbia to Baja Calif. Also Peru. Type locality: Long I., New York.

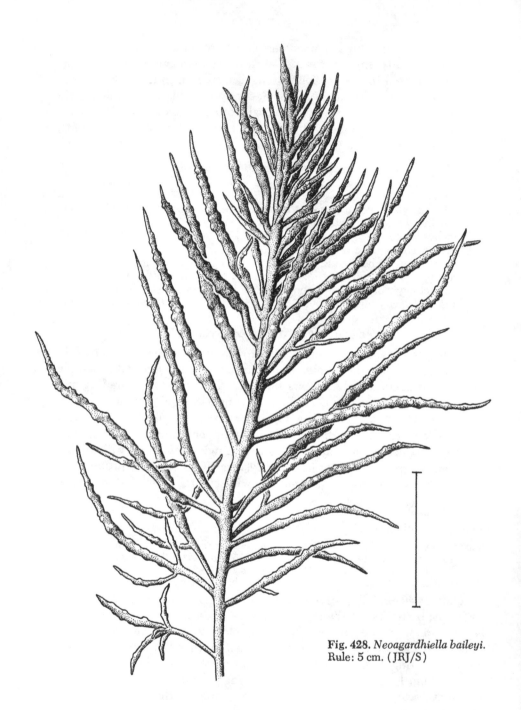

Fig. 428. *Neoagardhiella baileyi*.
Rule: 5 cm. (JRJ/S)

Gardneriella Kylin 1941

Thalli minute, "parasitic" on *Neoagardhiella*, mostly sessile, hemispherical to irregularly globular, the surfaces tubercular; thalli of branched, radiating filaments. Cells ellipsoidal to spherical, becoming progressively smaller outward. Tetrasporangia irregularly cruciately divided, distributed in cortical layers. Spermatangia in small, superficial sori. Carpogonial filaments as in *Neoagardhiella*. Carposporangia in short rows, peripheral on globose gonimoblast.

Gardneriella tuberifera Kyl.

Kylin 1941: 18; Smith 1944: 261.

Thalli whitish or pinkish, 2–6 mm diam.; with characters of the genus.

Host-specific, frequent in sheltered low-intertidal localities, Bodega (Sonoma Co.), Calif., to Pta. Baja, Baja Calif. Type locality: Pacific Grove, Calif.

It has been reported (Kugrens, 1971) that these growths may be galls of uncertain cause rather than parasitic algae.

Opuntiella Kylin 1925

Thallus with single, undivided primary blade arising from fleshy, discoid base; secondary blades proliferating from margin of primary blade, occasionally with additional orders of branching. Medulla of densely interwoven filaments. Cortex of compact, branching, anticlinal cell rows, the cells gradually smaller outward; large gland cells scattered among cortical filaments. Tetrasporangia zonately divided, distributed in outer cortex. Spermatangia unknown. Carpogonial branches usually 6-celled. Gonimoblast filaments developing toward medulla, most cells forming carposporangia.

Opuntiella californica (Farl.) Kyl.

Callymenia californica Farlow 1877: 241. *Opuntiella californica* (Farl.) Kylin 1925: 23; Smith 1944: 262.

Thallus to 20 cm tall, 30 cm broad, deep dark red; primary and secondary blades fan-shaped to broadly obovate, with stipes 4–8 mm long, 2–3 mm broad; base of blade frequently longitudinally thickened; proliferous blades 2–3 mm thick, commonly larger than primary blade.

Frequent on rocks, mostly subtidal (to 30 m), Alaska to Pta. Santo Tomás, Baja Calif. Type locality: Santa Cruz, Calif.

Reticulobotrys Dawson 1949

Thalli erect, branched, with discoid attachments. Branches cylindrical, multiaxial, solid below, saccate-fistulose above, the cavity filled with loose network of filaments. Inner cortex of large, vacuolate cells; outer cortex

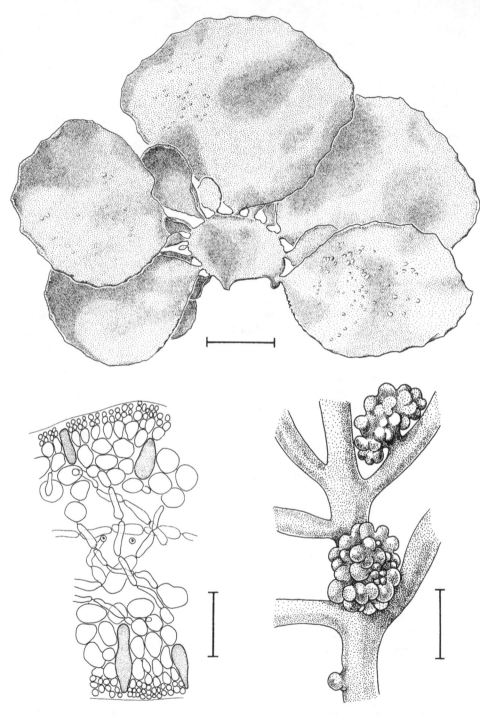

Fig. 429 (below right). *Gardneriella tuberifera*. Rule: 5 cm. (SM)   **Fig. 430** (above & below left). *Opuntiella californica*. Rules: habit, 5 cm; cross section, 100 μm. (JRJ/S & DBP/A)

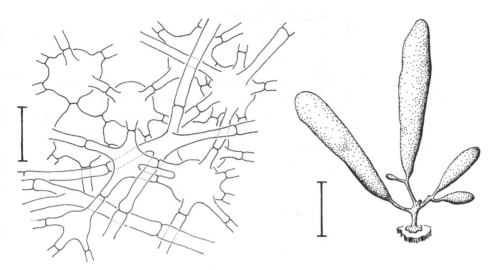

Fig. 431. *Reticulobotrys catalinae.* Rules: habit, 1 cm; inner wall of bladder, 100 μm. (both NLN)

of small pigmented cells. Tetrasporangia zonately divided, scattered in outer cortex. Spermatangia in extensive surface sori. Cystocarps slightly protuberant, with carpostome and filamentous envelope. Gonimoblast filaments radiating from very large fusion cell, bearing terminal carposporangia.

Reticulobotrys catalinae Daws.

Dawson 1949b: 12.

Thalli 6 cm tall, dark red, with several elongate-saccate branches arising from short, solid, branched axes 10–12 mm tall; saccate portions of branches 20–35 mm long, 3–6 mm diam., with rounded apices and abruptly or gradually narrowed bases; gonimoblasts in terminal parts of branches.

Infrequent to rare, subtidal (to 100 m), White Cove (Catalina I.; type locality) and La Jolla Submarine Canyon, Calif., and I. Guadalupe, Baja Calif. The type specimens were growing on brachiopods.

Sarcodiotheca Kylin 1932

Thalli complanate, with dichotomous branching. Medulla a loose meshwork of thin filaments. Inner cortex of large cells, with gradually smaller and pigmented cells toward surface. Tetrasporangia zonately divided, distributed in outer cortex. Spermatangia covering extensive mottled areas on branches. Cystocarps scattered, protuberant on 1 surface of branch, with prominent carpostome. Gonimoblast with central mass of sterile cells; carposporangia in peripheral rows.

Sarcodiotheca furcata (S. & G.) Kyl.

Anatheca furcata Setchell & Gardner 1903: 310. *Sarcodiotheca furcata* (S. & G.) Kylin 1925: 36; 1932: 16; Dawson 1961a: 299 (incl. synonymy).

Thalli to 25 cm tall, dull red, complanate, dichotomously to irregularly branched 5–7 times, narrowly cuneate below, arising from slender, cylindrical stipes, these 10–20 mm long, with small, discoid holdfasts; primary blades 10–20(30) mm broad, to 1 mm thick, the margins entire or proliferous; terminal divisions with sharply acute apices; cells of inner cortex to 70 μm diam., the pigmented cells of outer cortex 7–10 μm diam., in 2 layers; tetrasporangia distributed in cortex; spermatangia forming mi-

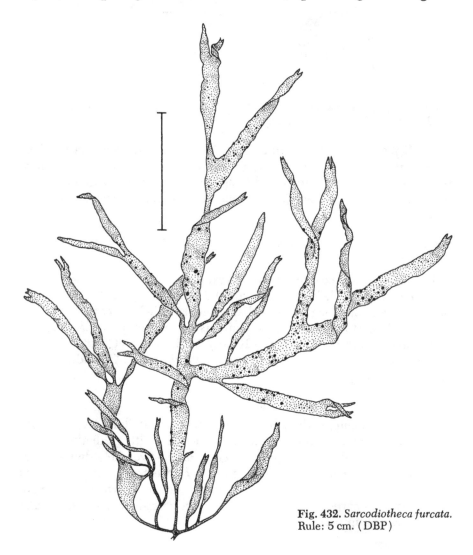

Fig. 432. *Sarcodiotheca furcata.*
Rule: 5 cm. (DBP)

nutely mottled areas over large portions of blades; cystocarps distributed over blades, 1–2 mm diam., prominently protuberant, commonly with distinctly raised carpostome.

On rocks, mostly deeply subtidal, S. Br. Columbia to Costa Rica and Galápagos Is.; mostly infrequent in Calif., but common subtidally at Bluebank Anchorage (Santa Cruz I.), Calif. Type locality: Whidbey I., Wash.

Family HYPNEACEAE

Thalli erect, radially branched. Branches typically cylindrical, commonly beset with numerous thornlike branchlets, obscurely uniaxial, the apical cell cutting off new increments alternately. Medulla pseudoparenchymatous with large inner cells, grading to small, pigmented cortical cells. Tetrasporangia zonately divided, in nemathecioid, thickened areas on short, terminal branchlets (stichidia). Spermatangia in slightly swollen nemathecia at base of similar branchlets. Plants procarpial; daughter cell of supporting cell functioning as auxiliary cell, identifiable before fertilization. Gonimoblast filaments developing first toward medulla and later outward, forming connections with pericarp and producing terminal carposporangia. Cystocarps nearly globular, on terminal branchlets; carpostome indistinct or lacking.

Hypnea Lamouroux 1813

Thalli free-living, with the characters of the family.

1. Main branches slightly to strongly compressed . . . . .H. variabilis
1. Main branches cylindrical 2
   2. Thalli bushy and caespitose or with caespitose basal portions, forming
     coarse mats . . . . H. johnstonii
   2. Thalli slender, erect, loosely branched or slightly entangled . . . .
    . . . . H. valentiae

Hypnea johnstonii S. & G.

Setchell & Gardner 1924a: 758; Dawson 1961a: 236.

Thalli bushy and caespitose, or with caespitose basal portions, forming coarse greenish to purple mats, to 20+ cm broad; main axes 7–10(14) cm tall, 1.5–2.5 mm diam.; ultimate branchlets numerous, aculeate, gradually shorter above; tetrasporangial branchlets simple or compound; sexual plants unknown.

Occasional, intertidal, on gravel, rocks or shells, Newport Harbor (Orange Co.), Calif., to Nayarit, Mexico. Type locality: Gulf of Calif.

Hypnea valentiae (Turn.) Mont.

Fucus valentiae Turner 1809: pl. 78. Hypnea valentiae (Turn.) Montagne 1841: 161; Dawson 1961a: 238 (incl. synonymy). H. californica Kylin 1941: 20.

Thalli erect to slightly decumbent, bushy, the branches numerous, with acute apices, commonly with small stellate groups of branchlets scattered over thallus; determinate ramuli simple to compound, bearing tetrasporangial and spermatangial nemathecia at or near their bases; sexual plants rare; cystocarps somewhat globose, conspicuous.

Hypnea valentiae var. valentiae

Thalli loosely branched, 10–25 cm tall, brownish-red, with cylindrical branches; main erect branches to 2 mm diam., gradually reduced in lateral branches; lesser branchlets straight to uncinate, 3–10 mm long.

On rocks, shells, etc., low intertidal, Santa Barbara, Calif. to Peru. Type locality: Red Sea. Widely distributed in warmer seas. A frequent summer annual in S. Calif.

Hypnea valentiae var. gardneri Hollenb.

Hollenberg 1972: 44.

Thalli to 8 cm tall, dark red, divaricately and laxly branched, with major axes cylindrical, moderately distinct to 1 mm diam.; tetrasporangial stichidia globular to stellate, with nemathecia involving entire stichidium; cystocarps sessile, globular, about 600 μm diam.; spermatangia unknown.

Seemingly rare, epiphytic, on drift and in unspecified habitats near San Pedro (type locality), and La Jolla, Calif.

Hypnea variabilis Okam.

Okamura 1907–42 (1909): 21; Dawson 1961a: 240.

Thalli bushy, to 10 cm tall, dull red, drying to nearly black; primary axes ill-defined, arising from stoloniferous base; older parts mostly strongly compressed or flattened, 1.2–2 mm wide, 0.7–1 mm thick; branching irregularly divaricate, in part pinnate, the branches gradually reduced to cylindrical ultimate branchlets, these 1–3 mm long; medulla with central core of smaller cells surrounded by outer medulla of large cells; tetrasporangial nemathecia partly to completely encircling base of determinate branchlets; tetrasporangia 15–16 μm diam., 30–40 μm long; spermatangia unknown; cystocarps globular, 0.5–1 mm diam.

Occasional, on rocks, intertidal, Santa Catalina I. and La Jolla, Calif., to I. San Roque, Baja Calif. Type locality: S. Japan.

Family PLOCAMIACEAE

Thalli erect, the fronds flattened, branched bilaterally; or thalli forming small cushions with branching obscure. Branching in erect thalli distichous, sympodial, alternate, pectinate and compressed, the branches uniaxial; apical cell and percurrent axial filament distinct. Cortex of small cells, progressively larger inward. Tetrasporangia zonately divided, in stichidia. Spermatangia superficial on terminal branchlets. Plants procarpial; carpogonial branches 3-celled, the supporting cell functioning as auxiliary

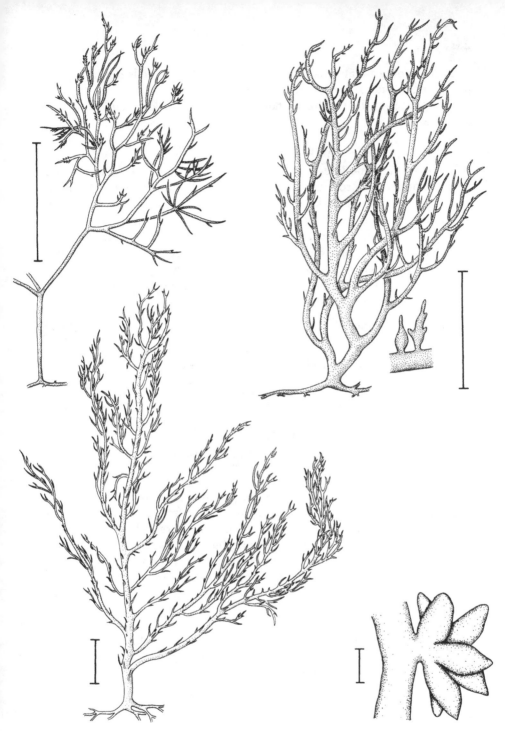

Fig. 433 (two above right). *Hypnea johnstonii*. Rule: 3 cm. (DBP)   **Fig. 434** (below left). *H. valentiae* var. *valentiae*. Rule: 3 cm. (DBP)   **Fig. 435** (below right). *H. valentiae* var. *gardneri*. Rule: 100 μm. (GJH)   **Fig. 436** (above left). *H. variabilis*. Rule: 3 cm. (DBP)

cell. Gonimoblast filaments developing outward, most cells becoming car-posporangia. Mature cystocarps protuberant.

Two genera in the California flora, *Plocamium* (below) and *Plocamio-colax* (p. 494).

Plocamium Lamouroux 1813

Thalli erect, sometimes arising from discoid attachments, most often from stoloniferous, prostrate branches with small attachment disks. Erect branches freely branched, compressed to strongly flattened. Branching sympodial, alternately distichous, pectinate, each pectination with 2–5 branchlets. Branches uniaxial, each cell of axial filaments bearing 2 lateral filaments. Cells cut off from lateral filaments progressively smaller outward, forming compact cortex. Tetrasporangia in groups in small, usually compound, special branchlets. Spermatangia covering ultimate branchlets; cystocarps without carpostome.

1. Some ultimate branchlets strongly incurved . . . . *P. violaceum*
1. Branchlets straight or curving outward 2
   2. Branching relatively sparse throughout, especially below; main axes appearing naked below owing to fewer and smaller branches . . . .
   . . . . *P. oregonum*
   2. Branching usually dense throughout, or especially above; main axes mostly not appearing naked below . . . . *P. cartilagineum*

Plocamium cartilagineum (L.) Dix.

Fucus cartilagineus Linnaeus 1753: 1161. *Plocamium cartilagineum* (L.) Dixon 1967: 58. *P. pacificum* Kylin 1925: 42; Smith 1944: 264. *P. coccineum* var. *pacificum* (Kyl.) Dawson 1961a: 220.

Thalli 4–25 cm tall, deep red or purplish-red to pink, with several to many erect, plumose axes 1–2 mm diam., arising from stoloniferous, prostrate branches; branches of various orders, compressed, the branchlets straight, or slightly incurved, or curving outward in pectinate groups of 3 or 4, the lowermost of each group usually simple and determinate, the others usually indeterminate and successively pectinate; tetrasporangia borne in 2 rows in swollen tips of mostly compound stichidia; sexual structures as for genus.

Common intertidally to subtidally (to 40 m), often on rocks buried in sand, S. Br. Columbia to Is. Revillagigedo, Mexico. Type locality uncertain.

Much smaller and more delicate forms of this species occur along with coarser forms throughout Calif.

Plocamium oregonum Doty

Doty 1947b: 177.

Thalli forming hemispherical tufts, dull red, to 13 cm tall; major axes mostly less than 1 mm broad, somewhat pinnately branched above; branch-

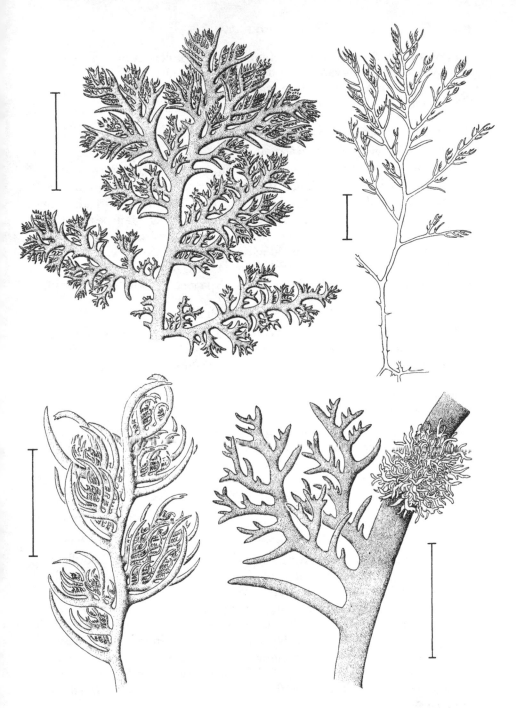

Fig. 437 (above left). *Plocamium cartilagineum.* Rule: 1 cm. (JRJ/S) **Fig. 438** (above right). *P. oregonum.* Rule: 1 cm. (DBP) **Fig. 439** (below left). *P. violaceum.* Rule: 1 cm. (JRJ/S) **Fig. 440** (below right). *Plocamiocolax pulvinata.* Rule: 5 mm. (JRJ/S)

lets few and short on lower parts of main branches, slender in upper parts, the apices curving outward; reproduction as for genus.

Occasional, saxicolous, lower intertidal to subtidal (3 m), S. Br. Columbia to Sonoma Co., Calif. Type locality: Brookings, Ore.

Plocamium violaceum Farl.

Farlow 1877: 240; Smith 1944: 264; Dawson 1961a: 221.

Thalli mostly 4–5 cm tall, reddish-violet, usually short and congested, with relatively few long, primary branches arising from mat of stoloniferous branches; main axes 1–1.5 mm diam.; branches subcylindrical, with 3–5 orders of pectination; lowermost of each order of branches usually simple, 3–6 mm long, commonly strongly curved toward parent branch; tetrasporangia in compound stichidia; sexual structures as for genus.

Common, mostly on vertical rock faces in areas of heavy surf, midtidal, S. Br. Columbia to Is. San Benito, Mexico. Type locality: Santa Cruz, Calif.

Plocamiocolax Setchell 1923

Thalli pulvinate, "parasitic," composed of cylindrical, pectinately branched, radiating branches with blunt apices. Tetrasporangia zonately divided. Spermatangia unknown. Cystocarps protuberant, with carpostome.

Plocamiocolax pulvinata Setch.

Setchell 1923a: 396; Smith 1944: 265.

Thalli to 5 mm diam., white to light tan; branches to 2.5 mm long, 0.25 mm diam., irregularly or pectinately branched, straight or twisted, cylindrical, tapering to broadly rounded apex.

Infrequent, usually subtidal, "parasitic" on *Plocamium cartilagineum*, San Juan I., Wash., to Cabo Colnett, Baja Calif. Type locality: Carmel Bay, Calif.

Family GRACILARIACEAE

Thalli mostly erect, free-living, cylindrical to complanate. Medulla of large cells, lacking central filament. Cortical cells progressively smaller outward. Tetrasporangia cruciately divided, scattered in outer cortex or borne in nemathecia. Spermatangia in superficial areas or in small pits. Carpogonial branches of 2 or 3 cells; supporting cell bearing sterile filaments. Gonimoblast filaments developing outward from mass of sterile cells, most gonimoblast cells forming carposporangia. Mature cystocarp protuberant, with thick pericarp and distinct carpostome.

Two genera in the California flora, *Gracilaria* (below) and *Gracilariophila* (p. 500).

Gracilaria Greville 1830

Thalli saxicolous, with several to numerous erect branches arising from mostly discoid base. Erect branches cylindrical to complanate, fleshy to cartilaginous. Gonimoblast with or without special filaments connecting with pericarp (Papenfuss, 1967, has found filaments present or lacking in the type species of *Gracilaria*).

1. Branches essentially cylindrical throughout or only slightly flattened 2
1. Branches distinctly flattened 6
   2. Erect branches mostly less than 15 cm tall 3
   2. Erect branches mostly 30+ cm tall 5
3. Erect branches stiff but not turgid, arising from compact, semistoloniferous base, 1–1.3 mm diam. . . . .*G. andersonii*
3. Erect branches turgid, arising from discoid base, to 5 mm diam. in thickest part 4
   4. Branches pale, irregularly branched, crisp to brittle . . . . *G. turgida*
   4. Branches deep burgundy red, 4–5 times subdichotomously branched, turgid but not brittle . . . . *G. robusta*
5. Branching pyramidal, of 3 or 4 orders; spermatangia in small, nonconfluent pits . . . . *G. verrucosa*
5. Branching mostly basal, mostly of 2 orders; spermatangia in continuous, superficial layer . . . . *G. sjoestedtii*
   6. Plants 6–8 cm tall; tetrasporangia distributed in cortex . . . . . . *G. veleroae*
   6. Plants 10–20 cm tall; tetrasporangia in nemathecioid areas. . . .*G. textorii*

Gracilaria andersonii (Grun.) Kyl.

Cordylecladia andersonii Grunow 1886: 62. *Gracilaria andersonii* (Grun.) Kylin 1941: 21. *Gracilariopsis andersonii* (Grun.) Dawson 1949a: 43.

Thalli usually with numerous erect branches, 6–18 cm tall, arising from semistoloniferous bases, dark red below, grading to straw-colored apices; branches terete, 1–1.3 mm diam., irregularly and abundantly multifariously branched, wiry and stiff; medulla of large cells; cortex of about 3 layers of smaller cells somewhat anticlinally arranged; tetrasporangia oblong to ovoid, 30–35 µm long, densely grouped in cortex of upper branches; spermatangia in deep, closely grouped pockets; cystocarps somewhat globose, slightly rostrate, mostly 900+ µm diam., without special filaments between gonimoblast and pericarp.

Frequent on rocks usually shallowly embedded in sand, low intertidal, Santa Barbara, Calif., to Pta. María, Baja Calif. Type locality: "California."

Gracilaria robusta Setch.

Setchell, P.B.-A. 1895–1919 [1899]: no. 635; Smith 1944: 267. *Gracilariopsis robusta* (Setch.) Dawson 1949a: 42. *Gracilariopsis claviformis* Daws. 1961a: 340. *Gracilaria claviformis* (Daws.) Papenfuss 1967: 100.

Thalli deep red, to 15 cm tall, several erect branches arising from discoid holdfasts; stipes weak, 1–3 mm long, 1 mm thick; erect branches usually 4–5 times dichotomously branched, usually in 1 plane, occasionally with less branching, each dichotomous order shorter and broader than the preceding; lower branches compressed, upper branches terete to compressed, stiff and turgid; tetrasporangia grouped in a somewhat nemathecioid cor-

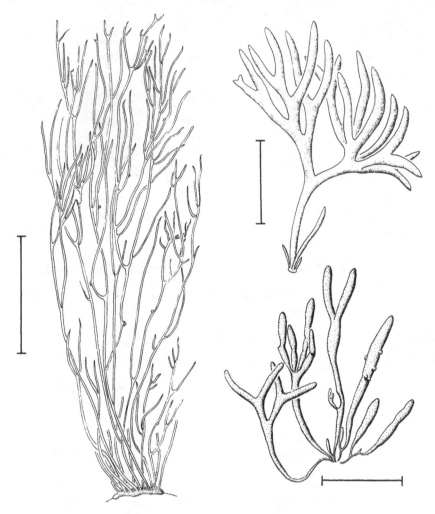

Fig. 441 (left). *Gracilaria andersonii*. Rule: 2 cm. (DBP) **Fig. 442** (two at right). *G. robusta*: above, habit of Monterey specimens; below, habit of Channel Is. specimens. Rules: both 3 cm. (JRJ/S & NLN)

tex of anticlinal cell rows; spermatangial pockets deep, ovoid, closely spaced, with small ostioles; cystocarps 2–4 mm diam., bulging prominently, nonrostrate, without special filaments between gonimoblast and pericarp.

Locally frequent, saxicolous, low intertidal to subtidal (9 m), Monterey Peninsula; common subtidally off Santa Cruz I., Calif. Type locality: Monterey, Calif.

Numerous specimens dredged off Santa Cruz I., referred to this species, differ in having mostly unbranched erect branches.

Fig. 443. *Gracilaria sjoestedtii.*
Rule: 3 cm. (JRJ/S)

Gracilaria sjoestedtii Kyl.

Kylin 1930: 55; 1941: 21; Smith 1944: 267. *Gracilariopsis sjoestedtii* (Kyl.) Dawson 1949a: 40.

Thalli to 2 m tall, brownish-red, few to many erect axes arising from discoid holdfast and associated semistoloniferous branches; branching sparse and irregular, the branches mainly basal, mostly simple, 0.5–1.5 (3.5) mm diam.; tetrasporangia ellipsoidal, ±38 μm long, distributed in outer cortex; spermatangia in continuous surface layer over major portion of surface of branches; cystocarps protuberant, 0.8–1.8 mm diam., without special filaments between gonimoblast and pericarp.

Locally common, on rocks half-buried in sand, midtidal to subtidal (to 15 m), southern British Columbia to Costa Rica. Type locality: Pacific Grove, Calif.

Gracilaria textorii (Sur.) J. Ag.

Sphaerococcus textorii Suringar 1870: 36. *Gracilaria textorii* (Sur.) J. Agardh 1876: 449.

The type variety is not known in California.

Gracilaria textorii var. cunninghamii (Farl.) Daws.

Gracilaria cunninghamii Farlow 1901: 93. *G. textorii* var. *cunninghamii* (Farl.) Dawson 1961a: 213.

Thalli complanate, ligulate, 10–20 cm tall, dark red, openly dichotomously branched, sparsely proliferous, subpalmate above, narrowed below to small, discoid holdfasts; branches 3–8 mm broad, 400–600 μm thick, the pigmented cortex with 1 or 2(3) layers of subcubical cells; tetrasporangia abundant, in confluent nemathecia; spermatangia in relatively deep, somewhat confluent, pitlike cavities, these in distinct patches separated by highly modified cortical cells; cystocarps globose, slightly basally constricted, nonrostrate, with special filaments between gonimoblast and pericarp.

Occasional on open coast, saxicolous, low intertidal, San Luis Obispo Co. and Santa Barbara (type locality), Calif., to I. Magdalena, Baja Calif.

Gracilaria turgida Daws.

Dawson 1949a: 14; 1961a: 213.

Thalli erect, cylindrical, 8–15 cm tall, 2–5 mm diam., robust, turgid, almost brittle when fresh, dull brownish-red to very pale, with 1 to several erect axes, these irregularly or subsecundly branched 1 or 2 times, arising from discoid holdfasts; apices of branches blunt; outer cortical cells slightly elongate anticlinally, with numerous deciduous hairs intermixed; tetrasporangia distributed in outer cortex; spermatangia in cuplike cavities separated by elongate cortical cells; cystocarps 700–900 μm diam., globose,

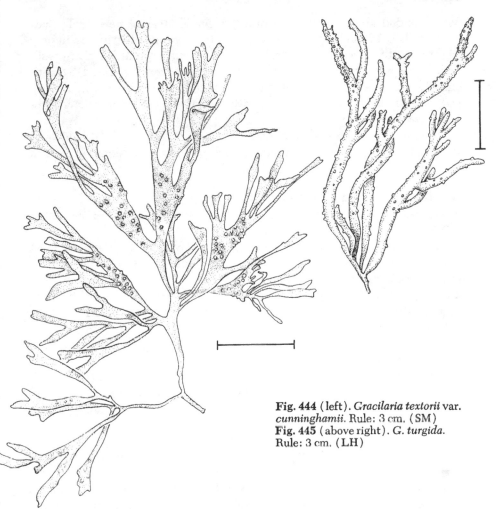

Fig. 444 (left). *Gracilaria textorii* var. *cunninghamii*. Rule: 3 cm. (SM)
Fig. 445 (above right). *G. turgida*. Rule: 3 cm. (LH)

slightly rostrate and slightly basally constricted, with special filaments connecting gonimoblast and pericarp.

Infrequent to rare, on rocks in bays or estuaries, midtidal to low intertidal, upper Newport Harbor (Orange Co.; type locality), Calif., to Bahía Magdalena, Baja Calif.

Gracilaria veleroae Daws.

Dawson 1944a: 297; 1961a: 214.

Thalli erect, membranous, dark red, 6–8 cm tall; blades with cuneate bases, arising from small, discoid holdfasts; branching irregularly dichotomo-flabellate; divisions mostly 5–7 mm broad, less than 250 µm thick, with rounded apices; cortex essentially a single layer of periclinally flattened cells ±12 μm long, ±5 μm diam.; tetrasporangia distributed in only slightly modified cortex, spherical, 25–35 μm diam.; spermatangia in well-defined depressions, these distributed over branches, separated by anticlinally elongate cells; cystocarps protuberant, ±1 mm diam., slightly rostrate, mostly near branch margins, with special filaments connecting gonimoblast and pericarp.

Predominantly subtidal, saxicolous, La Jolla, Calif., to Baja Calif. and Is. Revillagigedo, Mexico; common in Gulf of Calif., occasional in drift in S. Calif. Type locality: Gulf of Calif.

Gracilaria verrucosa (Huds.) Papenf.

Fucus verrucosus Hudson 1762: 470. *Gracilaria verrucosa* (Huds.) Papenfuss 1950: 195; Dawson 1961a: 214.

Thalli erect, pale reddish to pinkish, sparsely radially branched from percurrent axes 30–50 cm tall, 1–2 mm diam., sometimes beset below with numerous simple or branched spines; branches cylindrical, branching 3 or 4 times; cortex of mostly 2 layers of small cells; tetrasporangia distributed in unmodified cortex; spermatangial sori forming rounded and slightly elevated surface patches, mostly toward branch apices; cystocarps hemispherical to globose, with or without abundant filaments connecting pericarp and gonimoblast.

Infrequent, discontinuous, in coarse sand or on rocks, mostly in sheltered water, midtidal to upper subtidal, S. Br. Columbia to Baja Calif. and Gulf of Calif. Of worldwide distribution. Type locality: England.

Gracilariophila Setchell & Wilson 1910

Plants minute, colorless, "parasitic" on *Gracilaria* species. External portion globose. Medulla of angular, thick-walled cells. Cortex of small, angular cells. Tetrasporangia distributed in outer cortex. Spermatangia covering outer surface. Cystocarps solitary or in groups, similar to those of *Gracilaria*.

Gracilariophila gardneri Setch.

Setchell 1923a: 393; Dawson 1949a: 51.

Thalli depressed-globose or irregularly tuberculate, 3–3.5 mm diam., "parasitic" on *Gracilaria textorii* var. *cunninghamii*, with cellular processes penetrating hosts; tetrasporangia 28–34 μm long, borne in cortex of entire external portion of thallus, very irregularly cruciate to nearly zonate; spermatangia in dense peripheral layer; cystocarps solitary or in groups, hemispherical, nonrostrate, obscurely ostiolate.

Infrequent, Santa Monica, Calif. (type locality), to Bahía Bocochibampo (Sonora), Mexico.

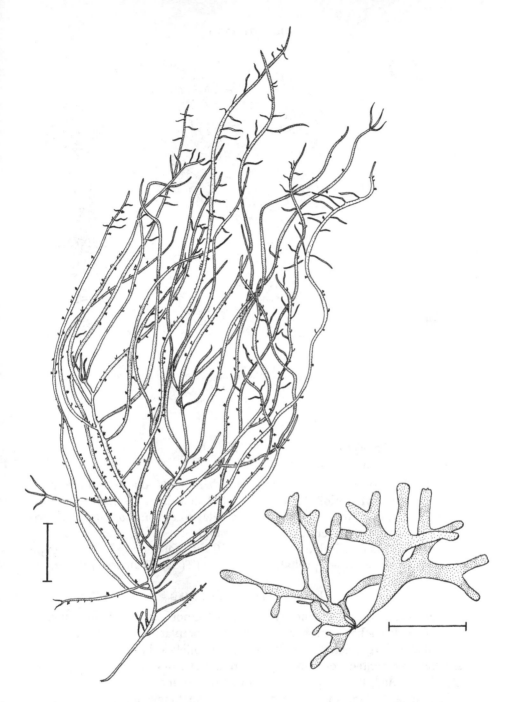

Fig. 446 (below right). *Gracilaria veleroae*. Rule: 3 cm. (DBP) **Fig. 447** (left). *G. verrucosa*. Rule: 3 cm. (DBP)

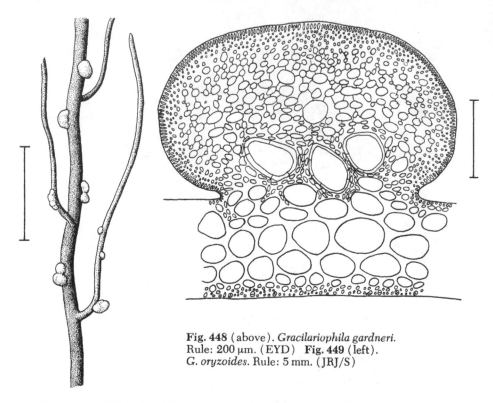

Fig. 448 (above). *Gracilariophila gardneri.*
Rule: 200 μm. (EYD) **Fig. 449** (left).
G. oryzoides. Rule: 5 mm. (JRJ/S)

Gracilariophila oryzoides Setch. & Wils.

Setchell & Wilson 1910: 81; Smith 1944: 268.

Thalli "parasitic" on *Gracilaria sjoestedtii* and *G. andersonii*, forming colorless, somewhat globular tubercles 1–2 mm diam. on lower parts of hosts; structurally and reproductively similar to *Gracilariophila gardneri*, but with more strictly cruciate tetrasporangia.

Smith I., Wash., to Bahía Rosario, Baja Calif.; common in C. Calif. Type locality: not designated, probably "Fort Point" (San Francisco), Calif.

Family PHYLLOPHORACEAE

Thalli erect, saxicolous, more or less cartilaginous, mostly dichotomously branched. Branches cylindrical to broadly complanate. Thalli multiaxial. Medulla of large, compact cells, typically thick-walled. Cortex firm, of smaller, pigmented cells, mostly in anticlinal rows. Tetrasporangia cruciately divided, typically borne in anticlinal catenate series in superficial nemathecia, lacking in certain species. Spermatangia usually superficial, variously grouped. Plants procarpial, with supporting cells serving as auxiliary cells. Cystocarps typically in groups separated by sterile filaments, forming carposporangia or carpotetrasporangia. Monosporangia in a few species.

1. Thalli primarily crustose; erect axes less than 2 mm tall......*Besa* (p. 515)
1. Thalli not primarily crustose 2
  2. Procarps numerous, in conspicuous, interrupted line resembling a midrib .... *Stenogramme* (p. 512)
  2. Procarps variously distributed, the branches without interrupted median line 3
3. Branches cylindrical or subcylindrical; medulla partly of slender, parallel, longitudinal filaments .... *Ahnfeltia* (below)
3. Branches distinctly flattened to complanate; medulla of large, more or less isodiametric cells 4
  4. Spermatangia and cystocarps developing in small, specialized leaflets on surface of blades .... *Ozophora* (p. 511)
  4. Spermatangia not developing in specialized surface leaflets.......... 5
5. Plants with separate gametangial and sporangial stages; tetrasporangia typically cruciately divided, in anticlinal chains in nemathecia........
.... *Petroglossum* (p. 508)
5. Plants without separate sporangial stage; tetrasporangia mostly imperfectly divided, arising from gonimoblast filaments in nemathecial pustules in certain species, unknown in other species....... *Gymnogongrus* (p. 505)

Ahnfeltia Fries 1835

Thalli erect, with numerous cylindrical to slightly compressed branches arising from cylindrical rhizomes. Branches repeatedly dichotomous or irregularly branched, rigid to wiry. Medulla partly of narrow, parallel, longitudinal filaments. Outer cortex of radially arranged, firmly adjoined cell rows. Monosporangia produced in nemathecioid enlargements of branches, these the only known reproductive structures.

Ahnfeltia gigartinoides J. Ag.

J. Agardh 1847: 12; Smith 1944: 272.

Thalli tufted, deep red to purplish-black, 10–30 cm tall, 10–15 times dichotomous, somewhat distichous; branches rigid, cylindrical to subcylindrical, 0.5–1 mm diam.; sometimes with short, simple, proliferous branches on basal parts. Reproduction imperfectly known.

Frequent to uncommon on intertidal or subtidal rocks, Vancouver I., Br. Columbia, to Pta. Santo Tomás, Baja Calif. and southward. Type locality: St. Augustine (Oaxaca), Mexico.

Ahnfeltia plicata (Huds.) Fries

Fucus plicatus Hudson 1762: 470. *Ahnfeltia plicata* (Huds.) Fries 1835: 310; Smith 1944: 271.

Thalli purplish-black, 5–14 cm tall; branches numerous, cylindrical, wiry, densely irregularly branched, the branches 250–500 µm diam., with

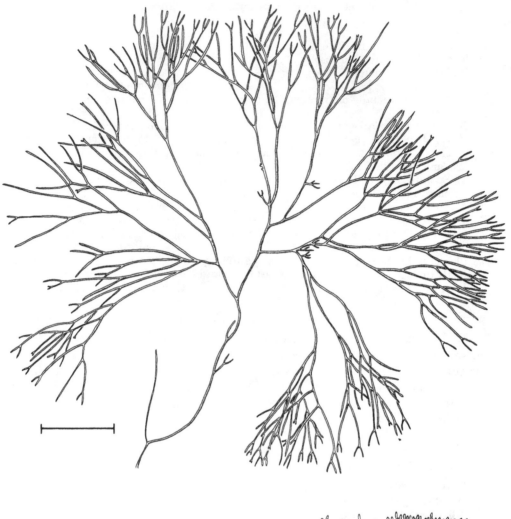

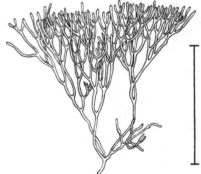

Fig. 450 (center). *Ahnfeltia gigartinoides.*
Rule: 3 cm. (JRJ/S) **Fig. 451** (below
right). *A. plicata.* Rule: 3 cm. (JRJ/S)

proliferous branches arising mostly on mature lower parts; reproduction not observed in Calif. specimens.

Common in restricted localities, usually half-buried in sand, midtidal to lower intertidal, Bering Sea to Baja Calif. Widely distributed. Type locality: England.

Gymnogongrus Martius 1828

Thalli erect, saxicolous, several to many main branches arising from discoid holdfasts. Branches rigid, repeatedly dichotomous, mostly cylindrical below, compressed or distinctly complanate above, the branching primarily in 1 plane. Medulla of large, angular cells. Cortex of smaller cells in anticlinal rows. Tetrasporangial phase lacking in certain species; in other species reduced nemathecioid or pustular tetrasporophytes developing from gonimoblast and typical carposporophytes lacking. Plants dioecious. Spermatangia in irregular surface areas. Carpogonial branches borne on inner cortical cells. Mature cystocarps deeply embedded, globose, without carpostome; carposporangia in groups separated by sterile cells.

1. Erect branches mostly less than 20 mm tall, arising from thin, crustose base .... *G. crustiforme*
1. Erect branches mostly 70+ mm tall 2
   2. Upper branches less than 2 mm broad ....*G. leptophyllus*
   2. Upper branches commonly 4+ mm broad 3
3. Plants red to reddish-purple, not tufted; branches above first dichotomy complanate, uniformly broad .... *G. platyphyllus*
3. Plants light tan to brownish-purple, tufted; branches above first dichotomy not uniformly broad 4
   4. Erect axes 50+, arising from crustose base; ultimate branches not complanate .... *G. linearis*
   4. Erect axes few, arising from crustose base; ultimate branches distinctly to markedly complanate ....*G. martinensis*

Gymnogongrus crustiforme Daws.

Dawson 1961a: 248.

Thalli dull red, with ligulate, erect branches, arising from closely adherent, crustose bases, these 10+ mm broad; erect branches usually less than 1 cm tall, slightly stipitate, with blunt apices, 600–800 μm broad, ±150 μm thick, irregularly forked 1 or 2 times near apices, rarely with 1 or more lateral branches; medullary cells elliptical, 20–30 μm maximum diam., grading into anticlinal cortical cell rows; tetrasporangia unknown; spermatangial nemathecia in upper part of erect branches, longitudinally elongate, 300–400 μm broad, sharply delimited; cystocarps in upper parts of erect branches, 300–500 μm diam., prominently protuberant, mostly on 1 side only.

Infrequent on rocks in surf, Corona del Mar (Orange Co.) and Laguna Beach, Calif., and Salina Cruz, Mexico (type locality).

Gymnogongrus leptophyllus J. Ag.

J. Agardh 1876: 211; Smith 1944: 273.

Thalli erect, the branches forming hemispherical tufts 4–7 cm tall, to 10 cm diam., arising from discoid holdfasts, deep dull carmine; erect branches with 5–8 dichotomies, frequently with a few short, proliferous branches; branches cylindrical or subcylindrical below, mostly markedly flattened and 0.75–1.25 mm broad in terminal divisions; tetrasporangia unknown; spermatangia forming dense, continuous layer 20–25 μm thick over upper branches; cystocarps ellipsoidal, to 1.5 mm long, bulging on either side of upper branches.

Frequent on sand-swept rocks in both open and protected localities, intertidal to subtidal (30 m), Alaska to Pta. Baja, Baja Calif.; common on Monterey Peninsula and Channel Is., Calif. Type locality: "ad oras Californiae," probably Santa Cruz, Calif.

Gymnogongrus linearis (C. Ag.) J. Ag.

Sphaerococcus linearis C. Agardh 1822a: 250. Gymnogongrus linearis (C. Ag.) J. Agardh 1851: 325; Doubt 1935: 294; Smith 1944: 274.

Thalli in dense tufts or patches, these to 40 cm broad, with 50–150 erect branches, 10–18 cm tall, arising from a crustose base; erect branches light tan to deep brownish-purple, with 3–6 successive dichotomies at wide angles, all approximately in same plane, with cylindrical, stipelike basal part 4–7 cm long; lesser branches mostly 5–8 mm broad, oval to elliptical in cross section; tetrasporangia unknown; plants dioecious; spermatangia forming extensive surface areas; carposporangial nemathecia abundant, on all but final dichotomies, 1–2 mm diam., conspicuously protuberant.

Common to abundant on sand-swept rocks, midtidal to lower intertidal, S. Br. Columbia to Pt. Conception, Calif. Type locality: Trinidad (Humboldt Co.), Calif.

Gymnogongrus martinensis S. & G.

Setchell & Gardner 1937a: 78; Dawson 1961a: 251.

Thalli caespitose, to 16 cm tall, dark red, with few to many erect axes arising from thin, discoid holdfasts; erect axes subcylindrical below, complanate above, 3–6 times irregularly dichotomously branched, with stipe to 45 mm long and with a few lateral proliferations; branches 2–6(10) mm broad above, mostly 0.5–2.5 mm thick; tetrasporangia and spermatangia unknown; cystocarps scattered or grouped on subultimate branches, ±1 mm diam., projecting mostly on 1 side of branches.

Rare, on rocks, Ventura Co., Calif., to I. Magdalena, Baja Calif.; in Calif., known only from a single collection at Hobson State Park, Ventura Co. Type locality: I. San Martín, Baja Calif.

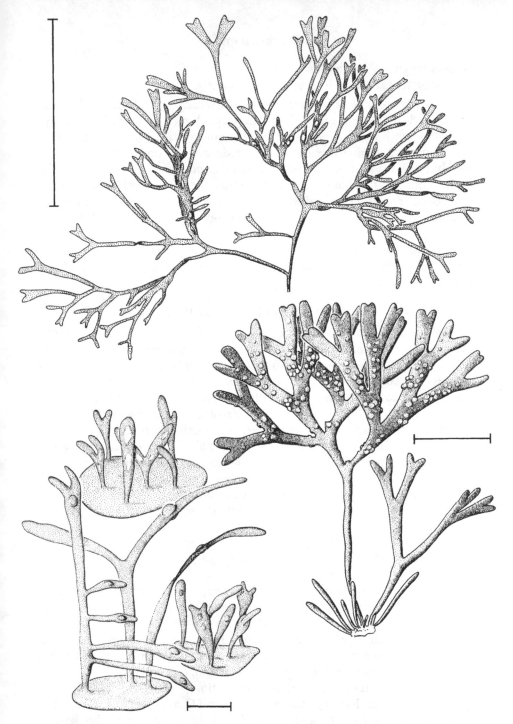

Fig. 452 (below left). *Gymnogongrus crustiforme*. Rule: 2 mm. (DBP/H) Fig. 453 (above). *G. leptophyllus*. Rule: 3 cm. (JRJ/S) Fig. 454 (below right). *G. linearis*. Rule: 3 cm. (JRJ/S)

Gymnogongrus platyphyllus Gardn.

Gardner 1927a: 247; Doubt 1935: 299; Smith 1944: 274.

Thalli with relatively few erect branches arising from discoid holdfasts; branches 8–15(25) cm tall, dull red to purplish-red, very briefly cylindrical below, with arching parallel fans of branches, these of 3–5 successive, overlapping, flattened dichotomies at wide angles; upper branches uniformly (3)4–6 mm wide, mostly less than 1 mm thick, the apices mostly blunt; spermatangia and free-living tetrasporophytes unknown; "tetrasporoblast" nemathecia (reduced "parasitic" tetrasporophytes) abundant, prominently protuberant, 2–3 mm diam., produced on all but final dichotomies.

Occasional on rock faces, low intertidal to subtidal (to 18 m), S. Br. Columbia to Pta. Santa Rosalía, Baja Calif. Type locality: Duxbury Reef (Marin Co.), Calif.

Petroglossum Hollenberg 1943

Thalli with several to numerous erect branches arising from expanded-crustose or somewhat stoloniferous bases. Erect branches ligulate, simple to irregularly dichotomously flabellate, commonly stipitate or considerably narrowed below, with or without lateral or terminal proliferations; texture firm to cartilaginous. Medulla of large, thick-walled cells. Cortex with mostly 1 or 2 layers of small cells. Tetrasporangia cruciately divided, in catenate, anticlinal rows, aggregated in nemathecia, these when mature usually occurring singly in the center of a thallus division. Spermatangia in well-defined nemathecia, these arising singly in center or toward apices of branches. Cystocarps borne singly in center of blade or in short lateral proliferations, protuberant on both sides of branches, with or without carpostome. Carpospores arrayed in groups, the groups separated by sterile cells.

Petroglossum pacificum Hollenb.

Hollenberg 1943a: 571.

Thalli of erect branches, to 5 cm tall, purplish-red, with few to many irregular to dichotomously flabellate divisions, these 1–3 mm broad, 100–135 μm thick, usually slightly crisped; base crustose; medullary cells 30–50(80) μm diam.; cortical cells 5–7 μm diam.; tetrasporangia unknown; spermatangial nemathecia ±1 mm broad; cystocarps on numerous short, lateral proliferations, 500–600 μm diam., without carpostome.

Infrequent on rocks exposed to wave action, Corona del Mar and La Jolla, Calif. Type locality: Corona del Mar (Orange Co.), Calif. Additional specimens from C. and N. Calif. may be referrable to this species but are mostly sterile.

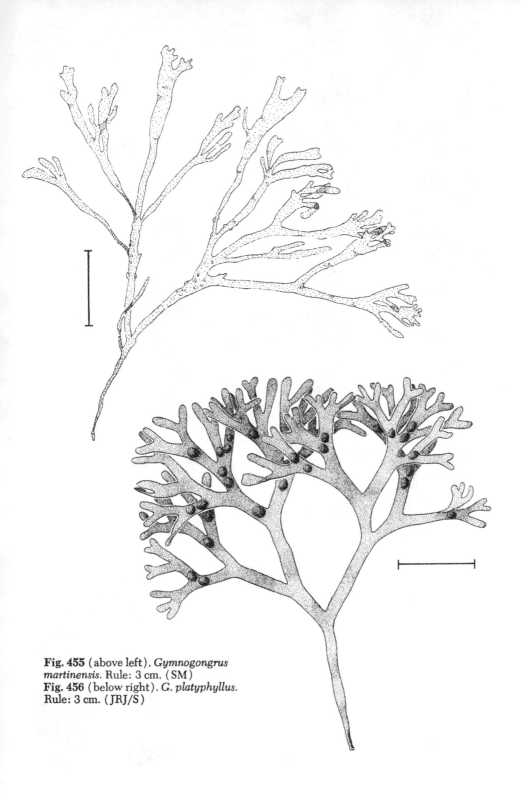

Fig. 455 (above left). *Gymnogongrus
martinensis.* Rule: 3 cm. (SM)
Fig. 456 (below right). *G. platyphyllus.*
Rule: 3 cm. (JRJ/S)

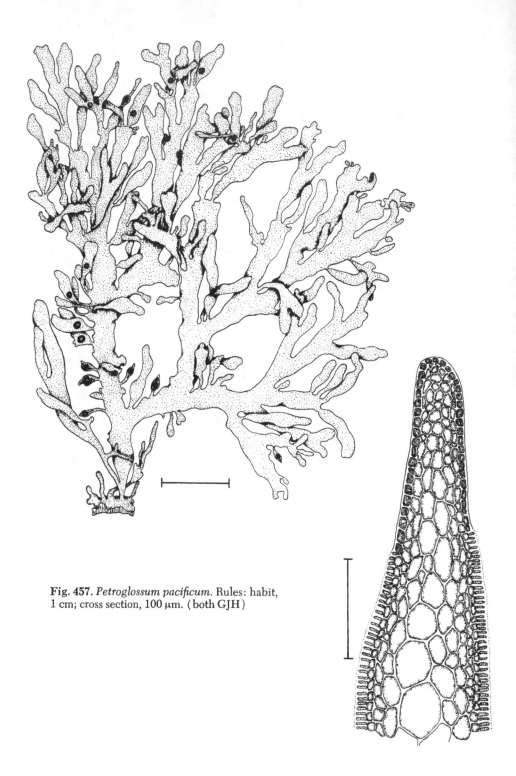

Fig. 457. *Petroglossum pacificum.* Rules: habit, 1 cm; cross section, 100 μm. (both GJH)

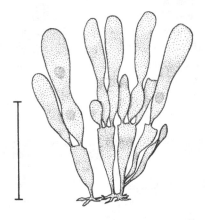

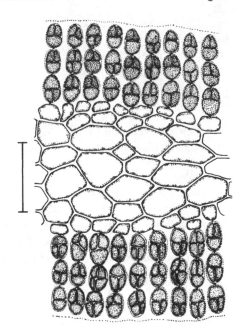

Fig. 458. *Petroglossum parvum.*
Rules: habit, 1 cm; cross section,
100 μm. (both GJH)

Petroglossum parvum Hollenb.

Hollenberg 1945a: 450; Dawson 1961a: 244.

Thalli 10–18 mm tall, arising from spreading, stoloniferous branches; erect branches simple, or mostly with simple divisions from cylindrical stipe, this 1 or 2 times dichotomous; branches ligulate to spatulate, 0.8–1.5 (2) mm wide, 60–120 μm thick, commonly producing several new branches from an eroded or grazed apex; tetrasporangia in anticlinal series of 3–5, grouped in small circular nemathecia, these occurring singly in center of blades, protruding on both surfaces; new nemathecium commonly arising distally before older one completely disappears; spermatangial nemathecia and gonimoblasts occurring singly in centers of blade divisions; cystocarps with small carpostome.

Frequent on rocks, low intertidal to subtidal (to 18 m), Santa Barbara, Calif., to I. Cedros, Baja Calif. Type locality: Laguna Beach, Calif.

Specimens represented by fig. 38 in Hollenberg & Abbott (1966) and as described by Doty (1947b) are very doubtfully referrable to this species. Their further classification awaits collection of fertile material.

Ozophora J. Agardh 1892

Thalli saxicolous, 1 to several erect branches arising from discoid holdfasts, with or without cylindrical stipes. Erect branches complanate, with linear to cuneate divisions, these commonly stipitate, the broadly round-tipped ultimate bladelets frequently arising as regenerative growths. Medulla of large, colorless cells. Cortex with 2–4 layers of small, pigmented cells. Monospores (undivided tetrasporangia?) borne in superficial or marginal pustules on asexual plants, prominently protuberant. Tetrasporangia cruciately divided, in anticlinal chains in surface nemathecia. Spermatangia in superficial areas on thin, cordate leaflets, these clustered on blade surface. Cystocarps protuberant, in short, simple-cylindrical or fusiform, surface proliferations.

Ozophora clevelandii (Farl.) Abb.

Phyllophora clevelandii Farlow 1875: 368. *Ozophora clevelandii* (Farl.) Abbott 1969: 48 (incl. synonymy). *O. californica* J. Agardh 1892: 82. *P. californica* (J. Ag.) Kylin 1931: 34.

Thalli deep rose, with several erect branches arising 10–28 cm from woody, discoid holdfasts, with cylindrical, wiry stipes to half the total height; blades mostly linear or spatulate, 5–20 mm wide, to 15 cm long, on short stipes of 2 or 3 orders of branching, occasionally regeneratively proliferous; cortex of 2 or 3 cell layers; monosporangia (undivided tetrasporangia?) in anticlinal chains in blisterlike pustules on blade surface; tetrasporangia in anticlinal chains in superficial nemathecia; spermatangial leaflets clustered on both median surfaces of blades; cystocarps in papillate leaflets, these in similar clusters or marginal.

Infrequent, on rocks, subtidal, Duxbury Reef (Marin Co.), Calif., to Bahía de San Quintín, Baja Calif. Type locality: San Diego, Calif.

Ozophora latifolia Abb.

Abbott 1969: 49.

Thalli erect, branched, to 30 cm tall, brick red, drying to rust, with or without short, cylindrical stipes, these when present less than 20 mm long; main axes bladelike, 2.5–3 cm broad, irregularly to regularly dichotomously flabellate, or with stipitate, proliferous marginal blades, these 7–10 cm long, 1–2 cm broad; cortex of 3 or 4 cell layers; monosporangia (undivided tetrasporangia?) in nemathecia on blade segments or marginal outgrowths; spermatangia in continuous surface areas, these on pink, cordate, leaflike proliferations, 1–2 mm long, 1–3 mm broad; cystocarps in simple or once-branched, cylindrical to fusiform, papillate outgrowths ±1 mm tall, mostly arising from blade surface.

Infrequent, on rocks, low intertidal to mostly subtidal, Wash. to San Diego, Calif. Type locality: Monterey, Calif., dredged in 15–18 m.

Stenogramme Harvey 1840

Thalli saxicolous, with several erect branches arising from discoid holdfasts. Erect branches complanate, dichotomously divided. Medulla with 1–3 layers of large, colorless, isodiametric cells. Cortex with 1 or 2 layers of much smaller, pigmented cells. Tetrasporangia cruciately divided, in anticlinal, catenate series in small, irregularly shaped nemathecia, these

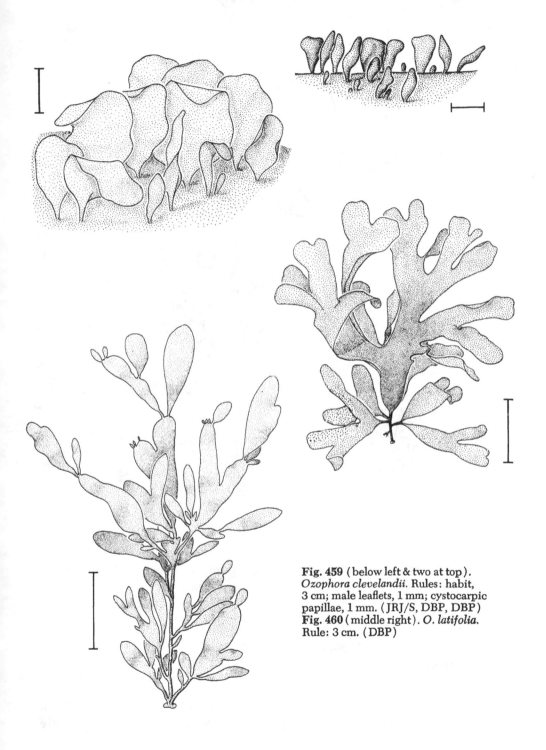

Fig. 459 (below left & two at top).
Ozophora clevelandii. Rules: habit,
3 cm; male leaflets, 1 mm; cystocarpic
papillae, 1 mm. (JRJ/S, DBP, DBP)
Fig. 460 (middle right). *O. latifolia.*
Rule: 3 cm. (DBP)

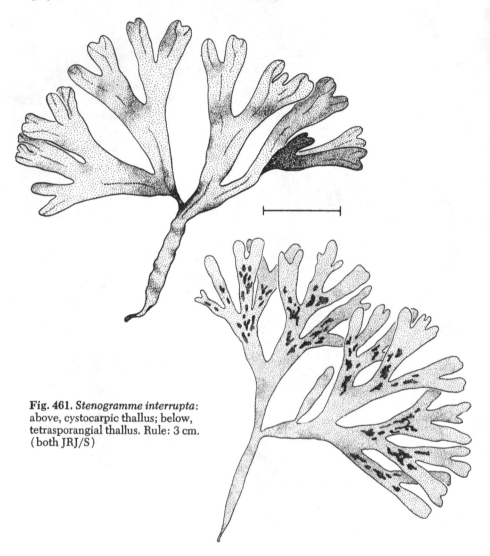

Fig. 461. *Stenogramme interrupta*:
above, cystocarpic thallus; below,
tetrasporangial thallus. Rule: 3 cm.
(both JRJ/S)

distributed over blades. Spermatangia borne in microscopic pockets in
large, irregularly shaped sori. Cystocarpic plants with axial longitudinal
soral areas resembling an interrupted midrib; mature cystocarps small,
globose, deeply embedded, without carpostome.

Stenogramme interrupta (C. Ag.) Mont.

Delesseria interrupta C. Agardh 1822a: 179. *Stenogramme interrupta* (C. Ag.) Montagne 1846b: 483; Dawson 1961a: 253. *S. californica* Harvey 1841a: 408; Smith 1944:
276.

Thalli deep red, 6–20 cm tall; erect branches cylindrical below, dichotomously branched and ligulate above; ultimate dichotomies linear, mostly

5–10 mm broad, occasionally with proliferous terminal blades, but apices commonly spatulate. Tetrasporangial thalli with broader blades than sexual thalli; sori irregularly spaced, producing mottled surface.

Frequent on rocks, low intertidal to subtidal (to 30 m), S. Br. Columbia to Cabo de San Lucas, Baja Calif. Also Atlantic Europe. Type locality: Spain.

Besa Setchell 1912

Thalli primarily crustose, few to many cells thick, noncalcareous, composed of adjoined erect filaments arising from ill-defined basal layers. Medulla of somewhat elongate cells, these loosely joined by slender connections. Cortex of anticlinal cell rows. Fertile branches papillate to club-like. Tetrasporangia unknown; spermatangia superficial in separate branchlets from cystocarps. Cystocarps terminal in fertile branches, without carpostome; carposporangia in groups separated by sterile cells.

Besa papillaeformis Setch.
Setchell 1912: 237.

Crust dark red, fleshy-cartilaginous, to 3 cm broad, many cells thick, of firmly adjoined, erect filaments; fertile branches papillate, 250–500 μm tall, to 400 μm diam., simple or with 1 or 2 branches; cystocarps relatively large, globose, terminal; spermatangia and tetrasporangia unknown.

Rare, on rocks, intertidal to subtidal (to 14 m); known only from Land's

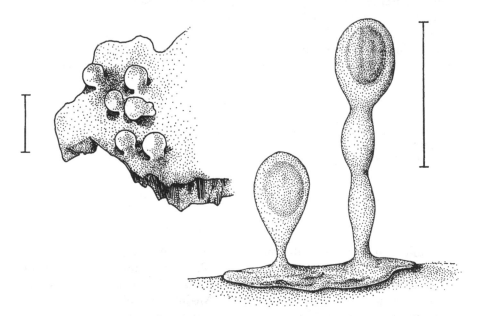

Fig. 462 (left). *Besa papillaeformis*. Rule: 1 mm. (SM).   Fig. 463 (right). *B. stipitata*. Rule: 1 cm. (DBP/H)

End (San Francisco: type locality), Pacific Grove, and Cypress Pt. (Monterey Co.), Calif.

Besa stipitata Hollenb. & Abb.

Hollenberg & Abbott 1968: 1249.

Crust dark red, delicate, 20+ mm broad, mostly 3 cells thick, with lower layer of radiating cell rows bearing erect filaments mostly 2 cells long; erect fertile cystocarpic branches stipitate, clavate, to 1.5 mm tall, with 1 terminal, protuberant cystocarp, this to 365 μm diam.; spermatangial branches slightly flattened, with short stipe, the spermatangia covering most of branch; tetrasporangia unknown.

Rare, on rocks, subtidal (to 14 m), San Juan Is., Wash., and Monterey Bay (type locality), Calif.

Family GIGARTINACEAE

Thalli erect, saxicolous, fleshy, mostly foliaceous or compressed, many of large size, branched or unbranched, multiaxial. Medulla of meshlike elongate anastomosing filaments, or medullary filaments of irregular shape and size, not meshlike. Cortex of branched filaments at right angles to surface. Tetrasporangia cruciately divided, in rows or globose masses, adjacent to or embedded in medulla, developing from subcortex or cortex, or from special filaments arising from medulla. Spermatangia in extensive patches, these superficial or partially embedded. Plants procarpial; carpogonial branch 3-celled, the supporting cell very large, enlarging further after fertilization of carpogonium and forming auxiliary cell. Gonimoblast filaments initiated toward medulla; nearly all cells becoming carposporangia, with or without dense ring of sterile filaments around gonimoblast.

1. Surfaces or margins of thalli covered with elongate or wartlike papillae . .
   . . . . *Gigartina* (below)
1. Surfaces or margins usually smooth, sometimes proliferous but not
   spiny 2
   2. Inner medullary filaments loosely arranged; tetrasporangia on outer
      medullary filaments . . . . *Iridaea* (p. 529)
   2. Inner medullary filaments closely knit; tetrasporangia in modified
      cortical cells . . . . *Rhodoglossum* (p. 538)

Gigartina Stackhouse 1809

Thalli with peglike or discoid holdfasts; fronds gregarious, relatively elongate, cylindrical or compressed to foliaceous, multiaxial, thickly or sparsely beset with papillate outgrowths, some of these clearly vegetative, others carrying reproductive structures. Medulla of delicate, cobwebby, anastomosing filaments, or with cells of various shapes and sizes. Cortex uniform, relatively thin, the cells small, oval to spherical, in 4–6(15)-celled anticlinal rows. Most species with tetrasporangial thalli similar in size and shape to gametangial thalli; in 2 species (*G. agardhii* and *G. papillata*) tetrasporangial plant crustose, not resembling erect thalli. Tetrasporangia developed from innermost cortical cells, cruciately divided, in globose to flattened sori on fertile papillae, or in chains in separate crustose stage. Spermatangia in irregularly shaped, continuous, confluent, superficial patches on papillae, sometimes spreading to flat surface of thallus; in some species entire thallus smooth, mostly without papillae, and lighter in color when spermatangial. Cystocarps making prominent bulges on papillae, in some species changing appearance of sterile thallus completely; firm, dense, internal ring of sterile tissue surrounding carposporangia.

Different species of *Gigartina* tend to occur in characteristic tidal zones.

1. Main branches commonly 4+ cm broad to broadly foliose, mostly little-branched 2
1. Main branches commonly narrower, plants repeatedly branched 4
  2. Blades large, usually divided, frequently 10 times as long as broad... .... *G. harveyana*
  2. Blades large, simple, rarely divided, 20+ cm long 3
3. Blades broad-obovate with rounded apices; papillae coarse and tapering .... *G. corymbifera*
3. Blades lanceolate with tapering apices; papillae crisp and hemispherical .... *G. exasperata*
  4. Thalli with few or no papillae; superficial and marginal earlike leaflets common .... *G. volans*
  4. Thalli with few to many crowded papillae; leaflets lacking 5
5. Axes and branches nearly same width throughout; papillae inconspicuous until fertile 6
5. Axes and branches of different widths; papillae conspicuous, well-developed before fertilization 7
  6. Branches intricate, complanate, less than 1 mm wide.......... *G. tepida*
  6. Lower branches arching, complanate, 2–3 mm wide ..... *G. canaliculata*
7. Fronds densely covered with long, flexible, rodlike papillae .... *G. leptorhynchos*
7. Fronds with short, stiff papillae 8
  8. Fronds pendant, narrow to slightly expanded; low intertidal; large papillae continuously distributed .... *G. spinosa*
  8. Fronds erect, crisp; high midtidal; papillae in irregularly distributed groups 9
9. Fronds narrow, broadening at apices; margins raised .... *G. agardhii*
9. Fronds usually wider throughout than *G. agardhii*; penultimate divisions mostly wider, but highly variable; margins usually not raised .... *G. papillata*

Gigartina agardhii S. & G.

Setchell & Gardner 1933: 290; Smith 1944: 284; Abbott 1970: 2.

Thalli reddish-brown, 5–10(15) cm tall, growing in small, occasional clumps. Erect portions narrow, attenuated, crisp; blades 2–5 mm wide, divided dichotomously several times, the apices commonly flaring; lateral margins with thickened ridge; blades furrowed, with occasional papillae in small groups; medulla filamentous, irregular; cortex of many branched filaments, closely apposed; fertile papillae restricted to flattened faces of blades, acutely pointed, at times bifurcate, sometimes flattened; tetrasporangia in separate crustose stage (*"Petrocelis"*).

Frequent to common, saxicolous, growing conspicuously among *Gigartina papillata*, high to midtidal, S. Br. Columbia to San Luis Obispo Co., Calif.

The sporangial stage has been known as *Petrocelis franciscana* (West, 1972).

Gigartina canaliculata Harv.

Harvey 1841a: 409; 1853: 174; Smith 1944: 278. *Gigartina serrata* Gardner 1927a: 334; Dawson 1961a: 274.

Thalli annual, from perennial bases, 10–25(44) cm tall; lower portions commonly greenish, upper portions purplish, drying to black; thalli compressed when young and appressed in flat clumps; mature reproductive thalli erect, the upper portions compressed to terete, growing in extensive clumps with entangled basal portions; lower branches unbranched or little branched, 2–3 mm wide; upper third of erect branches irregularly dichotomously divided, distichous in terminal portions. Medulla filamentous, the filaments irregular; cortex firm, the cells in thick rows; tetrasporangial thalli densely branched upward, bearing short, pointed, papillate, terminal branches, these with sori in slipper-shaped, internal nemathecia lateral to apices; spermatangial thalli also densely branched, the papillae pale, with sori encircling papillae subterminally; cystocarpic thalli with papillae and ultimate branches mostly more widely spaced than in other thalli, the gonimoblasts appearing beadlike to unaided eye, several on each papilla, surrounded by and partly enclosed by sterile branches.

Locally abundant, saxicolous, midtidal to lower intertidal, rarely subtidal, S. Ore. to I. Magdalena, Baja Calif. Type locality: San Francisco, Calif.

Gigartina corymbifera (Kütz.) J. Ag.

Mastocarpus corymbiferus Kützing 1847: 24. *Gigartina corymbifera* (Kütz.) J. Agardh 1876: 202; Setchell & Gardner 1933: 275; Smith 1944: 281 (incl. synonymy). *G. binghamiae* J. Ag. 1899: 33; Dawson 1961a: 267.

Thalli in clumps of large, crisp blades, to 1+ m long, to 30 cm wide; blades generally with stout holdfasts and strap-shaped apophyses, these 4–5 cm wide, somewhat flared upwardly at bases of blades; holdfasts and apophyses usually smooth; blades of various sizes and ages in clumps; young blades and basal portion of old blades iridescent; older blades yel-

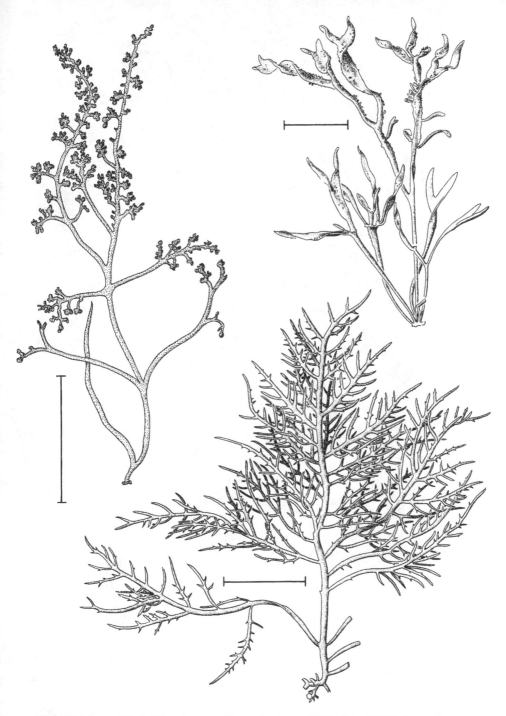

Fig. 464 (above right). *Gigartina agardhii.* Rule: 3 cm. (FT) **Fig. 465** (below right & above left). *G. canaliculata*: below right, habit of sterile thallus; above left, habit of fertile thallus. Rules: both 3 cm. (JRJ/S & CS)

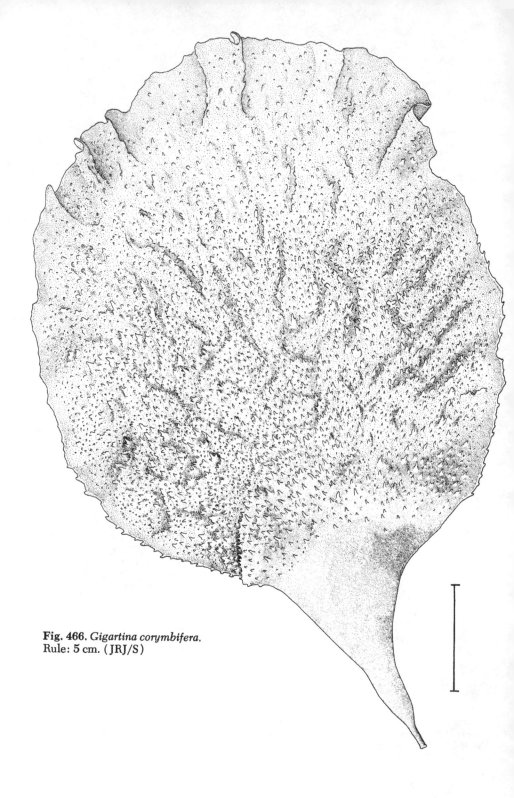

Fig. 466. *Gigartina corymbifera.*
Rule: 5 cm. (JRJ/S)

lowish-pink when plants in low intertidal, the plants uniformly bluish-rose when subtidal; medulla with filaments of variety of sizes and shapes; papillae very coarse, commonly 1–2 cm long, obtuse and occasionally bifurcate if tetrasporangial, acuminate and slender if spermatangial, short and mammillose (often corymbose) if cystocarpic, commonly with 2 or 3 terminally branched cystocarps per papilla.

Common, saxicolous, low intertidal (in areas exposed to surf and surge) to subtidal (30 m), Whidbey I. and Cape Flattery, Wash., to Cabo San Quintín, Baja Calif. Type locality: probably Monterey, Calif.

Gigartina exasperata Harv. & Bail.

Harvey & Bailey 1851: 371; Scagel 1957: 185 (incl. synonymy). *Gigartina californica* J. Agardh 1899: 39; Setchell & Gardner 1933: 272; Smith 1944: 280.

Thalli deep brownish-red, commonly 30–50 cm tall but frequently to 1+ m, usually 2 or 3 blades 15–30 cm wide growing from single holdfast, sometimes divided from common stipe; stipes short, cylindrical, merging into flattened apophyses, these gradually flaring to 2–3 cm wide; medulla filamentous, anastomosing; base of blade broadly lanceolate with gradually narrowing upper portion; apex continuous or divided, blunt to pointed; northern specimens with irregular margins, some almost dentate, the C. Calif. specimens with crisp, sometimes ridged margins; papillae of most specimens short, nearly hemispherical, crisp, those in northern specimens scattered randomly, elongate, the blades appearing thin, those in C. Calif. specimens closely placed, sometimes overlapping, the blades appearing thick, those in S. Calif. specimens more widely spaced; tetrasporangial papillae usually upwardly decurrent, with sharp apex, the tetrasporangial sori subterminal; spermatangial papillae with rounded apex; cystocarpic papillae broadly rounded at apex, frequently ornamented by 3 or 4 short, acuminate spines, usually only 1 cystocarp per papilla.

Common on all rocky headlands, saxicolous, subtidal (to 20 m), Vancouver I., Br. Columbia, to Pta. María, Baja Calif. Type locality: Strait of Juan de Fuca.

Gigartina harveyana (Kütz.) S. & G.

Mastocarpus harveyanus Kützing 1849: 734. *Gigartina harveyana* (Kütz.) Setchell & Gardner 1933: 276; Smith 1944: 282 (incl. synonymy); Dawson 1961a: 268. *G. boryi* S. & G. 1933: 269.

Thalli rusty red, 25–30(90) cm tall, in clumps, lax, with or without linear to lanceolate proliferations crowded in upper portions of blades; proliferations 2–8 cm long, 1–3 cm wide, with attenuated apices; blades linear-lanceolate to lanceolate, with attenuated apices, occasionally simple, or divided several times at base; short stipe sometimes present, usually dissected; medulla filamentous, with irregular cells; papillae numerous, nar-

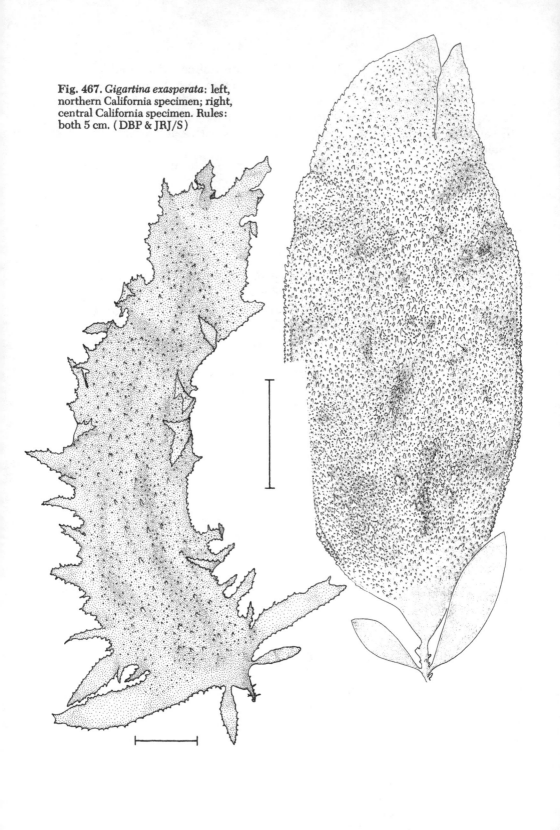

Fig. 467. *Gigartina exasperata*: left, northern California specimen; right, central California specimen. Rules: both 5 cm. (DBP & JRJ/S)

row, soft, frequently wide and flattened, on blade surfaces and margins; tetrasporangial papillae elongate, 3–9 mm long, the sori crowded, producing rugose surfaces; spermatangial thalli mostly smooth, with rudimentary papillae on blade surfaces; cystocarpic papillae usually marginal on proliferations and edges of blades, the papilla round and swollen at apex by single cystocarp.

Locally abundant in areas of fine sand covering low rocks, low intertidal (but higher than G. *corymbifera*), occasionally subtidal (to 5 m), Wash. to Pta. María, Baja Calif. Type locality: Monterey, Calif.

At one time, it seemed practical to recognize both G. *harveyana* and G. *boryi*; but in extensive collections the characters that had been used to separate them are found on the same plants as well as in different populatio..s.

Gigartina leptorhynchos J. Ag.

J. Agardh 1885: 28; Setchell & Gardner 1933: 267; Smith 1944: 279. *Gigartina multidichotoma* Dawson 1945b: 76; 1961a: 271. *G. leptorhynchos* f. *caespitosa* Daws. 1949b: 8.

Thalli dark brown to blackish, 10–15(40) cm tall, slender, woolly or soft, irregularly branched; main axes 0.5–1 cm diam., cylindrical to compressed; medulla delicate, anastomosing; terminal branchlets usually compressed, the last orders of branchlets congested; spines of irregular length and width, cylindrical, on margins and flattened faces of branches; tetrasporangial and spermatangial papillae acutely pointed; cystocarpic papillae tending to be flattened and foliose, occasionally dentate, each with 1–3 cystocarps, these occasionally surrounded by spines.

Common locally, saxicolous, usually below G. *papillata* and mixed with *Cryptosiphonia woodii*, lower midtidal, Humboldt Co., Calif., to I. Cedros, Baja Calif. Type locality: Santa Barbara, Calif.

N. Calif. specimens tend to be fairly uniform in size and branching pattern, and are distinguished by more or less flattened thalli; those from S. Calif. are larger and more cylindrical, with fewer laterals and with dense, long papillae and branchlets.

Gigartina papillata (C. Ag.) J. Ag.

Sphaerococcus papillatus C. Agardh 1821: pl. 19. *Gigartina papillata* (C. Ag.) J. Agardh 1846: pl. 19; Setchell & Gardner 1933: 287; Smith 1944: 283 (incl. synonymy); Abbott 1972b: 261. *Chondrus mamillosus* var. *sitchensis* Ruprecht 1851: 318. *G. sitchensis* (Rupr.) Kjellman 1889a: 31; Yendo 1916: 57. *G. dichotoma* Gardner 1927a: 333. *G. mamillosa sensu* Doty 1947b: 180; Scagel 1957: 187. *G. papillata* f. *cristata* Setchell, P.B.-A. 1895–1919 [1898]: no. 426. *G. cristata* (Setch.) S. & G. 1933: 289; Smith 1944: 283.

Thalli rarely more than 15 cm tall, with several erect, complanate to foliose branches, these linear, lanceolate, palmately divided, with dichotomous apices, or fimbriate, or apices not divided; thalli of at least 4 major

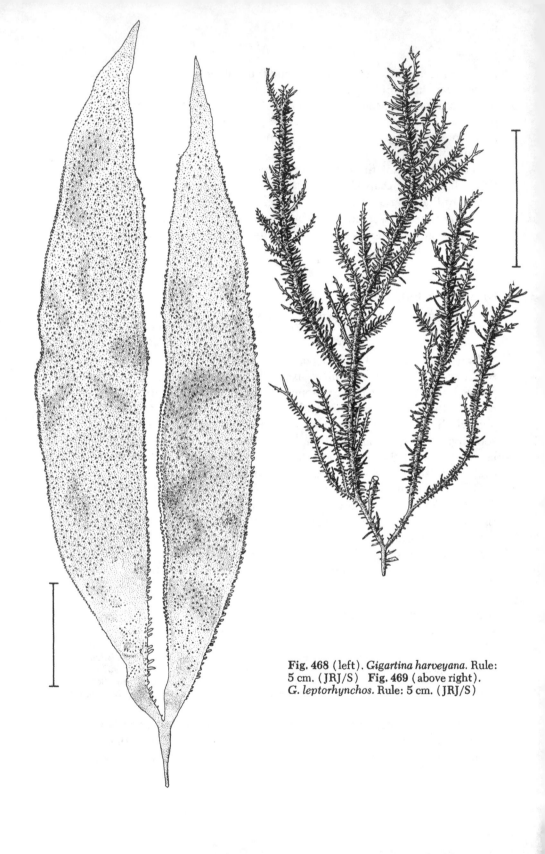

Fig. 468 (left). *Gigartina harveyana*. Rule:
5 cm. (JRJ/S)   **Fig. 469** (above right).
G. leptorhynchos. Rule: 5 cm. (JRJ/S)

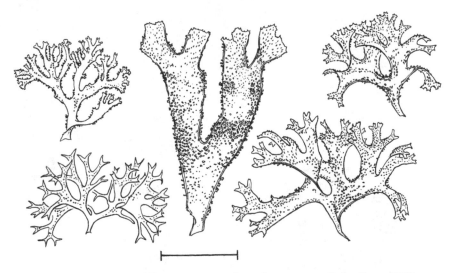

Fig. 470. *Gigartina papillata*: variation in branching pattern. Rule: 5 cm. (FT)

growth forms differing in width and shape of portions of thallus, and by distribution and size of papillae; tetrasporangia occurring in crustose stage known as the independent genus "*Petrocelis*"; spermatangial thalli lacking papillae, light rose to yellowish; cystocarpic thalli dark brown.

Common, saxicolous, high to middle intertidal, Alaska to Pta. Baja, Baja Calif.; common in N. and C. Calif. as the top 0.5–1.0 m band of brownish-red algae throughout the middle and high intertidal; less dominant south of Santa Barbara. Also known from Japan. Type locality: probably San Francisco, Calif.

The large number of synonyms is a reflection of the great variability shown by plants of this species, probably the most common red alga on the Pacific Coast.

Gigartina spinosa (Kütz.) Harv.

Mastocarpus spinosus Kützing 1847: 24. *Gigartina spinosa* (Kütz.) Harvey 1853: 177; Smith 1944: 168. *G. eatoniana* J. Agardh 1899: 15; Dawson 1961a: 254. *G. echinata* Gardner 1927a: 335. *G. armata* var. *echinata* (Gardn.) Daws. 1961a: 264. *G. armata* J. Ag. 1899: 15; Daws. 1961a: 262. *G. asperifolia* J. Ag. 1899: 15; Daws. 1961a: 264.

Thalli heavy, coarse, purplish to black or brownish to red, growing in thick isolated clumps, 20–30(40) cm tall, (1)4–6 cm broad, arising from short stipe, frequently divided dichotomously or unilaterally near top of stipe, irregularly divided upward, occasionally pinnately divided, sometimes remaining a simple undivided blade, or apices unequally dichotomously divided; superficial papillae often upwardly decurrent; tetrasporangial and spermatangial papillae simple; cystocarpic papillae divided

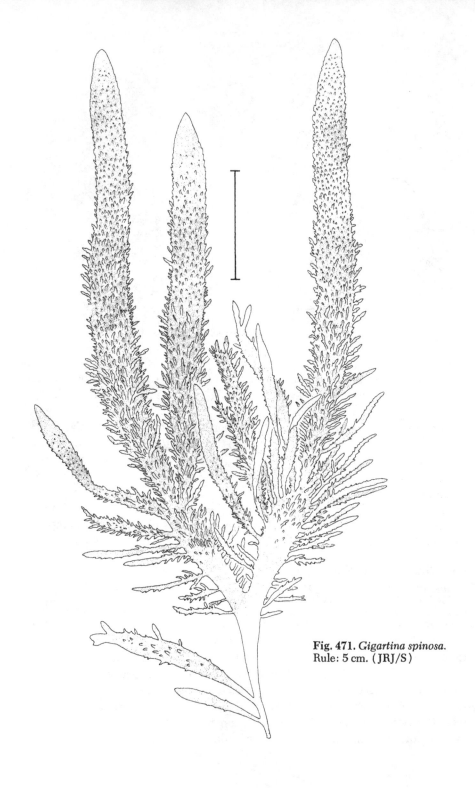

Fig. 471. *Gigartina spinosa.*
Rule: 5 cm. (JRJ/S)

many times, to 1 cm long, borne along margins or on flattened faces of blades; 3–5 cystocarps borne irregularly to alternately along papillae.

Frequent, saxicolous, low intertidal areas exposed to surge and surf, Monterey, Calif. (type locality), to I. Magdalena, Baja Calif.

Gigartina tepida Hollenb.

Hollenberg 1945a: 449; Dawson 1961a: 275.

Thalli 8–10 cm tall, purplish, loosely tufted; main branches intricate, 0.5–1 mm diam., terete to mostly compressed; laterals frequent basally, fewer upwards, at times opposite but mostly irregular to pinnately branched; branchlets on upper portions of thallus spinous, acute, scarcely papillate; apices of branches occasionally establishing secondary, adventitious, disklike attachments; reproductive structures inconspicuous in papillae along branch margins.

Infrequent, on shells and rocks in sheltered water, low intertidal, Wash., Morro Bay (San Luis Obispo Co.), mouth of San Gabriel R. (Los Angeles Co.), and Newport Bay (Orange Co.; type locality), Calif., and Baja Calif. through Gulf of Calif. to Guaymas, Mexico.

Gigartina volans (C. Ag.) J. Ag.

Sphaerococcus volans C. Agardh 1821: pl. 18. *Gigartina volans* (C. Ag.) J. Agardh 1846: pl. 18; Setchell & Gardner 1933: 279; Smith 1944: 282. *G. velifera* J. Ag. 1899: 41; S. & G. 1933: 280. *Rhodoglossum linguiforme* Dawson 1958: 73; 1961a: 259.

Thalli of two strongly dimorphic shapes: tetrasporangial thalli without papillae, the surfaces of thalli plane, thick (rather like *Iridaea cordata* var. *splendens* except for presence of lateral branchlets or their remnants near lower margins), branched more or less dichotomously 2 or 3 times from simple or branching compressed stipes, the blades lanceolate, tapering at apices; other thalli with simple or lobed blades, the apices broadly rounded, the margins and surfaces crowded with ovate, obovate, sometimes stipitate, randomly occurring, foliar proliferations, with or without papillae; all thalli in loose clumps, with several erect blades from common holdfasts, 10–20(40) cm tall, divided dichotomously once or twice or not divided; dark purple with brown, the stipe branched, compressed, 4–20 cm long, the apophyses sometimes clearly marked; fertile blades of all 3 kinds of thalli differing, the tetrasporangial blades without proliferations, or papillae, the sori embedded in surface layers, the spermatangial blades with obovate-ovate superficial leaflets, these bearing spermatangial sori, the cystocarpic blades with predominantly marginal papillae, 0.5–1 cm long, bearing 3–6 cystocarps terminally and occasionally laterally.

Locally abundant in sand-scoured areas, low intertidal and subtidal (to 10 m), N. Ore. to Pta. María, Baja Calif. Type locality: probably W. coast of N. America, possibly Monterey, Calif.

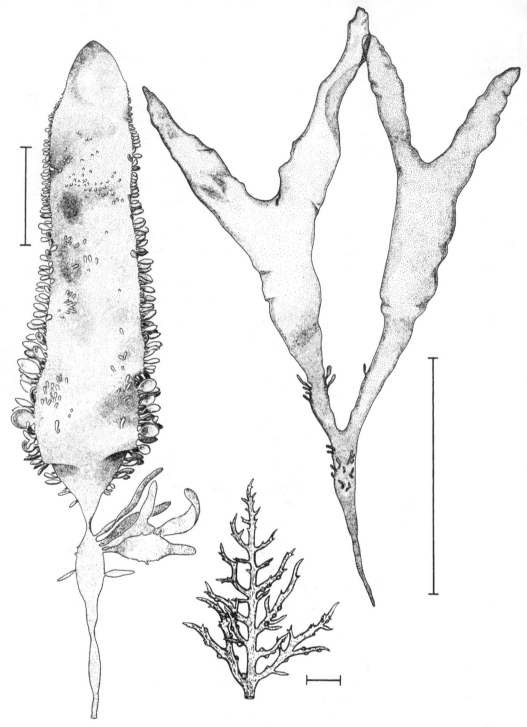

Fig. 472 (below center). *Gigartina tepida.* Rule: 5 mm. (GJH)   **Fig. 473** (two above).
G. volans: below, proliferous intertidal specimen; above, tetrasporangial subtidal
specimen. Rules: both 5 cm. (JRJ/S & SM)

Iridaea Bory 1826

Thalli in clumps, usually in dense horizontal bands 1–2 m deep; of long, lanceolate to cordate blades, these on short, cylindrical or compressed stipes arising from perennial discoid holdfasts. Stipes frequently giving rise to short, smooth, flaring apophyses bearing simple, cleft, dissected, or irregular blades, these mostly longer than wide, but size and shape unreliable because of shrinkage on drying. Thalli in certain species markedly iridescent when submerged. All thalli multiaxial. Medulla of interwoven meshwork of colorless, slender filaments, with much intercellular space. Cortex of short, branched rows of anticlinal filaments. Tetrasporangia cruciately divided, in globular to flattened internal sori, formed from specialized short laterals of medullary filaments. Spermatangia in extensive sori on blade surfaces. Plants procarpial; carpogonial branch 3-celled, borne on supporting cell also functioning as auxiliary cell. Gonimoblast filaments produced toward interior of blade; nearly all cells becoming carposporangia. Cystocarp with absorbing filaments radiating from basal portions of gonimoblast to periphery of gonimoblast, and with ring of medullary filaments encircling developing spore mass.

1. Blades narrower than 2 cm, less than 15 cm tall . . . . *I. cornucopiae*
1. Blades wider than 2 cm, 15+ cm tall, mostly longer than wide 2
   2. Blades purple, brownish, or greenish, with lavender or purple bases, iridescent 3
   2. Blades deep rose red, usually not iridescent, the bases to twice as broad as apex of blade 7
3. Margins entire, sometimes undulate; blades yellowish-green; midtidal to upper intertidal levels . . . . *I. flaccida*
3. Margins mostly scalloped, fringed, or irregular 4
   4. Blades never simple, variously lobed and irregular; cystocarpic sori conspicuous . . . . *I. heterocarpa*
   4. Blades usually simple except for occasionally divided apex; cysto- carpic sori inconspicuous 5
5. Bases of blades wider than apex; apophysis prominent . . . .*I. cordata*
5. Bases of blades linear to lanceolate; apophysis small or lacking 6
   6. Blades broadly lanceolate, variously divided at apex or undivided, few per clump; stipes usually flat, broad . . . . *I. cordata*
   6. Blades linear-lanceolate, spirally twisted, the apex undivided or divided once, many per clump; stipes long, terete . . . . *I. lineare*
7. Blades thick, mostly wider than long; outer medullary filaments thinner than others . . . . *I. punicea*
7. Blades longer than wide; medullary filaments more uniform in diam. . . . . *I. sanguinea*

Iridaea cordata (Turn.) Bory

Fucus cordatus Turner 1809: 118. *Iridaea cordata* (Turn.) Bory 1826a: 16.

Thalli in loose groups forming bands 1–2 m deep, with conspicuous blades 20–40+ cm tall, 12–24+ cm wide, borne on short or long stipes, with or without apophyses; blades cordate to broadly lanceolate, purple to blackish, some with brown overtones, often with bluish sheen; margins of blades entire, occasionally proliferous, often cleft or lobed, plane or broadly ruffled; blades mostly smooth, sometimes rubbery, bumpy when fertile; tetrasporangial sori numerous, small, closely grouped, extending to margins; spermatangial blades pale in color with spermatangia continuous; cystocarpic blades usually thick, mostly dark purple and drying to black.

A common, extremely variable species, better judged on a population basis than on a few specimens. The varieties recognized below may be ecological variants only; continuing field studies are necessary.

Iridaea cordata var. cordata

Abbott 1971b: 54 (incl. synonymy); 1972a: 253.

Blades 20–40(to 2 m) tall, 12–24(40) cm wide, liver-brown, purple, or violet, with blue iridescent sheen lost on drying; blades typically cordate but frequently noncordate to broadly lanceolate as well, sometimes lobed, cleft or split; margins frequently with pinnules; holdfasts fleshy, perennial; stipe typically present but usually short, less than 3 cm long.

Common to abundant northward, saxicolous, mostly lower intertidal to subtidal (5 m), Pribilof Is., Alaska, to Ventura, Calif.; occasional south of Monterey Peninsula. Also Japan and Kurile Is. Type locality: Banks I., Br. Columbia.

Iridaea cordata var. splendens (S. & G.) Abb.

Iridophycus splendens Setchell & Gardner 1937b: 170. *Iridaea cordata* var. *splendens* (S. & G.) Abbott 1971b: 55 (incl. synonymy). *Rhodoglossum coriaceum* Dawson 1945b: 75.

Thalli with blades 40–120+ cm tall, dark purple to blackish; cystocarpic blades drying to brown, others drying to dark purple or black; stipes 4–6 cm long, with short, cuneate apophyses, these sometimes lacking; blades thick, lanceolate to broadly ovate, entire, with tapering apex or terminally cleft or divided; stipes typically longer than in var. *cordata*, the blades broad-lanceolate and thick.

Abundant on exposed coasts, saxicolous, low intertidal to subtidal (7 m), Queen Charlotte Is., Br. Columbia, to N. Baja Calif. Type locality: Carmel, Calif.

Iridaea cornucopiae Post. & Rupr.

Postels & Ruprecht 1840: 18; Mikami 1965: 259; Abbott 1971b: 62. *Iridophycus parksii* Setchell & Gardner 1937b: 172; Abb. 1971b: 62.

Thalli with blades clustered on discoid holdfasts, the stipes short, flaring into furrowed apophyses; blades thick, obovate, round to spatulate, divided once or twice into narrow lobes, or not divided, 2–4(15) cm tall, 1–2 cm

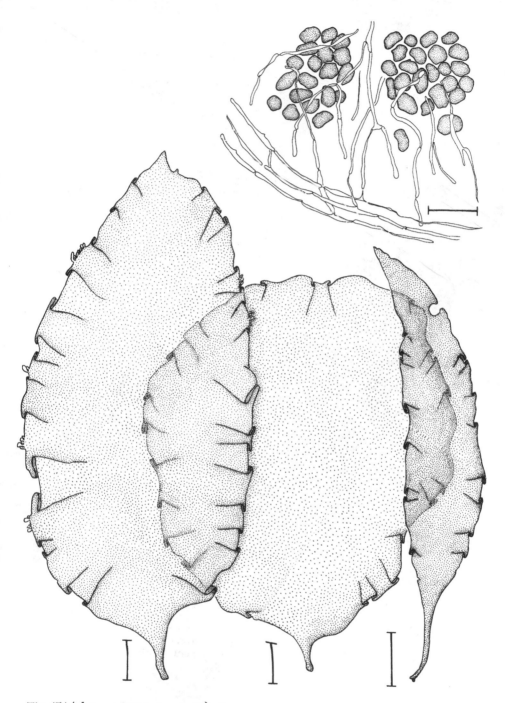

Fig. 474 (above right & below center). *Iridaea cordata* var. *cordata*. Rules: cross section, 50 μm; habit, 3 cm. (DBP & DBP) Fig. 475 (below left & below right). *I. cordata* var. *splendens*: variation in branching pattern. Rules: both 3 cm. (DBP)

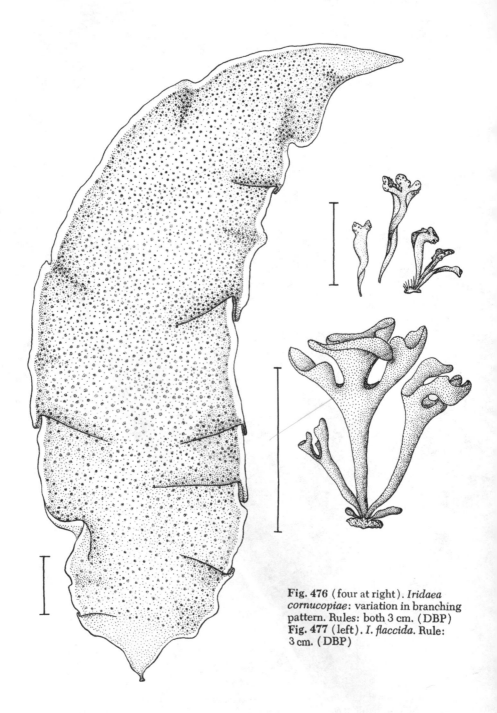

Fig. 476 (four at right). *Iridaea cornucopiae*: variation in branching pattern. Rules: both 3 cm. (DBP) **Fig. 477** (left). *I. flaccida*. Rule: 3 cm. (DBP)

wide, dark purple with brown; tetrasporangia in flat sori; cystocarps large, protuberant.

Infrequent, high intertidal to midtidal atop rocks, Alaska to Palmer's Pt. (Mendocino Co.), Calif. Also Hokkaido (N. Japan) and Kurile Is. Type locality: N. Pacific.

Very similar in habit to *Chondrus ocellatus* f. *parvus*; the gonimoblasts differ, however, those of *Chondrus* lacking the encircling ring of sterile tissue and the absorbing filaments. This form is reported from N. Japan and the Aleutians.

Iridaea flaccida (S. & G.) Silva

Iridophycus flaccidum Setchell & Gardner 1937b: 171; Smith 1944: 288. *Iridaea flaccida* (S. & G.) Silva 1957a: 328; Abbott 1971b: 64; 1972a: 253.

Blades narrow in proportion to length, 20–30(140) cm long, 8–20 cm wide, broadly lanceolate or cordate-obovate, the apices tapering, or apices forked or cleft, yellowish-green, with iridescent purples and browns in basal portion and in portions shaded from light; stipes less than 2 cm long, to 0.75 cm wide; apophyses cuneate to flaring; tetrasporangial blades with continuous or interrupted nonsporangial margin; spermatangial thalli pale yellow; cystocarps inconspicuous, externally resembling tetrasporangial sori.

Locally abundant, saxicolous, midtidal to low intertidal, Alaska to N. Baja Calif., including Channel Is.; in C. Calif., the most conspicuous and common of bladelike red algae in midtidal levels. Type locality: Carmel, Calif.

Iridaea heterocarpa Post. & Rupr.

Postels & Ruprecht 1840: 18; Abbott 1971b: 65 (incl. synonymy). *Iridophycus heterocarpum* (Post. & Rupr.) Setchell & Gardner 1937b: 170; Smith 1944: 201.

Thalli limp in habit; blades 10–15(40) cm tall, 1–3 cm wide below, wider above, usually irregularly divided, the margins frequently proliferous, reddish-brown, drying to brown and light tan; stipes less than 3 mm long or blades nonstipitate; apophyses quickly flaring into blades; tetrasporangial sori very small and closely set; cystocarps large, bulging prominently, to 4 mm diam.; spermatangial thalli resembling those of broad *Gigartina papillata*, pale in color, entire.

Common, in isolated clumps, saxicolous, midtidal, Alaska to Ventura, Calif. Also reported from Sakhalin, U.S.S.R. Type locality: N. Pacific.

Iridaea lineare (S. & G.) Kyl.

Iridophycus lineare Setchell & Gardner 1937b: 171; Smith 1944: 290. *Iridaea lineare* (S. & G.) Kylin 1941: 23; Abbott 1970: 2; 1971b: 68.

Thalli in dense tufts, to 0.5 m diam., bases of blades laterally cohering

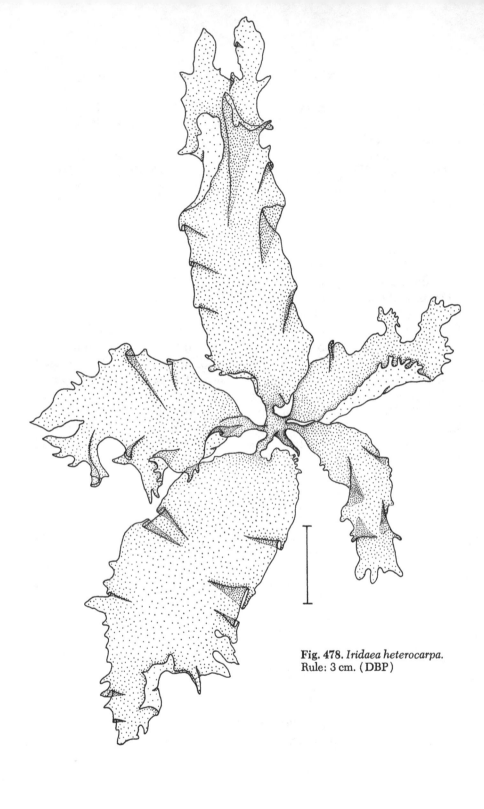

Fig. 478. *Iridaea heterocarpa.*
Rule: 3 cm. (DBP)

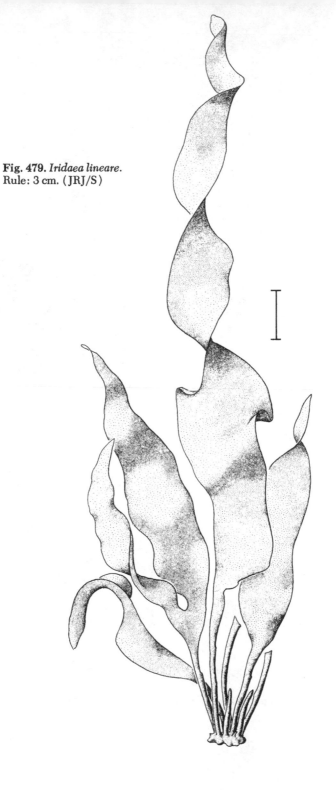

Fig. 479. *Iridaea lineare.*
Rule: 3 cm. (JRJ/S)

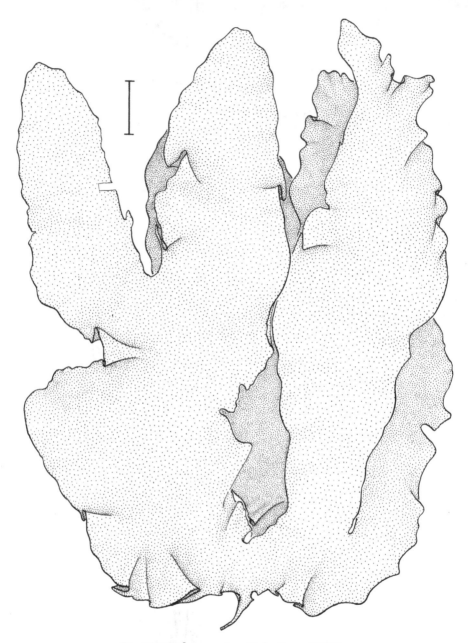

Fig. 480. *Iridaea punicea*. Rule: 3 cm. (DBP)

and sometimes furrowed; blades heavy, linear-lanceolate, plane or mostly spiraled, with attenuate apices, these sometimes divided, 0.25–1 m long, 3.5–7 cm broad, entire or with divisions of 2 or 3 lobes, the margins mostly crisped, purplish-brown to dark purple, drying with brown undertone; stipe distinct, 2–5 cm long, subcylindrical, occasionally with fine lateral papillae; tetrasporangial sori small, very closely set; spermatangial blades smooth, paler than other blades; cystocarps inconspicuous.

Locally abundant on exposed coasts, saxicolous, low intertidal, Alaska to Port Hueneme (Ventura Co.), Calif., including Channel Is. Type locality: Carmel, Calif.

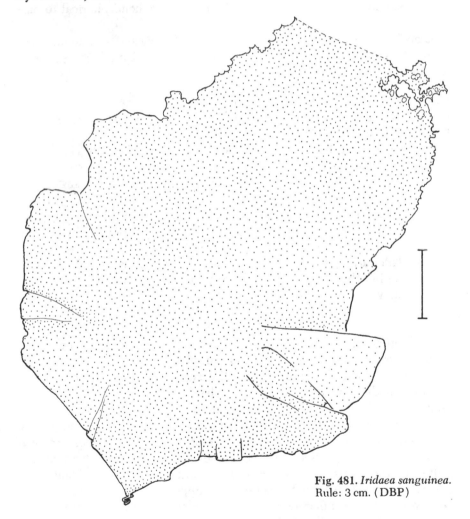

Fig. 481. *Iridaea sanguinea.*
Rule: 3 cm. (DBP)

Iridaea punicea Post. & Rupr.

Postels & Ruprecht 1840: 18; Abbott 1971b: 68. *Iridophycus whidbeyana* Setchell & Gardner 1937b: 172.

Thalli large, coarse, orbicular to somewhat broad-cordate, dark red with brownish tones, 20–40 cm tall, 20–30 cm broad, with short stipes; apophyses inconspicuous; apices sometimes cleft or lobed; medullary filaments distinctive for species, of 2 sizes, those immediately subcortical of 5–7 μm diam., the filaments internal to these thicker, to 3 times as thick; tetrasporangial sori small, uniformly distributed; spermatangial thalli unknown; cystocarps in irregular clusters, first forming along upper lateral margins, then spreading inward of blade, remaining discrete, hemispherical to umbonate.

Saxicolous, subtidal: common in Puget Sd., Wash.; occasional from Alaska to Shell Beach (San Luis Obispo Co.), Calif. Type locality: Sitka, Alaska.

Iridaea sanguinea (S.& G.) Hollenb. & Abb.

Iridophycus sanguineum Setchell & Gardner 1937b: 172. *Iridaea sanguinea* (S. & G.) Hollenberg & Abbott 1965: 1184; Abbott 1971b: 69.

Blades deep red, thick, to 40 cm tall, 20 cm wide, broadly lanceolate, entire, with triangular to rectangular bases, apophyses and stipes very short; medullary filaments 8–14 μm diam.; other characters suggesting *I. punicea*.

Infrequent, saxicolous, subtidal (to 30 m), Alaska to Carmel Submarine Canyon and San Luis Obispo Co., Calif. Type locality: Duxbury Reef (Marin Co.), Calif.

The blades are up to twice as thick as those of *I. punicea* and are more lanceolate; but these observations have been made on too few specimens to serve as valid generalizations.

Rhodoglossum J. Agardh 1876

Thalli in groups, 1 or more erect blades arising from discoid holdfasts. Blades sessile or stipitate, multiaxial, simple or dichotomously divided many times; margins usually entire, the surfaces usually smooth. Inner medulla of densely interwoven, very slender, colorless filaments; outer medulla of short, broad cells arranged in scalariform rows. Cortex of laterally adjoined, erect, branching filaments with very small, dense cells. Tetrasporangia cruciately divided, developing from innermost cortical cells, borne in compressed, globose masses at juncture of cortex and medulla. Spermatangia in colorless, nearly continuous patches. Plants procarpial; carpogonial branch 3-celled, on supporting cell serving as auxiliary cell. Cystocarps deeply embedded in blade, surrounded by nutritive layer of medullary cells; sterile cells separating sporangial masses.

1. Blades dichotomously divided . . . . *R. affine*
1. Blades simple 2
    2. Blades with proliferations on margins and surfaces . . . . *R. oweniae*
    2. Blades usually without proliferations 3
3. Blades lanceolate; cystocarps less than 1 mm diam. . . . .*R. californicum*
3. Blades elliptical to subcordate; most cystocarps more than 1 mm diam. . .
  . . . . *R. roseum*

Rhodoglossum affine (Harv.) Kyl.

Chondrus affinis Harvey 1841a: 408. *Rhodoglossum affine* (Harv.) Kylin 1928: 49.

Thalli in small, bushy, decumbent or erect tufts, 4–15 cm tall, often in bands ±0.5 m wide near tops of rocks, greenish-olive or reddish-purple to blackish; blades usually smooth, with 1 side concave, other side convex; two vegetative forms common, one with thalli 4–5 cm tall, having blades less than 2 mm wide, these repeatedly branched 6–8 times, the penultimate and ultimate dichotomies close together, the ultimate branches appearing fimbriate and congested (such plants most common in high midtidal), the other with thalli 12–15 cm tall, blades to 3 cm wide and usually more sparsely branched, occasionally with superficial and a few lateral proliferations (these plants common in low intertidal); reproductive structures as for genus, the cystocarps conspicuous, 1–2.5 mm diam.

Locally abundant, saxicolous, midtidal, occasionally lower, often associated with *Gigartina papillata*, Br. Columbia to I. Cedros, Baja Calif.; not as common south of San Luis Obispo Co. as northward. Type locality: Monterey, Calif.

Rhodoglossum californicum (J. Ag.) Abb.

Collinsia californica J. Agardh 1899: 70. *Rhodoglossum californicum* (J. Ag.) Abbott 1971b: 70. *R. americanum* Kylin 1941: 24; Smith 1944: 285.

Thalli mostly 25–40(300) cm tall, 2–10(20) cm wide, 1 to several thin blades arising from small, discoid holdfasts, deep pink to rust red; blades with short, cylindrical stipes 1–2 cm long, occasionally dichotomously branched once or twice, each branch terminating in 1 blade; blades linear-lanceolate to broadly lanceolate, with tapering apices, the margins usually entire, occasionally cleft and slightly ruffled, rarely with proliferous outgrowths near base; reproductive structures as for genus, the tetrasporangial sori unusually small and inconspicuous.

Locally common on rocks in sandy areas protected from surf, low intertidal to subtidal, Br. Columbia to N. Baja Calif. Type locality: Santa Barbara, Calif.

Rhodoglossum oweniae Hollenb. & Abb.

Hollenberg & Abbott 1968: 1249.

Thalli with 1 to several erect deep-rose to red blades, 50–70 cm tall,

Fig. 482 (below left). *Rhodoglossum affine.*
Rule: 3 cm. (JRJ/S) **Fig. 483** (right).
R. californicum. Rule: 3 cm. (JRJ/S)

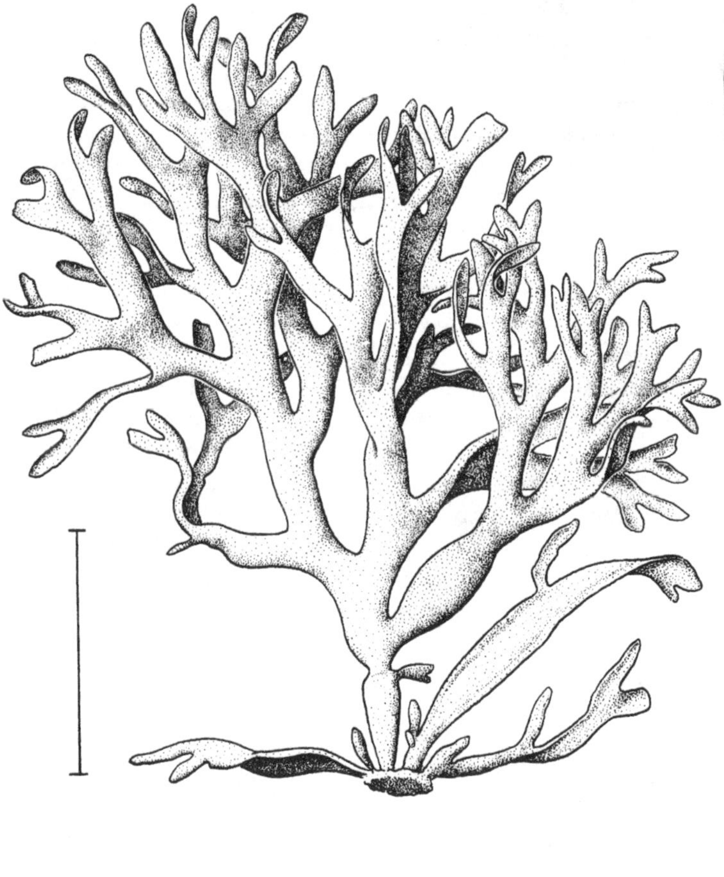

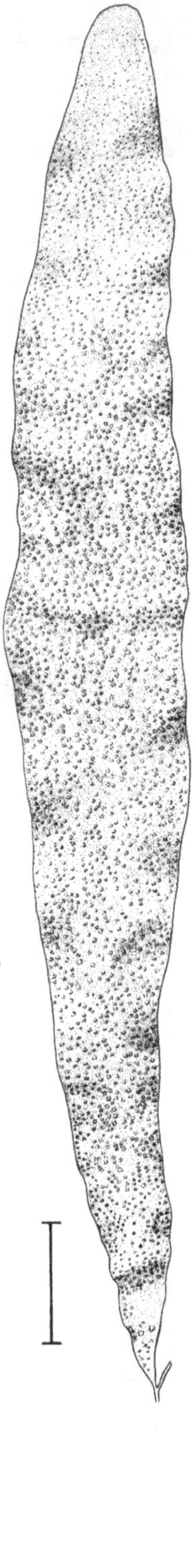

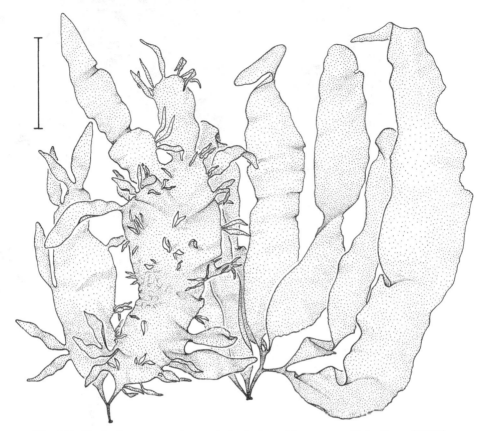

Fig. 484. *Rhodoglossum oweniae*: variation in branching pattern. Rule: 3 cm. (DBP)

4–12 cm wide, broadly lanceolate, ruffled, arising from branched **stipes**; older thalli characterized by marked proliferations, 2–20 cm long, 1–10 cm wide, in some cases resembling branches of second order; proliferations on margins and surfaces of blades, frequently stipitate; reproductive structures as for genus, the tetrasporangial sori fewer and larger than those of *R. californicum*.

Locally abundant, saxicolous, low intertidal, Humboldt Co., Calif. Type locality: Buhne Pt. (Humboldt Co.), Calif.

Rhodoglossum roseum (Kyl.) Smith

Iridaea rosea Kylin 1941: 24. *Rhodoglossum roseum* (Kyl.) Smith 1943: 216; 1944: 268; Abbott 1971b: 70. *R. parvum* Smith & Hollenberg 1943: 216.

Thalli with 1 or more erect blades arising from small discoid holdfasts, 8–15(20) cm tall, oval or obovate to subcordate; blades with rounded apices, sometimes with ruffled margins, occasionally with stipes 2–3 mm long; cystocarps conspicuous, usually more than 1 mm diam.

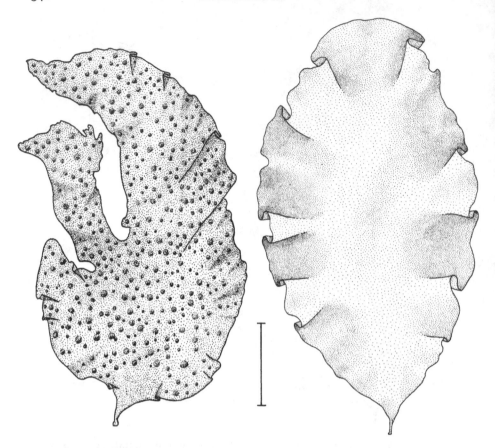

Fig. 485. *Rhodoglossum roseum*: left, cystocarpic thallus; right, tetrasporangial thallus. Rule: 3 cm. (CS & JRJ/S)

Frequent on exposed coasts, saxicolous, low intertidal to shallow subtidal, Alaska to N. Baja Calif. Type locality: Pacific Grove, Calif.

Because of their roughness, the cystocarpic thalli of this species almost never adhere to mounting paper after drying.

Small, depauperate plants with relatively long stipes and congested, sometimes branching upper blades, were previously known as *R. parvum*.

Order RHODYMENIALES

Thalli multiaxial, the axes formed from terminally grouped apical cells, the plants erect to decumbent or repent, cylindrical to foliose and blade-like, branched or unbranched. Branches solid or hollow. Cortex mostly of 2 to several cell layers, with cells and filaments of varied arrangement, frequently with diaphragms and internal gland cells. Medulla pseudoparenchymatous, or filamentous, or hollow. Tetrasporangia embedded just beneath surface, scattered or in sori, cruciately or tetrahedrally divided. Spermatangia scattered or in sori; auxiliary-cell filament of 2 or 3 cells, borne on supporting cell of carpogonial filament; auxiliary cell formed before fertilization. Gonimoblast filaments growing toward surface. Mature cystocarp projecting; pericarp with or without ostiole.

Two families in the California flora, Rhodymeniaceae (below) and Champiaceae (p. 562).

Family RHODYMENIACEAE

Thalli cylindrical to foliose, mostly branched. Medulla continuous, composed of angular cells or becoming hollow through gelatinization of initial medullary layers. Cortex with anticlinal rows of 3 or 4 cells. Gland cells secretory in function, isolated to clustered. Tetrasporangia cruciately divided. Carpogonial filament 3-celled. Auxiliary cell usually not fusing with other cells after fertilization. Most cells of gonimoblast becoming carposporangia.

```
1. Plants "parasitic" ............................................. 2
1. Plants free-living ............................................. 3
   2. Plants "parasitic" on Fauchea ................Faucheocolax (p. 546)
   2. Plants "parasitic" on Rhodymenia ...........Rhodymeniocolax (p. 561)
3. Thallus wholly or partly hollow or with hollow, bulbous branch apices .... 4
3. Thallus solid, without bulbous apices ............................. 6
   4. Thallus saccate, mainly unbranched .............Halosaccion (p. 550)
   4. Thallus primarily not saccate ................................... 5
5. Axes and branches solid, cylindrical, with hollow, bulbous branch apices
   .......................................... Botryocladia (p. 550)
5. Thalli hollow, bladelike, dichotomous, with curved internal diaphragms
   ............................................. Fryeella (p. 548)
   6. Blades peltate ............................................. 7
   6. Blades usually dichotomously or palmately branched .............. 8
7. Blades circular, entire ...........................Maripelta (p. 548)
7. Blades stellate, developing secondary attachments from apices........
   ............................................ Sciadophycus (p. 547)
   8. Cortical cells in distinct, anticlinal filaments; cystocarps often coronate
      .......................................... Fauchea (below)
   8. Cortical cells showing no filamentous arrangement; cystocarps often
      rostrate, but not coronate ..................................... 9
9. Cortex essentially monostromatic; cystocarps chiefly marginal ........
   ......................................... Leptofauchea (p. 547)
9. Cortex of 2 or 3 cell layers; cystocarps scattered or restricted to blade
   apices, but not chiefly marginal ..................Rhodymenia (p. 553)
```

Fauchea Montagne 1846

Thalli erect or repent, arising from small, discoid bases. Blades usually sessile, irregularly flabellately divided; upper divisions short, frequently fringed. Medulla of large cells in 2–4 layers. Cortex of small cells, usually in distinct anticlinal filaments. Tetrasporangia cruciately divided, in small, irregular, nemathecioid sori, appearing as streaks or patches. Cystocarps marginal or scattered, protuberant, smooth or coronate, with network of filaments between gonimoblast and pericarp. Spermatangia in single, continuous superficial sorus.

1. Cystocarps scattered on surface and margins of blades . . . .*F. laciniata*
1. Cystocarps on margins of blades only 2
   2. Thalli to 34 cm tall, straw-colored to light rose, with branches to 20
     mm broad . . . .*F. fryeana*
   2. Thalli less than 8 cm tall, brownish-red to deep red, the widest
     branches less than 10 mm broad . . . . *F. galapagensis*

Fauchea fryeana Setch.

Setchell 1912: 239; Hollenberg & Abbott 1966: 93.

Thalli to 34 cm tall, straw-colored to light rose; branches to 2 cm broad; tetrasporangia in irregularly elongate, flabellately radiating nemathecia; cystocarps marginal only, closely set, giving margin a fimbriate appearance, mostly not coronate or papillate.

Occasional on subtidal rocks, S. Br. Columbia to Monterey Peninsula, Calif. Type locality: Friday Harbor, Wash.

Fauchea galapagensis Tayl.

Taylor 1945: 246; Hollenberg & Abbott 1966: 93.

Thalli forming small, rounded tufts, 5–8 cm tall, brownish-red to deep red; blades mostly regularly dichotomous, branching 3 or 4 times; penultimate branches to 10 mm broad, the ultimate branches to 5 mm broad; tetrasporangia in mottled sori, covering surface of blades except on ultimate branches; cystocarps mostly coronate, marginal, at times dense and forming marginal fringe.

Occasional on rocks, subtidal (8–20 m), Monterey Bay and Santa Catalina I., Calif. Type locality: Galápagos Is.

Fauchea laciniata J. Ag.

J. Agardh 1884–85 (1885): 40; Kylin 1931: 9. *Fauchea laciniata* f. *pygmaea* Setchell & Gardner 1912b: 238; Dawson 1963a: 440. *F. pygmaea* (S. & G.) Kyl. 1941: 27. *F. media* Kyl. 1941: 27.

Thalli mostly 3–12 cm tall, forming loose clusters of blades, deep red with bluish cast; blades dichotomo-flabellately branched, membranous, lubricous; branches 6–15 mm wide below, reduced to 2–6 mm wide above; medulla of 3 or 4 cell layers; tetrasporangial nemathecia elongate, irregularly anastomosing; cystocarps mostly superficial and marginal, usually markedly coronate.

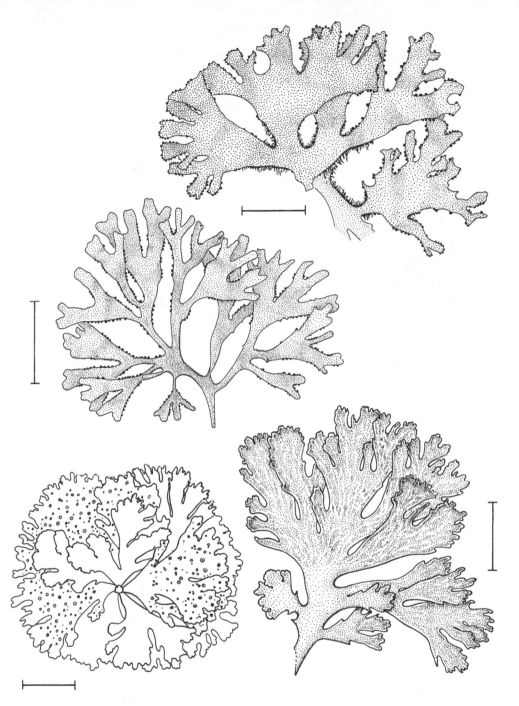

Fig. 486 (above right). *Fauchea fryeana.* Rule: 2 cm. (DBP) **Fig. 487** (middle left). *F. galapagensis.* Rule: 2 cm. (DBP) **Fig. 488** (two at bottom). *F. laciniata*: left, cysto-carpic thallus; right, tetrasporangial thallus. Rules: left, 2 cm; right, 10 cm. (both DBP)

Occasional to frequent on rocks, intertidal to subtidal, Masterman I., Br. Columbia, to N. Baja Calif.; in Calif., from Monterey Peninsula, Santa Barbara (type locality), Channel Is., and Laguna Beach.

We consider *F. media* and *F. pygmaea* as merely small representatives of the species.

Faucheocolax Setchell 1923

Plants "parasitic" on *Fauchea*. Thalli with very short basal stalks bearing many short, wartlike branches. Tetrasporangial specimens more distinctly branched than cystocarpic specimens. Vegetative structure and reproduction as in *Fauchea*.

Faucheocolax attenuata Setch.

Setchell 1923a: 394; Sparling 1957: 345.

Thalli mostly 1–3 mm diam., whitish to pinkish; tetrasporophyte with fertile branches inflated below and attenuated at apices; cystocarpic plants with a few short lobes, the lobes soon becoming obscured by the developing pericarp.

Occasional on *Fauchea*, subtidal (to 10 m), Wash. to Carmel Bay, Calif. (type locality).

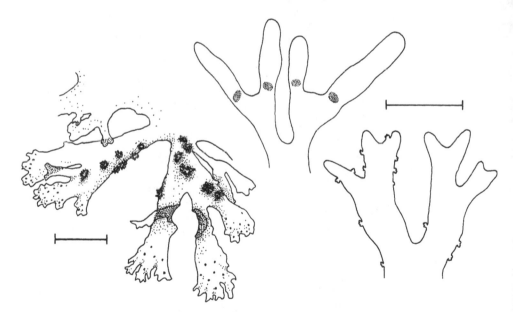

Fig. 489 (left). *Faucheocolax attenuata*. Rule: 2 cm. (LH) **Fig. 490** (center & right). *Leptofauchea pacifica*: center, spermatangial thallus; right, cystocarpic thallus. Rule: 1 cm. (both SM, after Dawson)

Leptofauchea Kylin 1931

Thalli erect, dichotomously branched, ligulate. Medulla of large, thin-walled cells. Cortex mostly single layer of small cells. Tetrasporangia cruciately divided, arising laterally on nemathecial paraphyses. Spermatangia unknown; cystocarps marginal, strongly projecting, not coronate, with network of slender filaments between gonimoblast and pericarp.

Leptofauchea pacifica Daws.

Dawson 1944e: 104; 1963a: 444; Dawson & Neushul 1966: 178.

Thalli usually 3–7 cm tall, arising from small, discoid bases, pale red; blade divisions to 4 mm broad, 90–130(250) μm thick, the ultimate divisions reduced and sometimes with secondary attachment disks; medulla of 2 or 3 irregular cell layers; cortical cells 7–14 μm diam.; tetrasporangia in horizontally elliptical nemathecia, mostly borne just above points of forking of upper divisions and nearly as broad as division; cystocarps marginal, sessile, hemispherical, 750–800 μm diam., rostrate.

Infrequent, on subtidal rocks (16–40 m), Puget Sd., Wash., and off Anacapa I. and Santa Monica Bay, Calif. Type locality: I. Cedros, Baja Calif.

Sciadophycus Dawson 1944e

Thalli peltate, with short central stipes. Blades at first circular and entire, then stellate. Branching sympodial from apices of points of blade, the branches forming secondary attachments. Medulla of blade monostromatic, cells large. Cortex of 1 or 2 layers of very small cells. Tetrasporangia cruciately divided, in small, blisterlike nemathecia scattered over blade. Spermatangia unknown; cystocarps superficial, hemispherical, smooth and ostiolate, with network of sterile filaments between gonimoblast and pericarp.

Sciadophycus stellatus Daws.

Dawson 1944e: 105; Neushul, Scott, Dahl & Olsen 1967: 195. *Fauchea rhizophylla* Taylor 1945: 247.

Thalli with primary stipes simple or once-forked, 5–16 mm tall; primary blade deep rose, 4–6 cm broad, with 10–14 radiating points; old plants becoming eroded, losing primary blade; secondary and tertiary blades prostrate, spreading, each subsequent one becoming rounded, 10–15 mm broad, developing 3–5 stellate, stoloniferous attachments.

Infrequent, on subtidal rocks (13–55 m), Channel Is. to San Diego, Calif., to I. Natividad, Baja Calif. and Galápagos Is. Type locality: I. Cedros, Baja Calif. (40–50 m). Usually found in kelp beds (10–30 m) in S. Calif.

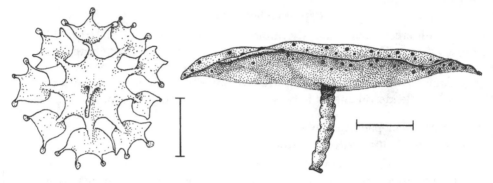

Fig. 491 (left). *Sciadophycus stellatus.* Rule: 1 cm. (LH, after Dawson) **Fig. 492** (right). *Maripelta rotata.* Rule: 1 cm. (DBP)

Maripelta Dawson 1963

Thalli symmetrically peltate, arising from simple or rarely branched, cylindrical stipes. Blades membranous, simple, rotate, expanding from apices of stipes, periodically deciduous from each sympodially produced increment. Medulla of large, vacuolate cells, merging abruptly into cortex with 2 or 3 layers of much smaller cells. Tetrasporangia in small, superficial nemathecia on upper blade surface, cruciately divided. Spermatangia superficial, in pale patches. Cystocarps scattered, superficial, subhemispherical, ostiolate, lacking filamentous network between gonimoblast and pericarp.

Maripelta rotata (Daws.) Daws.

Drouetia rotata Dawson 1949b: 9. *Maripelta rotata* (Daws.) Daws. 1963a: 446; Hollenberg & Abbott 1966: 97.

Thalli deep rose red, with bluish sheen in life; stipes 1–4.5 cm long, 2–4 mm diam., arising from discoid holdfasts, with up to 22 successive growth increments; blade single, 3–8(12) cm diam., 300–325 μm thick, slightly depressed at center, entire; blade with single medullary layer of very large cells, this nearly equal to thickness of blade; cortex with 2 or 3 layers of much smaller cells; cystocarps 1.2–1.4 mm diam., slightly rostrate.

Rare, on subtidal rocks (to 30 m), Carmel and Diablo Submarine Canyons and Santa Catalina I. (type locality), Calif., to Pta. Eugenio, Baja Calif.; frequent at 15–30 m under kelp canopy off La Jolla, Calif.

Fryeella Kylin 1931

Thalli erect to decumbent, complanate, regularly or irregularly dichotomously branched, hollow, the thin median cavity crossed by arching septa visible from exterior; cortex of (2)3–4 cell layers, the inner cells large and thin-walled, the outer cells small; gland cells abundant, mainly on septa;

tetrasporangia cruciately divided, in small irregular, frequently anastomosing, pustular sori; spermatangia unknown; cystocarps prominent, superficial.

Fryeella gardneri (Setch.) Kyl.

Fauchea gardneri Setchell 1901: 125. *Fryeella gardneri* (Setch.) Kylin 1931: 16; Hollenberg & Abbott 1966: 96. *Rhodymenia gardneri* (Setch.) Kyl. 1925: 41. *Fryeella gardneri* var. *prostrata* Dawson & Neushul 1966: 178.

Thalli to 16 cm tall, repent and in thick clusters, or erect and isolated, deep rose red to orange-red, with strong bluish cast to lower surface when fresh; branching 3 or 4 times, the second and third dichotomies to 4 cm

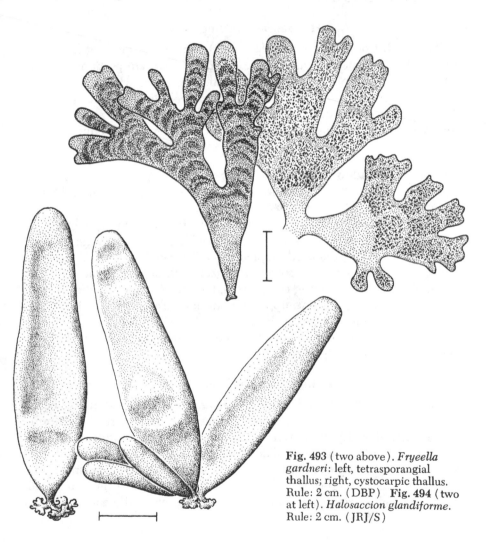

Fig. 493 (two above). *Fryeella gardneri*: left, tetrasporangial thallus; right, cystocarpic thallus. Rule: 2 cm. (DBP) Fig. 494 (two at left). *Halosaccion glandiforme*. Rule: 2 cm. (JRJ/S)

wide; ultimate divisions to 1 cm wide, with blunt apices; tetrasporangia in irregular, grapheiform sori, scattered over all except basal 1–2 cm of blade, produced on 1 side only (lower surface if repent); cystocarps usually 1+ mm diam.

Frequent, subtidal (7–50 m), S. Br. Columbia to Bahía Papalote, Baja Calif.; in Calif., Humboldt Bay, Monterey Peninsula (common in Carmel Submarine Canyon), Morro Bay, and Anacapa I. Type locality: Whidbey I., Wash.

Since prostrate forms are commonly found in company with more erect specimens, we do not regard var. *prostrata* Dawson & Neushul (1966: 178) as worthy of recognition.

Halosaccion Kützing 1843

Thalli with 1 or more erect, hollow, short-stipitate sacs arising from discoid holdfasts. Stipes inconspicuous, cylindrical. Saccate portions usually cylindrical, sometimes sparingly branched, sometimes with hollow, proliferous outgrowths. Cells next to cavity large, colorless, the cells progressively smaller outward. Tetrasporangia cruciately divided, scattered remotely over entire surface, embedded in cortex. Sexual reproduction unknown.

Halosaccion glandiforme (Gmel.) Rupr.

Ulva glandiformis Gmelin 1768: 232. *Halosaccion glandiforme* (Gmel.) Ruprecht 1851: 279; Smith 1944: 298 (incl. synonymy).

Thalli 5–15(25) cm tall, yellowish-brown to reddish-purple, usually with several simple sacs arising from single holdfasts; sacs of mature plants cylindrical, to 3 cm diam., the apices broadly rounded; sacs filled with water, this under pressure sometimes ejected through several terminal pores; older plants becoming compressed, the rounded apex eroding away.

Abundant in certain areas, upper to midtidal rocks, northwest Pacific to Pt. Conception, Calif. Type locality: Kamchatka, U.S.S.R.

The small sacs frequently found on older plants may represent growth from germinated tetrasporangia.

Botryocladia Kylin 1931

Thalli simple or branched, with solid, cylindrical, monopodial axes bearing 1 or more erect saccate branches. Saccate branches spherical, pyriform, or elongate, often 10–20 times diam. of bearing branch or stipe, with wall of large cells adjoining mucus-filled cavity and smaller cells toward surface. Gland cells single or clustered, projecting into cavity. Tetrasporangia scattered in cortex, cruciately divided. Spermatangia in surface areas on saccate branches; cystocarps ostiolate, projecting outwardly and inwardly.

1. Saccate branches with lobes and protuberances .... *B. hancockii*
1. Saccate branches ellipsoidal, without lobes or protuberances 2
   2. Axes prominent, bearing small saccate branches mostly 2–3 mm long
   .... *B. neushulii*
   2. Axes less prominent than vesicles; saccate branches mostly 10–40 mm
   long .... *B. pseudodichotoma*

Botryocladia hancockii Daws.

Dawson 1944a: 305; 1963a: 452.

Thalli mostly 3–5 cm tall, 2 or more deep-red compound-saccate branch-es arising from short, terete stipes; primary vesicles elongate, 4–6 mm diam., bearing several simple or branched elongate lobes, these sometimes further branched; outermost cortical layer netlike in surface view; gland cells occasional, in groups of 2–8, pyriform to subspherical, sessile or on special rotund cell; reproduction as for genus.

Infrequent, on subtidal rocks (18–38 m), Anacapa I., Calif., to Bahía Viscaíno, Baja Calif., and lower Gulf of Calif. Type locality: Bahía Agua Verde, Baja Calif., dredged in 20–40 m.

Botryocladia neushulii Daws.

Dawson 1958: 75.

Thalli mostly 8–20(32) cm tall, 1 to many erect axes arising from exten-sive basal disks; erect axes 1.5–2 mm diam., distantly irregularly branched, provided above with dark-red, sparse, small, multifarious, saccate vesicles, the lower 6–7 cm of axis usually bare; vesicles ovoid to long-ellipsoidal, 2–3(5) mm long, on short pedicels, usually solitary; gland cells usually solitary, but often with 1 or more branch cells from basal cell; reproduc-tion as for genus.

Occasional, on subtidal rocks (7–15 m), Santa Barbara, Channel Is., and La Jolla (type locality), Calif. to Bahía San Quintín, Baja Calif.

Botryocladia pseudodichotoma (Farl.) Kyl.

Chrysymenia pseudodichotoma Farlow 1889: 1. *Botryocladia pseudodichotoma* (Farl.) Kylin 1931: 18; Smith 1944: 297; Dawson 1963a: 453.

Thalli 10–15(30) cm tall with single, solid axes divergently branched, bearing numerous, stipitate, elongate-pyriform saccate vesicles, these 1–4(7) cm long at maturity, 6–25 mm diam.; gland cells sessile, produced in groups of 10–20 from some smaller inner cells; tetrasporangia abun-dant, scattered in upper half of saccate branches, often forming distinct sorus; cystocarps protuberant.

Frequent to locally abundant, on rocks, low intertidal to subtidal (to 25 m), Vancouver I., Br. Columbia, to I. Magdalena, Baja Calif.; usually at 6–18 m in Calif., occasionally to 38 m. Type locality: Santa Cruz, Calif.

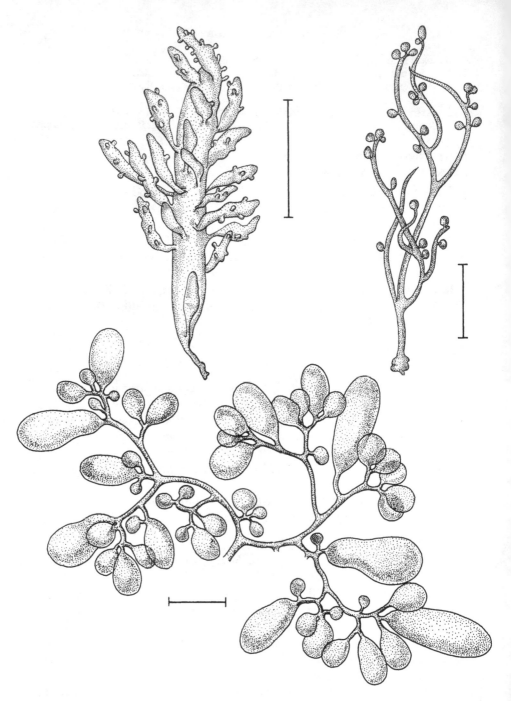

Fig. 495 (above left). *Botryocladia hancockii*. Rule: 2 cm. (SM) Fig. 496 (above right). *B. neushulii*. Rule: 2 cm. (SM) Fig. 497 (below). *B. pseudodichotoma*. Rule: 2 cm. (SM)

Rhodymenia Greville 1830

Thalli erect or spreading, 2–40 cm tall, usually stipitate from discoid holdfasts, with or without cylindrical stolons; branches firm, bladelike, usually dichotomously or irregularly divided; axial branching sometimes sympodial; medulla of large isodiametric cells grading outwardly to small-celled cortex of 2 or 3 cell layers. Tetrasporangia cruciately divided, scattered over entire blade or branch, or in subterminal or terminal sori, sometimes on small marginal or subterminal bladelets, embedded beneath little-modified cortex or in nemathecia. Spermatangia in small superficial sori; carpogonial branches 3- or 4-celled; base of carposporophyte with small fusion cell; cystocarps prominent, scattered or restricted to branch apices, with thick ostiolate pericarp.

1. Blades mostly 3–6 cm broad, the marginal proliferations frequently numerous .... *R. palmata* var. *mollis*
1. Blades mostly less than 1.5 cm broad, the marginal proliferations usually lacking 2
  2. Base with simple-discoid holdfast, the stolons inconspicuous, or basal attachment poorly known 3
  2. Base with stolons more or less prominent 7
3. Stipe essentially lacking .... *R. callophyllidoides*
3. Stipe prominent, sympodially branched 4
  4. Thallus horizontally disposed, 1–3 cm long; blades simple or once-forked .... *R. sympodiophyllum*
  4. Thallus erect, 10–25 cm tall; blades several times dichotomous ... 5
5. Medulla of many cell layers; stipes terete and relatively thick... *R. hancockii*
5. Medulla mostly of 2–6 cell layers; stipes not particularly thick 6
  6. Blades 120–500 μm thick, with 3–6 medullary cell layers; holdfast prominent, subconical, without stolons .... *R. arborescens*
  6. Blades 100–250 μm thick, with 2 or 3 medullary cell layers; holdfast simple, discoid, with some branched stolons .... *R. lobata*
7. Thalli mostly under 7 cm tall; divisions mostly less than 3 mm broad, regularly dichotomous .... *R. californica*
7. Thalli mostly 8–13 cm tall; divisions 4–13 mm broad 8
  8. Stipes sympodially branched; blades usually terminally narrowed; stolons prominent and numerous, providing many supplemental attachments .... *R. rhizoides*
  8. Stipes not sympodially branched; blades usually terminally broadened and blunt; stolons less prominent .... *R. pacifica*

Rhodymenia arborescens Daws.

Dawson 1941: 149; 1963a: 456 (incl. synonymy).

Mature thalli 10–15(24) cm tall, arising from solid, subconical holdfasts of 5–10 mm diam.; blades erect, deep rose, stipitate, arising sympodially from upper part of stipe of each preceding blade; blades dichotomo-flabellate, 60–100 mm broad, the divisions mostly 1–2(3) cm long, 3–4(6)

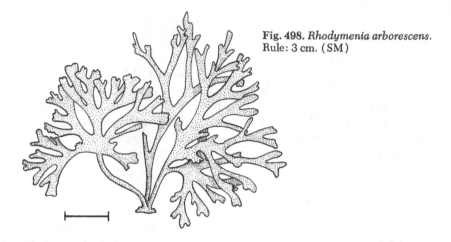

Fig. 498. *Rhodymenia arborescens.*
Rule: 3 cm. (SM)

mm broad, 120–320(500) μm thick, with 3–6 layers of large medullary cells grading through 3–8 layers of cortex and subcortex; tetrasporangia in terminal nemathecia; spermatangia in terminal sori; cystocarps unknown.

Frequent on subtidal rocks (9–40 m), commonly associated with *Macrocystis* under kelp canopy and into deeper water, Anacapa I., S. Calif. mainland, and throughout N. Baja Calif. to Sinaloa, Mexico. Type locality: Laguna Beach, Calif.

Rhodymenia californica Kyl.

Kylin 1931: 21; Dawson 1941: 135; Smith 1944: 301.

Thalli bushy, clumped, 1 to many erect or spreading blades dichotomously or flabellately branched on short stipes arising from basal disks, early producing irregular horizontal subcylindrical stolons; medulla of 2 or 3 cell layers grading abruptly through 2 or 3 cortical layers of smaller cells; tetrasporangia on 1 or both sides of blades, in rounded, terminal, usually nemathecial sori; spermatangia in superficial sori, covering terminal blades; cystocarps scarcely rostrate, usually aggregated on terminal divisions.

Rhodymenia californica var. californica

Thalli in fan-shaped clumps, spreading by several horizontal stolons arising from stipes; blades 2–5 times dichotomously divided, rigid, narrowly linear, 2.5–7(11) cm tall, 1.5–3 mm broad, 70–150 μm thick; terminal divisions blunt to pointed.

On shaded rocks, lower intertidal to subtidal, S. Br. Columbia to Nayarit, Mexico; frequent to common throughout Calif. Type locality: Pacific Grove, Calif.

Rhodymenia californica var. attenuata (Daws.) Daws.

Rhodymenia attenuata Dawson 1941: 139. *R. californica* var. *attenuata* (Daws.) Daws. 1963a: 459.

Thalli as for the species, but with the terminal divisions 1.5–2 times longer, the apices attenuate, the lower branching at longer intervals and more lax.

Occasional on rocks throughout range of species, characteristically in deeper water. Type locality: San Pedro, Calif.

Rhodymenia callophyllidoides Hollenb. & Abb.

Hollenberg & Abbott 1965: 1184.

Thalli closely tufted, erect, without apparent stipes or rhizomatous bases, orange-brown, 6–10 cm tall; blades irregularly dichotomous to fifth order, the divisions mostly less than 1 cm wide through broadest part, divaricately

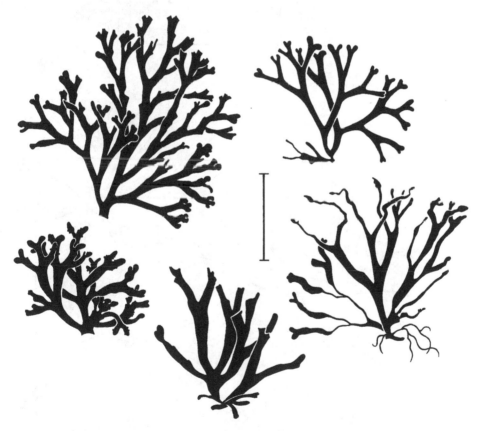

Fig. 499. *Rhodymenia californica* var. *californica*: variation in branching pattern. Rule: 3 cm. (FT)

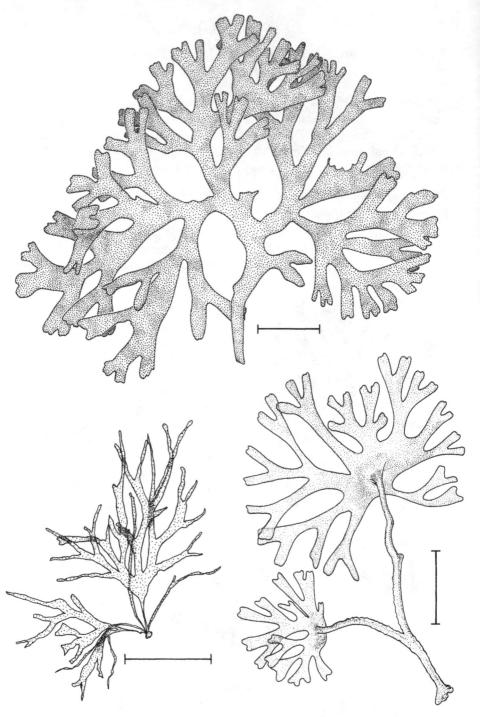

Fig. 500 (below left). *Rhodymenia californica* var. *attenuata*. Rule: 3 cm. (LH)
Fig. 501 (above). *R. callophyllidoides*. Rule: 3 cm. (DBP) **Fig. 502** (below right).
R. hancockii. Rule: 3 cm. (SM)

branched at apices; medulla of 3–5 layers of large cells grading through 2 or 3 cortical layers of very small cells. Tetrasporangia in mottled nemathecia, these submarginal and spreading toward center of branches; spermatangia unknown; cystocarps marginal or superficial on upper branches, protuberant, smooth or slightly rostrate.

Abundant on rocks subtidally (to about 20 m), Monterey Bay (type locality); infrequent at Carmel Bay and Channel Is., Calif., and at Bahía Papalote, Baja Calif.

Rhodymenia hancockii Daws.

Dawson 1941: 146; 1963a: 461.

Thalli 8–10 cm tall, sympodially branched, the basal attachments incompletely known; blades sometimes peltate, or more frequently dichotomo-flabellate, deep red, the lower divisions extending at right or acute angles to terete, relatively thick stipe; blade divisions 4.5–7 mm broad, with rounded apices, 150–400 µm thick; medulla many cells thick; reproduction unknown.

Rare, saxicolous, subtidal (to 42 m); known only from Santa Catalina I., Calif., and Gulf of Calif. (type locality).

Rhodymenia lobata Daws.

Dawson 1941: 147.

Thalli deep rose, caespitose, 20–25 cm tall, arising from simple, discoid bases with few or no stolons; with 1 to many compressed, branched stipes, 10–15 cm long, expanding to dichotomous blades; divisions 3–10 mm broad, the ultimate divisions 10–30 mm long; divisions often proliferous below, 100–250 µm thick, with 2 or 3 medullary layers and ±2 cortical layers; tetrasporangia in rounded nemathecia at expanded apices of ultimate divisions, sometimes in sori in sharply expanded, short-stipitate bladelets. Spermatangia and cystocarps unknown.

Known only from drift, Carmel Bay, Calif. (type locality).

Rhodymenia pacifica Kyl.

Kylin 1931: 21; Dawson 1963a: 461. R. lobulifera Daws. 1941: 137. R. palmettiformis Daws. 1941: 140.

Thalli deep rose, 3–13 cm tall, attached by small, primary discoid holdfasts, these supplemented by spreading, more or less prominent stolons; with 1 to several dichotomo-flabellate blades on compressed stipes 1–3 cm long; blades 2–6 times dichotomous at irregular intervals of 1–2.5 cm; divisions 3–13 mm wide, 140–230 µm thick, often terminally broadened and blunt; medulla of ±3 layers of large cells, grading through 3 subcortical layers to small cells of outer cortex; tetrasporangial sori in nemathecia, these borne terminally in somewhat expanded blade apices or on lateral or marginal bladelets; spermatangial sori superficial, discoloring the

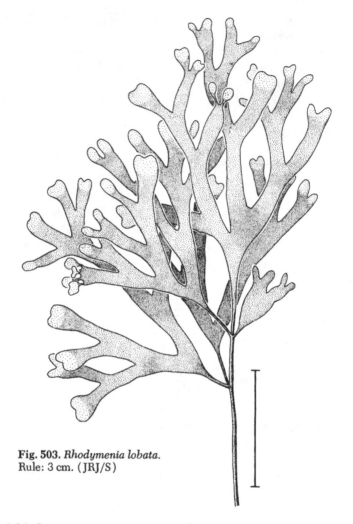

Fig. 503. *Rhodymenia lobata.*
Rule: 3 cm. (JRJ/S)

terminal blades; cystocarps aggregated marginally on terminal divisions.

Frequent to common on rocks, lower intertidal and below, N. Br. Columbia to N. Baja Calif.; in Calif., from Humboldt Co. to San Diego. Type locality: Pacific Grove, Calif.

Extremely variable in height, branching pattern, and shape of divisions, but generally broader in branch width than most other Calif. species (except *R. palmata* var. *mollis*).

Rhodymenia palmata (L.) Grev.

Fucus palmatus Linnaeus 1753: 630. *Rhodymenia palmata* (L.) Greville 1830: 93.

The type variety does not occur in California.

Rhodymenia palmata var. mollis S. & G.

Setchell & Gardner 1903: 315; Smith 1944: 301.

Thalli dull reddish, gregarious, 20–92 cm tall; stipes inconspicuous; blades complanate, somewhat flaccid, simple, or dichotomously to palmately lobed or cleft, the divisions linear-lanceolate to broadly ovate in outline, 3–6(20) cm wide, commonly with marginal proliferations; tetrasporangia associated with paraphyses, in indistinct sori scattered over surface of blade; spermatangia in sori; female plants unknown.

Occasional on rocks in groups (i.e. never isolated), lower intertidal to subtidal, Agattu I., Alaska, to San Luis Obispo Co., Calif. Type locality: not designated, probably Puget Sd., Wash.

Pressed specimens frequently resemble certain species of *Callophyllis*.

Rhodymenia rhizoides Daws.

Dawson 1941: 146.

Thalli of several erect, stipitate blades, 12–15 cm tall, arising from discoid bases; bases often obscured by loose tangle of abundantly branched

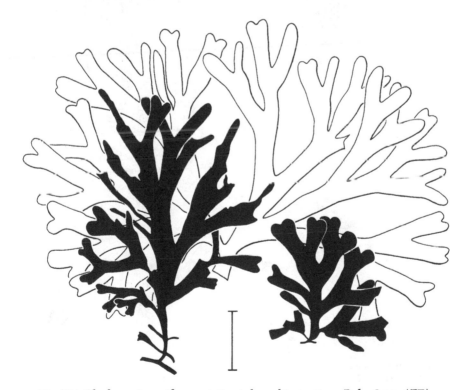

Fig. 504. *Rhodymenia pacifica*: variation in branching pattern. Rule: 3 cm. (FT)

stolons, these arising sympodially in all directions from lower parts of stipes, frequently forming secondary attachments and secondary erect branches and blades; blades deep rose, broadly flabellate, with divisions 4–6 mm wide, ±140 µm thick; tetrasporangia in rounded sori, nearly covering rotund, stipitate, proliferous bladelets, these emerging from eroded or grazed ends of main blades; sexual reproduction unknown.

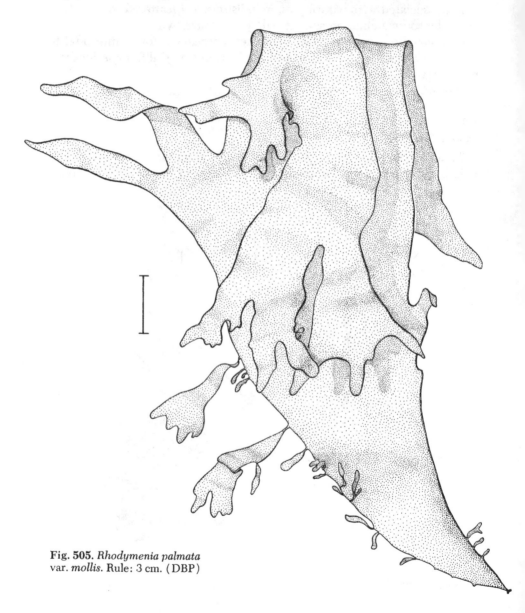

Fig. 505. *Rhodymenia palmata*
var. *mollis*. Rule: 3 cm. (DBP)

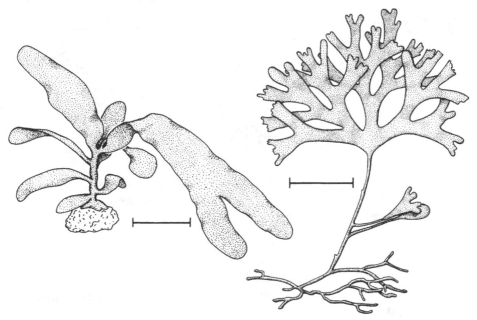

Fig. 506 (right). *Rhodymenia rhizoides*. Rule: 3 cm. (DBP) **Fig. 507** (left).
R. sympodiophyllum. Rule: 1 cm. (EYD)

Occasional on rocks, lowermost intertidal and below, Puget Sd., Wash.,
and Santa Cruz I. to San Diego (type locality), Calif.

Dawson (1963a) calls attention to the closeness of this insufficiently
known species to *R. pacifica*.

Rhodymenia sympodiophyllum Daws. & Neush.

Dawson & Neushul 1966: 179.

Thalli erect, 1–3 cm tall, stiff, attached by broad, low, conical holdfasts;
axes solitary, cylindrical, 500–700 µm diam., consisting of short-stipitate
bases of multifarious blades, these arising sympodially; blades obovate to
oblanceolate, simple, entire, broadly rounded, 8–15 mm long (sometimes
1 large blade to 25 mm long, and once-forked), to 6 mm wide, plane or
slightly undulate, somewhat horizontally disposed and reflexed, ±100 µm
thick, thinner toward margins; medulla of large rotund cells, the subcortex
of flattish cells, the cortex of 1 cell layer; reproduction unknown.

Known only from the type collection from 38 m, Anacapa I., Calif.

Fertile material is needed to establish the position of this entity with
confidence, since it is vegetatively similar to the Phyllophoraceae.

Rhodymeniocolax Setchell 1923

Plants "parasitic" on *Rhodymenia*, tuberculate, with solid bases penetrating hosts. Medullary cells rounded to slightly elongate. Cortical cells

Fig. 508. *Rhodymeniocolax botryoides.*
Rule: 3 mm. (SM)

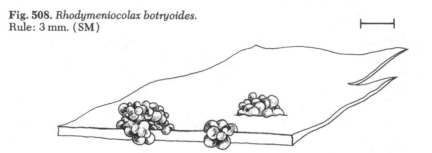

forming short, compact filaments. Tetrasporangia cruciately divided, produced over most of plant surface. Spermatangia unknown. Carpogonial branch of 4 cells. Cystocarps, much as in *Rhodymenia*, terminal.

Rhodymeniocolax botryoides Setch.

Setchell 1923a: 394; Sparling 1957: 362.

Thalli whitish, "parasitic," 3–4 mm diam., with many short, thick branches to 0.75 mm long, sometimes completely encircling host, without evident stalk.

Occasional "parasite" on *Rhodymenia*, mostly subtidal and drift, especially on *Rhodymenia pacifica*, Puget Sd., Wash., to C. and S. Calif. Type locality: San Pedro, Calif.

Family CHAMPIACEAE

Branches cylindrical to narrowly foliose, sometimes with medulla of angular cells, but usually hollow, interrupted at regular intervals by transverse diaphragms (septa). Sporangia mostly tetrahedrally divided, sometimes forming polyspores, usually distributed in outer cortex. Spermatangia in colorless superficial patches. Carpogonial branches mostly 4–celled; auxiliary cell fusing with other cells after fertilization. Most cells or only terminal cells of gonimoblast becoming carposporangia.

1. Branches without septa in hollow parts . . . . *Lomentaria* (p. 567)
1. Branches with septa dividing hollow parts 2
   2. Branches flattened; septa multistratose; plants bearing tetrasporangia
   . . . . *Binghamia* (p. 563)
   2. Branches cylindrical to moderately compressed; septa unistratose;
      plants bearing tetrasporangia or polysporangia 3
3. Plants mostly 10–26 cm tall, only the ultimate branchlets hollow and
   septate; tetrasporangia usually present. . . . . *Gastroclonium* (p. 567)
3. Plants less than 4 cm tall, tubular and septate throughout or with limited
   solid base 4
   4. Axes and branches tubular and septate throughout; plants with tetra-
      sporangia . . . . *Champia* (p. 564)
   4. Limited basal portion of axis or stipe solid; plants with polysporangia
   . . . . *Coeloseira* (p. 565)

Binghamia J. Agardh 1894

Thalli lubricous, complanate, irregularly dichotomously branched, sometimes with pinnate secondary branchlets at margins. Branches at first solid, later lacunose, with longitudinally anastomosing septa. Tetrasporangia tetrahedrally divided, in sori in shallow depressions in surface tissue. Spermatangia in superficial sori. Carpogonial branches unknown. Cystocarps prominent, urceolate, strongly rostrate. Gonimoblast pedicellate, most cells forming carposporangia.

Binghamia californica J. Ag.

J. Agardh 1894: 63; 1899: 158. *Binghamiella californica* (J. Ag.) Setchell & Dawson 1941: 380.

Thalli saxicolous, erect, dark red, 3–6 cm tall, 6–8 times dichotomous,

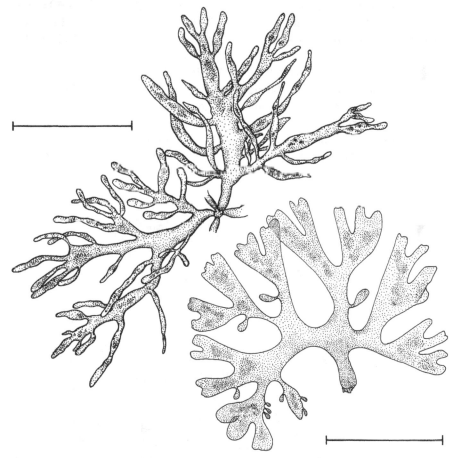

Fig. 509 (right). *Binghamia californica.* Rule: 1 cm. (SM) Fig. 510 (left). *B. forkii.* Rule: 1 cm. (DBP)

divisions divaricate, 2–3 mm wide, ±300 µm thick, the margins entire or with a few lobate proliferations, the apices obtuse, bilobate; cystocarps about 900 µm diam., 1 mm long; tetrasporangia and spermatangia as for genus.

Infrequent in drift from Santa Barbara (type locality) to Laguna Beach, Calif. Reported from Japan.

Binghamia forkii (Daws.) Silva

Binghamiella forkii Dawson 1944d: 96. *Binghamia forkii* (Daws.) Silva 1952: 307.

Thalli epiphytic, dark rose, 1–2 cm tall, adherent at most points of contact by small disks; main divisions 3 or 4 times dichotomous, divaricate, 1–2 mm wide, often with secondary pinnate branchlets, these conspicuous, irregular, 1–5 mm long, basally narrowed; cystocarps 450–600 µm diam., 500–700 µm long.

Occasional on other algae, or in low epiphytic turf, low intertidal (to 20 m), San Nicolas I., Santa Catalina I., Pt. Vicente (south of Redondo Beach), and La Jolla (type locality), Calif., to Pta. Banda, Baja Calif.

Champia Desvaux 1809

Thalli low-caespitose, gelatinous, tender, tubular-septate throughout, contracted at septa. Growth from apical group of cells. Branches with 1 or 2 layers of peripheral cells and with series of internal, longitudinal, gland-bearing filaments adjacent to walls. Tetrasporangia tetrahedrally divided, numerous, distributed in peripheral cell layers. Spermatangia in irregular

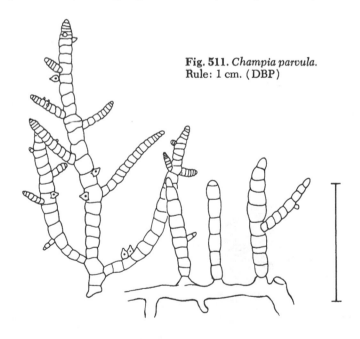

Fig. 511. *Champia parvula.*
Rule: 1 cm. (DBP)

surface areas. Cystocarps prominently protuberant with wide ostioles. Gonimoblasts of richly branched filaments, forming terminal carposporangia.

Champia parvula (C. Ag.) Harv.

Chondria parvula C. Agardh 1824: 207. *Champia parvula* (C. Ag.) Harvey 1853: 76.

Thalli caespitose to turf-forming, 1–2(6) cm tall, pink to straw-colored, attached by accessory disks; branching irregular, the branches cylindrical to slightly compressed, mostly 0.7–1.2 mm diam., the apices blunt; wall of tubular branches mostly of 1 cell layer; tetrasporangia 50–60 µm diam.; cystocarps 0.8–1 mm diam., somewhat rostrate.

Occasional on rocks or other algae, low intertidal to shallow subtidal (6 m), Channel Is. and La Jolla, Calif., to Baja Calif., Costa Rica, and Ecuador. Common in tropical and subtropical N. Atlantic and Pacific. Type locality: Spain.

Coeloseira Hollenberg 1940

Erect axes initially or ultimately solid, short, cylindrical, bearing hollow, septate branches with grouped apical cells. Branch walls 1–3 cells thick,

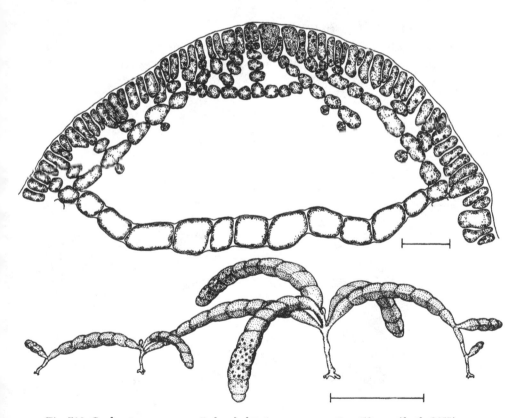

Fig. 512. *Coeloseira compressa.* Rules: habit, 1 cm; cross section, 70 µm. (both GJH)

the innermost layer filamentous in part and bearing gland cells. Polysporangia globose, with 12–16 spores, embedded in distal parts of branches. Spermatangia superficial, forming continuous areas over branches. Cystocarps strongly protuberant, without ostiole.

Coeloseira compressa Hollenb.

Hollenberg 1940: 874; Dawson 1963a: 469.

Thalli 3–7 cm tall, dark red, tubular and septate throughout, spreading; stipes initially tubular but secondarily solid, developing at apices of stoloniferous branches; branches distinctly compressed, ±1 mm diam., commonly with downward-curving apices frequently becoming stoloniferous; branch walls of 1 continuous outer cell layer and incomplete inner layer partly composed of longitudinal filaments; sporangia scattered, 70–100 μm diam.; cystocarps 300–350 μm diam., grouped near branch apices.

Locally frequent, saxicolous or sometimes epiphytic, midtidal to subtidal (18 m), Monterey Peninsula, Calif., to Baja Calif. Type locality: Corona del Mar (Orange Co.), Calif.

Coeloseira parva Hollenb.

Hollenberg 1940: 871; Dawson 1963a: 470.

Thalli mostly 2–3.5 cm tall, dark red, with tubular, septate branches borne on erect, solid, occasionally branched stipes arising from discoid or

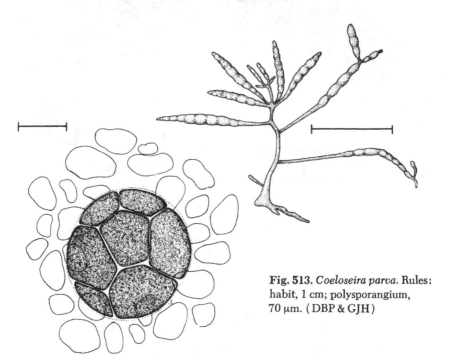

Fig. 513. *Coeloseira parva*. Rules: habit, 1 cm; polysporangium, 70 μm. (DBP & GJH)

partly stoloniferous holdfasts; axes solitary or mostly clustered or forming mats to 10 cm broad, bearing multifarious, simple or compound tubular branches, these 1–1.5 mm diam., prominently constricted at septa; branch walls of 2 or 3 cell layers; sporangia 80–120 μm diam., scattered mostly in subterminal segments of branches; cystocarps 280–300 μm diam., grouped near branch apices.

Occasional, saxicolous, low intertidal to subtidal (18–30 m) at Pacific Grove, subtidal (10–15 m) at Redondo Beach (type locality), Calif., and to Baja Calif.

Gastroclonium Kützing 1843

Erect axes solid, cylindrical, irregularly or subdichotomously branched, bearing many short branchlets. Branchlets and sometimes upper parts of major branches hollow, transversely septate, constricted at septa; with group of apical cells. Tubular branches with inner layer of elongate cells bearing gland cells and with progressively smaller cells toward surface. Tetrasporangia in groups encircling ultimate branchlets above several successive septa near branch apices, tetrahedrally divided or sometimes polysporangiate. Spermatangia in irregular surface patches on ultimate branchlets. Carpogonial filaments 4-celled. Fertilization followed by formation of bilobate placental cell. Mature cystocarp protuberant, globose, ostiolate.

Gastroclonium coulteri (Harv.) Kyl.

Lomentaria ovalis var. *coulteri* Harvey 1853: 78. *Gastroclonium coulteri* (Harv.) Kylin 1931: 30; Smith 1944: 303 (incl. synonymy).

Thalli 10–26(39) cm tall, with reddish bases and greenish tops; solid axes 1.5–2.5 mm diam.; septa of hollow axes at intervals of 1–2 mm; cystocarps 500–800 μm diam.; other characters as for genus.

Common on rocks, intertidal to subtidal (to 14 m), S. Br. Columbia to Baja Calif.; very common throughout Calif. Type locality: Monterey, Calif.

In winter, the plants are commonly eroded and bear only the lower axial portions.

Lomentaria Lyngbye 1819

Thalli erect, cylindrical or compressed, variously branched, hollow, without septa but closed at branch bases, with group of apical cells. Gland cells formed on internal longitudinal filaments. Sporangia scattered or somewhat grouped. Spermatangia formed in extensive surface areas. Carpogonial filaments 3-celled. Mostly gonimoblast cells forming carposporangia. Cystocarps protuberant, ostiolate.

Lomentaria caseae Daws.

Dawson 1945b: 80; 1963a: 465.

Thalli epiphytic on *Phyllospadix*, rose red, to 2.4 cm tall; branching chiefly distichous, subopposite to alternate; branches compressed, 0.3–0.4

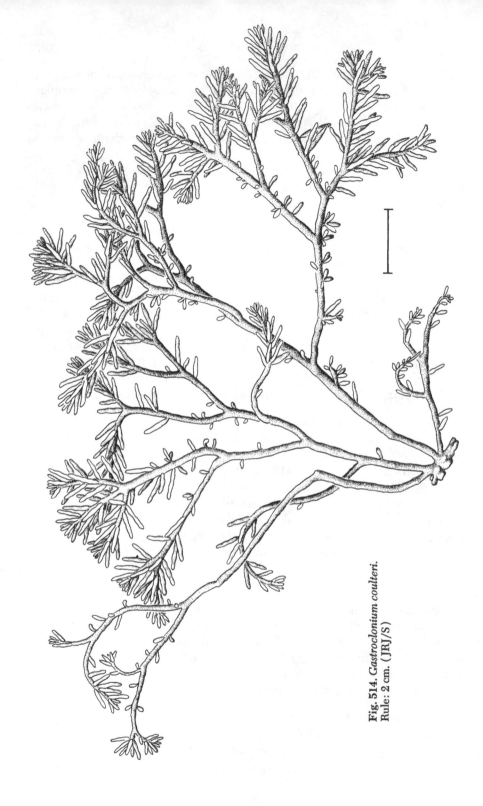

Fig. 514. *Gastroclonium coulteri.*
Rule: 2 cm. (JRJ/S)

Fig. 515 (left). *Lomentaria caseae*. Rule. 1 cm. (SM) **Fig. 516** (right). *L. hakodatensis*. Rule: 1 cm. (SM)

mm wide, somewhat clavate, with blunt, rounded apices; tetrasporangia ±50 μm diam., in irregular sori along primary branches of axis; spermatangia unknown; cystocarps ±350 μm diam., basally contracted, slightly rostrate.

Known only from old *Phyllospadix* leaves in drift at Del Mar (San Diego Co., type locality), Calif., and subtidal on *Zostera*, I. Guadalupe, Baja Calif.

Lomentaria hakodatensis Yendo

Yendo 1920: 6; Dawson 1963a: 466.

Thalli deep red, caespitose to bushy, 3–10 cm tall; branching chiefly radial, the branches cylindrical to compressed, somewhat terminally tapered to occasionally uncinate; tetrasporangia 100–140 μm diam., grouped in somewhat swollen, specialized, fusiform branchlets; spermatangia in reticulate sori in similar branchlets; cystocarps ±700 μm diam., prominently rostrate.

On rocks, subtidal (8–36 m), S. Calif. to Costa Rica; in Calif., known from tank-culture collection from 20 m, off Anacapa I., and saxicolous, low intertidal, La Jolla. Type locality: Japan.

Order CERAMIALES

Thalli uniaxial, basically filamentous, wholly erect, or erect filaments arising from prostrate portion, or prostrate with few erect portions, some of these dorsiventrally organized. Plants with no cortications, with only major branches corticated, or wholly corticated. Some thalli polysiphonous or foliaceous, uncorticated, partly corticated, or wholly corticated. Cells uninucleate or multinucleate, mostly with many discoid chloroplasts lacking pyrenoids. Filaments uniseriate, or multiseriate, or expanded delicate blades, all forms with 1 or more conspicuous apical cells. Growth entirely apical (in some Delesseriaceae also intercalary). Tetrasporangia tetrahedrally or cruciately divided, freely exposed or embedded in thallus, borne singly, in sori, or in stichidia. Reproduction also by polysporangia or parasporangia. Spermatangia in corymbose clusters, in compact masses on trichoblasts, or in sori. All female reproductive structures procarpic; supporting cell with or without sterile cells. Carpogonial branch 4-celled, fusing directly or indirectly with 1 or more auxiliary cells after fertilization of carpogonium; fertilization initiating formation of auxiliary cells. Gonimoblast filaments with all or only terminal cells producing carposporangia, naked or surrounded by loose sterile filamentous involucre or tightly woven pericarp of sterile tissue.

1. Thalli uniaxial, uncorticated to wholly corticated, the cortications rarely as long as bearing cell; most reproductive structures lacking elaborate sterile covering ....CERAMIACEAE (below)
1. Thalli at first uniaxial, soon foliaceous or polysiphonous, frequently corticated; most reproductive structures enclosed or surrounded by sterile tissue 2
   2. Thalli bladelike, mostly delicate, frequently with veins, not polysiphonous .... DELESSERIACEAE (p. 634)
   2. Thalli mostly not bladelike, the branches composed of pericentral cells cut off from axial cells and as long as axial cells, polysiphonous 3
3. Growth sympodial; tetrasporangia on stichidia........DASYACEAE (p. 674)
3. Growth not sympodial; tetrasporangia not on specialized branchlets....
   .... RHODOMELACEAE (p. 681)

Family CERAMIACEAE

Thalli mostly small, delicate, uniaxial, with branched, filamentous laterals, frequently epiphytic, the axis monosiphonous or rarely polysiphonous, uncorticated, partly corticated, or wholly corticated, the corticating cells always shorter than cell they encircle. Cells mostly uninucleate, in some species multinucleate. Tetrasporangia cruciately or tetrahedrally divided, borne singly or in open clusters; some species with polysporangia of 8, 16, 32, or 64 spores. Spermatangia in marginal patches or in spherical to elongate heads or clusters. Fertile female axes, including those of wholly uncorticated species, nearly always shortened determinate branches bearing 1–3 pericentral cells, 1 or more of these serving as supporting cells of 1 or more 4-celled carpogonial branches. Postfertilization events involving 1 or 2 connecting cells between carpogonium and auxiliary cell; after fertilization, usually 1 or more auxiliary cells formed from supporting cell or adjacent pericentral cells; fusion with adjacent cells frequent; all or most cells or only terminal cells of gonimoblast producing carposporangia. Loose covering of sterile filaments sometimes surrounding gonimoblast as involucre (but in *Lejolisea* the enclosing structure thin, parenchymatous, pericarplike), or sterile filaments lacking.

Antithamnion Nägeli 1847

(Contributed by Elise M. Wollaston)

Thallus usually of erect branches arising from prostrate base, rose red in all Calif. species. Axes filamentous, uncorticated, bearing opposite determinate branchlets on each axial cell, initiated in regular sequence at branch apices. Basal cell of branchlet usually of length equal to diam., always lacking determinate side branches. Cells uninucleate. Gland cells cut off from vegetative cell, this dividing further to form chain of 2–5 small cells extending beyond gland cell (gland-cell branch); gland-cell branches borne laterally or terminally. Tetrasporangia ovoid, cruciately divided, sessile or pedicellate on inner cells of opposite branchlets. Spermatangia appearing clustered on specialized branches borne on adaxial side of inner cells of branchlets. Carpogonial branches 4-celled, borne on basal cells of successive branchlets at branch apices. After fertilization, branch apex ceasing to elongate and deflected to 1 side as gonimoblast matures. Most cells of gonimoblast forming rounded gonimolobes successively, protected by surrounding branchlets.

Antithamnion pulchellum Gardner, 1927a: 412, and *A. gardneri* DeToni, 1936: [1], are known only from type specimens or very limited material and are doubtful species in the California flora.

1. Thallus mostly prostrate .... *A. hubbsii*
1. Thallus mostly erect 2
  2. Opposite branchlets simple, abruptly tapering apically; gland-cell branches adaxial on branchlets .... *A. dendroideum*
  2. Branchlets with side branches on adaxial side; gland-cell branches replacing or terminal on side branches 3
3. Lateral branches replacing 1 member of a pair of opposite branchlets.. .... *A. kylinii*
3. Lateral branches without an opposite branchlet .... *A. defectum*

Antithamnion defectum Kyl.

Kylin 1925: 46; Smith 1944: 308; Wollaston 1971: 75. *Antithamnion setaceum* Gardner 1927a: 373. *A. pygmaeum* Gardn. 1927b: 413; Smith 1944: 309.

Erect branches 2(4) cm tall, arising from prostrate base; branchlets opposite on each axial cell, pectinately branched adaxially; lateral branches replacing branchlets, each of these without opposite branchlet; gland cells near outer ends of lateral branches; tetrasporangia ovoid, to 80 μm long, usually pedicellate in clusters; spermatangial branches adaxial; development of carposporophyte as for genus.

Common on other algae, intertidal to subtidal (10 m), S. Br. Columbia to Baja Calif. Type locality: Friday Harbor, Wash.

Antithamnion dendroideum Smith & Hollenb.

Smith & Hollenberg 1943: 217; Smith 1944: 309; Wollaston 1971: 78.

Erect branches to 2 cm tall, arising from prostrate base; branchlets opposite on each axial cell, usually unbranched but occasionally with a few short side branches adaxially; terminal several cells of each branchlet tapering abruptly to acute apex; gland cells borne on short gland-cell branches on adaxial side of inner cells of branchlets; tetrasporangia ovoid, to 80 μm long, usually pedicellate on short adaxial branches; spermatangial branches adaxial; carpogonial branches and carposporophyte unknown.

Occasional, saxicolous or on *Cryptochiton stelleri*, subtidal (5–15 m), Monterey Peninsula, Calif., to I. Guadalupe, Baja Calif. Type locality: subtidal near Monterey, Calif.

Antithamnion hubbsii Daws.

Dawson 1962: 16.

Axes less than 1.5 cm long, partly to mostly prostrate; apical portions of thallus complanate, densely branched; branchlets opposite on each axial cell, curving upward, branched adaxially, occasionally abaxially as well; uniseriate rhizoids borne on most inner cells of vegetative branchlets, terminating in irregular disk; gland cells on short branchlets on adaxial side of upper branchlets, occasionally in series; reproduction unknown.

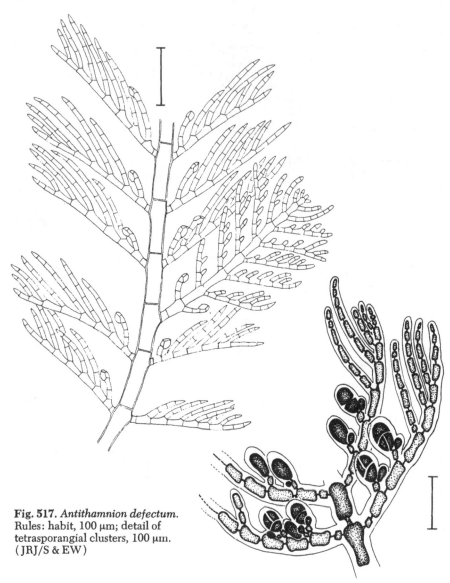

Fig. 517. *Antithamnion defectum.*
Rules: habit, 100 μm; detail of
tetrasporangial clusters, 100 μm.
(JRJ/S & EW)

Rare, epiphytic, subtidal, in 25 m off Isthmus Cove (Santa Catalina I.),
Calif., and in 70 m off I. Guadalupe, Baja Calif. (type locality).

Antithamnion kylinii Gardn.

Gardner 1927a: 411; Smith 1944: 307; Wollaston 1971: 78. *Antithamnion secundatum*
Gardn. 1927a: 413.

Erect branches 4(5) cm tall, without prostrate base; branchlets opposite on each axial cell, pectinately branched adaxially; lateral branches

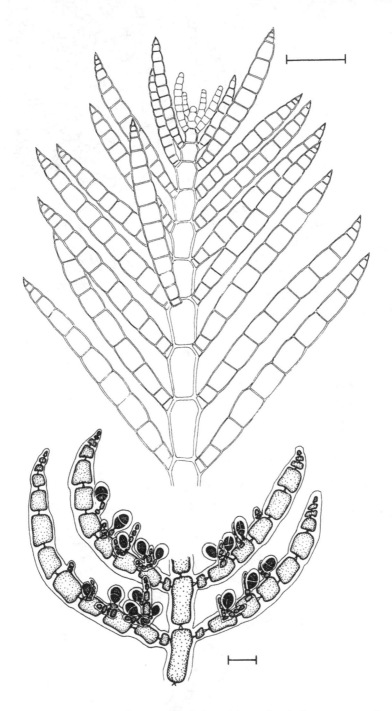

Fig. 518. *Antithamnion dendroideum.* Rules: habit, 100 μm; detail of tetrasporangial formation, 100 μm. (JRJ/S & EW)

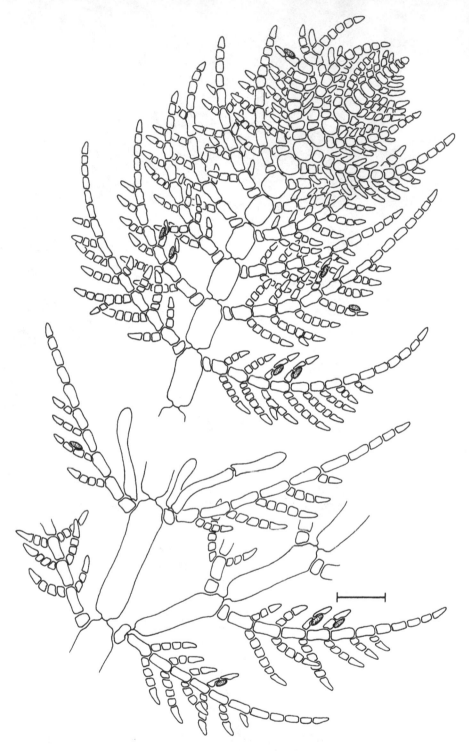

Fig. 519. *Antithamnion hubbsii.* Rule: 100 μm. (NLN)

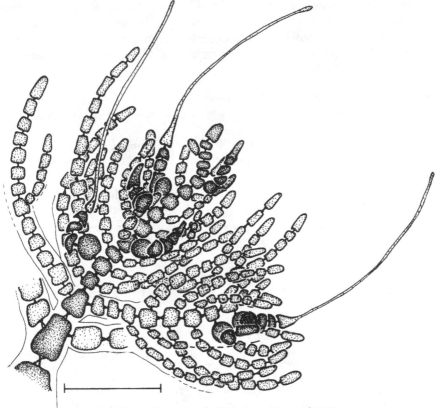

Fig. 520. *Antithamnion kylinii* Rule. 30 μm. (EW)

each with a branchlet opposite; gland cells near outer ends of side branches; tetrasporangia ovoid, to 70 μm long, pedicellate on inner cells of branchlets; spermatangia and gonimoblast development as for genus.

Common on other algae, low intertidal to subtidal (20 m), S. Br. Columbia to I. Magdalena, Baja Calif. Type locality: Vancouver I., Br. Columbia.

Hollenbergia Wollaston 1971
(Contributed by Elise M. Wollaston)

Thallus of erect, filamentous, uncorticated axes bearing 2–4 usually rebranched branchlets from each axial cell, these initiated in regular sequence at branch apices. Branchlets often unequal in each whorl and with basal cell equivalent in size to adjacent cells. Cells uninucleate. Gland cells near apices of branchlets, initiated laterally or terminally, always with tapering branch apex extending beyond gland cell. Tetrasporangia ovoid or subspherical, cruciately or tetrahedrally divided, borne on short branches on inner cells of branchlets. Spermatangia appearing clustered on specialized branches on inner cells of branchlets. Carpogonial branches 4-celled, borne on basal cells of successively formed young branchlets at branch apices. Branch apex after fertilization ceasing to elongate and deflecting to one side as gonimoblast matures. Carposporophyte development as in *Antithamnion*.

Hollenbergia nigricans (Gardn.) Woll.

Antithamnion nigricans Gardner 1927a: 409. *Hollenbergia nigricans* (Gardn.) Wollaston 1971: 83.

Thalli dark red, drying to blackish, erect branches to several cm tall, each axial cell bearing a whorl of (1)2(4) branchlets, each of these with acute-tipped side branches singly or in pairs; gland cells, when present, lateral on cells near apices of branchlets; tetrasporangia ovoid or subspherical, to 60 µm long, cruciately or tetrahedrally divided, on short branches borne on inner cells of branchlets; spermatangia unknown; carpogonial branches borne on basal cells of young branchlets near branch apices; gonimoblast development unknown.

Rare, epiphytic on *Lessoniopsis* or *Pleurophycus*, low intertidal, S. Br. Columbia, Whidbey I., Wash., and Mendocino Co., Calif. Type locality: Vancouver I., Br. Columbia.

Hollenbergia subulata (Harv.) Woll.

Callithamnion subulatum Harvey 1862: 175; J. Agardh 1876: 23. *Hollenbergia subulata* (Harv.) Wollaston 1971: 81. *Antithamnion subulatum* (Harv.) J. Ag. 1892: 20; Kylin 1925: 50; Smith 1944: 312. *A. densiusculum* Gardner 1927a: 374; Smith 1944: 313. *A. baylesiae* Gardn. 1927a: 375; Smith 1944: 313.

Thalli dark rose, erect, 3–10(25) cm tall, each axial cell bearing whorl

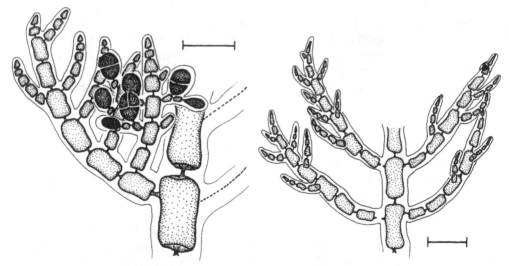

Fig. 521. *Hollenbergia nigricans*: left, tetrasporangia on short branchlets; right, detail of habit. Rules: both 100 µm. (both EW)

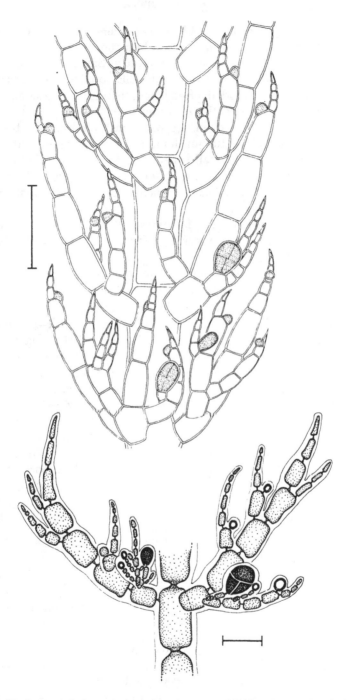

Fig. 522. *Hollenbergia subulata*: above, habit of portion of thallus; below, detail of tetra-sporangial formation. Rules: both 100 μm. (JRJ/S & EW)

of (2)3 or 4 rebranched branchlets, these sometimes differing in length and branching in a given whorl; lateral branches initiated from basal cells of branchlets, these later seeming to replace branchlets; branches bearing reproductive structures short, with cells smaller than those of other branches; gland cells rounded, often longer than broad, initiated terminally on end cells of branchlets, the branchlets continuing growth to form tapered, acute apices of 3–6 cells beyond final gland cells; tetrasporangia ovoid or subspherical, to 100 µm long, cruciately divided but sometimes appearing tetrahedral, on small-celled branches borne on basal and inner cells of branchlets; spermatangia as for genus; carpogonial branches on basal cells of young branchlets at branch apices; development of gonimoblast as for genus.

Infrequent on shells or rocks, low intertidal to subtidal, S. Br. Columbia to Pacific Grove, Calif. Type locality: Vancouver I., Br. Columbia.

Antithamnionella Lyle 1922
(Contributed by Elise M. Wollaston)

Thalli rose red, erect or of erect branches arising from prostrate bases; axes monosiphonous, uncorticated, bearing whorls of 1–4 branchlets on each axial cell, these initiated in irregular or unilateral sequence from curved branch apices. Branchlets rebranched or unbranched, with basal cells similar in size to adjacent cells. Cells uninucleate. Gland cells cut off laterally, usually from central or inner cells of branchlets. Tetrasporangia ovoid or subspherical, cruciately or tetrahedrally divided, sessile or pedicellate on inner cells of branchlets. Spermatangia on branched spermatangial branches borne adaxially on branchlets. Carpogonial branches 4-celled, borne singly at branch apices on basal cells of immature branchlets, these not elongating more than 1 or 2 cells beyond supporting cell; after fertilization, branch apex ceasing to elongate and deflecting to one side as gonimoblast matures. Most cells of gonimoblast becoming carposporangia, these formed successively in lobed or somewhat elongate groups.

1. Thalli prostrate, with erect branches less than 1 cm tall . . . . . A. breviramosa
1. Thalli erect, over 1 cm tall 2
   2. Mature thalli with axes not twined together; branchlets 2(4) per axial cell; gland cells frequent; tetrasporangia sessile . . . . A. glandulifera
   2. Mature thalli usually with axes twined together; branchlets 0–2 per axial cell; gland cells lacking (or very rare); tetrasporangia pedicellate . . . . A. pacifica

Antithamnionella breviramosa (Daws.) Wom. & Bail.

Antithamnion breviramosus Dawson 1949b: 14; 1962: 14; Itono 1969: 30. *Antithamnionella breviramosa* (Daws.) Womersley & Bailey 1970: 322; Wollaston 1971: 84.

Thallus with erect branches to 3 mm tall, arising from prostrate base;

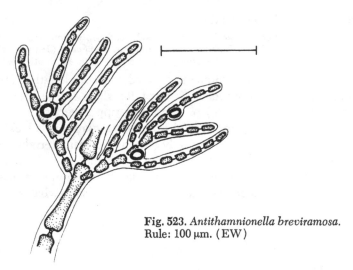

Fig. 523. *Antithamnionella breviramosa.*
Rule: 100 μm. (EW)

branchlets usually in whorls of 3 per axial cell, branched several times
or sometimes unbranched in lower thallus; gland cells lateral on cells of
branchlets; tetrasporangia ovoid to subspherical, to 40 μm long, cruciately
or tetrahedrally divided, sessile on inner cells of branchlets; spermatangial
branches borne adaxially on inner cells of branchlets; carpogonial branches
poorly known; carposporangia in lobed or ovoid groups.

Rare in S. Calif., occasionally on rocks but usually epiphytic on other
algae, low intertidal to subtidal, Gulf of Calif. to Colima, Mexico. Widespread in warmer seas. Type locality: Pebbly Beach (near Avalon, Santa
Catalina I.), Calif.

Antithamnionella glandulifera (Kyl.) Woll.

Antithamnion glanduliferum Kylin 1925: 47; Smith 1944: 310; Dawson 1962: 15. *Antithamnionella glandulifera* (Kyl.) Wollaston 1971: 85. *Antithamnion scrippsiana* Daws.
1949b: 15.

Thallus erect, to 5 cm tall; length of lower axial cells to 8 times diam.;
branchlets unbranched, usually 2 opposite on each axial cell, or occasionally 3 or 4 per whorl in lower thallus; lateral branches formed in place of
branchlets; gland cells frequent, lateral, usually on central to inner cells of
branchlets; tetrasporangia ovoid to subspherical, to 60 μm long, cruciately
or tetrahedrally divided, sessile, adaxial, commonly seriately developed
on axis of branchlets; spermatangial branches borne adaxially on inner
cells of branchlets; carpogonial branches borne singly at branch apices
on basal cells of young branchlets; carposporangia in lobed or somewhat
cone-shaped groups.

Common, low intertidal, usually epiphytic on other algae, S. Br. Columbia to Baja Calif. Type locality: Friday Harbor, Wash.

Antithamnionella pacifica (Harv.) Woll.

Callithamnion floccosum var. *pacificum* Harvey 1862: 176. *Antithamnionella pacifica* (Harv.) Wollaston 1971: 87. *Antithamnion floccosum* var. *pacificum* (Harv.) Setchell & Gardner 1903: 341. *Antithamnion pacificum* (Harv.) Kylin 1925: 47; Smith 1944: 310.

Thallus erect, the lower axes often intertwined, with cell length to 20(30) times diam.; branchlets unbranched (except for short fertile branches), usually 2 opposite (sometimes 1) on central or upper part of each axial cell; lateral branches arising in place of branchlets; gland cells lacking or very rare; tetrasporangia ovoid, cruciately (often appearing tetrahedrally) divided, terminal or lateral on short branches on adaxial side of inner cells of branchlets; carpogonial branches arising singly at branch apices on basal cells of young branchlets; carposporangia in somewhat elongate groups.

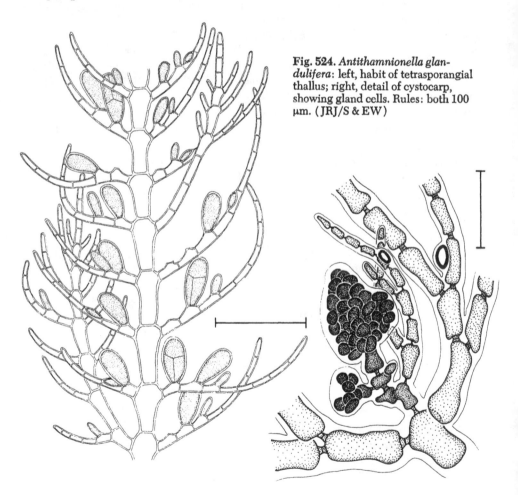

Fig. 524. *Antithamnionella glandulifera*: left, habit of tetrasporangial thallus; right, detail of cystocarp, showing gland cells. Rules: both 100 μm. (JRJ/S & EW)

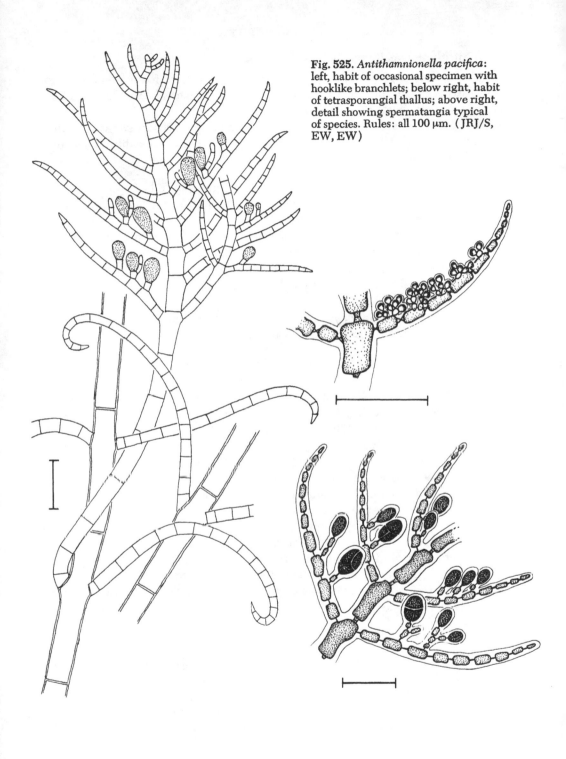

Fig. 525. *Antithamnionella pacifica*:
left, habit of occasional specimen with
hooklike branchlets; below right, habit
of tetrasporangial thallus; above right,
detail showing spermatangia typical
of species. Rules: all 100 μm. (JRJ/S,
EW, EW)

Antithamnionella pacifica var. pacifica

Thallus to 10 cm tall; vegetative and reproductive features much as for var. *uncinata*, but characterized by straight apices of branchlets and the axial cells much longer (20–30 times) than diam.

Common, subtidal and epiphytic on upper portions of stipes of *Nereocystis* or other large algae, Alaska to Baja Calif. Type locality: Vancouver I., Br. Columbia.

Antithamnionella pacifica var. uncinata (Gardn.) Woll.

Antithamnion uncinatum Gardner 1927a: 408; Smith 1944: 311. *Antithamnionella pacifica* var. *uncinata* (Gardn.) Wollaston 1971: 88.

Vegetative and reproductive features much as for var. *pacifica*, but characterized by outwardly curved circinnate apices of branchlets, especially in lower parts of thallus, and often with relatively shorter (less than 8 times diam.) axial cells (especially in less actively growing plants and in mature parts of thallus).

Common on *Nereocystis* or other large algae, Alaska to Baja Calif. Type locality: Dillon Beach (Marin Co.), Calif.

Scagelia Wollaston 1971
(Contributed by Elise M. Wollaston)

Thallus erect, filamentous. Axes uncorticated, bearing whorls of 2–4 branchlets on each axial cell, these initiated in irregular or unilateral sequence from curved branch apices. Branchlets rebranched, unequal in length in each whorl and each with basal cell similar in size to adjacent cells. Cells uninucleate. Gland cells cut off laterally from cells of branchlets. Tetrasporangia ovoid or subspherical, cruciately or tetrahedrally divided, on inner cells of branchlets. Spermatangia on rebranched spermatangial branches borne on inner cells of branchlets. Carpogonial branches 4-celled, borne singly at branch apices on basal cells of young branchlets, these continuing growth to their mature length; fertile branch apex also continuing to elongate after fertilization of carpogonium, and a series of several gonimoblasts sometimes occurring at intervals along each branch axis. Most cells of gonimoblast becoming carposporangia, forming successively in rounded groups.

Scagelia occidentale (Kyl.) Woll.

Antithamnion occidentale Kylin 1925: 47 (incl. synonymy); Smith 1944: 312. *Scagelia occidentale* (Kyl.) Wollaston 1971: 89. *Callithamnion americanum sensu* Harvey 1862: 175.

Thallus to 8 cm tall; branchlets in whorls of (2)3 or 4 per axial cell, 1 or 2 branchlets of each whorl usually longer than others; lateral branches replacing 1 of 2 smaller branchlets in mature whorls and usually occurring

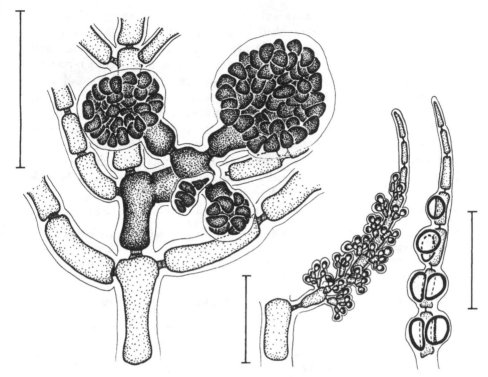

Fig. 526. *Scagelia occidentale*: left, carposporophyte with terminal gonimolobe first-formed; center, stalked clustered spermatangia; right, sessile gland cells. Rules: all 100 µm. (all EW)

at intervals of 1–4 axial cells; gland cells numerous, lateral on cells of branchlets, commonly with more than 1 gland cell per parent cell; tetrasporangia ovoid, to 70 µm long, cruciately (sometimes appearing tetrahedrally) divided, sessile, borne successively on adaxial side of inner cells of branchlets, often several per cell; spermatangial branches in whorls on inner cells of branchlets; carpogonial branches borne singly near branch apices on basal cells of young branchlets; carposporangia in rounded groups.

Common, saxicolous or epiphytic, low intertidal to subtidal, S. Br. Columbia to S. Calif. Type locality: Friday Harbor, Wash.

Platythamnion J. Agardh 1892
(Contributed by Elise M. Wollaston)

Thalli erect, attached at bases by rhizoids, rose red, the erect portions branched, distichous, completely lacking cortication, each axial cell bearing on upper part a whorl of 4 branchlets, consisting of 2 opposite major branchlets and, at right angles to these, 2 opposite minor branchlets. Each major branchlet with primary axis (rachis) bearing secondary branches (ramuli), these usually rebranched. Minor branchlets with short, branched ramuli usually arising from single basal cell, this immediately forming 1–3 upwardly directed ramuli. Indeterminate lateral branches replacing major branchlets in regular, alternate sequence at intervals of 2–4 axial cells (closer in newly formed branchlet areas). Cells uninucleate. Gland cells lateral on cells of branchlets. Tetrasporangia subspherical, cruciately divided, most abundant on inner ramuli of branchlets. Spermatangia on short, rebranched, spermatangial branches borne on inner cells of rachis and sometimes on ramuli of branchlets. Carpogonial branches 4-celled, 1 to several borne on basal cells of successive branchlets near branch apices. Gonimoblasts developed singly at branch apices, nearly all cells becoming rounded groups of carposporangia forming successively.

1. Mature major branchlets with 4 ramuli (2 long, 2 short) on each inner
   cell of rachis . . . . *P. heteromorphum*
1. Mature major branchlets with 1–3 ramuli on each inner cell of rachis 2
   2. Mature major branchlets with (2)3 ramuli, (1)2 above and 1 below,
      on each inner cell of rachis . . . . *P. villosum*
   2. Mature major branchlets with (0)1 or 2 ramuli on upper side only of
      each inner cell of rachis 3
3. Ramuli of mature major branchlets pectinately branched; young minor
   branchlets usually with 3 short, upwardly directed ramuli; lower axes of
   thallus with whorls of 2 long and 2 short branchlets . . . . *P. pectinatum*
3. Ramuli of mature major branchlets not consistently pectinately branched;
   young minor branchlets with single axis, approaching major branchlets
   in length; lower axes of thallus with whorls of 4 almost equal recurved
   branchlets . . . . *P. recurvatum*

Platythamnion heteromorphum (J. Ag.) J. Ag.

Callithamnion heteromorphum J. Agardh 1876: 23. *Platythamnion heteromorphum* (J. Ag.) J. Ag. 1892: 23; Kylin 1925: 51; Smith 1944: 315; Dawson 1962: 9 (incl. synonymy); Wollaston 1972: 46.

Thallus to 3 cm tall; axial cell length to 1.5 times diam.; mature major branchlets with 4 ramuli per cell of rachis, these in opposite pairs of long ramuli branching in plane of thallus and pairs of shorter ramuli at right angles to them; long ramuli usually branched equally on each side of rachis; short ramuli often remaining simple (sometimes lacking); mature minor branchlets radially branched from common basal cell, this somewhat rounded in form; branchlets of lower thallus similar to those above; tetrasporangia to 36 μm long, borne on inner cells of specialized branchlets that replace outer portions of axes of ramuli or branches of ramuli; spermatangia unknown; carposporangia in rounded groups near branch tips.

Occasional to locally frequent, mostly saxicolous, occasionally epiphytic, low intertidal to subtidal (30 m), Sunset Bay, Ore., to I. Guadalupe, Baja Calif. Type locality: Santa Cruz, Calif.

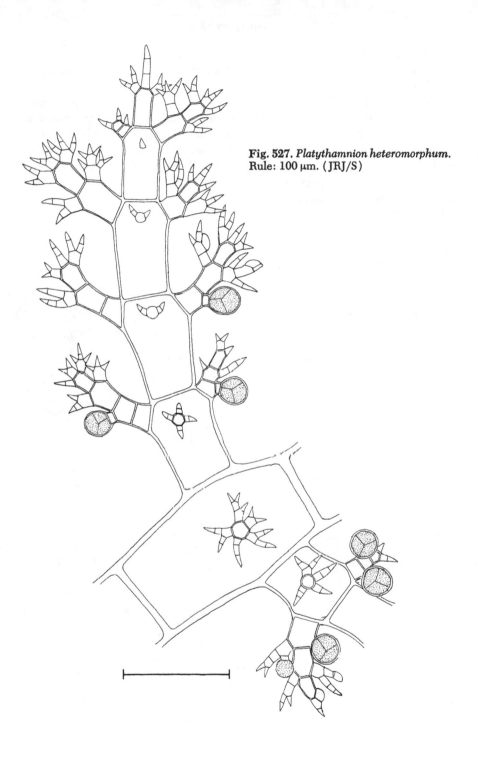

Fig. 527. *Platythamnion heteromorphum.*
Rule: 100 μm. (JRJ/S)

Platythamnion pectinatum Kyl.

Kylin 1925: 53; Smith 1944: 316; Dawson 1962: 6; Wollaston 1972: 51 (incl. synonymy).

Thallus to 5 cm tall; axial cell length to 2.5 times diam., but commonly much shorter; mature major branchlets with (1)2 ramuli on upper sides of inner cells of rachis, the outer rachis cells bearing only 1 ramulus on upper side; ramuli pectinately and abaxially branched; mature minor branchlets radially branched from common basal cell, the branchlets of lower areas similar to those above; tetrasporangia to 45 μm long, borne successively and often abundantly on lower cells of ramuli of branchlets, or sometimes on basal cells of rachis; spermatangial branches borne in place of branches

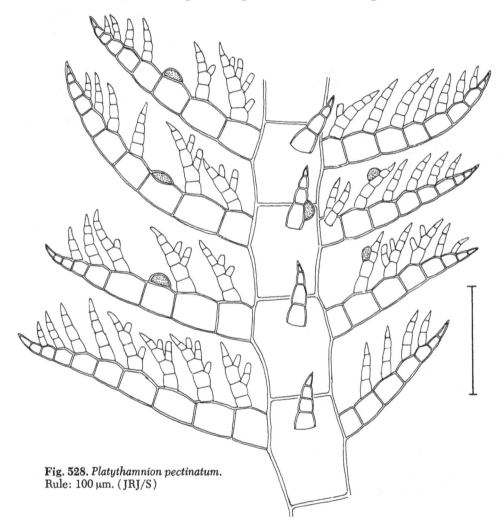

Fig. 528. *Platythamnion pectinatum.*
Rule: 100 μm. (JRJ/S)

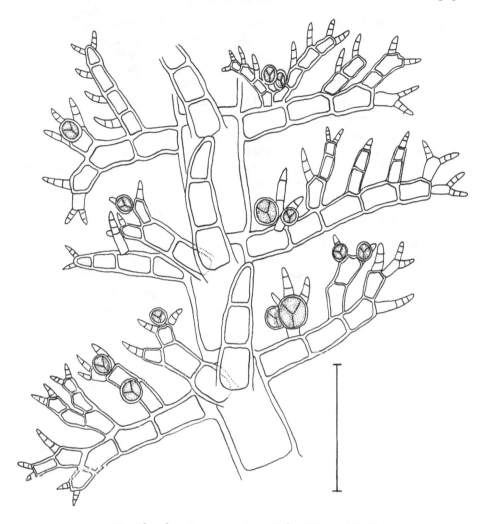

Fig. 529. *Platythamnion recurvatum.* Rule: 100 μm. (SM/Λ)

on ramuli of branchlets; usually only 1 carpogonial branch at each fertile apex.

Common and widespread, saxicolous, low intertidal to subtidal (40 m), S. Br. Columbia to Bahía San Lucas, Baja Calif., and Is. Revillagigedo, Mexico. Type locality: Friday Harbor, Wash.

Platythamnion recurvatum Woll.

Wollaston 1972: 51.

Thallus to 6 cm tall, often procumbent at base; axial cell length to twice diam.; mature major branchlets with 1 or 2 ramuli on upper side of inner cells of rachis and occasionally with 1 branched ramulus on lower side of basal cell; outer rachis cells each with 1 ramulus on upper side; branching of ramuli irregularly pectinate and abaxial, each branch tapering abruptly to acute tip; minor branchlets with single axes approaching major branchlets in length; whorls of lower axes with major branchlets recurved and often differing little in form from much-branched minor branchlets; tetrasporangia to 50 μm long, borne on lower cells of inner ramuli of branchlets; spermatangia unknown; carpogonial branch and carposporophyte development as for the genus, several carpogonial branches forming at each fertile branch apex.

Occasional, saxicolous, low intertidal, Monterey Peninsula, Calif. Type locality: middle reef of Moss Beach (near Pacific Grove), Calif.

Platythamnion villosum Kyl.

Kylin 1925: 51; Smith 1944: 315; Dawson 1962: 9; Wollaston 1972: 49 (incl. synonymy). *Platythamnion heteromorphum* f. *typicum* Setchell & Gardner 1903: 344.

Thallus to 6(8) cm tall; axial cell length to 3 times diam.; mature major branchlets with whorls of 3 irregularly branched ramuli on inner cells of

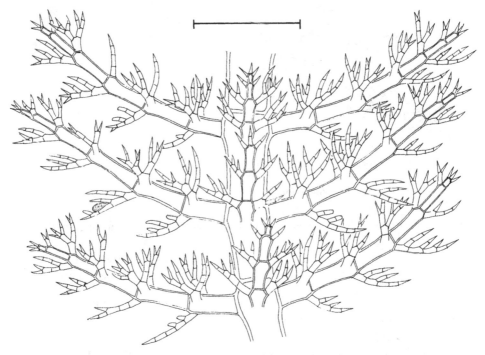

Fig. 530. *Platythamnion villosum.* Rule: 100 μm. (JRJ/S)

rachis, 2 above and 1 below per cell; mature minor branchlets radially branched from common basal cell; lower axes with markedly increased branching of both major and minor branchlets; tetrasporangia to 40(45) μm long, borne on lower cells of inner ramuli of branchlets or sometimes on basal cell of rachis; spermatangial branches borne on upper and occasionally on lower side of inner cells of branchlets; 1 to several carpogonial branches produced successively at each fertile branch apex.

Common, on pilings or saxicolous, low intertidal to subtidal, Sitka, Alaska, to Baja Calif., including Santa Catalina I., Calif. Type locality: Friday Harbor, Wash.

Ceramium Roth 1797

Thalli usually erect, sometimes partially or wholly prostrate, saxicolous or epiphytic; axes cylindrical, freely branched. Branches of all orders corticated in transverse bands at nodes, in continuous layer throughout, or continuously in main axis and intermittently in younger branches. Branching alternate, predominantly unilateral, or irregular, frequently appearing dichotomous. Branch apices usually forcipate, sometimes straight. Axial filament of large cells, each cutting off transverse band of corticating cells at upper end (node), with or without downward or upward growth of smaller, usually angular cells from corticating band. Chloroplasts many, discoid to fusiform to elongate, denser in smaller cells. Cells uninucleate. Tetrasporangia embedded in cortical bands or projecting from them, with or without surrounding sterile cells. Procarps with 1 or 2 4-celled carpogonial branches arising from fertile pericentral cell, this forming auxiliary cell after fertilization; connecting cell present or lacking. Gonimoblasts borne at nodes, usually terminating further growth there but initiating several sterile branchlets from below, these serving as an involucre; gonimoblast rounded, with 1–3 gonimolobes, surrounded by gelatinous envelope. All cells of gonimoblast becoming carposporangia.

1. Thalli corticated only at nodes 2
1. Thalli corticated almost continuously 13
  2. Thalli coarse; branches 120+ μm diam. in midsections 3
  2. Thalli delicate, details microscopic; branches less than 100 μm thick in midsections 6
3. Apices forcipate; tetrasporangia 1 or 2, at nodes, conspicuous ..... *C. zacae*
3. Apices straight; tetrasporangia numerous, in whorls 4
  4. Filaments with conspicuous gland cells irregularly scattered along lateral edges .... *C. clarionense*
  4. Filaments lacking gland cells 5
5. Thalli saxicolous; nodal bands discrete .... *C. gardneri*
5. Thalli epiphytic on various algae, especially *Gracilaria*; banding extending well beyond nodes .... *C. californicum*
  6. Tetrasporangia surrounded by sterile filaments .... *C. camouii*
  6. Tetrasporangia naked 7

7. Nodal bands divided into upper group of angular cells and lower group of pendulous, rounded cells .... *C. caudatum*
7. Nodal bands not divided, or if divided the lower cells transversely elongate 8
    8. Thalli mostly prostrate and creeping 9
    8. Thalli mostly erect 10
9. Filaments irregularly branched .... *C. serpens*
9. Filaments oppositely branched .... *C. procumbens*
    10. Tetrasporangia immersed 11
    10. Tetrasporangia emergent 12
11. Nodal bands close-set in terminal portion and in laterals; gland cells lacking .... *C. equisetoides*
11. Nodal bands not close-set; gland cells frequent among corticating filaments .... *C. gracillimum*
    12. Nodes complex, divided into 3 groups of cells, 2 groups set off from other group by clear space .... *C. taylorii*
    12. Nodes simple, with 1 group of 2 or 3 rows of rounded cells .... *C. affine*
13. Thalli attached by large, dense, basal tuft of penetrating rhizoids 14
13. Thalli attached by several tufts of nonpenetrating rhizoids issuing from nodes 15
    14. Rhizoids inflated, pigmented .... *C. codicola*
    14. Rhizoids not inflated, colorless .... *C. sinicola*
15. Erect axes heavily clothed with short, proliferous branches; gland cells lacking .... *C. pacificum*
15. Erect axes with occasional proliferous branches; gland cells at nodes.. .... *C. eatonianum*

Species Corticated Only at Nodes

Ceramium affine S. & G.

Setchell & Gardner 1930: 172; Dawson 1950e: 132.

Thalli dark red, 1–24 mm tall, 30–38(60) μm diam., not tapering perceptibly until 2 or 3 nodes from apex, dichotomously branched throughout, attached by unbranched rhizoids in lower portions, the apices straight to slightly incurved; internodal cells cylindrical with short conical ends, the length 4–6 times diam.; corticating bands narrow, with 2 or 3 rows of rounded cells, these larger below node than above; tetrasporangia 30–40 μm diam., adaxial on upper branches, 1–3 at each node; tetrasporangial wall remaining after release of spores; spermatangia surrounding nodal cortical cells; gonimoblasts 100–120 μm diam., with a few large carpospores. (After Dawson.)

Rare, shallow subtidal, on *Codium simulans* or *Sargassum* sp., Corona del Mar (Orange Co.), Calif., to Scammon Lagoon, Baja Calif. Type locality: I. Guadalupe, Baja Calif.

Judging from the small number of specimens of this taxon, we do not consider it necessary to recognize the two varieties of Dawson (1950e: 132).

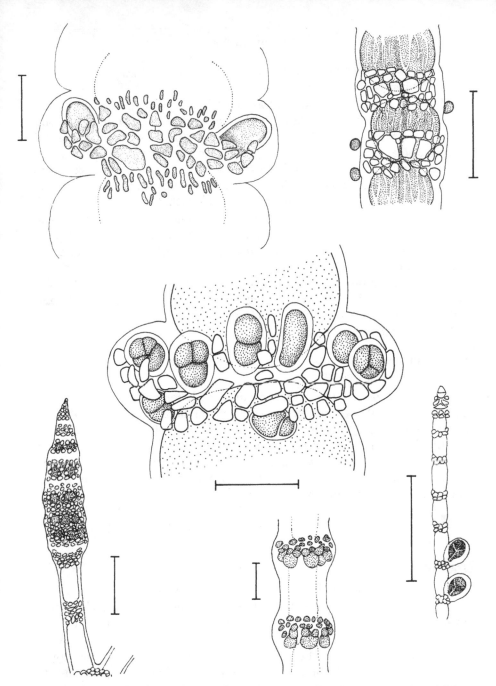

Fig. 531 (lower right). *Ceramium affine.* Rule: 200 μm. (LH) **Fig. 532** (above left).
C. californicum. Rule: 50 μm. (SM/A) **Fig. 533** (center). *C. camouii.* Rule: 50
μm. (DBP/A) **Fig. 534** (below center). *C. caudatum.* Rule: 50 μm. (LH/A)
Fig. 535 (above right). *C. clarionense.* Rule: 50 μm. (SM/A) **Fig. 536** (below left).
C. equisetoides. Rule: 200 μm. (LH/A)

Ceramium californicum J. Ag.

J. Agardh 1894: 45; Dawson 1962: 51.

Thalli dark rose, 2–7 cm tall, (150)400–600 µm diam., coarse, with 7 or 8 orders of branching, the last 4 orders corymbose or pyramidal; beset throughout with short, slender, proliferous branches of 1 or 2 orders; branch apices forcipate or nodding; cortications of branches ringlike, not spreading upward or downward, those on lower main axes with secondary corticating threads spreading close to next cortical band; internodal diam. of upper branches 0.5 to 0.8 times internodal length, that of main axes and lower branches about equal to length, to 1 mm diam.; tetrasporangia slightly emergent from upper portions of cortical band; spermatangia arising from marginal cells of cortical band; gonimoblasts axial, protected by arching sterile filaments.

Occasional, epiphytic on *Gracilaria* or other algae, low intertidal to subtidal (10 m), N. Wash. to Bahía Magdalena, Baja Calif., including Channel Is., Calif. Type locality: Santa Cruz, Calif.

Closely related to *C. gardneri.*

Ceramium camouii Daws.

Dawson 1944a: 319; 1962: 52.

Thalli dark rose, 3–10 mm tall, sparingly dichotomously branched, with occasional lateral branches; branches 30–50 µm diam.; apices straight, with many slender, early-deciduous hairs; cortications only at nodes, the corticating bands regular, not conspicuous, much shorter than internodes, of 2 or 3 rows of cells; tetrasporangia at first lateral, then in whorls, emergent, surrounded by sterile filaments; spermatangia on upper abaxial surfaces; gonimoblasts unknown. (After Dawson.)

Rare, saxicolous or epiphytic, subtidal (to 4 m), Santa Rosa I., Calif., to Baja Calif. and Gulf of Calif. Also Mauritius. Type locality: I. Turner, Gulf of Calif.

Ceramium caudatum S. & G.

Setchell & Gardner 1924a: 776; Dawson 1962: 52 (incl. synonymy).

Thalli small, rose red, creeping, attached by rhizoids from nodes; erect axes 4–15 mm tall, 70–100 µm diam. at upper nodes, to 180–200 µm diam. at lower nodes; apices forcipate; internodes twice as long as nodes; nodal bands projecting only slightly, divided into 2 groups of cells, the upper of 2 or 3 rows of small, angular cells, the lower of 1 or 2 irregular rows of rounded, larger cells, these appearing somewhat pendulous; tetrasporangia broadly elliptical to ovoid, 55–70 µm diam., projecting, few per node; spermatangia unknown; gonimoblasts in subterminal axils of branches, with few enveloping sterile branches. (After Dawson.)

Infrequent, epiphytic, low intertidal to subtidal, Anacapa I. and Santa

Catalina I., Calif., to Baja Calif. and Gulf of Calif. Type locality: Eureka (near La Paz), Gulf of Calif.

Ceramium clarionense S. & G.

Setchell & Gardner 1930: 170; Dawson 1950e: 134; 1962: 53.

Thalli dark red, to 1 cm tall, partially creeping, the erect axes 160–250 µm diam.; branches of 4 or 5 orders, regularly dichotomous, relatively straight, with circinnate apices; cortications in upper branches discrete; internodes shorter than nodes in upper parts, becoming 1.5 times as long as adjoining nodes in lower parts; nodal corticating cells with 1 irregular central row of large, rounded cells and 4–6 adjoining rows of smaller angular cells; internodes in lower parts occasionally tumid; cortical cells frequently with globular, shiny, deeply staining gland cells, these external to cortical band or included, irregularly arranged, few but conspicuous; tetrasporangia in whorls at nodes, emergent, twice as long as diam.; spermatangia in continuous, colorless clusters at nodes; gonimoblasts in subterminal axils of branches, with few enveloping sterile branches. (After Setchell & Gardner.)

Infrequent, epiphytic on variety of algae; subtidal (to 4–5 m) off Channel Is. and low intertidal at Corona del Mar (Orange Co.), Calif., to Oaxaca, Mexico, including Gulf of Calif. Type locality: I. Clarion (Is. Revillagigedo), Mexico.

Ceramium equisetoides Daws.

Dawson 1944a: 320; 1962: 55.

Thalli 8–15 mm tall, primarily dichotomously branched; branches 80–100 µm diam., corticated only at nodes, the young apices usually forcipate; cortical bands twice as long as wide, with 1 median row of large cells and 1 or 2 adjoining rows of irregularly arranged angular cells, these mostly under 10 µm diam.; internodes very short above, to 200 µm long below; tetrasporangia immersed in swollen terminal portions of main axes, or in special lateral branches, usually 1 whorl of these per node; tetrasporangial branches with close-set nodes, nonforcipate; spermatangia tufted, covering terminal branchlets; gonimoblasts unknown. (After Dawson.)

Rare, epiphytic, low intertidal, Newport Bay (Orange Co.), Calif.; rare in Gulf of Calif. and to southward in Mexico. Type locality: Bahía San Carlos (near Guaymas), Mexico.

Ceramium gardneri Kyl.

Kylin 1941: 29; Smith 1944: 325. *Ceramium californicum sensu* Gardner, P.B.-A. 1895–1919 [1917]: no. 2248.

Thalli dark rose, 2–3 cm tall, coarse, tufted, with 5 or 6 orders of branching, the last orders short, crowded, irregular, almost proliferous; apices usually straight, rarely forcipate; cortications throughout discrete at nodes, not spreading; central rows of nodal band irregular, 4–6 rows of smaller angular cells on each side of central row; internodal diam. in upper portions nearly twice length; internodes 3 or 4 times as long as diam. near base, 500–700 μm diam. in lower main axes (150–200 μm in S. Calif. specimens); tetrasporangia emergent, usually lateral at first, then arising from top portion of cortical band; spermatangia marginal on band; gonimoblasts axial, naked or protected by slender, sterile vegetative filaments.

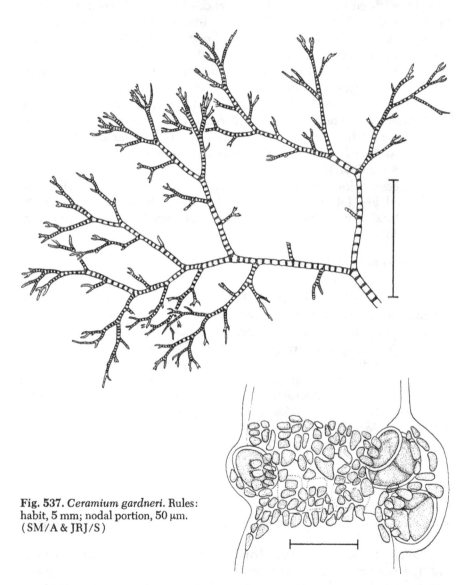

Fig. 537. *Ceramium gardneri.* Rules: habit, 5 mm; nodal portion, 50 μm. (SM/A & JRJ/S)

Occasional, saxicolous, low intertidal on exposed coasts, Br. Columbia to Channel Is., Calif. Type locality: Pescadero Pt. (Monterey Co.), Calif.

Except for its discrete cortical bands and its slightly smaller and saxicolous habit, this species scarcely differs from *Ceramium californicum*.

Ceramium gracillimum Griff. & Harv.

Griffiths & Harvey 1849: 206.

The type variety does not occur in California.

Ceramium gracillimum var. byssoideum (Harv.) Maz.

Ceramium byssoideum Harvey 1853: 218. *C. gracillimum* var. *byssoideum* (Harv.) Mazoyer 1938: 323; Feldmann-Mazoyer 1941: 293; Dawson 1962: 57 (incl. synonymy).

Thalli light pink, epiphytic, attached by simple rhizoids arising from nodes of semiprostrate or entangled lower filaments, 4–10 mm tall, 40–50 µm diam. above, 60–80 µm diam. below, corticated only at nodes; branching alternate, the apices nonforcipate but slightly incurved; internodes to 5 times length of nodes below; cortical bands of 2 groups, the upper group of large, angular cells cutting off a few smaller, superficial cells, the lower of 1 or 2 tiers of horizontally elongate cells sometimes with gland cells; tetrasporangia solitary to whorled, nonemergent, surrounded by sterile filaments; spermatangia in tufts at nodes; gonimoblasts not reported. (After Dawson.)

On other algae, rare in Calif.: subtidal (10–16 m) off Anacapa I., intertidal at Corona del Mar (Orange Co.); also Baja Calif. and Gulf of Calif. to Acapulco, Mexico, various Pacific islands, Mauritius, and W. Mediterranean. Type locality: Key West, Florida.

A critical examination of small, nodally corticated *Ceramium* species in the subtropics and tropics may ally many described taxa, including *C. affine*, with this variable species.

Ceramium procumbens S. & G.

Setchell & Gardner 1924a: 772; Dawson 1962: 63.

Thalli pink to rose, minute, 1–3 mm long, wholly repent, or creeping and partly erect, attached basally by many rhizoids from nodal cells; main axes 45–90 µm diam.; branching mostly opposite but in erect portions frequently alternate; cortication only at nodes; tetrasporangia immersed in nodal band, few but large, 0.25–0.5 times diam. of filament; spermatangia encircling nodal bands; gonimoblasts subterminal, surrounded by a few slender, sterile filaments. (After Dawson.)

Infrequent, epiphytic on a variety of algae, low intertidal to subtidal (3–15 m), Santa Cruz, Santa Catalina, and Anacapa Is., Calif., to Baja Calif. and Gulf of Calif. Type locality: I. Partida, Gulf of Calif.

Ceramium serpens S. & G.

Setchell & Gardner 1924a: 775; Dawson 1962: 64.

Thalli pink, microscopic; creeping filaments fastened by 1 unicellular rhizoid per node and giving rise to erect filaments, these (30)60–80 μm diam., 3–10 mm tall, the branching slight, the apices slightly forcipate; nodes inconspicuous, of 3 or 4 cell rows, 1 middle row larger than outer irregular rows; internodes spherical in upper thallus, cylindrical and 2 or 3 times as long as nodes in lower thallus; tetrasporangia 1 per node, 20–40 μm diam., conspicuous; sexual plants unknown. (After Dawson.)

Rare, epiphytic, low intertidal, on *Corallina* at Santa Catalina I., Calif.; on *Laurencia* at La Paz, Baja Calif. (type locality). Also Marshall Is., C. Pacific.

Ceramium taylorii Daws.

Dawson 1950e: 127; 1962: 65.

Thalli dark rose to red, of prostrate filaments giving rise to erect filaments, these 5–16 mm tall, 60–80 μm diam. above to 180 μm diam. below, alternately branched; apices nonforcipate, divergent when mature; nodes swollen, broader than long; internodes short above, 4–10 times as long as nodes below; mature nodes of 3 definite cell rows, upper 1 or 2 rows of small, angular cells, followed below by angular cells ±4 times as large as those of top rows, these separated by clear space from lowest 1 or 2 cell rows, these in turn at first horizontally divided, then divided again into small, angular cells; tetrasporangia 30 μm diam., whorled, 2–6 at divided cortical band of node, each tetrasporangium protected by modified nodal cells forming an involucre; spermatangia borne in more or less continuous superficial layer at nodes; gonimoblasts well covered by 5 or 6 long, arching involucral branches. (After Dawson.)

Occasional, epiphytic or saxicolous, low intertidal to shallow subtidal, Santa Catalina I. and San Pedro to La Jolla, Calif., and Pta. Descanso, Baja Calif., through Gulf of Calif. to Oaxaca, Mexico. Type locality: Cabeza Ballena (near Cabo San Lucas), Baja Calif.

Ceramium zacae S. & G.

Setchell & Gardner 1937a: 89; Dawson 1962: 67.

Thalli dark rose, of coarse filaments with penetrating rhizoids; erect portions 3–6 mm tall, the branches 100–300 μm diam., dichotomously divided, corticated only at nodes; corticating bands of 5 or 6 transverse, irregular cell rows; internodal cells subspherical, slightly longer than broad; tetrasporangia 35–40 μm diam., protruding on both adaxial and abaxial sides of nodes, sometimes in whorls, conspicuous; spermatangia in low, pulvinate, adaxial patches at nodes of upper segments; gonimoblasts lateral on upper branches, with sterile involucral branchlets. (After Dawson.)

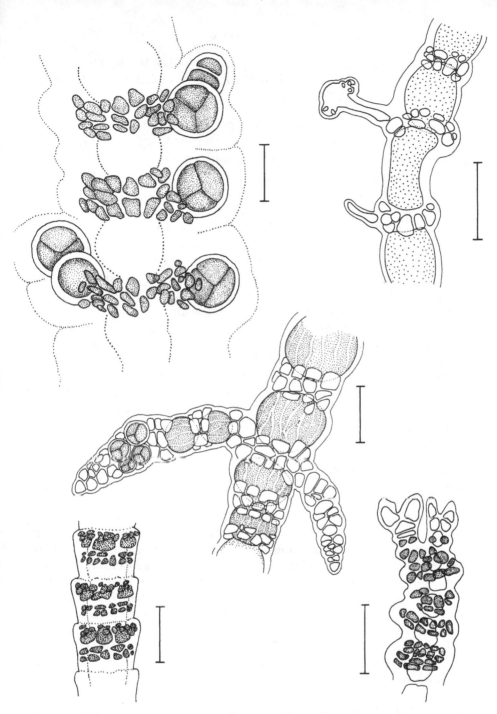

Fig. 538 (below right). *Ceramium gracillimum* var. *byssoideum*. Rule: 200 μm. (SM/A)
Fig. 539 (below center). *C. procumbens*. Rule: 50 μm. (SM/A)   **Fig. 540** (above right).
C. serpens. Rule: 50 μm. (SM/A)   **Fig. 541** (below left). *C. taylorii*. Rule: 50 μm.
(LH/A)   **Fig. 542** (above left). *C. zacae*. Rule: 50 μm. (SM/A)

Rare, epiphytic on *Codium fragile*, low intertidal, Wash., Corona del Mar (Orange Co.), Calif., and Baja Calif. to Gulf of Calif. Type locality: Bahía San Bartolome, Baja Calif.

Corticated Species

Ceramium codicola J. Ag.

J. Agardh 1894: 23; Smith 1944: 326; Dawson 1962: 54.

Thalli dull reddish-brown, forming coarse, fringing tufts along upper dichotomies of *Codium fragile*, 1–2.5 cm tall (late summer specimens to 5 cm), 150–300 μm diam.; branches predominately dichotomous, at times pectinate, the apices usually nonforcipate; attached to host by many bulbous, penetrating, pigmented rhizoids; completely corticated except for thin internodal spaces, with cells irregular throughout; tetrasporangia 35–40 μm diam., embedded at nodes, irregular or in whorls; spermatangia completely covering upper branches below ultimate dichotomies; gonimoblasts in upper axes, with inconspicuous sterile filaments.

Common, epiphytic wherever host occurs, low intertidal to subtidal (8 m), Alaska to Is. San Benito, Baja Calif., including all major offshore islands. Type locality: Santa Cruz, Calif.

Most easily separated from *Ceramium sinicola*, which it strongly resembles, by the bulbous, pigmented rhizoids, those of *C. sinicola* being noninflated and colorless. As far as is known, *C. codicola* is a strict epiphyte of *Codium*, whereas *C. sinicola* epiphytizes a large number of algae, including *Codium*.

Ceramium eatonianum (Farl.) DeToni

Centroceras eatonianum Farlow 1875: 373. *Ceramium eatonianum* (Farl.) DeToni 1903: 1493; Smith 1944: 327; Doty 1947b: 187 (incl. synonymy).

Thalli purplish-brown, 4–10(15) cm tall, in loose tufts; main axis 200–250 μm diam., dichotomously or flabellately branched, the apices usually divergent; lower axes appearing banded owing to greater concentration of cortical cells at nodes; internodal corticating cells in longitudinal rows; gland cells common among nodal cells; tetrasporangia embedded at nodes, 40–70(80) μm diam., making nodes torulose; spermatangia not seen in Calif. material; gonimoblasts lateral, terminating penultimate branchlets, without sterile subtending filaments.

Rare to common, on rocks in upper intertidal, rare subtidally (to 8 m), Wash. to I. Magdalena, Baja Calif.: common from Ore. to S. Calif.; less common to south. Probable type locality: Clatsop Co., Ore.

Ceramium pacificum (Coll.) Kyl.

Ceramium rubrum var. *pacificum* Collins 1913: 125. *C. pacificum* (Coll.) Kylin 1925: 61; Smith 1944: 326.

Thalli deep carmine, 5–18 cm tall; branches 200–300 μm diam. with long

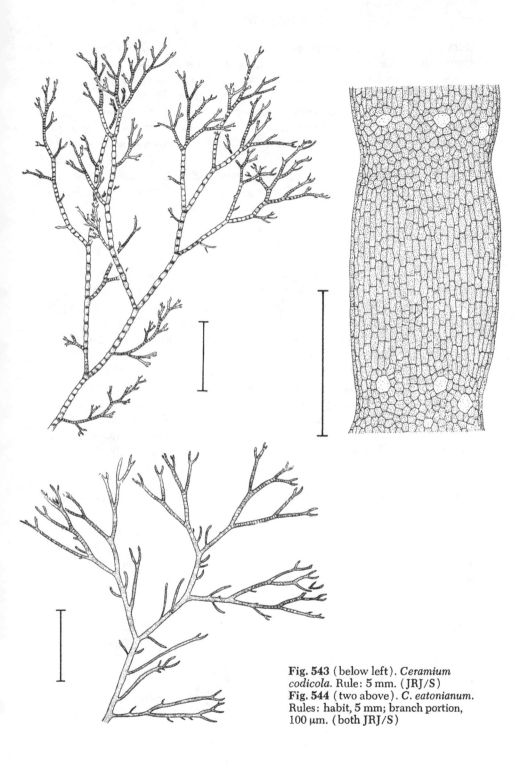

Fig. 543 (below left). *Ceramium codicola*. Rule: 5 mm. (JRJ/S)
Fig. 544 (two above). *C. eatonianum*. Rules: habit, 5 mm; branch portion, 100 μm. (both JRJ/S)

intervals between dichotomies, heavily proliferous with short radial branchlets; apices nodding or forcipate, occasionally with 1 or 2 short spines; cortications of small, angular cells, these closely packed above, becoming somewhat elongate below; tetrasporangia 35–40 μm diam., immersed in cortex of lateral branches, irregularly distributed; spermatangia in continuous patches in upper portions of branches, but not in youngest parts; gonimoblasts terminal on branches, surrounded by 5–7 stout, pointed sterile branches.

Locally abundant, epiphytic or saxicolous, midtidal to subtidal (10 m), Vancouver I., Br. Columbia, to Baja Calif. and Gulf of Calif. Type locality: Monterey, Calif.

Ceramium sinicola S. & G.

Setchell & Gardner 1924a: 773; Dawson 1950e: 118; 1962: 64.

Thalli dull red, epiphytic, 1–3 cm tall, attached basally by slender, branched, colorless rhizoids; branching dichotomous, the branches 150–250 μm diam., tapering gradually from base to apex; apices usually straight on older plants, forcipate or nodding on younger portions; completely corticated by angular cells except in younger portions, where internodes may be seen between developing cortical bands; occasionally with adventitious laterals in lower portions; tetrasporangia 35–50 μm diam., embedded in cortical bands, irregular or in rows, conspicuous when fully developed, marking cortical bands throughout; spermatangia encircling nodes of all but ultimate and lowest portions; gonimoblasts terminal, naked or with 4 or 5 stout, sterile branchlets; female gametangial plants shorter than tetrasporophytes, with branches shorter and apices usually blunt.

Frequent on a variety of algae, low intertidal to subtidal (8 m), Santa Barbara through Channel Is. to La Jolla, Calif., and Baja Calif. through Gulf of Calif. to Guaymas, Mexico. Type locality: Bahía Todos Santos, Baja Calif.

It may be useful to recognize the three varieties of Dawson (1950e: 119) for isolated specimens, but large numbers of specimens will demonstrate a complete range of the cortication characters by which these varieties are distinguished.

Centroceras Kützing 1841

Thalli small, purplish, erect or prostrate, with cylindrical, wholly corticated branches. Branching dichotomous, or irregular; apices forcipate. Cortical cells rectangular and arranged in longitudinal rows, cut off from upper end of large, cylindrical axial cells comparable to internodal cells of *Ceramium*; corticating cells dividing transversely and covering entire axial cell, forming 1 to many spines at upper portion of each node. Cells uninucle-

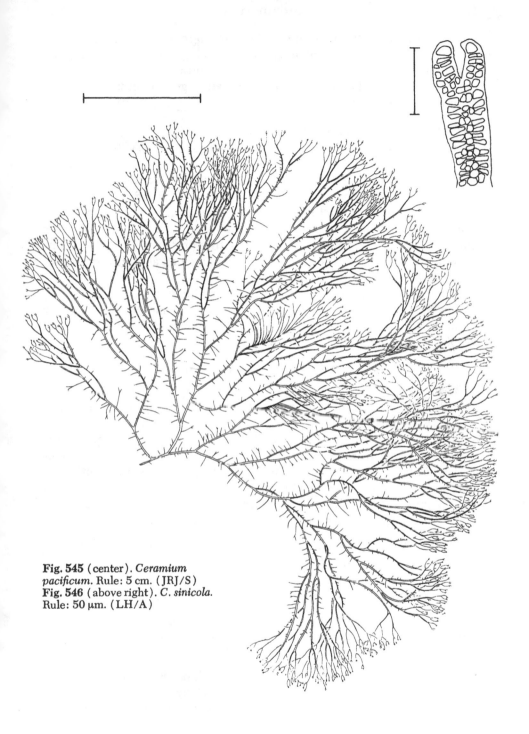

Fig. 545 (center). *Ceramium pacificum.* Rule: 5 cm. (JRJ/S)
Fig. 546 (above right). *C. sinicola.* Rule: 50 μm. (LH/A)

ate. Tetrasporangia tetrahedrally divided, in rings at nodes, sometimes on specialized branches. Spermatangia in terminal clusters from tufted, adventitious branchlets, these arising from pericentral cells at nodes. Development of gonimoblast as in *Ceramium*; mature gonimoblast surrounded by several sterile branches.

Centroceras clavulatum (C. Ag.) Mont.

Ceramium clavulatum C. Agardh 1822b: 2. *Centroceras clavulatum* (C. Ag.) Montagne 1846a: 140; Smith 1944: 328.

Thalli tufted, dark red, 1–3 cm tall in subtropical and tropical portions of range, 4–5(15) cm tall in northern portion of Pacific Coast range; branching alternate, sometimes almost ternate, each dichotomy of equal length; ultimate branches incurved in pairs; each segment in younger portions with several spines of 1 or 2 cells, the spines not always found in older portions; thalli brittle on drying, easily loosened at nodes; reproductive structures as described for genus.

Locally abundant on sand-swept rocks, midtidal to lower intertidal, throughout Calif. from Santa Cruz to San Diego, and throughout Baja Calif. to Callao, Peru (type locality). Common in most oceans; in the tropics mostly epiphytic or forming parts of turf.

Microcladia Greville 1830

Thalli erect, without creeping basal portions. Branching of 5–7 orders, usually in 1 plane, distichous, regularly alternate or appearing pectinate. Branches terete to strongly compressed, the apices forcipate; cells large, uniseriate, uninucleate, the plant corticated throughout by irregular cells of various sizes. Tetrasporangia tetrahedrally divided, embedded in cortical cells, borne on branches of last 3 orders. Spermatangia in continuous layers on ultimate branches. Procarps with 2 carpogonial branches on each supporting cell. Mature cystocarps globose, all cells of cystocarp becoming carposporangia; cystocarps with or without involucre.

1. Branching unilateral and pectinate . . . . *M. borealis*
1. Branching distichous and alternate 2
  2. Alternating branches bearing pinnate branchlets; cystocarps with involucre . . . . *M. coulteri*
  2. Alternating branches bearing irregular branchlets; cystocarps lacking involucre . . . . *M. californica*

Microcladia borealis Rupr.

Ruprecht 1851: 259; Smith 1944: 330.

Thalli 8–20 cm tall, deep red, with entangled, partly rhizomatous bases. Erect branches to 1 mm diam., tufted, branching from middle and top portions, the branches unilateral, pectinate, with pectination in successive

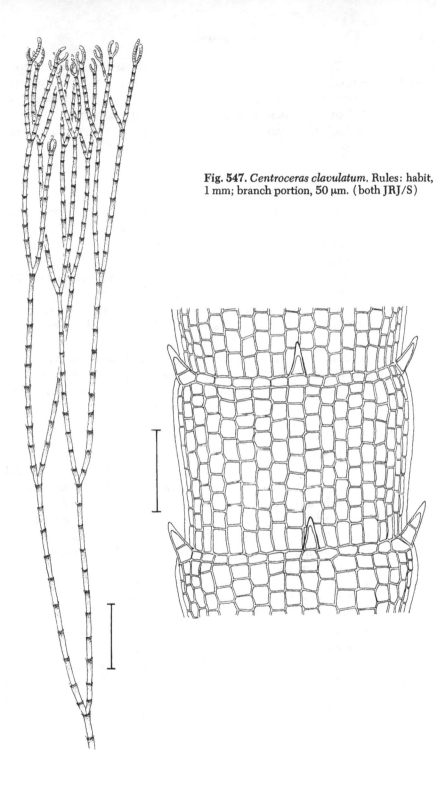

Fig. 547. *Centroceras clavulatum*. Rules: habit, 1 mm; branch portion, 50 μm. (both JRJ/S)

orders alternate. Branches of first and second orders arcuate and curving away from apex of axis or branch bearing them; branches of higher orders incurved and with forcipate tips. Tetrasporangia and spermatangia as for genus. Fertile female branches congested; cystocarps involucrate.

Common on high rocks exposed to surf, occasionally epiphytic, Alaska to San Luis Obispo Co., Calif. Type locality: Unalaska, Alaska.

Microcladia californica Farl.

Farlow 1875: 372; Smith 1944: 330.

Thalli dark red, 20–25 cm tall, the filaments coarse; lower branches entwined, the upper branches divaricate and spiny or arcuate and smooth, alternately branched; main axes compressed, the apices forcipate; lower portions of main axes with internal rhizoids produced by cortical cells; cystocarps lacking involucre, on short branches, borne close to bearing axis and there terminating growth of short branch.

Common at isolated localities, low intertidal, especially epiphytic on *Egregia menziesii*, San Francisco to San Diego, Calif. Type locality: Santa Cruz, Calif.

The spiny forms with divaricate branching differ from most specimens of *M. coulteri* (below) in having no involucres around the cystocarps, but otherwise the smooth forms resemble that species; thus the vegetative limits of the two species are confused.

Microcladia coulteri Harv.

Harvey 1853: 209; Smith 1944: 329.

Thalli 35–40 cm tall, deep rose, usually drying to black, with embedded holdfasts; branches with percurrent axis bearing 5–7 orders of branching; branching regularly alternate, distichous; cortical cells with internal rhizoids in lower portions; cystocarps on final orders of branches, several frequently on 1 branch, appearing seriately, maturing from base to apex, each surrounded by involucre of 3–6 corticated, sterile branchlets, these developing as in *Ceramium*.

Abundant epiphyte on large red algae, but occasionally on large brown algae, midtidal to subtidal (10 m), Vancouver I., Br. Columbia, to Baja Calif. Type locality: probably Monterey, Calif.

The Pacific Coast species of *Microcladia* share some characters, including the number of carpogonial branches, with the type species of the genus, *M. glandulosa*, and other characters (e.g. cortical rhizoids) with *Campylaephora* from the Northwest Pacific. Until details in these genera are studied, critically evaluated, and compared with the same characters in *Ceramium*, no definite position can be taken on the assignment of the Pacific Coast species.

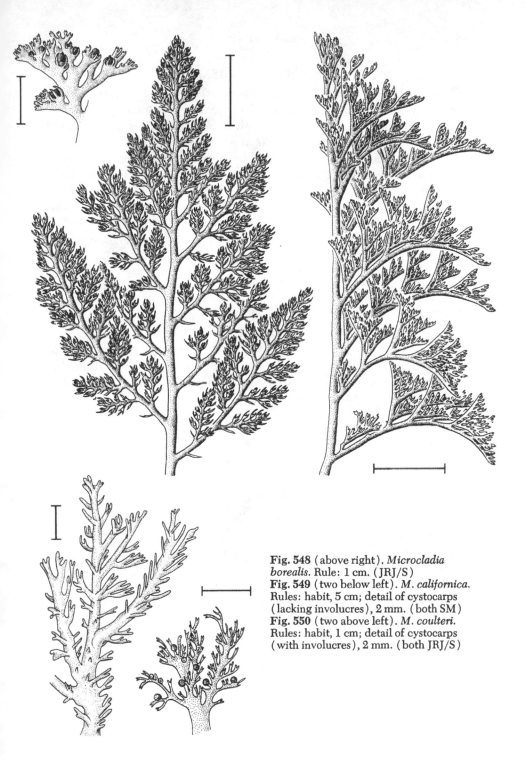

Fig. 548 (above right). *Microcladia borealis*. Rule: 1 cm. (JRJ/S)
Fig. 549 (two below left). *M. californica*. Rules: habit, 5 cm; detail of cystocarps (lacking involucres), 2 mm. (both SM)
Fig. 550 (two above left). *M. coulteri*. Rules: habit, 1 cm; detail of cystocarps (with involucres), 2 mm. (both JRJ/S)

Spyridia Harvey 1833

Thalli erect, terete, bushy, richly branched, with central axes of large cells, wholly corticated by elongate cells in main axes. Branches irregular, attenuate, corticated only at nodes, with or without spines on corticated areas, with or without spines or hooks at apex. Cortical cells in lower portions of main axis and in larger lateral branches having internal rhizoids; short, spinelike branches frequently covering larger lateral branches. Cells uninucleate. Tetrasporangia tetrahedrally divided, borne at nodes of lateral branches. Spermatangia in colorless patches at nodes. Gonimoblasts with 1–3 gonimolobes, surrounded by involucre of branched, sterile filaments; nearly all cells of gonimoblast becoming carposporangia.

Spyridia filamentosa (Wulf.) Harv.

Fucus filamentosus Wulfen 1803–5: 64. *Spyridia filamentosa* (Wulf.) Harvey 1833a: 336; Dawson 1962: 69.

Thalli rose red, tufted, 5–20 cm tall, variously branched, thickly beset with short, spinelike branches, giving fuzzy appearance; Calif. specimens all sterile and showing attenuate laterals without terminal hooks or spines.

Epiphytic; infrequent in Calif. (La Jolla); occasional in Baja Calif.; abundant in lower Gulf of Calif. (near La Paz). Widely distributed in subtropics and tropics. Type locality: Adriatic.

Fig. 551. *Spyridia filamentosa.*
Rule: 1 cm. (SM, after Børgesen)

Callithamnion Lyngbye 1819

Thalli filamentous, erect, usually short, delicate, tufted, saxicolous or epiphytic. Main axis and major laterals alternately branched; secondary laterals alternate or unilateral, sometimes distichous for short distances. Cells multinucleate. Occasional rhizoids produced by lower cells of branchlets and axis sometimes fusing laterally, sometimes penetrating host if epiphytic. Tetrasporangia tetrahedrally divided, borne singly or in pairs on upper side of branchlets, usually sessile. Spermatangia in flat, colorless tufts, adaxial on laterals. Procarps of 2 fertile pericentral cells serving as supporting cells opposite each other on main branches near apices, 1 of each pair bearing 4-celled carpogonial branch. Each supporting cell after fertilization producing 1 auxiliary cell, this fusing with carpogonium through a connecting cell. Nearly all gonimoblast cells but basal ones forming carposporangia. Gonimoblast globular or irregular, with 2–4 gonimolobes formed from the 2 auxiliary cells; involucre usually lacking.

1. Axis and main branches densely corticated .... *C. pikeanum*
1. Axis and main branches uncorticated 2
   2. Thallus strongly pinnate, the branching in 1 plane 3
   2. Thallus not pinnate, the branching in several planes 5
3. Nearly all cells as long as diam.; second-order branches infrequently re-branched .... *C. rupicolum*
3. Cells elongate; second-order branches pinnately divided 4
   4. Branchlet apices incurved and abruptly terminating in acute tip....
     .... *C. catalinense*
   4. Branchlet apices spreading, not abruptly terminated, the terminal cells blunt .... *C. biseriatum*
5. Terminal branchlets forming a corymb with axial branches ..... *C. paschale*
5. Terminal branchlets lax .... *C. acutum*

Callithamnion acutum Kyl.

Kylin 1925: 55. *Callithamnion californicum* Gardner 1927a: 378; Smith 1944: 318.

Thalli in small, soft, rose-colored tufts, uncorticated; bases of branched, slender, penetrating rhizoids; erect axes 1–3(5) cm tall, mostly alternately branched, occasionally unilateral or irregular in lower portion of axis; branchlets on basal cells of laterals frequently uniseriate, of 8–10 cells; branchlet apices acuminate; tetrasporangia on third-order branches, adaxial, sessile, on lowest cells of laterals; spermatangia in low adaxial tufts; gonimoblasts globular, of 3 or 4 gonimolobes, with a few involucral filaments arising from vegetative cell below fertile axial cell.

Occasional, on rocks, on decorator crabs, or epiphytic, low intertidal to subtidal (to 30 m), Wash. and Monterey Peninsula to San Luis Obispo Co., Calif. Type locality: Cattle Pt., Wash.

Not nearly as common as *C. biseriatum*, which is found in the same

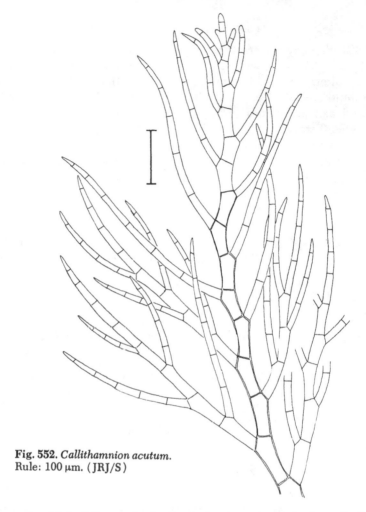

Fig. 552. *Callithamnion acutum.*
Rule: 100 μm. (JRJ/S)

habitats and which differs principally in having pinnately branched laterals and clustered tetrasporangia.

Callithamnion biseriatum Kyl.

Kylin 1925: 54; Smith 1944: 319.

Thalli erect, deep rose, 1–2 cm tall, uncorticated except at bases; rhizoids embedded in lateral walls of lower cells and penetrating host tissues; lower axial cells, including rhizoids, 60–78(150) μm diam.; erect portions with conspicuous percurrent axes, alternately branched in 1 plane, some branches more strongly developed than others, giving unilateral effect; 2 lowermost cells of lateral branches usually each with 1 short, strongly curved, uniseriate branchlet; remaining branchlets on lateral alternately

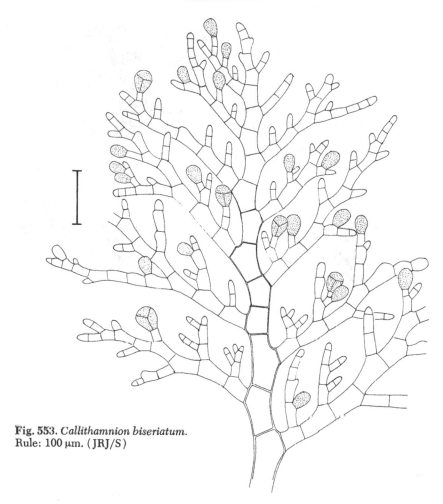

Fig. 553. *Callithamnion biseriatum.*
Rule: 100 μm. (JRJ/S)

branched in female thalli, irregularly branched and frequently unilateral in tetrasporangial thalli, unbranched in spermatangial thalli; apices of branchlets blunt; tetrasporangial thalli with adaxial branching only in fertile areas (alternate or irregular otherwise); tetrasporangia 42–46 μm diam., terminal or lateral on ultimate branchlets, in clusters of 1–3, distributed throughout thallus; spermatangia in low, flat clusters, crowded on second-order branches, then inhibiting third and higher orders of branching; gonimoblasts 65–80 μm diam., with 2–4 gonimolobes and 2–6 strongly developed but short involucral branchlets.

Locally frequent, epiphytic on foliose red algae, subtidal (to 30 m), Wash.; also Monterey Peninsula, Santa Cruz I., and La Jolla Submarine Canyon, Calif. Type locality: Friday Harbor, Wash.

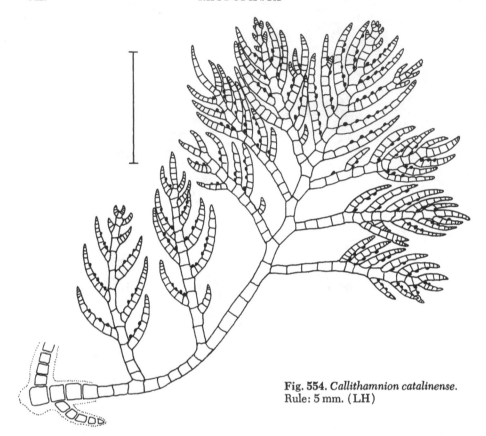

Fig. 554. *Callithamnion catalinense.*
Rule: 5 mm. (LH)

Callithamnion catalinense Daws.

Dawson 1962: 28.

Thalli dark rose, 11–14 mm tall; axial cells 165–200 μm diam., modified basally by simple or branched, uniseriate rhizoids arising from lateral portions of lowermost cells; branchlets simple, alternately placed, incurved, the lowest few cells without further branchlets, the next few cells with short adaxial branchlets, then alternately branched to apex, ending in 3 or 4 cells of reduced diam.; ultimate cells acute; tetrasporangia borne serially and adaxially on ultimate branches, 35–40 μm diam., 50–60 μm long, sessile, solitary; other reproductive structures unknown.

Rare, epiphytic or saxicolous, subtidal (15–30 m), Farnsworth Bank (off Santa Cruz I., Channel Is.; type locality) and Isthmus Cove (Santa Catalina I.), Calif., only known localities.

The stout thalli with acute branchlets resemble those of *Antithamnion dendroideum*, but differ vegetatively in their alternate branching.

Callithamnion paschale Børg.

Børgesen 1924: 294; Dawson 1962: 31.

Thalli rose pink, erect, to 2.5 cm tall, tufted, attached by rhizoids if on corallines, by broadened basal cells if on other algae, uncorticated; branching alternate, the upper portions of axes crowded, the upper branchlets of 4 or 5 orders, gradually becoming shorter upward; cells 1.5–2.5 times as long as diam.; in well-developed plants the ultimate cells of branchlets 6–10(20) μm diam.; tetrasporangia 44–55 μm diam., 45–60 μm long, sessile, adaxial; spermatangia in flat adaxial clusters; cystocarps small, without involucre, developing in pairs in apices of axes.

Rare; epiphytic, subtidal in Calif. (San Clemente and Santa Catalina Is.); intertidal to subtidal in Gulf of Calif. (I. San Benedicto). Type locality: Easter I.

Callithamnion pikeanum Harv.

Harvey 1853: 230; Smith 1944: 318 (incl. synonymy).

Thalli 10–20(40) cm tall, 1 to several conspicuous axes arising from common holdfasts, frequently intertwined; branches woolly, densely corticated, tannish-red to chocolate brown, dependent on epiphytes and degree of wetness for other colors; penultimate and ultimate branchlets uncorticated, closely alternately branched, overlapping and intertwined with each other, frequently with pointed, spinelike apices; tetrasporangia adaxial on ultimate branchlets, sessile, 70–90 μm diam.; spermatangia unknown; gonimoblasts globular, several gonimolobes occurring together.

Common atop rocks, high intertidal, Alaska to Pt. Dume (Los Angeles Co.), Calif. Type locality: San Francisco, Calif.

The several forms or varieties recognized by Gardner, Kylin, and others are an intergrading series scarcely worthy of mention.

Callithamnion rupicolum Anders.

Anderson 1894: 360; Smith 1944: 319; Abbott 1971a: 354. *Callithamnion breviramosum* Gardner 1927a: 403; Dawson 1962: 27. *C. rigidum* Daws. 1962: 33. *C. uncinatum* Daws. 1962: 35. *C. varispiralis* Daws. 1949b: 16.

Thalli forming soft, felted, deep rose to chocolate brown tufts 1–2(3.5) cm tall; branching alternate, occasionally unilateral; orders of branching and pattern highly variable, the simplest thalli pinnate, some bipinnate, others with mixture of patterns; apices of branchlets recurved or straight and with or without spines; cells almost uniformly as long as diam., in midsections (40)50–80 μm diam., becoming 2–4(8) times as long as diam. in basal portions; uniseriate to slightly branched rhizoidal filaments sometimes present, arising from basal cells of laterals and main axes; tetrasporangia variously placed on laterals of second or third order, but always adaxial; spermatangia usually restricted to second-order branches, but

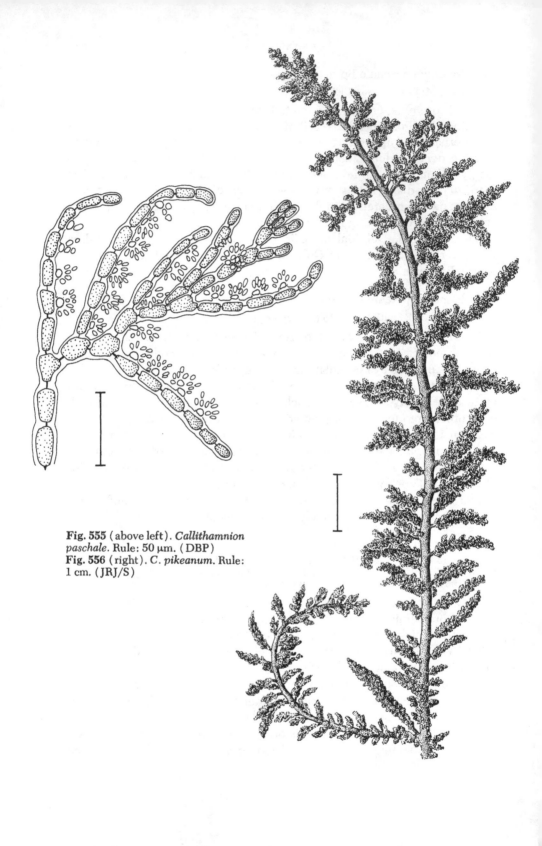

Fig. 555 (above left). *Callithamnion paschale*. Rule: 50 μm. (DBP)
Fig. 556 (right). *C. pikeanum*. Rule: 1 cm. (JRJ/S)

Fig. **557**. *Callithamnion rupicolum.*
Rule: 50 μm. (JRJ/S)

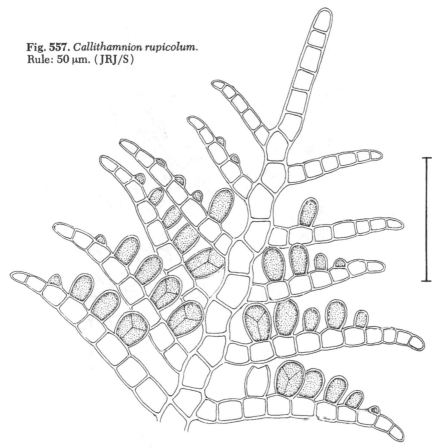

may occu In occasional isolated tufts on third-order branches; gonimoblasts rounded, the 2 or 3 gonimolobes of varying ages, mostly without
involucre, but occasionally with 2 or 3 rudimentary sterile branches.

Common to infrequent on a variety of algae or saxicolous, midtidal to
lower intertidal, Ft. Ross (Sonoma Co.), Calif., to I. Guadalupe, Baja
Calif., and Salina Cruz, Mexico, including Channel Is., Calif. Type locality: Santa Cruz, Calif.

Because he had no access to material from the type locality, Dawson's
concept of the species was limited; thus he segregated as species many
variants that those familiar with the taxon would not recognize as such.

On the Monterey Peninsula, gametangial thalli are found at a lower
tidal level than sporangial thalli.

Aglaothamnion Feldmann-Mazoyer 1941

Thalli deep rose, vegetatively resembling *Callithamnion* but with uninucleate cells. Reproductive structures, other than female structures, like

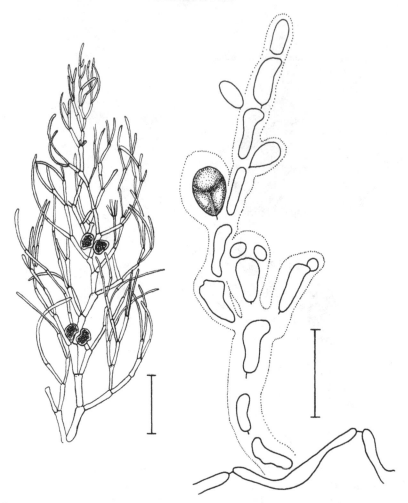

Fig. 558 (left). *Aglaothamnion cordatum*. Rule: 250 μm. (SM, after Børgesen)
Fig. 559 (right). *A. endovagum*. Rule: 50 μm. (SM/A)

those of *Callithamnion*. Procarps alternate along major axes, hence developing gonimoblasts describing a zigzag to right and left of axes. Mature gonimoblasts frequently cordate, reniform, or irregular, usually without involucre.

Aglaothamnion cordatum (Børg.) Feldm.-Maz.

Callithamnion cordatum Børgesen 1909: 10; 1915–20 (1917): 216. *Aglaothamnion cordatum* (Børg.) Feldmann-Mazoyer 1941: 459; Dawson 1962: 36.

Thalli tufted, 1–2(4) cm tall, with several strongly developed axes, basally attached by branched rhizoids; branches of higher orders rapidly diminishing in diam., the narrowing more gradual and not as clear in terminal portion of thalli; ultimate branches frequently becoming hairlike; rhizoids common from lower cell of main laterals, closely adpressed to bearing cell; tetrasporangia scattered through upper portions of thallus; spermatangia as in *Callithamnion*; gonimoblasts irregular, of 2–4 lobes.

Epiphytic and on pilings, subtidal (3 m), Santa Catalina I., Calif., to I. Guadalupe, Baja Calif.; rare in Calif., more common to south into tropics. Type locality: Virgin Is.

Aglaothamnion endovagum (S. & G.) Abb.

Callithamnion endovagum Setchell & Gardner 1924a: 771; Kylin 1941: 28; Dawson 1962: 29. *Aglaothamnion endovagum* (S. & G.) Abbott 1972b: 262. *Acrochaetium grateloupiae* Daws. 1950a: 153.

Thalli partly endophytic within bladelike red algae; creeping filaments irregular, rarely branched; erect, nonendophytic portions 200 µm tall; laterals rare, irregularly branched, opposite, alternate, or unilateral; cells 10–13 µm diam., to 20 µm at base of erect filaments; tetrasporangia 15–18 µm diam., 23–30 µm long, borne on main axis at position of lateral branchlets, lateral or terminal to axis, sessile; spermatangial plants unknown; procarps frequent, alternate on exposed upper axes; gonimoblasts elongate when young, twice as long as broad, assuming irregular globular shape when mature, all cells becoming carposporangia; involucre lacking.

Rare, low intertidal, in species of *Grateloupia* or *Prionitis*, Wash. and La Jolla, Calif., to Gulf of Calif. Type locality: I. San Esteban, Gulf of Calif.

Pleonosporium (Nägeli) Hauck 1885

Thalli filamentous, erect, frequently with adventitious rhizoids; main axes alternately branched; laterals pinnately or unilaterally branched; axis and primary branches partly corticated, or uncorticated. Cells multinucleate. Polysporangia sessile or pedicellate, restricted adaxially or borne alternately adaxially and abaxially along laterals. Spermatangial heads terminating ultimate branchlets, unilaterally placed or alternately adaxial and abaxial. Procarp on 2-celled axis, the lower cell producing 2 pericentral cells, 1 of these bearing 4-celled carpogonial branch and occasionally 1 sterile cell; after fertilization producing 1 auxiliary cell. Gonimoblast with nearly all cells becoming carposporangia, in 6–8 gonimolobes, surrounded by well-developed, arching involucral branches, these formed from vegetative cell basal to fertile axis and sometimes involving other lower vegetative cells.

1. Thalli with axis and major branches corticated, possibly only in basal parts . . . . *P. squarrulosum*
1. Thalli uncorticated 2

2. Branchlets blunt at apices; spermatangia borne alternately adaxially
   and abaxially on laterals .... *P. vancouverianum*
2. Branchlets attenuate at apices; spermatangia borne adaxially only on
   laterals .... *P. squarrosum*

Pleonosporium squarrosum Kyl.

Kylin 1925: 57. *Pleonosporium squarrosum* var. *obovatum* Gardner 1927a: 414.

Thalli 2–3 cm tall, dark red, alternately branched, distichous, with
simple laterals or the laterals pinnate; lowest cell of lateral bearing sparingly branched branchlet, the next order with simple, uniseriate filament;
cortications lacking; rhizoids descending from lower main axis and lower
cells of laterals; polysporangia borne laterally on second- or third-order
branchlets, ovate or obovate, with 16+ spores; spermatangial heads on
adaxial branchlets, short-pedicellate or nonpedicellate, seriately arranged,
the sterile branchlets bearing the heads curving and the apices nodding;
gonimoblasts terminating second-order laterals, with several gonimolobes,
involucrate.

Infrequent, intertidal at Friday Harbor, Wash. (type locality); subtidal
(10–13 m) at Monterey, Calif., on a variety of invertebrates (decorator
crabs, *Cryptochiton stelleri*, hydroids).

Differing from *P. vancouverianum* in having little-branched second-
order branchlets and adaxially arranged spermatangial heads. This species
resembles *Callithamnion acutum* in size, gross appearance, and microscopic details of branching, but differs in reproductive structures.

Pleonosporium squarrulosum (Harv.) Abb.

Callithamnion squarrulosum Harvey 1853: 232. *Pleonosporium squarrulosum* (Harv.)
Abbott 1972b: 262. *C. dasyoides* J. Agardh 1876: 31. *P. dasyoides* (J. Ag.) DeToni
1903: 1310; Smith 1944: 320. *P. polycarpum* Gardner 1927a: 378. *P. pygmaeum*
Gardn. 1927a: 379.

Thalli flaccid, soft pink, with strongly percurrent axes, 5–20 cm tall, alternately branched; upper thallus usually wider than lower portions, the
lower branches frequently eroded to axis; lower thallus heavily to lightly
corticated, attached by numerous rhizoids, arising from lower basal cells;
adventitious, uniseriate rhizoids common; laterals of lower-order branches
strongly developed, those of ultimate orders with only short, uniseriate
filaments; polysporangia with 32 or 64 spores, pedicellate, terminating
third-order branchlets; spermatangia terminating branchlets borne alternately along laterals, the basal cell of a lateral with abaxial spermatangial
branch; procarps borne terminally on short laterals of second- or third-
order branches, both adaxially and abaxially, forming large, conspicuous,
involucrate gonimoblasts of 4–6 gonimolobes.

Locally abundant, epiphytic on a variety of algae, low intertidal to subtidal (to 20 m), N. Br. Columbia, Wash., and Bodega Head (Sonoma Co.),
Calif., to N. Baja Calif. Type locality: San Francisco, Calif.

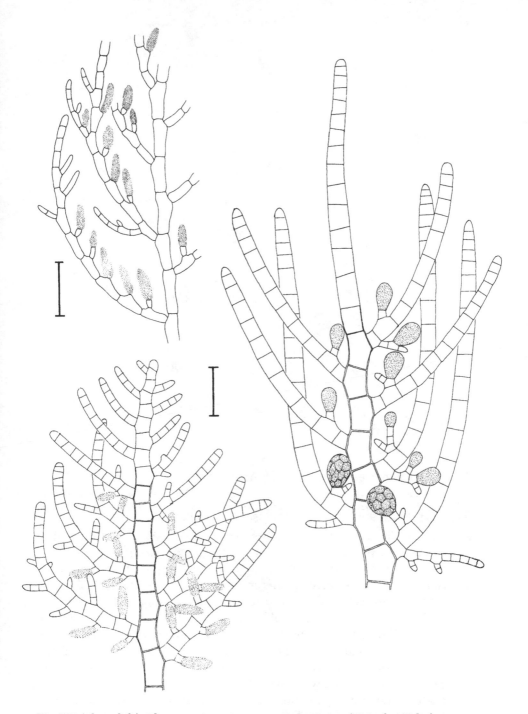

Fig. 560 (above left). *Pleonosporium squarrosum*. Rule: 1 mm. (SM, after Kylin)
Fig. 561 (right & below). *P. squarrulosum*: right, portion of polysporangial thallus; below, portion of thallus with spermatangial heads. Rule: 100 μm. (both JRJ/S)

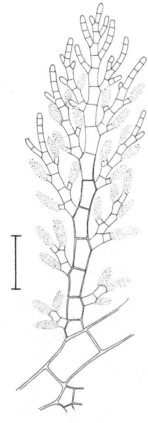

Fig. 562. *Pleonosporium vancouverianum.*
Rule: 100 µm. (JRJ/S)

Pleonosporium vancouverianum (J. Ag.) J. Ag.

Callithamnion vancouverianum J. Agardh 1876: 30. *Pleonosporium vancouverianum* (J. Ag.) J. Ag. 1892: 37; Kylin 1925: 57; Smith 1944: 32; Hollenberg & Abbott 1966: 101. *P. abysicola* Gardner 1927a: 380; Dawson 1962: 39.

Thalli 1–2.5 cm tall, deep rose, alternately and distichously branched to 2 or 3 orders; axis and laterals uncorticated; basal rhizoids few; cells of most axes from N. Calif. about one-fourth more elongate than those from S. Calif., resulting in longer plants in the north, more compact plants in the south; lowest cell of first-order branch abaxial, immediately abaxially rebranched, forming filaments of third order alternately; polysporangia ovate, of 16+ spores, borne on alternate branchlets, sessile or pedicellate; spermatangia in short-ovoid heads, pedicellate or nonpedicellate, terminating alternate branchlets along a lateral; gonimoblasts terminating second-order branchlets, with several gonimolobes, involucrate.

Locally abundant, epiphytic, lower intertidal to subtidal (to 20 m), Vancouver I. (type locality), Br. Columbia, Wash., and Monterey and Channel Is., Calif., to I. Magdalena, Baja Calif.

Of the three species of *Pleonosporium* recognized here, *P. vancouverianum* is the most constant in terms of branching pattern and *P. squarrulosum* the most varied, which may account in part for the number of names being placed in synonymy here. This is another example of a species previously described under various names because of material too limited in extent or restricted in habitat.

Griffithsia C. Agardh 1817

Thalli erect, monosiphonous, with subdichotomously branched filaments; cells usually very large, multinucleate, frequently visible to the unaided eye. Tetrasporangia and spermatangia in dense whorls, or tetrasporangia borne singly, with or without involucral cells. Trichoblasts associated with reproductive structures but soon shed in some species. Procarps 1 or 2, formed on the middle cell of 3-celled fertile axis, each middle cell bearing 4-celled carpogonial branch. Abaxial involucre of 2-celled branches formed from basal cell after fertilization and from large fusion cell; most cells of gonimoblast becoming carposporangia, enveloped in gelatinous matrix.

Griffithsia furcellata J. Ag.

J. Agardh 1842: 75; Hardy-Halos 1968: 523. *Neomonospora multiramosa* Setchell & Gardner 1937a: 87. *Griffithsia multiramosa* var. *minor* Taylor 1939: 14. *G. multiramosa* var. *balboensis* Hollenberg 1945a: 447. *Monospora tenuis* Okamura 1907–42 (1935): 23.

Thalli tufted, (2–4)7–15 cm tall, lubricous, dark rose red when dried, the filaments subdichotomously branched; cells cylindrical, (92)120–270 µm diam., 4–5 times as long basally; base entangled, with rhizoids produced by lower cells; tetrasporangia borne singly at nonforking nodes, on pedicels of 1 or 2 cells, without involucre or trichoblasts; spermatangia in small, terminal, oval heads; fertile procarpic thalli with colorless trichoblasts of 1 or 2 branches, these borne on the vegetative cell bearing fertile axis; procarp with 1 sterile cell and 1 carpogonial branch on fertile pericentral; postfertilization stages unknown; gonimoblast unknown.

Rare, epiphytic or unattached, subtidal (to 40 m), Newport Bay (Orange Co.), Calif., and San José del Cabo (Pta. Gorda), Baja Calif., to Gulf of Calif. Also Atlantic Coast of France, Canary Is., Japan. Type locality: Mediterranean.

Griffithsia pacifica Kyl.

Kylin 1925: 58; Smith 1944: 324.

Thalli tufted, 3–5 cm tall, orange-pink; branching regularly dichotomous in lower parts, irregular in terminal portions; bases frequently entangled, the rhizoids produced from lower ends of all entangled cells; rhizoids frequently unilateral on each cell from base to midportion of plant, these

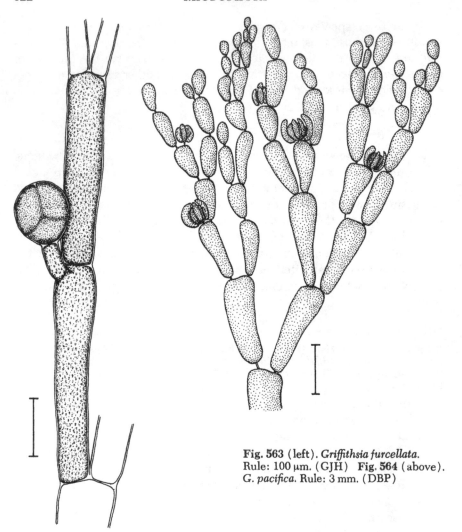

Fig. 563 (left). *Griffithsia furcellata.*
Rule: 100 μm. (GJH) **Fig. 564** (above).
G. pacifica. Rule: 3 mm. (DBP)

cells commonly fragmenting and regenerating; upper thallus of turgid, cylindroconical cells (sometimes cylindrical), these (0.3)1–2 mm diam. through middle portions of thallus; terminal cells often spherical; thalli usually sterile, occasionally tetrasporangial; tetrasporangia borne between articulations; plants rarely sexual; spermatangia without involucre; gonimoblasts terminating short laterals, involucrate.

Infrequent, saxicolous, or occasionally on sponges or holdfasts of large brown algae, in isolated tufts, lower intertidal to subtidal (to 20 m), S. Br. Columbia to Baja Calif. into Gulf of Calif. Type locality: Friday Harbor, Wash.

Tiffaniella Doty & Meñez 1960

Thalli filamentous, uncorticated, differentiated into basal portions bearing rhizoids with modified apices and erect portions of irregularly branched filaments with elongate multinucleate cells. Sporangial plants with polysporangia in irregular lateral clusters, each sporangium containing 8–32+ spores, or with tetrasporangia. Spermatangia in lateral, short to elongate heads, these variously arranged. Procarps formed on middle cell of 3-celled axis; 3 pericentral cells also produced (1 remaining functionless; 1 forming abaxial sterile cell, lateral, 4-celled carpogonial branch, and auxiliary cell after fertilization; and 1 forming second auxiliary cell). Both auxiliary cells fusing with fertilized carpogonium through connecting cell; fusion cell composed of fertile cells, axial cells, and 2 auxiliary cells, forming T-shaped gonimoblast; only terminal cells becoming carposporangia, each surrounded by gelatinous envelope; involucre lacking.

Tiffaniella phycophilum (Tayl.) Gord.

Spermothamnion phycophilum Taylor 1945: 263. *Tiffaniella phycophilum* (Tayl.) Gordon 1972: 124. *S. snyderiae* var. *attenuata* Dawson 1962: 45.

Thalli forming dense, soft turf 3–5 mm tall, dull brownish-red; creeping

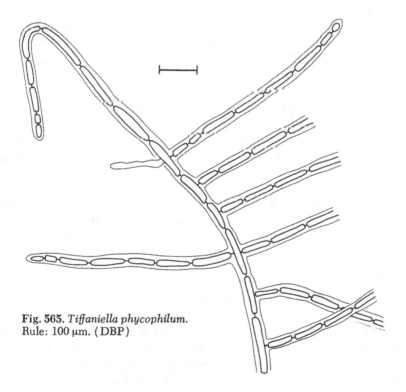

Fig. 565. *Tiffaniella phycophilum.*
Rule: 100 μm. (DBP)

filaments with unicellular rhizoids, these having expanded digitate haptera; erect filaments slender, 18–32 μm diam. in terminal portions, 57–78 μm below; cells cylindrical to slightly swollen; tetrasporangia 47 μm diam. by 79 μm long, in clusters of 5–7; spermatangia in solitary or seriate, dense adaxial heads, 40 μm diam., 65 μm long; gonimoblasts 180–200 μm diam., rounded, with erect, terminal carposporangia; involucre lacking.

Locally abundant, epiphytic on other algae, low intertidal to subtidal (to 6 m), Santa Catalina I., Calif., and I. Guadalupe, Mexico. Type locality: Galápagos Is.

Tiffaniella snyderiae (Farl.) Abb.

Spermothamnion snyderiae Farlow 1899: 74; Smith 1944: 322. *Tiffaniella snyderiae* (Farl.) Abbott 1971a: 349.

Thalli 2–5 cm tall, deep red; most branches unilateral or irregular, rarely with more than 3 orders of branching, all orders of nearly same diam.; basal rhizoids with expanded apices; cells in midsections (50)80–130 μm diam., 10–20 times as long; polysporangia in irregularly shaped, lateral clusters, broadly ellipsoidal, 75–90 μm diam., 95–110 μm long, with 8–32 spores; spermatangia dense in lateral branchlets, replacing adaxial branches; gonimoblasts rare, bilobate, 250–350 μm wide, flattened on top; first and second cells of fertile axis fusing with auxiliary cells; carposporangia ovate, 25–35 μm diam., 35–50 μm long, few per gonimoblast.

Locally abundant, saxicolous, low intertidal to subtidal (to 20 m), Wash. to Baja Calif., into Gulf of Calif. Type locality: Santa Cruz, Calif.

Ptilothamnion Thuret 1863

Thalli with prostrate and erect axes; prostrate axes little or much branched, with haptera; erect axes branched pinnately, unilaterally, or irregularly. Cells multinucleate. Tetrasporangia tetrahedrally divided, sessile or with short pedicels, frequently in small clusters; polysporangia in some species. Spermatangia in spherical to short-elongate heads, sessile, stalked or terminal. Procarp on 3-celled fertile axis, the subapical cell bearing 3 pericentral cells, only 1 of these fertile; 1 carpogonial branch and 1 auxiliary cell produced per procarp. Gonimoblast globular to irregular, of 1 lobe or of several lobes maturing in turn; involucre of bracts from basal cell of fertile row, formed before or after fertilization.

Ptilothamnion codicolum (Daws.) Abb.

Pleonosporium codicola Dawson 1962: 39. *Ptilothamnion codicolum* (Daws.) Abbott 1971a: 355.

Thalli rose red, minute, epiphytic or partly endophytic, with slender, erect filaments 3–5 mm tall, the lower portions penetrating hosts or entangled with turf-forming algae; rhizoids infrequent; branching sparse, mostly basal, irregular, the branches 20–30 μm diam.; polysporangia 90–

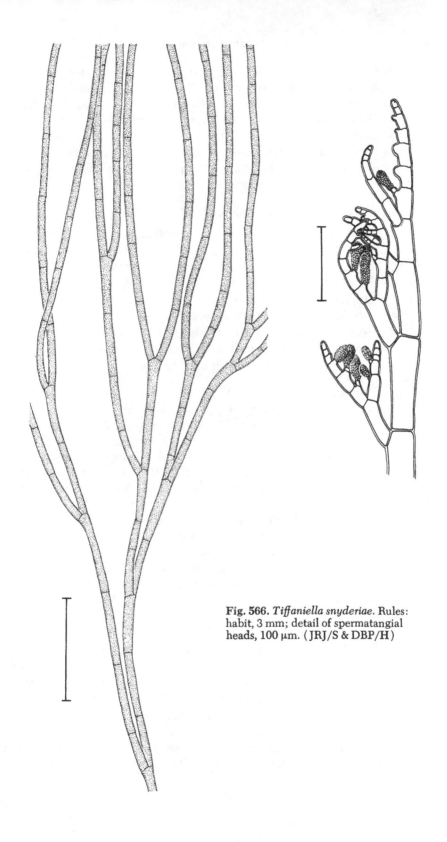

Fig. 566. *Tiffaniella snyderiae.* Rules:
habit, 3 mm; detail of spermatangial
heads, 100 µm. (JRJ/S & DBP/H)

100 μm diam., pedicellate, isolated or paired, borne basally; spermatangia in short, broad, oval heads, 40–60 μm tall, terminal on laterals; gonimoblasts near base, with 1–3 gonimolobes surrounded by common gelatinous envelope; involucre of unbranched or once-branched filaments of 6–8 cells, partly covering cystocarp, formed after fertilization.

Rare, partly within or entangled with small algae, subtidal, off Santa Rosa I. and Santa Catalina I., Calif.; and abundant on *Codium fragile*, I. Guadalupe (type locality), Mexico.

Ptilothamnionopsis Dixon 1971

Thalli filamentous, monosiphonous, noncorticated; differentiated into colorless, prostrate-creeping, much-branched portion and pigmented, erect, sparsely branched axes. Cells multinucleate. Tetrasporangia in clusters, tetrahedrally divided, terminal on short lateral filaments. Spermatangia in short, terminal, sometimes compound clusters. Procarps subterminal, of 3 pericentral cells, only 1 taking part in development of carposporophyte. Cystocarp globular to pyriform, formed of 2 principal and 1 or 2 subsidiary gonimolobes; nearly all cells becoming carposporangia, these in common gelatinous envelope; involucre lacking.

Ptilothamnionopsis lejolisea (Farl.) Dix.

Callithamnion lejolisea Farlow 1877: 244; Dawson 1962: 30; Hollenberg & Abbott 1966: 100. *Ptilothamnionopsis lejolisea* (Farl.) Dixon 1971: 58; Smith 1944: 183 (as *Rhodochorton amphiroae*).

Erect axes deep red, 1–2 mm tall, forming thick fur in articulations of erect corallines; filaments sparingly branched in upper portions, usually pectinate, tapering little toward apex; apical cells blunt; cells nearly as long as diam., 16–40 μm diam. in midsections; tetrasporangia if isolated or terminal in cluster 31 μm diam., 39 μm long, later ones smaller, 26–28 μm diam.; spermatangial heads 23–26 μm diam., 23–47 μm long, terminal or lateral; gonimoblasts spherical, 65–78 μm diam., low on thallus.

Locally common, particularly in articulations of *Calliarthron*, low intertidal to upper subtidal, Hope I., Br. Columbia, to San Diego, Calif. (type locality), and Baja Calif.

Lejolisea Bornet 1859

Thalli filamentous, noncorticated, with prostrate and erect portions; erect portions simple or irregularly branched, the cells uninucleate; prostrate portions with unicellular rhizoids bearing terminal, irregularly expanded disks. Tetrasporangia tetrahedrally divided, borne terminally or laterally, sessile or on short lateral branches; spermatangial heads short, borne on short laterals. Procarps on 3-celled, fertile axis, the subapical cell bearing 2 or 3 pericentral cells, 1 of these bearing persistent, 4-celled carpogonial branch. Fertile pericentral, after fertilization, giving rise to

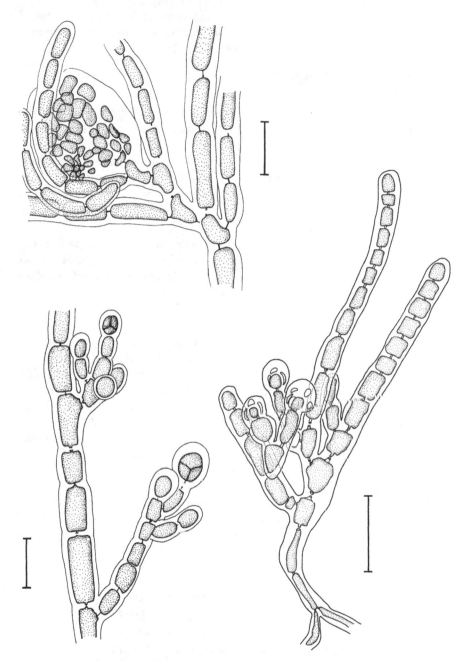

Fig. 567 (above left). *Ptilothamnion codicolum.* Rule: 50 µm. (DBP/A)   **Fig. 568**
(right & below). *Ptilothamnionopsis lejolisea*: right, habit of female thallus; below, habit
of tetrasporangial thallus. Rules: both 50 µm. (both DBP/A)

1 auxiliary cell, this forming small number of very large carposporangia. Gonimoblast surrounded by enclosed, filamentous involucre produced from cell below fertile cells and by internal divisions forming continuous envelope (like a pericarp) around gonimoblast.

Lejolisea pusilla Daws. & Neush.

Dawson & Neushul 1966: 181.

Thalli dark red, epiphytic on crustose corallines, tufted, about 1 mm tall, uniseriate throughout; basal parts of semiprostrate, spreading, stolon-like branches with frequent short ventral cells forming small, multicellular, rounded attachment disks; erect filaments irregularly and sparsely alternately branched; basal filaments and lower cells of erect filaments ± 12 μm diam., reduced above to ± 10 μm; tetrasporangia sessile on short protuberance from cell of erect branch, long-ovoid, 45–50 μm long, tetrahedrally divided. (After Dawson & Neushul.)

Found only once, subtidal (40 m) on *Lithothamnium*, Anacapa I., Calif.

Judging from the above description, there is no reason why this species was not placed in *Ptilothamnion*; but cystocarpic material is needed before an appropriate systematic position can be determined.

Bornetia Thuret 1855

Thalli tufted, of subdichotomous to unilaterally branched, uncorticated filaments. Cells elongate, cylindrical, multinucleate. Reproductive structures borne laterally on shortened branches. Tetrasporangia borne in clusters, sessile on adaxial or on both abaxial and adaxial surfaces of incurved, short, lateral branches, these terminated by 1 or 2 forcipate involucral cells. Spermatangia borne on similar branch systems, in oblong, densely branched heads placed in axils of dichotomy of sterile cells, involucrate. Fertile female axis of 5 or 6 cells. Procarp of 1 4-celled carpogonial branch and 1 or more pseudocarpogonia (resembling short, sterile, carpogonial branches) arising from sterile cells; fusion cell formed after fertilization and involving supporting cell, auxiliary cell, and lower cells of gonimoblast. Gonimoblasts with terminal, pyriform carpospores; involucre formed from basal and lower 2 cells of fertile axis, of large incurved cells surrounding and overtopping gonimoblast.

Bornetia californica Abb.

Abbott 1971a: 349. *Bornetia secundiflora sensu* Setchell 1901: 125.

Thalli dark carmine, tufted, saxicolous or epizoic; lowest cells horizontal and slightly modified as creeping filaments; erect filaments to 10 cm tall with few dichotomies; cells 350–500 μm (mostly 500 μm) diam. through midsections, 2 or more times as long, 2–3 mm long toward base; dichotomies frequently Y-shaped; tetrasporangia and spermatangial clusters on

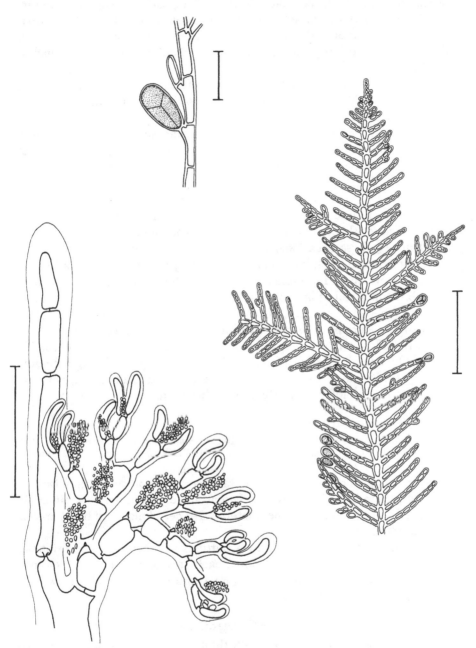

Fig. 569 (above left). *Lejolisea pusilla*. Rule: 50 μm. (EYD) **Fig. 570** (below left). *Bornetia californica*. Rule: 250 μm. (FT) **Fig. 571** (right). *Gymnothamnion elegans*. Rule: 2 mm. (DBP, after Okamura)

repeatedly branched, short laterals of 3 orders, each tetrasporangium sessile, 78–100 μm diam., attached to all 4 surfaces of branchlet cell; spermatangial heads oblong, densely branched, each spermatangium 2–3.5 μm diam.; fusion cell large and conspicuous; gonimoblasts large, with enclosing involucral cells.

Infrequent, saxicolous or epizoic, low intertidal to subtidal (to 6 m), C. Calif. from San Mateo Co. to San Luis Obispo Co. Type locality: Pescadero Beach (San Mateo Co.), Calif.

Very similar in external morphology to some specimens of *Griffithsia pacifica*, whose cells, however, are usually more barrel-shaped or are wider at the top than basally, and whose dichotomies are not Y-shaped.

Gymnothamnion J. Agardh 1892

Thalli small, uncorticated, usually epiphytic, differentiated into creeping portions with branched rhizoids and erect, pinnately branched tufts of filaments, these occasionally rebranching pinnately; cells uninucleate. Tetrasporangia tetrahedrally (rarely cruciately) divided, terminating laterals. Spermatangia in alternate clusters or whorled on laterals. Procarps on subterminal cell at apices of lateral branchlets, with 1 pericentral cell bearing 4-celled carpogonial branch and group of sterile cells; another pericentral cell remaining undivided or dividing into group of sterile cells, the terminal one developing a colorless hair; a sterile hair also usually formed by apical cell of fertile branchlet; connecting cell fusing with auxiliary cell after fertilization. Gonimoblast with 3–5 filamentous, compact to elongate gonimolobes; nearly all cells becoming carposporangia.

Gymnothamnion elegans (C. Ag.) J. Ag.

Callithamnion elegans C. Agardh 1828: 162. *Gymnothamnion elegans* (C. Ag.) J. Agardh 1892: 27; Feldmann-Mazoyer 1941: 354; Balakrishnan 1958: 138. *Plumaria ramosa* Yamada & Tanaka 1934: 346.

Thalli tufted, bright red, 1–1.5 cm tall; creeping portions with uniseriate rhizoids, these branched and modified at apices; erect portions pinnately and partly bipinnately branched; laterals 6–8 cells long, with blunt apices; reproduction as for genus.

Infrequent, on corallines or lower portions of other algae, occasionally saxicolous, low intertidal, Redondo Beach to Corona del Mar (Orange Co.), Calif. Widely distributed in tropics and subtropics; also New Zealand. Type locality: Algiers.

Ptilota C. Agardh 1817

Thalli large, feathery, several cells thick throughout; each primary axis bearing 2 primary laterals from alternate cells; 1 lateral frequently developing ahead of other, but each of indeterminate growth. All branches more or less compressed; ultimate branchlets resembling minute leaflets, often with serrate margins, each serration a potential fertile filament. Entire thallus corticated except at apices. Cells uninucleate. Tetrasporangia tetrahedrally divided, borne on margins of ultimate branchlets. Spermatangia rare; when present, in irregular clusters on determinate branchlets or arising from serrations of branchlets. Procarps borne in axils of ultimate branchlets, each procarp with 1 4-celled carpogonial branch and 2 or 3 groups of sterile cells, the terminal group bearing a colorless hair. Carpogonium after fertilization fusing with auxiliary cell through a connecting cell. Gonimoblast with fairly massive, filamentous involucre arising from lower vegetative portion of branch; nearly all cells becoming carposporangia.

Ptilota filicina J. Ag.

J. Agardh 1876: 76; Abbott 1972b: 264. *Ptilota plumosa* var. *filicina* Farlow 1875: 374 (in part). *P. filicina* (Farl.) J. Ag. 1885: 6 (in part). *P. tenuis* Kylin 1925: 60.

Thalli 10–35(60) cm tall, dark red, frequently drying to black, with 1 major axis or with several branches strongly developed; each order of branches with opposite pairs of branchlets, the branchlets alike in tetrasporangial thalli, unlike in spermatangial and cystocarpic thalli; ultimate branchlets leaflike, the margins sharply serrate on each side, the apices acute; tetrasporangia forming first on adaxial side, then involving entire leaflet, terminal on filaments, surrounded by sterile filaments; spermatangial clusters many, terminal on filaments arising from serrations of determinate branchlets; cystocarps surmounting apex of determinate branchlets, heavily covered by involucral filaments arising from lower portions of branchlet.

Locally abundant, saxicolous, low intertidal, Alaska to Punta Baja, Baja Calif. Also known from Japan. Type locality: San Francisco, Calif.

Neoptilota Kylin 1956

Thalli differing from those of *Ptilota* principally in having 1 of each pair of laterals remaining determinate and never bearing reproductive structures; in fertile thalli, every alternate branch thus sterile and opposite fertile branch. Tetrasporangia developing from short, indeterminate branchlets, in small clusters without sterile filaments; when mature, branchlet nearly entirely converted to bearing spores. Spermatangia rare, in continuous sori on adaxial side of branchlet. Cystocarps 1 to several, developed simultaneously on indeterminate branchlet.

1. Ultimate branchlets with smooth margins . . . . *N. hypnoides*
1. Ultimate branchlets with 1 or both margins scalloped, minutely or sharply serrate 2
   2. Abaxial margin strongly serrate; plants usually epiphytic on corallines . . . . *N. densa*
   2. Either abaxial or adaxial margins weakly serrate, frequently appearing smooth; plants usually epiphytic on Laminariales . . . . . . *N. californica*

Neoptilota californica (Harv.) Kyl.

Ptilota californica Harvey 1853: 222; Smith 1944: 332. *Neoptilota californica* (Harv.)
Kylin 1956: 392; Abbott 1972b: 263.

Thalli epiphytic, to 85 cm tall, with strongly developed and compressed
main axes and major laterals, dark red to reddish-brown, alternately
branched; branch apices drooping; branchlets of higher orders crowded,
lying close to axes, when dried giving thallus furry appearance; determinate branchlets simple, opposite each pinnately divided branchlet that
becomes fertile; ultimate branchlets with minute serrations on some abaxial margins, most margins appearing smooth; reproductive structures
as for genus.

Infrequent, epiphytic on *Laminaria sinclairii* or on some other Laminariales, low intertidal to shallow subtidal, Br. Columbia to San Luis
Obispo Co., Calif. Also known from Japan. Type locality: "California borealis rossica," probably near Fort Ross (Sonoma Co.), Calif.

The previously reported (Smith, 1944) wider distribution of this species
is not borne out by our examination of specimens. Most specimens have
proved to be *Ptilota filicina*; the gross habit of these thalli reminds one of
young, developing *Neoptilota densa*, but the branching resembles that of
P. filicina.

Neoptilota densa (C. Ag.) Kyl.

Ptilota densa C. Agardh 1822a: 387; Smith 1944: 333. *Neoptilota densa* (C. Ag.) Kylin
1956: 393.

Thalli coarse, dark red, to 30 cm tall; main axes 1–3 mm wide, with
densely fringed branches, irregularly branched; branchlets falcate, smooth
and concave on adaxial margin, strongly serrate and convex on abaxial
margin; tetrasporangia borne on adaxial side of strongly reflexed, abaxially serrate branchlets; procarps 1 to several on fertile branchlets; usually
2 gonimoblasts developing simultaneously and crowding branches.

Locally abundant, usually on articulated corallines, low intertidal to
subtidal (to 15 m), Tomales Bay (Marin Co.), Calif., to Bahía Rosario,
Baja Calif. Type locality: Monterey, Calif.

Neoptilota hypnoides (Harv.) Kyl.

Ptilota hypnoides Harvey 1833b: 164; 1853: 220; Smith 1944: 332. *Neoptilota hypnoides* (Harv.) Kylin 1956: 393; Abbott 1972b: 263.

Thalli erect, dark brownish-red, 10–25 cm tall, regularly branched, plumose, axes 1 to several, the central axes not strongly developed; branches
not crowded; fertile branchlets opposite spinelike (determinate) branch
and alternating on either side of axis; branchlets smooth, curving because
of stronger development of abaxial over adaxial margin; reproductive
structures as for genus.

Occasional, saxicolous or epiphytic on articulated corallines, low inter-

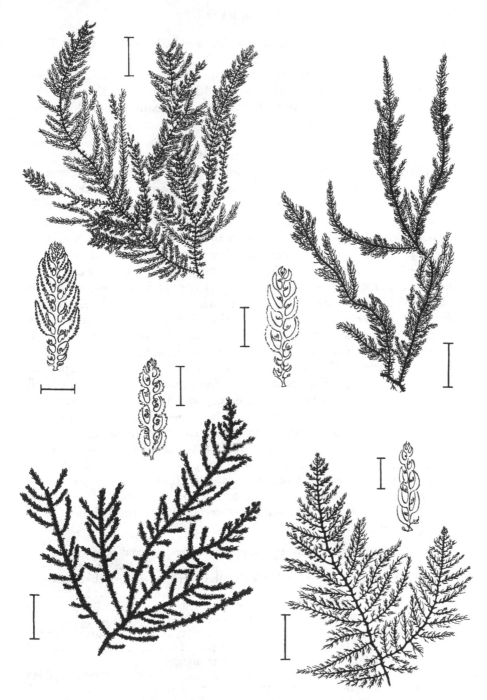

Fig. 572 (two above left). *Ptilota filicina*. Rules: habit, 2 cm; detail, 3 mm. **Fig. 573** (two above right). *Neoptilota californica*. Rules: habit, 2 cm; detail, 3 mm. **Fig. 574** (two below left). *N. densa*. Rules: habit, 2 cm; detail, 3 mm. **Fig. 575** (two below right). *N. hypnoides*. Rules: habit, 2 cm; detail, 3 mm. (all drawings NLN)

tidal to subtidal (to 10 m), Alaska to San Luis Obispo Co., Calif. Type locality: Monterey, Calif.

Northern specimens attributed to this species should be carefully studied, since *Neoptilota asplenioides*, reported from the N. Pacific, is similar except for serrate edges on the adaxial surface of leaflets, smooth abaxial surfaces, and longer thalli.

Family DELESSERIACEAE

Thalli flattened or compressed, mostly foliaceous, usually with conspicuous midribs, lateral veins, or both, or with network of macroscopic or microscopic veins; growth from single apical cell, some derivative cells showing intercalary divisions, all producing distichous internal laterals, these variously further divided; or growth from secondarily derived marginal apical cells, all cells joined and fused to form pseudoparenchymatous sheet 1 or more cells thick. Cells mostly uninucleate. Location and manner of branching of taxonomic importance. Reproductive structures embedded in blade surfaces, central or marginal, or in bladelets or proliferations, variously distributed on thallus. Tetrasporangia tetrahedrally divided, in sori mostly rounded or elongate and of various sizes. Spermatangia in sori, usually scattered over blade or close to margins. Procarps formed on pericentral cells with 1 or 2 groups of sterile cells associated with 4-celled carpogonial branch. After fertilization auxiliary cell cut off by pericentral cell and fusing with carpogonium, followed by fusion of various basal cells to form placenta, included in cystocarp; all, some, or only terminal cells of cystocarp producing carposporangia; large parenchymatous pericarp formed around spore mass, with conspicuous ostiole.

The taxonomic limits of some of the Pacific Coast genera in this family, especially *Botryoglossum*, *Cryptopleura*, and *Hymenena*, are poorly understood; nor are there clear limits to the species within these genera. They must be critically reevaluated, together with representatives from other parts of the world.

1. Thalli narrow, vegetatively appearing polysiphonous and compressed
.... *Platysiphonia* (p. 642)
1. Thalli foliose, ribbonlike, or bladelike, the surface cells bricklike, not polysiphonous 2
   2. Thalli delicate, filmy, with prominent midrib..................... 3
   2. Thalli not delicate, firm, the midrib inconspicuous or lacking.........11
3. Thalli branching from midrib .... *Delesseria* (p. 640)
3. Thalli branching from margins or from stipes........................ 4
   4. All terminal cells of third order reaching margin.................
   .... *Branchioglossum* (p. 634)
   4. Only some terminal cells of third order reaching margin............ 5
5. Branching sympodial, the blades on each division appearing pectinate
.... *Cumathamnion* (p. 640)
5. Branching monopodial, the divisions flabellate...................... 6

Branchioglossum Kylin 1924

Thalli membranous, ribbonlike, erect, of simple or lobed, branched or unbranched blades, or with secondary blades formed from margins. Blades with wide percurrent midrib several cells thick, with or without lateral veins, monostromatic except at midrib. Cells of thallus with 3 orders of branching; each axial cell bearing opposed pair of filaments (the cell rows of second order), each cell of these filaments producing cell rows of third order abaxially, the terminal cells of latter rows all reaching margin; intercalary divisions within cell rows not occurring. Tetrasporangial sori near margins, oval or linear, at first isolated, then confluent; sporangia tetrahedrally divided. Spermatangia formed in irregular, narrow sori on either side of midrib. Carpogonial branches formed on each side of blade, each associated with 2 groups of sterile cells. Mature cystocarps 1 or 2 per blade, borne along midrib, bulging toward 1 side of thallus, each with small, simple ostiole.

Branchioglossum undulatum Daws.

Dawson 1949b: 18; 1962: 76.

Thalli 1–2 cm tall, arising from small, spreading, discoid holdfasts; of

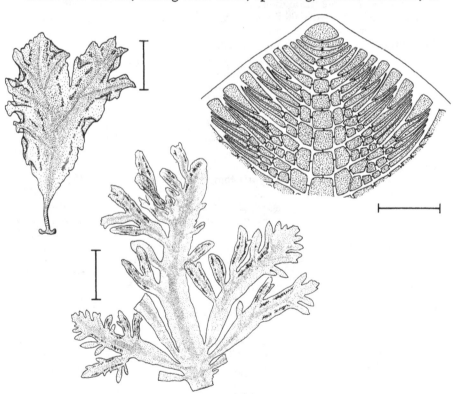

Fig. 576 (above left). *Branchioglossum undulatum.* Rule: 5 mm. (CS) **Fig. 577** (middle & right). *B. woodii.* Rules: habit, 5 mm; detail, 20 μm. (both JRJ/S)

1 or several rose pink, simple, oblanceolate blades, these in upper portions of several broad lobes with undulate to slightly ruffled margins, or branched basally into 1–3 lobes; cystocarps 1 per blade, median on midrib.

Saxicolous, low intertidal to subtidal at Carmel, Calif., subtidal (10 m) off Santa Rosa I. (type locality), and to San Diego, Calif.; also Gulf of Calif. Frequent in S. part of range, rare in north.

Branchioglossum woodii (J. Ag.) Kyl.

Delesseria woodii J. Agardh 1872: 54. *Branchioglossum woodii* (J. Ag.) Kylin 1924: 8; Smith 1944: 335. *B. macdougalii* Gardner 1927a: 103.

Thalli 1–2.5(7) cm tall, light pink, with percurrent axes profusely and suboppositely branched in upper portions; primary and secondary blades linear, 1–3 mm broad, with short terminal lobes, abruptly terminating in subacute to acute apices, the margins mostly smooth; reproduction as for genus, several cystocarps per blade.

Infrequent, saxicolous, low intertidal to subtidal (to 41 m), Vancouver I., Br. Columbia (type locality) and Ft. Bragg (Mendocino Co.), Monterey Peninsula, Santa Barbara, and San Diego, Calif., to Baja Calif. and Gulf of Calif.

Membranoptera Stackhouse 1809

Thalli erect or procumbent from small, conical holdfasts, alternately or subdichotomously divided into broad or linear blades. Blades with percurrent midrib more than 1 cell thick; portions lateral to midrib monostromatic and with or without diagonal parallel veins, occasionally with small proliferous blades from midrib and lateral veins. Internal cells with 3 orders of branching; terminal cells of third order not reaching margin; intercalary divisions within cell rows not occurring. Tetrasporangia tetrahedrally divided, in sori at blade apices, or scattered on either side of midrib and main lateral veins. Spermatangia in sori at blade apices, or scattered along midrib in upper halves of blades. Cystocarps borne on midrib, surrounded by hemispherical pericarp; carposporangia in chains.

1. Blades filiform, with inconspicuous midribs . . . .*M. weeksiae*
1. Blades broad, with conspicuous midribs 2
   2. Branches congested, the divisions nearly as broad as long. . . . .
. . . . *M. multiramosa*
   2. Branches not congested, the divisions linear, heavily proliferous. . .
. . . . *M. dimorpha*

Membranoptera dimorpha Gardn.

Gardner 1926: 211; Doty 1947b: 192; Scagel 1957: 218.

Thalli dark wine red, procumbent in groups, the blades linear-lanceolate, sparsely branching from margins; midribs conspicuous, traversing second- and third-order branches in upper portions; diagonal, lateral, and

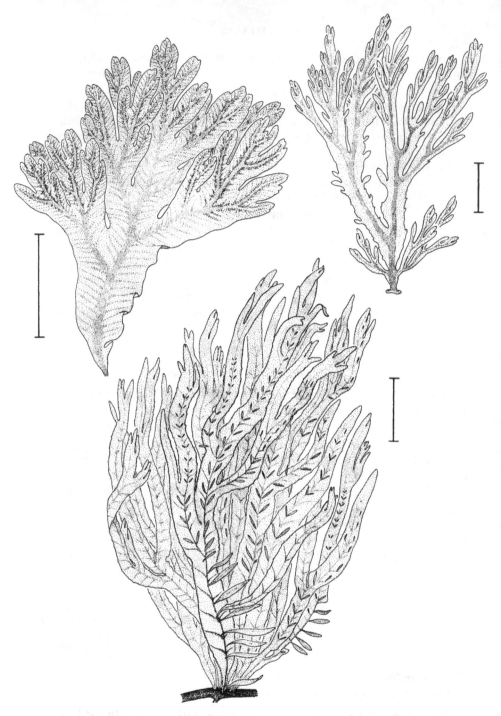

Fig. 578 (below). *Membranoptera dimorpha*. Rule: 2 cm. (DBP) **Fig. 579** (above left). *M. multiramosa*. Rule: 1 cm. (JRJ/S) **Fig. 580** (above right). *M. weeksiae*. Rule: 2 mm. (JRJ/S)

microscopic veins lying parallel throughout thalli; blades 8–15(25) cm tall, 5–8 mm wide, with smooth margins; midribs and laterals heavily proliferous, with small bladelets standing alternately at right angles to blades, these persisting on older plants otherwise eroded to midribs; bladelets oval on spermatangial and tetrasporangial plants, rounded to cordate on cystocarpic plants; tetrasporangia in sori on either side of midribs and larger lateral veins; spermatangia on main blades in sori similarly located; cystocarps basal to final dichotomies of lateral veins, or terminal on lateral veins in small bladelets, usually 1 or 2 per segment, with fimbriate ostioles.

Saxicolous or epiphytic: in drift at Amchitka I., Alaska, low intertidal from Queen Charlotte Is., Br. Columbia, to Carmel, Calif.; common in S. Ore. to Humboldt Bay, Calif., occasional to northward. Type locality: Neah Bay, Wash.

Membranoptera multiramosa Gardn.

Gardner 1926: 209; Smith 1944: 337. *Membranoptera edentata* Kylin 1941: 30.

Thalli deep red, 3–5 cm tall, alternately and closely flabellately branched from margin of blades, with 3–5 orders of branching; occasionally with small, proliferous blades from midribs; blades with conspicuous diagonal parallel veins; margins smooth; cystocarps with fimbriate ostioles.

Occasional, on rocks, low intertidal in shaded crannies to subtidal (to 10 m), Wash. to Channel Is., Calif. Type locality: Moss Beach (San Mateo Co.), Calif.

Membranoptera weeksiae S. & G.

Setchell & Gardner 1926a: 209; Smith 1944: 337.

Thalli ribbonlike, repeatedly branched, 3–5(9) cm tall in small, soft bushes, bright rose red, branching of 4–6 orders, usually alternate but occasionally subopposite or unilateral; with inconspicuous midribs and no lateral veins, the divisions remaining 0.5–1.5 mm broad throughout length if not eroded; margins smooth.

Epiphytic (on *Desmarestia latifrons* especially, but also on other algae) or saxicolous, low intertidal to subtidal (to 30 m), Wash. to San Diego, Calif.; especially common in spring near Pacific Grove, Calif. (type locality).

Holmesia J. Agardh 1890

Thalli erect, flattened, thick, bladelike, arising from compressed stipes attached to creeping, partly stoloniferous holdfasts. Branches occasional from lower portions, these with primary midrib only; lateral veins lacking; upper portions lacking midribs and veins. Reproductive structures on small, fertile, fusiform leaflets, these produced on surface of upper half of blades and giving that surface a rough, papillate aspect. Tetrasporangia cruciately divided. Cystocarps with gametophytic tissue stretching between inner pericarp wall and gonimoblast. Gonimoblast with large basal fusion cell; carposporangia terminal.

Holmesia californica (Daws.) Daws.

Loranthophycus californica Dawson 1944c: 655. *Holmesia californica* (Daws.) Daws. 1962: 80; Hollenberg & Abbott 1966: 103.

Thalli dark red, erect, membranous when young, when older and reproductive firm and leathery, 8–15(20) cm tall, the upper portions 4–7 cm broad, lobed once or twice; margins undulate; axes polystromatic throughout; tetrasporangia and spermatangia on fusiform papillae, these divided 1 or 2 times; cystocarps on broad, undivided papillae, to 1 mm diam., with beaked pericarp, this sometimes elongated to twice length of spore mass.

Saxicolous, occasional, Pine I., Br. Columbia, and N. Wash. to Pta. Santo Tomás, Baja Calif.; frequent subtidally (to 10–15 m) off several localities in Humboldt and Mendocino Cos., and from Monterey Peninsula to Channel Is. and Pt. Loma (San Diego Co.; type locality), Calif.

Delesseria Lamouroux 1813

Blades linear, arising from discoid holdfasts; axes with polystromatic, percurrent midribs; portions lateral to midrib monostromatic, with microscopic lateral veins. Axes freely branched 3 or 4 times from midribs; monostromatic portions with intercalary divisions in cell rows of second order, the terminal cells reaching margins. Tetrasporangial sori along midribs. Spermatangial sori mottled and irregular, distributed over thalli. Cystocarps borne on fertile leaflets along midribs, with conspicuous, ostiolate pericarps.

Delesseria decipiens J. Ag.

J. Agardh 1872: 58; Smith 1944: 338. *Apoglossum decipiens* J. Ag. 1898: 194.

Thalli pendant in clusters 10–25(61) cm long, straw-colored to dark purplish-red; branching usually alternate to 4 or 5 orders; axis 8–12 mm broad, frequently eroded, especially basally; monostromatic portions of axis and branches with inconspicuous veins diagonal to midrib; reproduction as for genus.

Common spring annual in C. Calif., saxicolous, low intertidal to subtidal (to 5 m), Alaska to San Luis Obispo Co., Calif. Type locality: Vancouver I., Br. Columbia.

Cumathamnion Wynne & Daniels 1966

Blades numerous, forming from sympodially developed axes, polystromatic, with prominent midribs and lateral veins; midrib with internal rhizoidal filaments; secondary blades produced along midrib; intercalary divisions present in cell rows of second order. Tetrasporangia in sori on unspecialized blades, tetrahedrally divided. Spermatangia in sori on both

Fig. 581 (center). *Holmesia californica.*
Rule: 3 cm. (DBP) **Fig. 582**
(below right). *Delesseria decipiens.*
Rule: 1 cm. (JRJ/S) **Fig. 583** (above
left). *Cumathamnion sympodophyllum.*
Rule: 3 cm. (SM)

surfaces of blade. Procarps on specialized fertile blades, developed from abaxial pericentral cell. Cystocarps 1 per blade, 500–800 µm diam., with beaked pericarp; carposporangia in chains.

Cumathamnion sympodophyllum Wynne & Dan.

Wynne & Daniels 1966: 13.

Thalli deep red, saxicolous, in bushy, robust clumps to 20 cm tall, with cartilaginous holdfasts; blade apices acute when young, obtuse when mature; reproduction as for genus.

Rare, low intertidal, Vancouver I., Br. Columbia; and on rocks exposed to heavy surf, Mendocino Bay, Calif. (type locality).

Platysiphonia Børgesen 1931

Thalli with prostrate, creeping axes bearing at frequent intervals erect, free branches, these alternate or unilateral; both prostrate and erect portions flattened, polysiphonous, with 4 pericentral cells in each segment. Prostrate axis with multicellular rhizoids arising from 1 or more marginal cells. Each lateral pericentral cell giving rise to pair of flanking cells, these either remaining undivided or cutting off 2 derivatives in row parallel to the long axis; branches arising endogenously. Tetrasporangia on short, stichidium-like branchlets, the several successive segments each producing an opposed pair of tetrasporangia, these cruciately or tetrahedrally divided. Spermatangia on short, flattened branchlets, near apex of main branches and covering flattened faces of branchlet. Cystocarps adaxial on ultimate branches, broad at base, borne singly; pericarp conspicuous; carposporangia terminal.

1. Erect axes with more than 1 cell row between pericentral cells and margin . . . . *P. decumbens*
1. Erect axes with only 1 cell row between pericentral cells and margin. 2
   2. Branches mostly alternate; plants saxicolous. . . . . *P. clevelandii*
   2. Branches mostly unilateral; plants epiphytic on *Codium* . . . . *P. parva*

Platysiphonia clevelandii (Farl.) Papenf.

Taenioma clevelandii Farlow 1877: 236; Smith 1944: 371. *Platysiphonia clevelandii* (Farl.) Papenfuss 1944: 206. *Platysiphonia* sp. Segawa 1949: 161.

Thalli 5–8 cm tall, deep pinkish-red; erect portions of main axes 200–350 µm broad, with 2 opposing, lateral pericentral cells each cutting off 1 row of corticating cells from every segment; rhizoids of prostrate portion multicellular, growing singly from marginal cells, each forming distally a flaring disk; tetrasporangia tetrahedrally divided; spermatangia and cystocarps as for genus.

Occasional, on rocks, low intertidal to subtidal (to 41 m), Puget Sd., Wash., and Pigeon Pt. (San Mateo Co.), Monterey Peninsula, San Pedro,

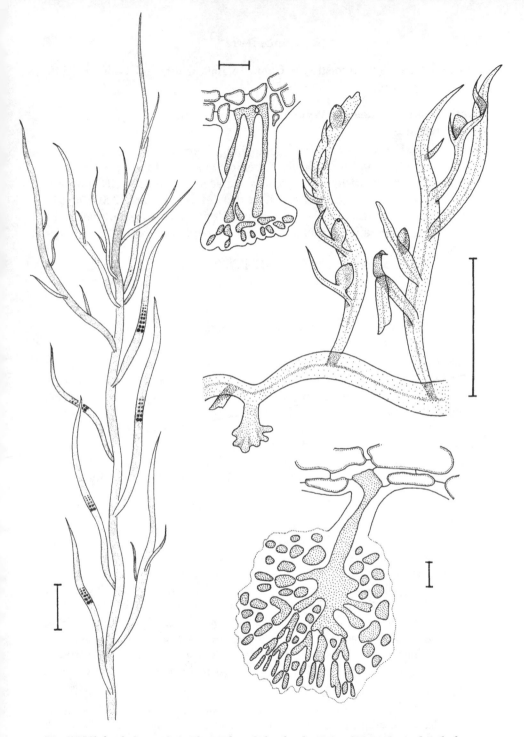

Fig. 584 (left & below right). *Platysiphonia clevelandii*. Rules: habit, 1 mm; detail of rhizoid formation, 100 μm. (JRJ/S & CS)   **Fig. 585** (two above right). *P. decumbens*. Rules: habit, 1 mm; detail of rhizoid formation, 100 μm. (SM & CS)

and San Diego (type locality), Calif., to Baja Calif. and Gulf of Calif. Also C. Japan.

Platysiphonia decumbens Wynne

Wynne 1969b: 190.

Thalli dark rose, 1–2 cm tall; branching frequent, the ends of some branches arching and secondarily attaching to substratum; prostrate portion with rhizoids arising from several adjacent marginal cells, forming bundle of cells terminating in modified disk; erect portions (200)350–450 µm wide, distinguished by 4 flanking cell derivatives on each margin of segment; tetrasporangia cruciately divided; spermatangia and cystocarps as for genus.

Occasional, on rocks, subtidal (3–15 m), Puget Sd.; Whidbey I., Wash. (type locality), and in Calif. at Santa Cruz, Monterey Peninsula, Santa Catalina I., and La Jolla.

Platysiphonia parva Silva & Cleary

Silva & Cleary 1954: 251.

Thalli deep rose, epiphytic on outer utricles of *Codium fragile*, to 1 cm tall, delicate; main axes 100–150 µm wide, the branching somewhat unilateral and sparse; prostrate portion with rhizoids arising from single marginal cells; rhizoids frequently little developed, but mature ones digitate and multicellular, forming small paddles; tetrasporangia tetrahedrally divided; spermatangia and cystocarps as for genus.

Rare but locally abundant, epiphytic, subtidal (25–40 m) at Anacapa I., Calif., and intertidal at I. Guadalupe, Baja Calif. (type locality).

Phycodrys Kützing 1843

Thalli bladelike, with discoid holdfasts, these developing entangled stolons. Blades elliptical to broadly lanceolate, with conspicuous midrib and paired lateral veins; axes monostromatic except for midrib and veins; margins smooth to strongly dentate; blades simple, lobed in upper portions, or producing bladelets along margins; older blades often with all of lower portion but midrib eroded away, and with new bladelets developing on old midrib. Tetrasporangia in rounded sori on proliferations from margin, or between lateral veins, or in both places, tetrahedrally divided. Spermatangia in continuous sori just within blade margin. Cystocarps scattered over blade, usually few and large, each with ostiole; carposporangia in chains.

1. Blades 1 or more, arising from creeping or entangled stolon......*P. setchellii*
1. Blades borne singly on discoid holdfast 2
   2. Primary blades membranous, with secondary bladelets on margins
     ....*P. isabelliae*
   2. Primary blades crisp, with rhizoids on margins............*P. profunda*

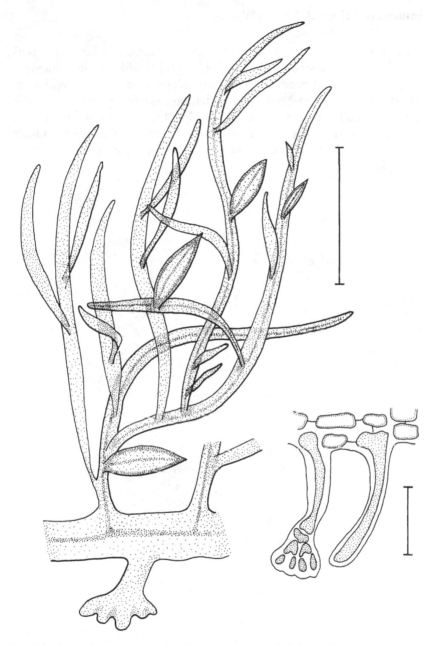

Fig. 586. *Platysiphonia parva*. Rules: habit, 1 mm; detail of rhizoid formation, 50 μm.
(SM & CS)

Phycodrys isabelliae R. Norr. & Wynne

R. Norris & Wynne 1968: 144.

Plants erect, 2–7 cm tall, arising from small discoid holdfasts, rose pink; blades simple or branched, 0.5–2 cm wide, oval to oblanceolate, membranous, with percurrent midrib and broken, irregular, opposite lateral veins; margins smooth or sometimes feebly dentate, rarely producing rhizoids; secondary bladelets to 1 cm long, 0.5 cm wide, the margins undulate; tetrasporangia in oval to irregular sori between lateral veins; spermatangial sori on each side of midrib; cystocarps to 1 mm diam., irregularly scattered over thallus.

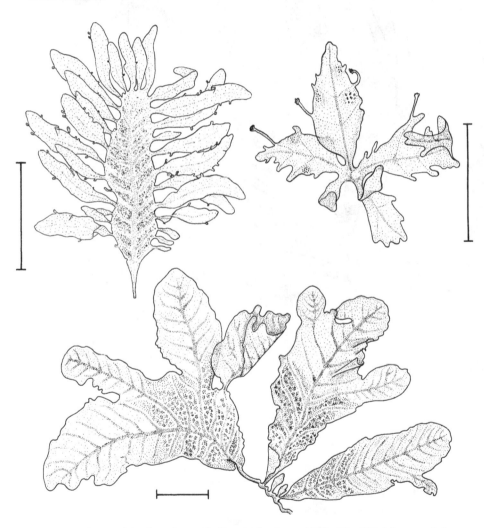

Fig. 587 (above left). *Phycodrys isabelliae*. Rule: 2 cm. (CS)   **Fig. 588** (above right). *P. profunda*. Rule: 2 cm. (SM)   **Fig. 589** (below). *P. setchellii*. Rule: 2 cm. (SM)

Occasional, on rocks, subtidal (to 20 m), Puget Sound, Wash. (type locality), Cape Arago, Ore., and Monterey Peninsula and Santa Cruz I., Calif.

Phycodrys profunda Daws.

Dawson 1962: 90.

Thalli rose red, decumbent, the bases entangled, irregular and somewhat creeping; stipe producing usually 1 major blade and several small ones; primary blade linear, 2–3 cm long; margins irregular, frequently dentate; midrib prominent, percurrent, with weak opposite veins; margins occasionally producing secondary, rudimentary bladelets in association with long, delicate, unicellular rhizoids; tetrasporangia in small sori between lateral veins; sexual reproduction unknown.

Frequent on rocks, low intertidal to subtidal (to 60 m), Monterey Peninsula, Santa Cruz I., Santa Catalina I. (type locality), and San Diego, Calif., and I. Guadalupe, Baja Calif.

Both *P. isabelliae* and *P. profunda* are probably more common than herbarium records would show, but until recently they have not been distinguished clearly from *Anisocladella*—which the latter species especially resembles—or from young *P. setchellii*. The occurrence of pinnate secondary blades in *P. isabelliae* is not shared with the other species; and the regular occurrence of marginal rhizoids in *P. profunda* seems to characterize that species.

Phycodrys setchellii Skottsb.

Skottsberg 1922: 433; Smith 1944: 342.

Plants to 20 cm tall, dark pink to brownish-red, drying darker; divisions to 4 cm wide, leaflike, dichotomously divided from base, the upper portions of elongate lobes with rounded apices, conspicuous percurrent midribs, and lateral veins; margins crisp and undulate; lateral blades occasionally produced as proliferations; tetrasporangial sori darker than remainder of blades, usually between lateral veins, occasionally marginal and on irregular marginal proliferations; sexual reproduction as for genus.

Occasional to frequent, on rocks, low intertidal to subtidal (to 40 m), Br. Columbia and Ore. to Baja Calif. and Gulf of Calif. Type locality: San Francisco, Calif.

Low-intertidal plants are usually short, with linear blades; subtidal plants are larger, the segments broader. The largest plants are from N. and C. Calif.

Erythroglossum J. Agardh 1898

Thalli erect, with very short, winged stipes grading into blades and branches. Blades flabellate, irregularly formed from margins but frequently divided to midrib; axes monostromatic with percurrent, polystromatic midrib; lateral veins lacking; margins toothed, especially in terminal portions. Tetrasporangia in row of sori parallel to margins. Spermatangia in elongate, submarginal sori. Cystocarps unknown.

Erythroglossum californicum (J. Ag.) J. Ag.

Delesseria californica J. Agardh 1884: 69. *Erythroglossum californicum* (J. Ag.) J. Ag. 1898: 176; Kylin 1924: 32; 1941: 31. *E. obcordatum* Dawson & Neushul 1966: 183.

Thalli tufted, to 10–12 cm tall, rose red and darker, arising from compacted bases; divisions erect, ribbonlike, linear, 2–3(5) mm wide, with percurrent, polystromatic midribs; divisions frequently eroded laterally, irregularly toothed where not eroded; apical margins irregular; apical cell inconspicuous; tetrasporangia at first in discrete, oval sori lying within margins and in row parallel to them, later becoming confluent in linear series. Spermatangia and cystocarps as for genus.

Rare, low intertidal nearly buried in sand at Santa Barbara (type locality), and subtidal off Anacapa I. and La Jolla, Calif.

Specimens reported from Ore. by Doty (1947b: 192) and from Pacific Grove by Smith (1944: 339) are to be identified with *Anisocladella pacifica*; those from I. Magdalena, Baja Calif., by Dawson (1962: 93), with *Sorella delicatula*.

Sorella Hollenberg 1943

Thalli small, delicate, flattened, erect but with basal branches somewhat prostrate, attached by hapterous branchlets from margins; axes monostromatic except for inconspicuous midrib, lacking lateral veins; branching from margins of about 4 orders; intercalary divisions in primary cell row. Tetrasporangial sori oval, occurring singly in center of branches near apices. Spermatangial sori oval to oblong, occurring singly or in pairs near branch apices. Cystocarps on 1 side of midrib.

Sorella delicatula (Gardn.) Hollenb.

Erythroglossum delicatula Gardner 1926: 208. *Sorella delicatula* (Gardn.) Hollenberg 1943a: 577; Dawson 1962: 83. *S. delicatula* var. *californica* Hollenb. 1943a: 577. *E. divaricata* Setchell & Gardner 1926b: 207; Smith 1944: 340. *S. divaricata* (S. & G.) Hollenb. 1943a: 578.

Thalli delicate, ribbonlike, dark rose, dichotomously and flabellately branched, the intervals usually long; flattened divisions 200–500(600) μm broad, the margins smooth; spermatangial sori in oval patches on ultimate divisions of branches; tetrasporangia and cystocarps as for genus.

Occasional, saxicolous, low intertidal to subtidal (to 30 m), El Jarro Pt. (Santa Cruz Co.) to Corona del Mar (Orange Co.), Calif., including Channel Is., to Gulf of Calif. Also known from Japan. Type locality: San Pedro, Calif.

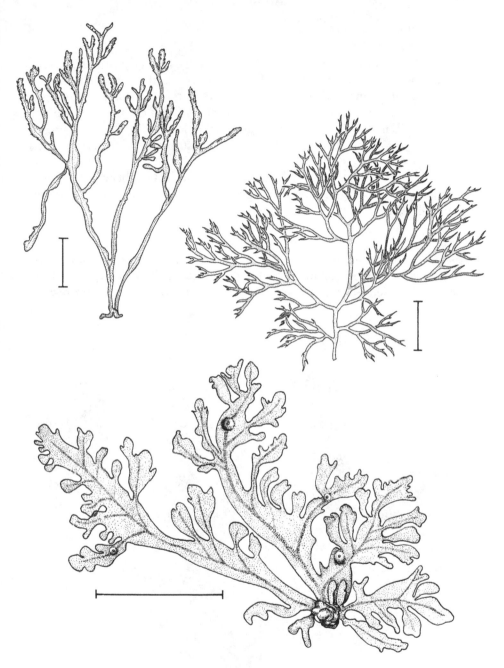

Fig. 590 (above left). *Erythroglossum californicum*. Rule: 2 cm. (SM) **Fig. 591** (above right). *Sorella delicatula*. Rule: 5 mm. (JRJ/S) **Fig. 592** (below). *S. pinnata*. Rule: 5 mm. (SM)

Sorella pinnata Hollenb.

Hollenberg 1943a: 578; Dawson 1962: 84.

Thalli semiprostrate to erect, 1–2(7) cm tall, pinnately branched, rose pink, monostromatic except at midrib and lower portions; branches membranous, lanceolate, 0.5–3.5 mm wide, the margins dentate, the midrib inconspicuous; tetrasporangial sori 300–400(800) μm diam., oval, solitary over midrib near apices; spermatangial sori oblong, 150–200 μm diam. on either side of midrib, occasionally in pairs, seriate along branch; cystocarps as for genus.

Occasional, epiphytic or saxicolous, low intertidal to subtidal (to 40 m), Channel Is. and Laguna Beach (type locality), Calif., to Baja Calif. and Gulf of Calif.

Polyneura Kylin 1924

Thalli erect, with 1 or more blades attached by discoid holdfasts or by flattened, irregularly divided, ribbonlike branches. Blades stipitate, ovate to broadly obcuneate, entire and monostromatic when young, later becom-

Fig. 593. *Polyneura latissima.* Rule: 2 cm. (JRJ/S)

ing lobed, incised, and polystromatic throughout, without midrib but with conspicuous, anastomosing veins. Tetrasporangia tetrahedrally divided, in sori scattered over entire blade, sometimes marginal. Spermatangial sori scattered over entire blade. Cystocarps with carposporangia in chains, with ostiolate pericarp.

Polyneura latissima (Harv.) Kyl.

Hymenena latissima Harvey 1862: 170. *Polyneura latissima* (Harv.) Kylin 1924: 37; Smith 1944: 341.

Blades 10–15(45) cm tall, sometimes as wide, deep pink to lake red, entire when young, later lobed unevenly, sometimes with thickened stipe; anastomosing veins forming conspicuous reticulum covering all but blade margins; reproduction as for genus.

Frequent on rocks, mean tide to subtidal (27 m), Esquimalt, Vancouver I., Br. Columbia (type locality), to Baja Calif.; one of commonest subtidal species.

Polyneurella Dawson 1944

Blade entire, from branched stipe, usually with 1 conspicuous blade and below it several smaller, subsidiary blades somewhat pinnately arranged. Blades monostromatic, with many microscopic veins more or less parallel but basally indistinctly anastomosing. Tetrasporangial sori small, scattered over blades. Sexual structures unknown.

Polyneurella hancockii Daws.

Dawson 1944a: 322.

Main blade rose red, ovate to oval, 2–6.5 cm tall, the subsidiary blades 1–10 mm long, short-stipitate, the stipes terete to compressed; reproduction as for genus.

Infrequent, subtidal in Carmel Submarine Canyon (15 m), and I. Angel de la Guarda, Gulf of Calif. (type locality, 22–40 m).

A poorly known taxon, owing to the limited number of collections.

Anisocladella Skottsberg 1923

Blades small, becoming erect from very narrow, flattened, branched, ribbonlike creeping portion; blades more or less ligulate, undivided, occasionally with proliferous bladelets at margins, with percurrent midrib and with unbranched parallel veins slightly diagonal to midrib; axes monostromatic except for midrib and veins; margins smooth to strongly dentate. Apex of blade with each cell of axial filament bearing 2 opposite lateral filaments, 1 more strongly developed; successive pairs with long and short filaments alternating. Tetrasporangia tetrahedrally divided, in sori between lateral veins. Spermatangial sori paired. Cystocarps with hemispherical pericarp, borne midway between midrib and margin.

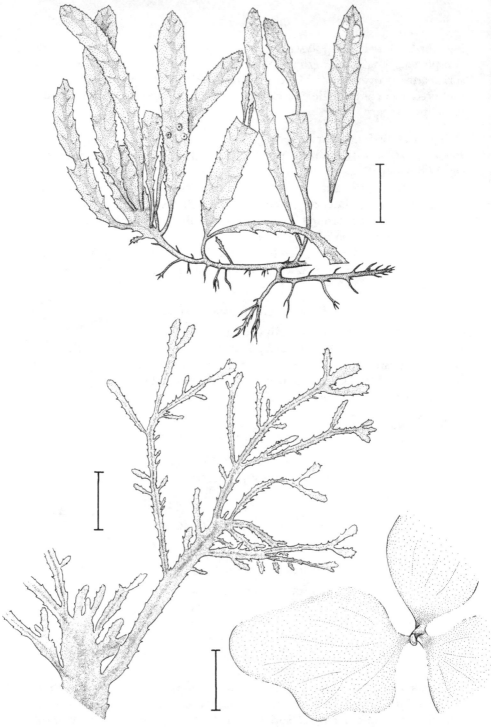

Fig. 594 (below right). *Polyneurella hancockii*. Rule: 5 mm. (SM) **Fig. 595** (above). *Anisocladella pacifica*. Rule: 1 cm. (DBP/H) **Fig. 596** (below left). *Nienburgia andersoniana*. Rule: 1 cm. (JRJ/S)

Anisocladella pacifica Kyl.

Kylin 1941: 31; Smith 1944: 343.

Blades erect, usually 1.5–3(5) cm tall, deep rose pink, forming entangled masses frequently buried in fine sand; blades linear, 2.5–5 mm broad, basally narrowed to short stipe, with opposite to subopposite pairs of diagonal, unbranched veins extending to or projecting beyond blade margins; prostrate portions freely branched, the branches sometimes reduced to creeping midribs; branching opposite, or alternate, or irregular and reduced to spinelike protuberances; reproduction as for genus.

Locally abundant in sandy places, low intertidal, Ore. to Bahía Asunción, Mexico. Type locality: Asilomar Pt. (Monterey Co.), Calif.

Herbarium specimens of this small alga are very similar to many specimens of *Phycodrys*, and except for size strongly resemble *Erythroglossum californicum*, from which they differ in having a clear midrib and lateral veins.

Nienburgia Kylin 1935

Thalli with prostrate and erect branches; prostrate systems entangled and with spinelike irregular branches; erect portions divided into linear blades with inconspicuous midrib in upper portions and with lateral veins; axes polystromatic. Secondary filaments from apical cells alternately developing into marginal teeth, occasionally initiating branches. Tetrasporangial sori scattered on blade proper, or on small, discrete, proliferous blades. Spermatangia in conspicuous, elliptical to linear sori at apices of branches or in small, terminally or laterally placed proliferations. Cystocarps scattered over both surfaces of blades, projecting beyond surface and surrounded by hemispherical, ostiolate pericarp; carposporangia terminal.

Nienburgia andersoniana (J. Ag.) Kyl.

Neuroglossum andersoniana J. Agardh 1876: 474. *Nienburgia andersoniana* (J. Ag.) Kylin 1935: 1; 1941: 32; Smith 1944: 345. *Heteronema andersoniana* (J. Ag.) Kyl. 1924: 46. *H. borealis* Kyl. 1924: 49. *Nienburgia borealis* (Kyl.) Kyl. 1935: 1.

Thalli tufted, bushy, the branches usually in 1 plane, rarely of more than 3 orders; bright rose red to dull carmine, drying to soft rose or greenish-black; main axes 1–16 mm broad, the secondary axes of varying width, or all nearly the same; branches alternate, irregular, flabellately branching from base, or branches crowded in upper portions only; margins regularly dentate, or if broadened sometimes edentate; basal portion often stemlike because of abrasion; tetrasporangia usually restricted to marginal proliferations; sexual reproduction as for genus.

Frequent, saxicolous, low intertidal to subtidal (to 20 m), Br. Columbia to Baja Calif.; common in region of type locality (Santa Cruz, Calif.), where usually sterile.

Myriogramme Kylin 1924

Thalli erect, with discoid holdfasts or prostrate systems of flattened branches. Blades more or less rounded or deeply and irregularly sinuate or lobed, without midrib or veins, monostromatic except basally. Apex with several marginal apical cells, the cells posterior to these irregularly arranged owing to intercalary cell divisions. Tetrasporangia in irregular or elliptical sori scattered over both surfaces of blades, tetrahedrally divided. Spermatangia in small, rounded sori scattered over entire blade except for base. Carpogonial branch with 2 groups of sterile cells. Cystocarps scattered over blade, with ostiolate, projecting pericarp; carposporangia in chains.

1. Thallus of deep lobes or divisions 2
1. Thallus with 1 undivided blade 3
   2. Branches divaricately, dichotomously divided, 2–4 mm wide.......
   .... *M. variegata*
   2. Branches pinnately divided, or irregularly palmately lobed, 1–3 cm
   wide .... *M. spectabilis*
1. Blades repent, overlapping, with ruffled terminal margins.....*M. caespitosa*
1. Blades erect or repent, not overlapping, the margins simple......*M. repens*

Myriogramme caespitosa Daws.

Dawson 1949b: 19; 1962: 96. *Myriogramme osoroi* Daws. 1950a: 158; 1962: 96.

Blades soft rose, repent, clustered, closely adhering to host or to rock, basally attached by small holdfasts, secondarily attached on ventral surface by numerous peglike disks; blades several, overlapping, with growing margins free, 0.5–1 cm long and wide, plane basally, scalloped and ruffled in upper lateral and terminal portions; tetrasporangia in small sori; spermatangial sori irregular, becoming confluent; cystocarps few, with beaked ostiole in pericarp.

Infrequent, locally abundant, on rocks or on *Macrocystis* or *Cystoseira*, lower intertidal to subtidal (13 m), Santa Rosa I. (type locality), Santa Cruz I., and mouth of San Gabriel R. (Los Angeles Co.), Calif., to Sonora, Mexico.

Material now available demonstrates that the distinguishing characteristics previously attributed to *M. caespitosa* and *M. osoroi* in fact intergrade; it seems advisable, therefore, to recognize a single taxon.

Myriogramme repens Hollenb.

Hollenberg 1945a: 449.

Thalli dark red, gregarious, erect or partly repent; blades oblanceolate to oval, occasionally lobed, 2 cm tall, 2–3 mm wide; from creeping bases, short-stipitate, monostromatic throughout; tetrasporangial sori ovate to

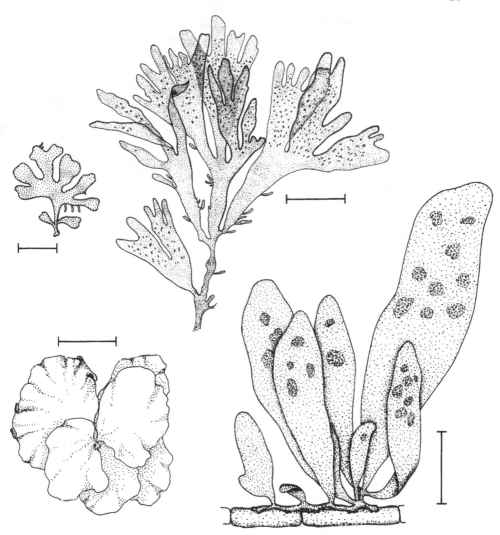

Fig. 597 (below left). *Myriogramme caespitosa*. Rule: 2 cm. (LH)   **Fig. 598** (below right). *M. repens*. Rule: 2 mm. (GJH)   **Fig. 599** (above center). *M. spectabilis*. Rule: 2 cm. (SM)   **Fig. 600** (above left). *M. variegata*. Rule: 5 mm. (CS)

lunate, scattered in middle and upper portions of blade; spermatangial sori rounded to irregular, in upper portions of blades, the cystocarps few, scattered, with short conical ostiole in pericarps.

Infrequent, low intertidal on rocks near Pt. Vicente (Los Angeles Co.; type locality), Calif.; subtidal (5–10 m) on variety of algae, Duncan Bay, Br. Columbia, and Santa Barbara I., Calif.

Myriogramme spectabilis (Eat.) Kyl.

Nitophyllum spectabile Eaton 1877: 245; Nott 1900: 21. *Myriogramme spectabilis* (Eat.) Kylin 1924: 58; Smith 1944: 346.

Thalli erect, foliose, 5–20 cm tall, bright rose red; blades irregularly ligulate, usually with narrow, palmate divisions, apically notched or rounded; divisions with entire margins, or appearing pinnate if proliferous; midribs and veins lacking. Tetrasporangial sori elliptical, regularly spaced over both surfaces of blades; spermatangia in light-colored, round sori; cystocarps numerous, irregularly distributed over both surfaces of blades.

Subtidal (to 30 m), Santa Cruz, Calif. (type locality), to Pta. Baja, Baja Calif.; frequently cast ashore on *Desmarestia latifrons* on Monterey Peninsula.

This species is frequently misidentified as *Nitophyllum* (in which the carposporangia are terminal) or as *Hymenena setchellii* (which possesses microscopic veins).

Myriogramme variegata Yam.

Yamada 1944: 22.

Thalli 1.5–2.5 cm tall, dark rose, attached in thick, soft clusters by small, irregularly shaped, discoid holdfasts; stipes short, simple or divided; upper portions bladelike, simple or irregularly dichotomously divided; margins simple or with proliferous outgrowths; reproductive structures borne in upper portions of blades; tetrasporangia in small sori; cystocarps rare; spermatangia unknown.

Rare, saxicolous, low intertidal, Pacific Grove, Calif. Type locality: Sagami Bay, C. Japan.

Nitophyllum Greville 1830

Thalli erect, with flat, lobed blades, or the blades with broad divisions, entire or ruffled, or expanded and ample with ruffled margins, monostromatic throughout. Tetrasporangia in oval sori scattered over blade. Spermatangial sori lunate to ovoid. Carpogonial branch single, with 1 group of sterile cells. Gonimoblasts with terminal carposporangia.

1. Blades narrow to ovate, stipitate .... *N. hollenbergii*
1. Blades dissected into flat divisions or ruffled lobes................... 2
   2. Blades dissected into cuneate, flat lobes; margins sometimes proliferous .... *N. northii*
   2. Blades with lobes strongly ruffled; thallus circular....... *N. cincinnatum*

Nitophyllum cincinnatum Abb.

Abbott 1969: 43. *Cryptopleura corallinara sensu* R. Norris & West 1967: 114.

Thalli gregarious, reddish-orange to reddish-brown, attached by small holdfasts, secondarily by occasional pegs from lower thallus, the lobes

crisp, densely ruffled, standing in free circular groups, bent back on themselves by ruffling, the margins undulate and fimbriate; tetrasporangial sori
oval, with center higher than edges; spermatangial sori ovate to irregular;
cystocarps numerous, scattered on outer portions of lobes; carposporangia terminal.

Locally abundant, epiphytic on articulated corallines, subtidal (to 12
m), C. Calif. from San Mateo Co. to Little Sur River (Monterey Co.).
Type locality: Carmel, Calif.

Somewhat similar to *N. nottii* R. Norris & Wynne (1968) from Washington, but more strongly ruffled.

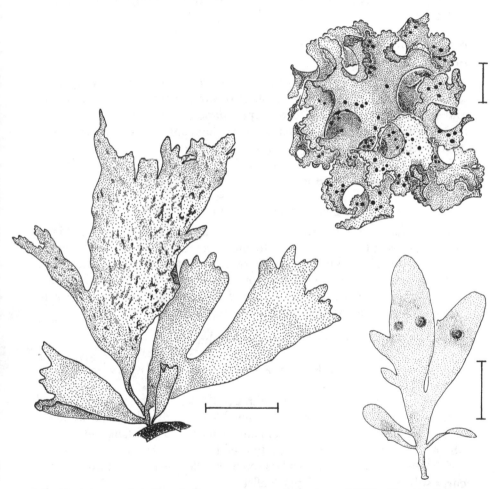

Fig. 601 (above right). *Nitophyllum cincinnatum.* Rule: 2 cm. (DBP)   Fig. 602 (below
right). *N. hollenbergii.* Rule: 5 mm. (JRJ/S)   Fig. 603 (left). *N. northii.* Rule: 2 cm.
(DBP)

Nitophyllum hollenbergii (Kyl.) Abb.

Myriogramme hollenbergii Kylin 1941: 32; Smith 1944: 346. *Nitophyllum hollenbergii* (Kyl.) Abbott 1969: 42.

Thalli erect, 1–2 cm tall, stipitate with terminal blades, occasionally with lateral blades; blades narrowly to broadly ovate, rounded or divided at apex, orange-red to bright red; tetrasporangial sori minute, round to irregular, irregularly scattered over thallus; spermatangial sori lunate on upper parts of blade, spreading downward in irregular patches; cystocarps few, with terminal carposporangia.

Occasional, saxicolous or epizoic, low intertidal to subtidal (to 35 m), Wash. to Baja Calif.; in Calif., from vicinity of Monterey (type locality) to La Jolla.

Nitophyllum northii Hollenb. & Abb.

Hollenberg & Abbott 1965: 1185.

Thalli gregarious, the blades dissected into cuneate divisions, frequently lacerated, sometimes proliferous, short-stipitate, orange-red; tetrasporangial sori elliptical, scattered over upper blades; spermatangial sori lunate to ovoid, frequently in linear series, becoming confluent; cystocarps infrequent, with terminal carposporangia.

Locally frequent, epiphytic, subtidal (8–35 m), Sonoma Co., Carmel Submarine Canyon (type locality), and Newport Bay, Calif.

Asterocolax Feldmann & Feldmann 1951

Thalli small, solitary, growing "parasitically" on Delesseriaceae. Major portion external to host, tuberculate and covered with erect, small, terete, needlelike leaflets. Young leaflets with transversely dividing apical cell, this obscured at maturity. Tetrasporangia scattered over nearly entire leaflet. Spermatangia covering surface of nearly entire leaflet. Cystocarps borne near apex; carposporangia in chains.

Asterocolax gardneri (Setch.) Feldm. & Feldm.

Polycoryne gardneri Setchell 1923a: 395; Smith 1944: 347; Wagner 1954: 317. *Asterocolax gardneri* (Setch.) Feldmann & Feldmann 1958: 59. *P. phycodricola* Dawson 1944e: 107.

Thalli 2–3 mm diam., whitish, with needlelike leaflets in low intertidal specimens to broader, somewhat flattened leaflets with clavate, acuminate, or obtuse apices in subtidal specimens; leaflets 1–3 mm long; tetrasporophytes with apical or median portion of fertile leaflets inflated; spermatangial plants smooth, with leaflets somewhat flattened when fertile; cystocarps subapical to apical, 1 per leaflet.

Occasional, commonly on bases of branches of *Nienburgia*, but also on *Phycodrys* or *Polyneura*, low intertidal to subtidal (to 20 m), N. Wash.;

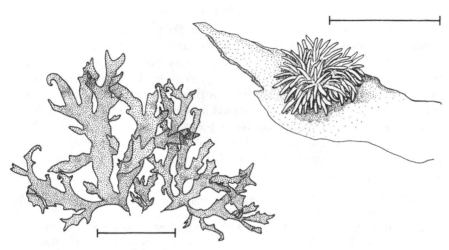

Fig. 604 (right). *Asterocolax gardneri*. Rule: 5 mm. (SM) Fig. 605 (left). *Acrosorium uncinatum*. Rule: 2 cm. (DBP)

in Calif. from Duxbury Reef (Marin Co.) to San Diego, and I. Cedros, Baja Calif. Type locality: Pt. Cavallo (Marin Co.), Calif.

Acrosorium Zanardini 1869

Thalli epiphytic or saxicolous, mostly small, entangled, flattened, with irregularly branched, ribbonlike blades 1 cell thick except in lower portions; with marginal initials. Macroscopic veins lacking, microscopic veins present. Tetrasporangia in 1 large sorus at each branch apex. Spermatangia in oval sori. Cystocarps distributed over thallus, with terminal carposporangia.

Acrosorium uncinatum (Turn.) Kyl.

Fucus laceratus var. *uncinatus* Turner 1808: 153. *Acrosorium uncinatum* (Turn.) Kylin 1924: 78; Dawson 1962: 94.

Thalli epiphytic or saxicolous, to 8 cm tall, deep rose red; erect portions of narrow blades 2–20 mm wide, the blades irregularly toothed, occasionally ending in hooks often functioning as tendrils; saxicolous thalli more robustly developed but shorter than epiphytic ones; rarely fertile in California, only the tetrasporangia seen.

Rare in north, subtidal (10–30 m) on worm tubes or rocks at Monterey Peninsula, and low intertidal to subtidal (to 30 m), commonly epiphytic on a variety of algae, especially *Pterocladia*, from San Luis Obispo Co., Calif., to Bahía San Lucas and Is. San Benito, Mexico, including Channel Is., Calif., and I. Guadalupe, Baja Calif. Type locality: England.

Hymenena Greville 1930

Thalli differentiated into prostrate, ribbonlike portions and erect blades. Blades undivided to much divided; divisions linear, spatulate, palmately or flabellately divided, monostromatic in upper portions and along edges, polystromatic through center and basally. Veins microscopic, or sometimes macroscopic, these several cells thick, without internal rhizoids. Tetrasporangia in sori, these small and scattered over blade, elliptical and in linear series, or in transverse bands across upper segments. Spermatangia, where known, in areolate sori scattered over blade. Cystocarps scattered over blade, or lateral to midline of branches, sometimes extending into marginal proliferations.

1. Blades flat, expanded, with irregular marginal proliferations....*H. setchellii*
1. Blades dissected into divisions, not flat 2
   2. Lateral margins irregular but without proliferous outgrowths; tetrasporangia in transverse bands ....*H. multiloba*
   2. Lateral margins smooth or with proliferous outgrowths; tetrasporangia not in transverse bands 3
3. Branches lanceolate, subdichotomously or palmately divided........... 4
3. Branches broad-spatulate, irregularly or subdichotomously divided...... 5
   4. Tetrasporangia in linear series covering more than half of blade.....
   .... *H. flabelligera*
   4. Tetrasporangia in linear series in upper third of blade.........*H. kylinii*
5. Divisions broadly cuneate, short; midrib inconspicuous; basal margins usually ruffled ....*H. cuneifolia*
5. Divisions irregularly divided; midrib lacking; margins plane to crenulate .... *H. smithii*

Hymenena cuneifolia Doty

Doty 1947b: 195.

Thalli deep purple-red, 8–20 cm tall, irregularly to regularly dichotomously divided; divisions short, cuneate, broadest through penultimate divisions; apices blunt, sometimes notched; divisions divided to near base; midribs short, in lowest 1–2 cm, usually inconspicuous, undivided; margins usually ruffled except on tetrasporangial plants; tetrasporangial sori linear in central and lower portions of plant; spermatangial sori discontinuous near ruffled margins; cystocarps sparsely placed along margins.

Locally abundant, saxicolous, low intertidal, Yaquina Head, Ore., to Bodega Head (Sonoma Co.), Calif.; frequent in Humboldt Co., Calif. Type locality: Coos Bay, Ore.

Hymenena flabelligera (J. Ag.) Kyl.

Nitophyllum flabelligerum J. Agardh 1876: 699. *Hymenena flabelligera* (J. Ag.) Kylin 1924: 83; Smith 1944: 348.

Thalli in deep salmon to dull rose clusters, 10–25(30) cm tall, subdi-

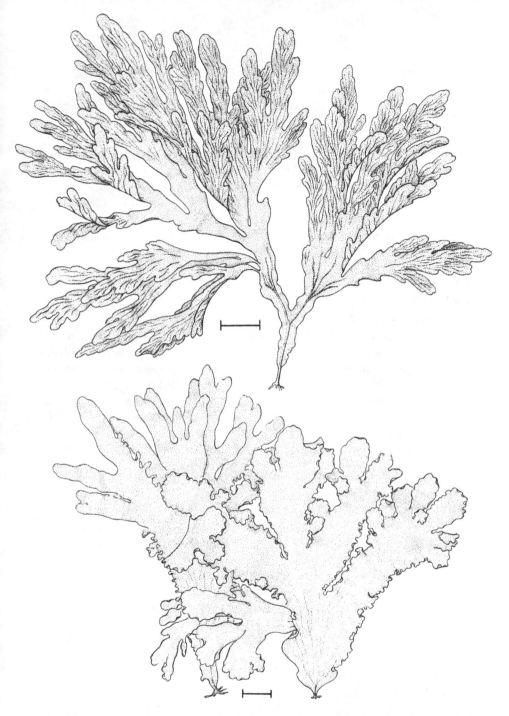

Fig. 606 (below). *Hymenena cuneifolia.* Rule: 1 cm. (CS)   **Fig. 607** (above). *H. flabelligera.* Rule: 1 cm. (DBP)

chotomously or palmately divided into many linear-lanceolate divisions; lower portions with distinct midribs in each division, these persisting upward and finally disappearing in upper lobes; midribs frequently abraded, appearing stipelike; margins of male plants ruffled, those of cystocarpic and tetrasporangial plants usually plane or slightly dentate; tetrasporangia in narrow, linear sori, these in parallel, long, longitudinally forking rows on upper half of blade; spermatangial sori not clearly shaped, giving divisions a pale, straw-colored cast; cystocarps scattered near margins.

Frequent in exposed places, saxicolous, low intertidal to subtidal (to 12 m), Vancouver I., Br. Columbia, to San Luis Obispo Co., Calif. Type locality: San Francisco, Calif.

A very common species; when very mature and eroded, it is difficult to distinguish from *Botryoglossum farlowianum.*

Hymenena kylinii Gardn.

Gardner 1927a: 242; Kylin 1941: 32; Smith 1944: 349.

Thalli in clusters, 6–8 cm tall, deep rose red; erect blades dichotomously to subdichotomously divided into a few branches 2–5 mm broad; blades with blunt apices and undulate, entire margins; lower portions of blades with inconspicuous midribs; tetrasporangial sori narrow, linear, running lengthwise on surface of blade, covering upper third of blade.

Infrequent but locally abundant on exposed open coast, saxicolous, low intertidal, Wash. and Ore. and in Calif. from San Francisco (type locality) to Monterey Peninsula.

Similar to *H. flabelligera*, but shorter and with less complex branching.

Hymenena multiloba (J. Ag.) Kyl.

Nitophyllum multilobum J. Agardh 1876: 698. *Hymenena multiloba* (J. Ag.) Kylin 1935: 3; Smith 1944: 350 (incl. synonymy).

Thalli short, densely branched, tufted, 4–8 cm tall, dark purplish-red to dull carmine, drying to black; blades subdichotomously to palmately divided, narrow, twisted, the margins entire or toothed; tetrasporangial sori broad, extending transversely across upper blades.

Locally abundant, in dense, often turfy clusters, intertidal on exposed rocks to subtidal (to 40 m), N. Br. Columbia to San Luis Obispo Co., Calif. Type locality: San Francisco, Calif.

Hymenena setchellii Gardn.

Gardner 1927a: 245; Smith 1944: 350 (incl. synonymy).

Thalli 10–20 cm tall, dark rose red to cherry red, mainly foliar; narrow, weak stipes sometimes present; foliose portions lobed several times, regularly, irregularly, or palmately; divisions 3–6+ cm wide, tapering toward

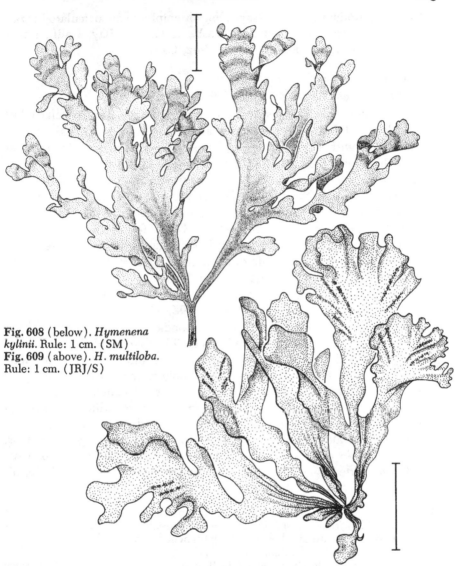

Fig. 608 (below). *Hymenena kylinii.* Rule: 1 cm. (SM)
Fig. 609 (above). *H. multiloba.* Rule: 1 cm. (JRJ/S)

apex; margins entire or toothed, ruffled, sometimes with minute, proliferous outgrowths; lower portions with dissected, inconspicuous midribs, the upper portions with microscopic veins; tetrasporangia in lunate sori, scattered over blades, sometimes appearing to be in lines; spermatangia in lunate, colorless sori in upper portions of blade; cystocarps widely spaced, in upper blades near margins; cystocarpic thalli more dissected than other thalli.

Locally abundant in season, saxicolous or epiphytic on articulated corallines, subtidal (8 m), Puget Sd., Wash., to Carmel Bay, Calif.; found mostly in drift. Type locality: Santa Cruz, Calif.

Hymenena smithii Kyl.

Kylin 1941: 33; Smith 1944: 351; Abbott 1970: 3.

Thalli 4–8 cm tall, dark reddish-brown, stipitate; blades irregularly but deeply lobed, lobes 1–2 cm wide, scarcely tapering to apices; margins crenulated or with small ruffles, or plane; blades monostromatic except at extreme base, with dissected midrib in lower portions, otherwise the blades with microscopic veins; reproductive structures as in *H. setchellii*.

Infrequent, epiphytic on stipes of *Pterygophora*, subtidal (6–10 m), Mukkaw Bay, Wash., Sunset Bay, Ore., and Carmel Bay (type locality) and Shell Beach (San Luis Obispo Co.), Calif.; found mostly in drift.

Specimens with narrowly tapering lobes, thought to be representative of the species, are not difficult to distinguish from *H. setchellii*, but the limits of the two species are not easy to define.

Cryptopleura Kützing 1843

Thalli differentiated into creeping, ribbonlike bases attached by hapterous pegs on ventral surfaces and erect, ribbonlike to bladelike tufts of branches divided flabellately, subalternately, or subdichotomously. Margins plane, undulate, or proliferous. Axes polystromatic except for upper and marginal portions; midribs a collection of macroscopic veins of main blade divisions, each vein dividing and gradually spreading upward, becoming narrower and disappearing in fine network of microscopic veins. Tetrasporangia along margins or in oval or lunate sori in proliferations. Spermatangia in elliptical to linear sori along margins. Cystocarps scattered over blades, each with an ostiolate pericarp; carposporangia terminal.

1. Blades with narrow divisions, mostly not broadening upward...........2
1. Blades with wide divisions, broadening toward apex of plant...........3
   2. Divisions repeatedly dichotomously branched, 1 mm wide.........
    ....*C. dichotoma*
   2. Divisions flabellately branched, 5–10 mm wide..............*C. rosacea*
3. Plants creeping, on other algae, closely adhering; reproductive sori relatively large ....*C. corallinara*
3. Plants erect; reproductive structures of relatively moderate size.........4
   4. Tetrasporangia in rounded marginal proliferations............*C. crispa*
   4. Tetrasporangia in marginal bands 5
5. Plants mostly more than 15 cm tall, with smooth margins; dark to light green with rose undertones ....*C. violacea*
5. Plants shorter than 15 cm, with ruffled margins; deep rose to purplish-red ....*C. lobulifera*

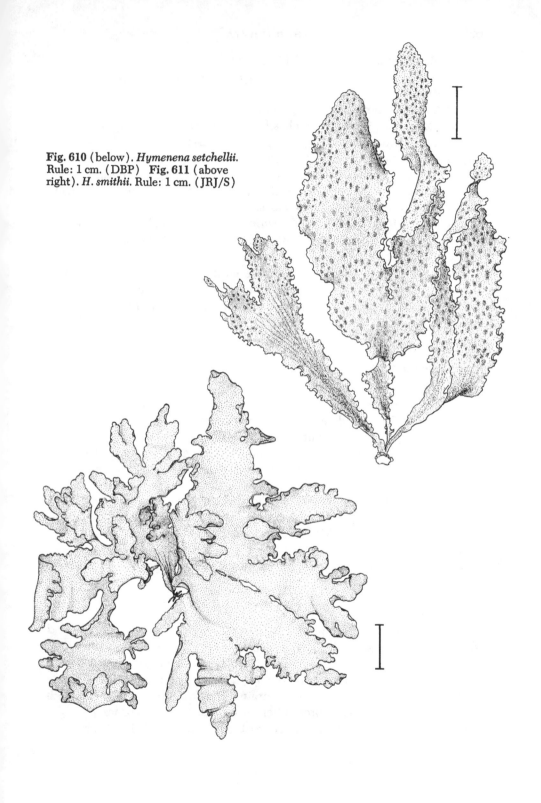

Fig. 610 (below). *Hymenena setchellii*. Rule: 1 cm. (DBP) **Fig. 611** (above right). *H. smithii*. Rule: 1 cm. (JRJ/S)

Cryptopleura corallinara (Nott) Gardn.

Nitophyllum corallinarum Nott 1900: 24. *Cryptopleura corallinara* (Nott) Gardner 1927a: 240; Dawson 1962: 98.

Thalli brownish-red, adhering closely to host by peglike attachments and rhizoids; margins and branches free, partly erect, the margins smooth to scalloped; erect blades (2)7(15) mm tall, 3–10 mm broad, occasionally irregularly divided into rounded lobes; tetrasporangial sori irregular in shape, occupying entire fertile lobes except for margins; spermatangial sori lunate, soon confluent with sori in adjacent lobes and becoming continuous just inside margin; cystocarps few.

Common in low intertidal on *Corallina* or other algae, less common subtidally, Monterey Peninsula, Calif., to I. Magdalena, Baja Calif. Type locality: San Diego, Calif.

Once it is distinguished as a separate entity from young *Cryptopleura violacea* in the northern and from young *C. crispa* in the southern part of its range (both occurring also on *Corallina*), *C. corallinara* will be found to be much more common than the present collections would indicate.

Cryptopleura crispa Kyl.

Kylin 1924: 90; Dawson 1962: 100.

Thalli rose red, essentially creeping but with erect, branched divisions 3–8 cm tall, 3–7 mm wide and apically wider; apices blunt-spatulate; margins at first undulate, becoming densely scalloped and crisp; blades with age basally eroded to midrib; tetrasporangial and spermatangial sori on isolated, rounded proliferations from margin; cystocarps scattered over outer marginal portions of blade.

Occasional, epiphytic on *Pterocladia*, *Gelidium*, or *Corallina*, low intertidal to subtidal (to 8 m), Santa Barbara through Channel Is., Calif., and mainland S. Calif. to I. Magdalena, Baja Calif. Type locality: La Jolla, Calif.

Closely related to, and perhaps not clearly distinct from, *C. violacea*.

Cryptopleura dichotoma Gardn.

Gardner 1927a: 242.

Thalli diminutive, light pink, 2–2.5 cm tall, ±1 mm wide, regularly and dichotomously branched, the distance between branchings usually equal to width of divisions, with widely divaricate divisions and blunt apices, but without midribs; macroscopic veins lacking; microscopic veins sparse but distinct, at times extending to margins of growing point; segments with smooth, entire margins; reproduction unknown. (After Gardner.)

Known only from type specimen, cast ashore at San Pedro, Calif.

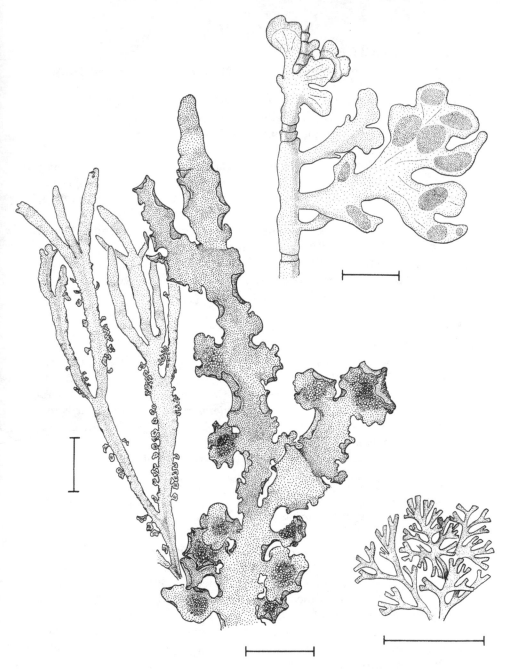

Fig. 612 (above right). *Cryptopleura corallinara*: habit on *Corallina*, showing tetra-sporangial sori. Rule: 5 mm. (DBP) **Fig. 613** (two at left). *C. crispa*: right, habit of young tetrasporangial thallus; left, habit of older thallus. Rules: both 2 cm. (CS) **Fig. 614** (below right). *C. dichotoma*: portion of (sterile) type specimen. Rule: 2 cm. (SM)

Cryptopleura lobulifera (J. Ag.) Kyl.

Neuroglossum lobuliferum J. Agardh 1898: 121. *Cryptopleura lobulifera* (J. Ag.) Kylin 1924: 90; Smith 1944: 352.

Thalli thickly branched, tufted, dark rose to purplish-red; erect branches subdichotomously divided, diverging widely, the divisions 8–15 mm broad, the margins ruffled or crisped; primary divisions with network of macroscopic veins in basal portions, none to few in midportions, terminally with minute or microscopic veins; tetrasporangial sori lunate to elliptical, lying within margins of blades on ruffled portions and within ruffles; spermatangial sori in similar locations; cystocarps 1.5–2 mm diam., submarginal, rarely in ruffles.

Frequent, on rocks or occasionally on other algae, low intertidal, Wash. to Baja Calif. Type locality: Pacific Grove, Calif.

Distinguished from *C. violacea* by the denser ruffled margins and purple coloring, and from narrow specimens of *Hymenena flabelligera* and *Botryoglossum* by the narrower blade divisions.

Cryptopleura rosacea Abb.

Abbott 1969: 45.

Thalli caespitose, in low, crisp, deep-rose clumps, flabellately branched in ribbonlike lobes 5–10 cm tall, bearing branches 2–3 cm broad; ultimate divisions with blunt to spatulate apices; lower portions with indistinct macroscopic veins, the upper portions with microscopic veins; cystocarps few in main divisions, 1–1.5 mm wide, somewhat flat. Tetrasporangia and spermatangia unknown.

Infrequent, cast ashore at Carmel, Calif. (type and only known locality).

Cryptopleura violacea (J. Ag.) Kyl.

Nitophyllum violaceum J. Agardh 1876: 700. *Cryptopleura violacea* (J. Ag.) Kylin 1924: 89; Smith 1944: 352. *C. brevis* Gardner 1927a: 241.

Thalli slender, 15–25 cm tall, deep rose to greenish-pink, drying to greenish, with long, ribbonlike central blades 10–25 mm broad, subdichotomously divided, the top portions commonly flabellately divided, 10–20 mm broad, with irregularly divaricate apices; midribs in lower portions inconspicuous, but sometimes strongly abraded and becoming stipelike; always some divisions of thalli with long intervals uninterrupted by branching; margins of divisions undulate, frequently with proliferous outgrowths; tetrasporangial sori marginal, at first continuous, then dissected by older sori dropping out, ultimately with spoon-shaped sori remaining; spermatangial thalli lighter in color than others, and with broader penultimate divisions; cystocarps distributed submarginally in upper portions of thallus.

Frequent, locally abundant, saxicolous, occasionally epiphytic, low in-

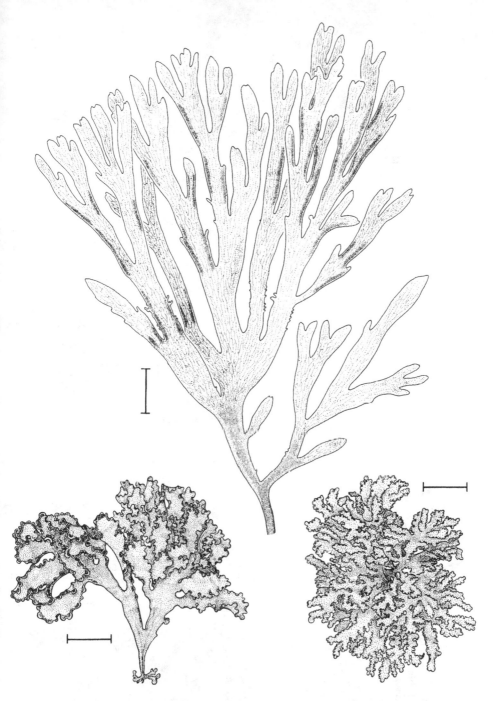

Fig. 615 (below left). *Cryptopleura lobulifera*: habit of tetrasporangial thallus. Rule: 2 cm. (DBP) **Fig. 616** (below right). *C. rosacea*: habit of cystocarpic thallus. Rule: 2 cm. (DBP) **Fig. 617** (above). *C. violacea*: habit of tetrasporangial thallus. Rule: 2 cm. (JRJ/S)

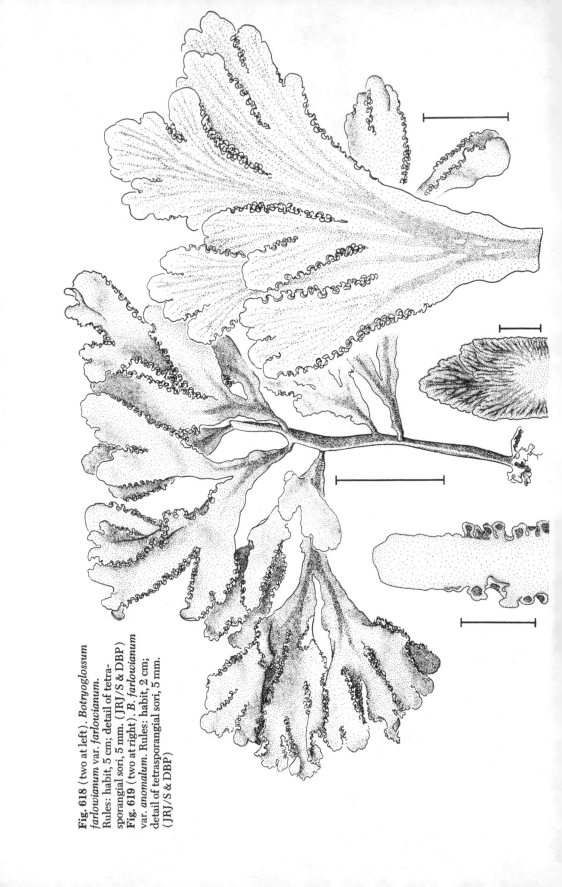

Fig. 618 (two at left). *Botryoglossum farlowianum* var. *farlowianum*. Rules: habit, 5 cm; detail of tetrasporangial sori, 5 mm. (JRJ/S & DBP) **Fig. 619** (two at right). *B. farlowianum* var. *anomalum*. Rules: habit, 2 cm; detail of tetrasporangial sori, 5 mm. (JRJ/S & DBP)

tertidal to subtidal (to 10 m), Vancouver I., Br. Columbia, to I. Magdalena, Baja Calif. Type locality: Golden Gate (San Francisco), Calif.

This is the most common species of *Cryptopleura*, and probably the most variable. The greenish cast in most specimens, the long uninterrupted divisions, and the marginal, ribbonlike tetrasporangial sori help to distinguish this species from *C. lobulifera*, which is typically thought to be shorter, with more congested branching and with strongly ruffled margins. These characteristics, however, intergrade to some degree.

Botryoglossum Kützing 1843

Thalli of 1 or more erect axes arising from ribbonlike bases, the axes polystromatic. Branching subdichotomous to flabellate, the divisions branched several times, each portion usually shorter than previous ones, the apices spatulate; margins plane, finely to coarsely crenulate, sometimes with congested proliferations. Midrib none, or 1–3 mm wide in basal portions, occasionally extending to terminal divisions, occasionally remaining, owing to erosion. Terminal divisions with microscopic veins; older thalli with scattered rosettes of tissue on blade surfaces. Tetrasporangia tetrahedrally divided in linear, longitudinal sori, or in discrete lunate sori along margins or in ruffles. Spermatangial thalli pale straw-colored; sori superficial, irregularly shaped. Cystocarps irregularly distributed submarginally, ostiolate.

Botryoglossum farlowianum (J. Ag.) DeToni

Nitophyllum farlowianum J. Agardh 1898: 95, *Botryoglossum farlowianum* (J Ag.) DeToni 1900: 676; Smith 1044: 354.

Thalli growing in thick clusters; branching of young thalli predominately flabellate, of older thalli predominately subdichotomous, owing to erosion of blades to midrib; margins with discrete proliferations, these often grading into closely packed marginal ruffles; tetrasporangia in discrete to confluent sori, these in lateral proliferations associated with or distinct from marginal ruffles; sexual reproduction as for genus.

Botryglossum farlowianum var. farlowianum

Thalli deep rose to brownish-red with some iridescence; erect axes 10–50 cm tall, blade segments 0.15–2(4) cm wide, with broadly rounded to blunt apices; margins with proliferations and ruffles; tetrasporangial sori first in proliferations, with age extending beyond.

Common on rocks, low intertidal to subtidal, Vancouver I., Br. Columbia, to Baja Calif.; commonly growing with *Hymenena flabelligera*, *Cryptopleura lobulifera*, and *Erythrophyllum delesserioides* in C. and N. Calif. Type locality: California, probably Monterey.

Botryoglossum farlowianum var. **anomalum** Hollenb. & Abb.

Hollenberg & Abbott 1965: 1185; 1966: 108.

Differing from type variety in having tetrasporangia in linear sori over microscopic veins in terminal segments, and often also in having sori in lateral proliferations.

Occasional, saxicolous, low intertidal, Duxbury Reef (Marin Co.), to Pismo Beach (San Luis Obispo Co.), Calif. Type locality: Mussel Pt. (Monterey Co.), Calif.

Botryoglossum ruprechtianum (J. Ag.) DeToni

Nitophyllum ruprechtiana J. Agardh 1872: 51. *Botryoglossum ruprechtianum* (J. Ag.) DeToni 1900: 676; Hollenberg & Abbott 1966: 110. *Cryptopleura spatulata* Gardner 1927a: 241; Dawson 1962: 100.

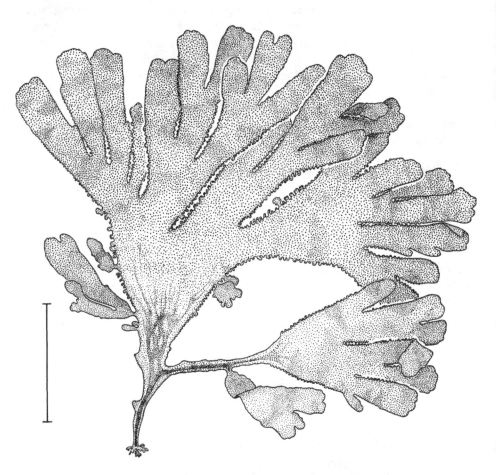

Fig. 620. *Botryoglossum ruprechtianum*. Rule: 5 cm. (DBP)

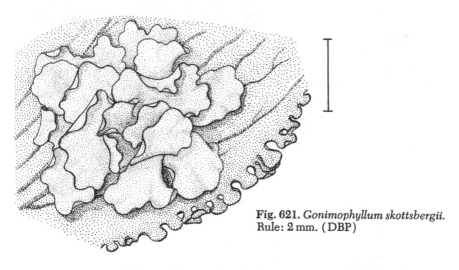

Fig. 621. *Gonimophyllum skottsbergii.*
Rule: 2 mm. (DBP)

Differing from *B. farlowianum* in having discrete proliferations on margins; ruffles lacking; and thallus delicate, not coarse.

Occasional, saxicolous, subtidal (to 30 m), Puget Sd., Wash., to I. Todos Santos, Baja Calif. Type locality: northwest coast of N. America.

Gonimophyllum Batters 1892

Thalli small, leafy, "parasitic" on various Delesseriaceae; major portions lying external to host. Blades rosette-like, expanded, simple or lobed, lying close to host; axes polystromatic throughout, with indistinct microscopic veins basally. Polysporangia in sori, nearly covering entire blade. Spermatangial sori covering most of both surfaces of blade. Cystocarps on lobed blades, several per blade; carposporangia terminal.

Gonimophyllum skottsbergii Setch.

Setchell 1923a: 394; Smith 1944: 355; Wagner 1954: 338.

Thalli to 1.5 cm diam., pinkish to soft rose, on lower portions of erect branches of host; blades 1–3 mm tall, with undulate margins; polysporangia with 30–50 polyspores; cystocarps bulging prominently toward 1 side of blade.

Occasional to rare, on stipes or laterally on lower blades of *Botryoglossum* spp. or *Hymenena* spp., low intertidal to subtidal (to 7 m), Wash. and C. Calif. to San Diego, Calif. Type locality: San Francisco, Calif.

Family DASYACEAE

Thalli with polysiphonous major axes and monosiphonous pigmented ultimate laterals. Growth sympodial, with main axis or apical cell repeatedly displaced to 1 side and lateral axis or lateral cell producing further axial growth. Branching distichous or radial, frequently with dorsiventral organization. Main axes corticated or uncorticated. Pericentral cells formed in clockwise sequence, the last-formed next to or nearly opposite first-formed. Tetrasporangia tetrahedrally divided, in cylindrical or compressed stichidia, several on each segment of stichidium. Spermatangia dense, terminal on lateral branchlets. Procarps on third-formed pericentral cell, or on last-formed cell, each procarp with 1 4-celled carpogonial branch and 2 groups of sterile cells. Fertilized carpogonium cutting off 1 or 2 connecting cells, these fusing with auxiliary cell cut off after fertilization. Fusion cell and pericarp formed, both relatively large.

1. Determinate laterals always polysiphonous in lower parts, with cells similar to those of bearing axis.................*Heterosiphonia* (p. 676)
1. Determinate laterals monosiphonous throughout, with basal cells quite unlike those of bearing axis.. 2
   2. Axes flattened, the branching distichous ....*Rhodoptilum* (p. 678)
   2. Axes cylindrical, the branching radial 3
3. Thalli parasitic on *Heterosiphonia*, in small whitish tufts...........
   .... *Colacodasya* (p. 680)
3. Thalli not parasitic 4
   4. Major portion of erect branches bare of laterals, the tufts of mono-siphonous laterals at tops of branches.........*Pogonophorella* (p. 678)
   4. Major portion of branches covered with laterals, leaving portions occasionally bare; dense woolly branchlets clumped along axes.......
   .... *Dasya* (below)

Dasya C. Agardh 1824

Thalli erect, radially branched, terete, with 4 or 5 pericentral cells cut off in clockwise sequence, each segment bearing monosiphonous, persistent, pigmented laterals, sometimes branched and occasionally polysiphonous basally; axes corticated. Tetrasporangia in stichidia, tetrahedrally divided. Spermatangia dense on stichidial branches. Procarps arising on third-formed pericentral cell, bearing 1 4-celled carpogonial branch and 2 groups of sterile cells; carposporangia in branching chains; pericarp formed after fertilization, large, usually elongate and beaked.

Dasya sinicola (S. & G.) Daws.

Heterosiphonia sinicola Setchell & Gardner 1924a: 770. *Dasya sinicola* (S. & G.) Dawson 1959: 32; 1963b: 408.

Thalli deep purple-red, several erect or partly procumbent axes arising from fleshy holdfasts; axes mostly bare of branches except at top, where tufts of monosiphonous laterals densely cover branches, the axes with 5 pericentral cells, heavily corticated; tetrasporangia in stichidia; spermatangial branches elongate, tapered, isolated or paired; the cystocarps large, scattered, urceolate, with beaked ostioles.

1. Thalli 10–13 mm tall, the laterals 40–50 μm diam. at bases. . . . .
. . . . var. *abyssicola*
1. Thalli 3–10 cm tall, the laterals 60+ μm diam. at bases. 2
   2. Determinate laterals 80–110 μm diam. near bases, tending to persist
   . . . . var. *californica*
   2. Determinate laterals usually 60–80 μm diam. near bases, tending to
   be deciduous, leaving much of axis bare. . . . . .var. *sinicola*

Dasya sinicola var. sinicola

Dawson 1963b: 408.

Thalli 3–10+ cm tall, frequently very lax when old, the axes 0.4–1.2 mm diam., the determinate laterals usually 60–80 μm near bases, tending to be deciduous, leaving much of axis bare.

Rare, subtidal (6–10 m), Santa Catalina I., Calif.; more common from I. Guadalupe, Mexico, through Baja Calif. and Gulf of Calif. into tropics. Type locality: La Paz, Gulf of Calif.

Vars. *abyssicola* and *californica* are smaller, and both occur outside the warmer areas favored by var. *sinicola*. Larger numbers of collections may

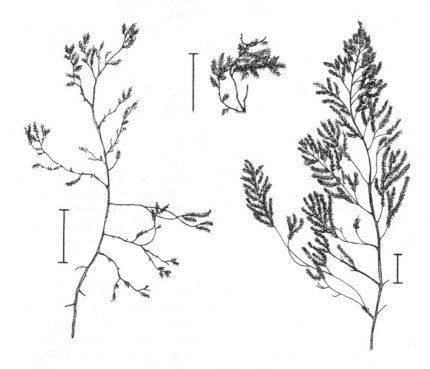

Fig. 622 (left). *Dasya sinicola* var. *sinicola*. Rule: 1 cm. (LH) **Fig. 623** (above). *D. sinicola* var. *abyssicola*. Rule: 1 cm. (LH) **Fig. 624** (right). *D. sinicola* var. *californica*. Rule: 1 cm. (SM)

show that there is an intergrading series, and thus no need to recognize varieties. Var. *sinicola* has been reported (Buggeln & Tsuda, 1969) from the mid-Pacific (Johnston I.).

Dasya sinicola var. abyssicola (Daws.) Daws.

Dasya abyssicola Dawson 1949b: 19. *D. sinicola* var. *abyssicola* (Daws.) Daws. 1963b: 410.

Thalli like the species but very small, the laterals 40–50 μm diam. at bases, 10–13 mm tall, the axes 200–300 μm diam.; laterals attenuate, less persistent than in var. *californica*.

Rare, saxicolous, subtidal (10–41 m), Santa Catalina I., Calif., to Pta. Frailes, Baja Calif. Type locality: San Clemente I., Calif.

Dasya sinicola var. californica (Gardn.) Daws.

Dasya californica Gardner 1927a: 340. *D. sinicola* var. *californica* (Gardn.) Dawson 1963b: 409.

Thalli like the species, but determinate laterals generally coarser, 80–100 μm at bases; laterals tending to persist, thus the upper branches more densely covered with laterals than in other varieties.

Frequent, saxicolous or entangled with other algae, low intertidal, Santa Catalina I., Newport Bay (Orange Co.), and La Jolla (type locality), Calif., to I. Magdalena, Baja Calif.

Heterosiphonia Montagne 1842

Thalli with discoid holdfasts giving rise to erect or prostrate axes. Erect axes distichous or subdichotomously branched, terete to compressed, corticated or uncorticated, with 4–12 pericentral cells. Laterals monosiphonous, pigmented, branched 1 to several times, separated by 2–9 internodes. Tetrasporangia tetrahedrally divided, in stichidia terminal on laterals, 4–6 on each segment of stichidium. Spermatangia on stichidial branchlets. Procarp on last-formed pericentral cell; carposporangia in rows or terminal, later-formed ones laterally produced; cystocarps with conspicuous pericarps.

Heterosiphonia erecta Gardn.

Gardner 1927a: 99; Setchell & Gardner 1937a: 84; Dawson 1963b: 403.

Thalli (1)2–5 cm tall, dark rose, partly prostrate; erect portions stout, branching closely every 2 segments; axes uncorticated; monosiphonous laterals short, with acute apices occasionally divergent; main axes of 4 pericentral cells; cystocarps sessile near bases of ultimate branches, with relatively long neck and definite ostioles.

Frequent, epiphytic on various algae, low intertidal to subtidal (to 10 m), San Clemente, San Nicolas, and Santa Catalina Is., mouth of San

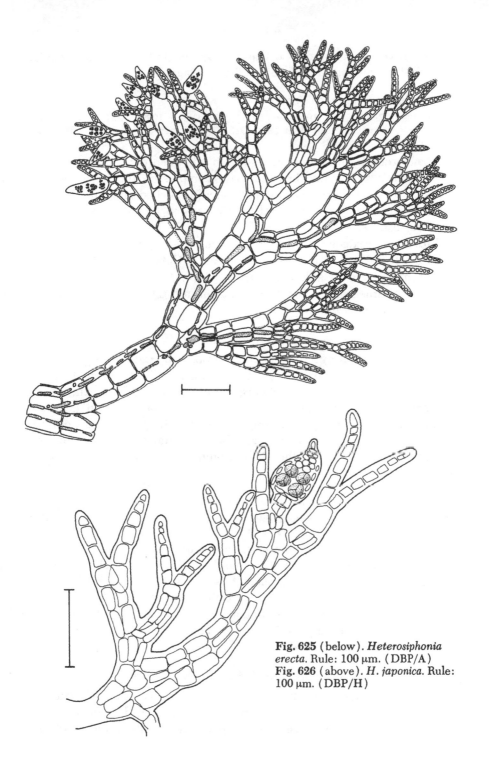

Fig. 625 (below). *Heterosiphonia
erecta*. Rule: 100 μm. (DBP/A)
Fig. 626 (above). *H. japonica*. Rule:
100 μm. (DBP/H)

Gabriel River (Los Angeles Co.), and Hyperion Outfall (Los Angeles Co.) to La Jolla (type locality), Calif., and to I. Magdalena, Baja Calif.

Heterosiphonia japonica Yendo

Yendo 1920: 8; Okamura 1907–42 (1921): 68; Abbott 1972b: 264. *Heterosiphonia densiuscula* Kylin 1925: 70; West 1970: 313. *H. asymmetria* Hollenberg 1945a: 449; Hollenberg & Abbott 1966: 112.

Thalli erect, from small disks, 2–20 cm tall, deep rose red, branched in 1 plane; main axes cylindrical near base, compressed above, composed of 5 pericentral cells; axes delicate, corticated, branched alternately, usually with 2 segments between laterals; cystocarps to 1 mm diam., with conspicuous ostiole.

Infrequent, saxicolous or on *Cryptochiton stelleri*, low intertidal to subtidal (to 15–70 m), S. Br. Columbia to Santa Catalina I. and Corona del Mar (Orange Co.), Calif. Type locality: Japan.

Most specimens from the eastern Pacific are tetrasporangial; gametophytes are very rare.

Pogonophorella Silva 1952

Thalli erect, sparingly to repeatedly branched, radially organized, terete, with 5 pericentral cells, heavily corticated. Each segment bearing branched, monosiphonous filaments, these mostly deciduous and persisting only in top portion of branches. Tetrasporangia in stichidia, tetrahedrally divided, regularly arranged, 3 per segment. Spermatangia covering the subterminal parts of the laterals, the spermatangial thalli more abundantly branched than the other thalli. Gonimoblasts sparse, solitary, broadly ovoid.

Pogonophorella californica (J. Ag.) Silva

Pogonophora californica J. Agardh 1890: 33; Kylin 1941: 34. *Pogonophorella californica* (J. Ag.) Silva 1952: 307; Dawson 1963b: 411.

Thalli 3–10 cm tall, forming loose tufts with creeping basal portions in fine sand, frequently only the terminal dark-red portions with tufted filaments projecting above sand; characters otherwise as for genus.

Locally frequent, saxicolous, in C. Calif. characteristically under *Phyllospadix* hummocks with loose sand, midtidal to subtidal (to 5 m), Bodega Bay (Marin Co.) to La Jolla, Calif., and Baja Calif. to Pta. Abreojos. Type locality: Santa Barbara, Calif.

Rhodoptilum (J. Agardh) Kylin 1956

Thalli erect, irregularly branched, with strongly compressed main axes; branches bilaterally arranged, short, or some as well developed as main axes; cells radially organized, with 5 pericentral cells heavily corticated

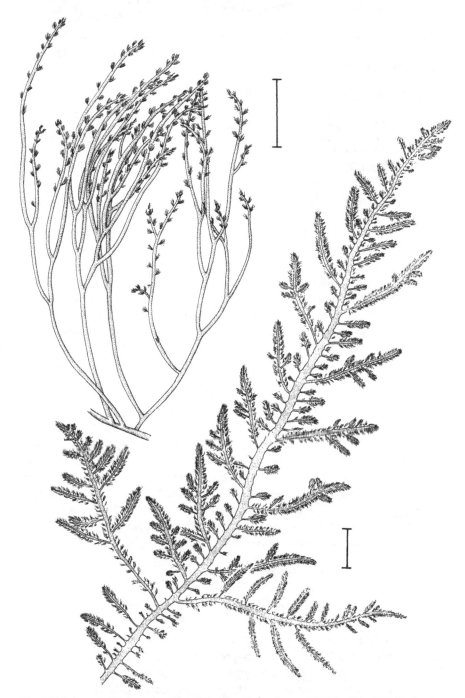

Fig. 627 (above). *Pogonophorella californica.* Rule: 1 cm. (DBP) **Fig. 628** (below).
Rhodoptilum plumosum. Rule: 1 cm. (JRJ/S)

by rhizoids; branches flattened, densely clothed on margins with mono-siphonous, branched, dark bluish-red filaments. Tetrasporangia, sperma-tangia, and gonimoblasts as in *Dasya*.

Rhodoptilum plumosum (Harv. & Bail.) Kyl.

Dasya plumosa Harvey & Bailey 1851: 371. *Rhodoptilum plumosum* (Harv. & Bail.) Kylin 1956: 461. *Dasyopsis plumosa* (Harv. & Bail.) Schmitz 1893: 231; Okamura 1907–42 (1910): 95. *Dasyopsis densa* Smith 1943: 217. *R. densum* (Smith) Dawson 1961b: 448; Daws. 1963b: 402.

Thalli 10–20(40) cm tall, usually with 1 to several leading axes, these strongly percurrent, flattened, to 4 mm broad; branching distichous; pri-mary laterals usually shorter than but occasionally as long as main axes; branching usually of 3 orders, occasionally 4; ultimate orders fringed along margins with branched filaments; northern plants with filaments to 1.5 mm long and forming silky fringe of approximately same length along a given axis, the southern plants with monosiphonous filaments less than 1 mm long, tending to be more tufted or interrupted in distribution along axis; filaments on young thalli and on those from N. Calif. and Wash. continu-ously distributed; tetrasporangial plants most commonly found.

Locally abundant, saxicolous, subtidal (to 30 m), Queen Charlotte Sd., Br. Columbia, to San Quintín Bay, Baja Calif.; in Calif., from Pt. Arena to La Jolla. Also reported from Japan. Type locality: Puget Sd., Wash.

Colacodasya Schmitz 1897

Thalli "parasitic" on *Heterosiphonia*, forming amorphous attachments to hosts. Erect branches tufted, free, richly branched, weakly dorsiventrally organized or radial, the lateral branches from every other segment cell; axes with 4–6 pericentral cells, corticated or uncorticated, occasionally with each pericentral divided by 2 or 3 crosswalls. Tetrasporangial sti-chidia unbranched, or branched, straight or weakly curved, with 4–6 spo-rangia per segment. Spermatangial stichidia oblong to lanceolate, com-monly curved, on short, monosiphonous pedicels. Cystocarps globular to slightly urceolate; carposporangia in short chains.

Colacodasya californica Hollenb.

Hollenberg 1970: 65.

Thalli parasitic on *Heterosiphonia erecta*, forming light pink to whitish, densely tufted growths in middle and upper portions of host, 1–2 mm tall, with 4 noncorticated pericentral cells; ultimate branchlets monosi-phonous, with acute apices; tetrasporangial stichidia cylindrical, with abruptly pointed apices, the sporangia in whorls of 4 in stichidia; sperma-tangial stichidia with sterile cells at apex; cystocarps globular to slightly urceolate, with slight basolateral spur.

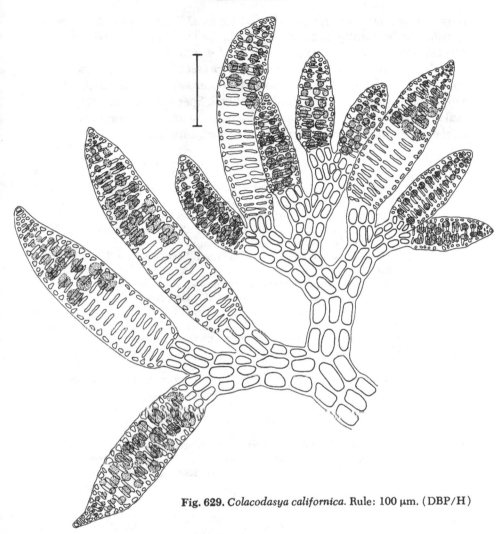

Fig. 629. *Colacodasya californica.* Rule: 100 μm. (DBP/H)

Rare, on *Heterosiphonia*, low intertidal, Pt. Vicente (Los Angeles Co.) to Laguna Beach (type locality), Calif.

Family RHODOMELACEAE

Thalli polysiphonous, mostly erect, with series of segments cut off from apical cells, each segment dividing longitudinally into central cell and 4+ pericentral cells; pericentral cells mostly remaining undivided but in certain species divided or partly or wholly obscured by 1 or more cortical layers of downwardly growing filaments, these initially cut off from pericentral cells. Branching mostly radial, sometimes bilateral or dorsiventral, mostly exogenous, but frequently endogenous. Branches mostly indeterminate but frequently of limited growth. Cells uninucleate, with numerous chloroplasts, mostly discoid. Branched or unbranched monosiphonous laterals, known as trichoblasts, commonly present in certain genera. Tetrasporangia mostly tetrahedrally but sometimes more or less cruciately divided, usually derived from pericentral cell. Spermatangia mostly on special branchlets of trichoblast origin. Carpogonia arising in connection with trichoblasts. Cystocarps with well-developed, ostiolate pericarp.

Polysiphonia Greville 1823

Thalli chiefly erect, reddish-brown to reddish-black, radially branched and polysiphonous, with usually limited prostrate portions, attached by mostly unicellular rhizoids. Trichoblasts commonly present, mostly in regular spiral positions near branch apices, never more than 1 per segment, mostly soon deciduous, leaving persistent scar cells at point of attachment. Pericentral cells 4+, relatively constant in number for given species; axes corticated or, usually, uncorticated. Tetrasporangia 1 per segment, mostly in spiral series. Spermatangial stichidia cylindrical, arising from entire trichoblast primordium or from primary branch of trichoblast. Cystocarps globular, briefly pedicellate.

1. Vegetative segments with 4 pericentral cells 2
1. Vegetative segments with more than 4 pericentral cells 11
2. Trichoblasts and scar cells commonly present, exogenous, mostly in regular spiral sequence 3
2. Trichoblasts and scar cells, when present, not in regular spiral sequence, except sometimes near branch apices 9
3. Trichoblast or scar cell separated from preceding branch by 1 or 2 naked segments . . . . *P. decussata*
3. Trichoblast or scar cell borne on each segment not bearing a branch 4
4. Branches arising at base of trichoblasts from common primordium . . .
. . . . *P. flaccidissima*
4. Branches wholly replacing trichoblasts in spiral sequence 5
5. Percurrent axes distinct; segments of main branches mostly shorter than broad . . . . *P. acuminata*
5. Percurrent axes not distinct; segments mostly as long as broad or longer 6
6. Plants epiphytic, mostly less than 10 mm tall . . . . *P. savatieri*
6. Plants epiphytic or saxicolous, larger 7

7. Segments of main axes commonly 2 or 3 times as long as diam......*P. mollis*
7. Segments of main axes mostly as long as diam...................... 8
   8. Thalli low, turf-forming, intertidal, mostly without cicatrigenous
     branches, never with cortical cells.......................*P. simplex*
   8. Thalli commonly 30–60 mm tall, not turf-forming, commonly with
     cicatrigenous branches and a few cortical cells below.........*P. bajacali*
9. Thalli mostly much larger, commonly richly branched..........*P. pacifica*
9. Thalli mostly less than 20 mm tall 10
   10. Thalli epizoic, growing on carapace of sea turtle............*P. carettia*
   10. Thalli mostly saxicolous.............................*P. scopulorum*
11. Major branches completely corticated......................*P. brodiaei*
11. All branches uncorticated 12
   12. Trichoblasts frequent to abundant 13
   12. Trichoblasts lacking or exceedingly rare 16
13. Numerous segments lacking trichoblast, scar cell, or branch......*P. hendryi*
13. All but lower segments bearing a trichoblast, scar cell, or branch....... 14
   14. Pericentral cells 5 or 6 ....*P. johnstonii*
   14. Pericentral cells 8–12 15
15. Thalli mostly under 20 mm tall...........................*P. confusa*
15. Thalli commonly 100–200 mm tall......................*P. paniculata*
   16. Thalli to 150 mm tall; branching alternate; pericentral cells mostly
     7–10 ....*P. nathanielii*
   16. Thalli mostly less than 50 mm tall; branching strictly dichotomous
     .... *P. indigena*

Species with Four Pericentral Cells

Polysiphonia acuminata Gardn.

Gardner 1927a: 100; Hollenberg 1942b: 782; Smith 1944: 360.

Thalli chiefly erect, 2–6 cm tall, attached by unicellular rhizoids from prostrate branches or from basal segments of erect branches; rhizoids cut off from pericentral cells as separate cells; erect branches with percurrent axes, 300–500 µm diam., with segments in median parts mostly shorter than broad. Pericentral cells 4, uncorticated. Trichoblasts in spiral sequence on each segment not bearing a branch, 2–6 segments between successive branches. Tetrasporangia 50–70 µm diam., spirally arranged in swollen segments of ultimate branches; spermatangial stichidia 100–150 µm long, 30–50 µm diam., arising as primary fork of trichoblast, sometimes with 1 or 2 sterile cells at apex. Cystocarps subglobose, 250–300 µm diam.

Occasional, mostly epiphytic, sometimes saxicolous, lower intertidal, Monterey to La Jolla, Calif. Type locality: White's Pt. (near San Pedro), Calif.

Polysiphonia bajacali Hollenb.

Hollenberg 1961: 347.

Thalli tufted, 3–6 cm tall, numerous erect branches arising from compact bases of prostrate branches attached by numerous rhizoids; main erect branches mostly 300–350 µm diam., the segments mostly shorter

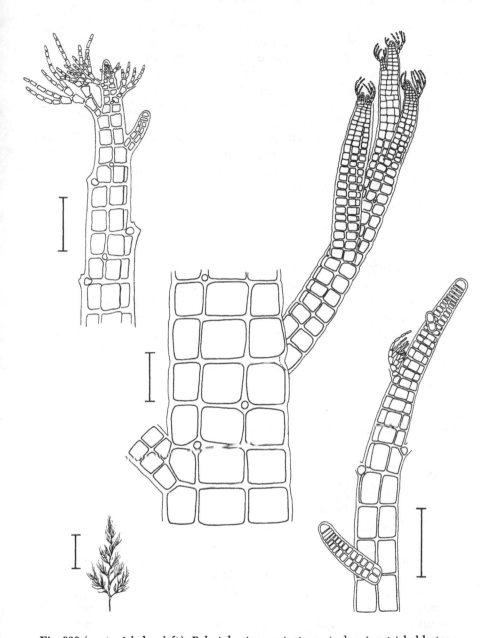

Fig. 630 (center & below left). *Polysiphonia acuminata*: main drawing, trichoblasts or scar cells on each segment not bearing a branch; inset, habit. Rules: main drawing, 100 µm; inset, 1 cm. (both DBP/H) **Fig. 631** (above left). *P. bajacali*: trichoblasts or scar cells on each segment. Rule: 100 µm. (DBP/H) **Fig. 632** (below right). *P. carettia*: trichoblasts or scar cells irregular. Rule: 50 µm. (DBP/H)

than diam.; pericentral cells 4, uncorticated or with a few cortical cells below; main axes indistinct, especially in densely branched upper parts; branches arising mostly from entire trichoblast primordium at irregular intervals; cicatrigenous branches common on lower axes; trichoblasts 1 per segment in spiral sequence, branching 3 or 4 times, soon deciduous; tetrasporangia 50–60 μm diam., in spiral series, slightly protuberant; spermatangial stichidia 200–240 μm long, comprising primary branch of trichoblast, lacking sterile apex; cystocarps urceolate to globular, 240–250 μm diam., with enlarged cells on ostiolar rim.

Infrequent, on rocks or other algae, low intertidal to subtidal (to 6 m), Anacapa I. and San Diego Bay, Calif., to I. Guadalupe, Baja Calif. (type locality).

Polysiphonia carettia Hollenb.

Hollenberg 1971a: 15.

Thalli in dense, soft mats on carapace of host; creeping filaments 50–70 μm diam., attached by unicellular rhizoids, these not cut off by crosswall from pericentral cells; erect filaments at least partly endogenous, to 1.5 cm tall, infrequently branched below, frequently to abundantly branched above; pericentral cells 4, uncorticated; segments in middle parts of erect branches 2–4 times as long as diam.; trichoblasts at irregular intervals, somewhat spirally arranged, mostly short, with 2–4 forks, quickly deciduous, leaving relatively large scar cells; lateral branches all or nearly all cicatrigenous; tetrasporangia 30–35 μm diam., in somewhat spiral arrangement; sexual plants unknown.

Known from single collection from carapace of sea turtle, Santa Catalina I., Calif.

Polysiphonia decussata Hollenb.

Hollenberg 1942b: 780; 1961: 351.

Thalli mostly 1–2 cm tall, deep brownish red; prostrate filaments 150–180 μm diam., attached by numerous unicellular rhizoids, these not cut off by crosswall from pericentral cells; erect branches 150–170 μm diam., the segments about as long as diam. in middle parts; lateral branches mostly exogenous, more or less determinate, replacing trichoblasts in part, somewhat distichous, but decussately arranged in relation to trichoblasts, the branches and trichoblasts occurring alternately in spiral sequence with one-fourth divergence and with 2 or 3 segments between branch and next trichoblast in spiral; pericentral cells 4, uncorticated; trichoblasts mostly with 1 or 2 dichotomies, soon deciduous; tetrasporangia 35–60 μm diam., in irregular spiral series; spermatangial stichidia 100–150 μm long, 30–40 μm diam., comprising primary branch of trichoblast; cystocarps ovoid, somewhat urceolate, 250–320 μm diam.

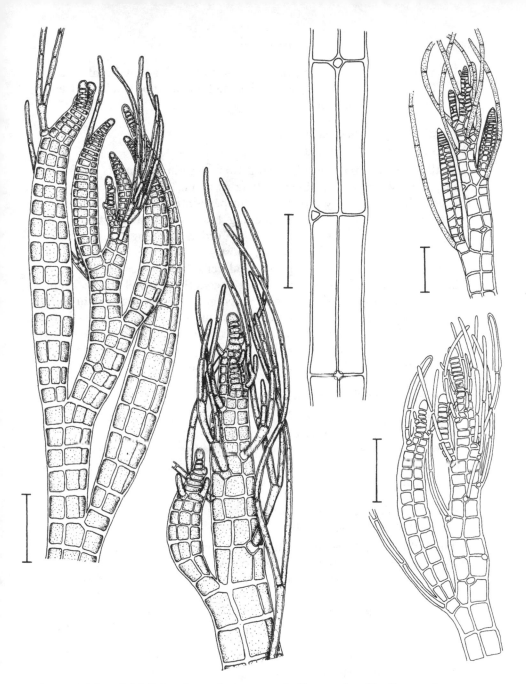

Fig. 633 (above left). *Polysiphonia decussata*: trichoblasts replaced by decussately arranged branches. Rule: 100 μm. (GJH)   **Fig. 634** (below right). *P. flaccidissima* var. *flaccidissima*: lateral branches and trichoblasts formed from same primordia. Rule: 50 μm. (SM/H)   **Fig. 635** (above right). *P. flaccidissima* var. *smithii*: coarser than var. *flaccidissima*. Rule: 200 μm. (LH, after JRJ/S)   **Fig. 636** (two in center). *P. mollis*: below, habit, showing abundant trichoblasts; above, detail, showing scar cells arranged in one-fourth-spiral sequence. Rule: 50 μm. (both GJH)

Infrequent on rocks among other turf-forming algae, midtidal to subtidal, Topanga Canyon area (north of Santa Monica; type locality), Calif., to I. Guadalupe, Baja Calif., and Galápagos Is.

Polysiphonia flaccidissima Hollenb.

Hollenberg 1942b: 783; 1961: 351.

Thalli very soft and flaccid, arising from creeping filaments 70–80 μm diam., the segments ±1.5 times as long as diam., attached by numerous rhizoids, these cut off by crosswall from pericentral cells; erect branches exogenous and assurgent from creeping branches with indistinct main axes; lateral branches of several to many orders, all exogenous and arising in connection with trichoblasts, mostly at intervals of 6 segments, gradually narrowed at bases; pericentral cells 4, uncorticated; cell walls thin and hyaline, perhaps accounting for flaccid thallus; trichoblasts simple or with 1 fork, tapering slightly to rounded apices, 1 per segment in spiral sequence, soon deciduous, leaving persistent scar cells or undeveloped trichoblast primordia resembling scar cells; tetrasporangia in more or less spiral series; spermatangial branches arising as primary branches of trichoblasts, with sterile apices of 1 or 2 cells; cystocarps globular.

Polysiphonia flaccidissima var. flaccidissima

Thalli mostly 1–2.5 cm tall, the erect branches 70–80 μm diam., the segments mostly 1 or 2 times as long as diam.

Infrequent to frequent in limited areas on other algae or on wave-swept rocks, Laguna Beach, Calif. (type locality), to Baja Calif. and Peru. Widely distributed in subtropical and tropical seas.

Polysiphonia flaccidissima var. smithii Hollenb.

Hollenberg 1942b: 784; Smith 1944: 361.

Thalli 2–5 cm tall, coarser and more laxly branched than in var. *flaccidissima*, the main axes 100–140 μm diam., the segments to 4–6 times as long as diam.; tetrasporangia 50–65 μm diam., in short, spiral series; sexual plants unknown.

Infrequent in sheltered water, growing on rocks or on wood, San Francisco to La Jolla, Calif. Type locality: Newport Harbor (Orange Co.), Calif.

Polysiphonia mollis Hook. & Harv.

Hooker & Harvey 1847: 43; Hollenberg 1961: 359. *Polysiphonia snyderiae* Kylin 1941: 35.

Thalli mostly 5–12 cm tall, arising from discoid base or mostly from creeping branches of limited extent, attached by numerous unicellular rhizoids, these cut off by crosswalls from pericentral cells; erect branches 300–400 μm diam. in median parts, the segments commonly 2 or 3 times as long, richly branched pseudodichotomously above, mostly relatively naked below; pericentral cells 4, uncorticated; trichoblasts with 2 or 3 dichotomies, 1 per segment in spiral sequence, soon deciduous, leaving persistent scar cells; branches commonly narrowed at base, at intervals of mostly 6–10 segments, replacing trichoblasts in spiral sequence; tetrasporangia 60–70 μm diam., in spiral series; spermatangial stichidia cylindrical, constituting primary branch of trichoblasts, mostly without sterile apex.

Frequent on rocks, wood, or shells, occasionally epiphytic, lower intertidal, mostly in sheltered water, S. Br. Columbia to Sinaloa, Mexico. Widely distributed in tropical and subtropical Pacific Ocean. Type locality: Tasmania.

Polysiphonia pacifica Hollenb.

Hollenberg 1942b: 777; 1961: 361; Smith 1944: 359.

Thalli medium to dark red, with inconspicuous creeping branches, attached by unicellular rhizoids, these mostly 1 per segment, arising from centers of pericentral cells, but not cut off as separate cells; erect branches to 15 cm tall, mostly endogenous; branching primarily alternate; main axes 60–200 μm diam., with (2)4–5 segments between successive branches; pericentral cells 4, uncorticated, with segments in main branches (2)3–10 times as long as diam.; trichoblasts and scar cells absent or exceedingly rare; tetrasporangia 50–70 μm diam., in long, straight series; spermatangial stichidia cylindrical, with short sterile apices of 1 to several cells, arising from entire trichoblast primordium; cystocarps slightly urceolate, mostly 200–500 μm diam.

Polysiphonia pacifica is a highly variable species, with several varieties recognized in Calif.

1. Thalli delicate, mostly under 2 cm tall; main branches 60–110 μm diam.
   . . . . .var. *delicatula*
1. Thalli coarse, mostly 5–10+ cm tall 2
   2. Ultimate branchlets relatively rigid, short, more or less determinate. . . . 3
   2. Ultimate branchlets indeterminate. 5
3. Thalli mostly under 7 cm tall; ultimate branchlets more or less distichous; main axes usually bearing determinate laterals throughout. . . . . . .var. *disticha*
3. Thalli mostly 10+ cm tall; ultimate branchlets mostly not distichous; main axes with long, naked portions below. 4
   4. Branching dense, the main axes indistinct. . . . . .var. *determinata*
   4. Branching lax, the main axes distinct. . . . . .var. *gracilis*
5. Branches mostly delicate and soft, deep red; branching dense above; plants to 20+ cm tall. . . . . .var. *pacifica*
5. Branches coarse, to 300 μm diam., brownish-red; branching very lax, especially below. . . . . .var. *distans*

Polysiphonia pacifica var. **pacifica**

Hollenberg 1942b: 777; 1961: 361.

Thalli 10–20(25) cm tall, the erect branches densely alternate near terminal portions, lax, irregularly branched, mostly naked below; main axes distinct or indistinct, 100–200(300) μm diam.; branches of many orders, typically not exceeding growing tip of main axes.

Frequent on rocks, pilings, etc., lower intertidal to subtidal (to 30 m), Alaska to Is. Coronados, Baja Calif., and Peru. Type locality: Santa Cruz, Calif.

Polysiphonia pacifica var. **delicatula** Hollenb.

Hollenberg 1942b: 778; 1961: 362.

Thalli under 4 cm tall, the main axes 60–110 μm diam., all branches indeterminate, short and blunt; branch apices often somewhat forcipate.

Occasional, saxicolous, mostly subtidal (to 9 m), Ft. Bragg (Mendocino Co.), Calif., to Baja Calif. and Galápagos Is. Type locality: Monterey, Calif.

Polysiphonia pacifica var. **determinata** Hollenb.

Hollenberg 1942b: 778.

Thalli mostly 10–15 cm tall, lax below, the main axes 170–340 μm diam., the ultimate branches coarse, more or less determinate, forming dense terminal tufts.

Occasional, saxicolous, mid- to low-intertidal, S. Br. Columbia to Malpaso Creek (Monterey Co.) and Government Pt. (Santa Barbara Co.), Calif.; common in C. Calif. Type locality: near Pebble Beach (Monterey Co.), Calif.

Polysiphonia pacifica var. **distans** Hollenb.

Hollenberg 1942b: 779.

Thalli 10–18 cm long, the main axes to 300 μm diam., the branches indeterminate, few, alternate, distant, not forming tufts; young branches strongly incurved.

Occasional on intertidal rocks, S. Br. Columbia to Monterey Peninsula, Calif. Type locality: Carmel, Calif.

Polysiphonia pacifica var. **disticha** Hollenb.

Hollenberg 1942b: 778.

Thalli 3–7 cm tall, the ultimate branches determinate, at first incurved, finally straight or incurved, the apices acute; branching alternate, frequently in pairs, these appearing distichous.

Occasional, saxicolous, upper intertidal, S. Br. Columbia to San Francisco, Calif. Type locality: near Cape Flattery, Wash.

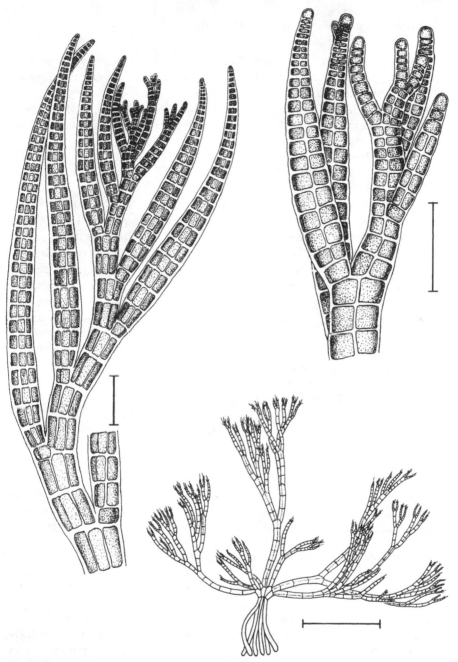

Fig. 637 (above right). *Polysiphonia pacifica* var. *pacifica*: branches crowded in upper portions of thallus. Rule: 100 μm. (GJH)   **Fig. 638** (left). *P. pacifica* var. *determinata*: branches throughout thallus, the main axes indistinct. Rule: 200 μm. (GJH)   **Fig. 639** (below right). *P. savatieri*. Rule: 2 mm. (DBP, after Dawson)

Polysiphonia pacifica var. **gracilis** Hollenb.

Hollenberg 1942b: 778.

Thalli 10–20 cm tall, very laxly branched; axes very distinct, commonly naked below, branches more or less determinate and mostly simple with 1 to several short branchlets near their tips.

Occasional, saxicolous, lower intertidal to upper subtidal, S. Br. Columbia to Crescent City (Del Norte Co.), Calif. Type locality: Orcas I., Wash.

Polysiphonia savatieri Har.

Hariot 1891: 226; Segi 1951: 202; Hollenberg 1961: 363. *Polysiphonia minutissima* Hollenb. 1942b: 781.

Thalli 3–10 mm tall, arising from tuft of rhizoids, these cut off by crosswalls from pericentral cells; erect branches spreading somewhat dichotomously, partly assurgent from brief, horizontal, radiating branches, resulting in bushlike aspect; branches of several orders, exogenous, arising independent of trichoblasts, with no constant interval between successive branches; main axes mostly indistinct, to 145 µm diam., the segments mostly shorter than diam.; pericentral cells 4, uncorticated; trichoblasts short, 1 per segment in spiral sequence, soon deciduous, leaving persistent scar cell; tetrasporangia about 80 µm diam., in short, spiral series, somewhat protuberant; spermatangial stichidia about 135 µm long, 40 µm diam., constituting a primary branch of trichoblast, mostly without sterile apex; cystocarps slightly urceolate, 225–290 µm diam.

Infrequent, epiphytic, low intertidal, Monterey Co., Calif., to I. Guadalupe, Baja Calif.; frequent on Channel Is., Calif. Also frequent in C. and W. tropical Pacific Ocean. Type locality: Japan.

Polysiphonia scopulorum Harv.

Harvey 1854: 540; Hollenberg 1968a: 79.

The type variety, not known from Calif., is distinguished by coarse erect branches, narrowing at the bases, and occurring mostly at every two or three segments of the prostrate branches.

Polysiphonia scopulorum var. **villum** (J. Ag.) Hollenb.

Polysiphonia villum J. Agardh 1863: 941. *P. scopulorum* var. *villum* (J. Ag.) Hollenberg 1968a: 81. *Lophosiphonia villum* (J. Ag.) Setchell & Gardner 1903: 329; Hollenb. 1942a: 535; Dawson 1963b: 421.

Thalli brownish-red, mostly 5–10 mm tall, arising from prostrate branches attached by unicellular rhizoids, these arising from centers of pericentral cells and not separated by crosswalls; erect branches simple or very sparingly branched, 40–60(80) µm diam., the segments 1–1.5 times as long as diam., arising at irregular intervals in strictly endogenous manner; lateral branches exogenous or endogenous, independent of trichoblasts, which they may replace; pericentral cells 4, uncorticated; trichoblasts in-

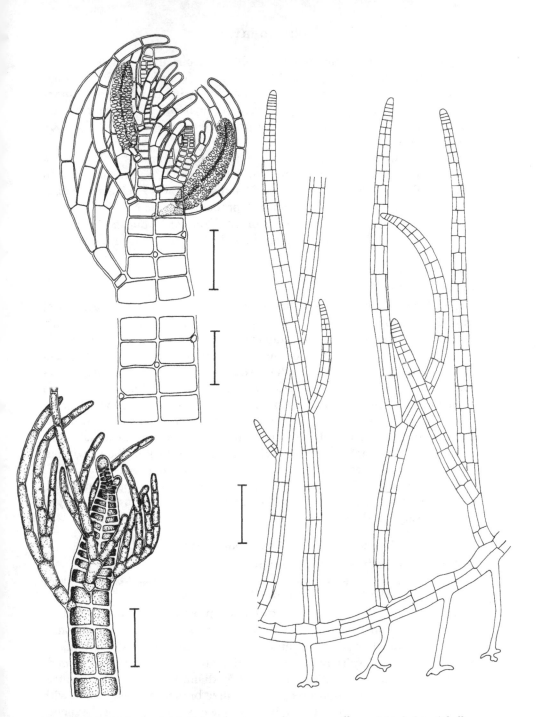

Fig. 640 (right & below left). *Polysiphonia scopulorum* var. *villum*: right, habit of thallus lacking trichoblasts; below left, detail of apex with trichoblasts. Rules: right, 100 μm; left, 200 μm. (JRJ/S & GJH) **Fig. 641** (two above left). *P. simplex*. Rules: habit, 50 μm; detail, 100 μm. (DBP/H)

frequent or lacking, sometimes relatively numerous and 1 per segment in spiral sequence, branching 1 or 2 times, relatively stiff, 250–480 µm long, soon deciduous; tetrasporangia 50–60 µm diam., in straight series, somewhat protuberant; spermatangial stichidia cylindrical, without sterile apices, arising from entire trichoblast primordium; cystocarps ovoid, 150–190 µm diam.

Frequent, with other algae or alone, forming continuous turf on rocks or occasionally on other algae, mid- to low-intertidal, S. Br. Columbia to Panama. Widely distributed in tropical and subtropical Pacific and Atlantic. Type locality: "Americae tropicae," probably Pacific Mexico.

This species has the growth habit of *Lophosiphonia*, but its radial construction and the branching of erect branches exclude it from that genus.

Polysiphonia simplex Hollenb.

Hollenberg 1942b: 782; 1961: 364.

Thalli forming dense mats often of considerable extent on rocks; basal branches creeping, tangled, 250–360 µm diam., mostly of short segments and attached by numerous unicellular rhizoids, these cut off by crosswalls from pericentral cells; erect branches exogenous, assurgent, mostly 2–3 cm tall, 160–250 µm diam. in median parts; main axes not prominent, sparingly branched, the branches radial and exogenous; trichoblasts with 1–3 dichotomies, 1 per segment in spiral sequence, soon deciduous, leaving scar cells; branches replacing trichoblasts; tetrasporangia to 70 µm diam., spirally arranged, somewhat protuberant; spermatangial stichidia a primary branch of trichoblast, 100–170 µm long, 35–40 µm diam., without sterile apices; cystocarps globose to ovoid, 300–350 µm diam.

Frequent, with other sand-binding algae, midtidal to upper intertidal, south of Redondo Beach, Calif., to Guerrero, Mexico, and Costa Rica. Type locality: Laguna Beach, Calif.

Species with More Than Four Pericentral Cells

Polysiphonia brodiaei (Dillw.) Spreng.

Conferva brodiaei Dillwyn 1802–9 (1809): pl. 109. *Polysiphonia brodiaei* (Dillw.) Sprengel 1827: 349; Hollenberg 1944a: 477; Smith 1944: 361.

Thalli mostly 15–25 cm tall, with limited prostrate system attached by numerous rhizoids; deep reddish-brown, soft, adhering closely to paper when dried; erect branches virgate, with percurrent axes 400–500 µm diam., bearing numerous spirally arranged penicillate laterals; segments in middle of chief branches 1–2 times as long as diam. with 6 or 7 pericentral cells; axes fully corticated except for smaller branches; trichoblasts with 1 or 2 forks, 1 per segment in spiral sequence, soon deciduous; branches arising as primary fork of trichoblast; tetrasporangia 70–80 µm diam., in spiral series in ultimate branches; spermatangial stichidia 160–200 µm

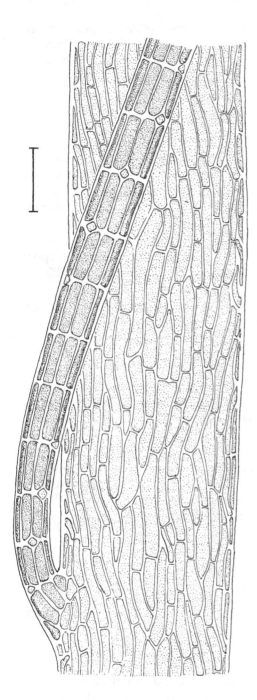

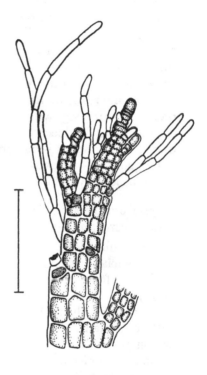

Fig. 642 (left). *Polysiphonia brodiaei*. Rule: 100 μm. (JRJ/S)
Fig. 643 (above). *P. confusa*. Rule: 100 μm. (DBP/H)

long, 35–40 µm diam., mostly without sterile apex, arising as primary fork of trichoblast; cystocarps ovoid to slightly urceolate, 400–430 µm diam.

Common on rocks or wood, low intertidal, in sheltered water of harbors, Seattle, Wash., and Eureka to Santa Monica, Calif. Common in N. Europe. Type locality: British Isles.

Polysiphonia confusa Hollenb.

Hollenberg 1961: 350. *Polysiphonia inconspicua* Hollenb. 1944a: 479.

Thalli 8–13(30) mm tall, arising from prostrate branches 100–170 µm diam., attached by unicellular rhizoids, these cut off as separate cells from pericentral cells; erect branches 60–140 µm diam., segments mostly 1–2.5 times as long, with relatively few lateral branches; lateral branches narrowed at base, arising exogenously in association with trichoblasts at irregular intervals; pericentral cells 8–10, uncorticated; trichoblasts abundant, to 1 mm long, with 1 or 2 dichotomies, 1 per segment, with spiral divergence, soon deciduous, leaving scar cells; tetrasporangia 60–80 µm diam., in long, spiral series; sexual plants unknown.

Frequent in turf on sand-swept rocks, midtidal, Corona del Mar (Orange Co.; type locality), Calif., to northern Baja Calif.

Polysiphonia hendryi Gardn.

Gardner 1927a: 101; Hollenberg 1944a: 482.

Thalli dull reddish-brown, 1.5–5 cm tall, arising from usually brief prostrate branches attached by unicellular rhizoids, these cut off from proximal end of pericentral cells, densely to laxly branched at intervals of mostly 3–5 segments; erect branches mostly 90–150 µm diam., uncorticated, with mostly 10–12 pericentral cells; branches arising in conjunction with trichoblasts; trichoblasts usually abundant, with 1 or 2 dichotomies, arising in spiral sequence, soon deciduous, leaving scar cells; tetrasporangia in spiral series in terminal branches; spermatangial stichidia 80–90 µm long, 25–30 µm diam., arising as primary fork of trichoblast, without sterile apex; cystocarps numerous, ovoid to globular.

1. Thalli mostly less than 1.5 cm tall, densely branched, usually epiphytic. . . . . .var. *hendryi*
1. Thalli mostly 3–10+ cm tall, saxicolous. 2
  2. Branching lax, the branches mostly at intervals of 4 or more segments; plants 5+ cm tall. . . . . .var. *deliquescens*
  2. Branching close, mostly at intervals of 2 or 3 segments; plants mostly less than 4 cm tall. 3
3. Branching very dense, forming compact, spongy, conical primary divisions; with prominent percurrent axes. .var. *compacta*
3. Branching moderately dense, not forming spongy, conical thalli or divisions; main axes less distinct. . . . . .var. *gardneri*

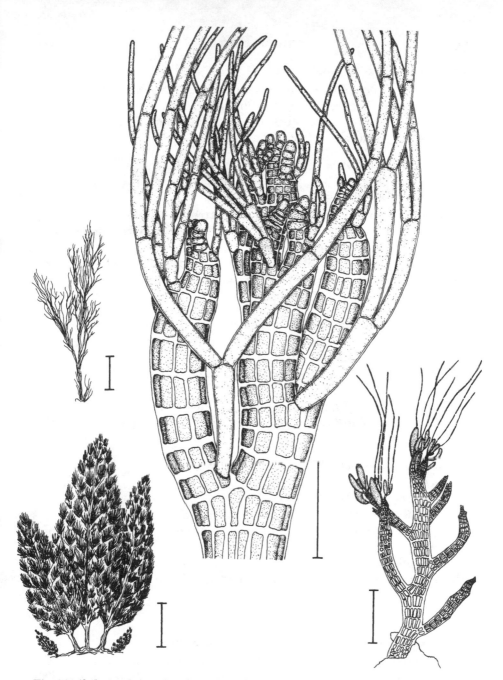

Fig. 644 (below right). *Polysiphonia hendryi* var. *hendryi*: less than 15 mm tall, densely branched. Rule: 250 μm. (DBP/H)　**Fig. 645** (below left). *P. hendryi* var. *compacta*: branches associated in spongy masses. Rule: 2 mm. (DBP/H)　**Fig. 646** (above left). *P. hendryi* var. *deliquescens*: branches lax, widely separated. Rule: 1 cm. (DBP/H)　**Fig. 647** (above center). *P. hendryi* var. *gardneri*: branches abundant, forming firm tufts. Rule: 100 μm. (GJH)

Polysiphonia hendryi var. hendryi

Thalli usually epiphytic; segments mostly about as long as diam.; mostly 2 or 3 segments between successive branches; main axes usually distinct.

Frequent on coralline or other algae, low intertidal, Ore. to I. Cedros, Baja Calif. Type locality: Santo Domingo, Baja Calif.

Polysiphonia hendryi var. compacta (Hollenb.) Hollenb.

Polysiphonia collinsii var. *compacta* Hollenberg 1944a: 481. *P. hendryi* var. *compacta* (Hollenb.) Hollenb. 1961: 356.

Thalli to 4 cm tall, with strongly percurrent axes, beset with densely branched laterals; conical divisions more or less spongy when out of water; mostly 3 segments between successive branches.

Occasional on rocks, midtidal to upper intertidal (along with var. *gardneri*), Pacific Grove, Calif., to northern Baja Calif. Type locality: Pt. Mugu (Ventura Co.), Calif.

Polysiphonia hendryi var. deliquescens (Hollenb.) Hollenb.

Polysiphonia collinsii var. *deliquescens* Hollenberg 1944a: 482. *P. hendryi* var. *deliquescens* (Hollenb.) Hollenb. 1961: 356.

Thalli to 5 cm tall, delicately and laxly branched, with mostly 5 segments between successive branches.

Occasional on rocks, probably mostly in sheltered water, low intertidal, Alaska to Humboldt Bay, Calif.; uncommon in Calif. Type locality: Sitka, Alaska.

Polysiphonia hendryi var. gardneri (Kyl.) Hollenb.

Polysiphonia gardneri Kylin 1941: 38. *P. hendryi* var. *gardneri* (Kyl.) Hollenberg 1961: 355. *P. collinsii* Hollenb. 1944a: 481.

Thalli 2–4 cm tall, abundantly branched, often forming firm tufts, but not densely compact; main axes distinct; branches mostly at intervals of 3 segments.

Relatively common on rocks in exposed areas, mostly high intertidal, Br. Columbia to Cabo San Lucas, Baja Calif. Type locality: San Pedro, Calif.

Polysiphonia indigena Hollenb.

Hollenberg 1958a: 63. *Polysiphonia dichotoma* Hollenb. 1944a: 477.

Thalli to 5 cm tall, forming loose tufts, with branches of several orders, strictly dichotomously branched at apices; main axes distinct below, 300–350 μm diam., the segments mostly half as long as diam. or shorter; pericentral cells mostly 22–24, uncorticated; trichoblasts and scar cells absent; tetrasporangia in straight series on smaller branches; sexual plants unknown.

Saxicolous, low intertidal, known only from Santa Cruz (type locality) and San Diego, Calif.

Polysiphonia johnstonii S. & G.

Setchell & Gardner 1924a: 767; Hollenberg 1961: 357.

Thalli dull brownish-red to nearly black, relatively rigid, attached by numerous rhizoids, these cut off as separate cells from pericentral cells of limited prostrate base; primary branches relatively distinct, alternately and closely branched at narrow angles; pericentral cells 5 or 6, uncorticated; trichoblasts branching 1 or 2 times, 1 per segment in spiral sequence, soon deciduous; branches replacing trichoblasts at irregular intervals; tetrasporangia 70–90 μm diam., in spiral series; spermatangial stichidia 100–140 μm long, 30–40 μm diam., arising from primary branch of trichoblasts; cystocarps globular, 450–500 μm diam.

Polysiphonia johnstonii var. johnstonii

Thalli larger than those of var. *concinna*, 5–12 cm tall, with 5–6 pericentral cells; main axes to 1 mm diam., the segments 1.5 times longer than diam.

Usually epiphytic, low intertidal, Santa Catalina I., Calif. (represented by single Calif. collection), to I. Margarita, Baja Calif., and Gulf of Calif. Type locality: I. San Esteban, Gulf of Calif.

Polysiphonia johnstonii var. concinna (Hollenb.) Hollenb.

Polysiphonia concinna Hollenberg 1944a: 474. *P. johnstonii* var. *concinna* (Hollenb.) Hollenb. 1961: 358.

Thalli similar to those of var. *johnstonii*, but smaller and shorter throughout, mostly less than 2 cm tall, with 5 pericentral cells; main axes 300–400 μm diam., the segments each about one-half as long as diam. and in some cases even stouter.

Occasional, epiphytic, low intertidal, Santa Catalina I., Calif., to Gulf of Calif. and Sinaloa, Mexico. Type locality: La Jolla, Calif.

Polysiphonia nathanielii Hollenb.

Hollenberg 1958a: 63. *Polysiphonia dictyurus sensu* Hollenb. 1944a: 479.

Thalli coarse, erect, 10–15 cm tall, with several main axes arising from basal mass of compact branches; main branches percurrent, sparingly branched, beset throughout with numerous short, closely set, spirally or sometimes somewhat distichously arranged branchlets of relatively limited growth, these arising exogenously at intervals of 2 or 3 segments and exhibiting one-fourth divergence; branchlets at first incurved, later straight or recurved; pericentral cells mostly 7–10, relatively large, uncorticated, the segments mostly shorter than diam. throughout; trichoblasts seemingly

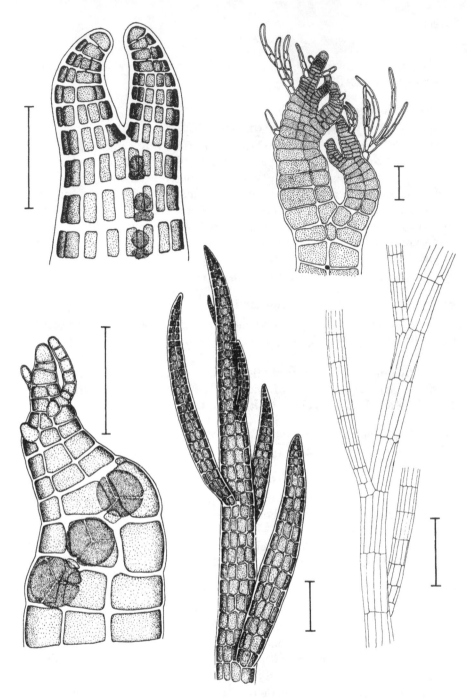

Fig. 648 (above left). *Polysiphonia indigena.* Rule: 100 μm. (GJH)   **Fig. 649** (above right). *P. johnstonii* var. *johnstonii.* Rule: 100 μm. (DBP/H)   **Fig. 650** (below left). *P. johnstonii* var. *concinna.* Rule: 100 μm. (GJH)   **Fig. 651** (below center). *P. nathanielii.* Rule: 250 μm. (GJH)   **Fig. 652** (below right). *P. paniculata.* Rule: 250 μm. (JRJ/S)

lacking; tetrasporangia in straight series in subultimate branchlets; spermatangial stichidia cylindrical, nearly or quite sessile, with 1-celled sterile apices; cystocarps unknown; dried specimens nearly black, of firm consistency, not adhering to paper.

Rare, on rocks, low intertidal, near Santa Monica, Calif. (type locality), and Rosarita Beach (near Tijuana), Baja Calif.

Polysiphonia paniculata Mont.

Montagne 1842: 254; Hollenberg 1944a: 480; 1961: 362. *Polysiphonia californica* Harvey 1853: 48; Smith 1944: 362.

Thalli densely tufted, soft, deep brownish-red, 10–20(30) cm tall, with densely matted but limited system of prostrate branches, attached by numerous rhizoids, these cut off from proximal end of pericentral cells as separate cells; erect branches endogenous, sparsely branched below, repeatedly and densely branched above; pericentral cells mostly 10–12, uncorticated; main axes distinct, 300–430 μm diam., the segments 1.5–2.5+ times as long; trichoblasts abundant, relatively coarse, to 800 μm long, with 1 or 2 forks, 1 per segment in one-fourth spiral sequence, soon deciduous, leaving relatively large scar cells; branches exogenous, arising in connection with trichoblasts; tetrasporangia 80–100 μm diam., in spiral series; spermatangial branches 130–250 μm long, 50–70 μm diam., without sterile apex, arising as primary branch of trichoblast; cystocarps globular-ovoid, 350–400 μm diam.

On sand-swept rocks, mostly lower intertidal, S. Br. Columbia to Baja Calif. and Gulf of Calif.; very common in C. and N. Calif. Type locality: Peru.

Murrayellopsis Post 1962

Thalli radially constructed, monopodial, with erect, uncorticated axes arising from decumbent primary branches. Erect branches bearing pigmented, monosiphonous, branched ramuli. Tetrasporangia tetrahedral, 2 per segment in spiral series in upper part of polysiphonous axes. Sexual plants unknown.

Murrayellopsis dawsonii Post

Post 1962: 1; Dawson 1963b: 415. *Veleroa subulata sensu* Hollenberg & Abbott 1966: 114.

Thalli flaccid, bright red, 7–10 cm tall; erect main axes ±120 μm diam.; pericentral cells 4 or 5; monosiphonous ramuli simple to mostly branched from first- and second-lowermost cells, 1 per segment in spiral sequence, mostly with sharp apices.

Occasional, locally abundant, tangled among other algae and sometimes forming "nests" of the ocean goldfish (*Hypsypops rubicunda*) commonly

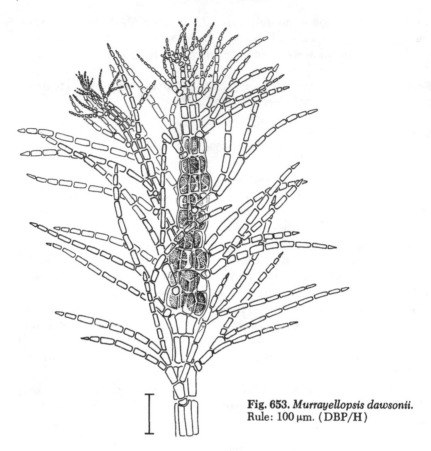

Fig. 653. *Murrayellopsis dawsonii.*
Rule: 100 µm. (DBP/H)

known as garibaldi, subtidal (5–11 m), Pacific Grove, Calif., to northern
Baja Calif.; common along Channel Is., Calif. Type locality: Pt. Loma
(San Diego Co.), Calif.

Veleroa Dawson 1944

Thalli erect, monopodial, sparingly branched, with large, persistent,
monosiphonous, pigmented ramuli. Main axes with 4 uncorticated pericentral cells. Monosiphonous ramuli 1 per segment, in spiral sequence.
Tetrasporangia 1 per segment, in spiral series. Sexual plants unknown.

Veleroa subulata Daws.

Dawson 1944a: 335; 1963b: 414.

Thalli pinkish-red, the main axes 10–15 mm tall, 50–70 µm diam., the
segments 1.5–2 times as long; colored ramuli 400–700 µm long, 25–40 µm
diam. at base, mostly simple, occasionally branched near base, subulate,

usually slightly curved upward, with sharp-pointed apex; tetrasporangia in series of 8–15 in main branches, slightly protuberant, ±50 μm diam.

Subtidal, known only from Santa Catalina I., Calif., and from 22 m, Bahía Tepoca, Sonora, Mexico (type locality).

Ophidocladus Falkenberg 1897

Thalli erect, uncorticated, the endogenous branches arising from pros-

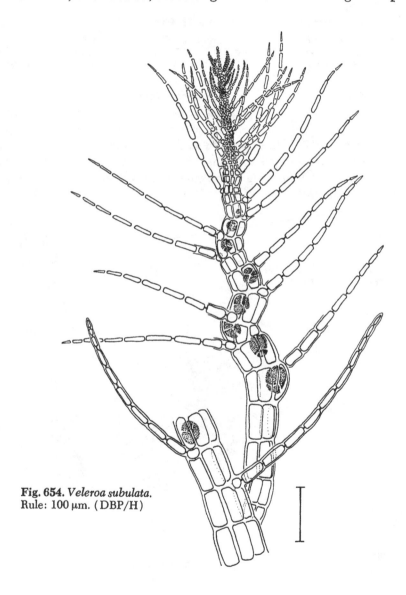

Fig. 654. *Veleroa subulata.*
Rule: 100 μm. (DBP/H)

trate branches attached by unicellular rhizoids; erect branches sparingly branched, at least in part exogenously, bearing distichous, branched, unpigmented trichoblasts; pericentral cells numerous, surrounding much larger central cell; tetrasporangia 2(3) per segment, arising in plane perpendicular to plane of trichoblasts, tetrahedrally divided; spermatangial stichidia cylindroconical, with sterile apices, wholly replacing trichoblasts; cystocarps globose, lateral on erect branches.

Ophidocladus simpliciusculus (Crouan & Crouan) Falk.

Polysiphonia simpliciuscula Crouan & Crouan 1867: 157. *Ophidocladus simpliciusculus* (Crouan & Crouan) Falkenberg 1897: 461; Saenger 1971: 291. *Rhodosiphonia californica* Hollenberg 1943a: 573; Dawson 1963b: 417. *O. californica* (Hollenb.) Kylin 1956: 542.

Thalli dark red, the prostrate branches attached by numerous rhizoids, usually several per segment, cut off as separate cells, mostly from distal end of pericentral cells; pericentral cells mostly 16–18; segments in erect branches mostly 0.5–1 times as long as diam., frequently twisted; chloroplasts commonly arranged in transverse bands; trichoblasts at intervals of 2 or 3 segments, branching 3 or 4 times; tetrasporangia ±60 µm diam., in long, nonspiraling series; spermatangial stichidia 90–300 µm diam., sim-

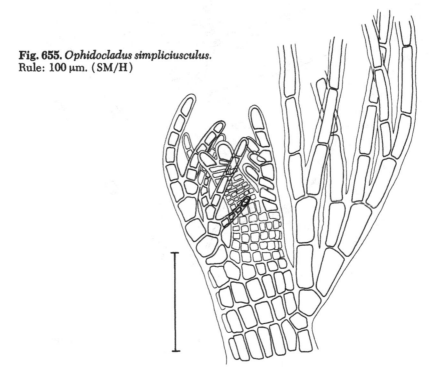

Fig. 655. *Ophidocladus simpliciusculus.*
Rule: 100 µm. (SM/H)

ple or forked, with sterile apices of 4–6 short cells; cystocarps 250–300 μm diam.

Occasional to common, on sand-swept rocks, midtidal to subtidal (to 11 m), Loon Pt. (Santa Barbara Co.), Calif., to Guerrero, Mexico. Type locality: Atlantic coast of France.

Pterosiphonia Falkenberg 1897

Thalli chiefly erect from prostrate branches; erect branches alternately distichous, with subordinate branches of 1 to several orders. Branches cylindrical to strongly compressed, with 4–20 pericentral cells, usually uncorticated. Ultimate branchlets determinate, typically without trichoblasts. Tetrasporangia 1 per segment in straight series, mostly in ultimate branchlets, tetrahedrally divided. Spermatangial stichidia clustered near apices of determinate branchlets or on spurlike branchlets from determinate branches, distichously arranged, with or without sterile apices, arising from entire trichoblast primordium. Cystocarps ovoid to globular, briefly pedicellate.

1. All but youngest branches completely corticated . . . . *P. baileyi*
1. All branches uncorticated 2
   2. Thalli mostly 10+ cm tall; branches cylindrical, not conspicuously distichous . . . .*P. bipinnata*
   2. Thalli mostly less than 10 cm tall; branches slightly compressed or flattened, conspicuously distichous 3
3. Ultimate branchlets strongly recurved . . . .*P. clevelandii*
3. Ultimate branchlets not strongly recurved 4
   4. Thalli mostly bipinnate or tripinnate; branches strongly flattened . . . .
   . . . .*P. dendroidea*
   4. Thalli once-pinnate, usually partly bipinnate; branchlets slightly or not compressed . . . .*P. pennata*

Pterosiphonia baileyi (Harv.) Falk.

Rytiphlaea baileyi Harvey 1853: 29. *Pterosiphonia baileyi* (Harv.) Falkenberg 1901: 270; Smith 1944: 367; Dawson 1963b: 426.

Thalli 8–25 cm tall, nearly black, several to many erect branches arising from common base; main axes percurrent, cylindrical, completely corticated, unbranched below, much branched above; branching mostly distichous, tripinnate, the ultimate branchlets subacute, 60–150 μm diam., uncorticated; pericentral cells 12–14; tetrasporangia 80–100 μm diam., in densely congested, incurved branchlets; spermatangial branchlets incurved, terete, blunt, 200–300 μm long, 60–70 μm diam.; cystocarps globular, 700–800 μm diam.

Frequent on rocks, intertidal to subtidal (to 6 m), Crescent City (Del Norte Co.), Calif., to I. San Roque, Baja Calif. Type locality: Monterey Bay, Calif.

Fig. 656. *Pterosiphonia baileyi.* Rule: 2 cm. (JRJ/S)

Pterosiphonia bipinnata (Post. & Rupr.) Falk.

Polysiphonia bipinnata Postels & Ruprecht 1840: 22. *Pterosiphonia bipinnata* (Post. & Rupr.) Falkenberg 1901: 273; Smith 1944: 366. *Pterosiphonia robusta* Gardner 1927a: 102. *Polysiphonia californica* var. *plumigera* Harvey 1862: 168.

Thalli dark brownish-red to bright red, 6–12(25) cm tall; erect branches cylindrical, uncorticated, laxly and repeatedly branched; final orders of branching distichous and sometimes dense, commonly with recurved branchlets near base of different orders of branches; major branches 180–200(260) μm diam., with mostly 3 segments between successive branches; pericentral cells 11–12(18); tetrasporangia in straight series in ultimate and subultimate branches; spermatangial stichidia 3 or 4, sometimes more, on short spur branchlets, these adaxial on lower third of ultimate branches, 400–480 μm long, 80–90 μm diam., with sterile apex of 3 or 4 short cells; cystocarps globular, to 450 μm diam.

Frequent on rocks, midtidal to upper subtidal, most common on ex-

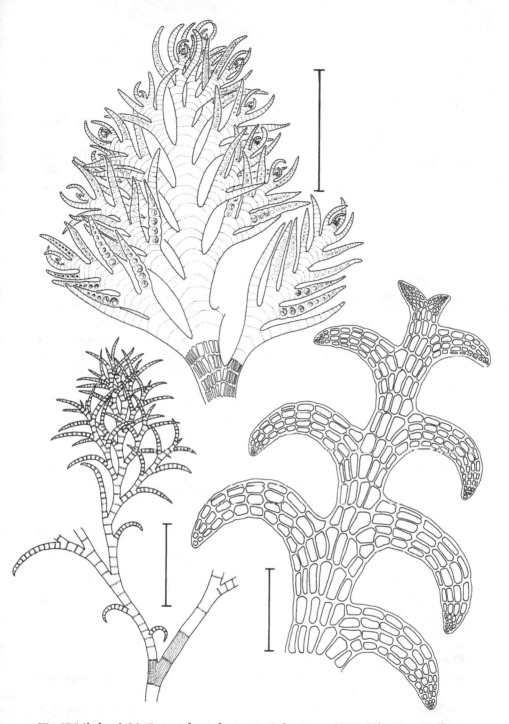

Fig. 657 (below left). *Pterosiphonia bipinnata*. Rule: 1 mm. (DBP/H)   **Fig. 658** (below right). *P. clevelandii*. Rule: 250 µm. (DBP/H)   **Fig. 659** (above). *P. dendroidea*. Rule: 1 mm. (DBP/H)

posed coast, Japan and Alaska to San Pedro, Calif. Type locality: Kamchatka Peninsula, U.S.S.R.

Pterosiphonia clevelandii (Farl.) Hollenb.

Polysiphonia dictyurus var. *clevelandii* Farlow 1877: 237. *Pterosiphonia clevelandii* (Farl.) Hollenberg 1970: 67.

Thalli erect, 2–3 cm tall, dark red, from prostrate branches; erect branches with mostly 3 orders of branching; main axes percurrent, compressed, 300–350 μm broad, the segments about one-half as long as broad, bearing lateral branches at intervals of 2 segments; ultimate branchlets strongly recurved, 300–350 μm long, at intervals of 2 segments; pericentral cells 10–12, uncorticated; reproduction unknown.

Known only from the type collection from drift, San Diego, Calif.

Pterosiphonia dendroidea (Mont.) Falk.

Polysiphonia dendroidea Montagne 1838: 353. *Pterosiphonia dendroidea* (Mont.) Falkenberg 1901: 268; Smith 1944: 366; Dawson 1963b: 426; André 1967: 37. *Pterosiphonia gracilis sensu* Hollenberg & Abbott 1966: 115.

Thalli brownish-red to deep red, mostly 2–10 cm tall, with numerous erect branches arising from creeping branches at intervals of mostly 3 segments; prostrate branches attached by groups of unicellular rhizoids, these cut off from pericentral cells of adjacent segments; erect axes percurrent, with branching of 2 or 3 orders; branches of all orders connate for 2–2.5 segments, arising at intervals of 2 segments; ultimate and subultimate branchlets mostly determinate; segments of main axes 200–500 μm broad, distinctly to strongly compressed, the segments mostly broader than long; tetrasporangia 60–70 μm diam.; spermatangial stichidia cylindroconical, incurved, 100–160 μm long, borne distichously in clusters of 1 per segment, on slender spur branchlets adaxial toward base of determinate branches, with sterile apices of 1 to several cells.

Common, mostly saxicolous, midtidal to lower intertidal and subtidal (10 m), northern Br. Columbia to I. Magdalena, Baja Calif., and Chile. Type locality: Peru.

This is a highly variable species. Variants include delicately branched forms mostly 2–3 cm tall, epiphytes, and an infrequent coarse form with prominent percurrent axes and bearing uniformly short laterals with short ultimate branchlets.

Pterosiphonia pennata (C. Ag.) Falk.

Hutchinsia pennata C. Agardh 1824: 146. *Pterosiphonia pennata* (C. Ag.) Falkenberg 1901: 263; Dawson 1963b: 427. *P. californica* Kylin 1941: 39.

Thalli saxicolous or occasionally epiphytic; main erect branches only slightly compressed, exogenous from creeping branches, mostly at intervals of 3 segments, 1–2.5 cm tall, once-pinnate or partly bipinnate; branch-

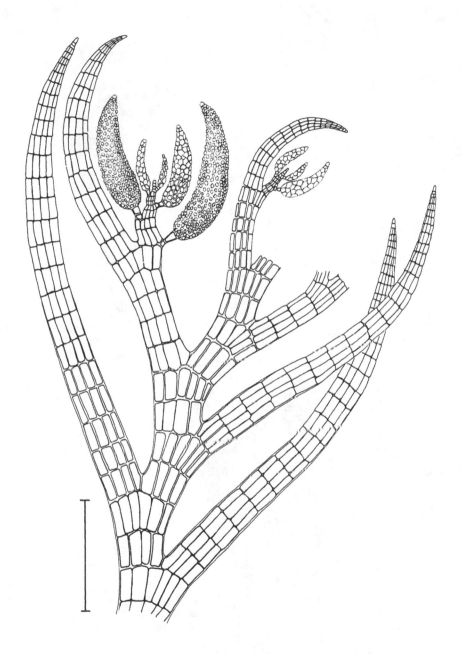

Fig. 660. *Pterosiphonia pennata.* Rule: 250 µm. (DBP/H)

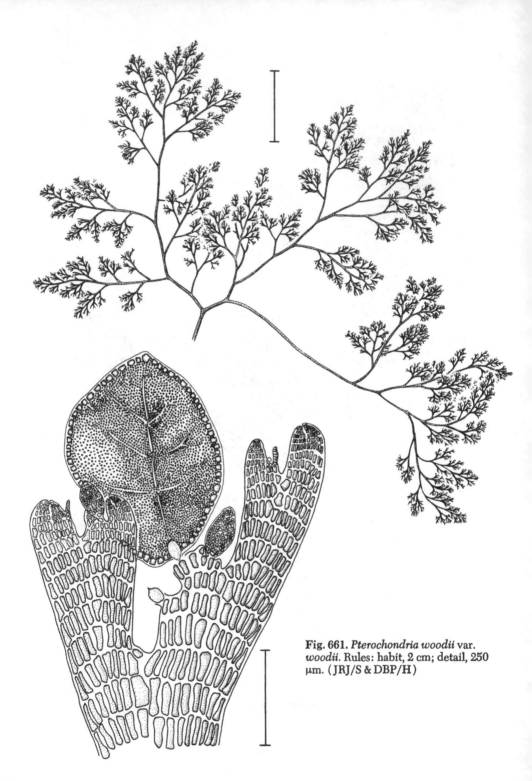

Fig. 661. *Pterochondria woodii* var. *woodii*. Rules: habit, 2 cm; detail, 250 μm. (JRJ/S & DBP/H)

lets cylindrical, mostly remaining simple, slightly incurved, mostly 1–1.5 mm long, at intervals of 2 segments, connate with axis for about 1 segment; segments in main parts of erect branches ±140 μm diam., 1–1.5 times as long; pericentral cells ±8; trichoblasts usually lacking; tetrasporangia in series of 8–11 in upper determinate branchlets; spermatangial stichidia several, borne distichously and adaxially on short, spurlike branchlets several segments from base of determinate branchlets; cystocarps not observed.

Occasional on rocks, low intertidal, Oakland, Calif., to Bahía San Lucas, Baja Calif. Also Japan and Europe. Type locality: Mediterranean.

Pterochondria Hollenberg 1942

Thalli epiphytic, wholly erect, attached basally by group of rhizoids. Branching of several orders, all alternate and distichous, mostly without percurrent axes; branches decurrent, strongly flattened, mostly uncorticated, with 12–30 pericentral cells; vegetative branches without trichoblasts. Tetrasporangia 1 per segment, in straight series on adaxial side of ultimate branches, tetrahedrally divided. Spermatangial stichidia discoid, oval, with sterile margin, borne near apices of smaller terminal branchlets. Cystocarps lateral on subultimate branches.

Pterochondria woodii (Harv.) Hollenb.

Polysiphonia woodii Harvey 1853: 52. *Pterochondria woodii* (Harv.) Hollenberg 1942a: 533; Dawson 1963b: 428.

Thalli strongly flattened, lying in ribbonlike clusters; pericentral cells ±12 above, increasing to ±30 below; major branches to 500 μm broad, with several segments between successive branches; axes sometimes corticated at lower indistinct nodes; tetrasporangia 80–100 μm diam.; spermatangial stichidia to 650 μm long, 390 μm broad, disk-shaped; mature cystocarps 400–800 μm diam.

Pterochondria woodii var. woodii

Thalli 8–15(25) cm tall, greenish-yellow to dull brownish-red, varying probably with light exposure; segments to 4 times as long as diam. below, shortening toward apices to as long as diam. or less; 3 or 4 segments between successive branches; cystocarps globular, 400–800 μm diam.

Common, growing on *Cystoseira* or occasionally on *Macrocystis* or other large brown algae, subtidal (5 m), S. Br. Columbia to Bahía del Rosario, Baja Calif. Type locality: N. California near Pt. Reyes (Marin Co.).

Pterochondria woodii var. pygmaea (Setch.) Daws.

Pterosiphonia woodii var. *pygmaea* Setchell, P.B.-A. 1895–1919 [1911]: no. 1744. *Pterochondria woodii* var. *pygmaea* (Setch.) Dawson 1963b: 429. *Pterochondria pygmaea* (Setch.) Hollenberg 1942a: 533.

Thalli like var. *woodii* but only 1–2 cm tall, compactly branched, the

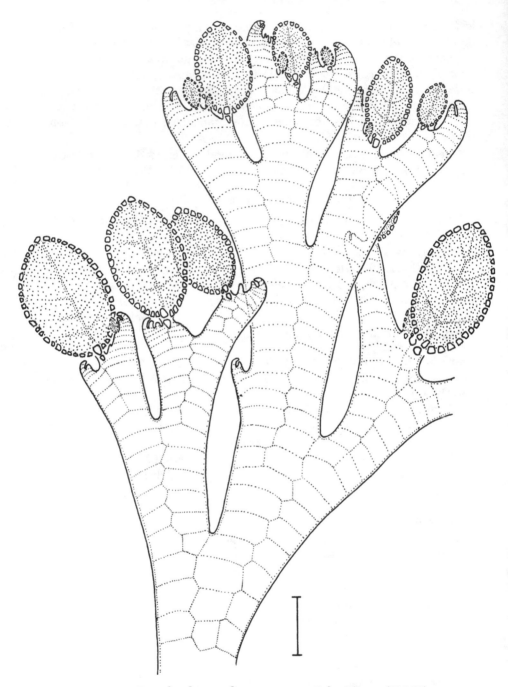

Fig. 662. *Pterochondria woodii* var. *pygmaea*. Rule: 250 μm. (SM/H)

branches narrowly divergent; pericentral cells 12–25, the segments mostly shorter than broad, the branch intervals shorter; cystocarps commonly ovate, 350–400 μm diam.

Frequent, mostly on *Cystoseira osmundacea*, shallow subtidal (1–5 m), Santa Barbara, Calif., to Bahía Asunción, Baja Calif.; common subtidally off Channel Is., Calif. Type locality: San Pedro, Calif.

Jantinella Kylin 1941

Thalli minute, whitish parasites, with nodular central part and very short, lax, polysiphonous free branches. Pericentral cells 7 or 8, each transversely divided into 2 or 3 cells. Vegetative branches without trichoblasts. Tetrasporangial stichidia lateral on free branches, lanceolate, with curved, sterile apices. Spermatangial stichidia cylindrical, with blunt apices. Cystocarps globular to urceolate, lateral on free branches.

Jantinella verrucaeformis (Setch. & McFadd.) Kyl.

Colacodasya verrucaeformis Setchell & McFadden 1911: 149. *Jantinella verrucaeformis* (Setch. & McFadd.) Kylin 1941: 40; Dawson 1963b: 438.

Thalli growing on *Chondria californica*, commonly causing slight bend at point of attachment, 0.5–1.5 mm diam.; free polysiphonous branches 180–360 μm long, 80–130 μm diam.; branches spirally inserted with approximately two-fifths divergence.

Frequent to common epiphyte, low intertidal, Government Pt. (Santa Barbara Co.), Calif., to Is. Revillagigedo, Mexico. Type locality: White's Pt. (San Pedro), Calif.

Levringiella Kylin 1956

Thalli minute, densely tufted, "parasitic" on *Pterosiphonia* species; polysiphonous, with dense monopodial branching. Branches spirally arranged; pericentral cells mostly 5–7, uncorticated; trichoblasts absent. Tetrasporangia 1 per segment, in straight series in determinate branches. Spermatangial stichidia cylindrical, on short pedicels. Cystocarps globular.

Levringiella gardneri (Setch.) Kyl.

Stromatocarpus gardneri Setchell 1923a: 395; Hollenberg 1948: 159. *Levringiella gardneri* (Setch.) Kylin 1956: 517.

Thalli small, tufted, pinkish-white, to 600 μm tall, the segments shorter than broad; branches radial, at approximately one-fourth divergence; tetrasporangia about 20 μm diam., in straight series; spermatangial stichidia cylindrical, arcuate, 140–160 μm long, 40–50 μm diam., with rounded, nonsterile apices; cystocarps 275–300 μm diam.

Rare, "parasitic" on *Pterosiphonia baileyi* or *P. dendroidea*, shallow subtidal (1–3 m), Malpaso Creek (Monterey Co.), San Luis Obispo Co., Santa Monica (type locality), and La Jolla, Calif.

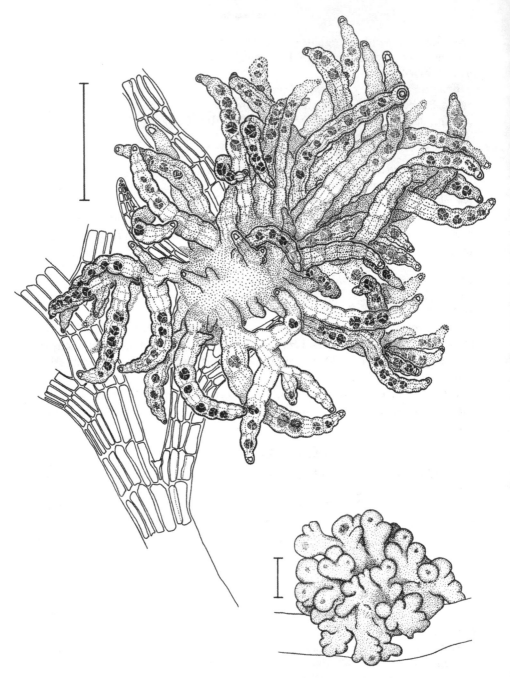

Fig. 663 (below right). *Jantinella verrucaeformis*. Rule: 250 μm. (EYD)   **Fig. 664** (above). *Levringiella gardneri*. Rule: 250 μm. (DBP/H)

Herposiphonia Nägeli 1846

Thalli of indeterminate prostrate branches attached by frequent unicellular rhizoids, these cut off as separate cells from distal end of pericentral cells. Primary indeterminate branches bearing determinate and indeterminate branches in more or less regular sequence, mostly with 3 determinate branches between successive indeterminate branches, these alternate on either side of axis. Mature determinate branches distichous or erect and parallel, with or without trichoblasts, mostly with characteristic number of pericentral cells for given species. Cells with numerous small, discoid chloroplasts usually randomly distributed. All branches uncorticated. Tetrasporangia in straight series, 1 per segment in determinate branches, tetrahedral or obliquely cruciate. Spermatangial stichidia cylindrical, on determinate branches, usually several per branch, without sterile apices and without accompanying trichoblasts. Cystocarps ovoid to somewhat urceolate, on determinate branches.

1. Mature determinate branches obviously distichous.................... 2
1. Mature determinate branches not distichous or obscurely so............ 3
    2. Determinate branches strictly distichous in 1 plane.........*H. plumula*
    2. Determinate branches distichous but at maturity standing at various
       angles with axis.....................................*H. verticillata*
3. Branches with frequent to regular bare segments..............*H. tenella*
3. Branches mostly with determinate or indeterminate branch on every
   segment.. 4
    4. Pericentral cells of determinate branches ±12; cystocarps terminal
       or subterminal .... *H. littoralis*
    4. Pericentral cells of determinate branches 6–8 in lower parts; cysto-
       carps usually on third segment from base..............*H. hollenbergii*

Herposiphonia hollenbergii Daws.

Dawson 1963b: 430.

Thalli minute, turf-forming, partly creeping, partly erect; branching with 3 erect, determinate branches between successive indeterminate branches, and with no bare nodes; determinate branches 1–4 mm long, of 40+ segments, the apices blunt; segments in lower part of determinate branches 40–50 µm diam., mostly 2–3 times as long, with 6–8 pericentral cells, those of upper parts 60–80 µm diam., gradually shortening to half as long as diam., with 10–12 pericentral cells; indeterminate branches, in some densely tufted plants, commonly elongate and erect, 4–5 mm tall, the indeterminate laterals on such branches mostly remaining rudimentary; trichoblasts short, sparse or mostly lacking; tetrasporangia ±50 µm diam., in continuous or broken series; spermatangial stichidia unknown; cystocarps mostly adaxial on third segment from base of determinate branches of erect indeterminate branches, urceolate, with prominent, slightly flared rostrum. (After Dawson.)

Fig. 665. *Herposiphonia hollenbergii.*
Rule: 250 µm. (EYD)

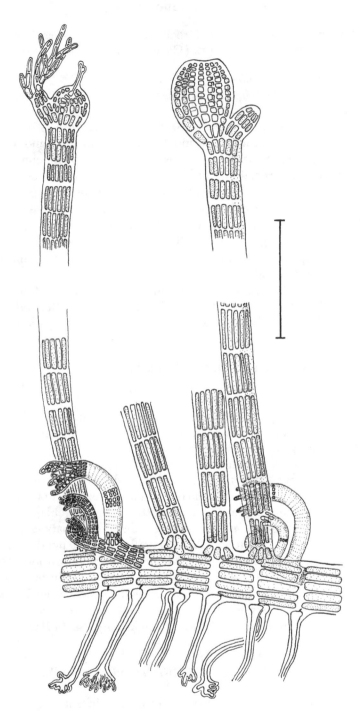

Fig. 666. *Herposiphonia littoralis*. Rule: 250 μm. (DBP/H)

Infrequent, on surf-dashed rocks, midtidal to lower intertidal, Santa Monica, Calif., to Salina Cruz (Oaxaca; type locality), Mexico.

Herposiphonia littoralis Hollenb.

Hollenberg 1970: 69. *Herposiphonia tenella sensu* Hollenb. 1948: 159; *sensu* Dawson 1963b: 435.

Thalli of prostrate branches 100–170 μm diam., with 3 erect determinate branches between successive indeterminate branches, and with no bare nodes; determinate branches simple, 2–5 mm tall, the indeterminate lateral branches mostly remaining rudimentary; determinate branches mostly 60–90 μm diam., of 20–27 segments, these 1.5–2 times as long as diam.; pericentral cells 10–12; chloroplasts commonly in transverse bands; trichoblasts 3 or 4, terminal to subterminal, 300–400+ μm long, branching 6 or 7 times; tetrasporangia 60–70 μm diam., in upper half or more of determinate branches; spermatangial stichidia not observed; procarps and young cystocarps subterminal.

Frequent to common, forming thin or felted patches 4–8 cm broad on rocks, midtidal to upper intertidal, Santa Catalina I., Redondo Beach, and Laguna Beach, Calif., to I. Magdalena, Baja Calif. Type locality: Corona del Mar (Orange Co.), Calif.

Herposiphonia plumula (J. Ag.) Hollenb.

Polysiphonia plumula J. Agardh 1884–85 (1885): 99. *Herposiphonia plumula* (J. Ag.) Hollenberg 1970: 68 (incl. synonymy).

Thalli creeping or partly erect, to 3 cm tall, brownish-red to bright red, strictly distichously branched, with 3 alternate determinate branches between successive alternate indeterminate branches, and with no bare nodes; determinate branches ±1.2 mm long, broadest at base, tapering to acute apices, with 10–12 segments; indeterminate branches progressively longer from apex to base of main axes; segments of axes 1–1.5 times as long as diam.; pericentral cells mostly 8–12; trichoblasts lacking; tetrasporangia obliquely cruciate, ±60 μm diam.; spermatangial stichidia cylindroconical and incurved, 160–360 μm long, 50–70 μm diam., 1 to several on adaxial side of determinate branchlets near bases, without sterile apices; cystocarps ovoid, 300–450 μm diam., occurring singly on adaxial side of determinate branchlets, usually on second segment from bases.

Frequent on rocks or other algae, midtidal to shallow subtidal (4 m), N. Br. Columbia to Oaxaca, Mexico. Type locality: Santa Barbara, Calif.

Herposiphonia tenella (C. Ag.) Ambr.

Hutchinsia tenella C. Agardh 1828: 105. *Herposiphonia tenella* (C. Ag.) Ambronn 1880: 195.

The type form is not known to occur in California.

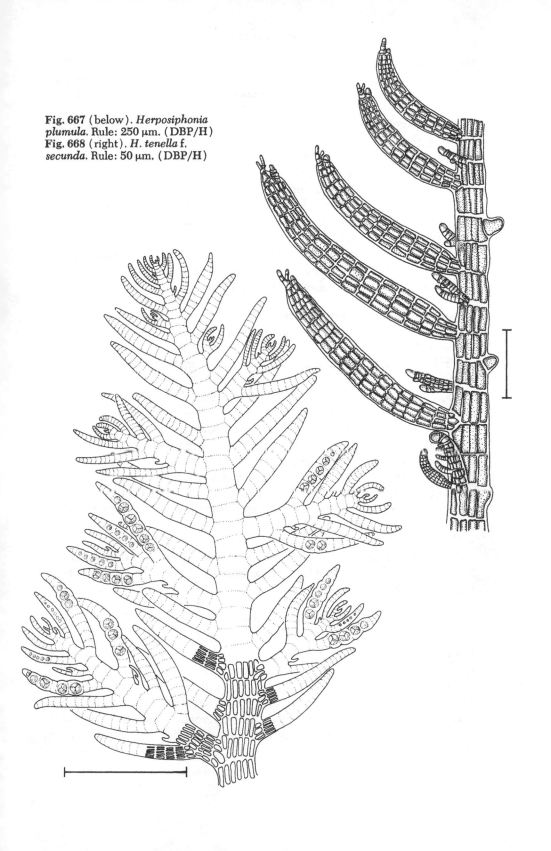

Fig. 667 (below). *Herposiphonia plumula*. Rule: 250 μm. (DBP/H)
Fig. 668 (right). *H. tenella* f. *secunda*. Rule: 50 μm. (DBP/H)

Herposiphonia tenella f. secunda (C. Ag.) Hollenb.

Hutchinsia secunda C. Agardh 1824: 149. *Herposiphonia tenella* f. *secunda* (C. Ag.) Hollenberg 1968b: 556. *Herposiphonia secunda* (C. Ag.) Ambronn 1880: 197; Dawson 1963b: 432.

Thalli creeping, to 15+ mm in extent; indeterminate branches (or primordia) alternating mostly on every third or fourth segment, with 1 or 2 (3) bare segments and sometimes 1 determinate branch between successive indeterminate branches; main axes 70–110 μm diam., with 8–10 pericentral cells; determinate branches simple, erect, circinnate when young, 40–70 μm diam., 0.5–1 mm tall, with 7–9 pericentral cells and usually several branched trichoblasts; reproduction of Pacific Coast specimens not observed.

Infrequent, on rocks, shells, or other algae, low intertidal, Santa Monica, Calif., to Panama. Type locality: Mediterranean.

Identity of sterile specimens very uncertain.

Herposiphonia verticillata (Harv.) Kyl.

Polysiphonia verticillata Harvey 1833a: 165. *Herposiphonia verticillata* (Harv.) Kylin 1925: 74; Smith 1944: 370.

Thalli primarily prostrate and somewhat matted, with some free branches 1–2 cm long, dark reddish-brown; 3 alternating determinate branches between successive indeterminate branches, and no bare nodes; determinate branches arising distichously but soon mostly curved and variously directed; indeterminate branches mostly remaining rudimentary; main axes 200–300 μm diam., with 14–16 pericentral cells, the segments mostly shorter than diam.; determinate branches 1–1.5 mm long, 100–200 μm diam., usually thickest in middle, terminally acute, frequently narrowed at base, the segments shorter than broad, with 10–12 pericentral cells; tetrasporangia ±80 μm diam., not distending segments; spermatangial stichidia 200–250 μm long, borne adaxially in groups of 3–8, usually with 1 sterile terminal cell.

Common epiphyte, low intertidal, Duncan Bay, Br. Columbia, and San Francisco, Calif. to Pta. Santa Rosalía, Baja Calif. Type locality: "California," probably Monterey.

Amplisiphonia Hollenberg 1939

Thalli prostrate, dorsiventral, expanded membranes, with rounded overlapping lobes, attached by numerous unicellular rhizoids from lower surface. Growth marginal from continuous row of apical cells, each cutting off derivatives and forming firmly adjoined polysiphonous branches, these branching dichotomously. Component branches with 3 dorsal and 2 ventral pericentral cells, uncorticated. Tetrasporangia obliquely cruciately divided, arising in narrowly attached, marginal, often ruffled lobes. Sper-

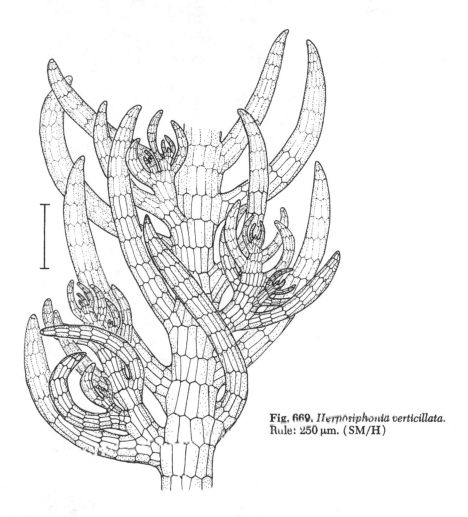

Fig. 669. *Herposiphonia verticillata.*
Rule: 250 µm. (SM/H)

matangial stichidia curved, borne spirally on numerous slender marginal branchlets. Cystocarps numerous, marginal, briefly pedicellate.

Amplisiphonia pacifica Hollenb.

Hollenberg 1939b: 382; Smith 1944: 372; Scagel 1953: 39; Hollenberg & Wynne 1970: 175.

Thalli in tightly adherent patches to 5.5 cm broad, but coalescing and overlapping others to 10–15 cm; lobes 140 µm thick, medium to dull red; tetrasporangial lobes marginal, ample, often ruffled at maturity, to 5 mm long, 6 mm broad; spermatangial stichidia cylindrical, 150–190(220) µm long, ±50 µm diam., narrowed apically, 1 per segment, on marginal radial

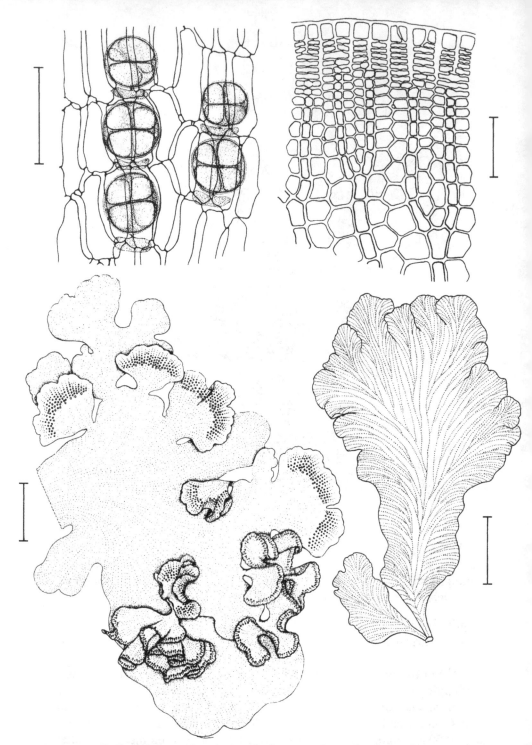

Fig. 670 (below left & two above). *Amplisiphonia pacifica*. Rules: habit of tetra-sporangial thallus, showing lobing, 1 cm; details of tetrasporangia and marginal area in sterile thallus, both 100 μm. (GJH) **Fig. 671** (below right). *Pollexfenia anacapensis*. Rule: 1 cm. (EYD)

branches, the stichidia curving inward, often with a few short, sterile apical cells; mature cystocarps ovoid to globular, 400–550 μm diam., 530–700 μm long.

Frequent on rocks, sometimes on other algae, midtidal to subtidal (to 25 m), N. Br. Columbia to Pta. Eugenio, Baja Calif. Type locality: Corona del Mar (Orange Co.), Calif.

Pollexfenia Harvey 1844

Thalli of erect, bladelike fronds arising from prostrate, dorsiventral bases with multicellular attachment organs. Blades of conjoined polysiphonous branches, these arising from single apical cell.

Pollexfenia anacapensis Daws. & Neush.

Dawson & Neushul 1966: 184.

Blades gregarious, erect, to 20 mm tall, arising from small, discoid holdfast; blades bilateral, not dorsiventral, 10–15 mm broad, sometimes deeply lobed, with cuneate to short-substipitate base, mostly 70–80 μm thick, to 150 μm thick below, the margins lobed and somewhat ruffled, the surface smooth; blades internally with pattern of delicate veins, these oppositely branched in lower parts, flabellately to laterally branched above and extending to margins; marginal initials soon forming central cells and 4 pericentral cells, 2 on either side; pericentral cells gradually becoming spherical while dividing and forming cortex of 2 or 3 layers; central cells elongating and appearing as veins of mature blade; reproduction unknown.

Known only from type collection growing in tank culture on worm tube collected at 40 m off Anacapa I., Calif.

The generic attribution of this material requires detailed examination, particularly of additional fertile collections. *Jeannerettia*, as interpreted by Papenfuss (1942), might possibly be a more appropriate assignment.

Chondria C. Agardh 1817

Thalli erect, slightly to very bushy, with several to many axes arising from common base; polysiphonous, with cortical layer thinly to completely obscuring 5 pericentral cells of polysiphonous axis. Branches cylindrical, mostly irregularly and radially alternate, sometimes opposite or whorled, with several orders of branching; smaller branches narrowed at base, with terminal tuft of branched trichoblasts, with or without terminal pit enclosing apical cell. Tetrasporangia restricted to ultimate branches, irregularly distributed, embedded just beneath cortical surface, tetrahedrally divided. Spermatangial stichidia discoid, with sterile margins, borne in clusters at branch apices. Cystocarps lateral on ultimate branches, covered by cortical layer.

1. Branches with terminal pit 2
1. Branches without terminal pit 4
   2. Thalli mostly 1–4 cm tall, spreading by stolons . . . . *C. arcuata*
   2. Thalli mostly 5–15 cm tall, chiefly erect, with few or no stolons 3
3. Branching frequently opposite or whorled . . . . *C. oppositiclada*
3. Branching irregularly alternate . . . . *C. dasyphylla*
   4. Axes all less than 700 µm diam.; plants delicate, mostly epiphytic,
      mostly less than 70 mm tall . . . . *C. californica*
   4. Axes 1–2 mm diam.; plants saxicolous, mostly 10+ cm tall 5
5. Thalli loosely and remotely branched, with short, determinate branches
   commonly in remote tufts on tetrasporic plants . . . . *C. nidifica*
5. Thalli densely branched, the ultimate branches not in remote tufts . . . . .
  . . . . *C. decipiens*

Chondria arcuata Hollenb.

Hollenberg 1945a: 447; Dawson 1963b: 442.

Thalli dull red, tufted, with stolonlike branches attached at frequent intervals by short, sturdy disks, and with erect cylindrical branches 3–4 cm tall, 300–400(700) µm diam.; branches conspicuously arched, narrowed at base, with frequent laterals, similarly curved; polysiphonous core of branches plainly evident through relatively thin cortex of 2 or 3 cell layers, resulting in segmented appearance of branches; pericentral cells 5, relatively large, in segments 1–1.5 times as long as diam.; growing apex of branches in terminal pit and bearing numerous branched trichoblasts, these protruding prominently; tetrasporangia 80–90 µm diam.; spermatangial disks oval, 180–215 µm broad; cystocarps ovoid, sessile, 360–450 µm diam.

Relatively rare, saxicolous, lower intertidal, reported from San Pedro, Corona del Mar (Orange Co.), and Laguna Beach (type locality), Calif.

Chondria californica (Coll.) Kyl.

Chondria tenuissima f. *californica* Collins, P.B.-A., 1895–1919 [1899]: no. 626. *C. californica* (Coll.) Kylin 1941: 41; Dawson 1963b: 443 (incl. synonymy). *Mychodea episcopalis sensu* Setchell, P.B.-A., 1895–1919 [1908]: no. 1497.

Thalli mostly 4–5 cm tall, but often taller, usually epiphytic, purplish, conspicuously iridescent in life, bushy and densely or openly branched, often attaching to hosts by slender, tendril-like branch tips; branches mostly cylindrical, 185–225 µm diam., or to 500+ µm broad if compressed, apically attenuated, with short tufts of trichoblasts near very slender apices; branching irregularly alternate, approximate or remote, multifarious, or somewhat distichous, with numerous sh llow depressions throughout, these representing branch primordia; tetrasporangia to 110 µm diam., in swollen, rugose branchlets; spermatangia unknown; cystocarps subspherical to slightly urceolate, 500–600 µm diam.

Common to very abundant locally, mostly on various algae, low inter-

Fig. 672 (two above right). *Chondria arcuata*. Rules: habit, 5 mm; cross section, 250 μm. (GJH)   **Fig. 673** (center). *C. californiea*. Rule: 1 cm. (SM)   **Fig. 674** (above left). *C. dasyphylla*. Rule: 5 mm. (SM)   **Fig. 675** (below middle). *C. decipiens*. Rule: 1 cm. (JRJ/S)   **Fig. 676** (two below right). *C. nidifica*. Rules: habit, 1 cm; detail of tetrasporangial apices, 100 μm. (both DBP)   **Fig. 677** (below left). *C. oppositiclada*. Rule: 1 cm. (DBP)

tidal to shallow subtidal (3 m), Santa Cruz I., Calif., to Gulf of Calif. and Panama. Type locality: La Jolla, Calif.

Chondria dasyphylla (Woodw.) C. Ag.

Fucus dasyphyllus Woodward 1794: 239. *Chondria dasyphylla* (Woodw.) C. Agardh 1817: xviii; Dawson 1963b: 444 (incl. synonymy).

Thalli mostly 5–15 cm tall, abundantly multifariously branched in irregular, alternate manner, usually with distinct main axes but with primary branches often pyramidally branched; branches 0.3–1.2 mm diam.; branch apices obtuse to truncate, with terminal pit and protruding trichoblasts; outer cortical cells narrowly oblong, in longitudinal rows; tetrasporangia 140–165 µm diam.; spermatangial disks circular to cordate, 500–600 µm broad; cystocarps subspherical to ovoid, the apex of bearing branch often with spur.

Frequent on rocks, low intertidal, Santa Barbara Co., Calif., to Agua Verde, Mexico. Widely distributed in N. Atlantic. Also Japan. Type locality: England.

Chondria decipiens Kyl.

Kylin 1941: 41; Dawson 1963b: 446.

Thalli mostly 8–16 cm tall, medium to deep brown, with several to many erect, much-branched axes arising from bases of compact, discoid attachments and stolons; erect axes ±1 mm diam., more or less percurrent, with many multifarious secondary and tertiary branches; ultimate branchlets fusiform, seldom exceeding 4 mm in length, terminally wrinkled or corrugated owing to small branch-initial depressions; trichoblasts often conspicuous on tapering branch apices; tetrasporangia developing acropetally in upper half of branchlets; spermatangial disks circular; cystocarps grouped near apices of branchlets, spherical to oval, 0.5–0.9 mm diam.

Occasional on rocks, pilings, etc., in somewhat sheltered locations, high to low intertidal, Santa Cruz, Calif., to Scammon Lagoon, Baja Calif.; frequent on Monterey Peninsula. Type locality: Pacific Grove, Calif.

Chondria nidifica Harv.

Harvey 1858: 125; Dawson 1963b: 447; Dawson & Tozun 1964: 286.

Thalli 15–20(40) cm tall, dark red, with few to many cylindrical axes arising from broad basal disks or from complex of discoid holdfasts and stolons; erect axes 1–2 mm diam., heavily corticated, mostly remotely and irregularly multifariously branched at intervals of 20–60 mm; branches terminally attenuate, the apices commonly cropped by grazing animals and regenerating new branches; smaller branchlets commonly in remote tufts along middle and upper parts of primary branches or arising from

broken ends; fertile branchlets of sexual plants less fasciculate; tetrasporangia in short, tufted, slender-fusiform branchlets; spermatangial disks
400–500 μm broad; cystocarps abundant, ±1 mm diam., multifarious on
ultimate and subultimate branchlets.

Occasional, saxicolous, midtidal to lower intertidal, Santa Cruz, Calif., to
Bahía Asunción, Baja Calif.; in S. Calif., common on midtidal reefs along
sand margins. Type locality: S. Calif. in vicinity of U.S.-Mexican border.

Chondria oppositiclada Daws.

Dawson 1945b: 78; 1963b: 447.

Thalli to 15 cm tall, abundantly branched; branches cylindrical, with 1
to several main erect axes from small, discoid bases supplemented by additional disks on ends of semiprostrate or arching stolons; branches 0.6–1
mm diam., of 2 or 3 orders; major branches alternate, radially disposed,
or in subopposite pairs or triplets at intervals of 3–4 mm; ultimate branchlets relatively short, clavate, 400–800 μm diam., in opposite or subopposite pairs or triplets, with contracted bases, obtuse apices, and apical pit
with small tuft of trichoblasts; tetrasporangia and spermatangial stichidia
unknown; cystocarps urceolate, 550–650 μm diam., sparse. (After Dawson.)

Frequent, epiphytic or saxicolous, in shallow tidepools or sand flats to
subtidal (to 18 m), San Diego Co., Calif., to Scammon Lagoon, Baja Calif.
Type locality: La Jolla, Calif. (in 10 m), on *Egregia*.

This species is very similar to *C. dasyphylla*, the two differing chiefly in
branching pattern.

Laurencia Lamouroux 1813

Thalli chiefly erect, solitary or tufted, epiphytic or saxicolous; erect axes
cylindrical to markedly flattened; branching pinnate or radial. Branch
apices blunt, with terminal pit containing single apical cell and rudimentary, evanescent trichoblasts; branches firm to cartilaginous, solidly parenchymatous, with obscure polysiphonous core surrounded by cortex of
usually isodiametric cells, these sometimes with lenticular thickenings in
walls. Tetrasporangia tetrahedrally divided, scattered in outermost cortical layers of branchlets. Spermatangial branches cylindrical, densely
crowded in conceptaclelike terminal pockets. Procarps arising in apical
pits; cystocarps morphologically terminal.

1. Erect branches compacted in cushionlike clump. . . . .*L. crispa*
1. Erect branches sometimes tufted but not forming cushionlike clump. 2
  2. Branches slightly to prominently flattened. 3
  2. Branches cylindrical. 7
3. Distichous branching evident throughout. 4
3. Distichous branching obscured by crowding of branchlets, or branching
   partly polystichous 6

4. Thalli saxicolous; lower portion of axes naked.............*L. spectabilis*
4. Thalli epiphytic; axes with numerous lateral branches throughout..... 5
5. Main axis prominently percurrent; medulla lacking lenticular thickenings...*L. splendens*
5. Main axis not prominently percurrent; lenticular thickenings abundant
in summer season.......................................*L. sinicola*
  6. Branches prominently flattened; crowded branchlets mostly shorter
than width of axis....................................*L. blinksii*
  6. Branches moderately flattened; branchlets not crowded, longer than
width of axis.......................................*L. subdisticha*
7. Hooked or tendril-like branches usually present in mature plants......
....*L. subopposita*
7. Hooked or tendril-like branches lacking........................... 8
  8. Main branches densely covered throughout with short-papillate
branchlets .... *L. snyderiae*
  8. Main branches not densely covered throughout with short-papillate
branchlets... 9
9. Thalli forming low turf or mats on rocks, mostly less than 30 mm tall..... 10
9. Thalli not forming low turf or mats, mostly 50+ mm tall.............. 11
  10. Erect branches mostly 20–30 mm tall; determinate branchlets not
crowded above; medullary cells lacking lenticular thickenings.. *L. lajolla*
  10. Erect branches less than 10 mm tall; determinate branchlets densely
crowded above, soon deciduous; medullary cells with frequent lenticular thickenings .... *L. decidua*
11. Thalli medium to deep red; branches nodding in 1 direction....*L. gardneri*
11. Thalli reddish-purple to brownish; branches not arcuate............. 12
  12. Medullary cells with numerous conspicuous lenticular thickenings
.... *L. masonii*
  12. Medullary cells with few or no lenticular thickenings.......*L. pacifica*

Laurencia blinksii Hollenb. & Abb.

Hollenberg & Abbott 1965: 1186; 1966: 117; Saito 1969: 87.

Thalli saxicolous, crisp, in dense, erect clumps 4–7(9) cm tall, arising from spreading basal attachments, these mostly 3–4 cm broad; erect fronds complanate and basically alternately distichous throughout, the distichous branching partly obscured by divergent growth and crowding of branches, these with laterals pyramidal in outline; subultimate branches lanceolate-conical, the axis 3–4 mm broad, irregularly divided or fungiform-lobed and congested; medullary cells 300–400 μm long, 60–150 μm diam.; cortical cells progressively smaller outward; lenticular thickenings lacking; surface cells 25–45 μm long, 13–15 μm wide in surface view, 15–30 μm deep anticlinally in smaller branches; tetrasporangial branches irregularly fungiform or alternately branched; spermatangial pockets occurring singly on short, subultimate divisions; cystocarps unknown.

Frequent on exposed rocky shores, low intertidal, Pigeon Pt. (San Mateo Co.), Monterey Peninsula, and near Malpaso Creek (south of Carmel; type locality), Calif.

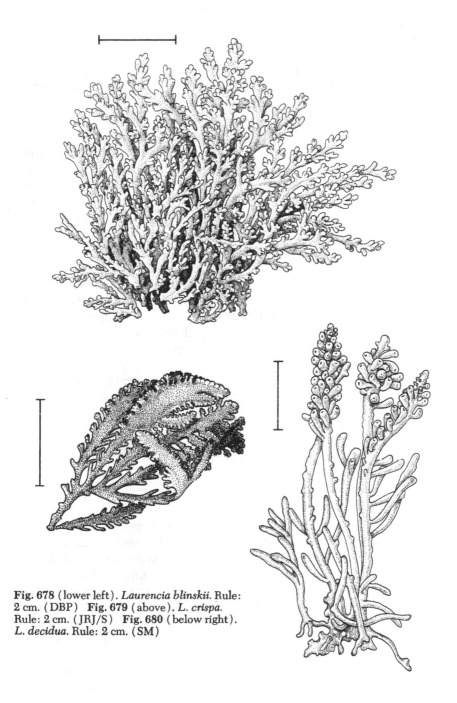

Fig. 678 (lower left). *Laurencia blinskii.* Rule: 2 cm. (DBP) **Fig. 679** (above). *L. crispa.* Rule: 2 cm. (JRJ/S) **Fig. 680** (below right). *L. decidua.* Rule: 2 cm. (SM)

Laurencia crispa Hollenb.

Hollenberg 1943c: 219.

Thalli 3–8(10) cm tall, crisp, greenish-brown, with many erect branches laterally compacted in cushionlike clumps, these to 30 cm broad; axes of erect branches 2 mm diam., subcylindrical to angular or flattened, with 3 or 4 orders of branching and with branchlets reduced to short, blunt projections; surface cells 15–25 µm diam., isodiametric, not forming palisade-like layer; medullary cells without lenticular thickenings; tetrasporangial plants taller and more compact than sexual plants; spermatangial pockets on ultimate branchlets or directly on surface of major branches; cystocarps rare, terminating lateral branches.

Occasional but locally abundant on rocks, midtidal to lower intertidal and subtidal; known only from type locality, Pacific Grove, Calif.

Laurencia decidua Daws.

Dawson 1954b: 8.

Thalli forming dense turf 6–10 mm thick, the color not recorded; creeping stoloniferous branches 300–400 µm diam., densely intergrown, attached to substratum and to one another by small disks; erect percurrent axes ±500 µm diam., sparingly branched below, bearing short, densely imbricated, determinate branchlets above, these ±1 mm long, soon deciduous, leaving axis denuded except for conical terminal group of branchlets; surface cells essentially isodiametric, not protuberant; medullary thickenings frequent; tetrasporangia in short, deciduous, determinate branchlets; sexual structures unknown. (After Dawson.)

Infrequent on exposed intertidal rocks, Laguna Beach, Calif., to I. Socorro (Is. Revillagigedo; type locality), Mexico.

Laurencia gardneri Hollenb.

Hollenberg 1943c: 218; Smith 1944: 379.

Thalli mostly 6–10 cm tall, rich red; erect fronds densely tufted, repeatedly branched radially from obscure axis; branches terete, firm to slightly cartilaginous, ±1 mm diam., those of common axis mostly curved, nodding in common direction; surface cells 25–40 µm broad, mostly elongate with length of branch, not forming palisade-like layer, each surface cell evidently containing single globular hyaline body resembling pyrenoid; lenticular thickenings rare or absent; tetrasporangia in elongate ultimate branchlets; spermatangia unknown; cystocarps inconspicuous, rarely swelling ultimate branchlets.

Frequent, on rocks or epiphytic, subtidal (to 20 m), Pacific Grove (type locality) and at various locations on Monterey Peninsula, Calif.; not known elsewhere.

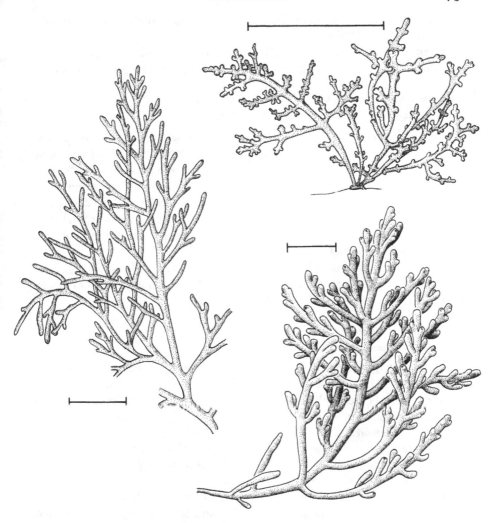

Fig. 681 (below right). *Laurencia gardneri*. Rule: 5 mm. (JRJ/S) **Fig. 682** (above right). *L. lajolla*. Rule: 2 cm. (DBP) **Fig. 683** (left). *L. masonii*. Rule: 5 mm. (SM)

Laurencia lajolla Daws.

Dawson 1958: 77; 1963b: 455.

Thalli dark red, densely tufted to turf-forming, 2–3 cm tall, with many erect, branched cylindrical axes arising from complex systems of tangled creeping branches attached by frequent irregular disks; erect axes sparsely branched at long intervals below, bearing frequent short, more or less determinate branches in upper parts, mostly 700–850 µm diam.; surface cells not forming palisade-like layer; medullary cells lacking lenticular thickenings; tetrasporangia in simple, short, lateral branches; sexual plants unknown. (After Dawson.)

Occasional component of coralline turf, or in midtidal or upper intertidal pools, La Jolla, Calif. (type locality), to Scammon Lagoon, Baja Calif.

Laurencia masonii S. & G.

Setchell & Gardner 1930: 155; Dawson 1963b: 455.

Thalli dark red, 6–20 cm tall, attached by small disks and secondary stolons; erect axes repeatedly and profusely branched, 1.5–2(3.5) mm diam.; branches usually divaricate, of 5+ orders, each successively shorter; medullary thickenings exceedingly abundant and prominent, preventing excessive shrinking of axes in drying; surface cells thin-walled, 25–30 µm diam.; tetrasporangia 90–120 µm diam., scattered in ultimate branchlets; spermatangia unknown; cystocarps ±1.3 mm diam., with prominent rostrum. (After Dawson.)

Frequent, on rocks or other algae, intertidal to subtidal (to 30 m), Santa Catalina I. and Laguna Beach, Calif., to I. Magdalena, Baja Calif. Type locality: I. Guadalupe, Baja Calif.

Laurencia pacifica Kyl.

Kylin 1941: 42; Smith 1944: 378; Dawson 1963b: 457 (incl. synonymy).

Thalli deep reddish-purple, usually with numerous radially branched, erect axes 6–16(30) cm tall, ±2 mm diam. below, arising from primary discoid holdfasts, generally obscured by numerous stolons, forming perennial basal cushion; erect axes of older plants mostly percurrent; branching irregularly alternate, subopposite or subverticillate, of 3 or 4 orders, the ultimate branchlets turbinate, 1–3 mm long, not dense or wartlike; cortex lacking lenticular thickenings; surface cells not forming palisade-like layer, each containing single spherical hyaline body; tetrasporangia in compound branchlets, these tending to form very dense, verticillate clusters around strongly percurrent upper parts of main axes; spermatangia in branched clusters in apical pits; cystocarps 1–1.2 mm diam., on upper branchlets.

Common to abundant in restricted localities, on rocks, low intertidal, usually in cool quiet water, Monterey Peninsula, Calif., to I. Magdalena, Baja Calif. Type locality: La Jolla, Calif.

The various forms of this highly variable species are treated by Dawson (1963).

Laurencia sinicola S. & G.

Setchell & Gardner 1924a: 764; Dawson 1963b: 461. *Laurencia scrippsensis* Daws. 1944b: 234.

Thalli dark rose red, 1.6–9 cm tall, with congested erect fronds arising

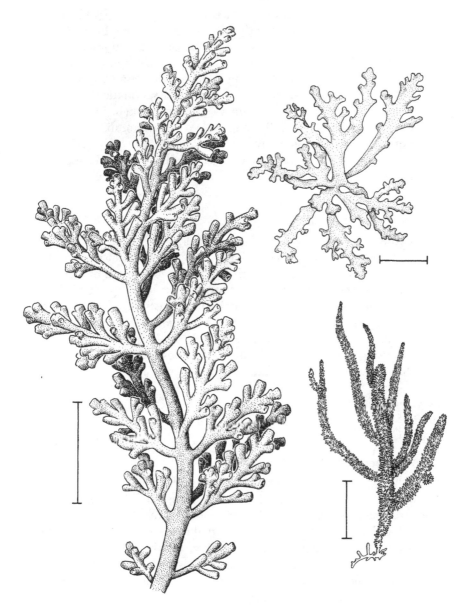

Fig. 684 (left). *Laurencia pacifica*. Rule: 1 cm. (JRJ/S) **Fig. 685** (above right). *L. sinicola*. Rule: 3 cm. (SM) **Fig. 686** (below right). *L. snyderiae*. Rule: 3 cm. (SM)

from common bases; main axes of erect fronds compressed, 1–1.5(4) mm broad, 400–500 μm thick, irregularly alternately pinnate, becoming decompound-pinnate; surface cells mostly isodiametric, 20–50 μm diam., not forming palisade-like layer; cortical cells mostly 70–90 μm diam., with abundant lenticular thickenings during summer season; tetrasporangia in outer ends of simple or digitate short pinnules; spermatangial pockets partly embedded or sessile-ovoid, solitary or in clusters on ultimate branchlets; cystocarps domoid, 600–800 μm diam., on distal ramuli.

Occasional on various algae, low intertidal, Santa Cruz I. and La Jolla, Calif.; frequent in Gulf of Calif. and to Is. Revillagigedo, Mexico. Type locality: Eureka (near La Paz), Gulf of Calif., on *Sargassum*.

Laurencia snyderiae Daws.

Dawson 1944d: 98; 1963b: 463.

Thalli rose red, with erect fronds in groups of 2–5 from low, primary stoloniferous clusters, each frond a limp, fleshy, cylindrical division with percurrent axis 8–12 cm long and with several indeterminate branches 2–4 cm long, arising irregularly or often as group from upper portion of main axis, or these sometimes abundant and obscuring distinctness of main axis; entire plant, except usually lower part of main axis, densely clothed with short, determinate, papillate branchlets 1.5–3 mm long; palisade-like surface cells lacking; medullary lenticular thickenings lacking; tetrasporangia and cystocarps borne on papillate branchlets; spermatangial plants not observed.

Occasional on midtidal rocks, Santa Catalina I. and Redondo Beach, Calif., to Scammon Lagoon, Baja Calif., and Gulf of Calif. Type locality: La Jolla, Calif.

Laurencia spectabilis Post. & Rupr.

Postels & Ruprecht 1840: 16; Smith 1944: 377; Dawson 1963b: 464.

Thalli to 30 cm tall, with several erect, decompound, compressed axes arising from conical base; branching pinnately alternate to subopposite, pyramidal or flabellate, with branches short above and progressively longer below; lower branches deciduous, or eroding, leaving lower axes more or less naked; branch apices rounded; tetrasporangia in short, dichotomously branched branchlets; spermatangial pockets adaxial on smaller branchlets; cystocarps ovoid on smaller branchlets.

1. Main axes more than 3 mm broad; apices of branchlets spatulate......
....var. *spectabilis*
1. Main axes less than 3 mm broad.................................. 2
   2. Axes ±1 mm broad; branching fastigiate, tending to be pinnate.....
....var. *tenuis*
   2. Axes 2–3 mm broad; branching divergent, irregular......var. *diegoensis*

Laurencia spectabilis var. spectabilis

Intertidal thalli deep purplish-red; northern and subtidal Calif. thalli rose red, to 30 cm tall; axes 3–5(6) mm broad; branching mostly symmetrical; branches markedly compressed with broadly rounded apices, borne at intervals of 2–8 mm above.

Frequent to common on rocks, midtidal to low intertidal and occasionally subtidal (to 15 m), Sitka, Alaska, to northern Santa Barbara Co., Calif.; infrequent to San Diego, Calif., and in upwelling cold water off Baja Calif. Type locality: Norfolk Sd., Alaska.

Laurencia spectabilis var. diegoensis (Daws.) Daws.

Laurencia diegoensis Dawson 1944b: 236. *L. spectabilis* var. *diegoensis* (Daws.) Daws. 1963b: 465.

Thalli mostly 8–28 cm tall, deep reddish-brown, with general features of var. *spectabilis* but with narrower branches, 2–3 mm broad, at relatively longer intervals of 6–10(17) mm; branches of different orders of variable length, resulting in asymmetrical appearance of frond; spermatangial pockets in prominent adaxial secund series on smaller branchlets.

Frequent on rocks, midtidal to mostly lower intertidal, Carpinteria (Santa Barbara Co.), Calif., to Pta. Baja, Baja Calif. Type locality: La Jolla, Calif.

Laurencia spectabilis var. tenuis Daws.

Dawson 1963b: 466.

Thalli similar to var. *diegoensis* but much more slender, with main axes 0.7–1.2 mm broad, less compressed than preceding varieties.

Infrequent, saxicolous or epiphytic, intertidal to subtidal, Santa Catalina I. and Pt. Loma (San Diego Co.), Calif., to I. Guadalupe, Baja Calif. (type locality).

Laurencia splendens Hollenb.

Hollenberg 1943c: 219; Smith 1944: 377; Dawson 1963b: 467. *Laurencia maxineae* Daws. 1944b: 233.

Thalli 3–20(30) cm tall, reddish-brown, with 1 to several erect, often branched axes arising from small basal attachments; erect axes strongly compressed except at base, bi- or tripinnate, bearing dense simple or compound branchlets 1–4 mm long throughout; surface cells mostly isodiametric, not forming palisade-like layer; cortical cells 50–60 μm diam., lacking lenticular thickenings; tetrasporangia in distal portion of little-modified ultimate branchlets; spermatangial pockets ovoid, solitary and terminal or variously positioned on ultimate branchlets; cystocarps ovoid, irregular in size and arrangement, mostly 1–3, lateral or terminal in pinnules.

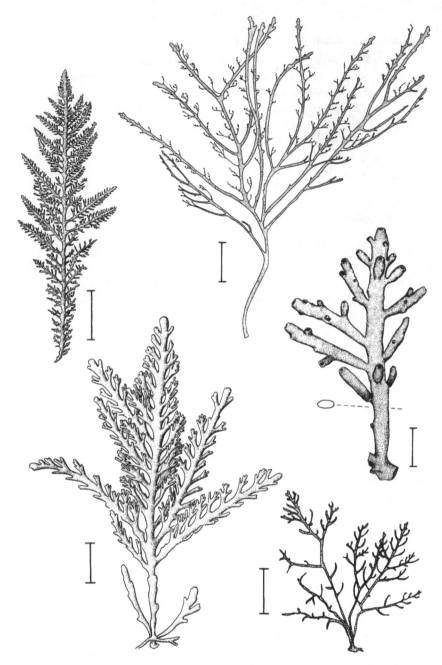

Fig. 687 (below left). *Laurencia spectabilis* var. *spectabilis*. Rule: 2 cm. (JRJ/S)
Fig. 688 (above right). *L. spectabilis* var. *diegoensis*. Rule: 2 cm. (SM) **Fig. 689**
(below right). *L. spectabilis* var. *tenuis*. Rule: 2 cm. (LH) **Fig. 690** (above left). *L. splendens*. Rule: 2 cm. (JRJ/S) **Fig. 691** (middle right). *L. subdisticha*. Rule: 2 mm. (SM, after Dawson)

Usually epiphytic, commonly on articulated corallines, low intertidal to subtidal (to 30 m), Santa Cruz, Calif., to I. Guadalupe, Baja Calif.; frequent in northern part of range, occasional in upwelling cool water of southern part of range. Type locality: Moss Beach (Monterey Co.), Calif.

Laurencia subdisticha Daws., Neush. & Wild.

Dawson, Neushul & Wildman 1960: 26; Dawson & Neushul 1966: 185.

Thalli to 6 cm tall, bright red, with several densely branched main axes

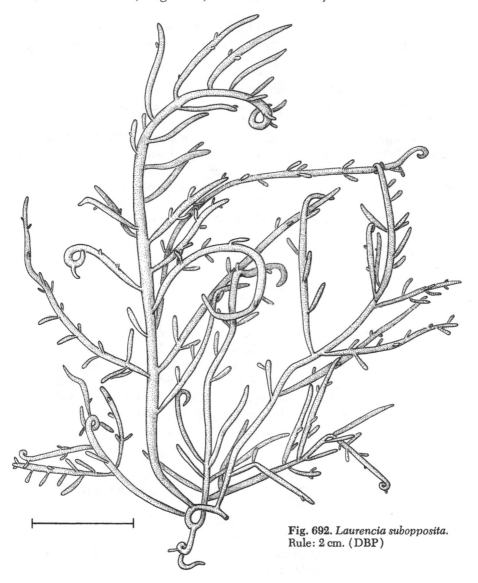

Fig. 692. *Laurencia subopposita.*
Rule: 2 cm. (DBP)

arising from nearly discoid bases; primary axes percurrent, (660)800–1500 μm broad, somewhat pyramidal in outline; branching mostly distichously alternate at intervals of 1–2 mm, but with occasional polystichous branches; branches compressed to flattened, especially below, (300)400–600 μm thick; medulla lacking lenticular thickenings, the surface cells mostly quadrangular, 35–40 μm diam., not protruding at apices; tetrasporangial branchlets simple, unmodified, ±2 mm long; spermatangial branches not observed; cystocarps scattered, shaped like inverted tops, ±600 μm diam.

Occasional, on rocks, mostly subtidal (to 18 m), northern Channel Is. and La Jolla, Calif., to Is. San Benito, Baja Calif. (type locality) and probably to Pacific Panama.

Laurencia subopposita (J. Ag.) Setch.

Chondriopsis subopposita J. Agardh 1892: 149. *Laurencia subopposita* (J. Ag.) Setchell 1914a: 9; Dawson 1963b: 469; Hollenberg & Abbott 1966: 18.

Thalli epiphytic, deep rose red, with branches arising 5–14 cm from entangled or entwined basal attachments, in quiet water occasionally forming globular masses to 30 cm diam., 25 cm tall; branches of several orders of varying lengths, with frequent terminal hooks coiling around host and other branches; individual branches terete, 0.8–2 mm diam.; tetrasporangia 60–80 μm diam., on terminal parts of short branchlets; sexual plants unknown.

Relatively infrequent, on articulated corallines or coarse algae, low intertidal to subtidal (to 15 m), Pebble Beach (Monterey Co.), Calif., to Pta. Eugenio, Baja Calif. Type locality: Santa Barbara, Calif.

Erythrocystis J. Agardh 1876

Thalli epiphytic, or semiparasitic on branch apices of several species of *Laurencia*, bright red to brownish, at first globular, attached by 1 large, cylindrical, rhizoidal cell penetrating deeply into branch apices of host tissue; early polysiphonous nature soon obscured by thick cortical layer. Tetrasporangia scattered, embedded in outer cortical layers. Spermatangial stichidia arising on trichoblasts, grouped in surface sori. Cystocarps conspicuous, with pericarp immersed in and united with cortical tissue.

Erythrocystis saccata (J. Ag.) Silva

Chylocladia saccata J. Agardh 1849: 89. *Erythrocystis saccata* (J. Ag.) Silva 1952: 308; Hollenberg & Abbott 1966: 119; Dawson 1963b: 441. *Ricardia saccata* (J. Ag.) Kylin 1928: 94; Smith 1944: 380.

Thalli to 3.5 cm long, obpyriform to ellipsoidal; with vegetative and reproductive characters of the genus.

Locally relatively common, epiphytic mostly on *L. pacifica* but occasionally on *L. subopposita* or other *Laurencia* spp., low intertidal, Pacific

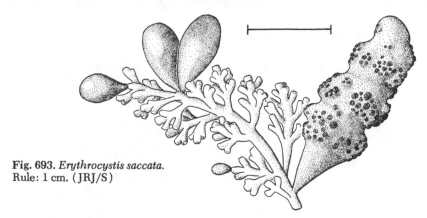

Fig. 693. *Erythrocystis saccata.*
Rule: 1 cm. (JRJ/S)

Grove, Calif., to Is. Revillagigedo, Mexico. Type locality: "California," probably vicinity of Monterey.

Janczewskia Solms-Laubach 1877

Thalli small, ivory white to pinkish, warty cushions "parasitic" on various Rhodomelaceae, especially *Laurencia* and *Chondria*, with strands of rhizoidal filaments penetrating hosts; external portion solidly parenchymatous. Tetrasporangia tetrahedrally divided, in outer cortex among paraphyses (trichoblasts?), in oval or irregular conceptacles. Spermatangia in plumose clusters lining conceptacles. Cystocarps with urn-shaped pericarps.

1. Thalli on *Laurencia*.. 2
1. Thalli on *Chondria*... 3
    2. Thalli on *L. spectabilis* and *L. splendens*, pinkish............*J. gardneri*
    2. Thalli on *L. subopposita*, ivory white ....*J. solmsii*
3. Thalli on *C. nidifica*, pink, burrlike........................*J. lappacea*
3. Thalli on *C. decipiens*, creamy white, flattened-reniform.......*J. moriformis*

Janczewskia gardneri Setch. & Guerns.

Setchell & Guernsey 1914: 12; Smith 1944: 381. *Janczewskia verrucaeformis sensu* Nott 1897: 83.

Thalli growing on *Laurencia spectabilis* or *L. splendens*, cushionlike, with free branches 2+ mm long, pinkish; tetrasporangia in outer cortical layers; spermatangial plumes slender and crowded, lining entire inner surface of conceptacle; cystocarps 1 or 2, at apices of free branches.

Occasional to frequent, parasitic on *Laurencia*, low intertidal to shallow subtidal, Vancouver I., Br. Columbia, to Pta. Baja, Baja Calif. Type locality: probably Duxbury Reef (Marin Co.), Calif.

Janczewskia lappacea Setch.

Setchell 1914a: 14; Dawson 1963b: 440.

Thalli growing on *Chondria nidifica*, pink, burrlike, 3–5 mm diam., the

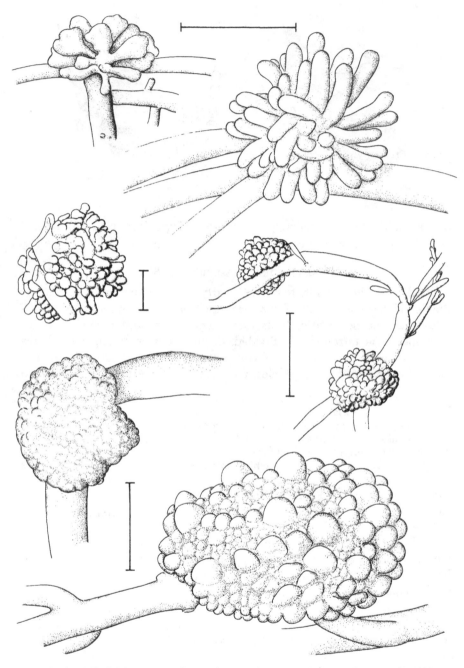

Fig. 694 (middle left). *Janczewskia gardneri.* Rule: 3 mm. (WAS) **Fig. 695** (two at top). *J. lappacea.* Rule: 3 mm. (WAS) **Fig. 696** (middle right). *J. moriformis.* Rule: 3 mm. (WAS) **Fig. 697** (below left & below right). *J. solmsii.* Rule: 3 mm. (WAS)

host usually bent at point of attachment; free tetrasporangial branches to 2 mm long, the tetrasporangia in outer cortex; spermatangial conceptacles terminal on branches, the spermatangial plumes relatively short, lining entire conceptacle cavity; cystocarps mostly single at apex of free branches.

Occasional to rare, parasitic on *Chondria*, low intertidal, San Luis Obispo Co. and San Pedro (type locality), Calif., to Pta. Baja, Baja Calif.

Janczewskia moriformis Setch.

Setchell 1914a: 11.

Thalli growing on *Chondria decipiens*, creamy white, flattened-reniform, 3–5 mm diam., the host not much bent at point of attachment; tetrasporangia and cystocarps on short free branches 0.6–0.8 mm long; tetrasporangia in outer cortical layers; spermatangial plants almost lacking branches, the spermatangial plumes lining walls and radiating toward center of conceptacles; cystocarps terminal, 1 or 2 per branch.

"Parasitic" on *Chondria*, low intertidal; known only from the type locality, Santa Monica, Calif.

Janczewskia solmsii Setch. & Guerns.

Setchell & Guernsey 1914: 9; Dawson 1963b: 440.

Thalli growing on *Laurencia subopposita* and *L. masonii*, pinkish, cushionlike, 8–10 mm diam., surface tessellated or botryoidal, without free branches; tetrasporangia in outer walls of apical pits; spermatangial plumes branched, arising in conical tufts from base of conceptacle; cystocarps globular, with narrow ostiole.

Occasional, "parasitic" on *Laurencia*, low intertidal, Redondo Beach, Calif., to I. Guadalupe, Baja Calif. Type locality: not specifically designated, specimens mentioned from Redondo Beach and San Pedro, Calif.

Rhodomela C. Agardh 1822

Thalli usually with several to many erect axes arising from a common base. Erect axes terete or slightly compressed, with several long branches divided into progressively shorter branches. Apices of young branchlets with branched trichoblasts; all branches polysiphonous, a condition soon obscured by division and redivision of pericentral cells. Tetrasporangia tetrahedrally divided, in 2 parallel rows in slightly inflated, stichidiumlike branchlets, 2 tetrasporangia per segment. Spermatangial stichidia in corymbose clusters. Cystocarps with ostiolate pericarp; only terminal cells of gonimoblast forming carposporangia.

Rhodomela larix (Turn.) C. Ag.

Fucus larix Turner 1819: 23. *Rhodomela larix* (Turn.) C. Agardh 1822a: 376; Smith 1944: 374.

Thalli annual, mostly 10–20 cm tall, brownish-black; ultimate branch-

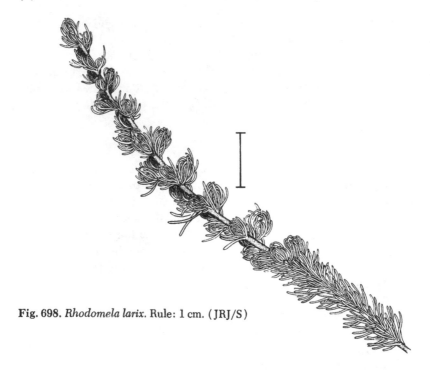

Fig. 698. *Rhodomela larix.* Rule: 1 cm. (JRJ/S)

lets 5–10 mm long, of mostly uniform length; branches and branchlets wiry; other characters as for genus.

Abundant in restricted areas on rocks, upper intertidal, mostly on sand-swept reefs, Bering Sea to Government Pt. (Santa Barbara Co.), Calif. Type locality: Nootka Sd. (Vancouver I.), Br. Columbia.

Odonthalia Lyngbye 1819

Thalli erect, freely branched, with several long, terete to markedly flattened branches, these with or without midrib. Branches basically polysiphonous, distichous, soon heavily corticated by repeated division of the 4 pericentral cells. Tetrasporangia in simple branchlets, these often borne in clusters; 2 opposite pericentral cells of each primary segment producing tetrahedrally divided tetrasporangium. Spermatangia covering surface of short, simple branchlets. Cystocarps clustered on tufted axillary branchlets or borne alternately and distichously on subultimate ramuli.

1. Branches distinctly to markedly flattened.............*O. washingtoniensis*
1. Branches terete or only slightly flattened......................... 2
   2. Thalli to 17 cm tall, terete throughout; cystocarps less than 0.5 mm
      diam. .... *O. oregona*
   2. Thalli to 40 cm tall, the ultimate branchlets flattened; cystocarps over
      0.5 mm diam. .... *O. floccosa*

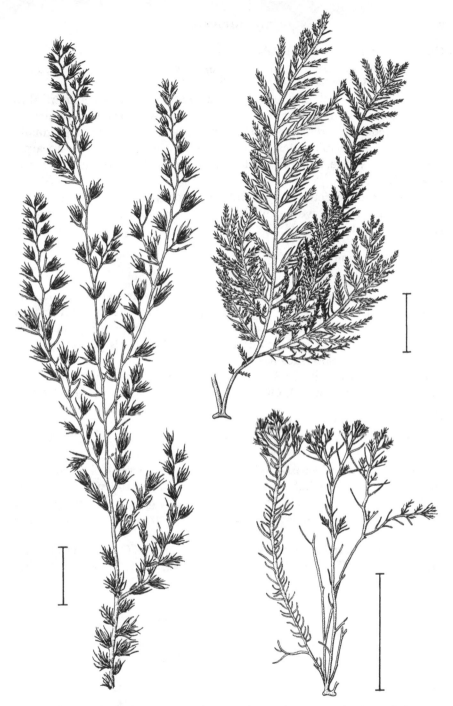

Fig. 699 (left). *Odonthalia floccosa*. Rule: 2 cm. (JRJ/S) **Fig. 700** (below right). *O. oregona*. Rule: 2 cm. (SM/H) **Fig. 701** (above right). *O. washingtoniensis*. Rule: 2 cm. (DBP)

Odonthalia floccosa (Esp.) Falk.

Fucus floccosus Esper 1802: 42, no. 115. *Odonthalia floccosa* (Esp.) Falkenberg 1901: 607; Setchell & Gardner 1903: 333; Smith 1944: 375.

Thalli to 40 cm tall, dark brown to black; major branches cylindrical to slightly compressed, ±1 mm diam., alternately distichous, with numerous short branchlets clustered on short laterals; branchlets flattened, gradually attenuate to acute apices; fertile branchlets, especially those of cystocarpic specimens, in dense, headlike clusters. Development of reproductive structures as for genus.

Abundant in restricted areas on intertidal rocks subjected to strong surf, S. Br. Columbia to Government Pt. (Santa Barbara Co.), Calif. Type locality: Trinidad (Humboldt Co.), Calif.

Odonthalia oregona Doty

Doty 1947b: 196.

Thalli to 17 cm tall, drying to black; slender branches terete throughout, less than 1 mm diam. below, ±250 μm diam. above; branches alternately distichous, gradually attenuated; tetrasporangia in densely grouped terminal branchlets; spermatangial structures unknown; cystocarps less than 0.5 mm diam., in close racemes.

Infrequent on subtidal rocks, Whidbey I., Wash.; mouth of Chetco River (Curry Co.; type locality), Ore., to Bodega Head (Sonoma Co.), Calif.

Odonthalia washingtoniensis Kyl.

Kylin 1925: 76; Scagel 1957: 245. *Odonthalia semicostata sensu* Setchell & Gardner 1903: 336.

Thalli to 25 cm tall, coarse and cartilaginous, drying to black, prominently distichous; branches slightly compressed below, flat and without definite midrib above; the ultimate branchlets subulate, alternate; tetrasporangia and cystocarps on densely tufted branchlets; spermatangial structures unknown.

Infrequent, on rocks, mostly subtidal (to 18 m), Br. Columbia to Bodega Head (Sonoma Co.) and San Luis Obispo Co., Calif. Type locality: San Juan I., Wash.

MASTER KEY TO GENERA

Master Key to Genera

ALGAE ARE "simple" plants; i.e., they have neither roots, nor stems, nor leaves, nor any of the elaborate cell types that characterize these structures in flowering plants. Indeed, some phycologists do not classify the algae with plants at all, because in these terms they are not plants (nor are they admitted to the Plant Kingdom by students of phylogeny, but are placed with the Protista, a repository of ancient and simple organisms that includes groups some of us place in the Protozoa, the Algae, or the Fungi). Algae do, however, possess chlorophyll *a*, which performs for them the same photosynthetic function it does in other plant groups. That fact suggests a relationship, but the inclusion of the algae in the Plant Kingdom is chiefly a matter of convenience.

Because algae are relatively simple in structure, there are few features that can be drawn upon for purposes of identification. Specimens must be studied closely, and attention must be given to details that may seem to reflect only minor differences. Care must also be taken to collect a number of specimens, in order that among them may be found an adequate characterization of the species.

The keys that follow are based on a series of choices, each choice presented as a dichotomy, or pair of opposed statements, usually called a *lead*; in each lead of the key, one choice, or *leg*, should more correctly or more fully characterize some feature of the plant in hand than the other leg does. Occasionally, both choices may seem to fit equally well. Where this happens, both legs should be followed on down to the names of genera, and the text descriptions of the genera and the species contained in them should be carefully read and compared with the plant in hand. No key can give this kind of comprehensive description.

The keys have been arranged in terms of "color groups" of algae. These groups correspond, in taxonomic terms, to the four divisions (or phyla) of the algae: the Chrysophyta, or yellow-brown algae (not keyed because only one macroscopic species has been recorded for the California coast); the Chlorophyta, or green algae (below); the Phaeophyta, or brown algae (p. 749), and the Rhodophyta, or red algae (p. 754). To use the keys effectively, then, one must start out in the correct color group. The greatest difficulty will be found with some of the red algae (notably *Iridaea*), which may appear to be green algae. Note, however, that the shade of green in these reds is yellowish or brownish, whereas the green of *Ulva*, which also has a flat blade, is the shade of a grassy lawn, and like a lawn could be dark green or light green. If still in doubt, cut a small section with a razor blade: green-algae blades are never more than two cells thick; red algae are mostly 10–30 or more cells thick in section. Some red algae are a dusky red, others rusty-red to blackish (especially the crustlike algae, such as *Petrocelis*), and could be mistaken for some brown algae, which may also be blackish. A section made with a razor blade, however, will usually show the golden-brown plastids of the brown algae, whereas red plastids do not show up well in sections of crusts. Red crusts are in fact nearly impossible to identify without tetrasporangial specimens; a good rule of thumb is that if specimens of red algae are sterile, they should be discarded, since they would prove difficult to identify, and occasionally more than one kind of fertile plant will be needed for positive identification. Finally, *Vaucheria*, the only chrysophyte included in this book, is a dark spinach green, which indicates that its chlorophyll masks the other pigments present; it lacks cross walls, and thus can be easily separated from other algae that are dark green.

The use of a microscope is mandatory for studies of the red algae, since the system of classification employed is based on internal reproductive and vegetative structures. However, a great deal can be learned with a hand lens or a magnifying glass.

The branching patterns of filamentous forms, whether the algae are green, brown, or red, are frequently very distinctive, and one can learn to associate these patterns with certain species. It is the branching pattern that makes so many seaweeds attractive, particularly in their natural habitats.

CHLOROPHYTA: THE GREEN ALGAE

1. Plants minute, bearing long nonseptate setae 2
1. Plants of various sizes, lacking long setae 3
   2. Setae terminal on erect filaments ....*Pilinella* (p. 56)
   2. Setae not terminal on erect filaments............*Bolbocoleon* (p. 58)
3. Plants unicellular coenocytes, either tubular or spherical.............. 4
3. Plants unicellular or multicellular coenocytes, neither tubular nor spherical 7
   4. Plants spherical coenocytes ....*"Halicystis"* (p. 115)
   4. Plants tubular coenocytes 5
5. Branches compacted, forming spongy thalli.............*Codium* (p. 116)
5. Branches not compacted, not forming spongy thalli 6
   6. Branching pinnate or radial, usually profuse........*Bryopsis* (p. 111)
   6. Branching irregular, dichotomous to unilateral, sparse to frequent
   .... *Derbesia* (p. 113)
7. Plants forming a globular or dendroid colony; habit saxicolous 8
7. Plants not forming a colony; habit various........................ 9
   8. Cells of colony radially distributed through common gelatinous matrix ....*"Collinsiella"* (p. 119)
   8. Cells of colony forming dendroid stalk system, not within common matrix .... *Prasinocladus* (p. 52)
9. Thalli discoid to pulvinate 10
9. Thalli not discoid to pulvinate 12
   10. Plants minute pulvinate epiphytes ....*Pseudopringsheimia* (p. 61)
   10. Plants minute to microscopic disks 11
11. Marginal cells commonly bifurcate ....*Ulvella* (p. 59)
11. Marginal cells not bifurcate......................*Pseudulvella* (p. 60)
   12. Plants simple or branched uniseriate filaments 13
   12. Plants multiseriate in irregular filaments or expanded sheets........23
13. Filaments unbranched 14
13. Filaments branched 17
   14. Chloroplast a transverse band, not reticulate........*Ulothrix* (p. 54)
   14. Chloroplast reticulate 15

15. Basal cell several times longer than other cells of filament. . . . .
. . . . *Chaetomorpha* (p. 100)
15. Basal cell not more than twice as long as other cells of filament. 16
    16. Attached by rhizoids from single basal cell. . . . .*Lola* (p. 92)
    16. Attached by rhizoids from several lowermost cells. . . .*Urospora* (p. 93)
17. Filaments endophytic or epiphytic 18
17. Filaments not endophytic or epiphytic 20
    18. Branch apices inflated when lying just beneath surface of host. . .
    . . . . *Endophyton* (p. 62)
    18. Branch apices not inflated 19
19. Branches endophytic, forming netlike arrangement. .*Pseudodictyon* (p. 63)
19. Branches endophytic or epiphytic, not forming netlike arrangement. . .
. . . . *Entocladia* (p. 64)
    20. Filaments cohering in ropelike strands by means of short curved
        branches or descending rhizoidal filaments. . . . .*Spongomorpha* (p. 96)
    20. Filaments not cohering in ropelike strands 21
21. Filaments unbranched, with occasional to frequent short rhizoidal
branches . . . . *Rhizoclonium* (p. 90)
21. Filaments repeatedly branched, lacking short rhizoidal branches. 22
    22. Lateral branches with crosswall at base . . . .*Cladophora* (p. 103)
    22. Lateral branches lacking crosswall at base. . . .*Cladophoropsis* (p. 110)
23. Thallus slender, composed mostly of 2 parallel rows of cells. . . . .
. . . . *Percursaria* (p. 69)
23. Thallus not composed of 2 parallel rows of cells. 24
    24. Plants bladelike or tubular 25
    24. Plants unicellular, mostly endophytic or endozoic 29
25. Blades 1 cell thick . . . . , , ...26
25. Blades 2 cells thick, or the thallus tubular. . . . . , . , ...27
    26. Blades less than 10 mm long, growing in spray zone, high on rocks
        coated with guano . . . .*Prasiola* (p. 89)
    26. Blades mostly over 15 mm long, usually intertidal . . . .
    . . . . *Monostroma* (p. 66)
27. Thallus bladelike . . . . *Ulva* (p. 77)
27. Thallus tubular, at least in part 28
    28. Erect tubular parts arising from common discoid base; rhizoidal
        processes from lower cells lacking . . . .*Blidingia* (p. 70)
    28. Erect tubular parts not arising from common discoid base; rhizoidal
        processes from lower cells numerous . . . .*Enteromorpha* (p. 73)
29. Plants commonly growing in matrix of mollusk shells. .*"Gomontia"* (p. 120)
29. Plants endophytic 30
    30. Plants growing among erect filaments of *Petrocelis.* . . . .
    . . . . *"Codiolum"* (pp. 95, 99)
    30. Plants growing in various algae, not in *Petrocelis.* . . . .
    . . . .*"Chlorochytrium"* (pp. 99, 100)

Phaeophyta: The Brown Algae

1. Thalli filamentous (sometimes with major branches corticated). 2
1. Thalli not filamentous, at least partly parenchymatous or pseudoparenchymatous 10
    2. Filaments wholly or primarily endophytic. . . . .*Streblonema* (p. 149)
    2. Filaments not wholly or primarily endophytic 3

3. Reproductive structures intercalary . . . . *Pilayella* (p. 146)
3. Reproductive structures mostly lateral or terminal. 4
    4. Hooked branchlets frequent; erect filaments cohering in strands. . .
    . . . . *Spongonema* (p. 148)
    4. Hooked branchlets lacking 5
5. Chloroplasts bandlike or ribbonlike, mostly few per cell. 6
5. Chloroplasts discoid, mostly many per cell. 7
    6. Growth zones present at base of hairlike branches. . . *Kuckuckia* (p. 140)
    6. Growth zones not evident, hairlike branches usually lacking . . . .
    . . . . *Ectocarpus* (p. 123)
7. Plurangia mostly in compact clusters . . . . *Sorocarpus* (p. 130)
7. Plurangia not in compact clusters, mostly not more than 1 per branch
  cell 8
    8. Filaments with frequent distinct growth zones between successive
    branches . . . . *Acinetospora* (p. 136)
    8. Filaments with few or no distinct growth zones. 9
9. Filaments commonly with unbranched hairlike extensions above distinct growth zone; plurangia mostly symmetrical and pedicellate. . . . .
  . . . . *Feldmannia* (p. 130)
9. Filaments lacking distinct growth zone; plurangia mostly asymmetri-
  cal and sessile . . . . *Giffordia* (p. 140)
    10. Thallus crustose, with erect filaments from basal layer 1 or more
    cells thick 11
    10. Thallus not crustose (though holdfast may be more or less crus-
    tose) 22
11. Crusts 5 mm or more broad, on rocks or shells. 12
11. Crusts minute, discoid, usually epiphytic 20
    12. Unangia borne laterally on cortical filaments, never in sori. . . . . . .
    . . . . *Cylindrocarpus* (p. 177)
    12. Unangia, if present, borne terminally on erect filaments, or in sori
    at base of paraphyses 13
13. Erect filaments loosely cohering, separated by gelatinous material. 14
13. Erect filaments firmly cohering 16
    14. Crusts less than 100 μm thick; sporangia terminal on erect fila-
    ments . . . . *Petroderma* (p. 174)
    14. Crusts mostly 500 μm or more thick 15
15. Unangia terminal; plurangia biquadriseriate. . . *Hapalospongidion* (p. 170)
15. Unangia unknown; plurangia mostly in terminal pairs. . . . *Diplura* (p. 172)
    16. Unangia borne at base of multicellular paraphyses; chloroplasts
    1-per cell 17
    16. Unangia unknown; chloroplasts many per cell 19
17. Crusts undivided, of variable extent. . . . . *Ralfsia* (in part; p. 164)
17. Crusts composed of overlapping branches or flattened lobes. 18
    18. Overlapping lobes broad and flat. . . . . *Ralfsia fungiformis* (p. 165)
    18. Overlapping branches linear, resembling hapteres . . . .
    . . . . *Hapterophycus* (p. 172)
19. Tip of plurangia of 2–4 sterile cells; crust light brown. . . . .
  . . . . *Endoplura* (p. 173)
19. Tip of plurangia not sterile, with frequent colorless unicellular para-
  physes among plurangia; crust dark brown to black. . . . .
  . . . . *Pseudolithoderma* (p. 174)

20. Erect filaments arising from distromatic basal layer.............
.... *Hecatonema* (p. 162)
20. Erect filaments arising from monostromatic basal layer...........21
21. Plurangia uniseriate ....*Myrionema* (p. 157)
21. Plurangia multiseriate ....*Composonema* (p. 160)
    22. Plant minute, tufted, to 7 mm high, with free paraphyses from
        pulvinate pseudoparenchymatous base ....*Elachista* (p. 178)
    22. Plant larger, lacking free paraphyses 23
23. Thallus primarily hollow throughout at maturity 24
23. Thallus not primarily hollow throughout at maturity 31
    24. Thallus erect, cylindrical or compressed 25
    24. Thallus globular or saccate, often flattened or collapsed..........27
25. Thallus branched, not tufted ....*Rosenvingea* (p. 202)
25. Thallus unbranched, tufted 26
    26. Thallus often constricted at intervals; unangia borne on minute
        crustose phase only .... *Scytosiphon* (p. 198)
    26. Thallus not constricted; unangia borne superficially among multi-
        cellular paraphyses ....*Melanosiphon* (p. 192)
27. Thallus fleshy, the outer cortex wholly of anticlinal filaments........
.... *Leathesia* (in part; p. 176)
27. Thallus not fleshy, wholly or primarily parenchymatous 28
    28. Surface with numerous conspicuous sori; plants globular saccate
        epiphytes .... *Soranthera* (p. 196)
    28. Surface lacking conspicuous sori 29
29. Plants cylindrical or flattened, saccate, mostly epiphytic............
.... *Coilodesme* (p. 188)
29. Plants globular or irregularly lobed, or convoluted..................30
    30. Thallus saccate or a folded net, with numerous irregular perfora-
        tions .... *Hydroclathrus* (p. 206)
    30. Thallus saccate, frequently with folds or tubercles, but lacking
        perforations .... *Colpomenia* (p. 202)
31. Plants globular epiphytes to 5 mm diam., seldom hollow...........
.... *Leathesia* (in part; p. 176)
31. Plants not globular epiphytes 32
    32. Pneumatocysts 1 or more 33
    32. Pneumatocysts lacking 39
33. Pneumatocyst single, large 34
33. Pneumatocysts numerous, small 35
    34. Blades borne on forking, antlerlike stalk........*Pelagophycus* (p. 251)
    34. Blades borne on 4 short, repeatedly forked stalks surmounting
        pneumatocyst .... *Nereocystis* (p. 253)
35. Pneumatocysts borne singly or in clusters 36
35. Pneumatocysts borne in series on smaller branches.................38
    36. Pneumatocysts spherical, mostly less than 8 mm diam., borne
        singly in axils of "leaves"...................*Sargassum* (p. 273)
    36. Pneumatocysts mostly ellipsoidal, mostly over 10 mm diam., borne
        singly along elongate axis 37
37. Pneumatocysts on cylindrical axis, each bearing relatively large terminal blade .... *Macrocystis* (p. 255)
37. Pneumatocysts on flattened axis, each commonly bearing relatively
  small terminal blade ....*Egregia* (p. 242)

38. Pneumatocysts spherical or slightly flattened, mostly in series lacking a flattened margin; stipe triangular in cross section. . . . . *Cystoseira* (p. 267)
38. Pneumatocysts distinctly flattened, in series forming a podlike structure with continuous flattened margin; stipe terete. . . . . *Halidrys* (p. 272)
39. Thallus unbranched, primarily bladelike (the blade sometimes divided) 40
39. Thallus branched, or with several blades from common base. 50
    40. Holdfast composed of evident hapterous branches 41
    40. Holdfast discoid 48
41. Blade with 5 raised longitudinal ribs . . . . *Costaria* (p. 237)
41. Blades lacking raised longitudinal ribs 42
    42. Blades with midrib, midvein, or median longitudinal fold. 43
    42. Blades lacking midrib, midvein, or median longitudinal fold. 46
43. Blade with numerous perforations . . . . *Agarum* (p. 234)
43. Blade lacking regular perforations 44
    44. Blade with reticulum of ridges on either side of smooth flattened midrib . . . . *Dictyoneuropsis* (p. 248)
    44. Blade lacking reticulations 45
45. Sori on special sporophylls borne laterally on stipe at base of blade. . . . . *Alaria* (p. 241)
45. Sori on median longitudinal fold . . . . *Pleurophycus* (p. 234)
    46. Blade with reticulum of ridges, proliferating by split in basal meristem . . . . *Dictyoneurum* (p. 246)
    46. Blade lacking reticulum of ridges, not proliferating by split in meristem 47
47. Blade with distinct stipe . . . . *Laminaria* (in part; p. 229)
47. Blade lacking stipe . . . . *Hedophyllum* (p. 233)
    48. Blade 45 cm long or longer; discoid base 1–3 cm broad. . . . . *Laminaria* (in part; p. 229)
    48. Blade mostly less than 25 cm long; discoid base less than 5 mm broad 49
49. Unangia and plurangia partly exserted, arising from transformed surface cells . . . . *Punctaria* (p. 192)
49. Unangia entirely above surface of blade, plurangia unknown. . . . . *Halorhipis* (p. 194)
    50. Branches cylindrical to slightly flattened 51
    50. Branches, at least ultimate blades or divisions, clearly to markedly flattened 59
51. Branch apices with microscopically conspicuous apical cell; lower cells in successive tiers . . . . *Sphacelaria* (p. 216)
51. Branch apices lacking conspicuous apical cell; lower cells not tiered 52
    52. Branches and/or veins opposite (chiefly alternate in *D. latifrons*) . . . . *Desmarestia* (in part; p. 220)
    52. Branches not opposite (somewhat opposite in *Pterygophora*). 53
53. Branches terminating in single filament. . . . . *Stictyosiphon* (p. 186)
53. Branches not terminating in single filament 54
    54. Branches with terminal tuft of hairs. . . . . *Sporochnus* (p. 184)
    54. Branches lacking terminal tuft of hairs 55
55. Branching chiefly dichotomous 56
55. Branching not chiefly dichotomous 57

56. Plants less than 15 cm high, in extreme upper intertidal.........
.... *Pelvetiopsis* (in part; p. 264)
56. Plants commonly 30–60 cm high, in upper intertidal...........
.... *Pelvetia* (in part; p. 261)
57. Branches bearing numerous short laterals; plants not gelatinous......
.... *Analipus* (p. 180)
57. Branches not bearing numerous short laterals; plants gelatinous 58
58. Main axes distinct, the plants uniaxial; cortical filaments 4–6 cells
long; axes usually with numerous unbranched hairs.............
.... *Haplogloia* (p. 180)
58. Main axes indistinct, the plants multiaxial; cortical filaments 8–11
cells long; axes lacking hairs .... *Tinocladia* (p. 183)
59. Plants with conspicuous stumplike, woody base, bearing numerous
linear blades .... *Lessoniopsis* (p. 246)
59. Plants lacking conspicuous stumplike base, the blades few, simple or
divided 60
60. Branches 3–10 cells thick 61
60. Branches many, more than 10 cells thick 66
61. Thalli fan-shaped or with fan-shaped divisions 62
61. Thalli not fan-shaped, dichotomously branched 64
62. Plants mostly less than 3 cm high, monostromatic at margins.....
.... *Syringoderma* (p. 210)
62. Plants mostly over 10 cm high, polystromatic throughout.........63
63. Thalli mostly 10–15 cm high, with numerous relatively narrow divisions
.... *Zonaria* (p. 215)
63. Thalli mostly 20 cm or more high, with few relatively broad divisions,
the divisions much longer than broad .... *Taonia* (p. 213)
64. Branches with evident midrib .... *Dictyopteris* (p. 211)
64. Branches lacking midrib ....,,65
65. Branches light brown, more or less delicate, 3 cells thick throughout
.... *Dictyota* (p. 207)
65. Branches medium to dark brown, coarser, 5 or 6 cells thick at margins
.... *Pachydictyon* (p. 209)
66. Branching dichotomous throughout 67
66. Branching, if present, not strictly dichotomous 70
67. Branches with midrib 68
67. Branches lacking midrib 69
68. Branches with line of whitish hair tufts on each side of midrib
.... *Hesperophycus* (p. 265)
68. Branches lacking line of hair tufts on each side of midrib........
.... *Fucus* (p. 259)
69. Thalli mostly over 200 cm high, with mostly more than 3 cm between
successive dichotomies; high intertidal......... *Pelvetia* (in part; p. 261)
69. Thalli mostly less than 10 cm high, with mostly less than 2 cm between
successive dichotomies; at extreme high tide level ....
.... *Pelvetiopsis* (in part; p. 264)
70. Blades with midrib and paired lateral veins or marginal teeth;
plants bleaching rapidly .... *Desmarestia* (in part; p. 220)
70. Blades lacking evident veins or paired marginal points; plants not
bleaching 71
71. Holdfast composed of a number of hapterous branches..............72
71. Holdfast discoid, bearing several to many blades 75

72. Stipe divided apically into 2 branches, each bearing numerous
    blades .... *Eisenia* (p. 242)
72. Stipe normally undivided 73
73. Stipe terminating in tuft of longitudinally grooved blades...........
    .... *Postelsia* (p. 251)
73. Stipe terminating in nongrooved blades 74
    74. Stipe bearing single blade with several straplike divisions.......
        .... *Laminaria* (in part; p. 229)
    74. Stipe bearing numerous lateral blades below flattened main blade
        .... *Pterygophora* (p. 241)
75. Plants dark brown, with cuneate blades, these broadest near apices;
    unangia abundant in surface areas.............. *Phaeostrophion* (p. 195)
75. Plants medium brown, with linear to spatulate blades; unangia lacking
    (though known for small crustose stage in *Petalonia*)................76
    76. Medulla of large rounded to angular cells......... *Petalonia* (p. 200)
    76. Medulla of thick-walled interwoven filaments.... *Endarachne* (p. 200)

RHODOPHYTA: THE RED ALGAE

1. Thallus usually not impregnated with calcium carbonate............. 2
1. Thallus impregnated with calcium carbonate, stony or semicalcified,
   pale pink to whitish 198
    2. Thallus branched, not filamentous, more than 1 cell broad......... 3
    2. Thallus a simple or branched filament (sometimes partly corticated)
      171
3. Thallus not parasitic, although sometimes epiphytic................ 4
3. Thallus parasitic(?), usually very small, pale or whitish; plants of each
   species usually limited to plants of single host genus.................160
    4. Thallus more or less erect 5
    4. Thallus prostrate, discoid or crustose 146
5. Cells not in regular transverse series 6
5. Cells in regular transverse series (polysiphonous), with 3 or more pericentral cells surrounding a central cell in each tier or segment........132
    6. Branches mainly cylindrical or of slightly flattened cylindrical form,
       oval to elliptical in cross section, not bladelike.................. 7
    6. Branches mostly bladelike or with bladelike divisions, linear in cross
       section of widest parts 60
7. Plant hollow, saccate, and mostly unbranched....... *Halosaccion* (p. 550)
7. Plant not hollow, or if hollow then branched....................... 8
    8. Lateral branches in transverse whorls... *Gloiosiphonia* (in part; p. 419)
    8. Lateral branches not in transverse whorls 9
9. Branches terminating in conspicuous elongate to obovoid vesicular enlargements 10
9. Branches lacking conspicuous enlargements 11
    10. Terminal enlargements simple, oblanceolate, several times as long
        as thick; inner cortex of interconnected stellate cells...........
        .... *Reticulobotrys* (p. 485)
    10. Terminal enlargements simple or compound, if simple usually obo-
        void but not oblanceolate; inner cortex lacking stellate cells......
        .... *Botryocladia* (p. 550)
11. At least ultimate branches with series of encircling cellular bands..... 12
11. Branches lacking series of encircling cellular bands 14

12. Nodal bands on smaller branchlets composed of a single row of
cells .... *Spyridia* (p. 608)
12. Nodal bands usually several to many cells wide................. 13
13. Cells of nodal bands in longitudinal rows...........*Centroceras* (p. 602)
13. Cells of nodal bands mostly not in longitudinal rows...*Ceramium* (p. 591)
    14. Branching dichotomous or mostly so 15
    14. Branching not primarily dichotomous 23
15. Branches uniaxial 16
15. Branches multiaxial 17
    16. Main axes flattened, commonly 10 cm or more tall, 4–5 mm wide
    .... *Leptocladia* (p. 363)
    16. Main axes cylindrical, mostly less than 4 cm tall, 1–2 mm wide
    .... *Gloiopeltis* (p. 422)
17. Outer cortex with numerous colorless utricles 18
17. Outer cortex lacking colorless utricles 20
    18. Branches conspicuously narrowing toward base......*Scinaia* (p. 331)
    18. Branches not conspicuously narrowing toward base.............. 19
19. Plants with pigmented cells among colorless utricles...............
    .... *Pseudogloiophloea* (p. 334)
19. Plants lacking pigmented cells among colorless utricles.............
    .... *Pseudoscinaia* (p. 333)
    20. Branches mostly cylindrical throughout...........*Ahnfeltia* (p. 503)
    20. Branches distinctly flattened in upper parts 21
21. Medulla of large cells, not filamentous..........*Gymnogongrus* (p. 505)
21. Medulla filamentous, at least in part.......................... 22
    22. Branches usually with a few short proliferous branchlets.........
    .... *Prionitis* (in part; p. 442)
    22. Branches lacking proliferous branchlets ....
    .... *Carpopeltis* (in part; p. 441)
23. At least the ultimate branches hollow 24
23. Branches not hollow 27
    24. Branches not transversely septate..............*Lomentaria* (p. 567)
    24. Branches hollow, transversely septate 25
25. Mature plants usually over 10 cm tall; major branches with many short
branchlets .... *Gastroclonium* (p. 567)
25. Mature plants mostly less than 4 cm tall 26
    26. Asexual plants forming polysporangia ....*Coeloseira* (p. 565)
    26. Asexual plants forming tetrasporangia...........*Champia* (p. 564)
27. Branch with core of longitudinal filaments...................... 28
27. Branch lacking core of longitudinal filaments (polysiphonous in *Rhodoptilum*) 32
    28. Plants firm, neither slimy nor elastic ....*Neoagardhiella* (p. 483)
    28. Plants gelatinous, or elastic and slippery...................... 29
29. Plants epiphytic, 1–2 cm tall; cortical filaments frequently with terminal hyaline hair ....*Helminthora* (p. 325)
29. Plants saxicolous, 10–20 cm tall or taller........................ 30
    30. Thallus sparingly branched or unbranched, wormlike; carpogonial
    branches terminal on cortical filaments...........*Nemalion* (p. 324)
    30. Thallus with few to numerous branches; carpogonial branches lat-
    eral toward base of cortical filaments......................... 31
31. Main axis with numerous short lateral branchlets.....*Cumagloia* (p. 329)
31. Main axis with few moderately long branches....*Helminthocladia* (p. 327)

32. Major branches with many short lateral branches all of approximately same length 33
32. Major branches with progressively shorter branches 41
33. Short lateral branches distichous 34
33. Short lateral branches not distichous (obscurely distichous in *Odonthalia*) 39
    34. Short branches filamentous ....*Rhodoptilum* (p. 678)
    34. Short branches not filamentous 35
35. Two branches of a subopposite pair alike................*Pikea* (p. 359)
35. Two branches of a subopposite pair not alike 36
    36. Alternate short branchlets flattened, resembling minute leaves..... 37
    36. No short branchlets markedly flattened 38
37. Short branchlet opposite larger branch sometimes bearing reproductive structures .... *Ptilota* (p. 630)
37. Short branchlet opposite larger branch never bearing reproductive structures .... *Neoptilota* (p. 631)
    38. Long and short lateral branchlets regularly alternate...........
    .... *Bonnemaisonia* (p. 336)
    38. Long and short lateral branchlets not regularly alternate........
    .... *Schimmelmannia* (p. 421)
39. Short branches obscurely distichous or in clusters................
    .... *Odonthalia* (in part; p. 742)
39. Short branches spirally or irregularly arranged 40
    40. Short branches mostly spirally arranged.........*Rhodomela* (p. 741)
    40. Short branches irregularly arranged.......*Gracilaria* (in part; p. 494)
41. Smaller branches unilaterally branched in groups (pectinate)......... 42
41. Smaller branches not unilaterally branched 43
    42. Unilateral branches incurved toward apex of bearing branch....
    .... *Plocamium* (p. 492)
    42. Unilateral branches curving away from apex of bearing branch...
    ....*Microcladia* (in part; p. 604)
43. Branching mostly distichous 44
43. Branching not distichous 49
    44. Branches tough, owing to numerous colorless thick-walled internal filaments 45
    44. Branches not particularly tough, lacking colorless thick-walled internal filaments 46
45. Gonimoblast in unilocular compartment...........*Pterocladia* (p. 349)
45. Gonimoblast in bilocular compartment...............*Gelidium* (p. 342)
    46. Branches with apical pit................*Laurencia* (in part; p. 727)
    46. Branches lacking apical pit 47
47. Plants epiphytic.......................*Microcladia* (in part; p. 604)
47. Plants not epiphytic 48
    48. Branches regularly and conspicuously distichous..............
    ....*Odonthalia* (in part; p. 742)
    48. Branches irregularly and often inconspicuously distichous.......
    .... *Gigartina* (in part; p. 516)
49. Smaller branches covered with minute spines........*Endocladia* (p. 422)
49. No branches covered with minute spines 50
    50. Plants densely and finely plumose...........*Asparagopsis* (p. 340)
    50. Plants not densely and finely plumose...................... 51

51. Apices of branches blunted by small terminal pit.................. 52
51. Apices of branches lacking terminal pit........................... 53
   52. Apical pit with tuft of protruding hairs......*Chondria* (in part; p. 723)
   52. Apical pit lacking evident protruding hairs..*Laurencia* (in part; p. 727)
53. Branch apices acute, usually with tuft of hairs...*Chondria* (in part; p. 723)
53. Branch apices lacking tuft of hairs............................... 54
   54. Branches with distinct central filament (usually evident in finer
      branches) 55
   54. Branches lacking central filament 58
55. Branches soft and gelatinous 56
55. Branches, if soft, not gelatinous 57
   56. Branches robust and clublike.................*Dudresnaya* (p. 354)
   56. Branches slender, attenuate..........*Gloiosiphonia* (in part; p. 419)
57. Branches cylindrical ....*Cryptosiphonia* (p. 361)
57. Branches slightly to conspicuously compressed...*Farlowia* (in part; p. 355)
   58. Medulla more or less solid, with large colorless globular cells.....
      .... *Hypnea* (p. 489)
   58. Medulla filamentous, or with elongated colorless cells........... 59
59. Branches densely covered on all sides with numerous short, sometimes
    spinose branchlets......................*Gigartina* (in part; p. 516)
59. Branches not densely covered with branchlets....*Prionitis* (in part; p. 442)
   60. Bladelike thallus normally undivided, but at times lobed or with
      proliferous blades 61
   60. Blades more or less divided 94
61. Blades discoid, 1 or more, borne terminally on thick erect stipe........ 62
61. Blades not discoid, lacking thick erect stipe....................... 64
   62. Blade stellate ....*Sciadophycus* (p. 547)
   62. Blade not stellate 63
63. Medulla of blade composed of thickly matted filaments.............
    .... *Constantinea* (p. 365)
63. Medulla of blade composed of relatively few large cells............
    .... *Maripelta* (p. 548)
   64. Blade subsessile, attached near center or margin of lower face...
      ....*Callophyllis* (in part; p. 459)
   64. Blade erect, basally attached 65
65. Blades usually with marginal proliferations 66
65. Blades usually lacking proliferations 67
   66. Blades longer than broad, relatively thin and slippery; prolifera-
      tions irregular in shape...............*Grateloupia* (in part; p. 430)
   66. Blades mostly broader than long, thick and not slippery; prolifera-
      tions similar to blades.......................*Opuntiella* (p. 485)
67. Entire blade either 1 or 2 cells thick............................. 68
67. Blade more than 2 cells thick, at least at base..................... 74
   68. Blade 2 cells thick.....................*Porphyra* (in part; p. 294)
   68. Blade 1 cell thick 69
69. Mature blades usually over 8 cm tall..........*Porphyra* (in part; p. 294)
69. Mature blades mostly less than 5 cm tall.......................... 70
   70. Blades mostly less than 12 cells broad..*Erythrotrichia* (in part; p. 284)
   70. Blades many cells broad 71
71. Plants initially saccate, mostly less than 5 mm tall...*Porphyropsis* (p. 291)
71. Plants not initially saccate, mostly 10 mm or more tall.............. 72

181. At least 1 branchlet in a whorl longer than others; branch continuing
    growth beyond developing gonimoblast. . . . .*Scagelia* (p. 584)
181. All branchlets of a whorl of approximately same length; branch not
    continuing growth beyond developing gonimoblast . . . .
    . . . . *Antithamnionella* (p. 580)
    182. Branching regularly alternate 183
    182. Branching not regularly alternate 185
183. Sporangia producing more than 4 spores (polysporangial) . . . .
    . . . . *Pleonosporium* (p. 617)
183. Sporangia producing 4 spores (tetrasporangial) 184
    184. Cells multinucleate; gonimoblast with rounded lobes. . . . .
    . . . . *Callithamnion* (p. 609)
    184. Cells uninucleate; gonimoblast with angular lobes. . . . .
    . . . . *Aglaothamnion* (p. 615)
185. Cells up to 0.5 mm diam., with upper end frequently distinctly larger
    than lower end 186
185. Cells less than 0.2 mm diam., cylindrical. 187
    186. Tetrasporangia and spermatangia borne on condensed lateral
    branch systems; upper end of cells frequently Y-shaped. . . . . . .
    . . . . *Bornetia* (p. 628)
    186. Tetrasporangia and spermatangia not borne on condensed
    branch systems; upper end of cells not Y-shaped . . . .
    . . . . *Griffithsia* (p. 621)
187. Branching predominantly unilateral 188
187. Branching not predominantly unilateral 190
    188. Sporangia producing 1 spore. . . . . . .*Acrochaetium* (in part; p. 308)
    188. Sporangia producing 4 or more spores. 189
189. Sporangia producing polyspores. . . . .*Tiffaniella* (in part; p. 623)
189. Sporangia producing mostly tetraspores. . .*Rhodochorton* (in part; p. 321)
    190. Filaments usually growing in matrix of mollusk shells. . . . .
    . . . . "*Conchocelis*" (p. 304)
    190. Filaments not growing in calcareous substratum. 191
191. Sporangia producing tetraspores, bispores, or polyspores. 192
191. Sporangia producing mostly monospores 197
    192. Sporangia (in our species) producing polyspores. 193
    192. Sporangia producing mostly tetraspores 194
193. Gonimoblast with involucral branchlets . . . .*Ptilothamnion* (p. 624)
193. Gonimoblast lacking involucre. . . . .*Tiffaniella* (in part; p. 623)
    194. Branches with numerous minute gland cells mostly at juncture
    of cells. . . . ."*Trailliella*" (p. 338)
    194. Branches lacking minute gland cells at juncture of cells. 195
195. Spermatangia in lateral or terminal clusters; gonimoblasts, so far as
    known, forming carpotetrasporangia. . . . .*Rhodochorton* (in part; p. 321)
195. Spermatangia in compact heads with gelatinous enclosing envelope;
    gonimoblasts forming carposporangia 196
    196. Gonimoblast with jacket of sterile cells (pericarp). . . . .
    . . . . *Lejolisea* (p. 626)
    196. Gonimoblast lacking sterile cells. . . . . . .*Ptilothamnionopsis* (p. 626)
197. Plants less than 50 μm tall, host-specific epiphytes; spermatangia
    separately stalked . . . . *Kylinia* (p. 323)
197. Plants mostly over 1 mm tall, epiphytic, epizoic, or endozoic; sperma-
    tangia clustered, several to many per stalk. .*Acrochaetium* (in part; p. 308)

REFERENCE MATERIAL

Glossary

MOST OF THE technical terms employed in this book may be found in any good dictionary, and this glossary is compiled with that fact in mind—not all of the terms used are defined here. Many descriptive terms, for example, apply equally well to algae and terrestrial plants; only the more unfamiliar of these are included here. Others, though, are often used somewhat differently when applied to algae, since their morphological parts are different from (and only apparently analogous to) those of most flowering plants; all of these, if used in the text, are defined here. Still other terms apply only to algae, and are not found in any general dictionary or in dictionaries of biology—which are few—or even in textbooks; all such terms, if used, are given here.

The great detail necessary to define certain chemicals, such as pigments, storage materials, and cell-wall materials, we believe to be beyond the scope of this book, and because of the active research in these topics much information of this kind may be out of date before this book is out of date. We refer the reader to a recent compilation that contains information up to 1974: W. D. P. Stewart, ed., *Algal physiology and biochemistry* (University of California Press, 989 pp.); the entries for Pigment, Storage products, and Wall materials draw upon this volume.

ABAXIAL. On the side away from the axis, turned toward the base; said of branches or reproductive structures. *See* Adaxial.

ABSCISSION. Natural separation and loss of a part, generally of a specialized separation layer.

ABSORBING FILAMENT. One of the sterile filaments extending from within the carposporophyte to the medullary tissue (Iridaea).

ACCESSORY. Additional or auxiliary; appended to a main or central structure.

ACCESSORY PIGMENT. Pigment other than chlorophyll *a*, but functioning in photosynthesis. *See* Pigment.

ACROCHAETIOID (or in an ACROCHAETIUM PHASE or STAGE). Resembling the genus *Acrochaetium* (Nemaliales) in being microscopic, filamentous, branched or unbranched; also a phase (in a life history) that resembles this genus.

ACRONEMATIC. Lacking mastigonemes; said of a flagellum.

ACROPETAL. Arising (maturing) from the base toward the apex; said of branches or reproductive structures.

ACULEATE. Prickly, spinelike; also, having processes resembling prickles or spines.

ACUMINATE. Tapering gradually to a point.

ACUTE. Sharp, ending abruptly in a point; describing less than a right angle; said of blades or branches.

ADAXIAL. On the side toward the axis, turned toward the apex; said of branches or reproductive structures. *See also* Abaxial.

ADJOINED. Lying in contact, but not fused.

AGAR. *See* Wall materials.

AKINETE. A single cell that separates from a vegetative cell to form a nonmotile reproductive body, or spore (Chrysophyta).

ALA (pl. ALAE). A flat lateral expansion; a wing.

ALATE. Winglike; said of blades or branches.

ALGINIC ACID. *See* Wall materials.

ALTERNATE. Arranged singly at regular intervals on an axis and alternating in orientation to the axis.

ALVEOLATE. Having numerous small cavities, pits, or depressions; honeycombed; said of surfaces or tissues.

AMPULLA (pl. AMPULLAE). A special cluster of filaments in the cortex, bearing either a carpogonial branch or an auxiliary cell (Cryptonemiaceae, Rhodophyta).

ANASTOMOSIS (pl. ANASTOMOSES). A joining at adjacent parts; said of blades or filaments.

ANDROPHORE. A stalk or structure bearing male reproductive organs.

ANISOGAMY. The union of two gametes (both usually motile) of unequal size (anisogametes). *See also* Isogamy.

ANNULAR. Ringed; arranged in a ring or circle.

ANTHERIDIUM (pl. ANTHERIDIA). A reproductive structure producing motile male gametes, the gametes bearing one or more flagella. *See also* Spermatangium.

ANTICLINAL. Perpendicular to the surface or circumference of a structure; said of medullary or cortical cells.

APEX (pl. APICES). Tip, distal end, or point.

APICAL CELL. An initial cell at the apex of a branch or stem; branch primordium. *See also* Meristem.

APICAL MARGIN. A plant margin composed entirely of specialized cells (apical cells) that initiate growth in some brown algae, and in many red algae.

APICAL MERISTEM. *See* Meristem.

APICULATE. Terminating in a short, somewhat flexible point.

APLANOSPORE. A nonmotile spore with its wall free from the sporangial wall (in green algae).

APOPHYSIS (pl. APOPHYSES). In foliose algae (e.g. *Iridaea*), the region of broadening between stipe and blade.

APPRESSED. Lying flat against something.

ARCUATE. Curved like a bow.

ARTICULATION. The uncalcified joint between segments of coralline algae. *See also* Geniculum.

ASCOCYST. A unicellular, hyaline, paraphysislike structure (brown algae).

ASSIMILATING (ASSIMILATORY) FILAMENT. An erect or free photosynthetic filament, usually branched, sometimes occurring with nonphotosynthetic tissue. Found in some brown and red algae.

ASSURGENT. Curving obliquely upward; ascending; said of essentially horizontal filaments that turn upward.

ATTENUATE. Narrow and gradually tapering toward the base or apex; drawn out; said of a blade or branch.

AURICULATE. Earlike; said of blades or proliferations.

AUXILIARY CELL. In the Florideophyceae, a specialized cell that receives the zygote nucleus or a division product of the zygote nucleus, usually initiating gonimoblast formation. This cell may be isolated, or may be borne in a branch, filament, or cluster.

AXIAL FILAMENT. A single series of cells extending longitudinally through the center of an axis or a branch.

AXIL. The angle between the axis and a lateral branch.

AXILE. Belonging to or in the axis of a cell; said of some plastids, these usually stellate in shape.

AXILLARY. Situated in the axil.

AXIS (pl. AXES). A main longitudinal structure, often bearing branches or laterals. Axes may be uniaxial or multiaxial (having one or many internal component parts).

BASIONYM. The original name of a taxon used in a new combination because of change in taxonomic position or rank; required in order to establish the nomenclatural validity of a taxon. *See also* Homonym, Synonym.

BASOLATERAL. Situated at the edge of the base.

BENTHIC. Growing or living on the sea bottom.

BIFID. Cleft in two parts.

BIFLAGELLATE. Having two flagella.

BIFURCATE. Forked. *See also* Dichotomous.

BILATERAL. Having two corresponding or complementary sides; arranged on opposite sides.

BILOBATE. Two-lobed.

BILOCULAR. Having two cavities.

BIPINNATE. Having pinnae of two orders

(i.e., each pinnate portion again pinnate).

BISPORANGIUM (pl. BISPORANGIA). A structure producing bispores (two spores), which are incompletely divided tetrasporangia and are diploid (certain red algae, especially the Corallinaceae).

BLADE. A flattened leaflike thallus or thallus part.

BLADELET. A small blade.

BOTRYOIDAL. Resembling a cluster of grapes.

BRANCH. A plant division subordinate to the main axis.

BRANCHING. See Dichotomous, Divaricate, Divergent, Order, Secund.

BRANCHLET. A small branch, usually the ultimate branch in a system of branching.

BULLA (pl. BULLAE). A local outward bulging, blistering, or puckering of the surface.

BULLATE. Having bullae.

BULLATION. An inflated or blisterlike swelling.

CAECOSTOMA (pl. CAECOSTOMATA). A cryptostoma without hairs or ostiole (Fucales).

CAESPITOSE. Growing in tufts; matted.

CALCAREOUS. Containing large amounts of calcium carbonate; stony.

CALCIFIED. Encrusted or impregnated with calcium carbonate.

CANALICULATE. Having or forming a channel or groove.

CAPITULIFORM. Having a dense terminal head or cluster.

CAROTENE, CAROTENOID. See Pigment.

CARPOGONIAL. Bearing or associated with carpogonia.

CARPOGONIUM (pl. CARPOGONIA). The female sex cell containing the egg (Rhodophyta); the terminal cell in a carpogonial branch or filament.

CARPOSPORANGIUM (pl. CARPOSPORANGIA). The reproductive cell of a carposporophyte, producing carpospores (most Rhodophyta).

CARPOSPORE. A spore produced in a carposporangium, usually diploid. In most red algae, the germinating carpospore produces a free-living plant, the tetrasporophyte.

CARPOSPOROPHYTE. A multicellular phase in the life history of Rhodophyta resulting from the development of the zygote and terminated by carpospore production. Carposporophytes are borne on the gametophyte and (in most cases investigated) have a different chromosome count (diploid) from the bearing plant, which is haploid. Most carposporophytes are filamentous.

CARPOSTOME (pl. CARPOSTOMATA). Opening in a little-modified cortex through which carpospores emerge (many Cryptonemiales, some Gigartinales). See also Pericarp, Ostiole.

CARPOTETRASPORANGIUM (pl. CARPOTETRASPORANGIA). A quartet of spores (carpotetraspores) borne in place of a carpospore in the carposporophyte or gonimoblast (Rhodophyta).

CARRAGEENAN. See Wall material.

CARTILAGINOUS. Firm, tough, and elastic.

CATENATE. Arranged in a single series (e.g. as in a chain of spherical cells).

CELLULOSE. See Wall material.

CHLOROPHYLL. See Pigment.

CHLOROPLAST. A cytoplasmic organelle (or plastid) in which chlorophyll is the predominant pigment. When other pigments (such as fucoxanthin or phycoerythrin) predominate, the structure is sometimes referred to as a chromatophore or chromoplast.

CHRYSOLAMINARIN. See Pigment.

CICATRIGENOUS. Arising from the scar cell that is the remnant of a trichoblast (Rhodomelaceae).

CIRCINATE. Coiled inward or downward from the tip, the tip emerging last during the uncoiling.

CLASS. A taxonomic division midway between phylum (or division) and order.

CLAVATE. Club-shaped; said of a filament or branch thickened toward the tip.

COALESCENT. Joined or fused together.

COCCOID. Rounded, pelletlike.

COENOCYTE. A multinucleate cell resulting from nuclear division without a corresponding cytoplasmic division, the cell having few or no crosswalls.

COMPLANATE. Flattened uniformly. See also Compressed, Dilated.

COMPOUND. Having two or more parts; having branching of the second or a higher order. See also Order.

COMPRESSED. Slightly flattened, the thickness varying in cross section. Never as evenly flattened as complanate. *See also* Dilated.

CONCEPTACLE. A fertile cavity, usually immersed but with a surface opening (ostiole). If deeply embedded, the tissues are modified to form a roof and a floor (Fucales and the crustose Corallinaceae).

CONCHOCELIS PHASE. Microscopic thalli, branched and filamentous, and usually bearing conchospores; a phase in the life history of *Porphyra* (Rhodophyta).

CONCHOSPORE. A fertile spore borne as in *Conchocelis* (Rhodophyta).

CONFLUENT. A growing together to form a merged mass.

CONJOINED. Lying together.

CONNATE. United or fused.

CONNECTING FILAMENT. A delicate colorless filament functioning in the transfer of a diploid nucleus from the fertilized carpogonium to a nutritive cell or auxiliary cell, or connecting several auxiliary cells to another (Cryptonemiales, Gigartinales).

CORALLINACEOUS. Resembling a coralline alga in structure.

CORALLINE. A calcareous red alga of the family Corallinaceae.

CORALLOID. Having slender, erect, nearly straight branches, like those of *Corallium*, the genus of precious red coral.

CORDATE. Heart-shaped, with the point upward or outward.

CORIACEOUS. Leathery.

CORONATE. Crowned; furnished with a crown.

CORRUGATE. Having small wrinkles or folds.

CORTEX. Tissue external to the medulla in bladelike thalli; external to the axial filament in narrow, erect thalli; external to the pericentral cells in the Rhodomelaceae.

CORTICAL. Relating to the cortex.

CORTICATED. Having a cortex.

CORTICATING BAND. A nodal band in some species of Ceramiaceae in which uncorticated internodes alternate with corticated nodes.

CORYMB. A flat-topped cluster.

COVER CELL. A special cell cut off by a pericentral cell (Polysiphonia, Rhodomelaceae). The term is sometimes used to describe the epithallium in crustose coralline algae.

CREEPING. Running along, at, or near the substrate.

CRENATE. Having a scalloped margin.

CRENULATE. Minutely crenate.

CROSSWALL. An internal cell partition.

CRUCIATE. Having the contents of a tetrasporangium divided in two planes at right angles to one another.

CRUSTOSE. Crustlike, thin, flattened against substratum; said of the vegetative habit of some green, brown, and red algae.

CRYPTOSTOMA (pl. CRYPTOSTOMATA). A sterile surface cavity containing hairs, especially prominent in *Hesperophycus* and present in other Fucales.

CUCULLATE. Hooded or hood-shaped; said usually of a large blade whose upper margin is curved over against the blade surface.

CUNEATE. Wedge-shaped, broader at one end than at the other (the narrow end is usually at the point of attachment); said of blades.

CYSTOCARP. Term loosely used for a gonimoblast, with or without enclosing tissue; generally equivalent to carposporophyte.

CYTOPLASM. Living cell material.

DECIDUOUS. Falling away, not persistent; usually said of blades or branches.

DECOMPOUND. Compound more than once; repeatedly divided.

DECUMBENT. Prostrate and curving upward. *See* Procumbent.

DECUSSATE. Lying in opposite pairs, each pair at right angles to pair above or below; said of branching or spore patterns.

DEFLECTED. Bent away from.

DENTATE. Toothed.

DENTICULATE. Minutely toothed.

DETERMINATE. Of limited growth; said of branches or branchlets. *See* Indeterminate.

DIAPHRAGM. A thin partition of cells across a hollow space (Rhodymeniales).

DICHOTOMOUS. Branching by forking in pairs; bifurcate. Dichotomies are described as equally or irregularly dichotomous.

DIGITATE. Shaped like a hand; compound, with the members arising at the same level.

DILATED. Expanded laterally; flattened. *See also* Complanate, Compressed.

DIMORPHIC. Occurring in two forms (e.g. differing generations).

DINOFLAGELLATE. A unicellular planktonic organism with two flagella (Pyrrophyta.).

DIOECIOUS. Having the male and female reproductive structures borne on separate individual plants; said of a species. See also Monoecious.

DIPLOID. Having a double set of chromosomes (2N) in each nucleus. See also Haploid.

DISCOID. Having the form of a disk; said of holdfasts.

DISSECTED. Divided into a number of slender segments; said of blades.

DISTAL. Remote from the point of attachment. See also Proximal.

DISTICHOUS. Arranged in two rows on opposite sides of an axis.

DISTROMATIC. Having two layers or sheets of cells (as in Ulva, Chlorophyta).

DIVARICATE. Spreading widely; said of branching.

DIVERGENT. Spreading moderately; said of branching.

DIVISION. The largest taxonomic unit employed in classification of the plant kingdom. See also Phylum.

DOLABRIFORM. Hatchet-shaped.

DOLIFORM. See Dolabriform.

DOMOID. Dome-shaped; said of the pericarp.

DORSAL. Morphologically, the upper surface of a dorsiventral structure; located on the upper surface.

DORSIVENTRAL. Having morphologically distinct upper and lower surfaces, the surfaces similar or dissimilar.

ECORTICATE. Lacking a cortex.

EGG. A female gamete without flagella, usually borne in an oogonium.

ELLIPSOID. Elliptical in outline or section.

ENCRUSTING. See Crustose.

ENDOGENOUS. Originating internally (as in the spores of Chrysophyta) or internally and below the surface, usually from a central cell (as in the development of branches in the Delesseriaceae and Rhodomelaceae).

ENDOPHYTE. A plant growing within the tissue of another plant but provided with chloroplasts, i.e. not a parasite. See also Epiphyte.

ENDOZOIC. Growing within part of an animal (often in a chitinous layer, or in the tunic of an ascidian). See also Epizoic.

ENTIRE. Continuous, without divisions, lobes, or marginal indentations; said of blade margins. See also Simple.

EPIPHYTE. A plant growing on the surface of another plant, usually not parasitic.

EPIPHYTIC. Having the habit of an epiphyte.

EPITHALLIUM (pl. EPITHALLIA; also EPITHALLUS, pl. EPITHALLI). A mono- or polystromatic structure of short cells covering most parts of corallinaceous (or some other crustose) algae; frequently forms the roof of conceptacles in crustose corallines.

EPIZOIC. Growing on the external surface of an animal.

ESTIPITATE. Having no stipe.

EVANESCENT. Of short duration.

EXCRESCENCE. A knobby, simple or branched, rounded, or dissected protuberance from the upper surface of some crustose coralline algae.

EXOGENOUS. Originating externally; in the Rhodomelaceae, said of a branch initially originating from an axial cell before pericentral cells are cut off.

EXSERTED. Projecting beyond a usual containing structure. See also Included.

EYESPOT. A microscopic, lenslike structure in unicellular green algae or in the motile cells of other green algae, functioning as a primitive light receptor. Also present in some dinoflagellates.

FACE. The side or flat surface of a flattened structure, usually a blade.

FALCATE. Sickle-shaped; said of a blade.

FASCICULATE. Lying in a small, dense cluster or bundle; said of filaments or branches.

FASTIGIATE. Having the branches erect and more or less appressed.

FENESTRATE. Having perforations or transparent areas; windowlike.

FERTILE. Bearing reproductive structures.

FILAMENT. A branched or unbranched row of cells joined end to end.

FILAMENTOUS. Generally, hairlike; specifically, having single rows of cells.

FILAMENTOUS PHASE. A plant body made up of microscopic branched or unbranched filaments, serving as one or more phases in a life history where the other phase or phases are not filamentous.

FILIFORM. Having the form of a thread or filament; very slender; said of branches.

FIMBRIA (pl. FIMBRIAE). A bordering fringe.

FIMBRIATE. Fringed.

FISTULOSE (also FISTULOUS). Hollow and cylindrical in structure; said of a thallus.

FLABELLATE. Fan-shaped.

FLACCID. Soft and flabby.

FLAGELLATE. Bearing one or more flagella.

FLAGELLUM (pl. FLAGELLA). A microscopic whiplike structure whose beating moves a cell.

FLEXUOUS. Repeatedly bent; zigzag; said of axes.

FLOOR. The inner surface of a deep conceptacle.

FLORIDEAN STARCH. See Storage material.

FOLIACEOUS (also FOLIAR, FOLIOSE). Leaflike; broad and flat.

FOOT. The penetrating part of a "parasitic" alga.

FORCIPATE. Forked and incurved, like pincers or crab claws; said of the branches of some red algae, especially *Ceramium*.

FORM (also FORMA, pl. FORMAE). A taxonomic category indicating minor morphological differences within a species, variety, or subspecies.

FROND. A single and commonly leaflike, erect part of the thallus.

FUCOXANTHIN. See Pigment.

FURCATE. Forked or cleft.

FUSIFORM. Spindle-shaped; swollen in the middle and narrowing toward the ends.

FUSION CELL. A cell that has enlarged following fertilization, often joining with neighboring cells to form a much larger common cell in which nuclei and cytoplasm are mixed. Common in red algae, especially the Corallinaceae. See *also* Placenta.

GAMETANGIUM (pl. GAMETANGIA). A cell or multicellular structure that produces gametes.

GAMETE. A sexual cell. Gametes are male or female, differing either physiologically or both physiologically and morphologically.

GAMETOPHYTE. The plant form or phase (in a life history) that bears gametes.

GELATINOUS. Slimy; jellylike.

GENICULATE. Bent abruptly, like a knee.

GENICULUM (pl. GENICULA). An uncalcified joint or articulation between the calcified segments of a jointed coralline alga. See *also* Intergeniculum.

GLAND CELL. A colorless, usually refractive (shiny) cell, often bearing distinctive chemicals. Frequent in red algae.

GONIMOBLAST. A structure of carposporophyte tissue developing from the fertilized carpogonium or from an auxiliary cell and usually composed of both sterile and fertile cells (carposporangia), sometimes almost entirely of fertile cells (Rhodophyta).

GONIMOLOBE. Carposporophyte tissue that is almost entirely fertile, in which the carposporangia mature in sequential groups (lobes).

GROWTH BAND (also GROWTH LINE, RIDGE, or ZONE). An area of cells in which cell division is more or less continual; a localized region functioning to expand the plant body. See *also* Meristem.

HAIR. A slender cell; an unbranched, nonpigmented filament; a surface extension without thickened walls. Usually deciduous.

HAPLOID. Having a single set of chromosomes (1N) per nucleus.

HAPTEROID (also HAPTEROUS). Relating to the basal attachment structure or holdfast; in general, cylindrical and much branched.

HAPTERON (pl. HAPTERA). The rootlike portion of a holdfast, more massive than a rhizoid. Attaching organ of the Laminariales.

HEMIPARASITE. A partial "parasite."

HETEROCYST. See Megacell.

HETEROMORPHIC. Having a life history in which one of the phases (gametophyte or sporophyte) is different in size or structure from the other phase. See *also* Isomorphic.

HOLDFAST. The structure by which an alga is attached to the substratum; a process similar to but larger than a rhizoid, or made up of a number of rhizoids.

HOMONYM. A generic or specific name the same as the one previously published. Later homonyms are rejected under the International Code of Botanical Nomenclature.

HYALINE. Colorless, transparent; said of hairs.

HYPOGYNOUS CELL. The cell directly below the cell carrying female reproductive structures (particularly the Ceramiales).

HYPOTHALLIUM (pl. HYPOTHALLIA; also HYPOTHALLUS, pl. HYPOTHALLI). The lowermost differentiated layer or layers of a crustose alga, usually parallel to the substratum.

IMBRICATED. Overlapping, like shingles on a roof.

IMMERSED. Deeply embedded.

IMPERFORATE. Unbroken; having no openings.

INCISED. Cut or slashed irregularly and more or less deeply.

INCLUDED. Contained entirely within another structure.

INDETERMINATE. Having essentially unlimited growth; said of a branch having the potential of resembling the main axis in structure and function.

INFLATED. Swollen; bladdery.

INITIAL. The earliest stage of a cell, tissue, or structure.

INSERTED. Attached to or growing out of.

INTERCALARY. Occurring anywhere along the length of a structure except at the apex; said of some meristems and sporangia.

INTERCELLULAR. Lying between cells; connecting cells.

INTERGENICULUM (pl. INTERGENICULA). A calcified segment between the uncalcified joints of a coralline alga. See also Geniculum.

INTERNODE. A segment of a jointed thallus, branch, or axis lying between two nodes.

INTERPOLATED. Inserted or introduced between other things or parts.

INTERTIDAL. Lying between high and low tide levels. In the Monterey area, between about +4.5 ft (1.4 m) and −1.5 ft (0.5 m).

INTRACELLULAR. Contained entirely within a cell.

INTRAMATRICAL. Lying within a matrix of enclosing material; said especially of some green algae.

INVESTING. Covering completely; enwrapping.

INVOLUCRATE. Having an involucre.

INVOLUCRE. A group of sterile branches (filaments) subtending a cystocarp and frequently arching over it.

INVOLUTE. Having the edges rolled inward or toward the upper side; said of blades or other flattened structures.

IRIDESCENT. Exhibiting a shimmering of rainbow colors.

ISOGAMOUS. Characterized by isogamy.

ISOGAMY. The union of two gametes that are similar in size and appearance (isogametes).

ISOMORPHIC. Having a life history in which the gamete-bearing and spore-bearing thalli look alike. See also Heteromorphic.

KARYOTIC. Having a nucleus.

LACERATE. Irregularly cleft or cut; torn along the margin.

LACINIATE. Slashed; cut into long narrow divisions.

LAMELLATE. Having thin plates or layers.

LAMINARIN. See Storage products.

LAMINATE. Platelike or sheetlike; sometimes said of plastids.

LANCEOLATE. Lance-shaped; narrow and tapering.

LEAFLET. A small, leaflike division (same as bladelet).

LENTICULAR. Resembling a biconvex lens or a lentil seed.

LENTICULAR THICKENING. An opaque transverse or longitudinal thickening in the internal walls of the medullary cells in Laurencia or Chondria (Rhodomelaceae).

LIGULATE. Straplike and short; tongue-shaped.

LINEAR. Long and narrow, with parallel margins; slightly broader than filiform.

LOBATE. Divided into lobes.

LOBE. A discrete segment or portion of a structure, and especially a rounded portion.

LOCULE (also LOCULUS, pl. LOCULI). A pocket or "cell" within a structure.

LUBRICOUS. Slippery.

MACROSCOPIC. Visible to the unaided eye.

MAMMILOSE (also MAMMILATE). Bearing teat-shaped lumps or processes; teat-shaped in overall structure.

MARGIN. The outer edge or periphery, as of a blade.

MARGINAL. Relating to the margin; situated on the margin.

MASTIGONEME. A fine, microscopic hair-like structure on a flagellum.

MEDIAN. Situated in or along the middle.

MEDULLA (pl. MEDULLAE). A central core of tissue in multicellular algae; usually colorless. In red algae, constructed of similar medullary filaments or pseudo-parenchyma; in the Laminariales, made up of cells of differing structure and functions.

MEDULLARY. Pertaining to the medulla.

MEGACELL. An enlarged internal hyaline cell in crustose coralline algae; also called heterocyst or trichocyst.

MERISTEM. A group or region of cells dividing rapidly and initiating growth. Occurs in several locations: apical, usually as a single actively dividing cell; marginal, with one or more rows of cells dividing; intercalary, with a cambium-like group of cells dividing on either or both sides of an initial located between the apex and base of a structure (in the Laminariales, between stipe and blade); basal, with dividing initials at the base of a terminal branch. *See also* Trichothallic growth.

MERISTODERM. A row of superficial (epidermal) cells that provide for increase in girth by active division (Laminariales).

MIDRIB. The thickened longitudinal axis of a flattened branch or blade.

MIDVEIN. A usually delicate median line of cells (Delesseriaceae), the blade thicker through this region than on either side.

MONILIFORM. Resembling a string of cylindrical beads; said of some filaments.

MONOCARPOGONIAL. Bearing a single carpogonial filament per auxiliary cell.

MONOECIOUS. Having the male and female reproductive structures borne on the same individual plant; said of a species. *See also* Dioecious.

MONOPODIAL. Having a distinct main axis initiating continuous growth and giving off branches.

MONOSIPHONOUS. Composed of a single row of cells. *See also* Uniseriate.

MONOSPORE. A spore formed singly in or from a parent cell (monosporangium), commonly repeating the parent phase in a life history (Rhodophyta).

MONOSTROMATIC. Composed of a single layer or sheet of cells.

MONOTYPIC. Having only one species in a genus.

MOTILE. Capable of movement, generally by flagella; said of reproductive cells.

MUCILAGINOUS. Slimy or mucilage-like.

MUCRONATE. Ending in a short, sharp point from a broad base.

MULTIAXIAL. Having a medulla with a central core of several parallel longitudinal filaments, each derived from an apical cell.

MULTIFARIOUS. Branching in irregular directions; commonly radially branched.

MULTIFID. Having many divisions.

MULTIPORATE. Having more than one pore per conceptacle for the discharge of spores (Corallinaceae).

MULTISERIATE. Consisting of several rows of cells in longitudinal series and in one or two planes.

MULTISTRATOSE. Having the cells multiseriate and repeatedly divided.

NEMATHECIOID. Bearing nemathecia; pertaining to nemathecia.

NEMATHECIUM (pl. NEMATHECIA). A wart-like elevation containing or bearing reproductive structures; a specialized sorus that may occur in bands.

NODAL. Situated at a node; pertaining to or like a node.

NODE. The regions of an axis that are points of insertion of branches.

NUTRITIVE CELL. A specific cell in the carpogonial branch with which the carpogonium fuses after fertilization (some Cryptonemiales).

NUTRITIVE FILAMENT. *See also* Connecting filament.

NUTRITIVE TISSUE. Sterile (nonsporangial) tissue associated with the gonimoblast, or a mixture of vegetative and nonsporangial cells that presumably furnish the developing carposporophyte with extra food.

OB-. Prefix signifying inversely or contrarily; upside-down. An obovate structure is broadest above the middle (i.e. toward the apex), an ovate structure broadest below the middle (i.e. toward the distal end). The rightside-up or "normal" position of the structure described is generally considered to have its broadest end toward the base or point of attachment.

OBLATE. Flattened or depressed at the

poles or on opposite sides; generally said of a spherical structure.

OBTUSE. Blunt or rounded at the end. *See also* Retuse.

OOBLAST. A connecting tube through which a zygote nucleus moves from a carpogonium to an auxiliary cell (Florideophyceae).

OOGAMY. A union of gametes in which a sperm fertilizes a relatively large, nonmotile egg.

OOGONIUM (pl. OOGONIA). A female reproductive cell containing one or more eggs (female gametes).

ORBICULAR. Approximately circular in outline.

ORDER. Degree of branching: in branching of the third order, for example, the main axis gives off branches (order 1), which rebranch (order 2) and then rebranch again to produce the third-order, or ultimate, branches. Also, a taxonomic grouping (suffix -ales) between class and family.

ORGANELLE. A specialized structure within a cell, e.g., a plastid, a nucleus, or a pyrenoid.

OSTIOLE. A regularly formed opening in a conceptacle or in a cystocarp with pericarp.

OVATE. Essentially ovoid, but with the distal end narrowed and pointed.

OVOID. Egg-shaped in outline.

PANTONEMATIC. Said of flagella that bear ultramicroscopic hairs (mastigonemes).

PAPILLA (pl. PAPILLAE). A short, nipplelike, superficial outgrowth on a surface or margin.

PAPILLATE. Having papillae; shaped like a papilla.

PARAPHYSIS (pl. PARAPHYSES). A short, slender nonreproductive filament or cell occurring among reproductive structures, usually in a sorus.

"PARASITE." A red alga much reduced in size and usually having little pigmentation; in some cases probably a hemiparasitic or host-specific epiphyte.

PARASPORANGIUM (pl. PARASPORANGIA). A reproductive cell bearing paraspores (rare; in red algae).

PARASPORE. A reproductive structure similar to a tetraspore but not produced in the same way and having triploid (3N) chromosomes.

PARENCHYMA (pl. PARENCHYMAE). Tissue composed of thin-walled cells mostly of equal diameters; in the Fucales, said of internal tissue produced by a manysided apical cell.

PARENCHYMATOUS. Composed of or containing parenchyma.

PARIETAL. Outer or peripheral in a cell or other structure; most frequently said of the position of a plastid in a cell.

PECTINATE. Having closely set unilateral branches, like the teeth of a comb, the branches usually of equal length.

PEDICEL. A narrow, stemlike support; in some brown and red algae, a single cell supporting sporangia or gametangia.

PEDICELLATE. Borne on or having a pedicel.

PELTATE. Shield-shaped; circular, having the stipe attached to the lower surface at the center.

PENICILLATE. Brushlike; having a terminal tuft of fine hairs.

PERCURRENT. Extending through the entire length of a structure; usually said of a persistent axis.

PERFOLIATE. Passing through a blade.

PERICARP. A projecting layer or layers of sterile (vegetative) cells enclosing and usually obscuring the carposporophyte.

PERICENTRAL. Surrounding and formed from a central axial cell (Rhodomelaceae especially). A fertile pericentral cell is usually the cell that supports the carpogonial branch in the Ceramiales.

PERICLINAL. Parallel and interior to the surface; longitudinal; said of medullary filaments.

PERIPHERAL. Pertaining to an outer region, surface, or margin.

PERITHALLIUM (pl. PERITHALLIA; also PERITHALLUS, pl. PERITHALLI). An erect layer in crustose algae, approximately perpendicular to the hypothallium.

PHAEOPHYCEAN HAIRS. Colorless, endogenous, uniseriate filaments having a basal meristem responsible for elongation; occurring in many orders of the Phaeophyta, especially the Sphacelariales.

PHYCOBILIN. *See* Pigment.

PHYCOLOGY. The study of algae.

PHYLUM. One of the primary taxonomic groupings within the plant or animal kingdoms. *See also* Division.

PIGMENT. A colored substance within cells; in algae, generally concentrated in plastids. In this volume the term refers to substances of several kinds. Chlorophylls form the photosynthetic green pigmentation in plants; all of the algal groups contain chlorophyll *a*; additionally, chlorophyll *b* characterizes the green algae (Chlorophyta), and chlorophyll *c* the yellow-browns (Chrysophyta) and brown (Phaeophyta). Phycobilins are protein-bound pigments found in the Rhodophyta, including *r*-phycoerythrin (reddish) and *r*-phycocyanin (bluish). Caretenoids include the oxygen-free carotenes (red-orange) and the oxygen-containing xanthophylls (golden brown). There are several kinds of carotenes, and eight or more xanthophylls (of which the photosynthetically active fucoxanthin is the best known).

PIGMENTED. Containing colored matter; said of either cells or tissues.

PILIFEROUS. Bearing hairs or setae.

PINNA (pl. PINNAE). A featherlike structure. A pinna consists of a central axis and branchlets (pinnules) that are arranged distichously (e.g. *Antithamnionella*) or alternately (e.g. *Pleonosporium*).

PINNATE. Featherlike, with distichous laterals on a central axis. In a bipinnate structure, the laterals are again pinnate.

PINNULE. The ultimate division of a pinnately branched structure.

PIT CONNECTION. A narrow cytoplasmic plug between adjacent cells (Rhodophyta).

PLACENTA. A large composite cell usually consisting of fused fertile tissue and bearing gonimoblast filaments or carposporangia (Rhodophyta).

PLASTID. A cell organelle containing pigments.

PLETHYSMOTHALLUS (pl. PLETHYSMOTHALLI). A microscopic filamentous phase in the life history of some brown algae, often bearing spores that repeat the phase.

PLICATE. Folded like a fan.

PLURANGIUM (pl. PLURANGIA). A plurilocular reproductive structure that may bear either gametes or sporangia (Phaeophyta).

PLURILOCULAR. Having several to many small compartments.

PNEUMATOCYST. A large air bladder or float (Phaeophyta); small floats are usually called vesicles.

POLYSIPHONOUS. Composed of tiers of usually vertically elongated cells, transversely arranged, the lateral (pericentral) cells surrounding a central axis (siphon) (some Ceramiales).

POLYSPORANGIUM (pl. POLYSPORANGIA). A sporangium producing more than four spores (polyspores) (Rhodophyta).

POLYSTICHOUS. Arranged in many ranks.

POLYSTROMATIC. Bladelike and having many cell layers (= multistratose).

PORATE. Filled with opening; porous.

PRIMORDIUM (pl. PRIMORDIA). A precursor or originating structure.

PROCARP. Carpogonial branch adjacent to one or more auxiliary cells; other special cells may be involved (Rhodophyta).

PROCUMBENT. Trailing; lying flat. See *also* Decumbent.

PROLIFERATION. An outgrowth, commonly smaller than and similar to the part bearing it; common in some red algae.

PROLIFEROUS. Bearing proliferations or outgrowths.

PROPAGULUM (pl. PROPAGULA). A vegetative structure of propagation, developed from a lateral modified by division of the apical cell to produce two or more arms or knobs, which may become elongated. An important structure in the taxonomy of *Sphacelaria*.

PROSTRATE. Lying along the substratum.

PROTANDROUS. Having male gametes produced before female gametes on the same plant. The plant may appear dioecious but is really monoecious.

PROTUBERANT. Bulging outward.

PROXIMAL. Toward the base or point of attachment. See *also* Distal.

PSEUDO-. A prefix applied to a structure or organ resembling another organ (e.g. hair, pseudohair) but different in internal structure or origin.

PSEUDOPARENCHYMA. Tissue resembling parenchyma but developmentally filamentous (most red algae).

PSEUDOPARENCHYMATOUS. Like pseudoparenchyma in structure.

PULVINATE. Cushion-shaped.

PYRENOID. An organelle usually occurring within a plastid, where it may be immersed or projecting; often associated with reserve food accumulation in starch plates; sometimes distinguishable when stained. Not present in all groups of algae, though frequent in simple green algae.

PYRIFORM. Pear-shaped, broader at bottom than at top.

RACEME. A simple elongate, indeterminate group of reproductive cells.

RACHIS. The axis of a compound or branching structure; the central axis of a pinna.

RAMULUS (pl. RAMULI). A secondary branch.

RECEPTACLE. The terminal portion of a branch, bearing numerous embedded conceptacles (Fucales).

RENIFORM. Kidney-shaped.

REPENT. Creeping.

RETICULATE. Forming a network.

RETICULUM (pl. RETICULI). A network.

RETUSE. Shallowly notched at the apex, the apex usually obtuse.

RHIZOID. A unicellular or filamentous attachment structure, usually colorless; smaller than a holdfast.

RHIZOIDAL. Pertaining to a rhizoid.

RHIZOMATOUS. Having one or more rhizomes.

"RHIZOME." A prostrate, thickened, rhizome-like axis (a true rhizome, in flowering plants, is an underground stem).

RIB. A thickened veinlike reinforcement in a bladelike structure.

ROOF. The outer surface of a deeply embedded conceptacle.

ROSTRATE. Beaklike.

ROSULATE. Having the form of a rosette; said of a thallus.

ROTATE. Wheel-shaped.

RUGOSE. Wrinkled or ridged.

SACCATE. Saclike.

SAGITTATE. Shaped like an arrowhead.

SAXICOLOUS. Growing on rocks.

SCALARIFORM. Resembling a ladder; having ladderlike transverse markings.

SCAR CELL. The persistent basal cell of a trichoblast (Rhodomelaceae). Scar cells are useful indicators of the branching pattern of trichoblasts.

SECUND. Arranged on one side only; unilateral. Branchlets may be equal or unequal in length.

SEGMENT. A portion of a branch between successive nodes or lateral branches.

SEGREGATIVE CELL DIVISION. The process of forming new vegetative cells by internal division of the protoplasm of a parent cell, the cells thus formed later increasing in size and shape as free-living plants. Occurs only in the Siphonocladales.

SEPTUM (pl. SEPTA). A partition, usually a wall, either transverse or longitudinal.

SERIATE. In a series, usually in a whorl, cycle, or row.

SERRATE. Toothed on the margins; sawlike, and with the teeth pointing toward the apex.

SESSILE. Directly attached; lacking a stipe or stalk.

SETA (pl. SETAE). A bristle; a stiff hairlike structure.

SHEATH. A tubular enveloping structure.

SIMPLE. Unbranched; undivided; said of a thallus or blade. See Entire.

SINUATE. Having a strongly wavy margin.

SIPHON. A tubular structure.

SIPHONOUS. Multinucleate and tubular; i.e. with few or no transverse septa (same as coenocytic); said of the Chlorophyta.

SORUS (pl. SORI). A group or cluster of reproductive structures not elevated above the surface. Internal sori occur in some red algae; sori are superficial in both brown and red algae.

SPATULATE. Spoon-shaped; oblong, or having a broad, rounded tip with narrowed basal end.

SPERMATANGIUM (pl. SPERMATANGIA). A cell producing a single spermatium.

SPERMATIUM (pl. SPERMATIA). A male gamete with no flagella (Rhodophyta).

SPORANGIUM (pl. SPORANGIA). A structure in which spores are formed. The sporangial phase of a life history produces spores (usually diploid).

SPORE. A motile or nonmotile asexual reproductive structure. If motile, spores are termed zoospores; if nonmotile, termed aplanospores, monospores, carpospores, tetraspores, polyspores, etc.

SPORIFEROUS. Bearing or producing spores.

SPOROPHYLL. A bladelike structure, usually modified and bearing sporangia (especially Laminariales).

SPOROPHYTE. The sporangial phase in a life history; the diploid phase.

SQUAMULOSE. Covered with small scales.

STATOSPORE. An endogenous cyst.

STELLATE. Star-shaped; said usually of plastids or medullary cells.

STERILE CELL. A cell produced during formation of the procarp. The number and location are often of taxonomic significance in the Delesseriaceae.

STICHIDIUM (pl. STICHIDIA). A specialized

branch, usually swollen, producing tetrasporangia (*Plocamium* and Dasyaceae).

STIPE. A thickened, stemlike structure bearing other structures such as blades. A stipe is usually short, but some stipes are very long (e.g. *Nereocystis, Pelagophycus*).

STIPITATE. Having a stipe.

STOLON. A branch or runner growing out from the base of a parent plant and capable of producing offshoots.

STOLONIFEROUS. Provided with stolons.

STORAGE PRODUCTS. Food reserves within cells, usually in plastids. Starch in the Chlorophyta is made up of glucose units and is identical in chemical structure to the starch of flowering plants. Floridean starch occurs in red algae outside the chloroplast and contains only one of the chemical fractions (amylopectin) found in the starch of green plants. Chrysolaminarin is a polysaccharide composed of glucose units in a special linkage, accumulating during photosynthesis and occurring in the Chrysophyta (as well as in the Bacillariophyta, the diatoms). Laminarin, a polysaccharide composed of glucose and mannitol in a special linkage, occurs in the Phaeophyta.

STRATIFIED. Arranged in layers.

SUB-. A prefix meaning less than, almost, approaching, etc.; subdichotomous = roughly dichotomous.

SUBSTRATUM (pl. SUBSTRATA). The surface or material on which an alga is growing.

SUBTENDING. Standing below.

SUBTIDAL. Below the lowest low-tide level.

SUBULATE. Awl-shaped.

SUPERFICIAL. On the surface.

SUPPORTING CELL. A cell bearing the carpogonial filament or carpogonium, and sometimes an auxiliary cell. A supporting cell occasionally functions as an auxiliary cell. In the Ceramiales, a specialized pericentral cell.

SYMPODIAL. Developing so that the apparent main axis does not extend by continuous terminal growth but sends off a secondary branch in the direction of growth, the former axis usually diverging and ending as a secondary branch, the apparent axis thus really a series of superposed branches.

SYNONYM. An additional name for the same taxon.

TAXON (pl. TAXA). A unit in a system of classification; a taxonomic group of any rank.

TERETE. Cylindrical in cross section.

TERMINAL. At the tip or apex.

TERNATE. Divided into three parts; arranged in groups of three.

TESSELLATED. Arranged in a checkerboard pattern.

TETRAHEDRAL. Said of tetrasporangia; divided obliquely so that only three of the four spores are normally visible in a given view.

TETRASPORANGIUM (pl. TETRASPORANGIA). A sporangium in which four spores are formed in a definite manner, as tetrahedral, decussate, cruciate, or zonate.

TETRASPORE. A spore produced in a tetrasporangium. In the Dictyotales (Phaeophyta) and in most red algae.

TETRASPOROBLAST. A gonimoblast producing tetrasporangia instead of the more usual carposporangia.

TETRASPOROPHYTE. A sporangial phase in a life history in which tetrasporangia are formed (Rhodophyta).

THALLUS (pl. THALLI). The plant body of an alga.

THECA (pl. THECAE). A sac or case.

TOMENTOSE. Woolly; covered with dense, matted hairs or filaments.

TONGUE CELL. A specialized initial cell formed during the development of a conceptacle.

TOOTHED. Having the margin dissected into a series of points. See also Dentate, Serrate.

TORTUOUS. Growing in repeated bends or twists; winding irregularly.

TORULOSE. Shaped like a torus; thickly ringlike with a central hollow.

TRICHOBLAST. A simple or branched, colorless, hairlike filament (Rhodomelaceae).

TRICHOCYST. See Megacell.

TRICHOTHALLIC GROWTH. A pattern of cell initiation in which girth especially is increased by a compounding of separate filaments that arise from numerous basal divisions in the laterals; in *Desmarestia*, for example, the silky hairs that occur seasonally are an external sign of trichothallic growth. See also Meristem.

TRICHOTOMOUS. Three-forked.

TRIPINNATE. Branching pinnately three times.

TUBERCULATE. Having irregular warty outgrowths.

TURBINATE. Top-shaped; inversely conical.

TYPE LOCALITY. The geographic locality where a species was first collected. Studying morphological variation in collections from a type locality is essential to understanding the limits of the described species.

TYPE SPECIMEN. In taxonomy, the specimen on which the original description of a taxon was based (often abbreviated to "type").

ULTIMATE. Final; apical; terminal.

UMBILICATE. Depressed in the center; navel-like.

UMBONATE. Having a central knob or convex elevation.

UNANGIUM (pl. UNANGIA). A unilocular reproductive structure; a unilocular sporangium (Phaeophyta).

UNCINATE. Hooklike.

UNDULATE. Wavy; said of a margin or surface.

UNIAXIAL. Having a single central longitudinal filament that forms the axis.

UNILOCULAR. Having a single compartment (Phaeophyta).

UNIPORATE. Having one pore per conceptacle for the discharge of spores (Corallinaceae).

UNISERIATE. Occurring in a single row or series.

UNISTRATOSE. See Monostromatic.

URCEOLATE. Urn-shaped; said of reproductive structures.

UTRICLE. An inflated portion of a tubular thallus (same as cortex) in some green algae (e.g. the Codiaceae).

VARIETY. A taxon subordinate to species or subspecies. See also Form.

VEGETATIVE. Not reproductive, and not associated with reproductive cells. Cells produced with reproductive cells but themselves not reproductive are termed sterile cells.

VEIN. A smaller branch from a midrib; a slightly thickened narrow line within the blade (especially in the Delesseriaceae). Microscopic veins are usually only one or two cells wide and must be viewed with some magnification.

VENTRAL. Morphologically, the lower surface of a dorsiventral structure; located on the lower surface.

VERRUCOSE. Covered with tubercles or warts; warty or wartlike.

VERTICIL. A whorl of branches, usually at a node.

VESICLE. A small air bladder or float (Phaeophyta). Large floats are called pneumatocysts.

WALL MATERIALS. Substances in the walls of algal cells, of several kinds. Agar and carrageenan are complex polysaccharides that form gels, derived from certain red algae. Alginic acids are linear polysaccharides occurring in the walls of many brown algae and some bacteria, widely used in industry as suspending agents, thickeners, and emulsifiers. Cellulose, the major wall material in most green, brown, and red algae, made up of glucose units that are variously oriented and capable of accommodating a variety of additional chemical groups in different arrangements.

WHORL. Three or more branches attached at a common level on the axis.

XANTHOPHYLL. See Pigment.

ZONAL. Arranged in layers or in concentric zones.

ZONATE. Divided on parallel planes (as in the division of tetrasporangia).

ZOOID. A motile reproductive cell with flagella. The term is used when the sex or chromosome number of the cell is unknown.

ZOOSPORE. A motile spore with flagella (Chlorophyta and Phaeophyta), produced by the thallus directly or by a zoosporangium.

ZYGOTE. A cell formed by the union of two gametes. Also a fusion nucleus resulting from the union of two gametic nuclei.

Literature Cited

Abbott, I. A. 1961. On Schimmelmannia from California and Japan. *Pacific Naturalist*, 2: 379–86, 2 pls.

———. 1962a. Structure and reproduction of Farlowia (Rhodophyceae). *Phycologia*, 2: 29–37. 12 figs.

———. 1962b. Some Liagora-inhabiting species of Acrochaetium. *B. P. Bishop Museum Occ. Pap.*, 23: 77–120. 17 figs.

———. 1965. Helminthora and Helminthocladia from California. *Hydrobiologia*, 25: 88–98. 8 figs.

———. 1967a. Studies in some foliose red algae of the Pacific coast. I. Cryptonemiaceae. *Jour. Phycol.*, 3: 139–49. 13 figs.

———. 1967b. Studies in the foliose red algae of the Pacific coast. II. Schizymenia. *Bull. So. Calif. Acad. Sci.*, 66: 161–74. 11 figs.

———. 1968. Studies in some foliose red algae of the Pacific coast. III. Dumontiaceae, Weeksiaceae, Kallymeniaceae. *Jour. Phycol.*, 4: 180–98. 38 figs.

———. 1969. Some new species, new combinations, and new records of red algae from the Pacific coast. *Madroño*, 20: 42–53. 14 figs.

———. 1970. On some new records of marine algae from Washington state. *Syesis*, 3: 1–4. 1 fig.

———. 1971a. On some Ceramiaceae (Rhodophyta) from California. *Pacific Science*, 25: 349–56. 2 figs.

———. 1971b. On the species of Iridaea (Rhodophyta) from the Pacific coast of North America. *Syesis*, 4: 51–72. 24 figs.

———. 1972a. Field studies which evaluate criteria used in separating species of Iridaea (Rhodophyta), in I. Abbott and M. Kurogi, eds., *Contributions to the Systematics of the Benthic Marine Algae of the North Pacific*. Japanese Soc. Phycol. Kobe. Pp. 253–64. 13 figs.

———. 1972b. Taxonomic and nomenclatural notes on North Pacific marine algae. *Phycologia*, 11: 259–65.

Abbott, I. A., and R. E. Norris. 1965. Studies on Callophyllis (Rhodophyceae) from the Pacific coast of North America. *Nova Hedwigia*, 10: 67–84. 14 pls.

Adey, W. H. 1966. The genera Lithothamnium, Leptophytum (nov. gen.) and Phymatolithon in the Gulf of Maine. *Hydrobiologia*, 28: 321–70. 112 figs. 16 pls.

———. 1970. A revision of the Foslie crustose coralline herbarium. *Kgl. Norske Vidensk. Selsk. Skr.*, 1970: 1–46.

Adey, W. H., and H. W. Johansen. 1972. Morphology and taxonomy of Corallinaceae with special reference to Clathromorphum, Mesophyllum, and Neopolyporolithon, gen. nov. (Rhodophyceae, Cryptonemiales). *Phycologia*, 11: 159–
80. 68 figs.

Agardh, C. A. 1817. *Synopsis algarum Scandinaviae* . . . Lund. 135 pp.

————. 1820. *Species algarum*. Lund. Vol. 1(1): 1–168.

————. 1821. *Icones algarum ineditae*, fasc. 2 Ed. Nov. Stockholm. 20 pls.

————. 1822a. *Species algarum*. Lund. Vol. 1(2): 169–531.

————. 1822b. Algae, in C. S. Kunth, *Synopsis plantarum* . . . Paris. Vol. 1: 1–6.

————. 1824. *Systema algarum*. Lund. 312 pp.

————. 1828. *Species algarum*. Griefswald. Vol. 2(1): 1–189.

Agardh, J. G. 1841. In historiam algarum symbolae. *Linnaea*, 15: 1–50, 443–57.

————. 1842. *Algae Maris Mediterranei et Adriatici* . . . Paris. 164 pp.

————. 1846. [Gigartina spp.], in C. Agardh, *Icones algarum* Ed. Nov. Stockholm, pls. 18, 19.

————. 1847. Nya alger från Mexico. *Öfvers. Kgl. Svensk. Vetensk. Ak. Förh.*,
4: 5–17.

————. 1848. *Species genera et ordines algarum* . . . Lund. Vol. 1. 363 pp.

————. 1849. Algologiska Bidrag. *Öfvers. Kgl. Svensk. Vetensk. Ak. Förh.*, 6:
79–89.

————. 1851. *Species genera et ordines algarum* . . . Lund. Vol. 2(2:1): 1–
504.

————. 1852. *Ibid.* Vol. 2 (3:1): 701–86.

————. 1863. *Ibid.* Vol. 2(3:2): 787–1291.

————. 1872. Till algernes systematik. Nya bidrag. *Lunds Univ. Årsskr.*, 9(8):
1–71.

————. 1876. *Species genera et ordines algarum* . . . Leipzig. Vol. 3(1): *Epi-
crisis systematis floridearum.* 724 pp.

————. 1883. Till algernes systematik. Nya bidrag. *Lunds Univ. Årsskr.*, 19(2):
1–182.

————. 1884–85. Till algernes systematik. Nya bidrag. *Ibid.*, 21(8): 1–117.

————. 1889. Species Sargassorum Australiae, descriptae et dispositae . . . *Kgl.
Svensk. Vetensk. Ak. Handl.* 23(3): 1–133. 31 pls.

————. 1889–90. Till algernes systematik. Nya bidrag. *Lunds Univ. Årsskr.*
26(3): 1–120.

————. 1892. Analecta algologica. *Ibid.*, 28(6): 1–182. 3 pls.

————. 1894. Analecta algologica. Cont. I. *Ibid.*, 29(9): 1–144. 2 pls.

————. 1896. Analecta algologica. Cont. III. *Ibid.*, 30(2): 1–140. 1 pl.

————. 1898. *Species genera et ordines algarum* . . . Lund. Vol. 3(3). 239 pp.

————. 1899. Analecta algologica. Cont. V. *Lunds Univ. Årsskr.*, 35(35):
1–160. 3 pls.

Ambronn, H. 1880. Über einige Fälle von Bilateralität bei den Florideen. *Bot.
Zeit.*, 38: 161–74, 177–85, 193–200, 209–32.

Anderson, C. L. 1892. Algae, in J. W. Blankinship and C. A. Keeler, On the
natural history of the Farallon Islands. *Zoe*, 3: 148–50.

————. 1894. Some new and some old algae but recently recognized on the
California coast. *Ibid.*, 4: 358–62. 2 figs.

Ardissone, F. 1883. Phycologia Mediterranea. *Mem. Soc. Critt. Italiana*, 1. 516 pp.

Ardré, F. 1959. Un intéressant Hildenbrandia du Portugal. *Rev. Algologique*, sér. 2, 4: 227–37. 3 pls.

————. 1967. Remarques sur la structure des Pterosiphonia (Rhodomélacées, Céramiales) et leurs rapports systématiques avec les Polysiphonia. *Ibid.*, 9: 37–77. 4 figs. 6 pls.

Ardré, F., and P. Gayral. 1961. Quelques Grateloupia de l'Atlantique et du Pacifique. *Ibid.*, 6: 38–48. 3 pls.

Areschoug, J. E. 1842. Algarum minus rite cognitarum pugillus primus. *Linnaea*, 16: 225–36. 2 figs. 1 pl.

————. 1847. Enumeratio phycearum in maribus Scandinaviae crescentium. Sectio prior, Fucaceas continens. *Nova Acta Reg. Soc. Sci. Upsaliensis*, 2d ser., 13: 223–382. 9 pls.

————. 1850. Enumeratio phycearum in maribus Scandinaviae crescentium. Sectio posterior, Ulvaceas continens. *Ibid.*, 2d ser., 14: 385–454. 3 pls.

————. 1852. Ordo XII. Corallineae, in J. Agardh, *Species genera et ordines algarum*... Lund. Vol. 2(2:3): 529–676.

————. 1866. Observationes phycologicae. I. De Confervaceis nonnulis. *Nova Acta Reg. Soc. Sci. Upsaliensis*, 3d ser., 6(2): 1–26. 4 pls.

————. 1876. De tribus Laminarieis et de Stephanocystide osmundaceae (Turn.) Trev. observationes praecursorias. *Bot. Notiser*, 1876: 65–73.

————. 1881. Beskrifning på ett nytt algslägte Pelagophycus, hörande till Laminarieernas familj. *Ibid.*, 1881: 49–50.

————. 1883. Observationes phycologicae. IV. De Laminariaceis nonnullis. *Nova Acta Reg. Soc. Sci. Upsaliensis*, 3d ser., 12(8): 1–23.

Balakrishnan, M. S. 1958. Notes on Indian red algae, I. *Jour. Indian Bot. Soc.*, 37: 138–46. 21 figs.

Batters, E. A. L. 1890. A list of the marine algae on Berwick-on-Tweed. *Hist. Berwickshire Nat. Club*, 12: 221–392.

————. 1896. Some new British marine algae. *Jour. Bot. Lond.*, 34: 6–11.

————. 1902. A catalogue of the British marine algae. *Ibid.*, 40 (Suppl. 1): 1–107.

Berkeley, M. J. 1833. *Gleanings of British Algae*... London. 50 pp. 20 pls.

Blackler, H. 1964. Some observations on the genus Colpomenia..., in A. Davy de Virville and J. Feldmann, eds., Proc. *4th Internat. Seaweed Symposium*. Oxford. Pp. 50–54.

Bliding, C. 1963. A critical survey of European taxa in Ulvales. I. Capsosiphon, Percursaria, Blidingia, Enteromorpha. *Bot. Notiser (Suppl.), Opera Botanica*, 8(3): 1–160. 92 figs.

————. 1968. A critical survey of European taxa in Ulvales. II. Ulva, Ulvaria, Monostroma, Kornmannia. *Bot. Notiser*, 121: 535–629. 47 figs.

Boillot, A., and F. Magne. 1973. Le cycle biologique de Kylinia rosulata Rosenvinge (Rhodophycées, Achrochaetiales). *Soc. Phycol. France Bull.*, 18: 47–53: 2 figs.

Børgesen, F. 1909. Some new or little known West Indian Florideae. *Bot. Tidsskr.*, 30: 1–19. 1 pl.

———. 1914. The marine algae of the Danish West Indies. II. Phaeophyceae. *Ibid.*, 2(2): 1–66. 44 figs.

———. 1915–20. The marine algae of the Danish West Indies. III. Rhodophyceae. *Ibid.*, 3: 1–504. 435 figs.

———. 1924. Marine Algae of Easter Island, vol. 2 of C. Skottsberg, ed., *The Natural History of Juan Fernandez and Easter Island*. Uppsala. Pp. 247–309.

Bornet, E. 1878. [Melobesia thuretii], in G. Thuret & E. Bornet, *Etudes phycologiques*, Paris. P. 96.

———. 1883. [Polysiphonia hillebrandii] in F. Ardissone, Phycologia Mediterranea. *Mem. Soc. Critt. Italiana.* 1: 376.

———. 1904. Deux Chantransia corymbifera Thuret, Acrochaetium et Chantransia. *Bull. Soc. Bot. France*, 51 (Suppl.): xiv–xxii.

Bornet, E., and C. Flahault. 1888. Note sur deux nouveaux genres d'algues perforantes. *Jour. de Bot.*, 2: 161–65.

Bornet, E., and G. Thuret. 1876. *Notes algologiques* . . . Paris. Fasc. 1. 70 pp. 25 pls.

Bory de Saint-Vincent, J. B. 1804. *Voyage dans les quatre principales îles des mers d'Afrique* . . . Paris. 3 vols.

———. 1826a. [Iridaea], Iridée, in J. B. Bory de Saint-Vincent, ed., *Dict. Class. Hist. Nat.* Paris. Vol. 9: 15–16.

———. 1826b. [Macrocystis integrifolia], in J. B. Bory de Saint-Vincent, ed., *ibid.*, vol. 10: 8–10.

Braun, A. 1855. *Algarum unicellularum genera nova et minus cognitae*. Leipzig. 111 pp. 6 pls.

Bravo, L. K. 1965. Studies on the life history of Prasiola meridionalis. *Phycologia*, 4: 177–94. 27 figs.

Buffham, T. H. 1896. On Bonnemaisonia hamifera Hariot. *Jour. Quekett Micr. Club*, 2d ser. 6: 177–82. 1 pl.

Buggeln, R., and R. Tsuda. 1969. A record of benthic marine algae from Johnson Atoll. *Atoll Res. Bull.*, no. 120: 1–20. 1 chart.

Cardinal, A. 1964. Etude sur les Ectocarpacées de la Manche. *Nova Hedwigia Beihefte*, 15: 1–86. 41 figs.

Chapman, A. R. O. 1972. Morphological variation and its taxonomic implications in the ligulate members of the genus Desmarestia occurring on the west coast of North America. *Syesis*, 5: 1–20. 24 figs.

Chapman, V. J. 1962. A contribution to the ecology of Egregia laevigata Setchell. *Bot. Marina*, 3(2): 33–55.

Chiang, Y.-M. 1970. Morphological studies of red algae of the family Cryptonemiaceae. *Univ. Calif. Publ. Bot.*, 58: 1–83. 34 figs. 10 pls.

Chihara, M. 1958. Studies on the life-history of the green algae in the warm seas around Japan, 7. *Jour. Japanese Bot.*, 33: 307–13. 2 figs.

———. 1960. Collinsiella in Japan, with special reference to the life-history. *Sci. Repts. Tokyo Kyoiku Daigaku*, 9(140): 181–98. 11 figs.

———. 1961. Life cycle of the Bonnemaisoniaceous algae in Japan, 1. *Ibid.*, 10: 121–54.

———. 1963. The life history of Prasinocladus ascus as found in Japan, with special reference to the systematic position of the genus. *Phycologia*, 3: 19–28. 3 figs.

————. 1969. Culture study of Chlorochytrium inclusum from the northeast Pacific. *Phycologia*, 8: 127–33. 3 figs.

Cienkowski, L. 1881. An account of the White Sea excursion for 1880. *Trudy imp. S-petersb. Obschch. Estest.*, 12: 130–71. (In Russian.)

Collins, F. S. 1903. [Cladophora columbiana], in W. A. Setchell and N. L. Gardner, Algae of northwestern America. *Univ. Calif. Publ. Bot.*, 1: 226.

———— 1906a. New species, etc., issued in the Phycotheca Boreali-Americana. *Rhodora*, 8: 104–13.

————. 1906b. Acrochaetium and Chantransia in North America. *Ibid.*, 8: 189–96.

————. 1909a. The green algae of North America. *Tufts College Studies, Sci. Ser.*, 2(3): 79–480. 18 pls.

————. 1909b. New species of Cladophora. *Rhodora*, 11: 17–20. 1 pl.

————. 1913. The marine algae of Vancouver Island. *Victoria Mem. Mus. Bull.*, 1: 99–137.

————. 1918. The green algae of North America. 2d suppl. paper. *Tufts College Studies, Sci. Ser.*, 4(7): 1–106. 3 pls.

Collins, F. S., and A. B. Hervey. 1917. The algae of Bermuda. *Proc. Amer. Acad. Arts & Sci.*, 53: 1–195.

Collins, F. S., I. Holden, and W. A. Setchell. 1895–1919. *Phycotheca Boreali-Americana. Exsiccatae.* Fasc. 1–46 and A–E. Malden, Mass.

Crouan, P. L., and H. M. Crouan. 1867. *Florule du Finistère*. Paris. 262 pp.

Dangeard, P. A. 1931. L'Ulvella lens de Crouan et l'Ulvella setchellii sp. nov. *Bull. Soc. Bot. France*, 78: 312–18. 7 figs.

————. 1969. Quelques Chlorophycées rares ou nouvelles. *Botaniste*, sér. 52, 1969: 29–58. 7 pls.

Dawson, E. Y. 1941. A review of the genus Rhodymenia with descriptions of new species. *Allan Hancock Pacific Expeditions*, 3: 123–80. 13 pls.

————. 1944a. The marine algae of the Gulf of California. *Ibid.*, 3: 189–453. 47 pls.

————. 1944b. Some new Laurenciae from Southern California. *Madroño*, 7: 233–40. 3 pls.

————. 1944c. A new parasitic red alga from Southern California. *Bull. Torrey Bot. Club*, 71: 655–57. 4 figs.

————. 1944d. Notes on Pacific coast marine algae, I. *Bull. So. Calif. Acad. Sci.*, 43: 95–101. 1 pl.

————. 1944e. Some new and unreported sublittoral algae from Cerros I., Mexico. *Ibid.*, 43: 102–12. 3 pls.

————. 1945a. Notes on Pacific coast marine algae, III. *Madroño*, 8: 93–97. 1 pl.

————. 1945b. New and unreported marine algae from Southern California and northwestern Mexico. *Bull. So. Calif. Acad. Sci.*, 44: 75–91. 6 pls.

————. 1949a. Studies of northeast Pacific Gracilariaceae. *Allan Hancock Foundation Publ. Occ. Pap.*, 7: 1–105. 25 pls.

————. 1949b. Contributions toward a marine flora of the Southern California Channel Islands, I–III. *Ibid.*, 8: 1–57. 14 pls.

————. 1950a. Notes on Pacific coast marine algae, IV. *Amer. Jour. Bot.*, 37: 149–58.

———. 1950b. Notes on some Pacific Mexican Dictyotaceae. *Bull. Torrey Bot. Club*, 77: 83–93. 3 figs.

———. 1950c. Notes on Pacific coast marine algae, V. *Amer. Jour. Bot.*, 37: 337–44. 6 figs.

———. 1950d. On the status of the brown alga Dictyota binghamiae J. G. Agardh. *Wasmann Jour. Biol.*, 8: 267–69.

———. 1950e. A review of Ceramium along the Pacific coast of North America with special reference to its Mexican representatives. *Farlowia*, 4: 113–38. 4 pls.

———. 1953a. Marine red algae of Pacific Mexico. I. Bangiales to Corallinaceae subf. Corallinoideae. *Allan Hancock Pacific Expeditions*, 17: 1–239. 33 pls.

———. 1953b. Notes on Pacific coast marine algae. VI. *Wasmann Jour. Biol.*, 11: 323–51. 7 pls.

———. 1954a. Marine red algae of Pacific Mexico. II. Cryptonemiales (cont.). *Allan Hancock Pacific Expeditions*, 17: 241–397. 44 pls.

———. 1954b. The marine flora of Isla San Benedicto following the volcanic eruption of 1952–1953. *Allan Hancock Foundation Publ. Occ. Pap.*, 16: 1–25.

———. 1957. Notes on eastern Pacific insular marine algae. *Los Angeles Co. Mus. Contrib. Sci.*, 8: 1–8. 4 figs.

———. 1958. Notes on Pacific coast marine algae. VII. *Bull. So. Calif. Acad. Sci.*, 57: 65–80. 5 pls.

———. 1959. Marine algae from the 1958 cruise of the Stella Polaris . . . *Los Angeles Co. Mus. Contrib. Sci.*, 27: 1–39. 9 figs.

———. 1960. Marine red algae of Pacific Mexico. III. *Cryptonemiales, Corallinaceae subf. Melobesioideae. Pacific Naturalist*, 2(1): 1–125. 50 pls.

———. 1961a. Marine red algae of Pacific Mexico. IV. Gigartinales. *Ibid.*, 2(5): 191–341. 61 pls.

———. 1961b. A guide to the literature and distributions of Pacific benthic algae from Alaska to the Galapagos Islands. *Pacific Science*, 15: 370–461.

———. 1962. Marine red algae of Pacific Mexico. VII. Ceramiales: Ceramiaceae, Delesseriaceae. *Allan Hancock Pacific Expeditions*, 26: 1–207. 50 pls.

———. 1963a. Marine red algae of Pacific Mexico. VI. Rhodymeniales. *Nova Hedwigia*, 5: 437–76. 19 pls.

———. 1963b. Marine red algae of Pacific Mexico. VIII. Ceramiales: Dasyaceae, Rhodomelaceae. *Ibid.*, 6: 401–81. 46 pls.

———. 1964. A review of Yendo's jointed coralline algae of Port Renfrew, Vancouver Island. *Ibid.*, 7: 537–43.

———. 1965. *Some marine algae in the vicinity of Humboldt State College.* Humboldt State College, Arcata, Calif. 76 pp.

Dawson, E. Y., C. Acleto, and N. Foldvik. 1964. The seaweeds of Peru. *Nova Hedwigia Beihefte*, 13: 1–111. 81 pls.

Dawson, E. Y., and M. Neushul. 1966. New records of marine algae from Anacapa Island, California. *Nova Hedwigia*, 12: 173–87. 3 pls.

Dawson, E. Y., M. Neushul, and R. D. Wildman. 1960. New records of sublittoral marine plants from Pacific Baja California. *Pacific Naturalist*, 1: 1–30. 4 pls.

Dawson, E. Y., and B. Tözün. 1964. The structure and reproduction of the red alga Chondria nidifica Harvey. *Trans. San Diego Soc. Nat. Hist.*, 13: 285–300. 8 figs.

Decaisne, J. 1842. Mémoire sur les Corallines ou Polypiers calcifères. *Ann. Sci. Nat. Bot.*, sér. 2, 18: 96–128.

———. 1847. [Corallina chilensis], in W. H. Harvey, *Nereis Australis* . . . London. P. 103.

———. 1864. Botanique, in A. Du Petit-Thouars, ed., *Voyage autour du monde sur la frégate la Vénus commandée par Abel Du Petit-Thousars.* Paris. 54 pp.

De la Pylaie, A. 1829. *Flora de Terre-Neuve* . . . Paris. 128 pp.

Delile, A. R. 1813. *Flore d'Egypte. . . . Description de l'Egypte* . . . Paris. Vol. 2: 145–320.

Denizot, M. 1968. Les algues Floridées encroutantes, à l'exclusion des Corallinacées. *Laboratoire de Cryptogamie, Mus. Natl. Hist. Nat.* Paris. 310 pp. 227 figs.

Derbès, A., and A. J. J. Solier. 1851. Algues, in L. Castagne, *Supplément au catalogue des plantes qui croissent naturellement aux environs de Marseille.* Aix. Pp. 93–121.

DeToni, G. B. 1936. *Noterella di nomenclature algologica, VII. Primo elenco di Floridee omonime.* Brescia. 8 pp. (Unnumbered.)

DeToni, J. B. 1895. *Sylloge algarum* . . . Padua. Vol. 3: 1–638.

———. 1900. *Ibid.* Vol. 4(2): 387–776.

———. 1903. *Ibid.* Vol. 4(3): 775–1525.

———. 1924. *Ibid.* Vol. 6(5): 1–767.

Dillwyn, L. W. 1802–9. *British Confervae.* London. 87 pp. 115 pls.

Dixon, P. S. 1959. Taxonomic and nomenclatural notes on the Florideae. I. *Bot. Notiser*, 112: 339–52.

———. 1964a. Taxonomic and nomenclatural notes on the Florideae. IV. *Ibid.*, 117: 56–78.

———. 1964b. Asparagopsis in Europe. *Nature*, 201: 902.

———. 1967. The typification of Fucus cartilagineus L. and F. corneus Huds. *Blumea*, 15: 55–62.

———. 1971. A study of Callithamnion lejolisea Farl. *Jour. Phycol.*, 7: 58–63. 5 figs.

Doty, M. S. 1947a. The marine algae of Oregon. I. Chlorophyta and Phaeophyta. *Farlowia*, 3: 1–65. 10 pls.

———. 1947b. The marine algae of Oregon. II. Rhodophyta. *Ibid.*, 3: 159–215. 4 pls.

Doubt, D. G. 1935. Notes on two species of Gymnogongrus. *Amer. Jour. Bot.*, 22: 294–310. 22 figs.

Drew, K. M. 1928. A revision of the genera Chantransia, Rhodochorton, and Acrochaetium. *Univ. Calif. Publ. Bot.*, 14: 139–224. 12 pls.

———. 1956. Reproduction in the Bangiophycidae. *Bot. Rev.*, 22: 553–611.

Druehl, L. D. 1968. Taxonomy and distribution of northeast Pacific species of Laminaria. *Can. Jour. Bot.*, 46: 539–47. 6 pls.

Eaton, D. C. 1877. Description of a new alga of California. *Proc. Amer. Acad. Arts & Sci.*, n.s., 4: 245.

Edelstein, T. 1970. The life history of Gloiosiphonia capillaris (Hudson) Car-
michael. *Phycologia*, 9: 55–59. 13 figs.

——. 1972. On the taxonomic status of Gloiosiphonia californica (Farlow)
J. Agardh (Cryptonemiales, Gloiosiphoniaceae). *Syesis*, 5: 227–34.

Edelstein, T., and J. McLachlan. 1969. Petroderma maculiforme on the coast of
Nova Scotia. *Can. Jour. Bot.*, 47: 561–63. 19 figs.

Edelstein, T., M. J. Wynne, and J. McLachlan. 1970. Melanosiphon intestinalis
(Saund.) Wynne, a new record for the Atlantic. *Phycologia*, 9: 5–9.

Esper, E. J. C. 1797–1808. *Icones fucorum* . . . Nürenberg. 139 pp. 177 pls.

Falkenberg, P. (with Fr. Schmitz). 1897. Rhodomelaceae, in A. Engler and K.
Prantl, eds., *Die natürlichen Pflanzenfamilien* . . . 1(2). Leipzig. 580 pp.

——. 1901. Die Rhodomelaceen des Golfes von Neapel und der angrenzenden Meeres-Abschnitte. *Fauna und Flora des Golfes von Neapel und der
angrenzenden Meeres-Abschnitte. Monographie* 26: 7–754. 10 figs. 24 pls.

Fan, K. C. 1959. Studies on the life histories of marine algae. I. Codiolum petrocelidis and Spongomorpha coalita. *Bull. Torrey Bot. Club*, 86: 1–12. 40 figs.

Fan, K. C., and G. F. Papenfuss. 1959. Red algal parasites occurring on members of the Gelidiales. *Madroño*, 15: 33–38. 10 figs.

Farlow, W. G. 1875. List of the marine algae of the United States, with notes
on new and imperfectly known species. *Proc. Amer. Acad. Arts & Sci.*, n.s., 2:
351–80.

——. 1877. On some algae new to the United States. *Proc. Amer. Acad. Arts
& Sci.*, n.s., 4: 235–45.

——. 1889. On some new or imperfectly known algae of the United States, I.
Bull. Torrey Bot. Club, 16(1): 1–12. 2 pls.

——. 1899. Three undescribed Californian algae. *Erythea*, 7: 73–76.

——. 1901. [Gracilaria cunninghamii] in J. G. Agardh, *Species genera et
ordines algarum* . . . Lund. Vol. 3(4): 93.

——. 1909. [Chaetomorpha clavata var. torta], in F. S. Collins, The green
algae of North America. *Tufts College Studies, Sci. Ser.*, 2(3): 323.

Farlow, W. G., C. L. Anderson, and D. C. Eaton. 1877–89. *Algae Americae
borealis exsiccatae.*, Fasc. 1–5. Boston.

Farlow, W. G., & W. A. Setchell. 1900. [Lessonia littoralis] in J. E. Tilden, 1894–
1900, *American Algae* (Exsiccatae). Minneapolis. No. 342.

Feldmann, J. 1937. Les algues marines de la côte des Albères. I–III. Cyanophy-
cées, Chlorophycées, Phéophycées. *Rev. Algologique*, 9: 141–235. 67 figs. 10
pls.

——. 1958. Le genre Kylinia Rosenvinge (Acrochaetiales) et sa reproduction.
Bull. Soc. Bot. France, 105: 493–500. 3 figs.

Feldmann, J., and G. Feldmann. 1958. Recherches sur quelques Floridées parasites. *Rev. Gén. Bot.*, 65: 1–78. 30 figs. 1 pl.

Feldmann J., and G. Hamel. 1936. Floridées de France. VII. Gélidiales. *Rev.
Algologique*, 9: 85–140.

Feldmann-Mazoyer, G. 1941. *Recherches sur les Ceramiacées de la Méditer-
ranée occidentale*. Algiers. 504 pp. 191 figs. 4 pls.

Fensholt, D. 1955. An emendation of the genus Cystophyllum (Fucales). *Amer.
Jour. Bot.* 42: 305–22. 51 figs.

Foslie, M. H. 1894. New or critical Norwegian algae. *Kgl. Norske Vidensk. Selsk. Skr.*, 1893(6): 114–44. 3 pls.

———. 1895. The Norwegian forms of Lithothamnion. *Ibid.*, 1894(2): 29–208. 23 pls.

———. 1897. On some Lithothamnia. *Ibid.*, 1897(1): 1–20.

———. 1898. List of species of the Lithothamnia. *Ibid.*, 1898(3): 1–11.

———. 1900a. Five new calcareous algae. *Ibid.*, 1900(3): 1–6.

———. 1900b. Revised systematical survey of the Melobesieae. *Ibid.*, 1900(5): 1–22.

———. 1901. New Melobesieae. *Ibid.*, 1900(6): 1–24.

———. 1902a. New species or forms of Melobesieae. *Ibid.*, 1902(2): 1–11.

———. 1902b. [Dermatolithon corallinae] in F. Børgesen, Marine Algae of the Faeroës. *Botany of the Faeroës.* Copenhagen. Pt. 1: 402.

———. 1903. Two new Lithothamnia. *Ibid.*, 1903(2): 1–4.

———. 1905a. Remarks on northern Lithothamnia. *Ibid.*, 1905(3): 1–138.

———. 1905b. New Lithothamnia and systematical remarks. *Ibid.*, 1905(5): 1–9.

———. 1906. Algologiske notiser, II. *Ibid.*, 1906(2): 1–28.

———. 1907a. Algologiske notiser, III. *Ibid.*, 1906(8): 1–34.

———. 1907b. Algologiske notiser, IV. *Ibid.*, 1907(6): 1–30.

———. 1909. Algologiske notiser, VI. *Ibid.*, 1909(2): 1–63.

———. 1929. [Lithothamnium microsporum] in H. Printz, ed., *Contribution to a monograph of the Lithothamnia.* Trondheim. P. 51.

Fries, E. 1835. *Corpus florarum provincalium Sueciae.* Uppsala. Vol. I: *Floram Scanicum.* 394 pp.

Ganesan, E. K. 1962. Notes on Indian red algae. II. Dermatolithon ascripticium (Foslie) Setchell et Mason. *Phykos*, 1: 108–14. 12 figs.

———. 1968a. Studies on the morphology and reproduction of the articulated corallines. II. Corallina Linnaeus emend. Lamouroux. *Bol. Inst. Oceanogr., Univ. Oriente*, 7: 65–97. 34 figs. 3 pls.

———. 1968b. Studies of the morphology and reproduction of the articulated corallines. IV. Serraticardia (Yendo) Silva, Calliarthron Manza and Bossiella Silva. *Bot. Marina*, 11: 10–30. 35 figs.

Gardner, N. L. 1909. New Chlorophyceae from California. *Univ. Calif. Publ. Bot.*, 3: 371–75. 1 pl.

———. 1910. Variations in nuclear extrusion among the Fucaceae. *Ibid.*, 4: 121–36. 2 pls.

———. 1913. New Fucaceae. *Ibid.*, 4: 317–74. 18 pls.

———. 1917. New Pacific coast marine algae, I. *Ibid.*, 6: 377–416. 5 pls.

———. 1918. [Cladophora hemisphaerica], in F. S. Collins, The green algae of North America. 2d suppl. paper. *Tufts College Studies, Sci. Ser.*, 4(7): 83.

———. 1919. New Pacific coast marine algae, IV. *Univ. Calif. Publ. Bot.*, 6: 487–96. 1 pl.

———. 1922. The genus Fucus on the Pacific coast of North America. *Ibid.*, 10: 1–180. 60 pls.

———. 1926. New Rhodophyceae from the Pacific coast of North America, I. *Ibid.*, 13: 205–26. 7 pls.

————. 1927a. New Rhodophyceae from the Pacific coast of North America, II–VI. *Ibid.*, 13: 235–72, 333–68, 373–402, 403–34; 14: 99–138. 62 pls.

————. 1927b. New species of Gelidium on the Pacific coast of North America. *Ibid.*, 13: 273–318.

————. 1940. New species of Melanophyceae from the Pacific coast of North America. *Ibid.*, 19: 267–86. 6 pls.

————. 1944. [Desmarestia linearis], in G. M. Smith, *Marine Algae of the Monterey Peninsula, California.* Stanford, Calif. P. 120.

Gmelin, S. G. 1768. *Historia Fucorum.* St. Petersburg. 239 pp. 33 pls.

Gordon, E. 1972. Comparative morphology and taxonomy of the Wrangelieae, Sphondylothamnieae, and Spermothamnieae (Ceramiaceae, Rhodophyta). *Austr. J. Bot.* (Suppl.), ser. 4. 180 pp. 63 figs.

Gray, J. E. 1867. Lithothrix, a new genus of Corallinae. *Jour. Bot. Lond.* 5: 33. 1 fig.

Greville, R. K. 1822–28. *Scottish Cryptogamic Flora* . . . Edinburgh. 6 vols. 360 pp., 360 pls.

————. 1826. Some account of a collection of plants from the Ionian Sea. *Trans. Linn. Soc.*, 15: 335–48.

————. 1830. *Algae brittannicae* . . . Edinburgh. 218 pp. 19 pls.

Griffiths, A., and W. H. Harvey (1846–51). 1849. [Ceramium gracillimum] in W. H. Harvey, *Phycologia Britannica*, p. 206.

Grunow, A. 1873. Algen der Fidschi-, Tonga-, und Samoa-Inseln . . . *[Hamburg] Mus. Godeffroy Jour.*, 3(6): 23–50.

————. 1886. [Cordylecladia andersonii] in A. Piccone, *Alghe del viaggio di circumnavigazione del Vittor Pisani.* Genoa. P. 62.

————. 1915. Additamenta ad cognitionen Sargassorum. *Verh. Kaiserl. Zool.-bot. Ges. Wien*, 65: 325–448.

Gunnerus, J. E. 1772. *Flora Norvegica, observationibus praesertim oeconomicis panosque Norvegici locupletata.* Vol. 2. Copenhagen. 148 pp.

Hamel, A., and G. Hamel. 1929. Sur l'hétérogamie d'une Cladophoracée, Lola (nov. gen.) lubrica (Setch. et Gardn.). *Compt. Rend. Acad. Sci. Paris*, 189: 1094–96.

Hamel, G. 1927. *Recherches sur les genres Acrochaetium Naeg. et Rhodochorton Naeg.* St.-Malo. 117 pp. 47 figs.

————. 1931–39. *Phéophycées de France.* Paris. 432 pp. + xlvii.

Hardy-Halos, M.-T. 1968. Observations sur la morphologie du Neomonospora furcellata (J. Ag.) G. Feldmann et Meslin (Rhodophyceae-Ceramiaceae) et sur sa position taxonomique. *Bull. Soc. Bot. France*, 115: 523–28.

Hariot, P. 1889. Algues, in *Mission scientifique du Cap Horn, 1882–1883.* Paris. Vol. 5 (Botanique): 3–109. 9 pls.

————. 1891. Liste des algues marines raportées de Yokosuka (Japan) par M. le Dr. Savatier. *Mém. Soc. Nat. Sci. et Math. Cherbourg*, 27: 211–30.

Harvey, W. H. 1833a. Confervoideae, in W. J. Hooker, *The English Flora of Sir James Edward Smith. Class XXIV. Cryptogamia.* London. Vol. 5(1): 322–85.

————. 1833b. Algae, in W. J. Hooker and G. A. W. Arnott, *The Botany of Captain Beechey's Voyage.* London. Pp. 163–65.

————. 1836. Algae, in J. T. Mackay, *Flora Hibernica* . . . Dublin. Pt. 3: 154–254.

————. 1841a. Algae, in W. J. Hooker and G. A. W. Arnott, *The Botany of Captain Beechey's Voyage*. London. Pp. 406–9.

————. 1841b. *A Manual of the British Marine Algae* . . . London. 229 pp.

————. 1846–7. Algae, in J. D. Hooker, *The Botany of the Antarctic Voyage of H.M. Discovery Ships Erebus and Terror* . . . I. Flora Antarctica. London. Pt. 2 (*Botany* . . .): 454–502. 30 pls.

————. 1846–51. *Phycologia britannica*. London. 3 vols. 360 pls.

————. 1847. *Nereis australis* . . . London. 124 pp. 50 pls.

————. 1852. Nereis boreali-americana. I. Melanospermeae. *Smithsonian Contr. to Knowledge*, 3 (art. 4): 1–150. 12 pls.

————. 1853. Nereis boreali-americana. II. Rhodospermeae. *Ibid.*, 5 (art. 5): 1–258. 24 pls.

————. 1854. Some account of the marine botany of Western Australia. *Trans. Roy. Irish Acad.*, 22: 525–66.

————. 1857. Algae, in M. C. Perry, *Narrative of the Expedition of an American Squadron* . . . Washington, D.C. Vol. 2: 331–32.

————. 1858. Nereis boreali-americana. III. Chlorospermeae. *Smithsonian Contr. to Knowledge*, 10 (art 2): 1–140. 14 pls.

————. 1859. Characters of new algae . . . *Proc. Amer. Acad. Arts & Sci.*, 4: 327–34.

————. 1862. Notice of a collection of algae made on the northwest coast of North America, chiefly at Vancouver's Island, by David Lyall, Esq., M.D., R.N., in the years 1859–61. *Jour. Linn. Soc. Bot.*, 6: 157–77.

————. 1863. *Phycologia Australica*. London. Vol. 5. 60 pls.

Harvey, W. H., and J. W. Bailey. 1851. Descriptions of seventeen new species of algae, collected by the United States Exploring Expedition. *Proc. Boston Soc. Nat. Hist.*, 3: 370–73.

Hauck, F. 1883–85. Die Meeresalgen Deutschlands und Österreichs, in L. Rabenhorst, *Kryptogamen-Flora von Deutschland, Österreich und der Schweiz*. Leipzig. 2d ed. Vol. 2. 575 pp. 236 figs. 5 pls.

Haupt, A. W. 1932. Structure and development of Zonaria farlowii. *Amer. Jour. Bot.*, 19: 239–54. 4 figs. 4 pls.

Heerebout, G. R. 1968. Studies on the Erythropeltidaceae (Rhodophyceae-Bangiophycidae). *Blumea*, 16: 139–57. 19 figs.

Hirose, H., and K. Yoshida. 1964. A review of the life history of the genus Monostroma. *Bull. Japan. Soc. Phycol.*, 12: 19–31.

Hoek, C. van den. 1963. *Revision of the European Species of Cladophora*. Leiden. 248 pp. 721 figs. 55 pls.

Hollenberg, G. J. 1935. A study of Halicystis ovalis. I. Morphology and reproduction. *Amer. Jour. Bot.*, 22: 782–812. 5 figs. 4 pls.

————. 1936. A study of Halicystis ovalis. II. Periodicity in the formation of gametes. *Ibid.*, 23: 1–3. 1 fig.

————. 1939a. Culture studies of marine algae. I. Eisenia arborea. *Ibid.*, 26: 34–41.

————. 1939b. A morphological study of Amplisiphonia, a new member of the Rhodomelaceae. *Bot. Gaz.*, 101: 380–90. 13 figs.

————. 1940. New marine algae from Southern California, I. *Amer. Jour. Bot.*, 27: 868–77. 17 figs.

————. 1941. Culture studies of marine algae. II. Hapterophycus canaliculatus S. & G. *Ibid.*, 28: 676:83. 16 figs.

————. 1942a. Phycological notes, I. *Bull Torrey Bot. Club*, 69: 528–38. 15 figs.

————. 1942b. An account of the species of Polysiphonia on the Pacific coast of North America. I. Oligosiphonia. *Amer. Jour. Bot.*, 29: 772–85. 21 figs.

————. 1943a. New marine algae from Southern California, II. *Ibid.*, 30: 571–79. 16 figs.

————. 1943b. [Porphyra pulchra], in G. M. Smith and G. J. Hollenberg, On some Rhodophyceae from the Monterey Peninsula, California. *Amer. Jour. Bot.*, 30: 211–22.

————. 1943c. [Laurencia spp.], in *ibid.*, pp. 218–19.

————. 1943d. [Porphyra schizophylla], in *ibid.*, p. 213.

————. 1944a. An account of the species of Polysiphonia on the Pacific coast of North America. II. Polysiphonia. *Ibid.*, 31: 474–83. 12 figs.

————. 1944b. [Ralfsia pacifica], in G. M. Smith, *Marine Algae of the Monterey Peninsula, California*. Stanford, Calif. P. 95.

————. 1945a. New marine algae from Southern California, III. *Amer. Jour. Bot.*, 32: 477–51. 9 figs.

————. 1945b. [Ralfsia occidentalis], in W. R. Taylor, Pacific marine algae of the Allan Hancock Expeditions to the Galapagos Islands. *Allan Hancock Pacific Expeditions*, 12: 81.

————. 1948. Notes on Pacific coast marine algae. *Madroño*, 9: 155–62.

————. 1957. Culture studies of Spongomorpha coalita. *Phycol. News Bull.*, 32: 76.

————. 1958a. Phycological notes, II. *Bull. Torrey Bot. Club*, 85: 63–69. 2 figs.

————. 1958b. Observations concerning the life cycle of Spongomorpha coalita (Ruprecht) Collins. *Madroño*, 14: 249–51. 7 figs.

————. 1959. Smithora, an interesting new algae genus in the Erythropeltidaceae. *Pacific Naturalist*, 1(8): 3–11. 5 figs.

————. 1961. Marine red algae of Pacific Mexico. V. The genus Polysiphonia. *Ibid.*, 2(5–6): 345–75.

————. 1968a. An account of the species of Polysiphonia of the Central and Western tropical Pacific. I. Oligosiphonia. *Pac. Sci.*, 22: 56–98. 43 figs.

————. 1968b. An account of the species of the red alga Herposiphonia occurring in the Central and Western tropical Pacific Ocean. *Ibid.*, pp. 536–59. 25 figs.

————. 1969. An account of the Ralfsiaceae (Phaeophyta) of California. *Jour. Phycol*, 5: 290–301. 30 figs.

————. 1970. Phycological notes. IV. Including new marine algae and new records for California. *Phycologia*, 9: 61–72. 23 figs.

————. 1971a. Phycological notes. V. New species of marine algae from California. *Ibid.*, 10: 11–16. 11 figs.

————. 1971b. Phycological notes. VI. New records, new combinations, and noteworthy observations concerning marine algae of California. *Ibid.*, 10: 281–89. 6 figs.

————. 1972. Phycological notes. VII. Concerning three Pacific coast species, especially Porphyra miniata (C. Ag.) C. Ag. (Rhodophyceae, Bangiales). *Ibid.*, 11: 43–46.

Hollenberg, G. J., and I. A. Abbott. 1965. New species and new combinations of marine algae from the region of Monterey, California. *Can. Jour. Bot.*, 43: 1177–88. 13 figs. 2 pls.

————. 1966. *Supplement to Smith's Marine Algae of the Monterey Peninsula.* Stanford, Calif. 130 pp. 53 figs.

————. 1968. New species of marine algae from California. *Can. Jour. Bot.*, 46: 1235–51. 14 figs.

Hollenberg, G. J., and M. J. Wynne. 1970. Sexual plants of Amplisiphonia pacifica (Rhodophyta). *Phycologia*, 9: 175–78. 5 figs.

Holmes, E. M. 1896. New marine algae from Japan. *Jour. Linn. Soc. Bot.*, 31: 248–60. 12 pls.

Hooker, W. J. 1833. *The English Flora of Sir James Edward Smith. Class XXIV. Cryptogamia.* London. Vol. 5(1). 432 pp.

Hooker, W. J., and W. H. Harvey. 1847. [Polysiphonia mollis] in W. H. Harvey, *Nereis Australis . . .* London. P. 43.

Hoppaugh, K. M. 1930. A taxonomic study of species of the genus Vaucheria collected in California. *Amer. Jour. Bot.*, 17: 329–47. 4 figs. 4 pls.

Howe, M. A. 1911. Phycological studies. V. Some marine algae of Lower California, Mexico. *Bull. Torrey Bot. Club*, 38: 489–514. 1 fig. 8 pls.

————. 1914. The marine algae of Peru. *Mem. Torrey Bot. Club*, 15: 1–185. 44 figs. 66 pls.

————. 1920. Algae, in N. L. Britton and C. F. Millspaugh, *The Bahama Flora.* New York. Pp. 553–618.

Hudson, P. R., and M. J. Wynne. 1969. Sexual plants of Bonnemaisonia geniculata (Nemaliales). *Phycologia*, 8: 207–13. 11 figs.

Hudson, W. 1762. *Flora Anglica . . .* London. 1st ed. 506 pp.

————. 1778. *Ibid.*, 2d ed. 680 pp.

Hus, H. T. A. 1902. An account of the species of Porphyra found on the Pacific coast of North America. *Proc. Calif. Acad. Sci.*, 3d ser., *Botany*, 2: 173–240. 3 pls.

Itono, H. 1969. The genus Antihamnion (Ceramiaceae) in southern Japan and adjacent waters, I. *Mem. Fac. Fish., Kagoshima Univ.*, 18: 29–45.

Jaasund, E. 1957. Marine algae from northern Norway, II. *Bot. Notiser*, 110: 223.

Jao, C. C. 1937. New marine algae from Washington. *Mich. Acad. Sci., Arts & Letters Pap.*, 22: 99–115. 3 pls.

Johansen, H. W. 1966. A new member of the Corallinaceae: Chiharaea bodegensis gen. et sp. nov. *Phycologia*, 6: 51–61. 11 figs.

————. 1969. Morphology and systematics of coralline algae with special reference to Calliarthron. *Univ. Calif. Publ. Bot.*, 49: 1–78. 19 pls.

————. 1971a. Changes and additions to the articulated coralline flora of California. *Phycologia*, 10: 241–49.

————. 1971b. Bossiella, a genus of articulated corallines (Rhodophyceae, Cryptonemiales) in the eastern Pacific. *Ibid.*, 10: 381–96.

Kjeldsen, C. K. 1972. Pleurophycus gardneri Setchell & Saunders, a new alga for Northern California. *Madroño*, 21: 416.

Kjellman, F. R. 1872. *Bidrag till Kännedomen om Skandinaviens Ectocarpeer och Tilopterider.* Stockholm. 112 pp. 2 pls.

――――. 1883. The algae of the Arctic Sea. *Kgl. Svensk. Vetensk. Ak. Handl.,* 20(5): 1–350. 31 pls.

――――. 1889a. Om Beringshafvets Algflora. *Ibid.,* 23(8): 1–58. 7 pls.

――――. 1889b. Undersökning of några till slägtet Adenocystis Hook. fil. et Harv. Hänförda Alger. *Bihang Kgl. Svensk. Vetensk. Ak. Handl.,* 15(1): 4–20. 28 figs.

――――. 1897. Marine Chlorophyceer fran Japan. *Ibid.,* 23(11): 1–44. 7 pls.

Kornmann, P. 1938. Zur Entwicklungsgeschichte von Derbesia und Halicystis. *Planta,* 28: 464–70. 4 figs.

――――. 1956. Zur Morphologie und Entwicklung von Percursaria percursa. *Helgol. wiss. Meeresunters.,* 5: 259–72. 7 figs.

――――. 1960. Ectocarpaceen-Studien. *Ibid.,* 7: 93–113. 6 figs.

――――. 1964. Die Ulothrix-Arten von Helgoland, I. *Ibid.,* 11: 27–38.

Kuckuck, P. 1894. Bemerkungen zur marinen Algenvegetation von Helgoland. *Wiss. Meeresunters.,* n.f., 1: 225–63. 29 figs.

――――. 1896. Bemerkungen zur marinen Algenvegetation von Helgoland. *Ibid.* 2: 371–400. 21 figs.

Kugrens, P. 1971. Comparative ultrastructure of vegetative and reproductive structures in parasitic red algae. Ph.D. dissertation, Univ. Calif., Berkeley. 308 pp.

Kuntze, O. 1891. *Revisio generum plantarum* . . . Würzburg. Vol. 2: 375–1011.

――――. 1898. *Ibid.* Vol. 3: 1–576.

Kützing, F. T. 1843. *Phycologia generalis.* Leipzig. 458 pp. 80 pls.

――――. 1845. *Phycologia Germanica.* Nordhausen. 340 pp.

――――. 1847. Diagnosen und Bemerkungen zu neuen order kritischen Algen. *Bot. Zeitg.,* 5: 1–5, 22–25, 33–38, 52–55, 164–67, 177–80, 193–98, 219–23.

――――. 1849. *Species algarum.* Leipzig. 922 pp.

――――. 1855. *Tabulae phycologicae.* Nordhausen. Vol. 5. 30 pp. 100 pls.

――――. 1856. *Ibid.* Vol. 6. 35 pp. 100 pls.

――――. 1858. *Ibid.* Vol. 8. 48 pp. 100 pls.

――――. 1859. *Ibid.* Vol. 9. 42 pp. 100 pls.

――――. 1866. *Ibid.* Vol. 16. 35 pp. 100 pls.

Kylin, H. 1906. Zur Kenntnis einiger schwedischen Chantransia-Arten, in *Botaniska Studier til F. R. Kjellman.* Uppsala. Pp. 113–26.

――――. 1924. Studien über die Delesseriaceen. *Lunds Univ. Årsskr.,* n.f., 20 (6): 1–111. 80 figs.

――――. 1925. The marine red algae in the vicinity of the biological station at Friday Harbor, Washington. *Ibid.,* 21(9): 1–87. 47 figs.

――――. 1928. Entwicklungsgeschichtliche Florideen-studien. *Ibid.,* 24(4): 1–127. 64 figs.

――――. 1930. Über die Entwicklungsgeschichte der Florideen. *Ibid.,* 26(6): 1–104. 56 figs.

――――. 1931. Die Florideenordnung Rhodymeniales. *Ibid.,* 27(11): 1–48. 8 figs. 20 pls.

――――. 1932. Die Florideenordnung Gigartinales. *Ibid.,* 28(8): 1–88. 22 figs. 28 pls.

————. 1935. Zur Nomenklatur einiger Delesseriaceen. *Förhandl. Kgl. Fysiografiska Sallsk. i Lund,* 5(23): 1–5.

————. 1940. Die Phaeophyceenordnung Chordariales. *Lunds Univ. Årsskr.,* n.f., 36(9): 1–67. 30 figs. 8 pls.

————. 1941. Californische Rhodophyceen. *Ibid.,* 37(2): 1–51. 7 figs. 13 pls.

————. 1947a. Die Phaeophyceen der schwedischen Westküste. *Ibid.,* 43(4): 1–99. 61 figs. 18 pls.

————. 1947b. Über die Fortpflanzungsverhältnisse in der Ordnung Ulvales. *Förhandl. Kgl. Fysiografiska Sallsk. i Lund,* 17 (19): 174–82.

————. 1956. *Die Gattungen der Rhodophyceen.* Lund. 673 pp. 458 figs.

Lagerheim, G. 1885. Codolium polyrhizum n. sp. *Öfvers. Kgl. Svensk. Vetensk. Ak. Förh.,* 42(8): 21–31. 1 pl.

Lamouroux, J. V. F. 1809. Mémoire sur trois nouveaux genres de la famille des algues marines. *Jour. de Bot.,* 2: 129–36.

————. 1813. Essai sur les genres de la famille des thalassiophytes non articulées. *Ann. Mus. Hist. Nat.* [Paris], 20: 21–47, 115–39, 267–93. 7 pls.

————. 1816. *Histoire des polypiers coralligènes flexibles, vulgairement nommés zoophytes.* Caen. 559 pp. 19 pls.

————. 1821. *Exposition méthodique des genres de l'ordre des Polypiers.* Paris. 115 pp. 84 pls.

Le Jolis, A. 1863. Liste des algues marines de Cherbourg. *Mém. Imp. Soc. Sci. Nat. Cherbourg,* 10: 1–168. 6 pls.

Leman, D. S. 1822. [Laminaria porra], in F. G. Levrault, ed., *Dictionnaire des sciences naturelles . . .* Paris. Vol. 25: 189.

Lemoine, M. 1924. Melobesieae, in F. Børgesen, Marine Algae of Easter Island. Vol. 2 of C. Skottsberg, ed., *The Natural History of Juan Fernandez and Easter Island.* Uppsala. Pp. 285–90.

Levring, T. 1939. Über die Phaeophyceengattungen Myriogloia Kuck. und Haplogloia nov. gen. *Bot. Notiser,* 1939: 40–52. 5 figs.

————. 1940. Die Phaeophyceengattungen Clanidophora und Syringoderma. *Förhandl. Kgl. Fysiografiska Sallsk. i Lund,* 10(20): 1–11. 5 figs.

————. 1956. [Pseudogloiophloea confusa], in N. Svedelius, Are the haplobiontic Florideae to be considered reduced types? *Bot. Tidsskr.,* 50: 8.

————. 1960. Contributions to the marine algal flora of Chile. *Lunds Univ. Årsskr.,* n.f., 56(10): 1–83. 22 figs.

Lewin, R. A., and J. A. Robertson. 1971. Influence of salinity on the form of Asterocytis in pure culture. *Jour. Phycol.,* 7: 236–38.

Lightfoot, J. 1777. *Flora Scotica . . .* London. 2 vols. 1,149 pp.

Link, H. F. 1820. Epistola de algis aquaticis in genera disponendis, in C. G. D. Nees, *Horae physicae berolinenses.* Bonn. pp. 1–8. 1 pl.

Linnaeus, C. 1753. *Species plantarum . . .* Stockholm. 1st ed. Vol. 2: 561–1,200.

————. 1755. *Flora Suecia . . .* Stockholm. 2d ed. 464 pp. 1 pl.

————. 1758. *Systema naturae . . .* 10th ed. Vol. 1. Stockholm. 823 pp.

————. 1767. *Systema naturae,* 12th ed. Vol. 2. Regnum vegetabile. Stockholm. 736 pp.

————. 1771. *Mantissa plantarum.* Stockholm. 2d ed. 587 pp.

Loiseaux, S. 1967a. Morphologie et cytologie des Myrionémacées. Critères taxonomiques. *Rev. Gén. Bot.,* 74: 329–47. 6 figs. 3 pls.

————. 1967b. Recherches sur les cycles de développement des Myrionémata-cées (Phéophycées). I-II. Hécatonématées et Myrionématées. *Ibid.*, 74: 529–76. 21 figs. 2 pls.

————. 1970a. Streblonema anomalum S. & G. and Compsonema sporangii-ferum S. & G., stages in the life history of a minute Scytosiphon. *Phycologia,* 9: 185–91. 4 figs.

————. 1970b. Notes on several Myrionemataceae from California using cul-ture studies. *Jour. Phycol.*, 6: 248–60. 5 figs.

Loomis, N. H. 1949. New species of Gelidium and Pterocladia with notes on the structure of the thalli in these genera. *Allan Hancock Foundation Publ. Occ. Pap.*, 6: 1–28. 10 pls.

————. 1960. New species of Gelidium and Pterocladia from the Pacific coast of the United States and the Hawaiian Islands. *Ibid.*, 24: 1–35.

Lyngbye, H. C. 1819. *Tentamen hydrophytologiae Danicae.* Copenhagen. 248 pp. 70 pls.

McFadden, M. E. 1911. On a Colacodasya from Southern California. *Univ. Calif. Publ. Bot.*, 4: 143–50. 1 pl.

Manza, A. V. 1937a. The genera of the articulated corallines. *Proc. Natl. Acad. Sci.*, 23: 44–48.

————. 1937b. Some North Pacific species of articulated corallines. *Ibid.*, 23: 561–67.

————. 1940. A revision of the genera of articulated corallines. *Philippine Jour. Sci.*, 71(3): 239–316. 20 pls.

Masaki, T. 1968. Studies on the Melobesioidae of Japan. *Mem. Fac. Fish., Hok-kaido Univ.*, 16: 1–80. 79 pls.

Masaki, T., and J. Tokida. 1960. Studies on the Melobesioidae of Japan, II–III. *Bull. Fac. Fish., Hokkaido Univ.*, 10: 285–90; 11: 37–42. 15 pls.

————. 1961. Studies on the Melobesioidae of Japan, V. *Ibid.*, 12: 161–65. 8 pls.

Mason, L. R. 1943a. [Lithothamnium aculeiferum], in W. A. Setchell and L. R. Mason, New or little known crustaceous corallines of Pacific North America. *Proc. Natl. Acad. Sci.*, 29: 94.

————. 1943b. [Lithothamnium crassiusculum], in *ibid.*, p. 93.

————. 1943c. [Lithothamnium giganteum], *ibid.*

————. 1953. The crustaceous coralline algae of the Pacific coast of the United States, Canada and Alaska. *Univ. Calif. Publ. Bot.*, 26: 313–89. 20 pls.

Mathieson, A. C. 1966. Morphological studies of the marine brown algae Taonia lennebackerae Farlow ex J. Agardh, I. *Nova Hedwigia*, 12: 65–79. 1 fig. 4 pls.

————. 1967. Morphology and life history of Phaestrophion irregulare S. et G. *Ibid.*, 13: 293–318. 15 figs.

Mazoyer, G. 1938. Les Céramiées de l'Afrique du Nord. *Bull. Soc. Hist. Nat. Afrique du Nord*, 29: 317–31.

Mertens, H. 1818. [Conferva wormskioldia], in J. W. Hornemann, *Icones plan-tarum . . . ad illustrandum, Flora Danicae.* Copenhagen. Vol. 9: 6.

————. 1829. Zwei botanisch-wissenschaftliche Berichte vom Dr. Heinrich Mer-tens . . . *Linnaea*, 4: 43–58.

Mikami, H. 1965. A systematic study of the Phyllophoraceae and Gigartinaceae from Japan and its vicinity. *Sci. Pap., Inst. Algol. Res., Hokkaido Univ. Fac. Sci.*, 5: 181–285. 55 figs. 11 pls.

Montagne, J. F. C. 1837–38. [Premiére] Centurie des plantes cellulaires exotiques nouvelles. *Ann. Sci. Nat. Bot.*, sér. 2, 8: 345–70.

———. 1839. Botanique, Cryptogamie. II. Florula Boliviensis stirpes novae et minus cognitae, in A. d'Orbigny, *Voyage dans l'Amerique meridionale*. Paris. Vol. 7: 13–39. 7 pls.

———. 1841. Plantes cellulaires, in P. B. Webb and S. Berthelot, eds., *Histoire naturelle des Iles Canaries*. Paris. Vol. 3(2): 161–76.

———. 1842. Troisième centurie de cellulaires exotiques nouvelles. *Ann. Sci. Nat. Bot.*, sér. 2, 18: 241–82.

———. 1846a. Phyceae, in M. C. Durieu de Maisonneuve, *Exploration scientifique de l'Algérie . . . Botanique. Flore d'Algérie. Cryptogamie*. Paris. Pp. 1–197. 16 pls.

———. 1846b. Note sur le genre Stenogramma Harv., de la famille des Floridées. *Duchatres Rev. Bot.* 1: 481–83.

Mower, A., and T. Widdowson. 1969. New records of marine algae from Southern California. *Bull. So. Calif. Acad. Sci.*, 68: 72–81. 3 figs.

Müller, O. F. 1775–82. *Icones plantarum . . . ad illustrandum, Florae Danicae . . .* Copenhagen. Vols. 4 and 5.

Nägeli, C. 1849. [Enteromorpha minima], in F. T. Kützing, *Species algarum*. Leipzig. P. 482.

———. 1862. Beiträge zur Morphologie und Systematik der Ceramiaceae. *Sitzungsber. k. b. Akad. Wiss. München*, 2: 297–415.

Nakamura, Y. 1944. The species of Rhodochorton from Japan, II. *Sci. Pap., Inst. Algol. Res., Hokkaido Univ. Fac. Sci.*, 3: 99–119. 13 figs.

Nardo, G. D. 1834. De novo genere algarum cui nomen est Hildenbrandia. *Isis*, 1834: 675–76.

Neushul, M. 1971. The species of Macrocystis with particular reference to those of North and South America, in W. J. North, ed., The biology of giant kelp beds (Macrocystis) in California. *Nova Hedwigia*, 32. 600 pp. 166 figs. 116 tables.

Neushul, M., J. Scott, A. L. Dahl, and D. Olsen. 1967. Growth and development of Sciadophycus stellatus Dawson. *Bull. So. Calif. Acad. Sci.*, 66: 195–200. 8 figs.

Nichols, H. W., and E. K. Lissant. 1967. Developmental studies of Erythrocladia Rosenvinge in culture. *Jour. Phycol.*, 3: 6–18. 103 figs.

Nichols, M. B. 1909. Contributions to the knowledge of the California species of crustaceous corallines, II. *Univ. Calif. Publ. Bot.*, 3: 349–70. pls. 10–13.

Norris, J. N. 1971. Observations on the genus Blidingia (Chlorophyta) in California. *Jour. Phycol.*, 7: 145–49. 4 figs.

Norris, R. E. 1957. Morphological studies on the Kallymeniaceae. *Univ. Calif. Publ. Bot.*, 28: 251–334. 25 figs. 13 pls.

Norris, R. E., and J. West. 1967. Notes on marine algae of Washington and southern British Columbia, II. *Madroño*, 19: 111–16.

Norris, R. E., and M. J. Wynne. 1968. Notes on marine algae of Washington and southern British Columbia, III. *Syesis*, 1: 133–46.

North, W. J. 1971. The biology of giant kelp beds (Macrocystis) in California. *Nova Hedwigia*, 32. 600 pp. 166 figs. 116 tables.

Nott, C. P. 1897. Some parasitic Florideae of the California coast. *Erythea*, 5: 81–84.

———. 1900. Nitophylla of California. *Proc. Calif. Acad. Sci.*, 3d ser., *Botany*, 2: 1–62. 9 pls.

Okamura, K. 1899. Contributions to the knowledge of the marine algae of Japan, III. *Bot. Mag. Tokyo*, 13: 2–10, 35–43. 1 pl.

———. 1903. Contents of the "Algae Japonicae Exsiccatae." Fasc. II. *Ibid.*, 17: 1–4.

———. 1907–42. *Icones of Japanese Algae.* Tokyo. 7 vols. 345 pls. (Published by the author.)

Papenfuss, G. F. 1942. Notes on algal nomenclature. I. Pollexfenia, Jeannerettia and Mesotrema. *Proc. Natl. Acad. Sci.*, 28: 446–51.

———. 1944. Structure and taxonomy of Taenioma, including a discussion on the phylogeny of the Ceramiales. *Madroño*, 7: 193–214. 17 figs. 2 pls.

———. 1945. Review of the Acrochaetium-Rhodochorton complex of the red algae. *Univ. Calif. Publ. Bot.*, 18: 299–334.

———. 1950. Review of the genera of algae described by Stackhouse. *Hydrobiologia*, 2: 181–208.

———. 1960. On the genera of the Ulvales and the status of the order. *Jour. Linn. Soc. Bot.*, 56: 303–16. 9 figs. 6 pls.

———. 1967. Notes on algal nomenclature. V. Various Chlorophyceae and Rhodophyceae, *Phykos, Prof. Iyengar Memorial Volume*, 1966, 5: 95–105.

Parker, B. C., and J. Bleck. 1965. A new species of elk kelp. *Trans. San Diego Soc. Nat. Hist.*, 14: 57–64.

Pease, V. A. 1920. Taxonomy and morphology of the ligulate species of the genus Desmarestia. *Puget Sound Biol. Sta. Publ.* 2: 313–67. 10 pls.

Post, E. 1962. *Murrayellopsis dawsonii gen. et. spec. nov.* Kiel. 4 pp. 3 figs. (Published by author.)

Postels, A., and F. Ruprecht. 1840. *Illustrationes algarum* . . . St. Petersburg. 22 pp. 40 pls.

Powell, H. T. 1957. Studies in the genus Fucus L. I. Fucus distichus L. emend. Powell. *Marine Biol. Assoc. U.K. Jour.*, 36: 407–32. 2 pls.

Pringsheim, N. 1863. Beiträge zur Morphologie der Meeresalgen. *Physikal. Abhandl. Kgl. Akad. Wissensch.*, 1862: 1–37.

Proskauer, J. 1950. On Prasinocladus. *Amer. Jour. Bot.*, 37: 59–66. 40 figs.

Ramus, J. 1969. The developmental sequence of the marine red alga Pseudogloiophloea in culture. *Univ. Calif. Publ. Bot.*, 52: 1–28. 12 pls.

Ravanko, O. 1970. Morphological, developmental and taxonomic studies in the Ectocarpus complex. *Nova Hedwigia*, 20: 179–252. 52 figs.

Reinke, J. 1879. Zwei parasitische Algen. *Bot. Zeitg.*, 37: 473–78. 1 pl.

———. 1889a. Algenflora der westlichen Ostsee deutschen Antheils. *Wiss. Meeresunters.*, 6: 1–101. 8 figs.

———. 1889b. *Atlas deutscher Meersalgen.* Berlin. Pt. 1. 34 pp. 25 pls.

———. 1892. *Ibid.* Pt. 2. 70 pp. 25 pls.

———. 1903. *Studien zur vergleichenden Entwicklungsgeschichte der Laminariaceen.* Kiel. 67 pp. 15 figs.

Reinsch, P. F. 1874–75. *Contributiones ad algologiam et fungologiam.* Nürenberg. 103 pp. 131 pls.

Rosenvinge, L. K. 1893. Grønlands Havalger. *Meddelelser om Grønland*, 3: 795–981. 57 figs. 2 pls.

———. 1900. Note sur une Floridée aérienne (Rhodochorton islandicum nov. sp.). *Bot. Tidsskr.*, 23: 61–81.

———, 1909–31. The marine algae of Denmark. Rhodophyceae. *Kgl. Danske Vidensk. Selsk. Skr.*, 7 Raekke, Afd., 7: 1–637. 618 figs. 8 pls.

Roth, A. G. 1797–1806. *Catalecta botanica* . . . Leipzig. Fasc. 1. 244 pp. 8 pls. Fasc. 2. 258 pp. 9 pls. Fasc. 3. 350 pp. 12 pls.

Ruprecht, F. J. 1851. Tange des Ochtskischen Meeres, in A. T. von Middendorff, ed., *Sibirische Reise. Botanik.* St. Petersburg. Vol. 1(2): 193–435. 10 pls.

———. 1852. Neue oder unvollständig bekannte Pflanzen aus dem nördlichen Theile des Stillen Oceans. *Mém. Acad. St.-Pétersb. Sci. Nat. Bot.*, 7: 55–82. 8 pls.

Saenger, P. 1971. On the occurrence of Ophidocladus (Rhodomelaceae) in South Africa. *Jour. S. Afr. Bot.*, 37: 291–94.

Saito, Y. 1969. On morphological distinctions of some species of Pacific North American Laurencia. *Phycologia*, 8: 85–90. 4 figs.

Sakai, Y. 1964. The species of Cladophora from Japan and its vicinity. *Ibid.*, 5: 1–104. 45 figs. 17 pls.

Saunders, De A. 1895. A preliminary paper on Costaria with description of a new species. *Bot. Gaz.*, 20: 54–58. 1 pl.

———. 1898. Phycological memoirs. *Proc. Calif. Acad. Sci.*, 3d ser., Botany, 1: 147–68. 21 pls.

———. 1899. New and little-known brown algae of the Pacific coast. *Erythea*, 7: 37–40. 1 pl.

———. 1901a. Papers from the Harriman Alaska Expedition. XXV. The algae. *Proc. Wash. Acad. Sci.*, 3: 391–486. 20 pls.

———. 1901b. A new species of Alaria. *Minn. Bot. Studies*, 2: 561–62. 1 pl.

Sauvageau, C. 1897. Sur quelques Myrionémacées. *Ann. Sci. Nat. Bot.*, sér. 8, 5: 161–288. 29 figs.

———. 1901. Remarques sur les Sphacelariacées. *Jour. de Bot.*, 15: 67–183. 38 figs.

———. 1927a. Sur les problèmes du Giraudya. *Bull. Sta. Biol. Arachon*, 24: 3–74. 18 figs.

———. 1927b. Sur le Colpomenia sinuosa Derb. et Sol. *Ibid.*, 24: 309–53. 8 figs.

Scagel, R. F. 1953. A morphological study of some dorsiventral Rhodomelaceae. *Univ. Calif. Publ. Bot.*, 27: 1–108. 20 figs.

———. 1957. An annotated list of the marine algae of British Columbia and northern Washington. *Bull. Natl. Mus. Can.*, 150: 1–289. 1 fig.

———. 1960. Life history studies of the Pacific coast marine alga Collinsiella tuberculata Setchell and Gardner. *Can. Jour. Bot.*, 38: 969–83. 48 figs.

————. 1966. Marine algae of British Columbia and northern Washington. I. Chlorophyceae (green algae). *Bull. Nat. Mus. Can.*, 207: 1–257. 49 pls.

Schmitz, C. J. F. 1889. Systematische Übersicht der bisher bekannten Gattungen der Florideen. *Flora*, 72: 435–56. 1 pl.

————. 1893. Die Gattung Lophothalia J. Ag. *Ber. deutsch. bot. Ges.*, 11: 212–32.

————. 1896–97. Rhodophyceae, in A. Engler and K. Prantl, eds., *Die natürlichen pflanzenfamilien.* Leipzig. Vol. 1(2): 298–544.

Scott, J. L., and P. S. Dixon. 1971. The life history of Pikea californica Harv. *Jour. Phycol.*, 7: 295–300.

Segawa, S. 1936. The development of the female reproductive organs in "Nagaobane." *Plants and animals*, 6: 1987–90. (In Japanese.)

————. 1947. Systematic anatomy of the articulated Corallines. XI. Lithothrix aspergillum Gray. *Seibutsu*, 2: 87–90. 4 figs.

————. 1949. Five Floridean genera new to Japan. *Jour. Japanese Bot.*, 24: 159–65.

————. 1955. Systematic anatomy of the articulated corallines. (Suppl. report.) The structure and reproduction of Yamadaea melobesioides Segawa. *Bot. Mag. Tokyo*, 68: 241–47. 6 figs.

Segi, T. 1951. Systematic study of the genus Polysiphonia from Japan and its vicinity. *Jour. Fac. Fish., Mie Prefecture Univ.*, 1: 169–272. 36 figs. 16 pls.

Setchell, W. A. 1893. On the classification and geographical distribution of the Laminariaceae. *Trans. Conn. Acad.* 9: 333–75.

————. 1901. Notes on algae, I. *Zoe*, 5: 121–29.

————. 1905. Parasitic Florideae of California. *Nuova Notarisia*, 16: 59–63.

————. 1906. A revision of the genus Constantinea. *Ibid.*, 17: 162–73.

————. 1908a. Critical notes on Laminariaceae. *Ibid.* 19: 90–101.

————. 1908b. Nereocystis and Pelagophycus. *Bot. Gaz.*, 45: 125–34.

————. 1912. Algae novae et minus cognitae, I. *Univ. Calif. Publ. Bot.*, 4: 229–68. 7 pls.

————. 1914a. Parasitic Florideae, I. *Ibid.*, 6: 1–34. 6 pls.

————. 1914b. The Scinaia assemblage. *Ibid.*, 6: 79–152. 7 pls.

————. 1923a. Parasitic Florideae, II. *Ibid.*, 10: 393–96.

————. 1923b. A revision of the west North American species of Callophyllis. *Ibid.*, 10: 397–401.

————. 1925. [Egregia laevigata f. borealis] in W. A. Setchell and N. L. Gardner, The marine algae of the Pacific coast of North America. III. Melanophyceae. *Ibid.*, 8: 649.

————. 1940. Fucus cordatus Turner. *Proc. Natl. Acad. Sci.*, 26: 643–51.

Setchell, W. A., and E. Y. Dawson. 1941. Binghamia, the alga versus Binghamia, the cactus. *Ibid.*, 27: 376–81. 1 pl.

Setchell, W. A., and W. G. Farlow. 1900. [Lessonia littorales] in J. E. Tilden, *American Algae* (Exsiccatae). 1900. No. 342.

Setchell, W. A., and M. H. Foslie. 1902a. [Melobesia marginata] in M. H. Foslie, *New* species or forms of Melobesieae. *Kgl. Norske Vidensk. Selsk. Skr.*, 1902(2): 10.

————. 1902b. [Lithothamnium conchatum] in *ibid.*, p. 5.

———. 1903. [Lithothamnium lamellatum] in M. H. Foslie, Two new Lithothamnia. *Kgl. Norske Vidensk. Selsk. Skr.*, 1903(2): 4.

———. 1907. [Lithothamnium parcum] in M. H. Foslie, Algologiske notiser, IV. *Ibid.*, 1907(6): 14.

Setchell, W. A., and N. L. Gardner. 1903. Algae of northwestern America. *Univ. Calif. Publ. Bot.*, 1: 165–419. 11 pls.

———. 1910. [Hesperophycus harveyanus] in N. L. Gardner, Variations in nuclear extrusion among the Fucaceae. *Ibid.*, 4: 127.

———. 1912a. [Hapterophycus canaliculatus] in W. A. Setchell, Algae novae et mius cognitae, I. *Ibid.*, 4: 233.

———. 1912b. [Fauchea laciniata] in *ibid.*, p. 238.

———. 1917a. [Pelvetia fastigiata] in N. L. Gardner, New Pacific coast marine algae, I. *Ibid.*, 6: 386.

———. 1917b. [Cystoseira neglecta] in *ibid.*, p. 388.

———. 1917c. [Sargassum dissectifolium] in *ibid.*, p. 136.

———. 1917d. [Cumagloia andersonii] in *ibid.*, p. 399.

———. 1917e. [Coriophyllum expansum] in *ibid.*, p. 396.

———. 1917f. [Petrocelis franciscana] in *ibid.*, p. 391.

———. 1917g. [Hildenbrandia occidentalis] in *ibid.*, p. 393.

———. 1917h. [Gayella constricta] in *ibid.*, p. 384.

———. 1917i. [Chlorochytrium porphyrae] in *ibid.*, p. 379.

———. 1917j. [Myelophycus intestinalis f. tenuis] in *ibid.*, p. 385.

———. 1919. [Rhizoclonium lubricum] in N. L. Gardner, New Pacific coast marine algae, IV. *Ibid.*, 6: 487–96. 1 pl.

———. 1920a. Phycological contributions, I. *Ibid.*, 7: 279–324.

———. 1920b. The marine algae of the Pacific coast of North America. II. Chlorophyceae. *Ibid.*, 8: 139–374. 25 pls.

———. 1922. Phycological contributions. II-VI. New species of: Myrionema; Compsonema; Hecatonema; Pylaiella and Streblonema; and Ectocarpus. *Ibid.* 7: 334–52, 353–76, 377–84, 385–402, 403–26. 18 pls.

———. 1924a. New marine algae from the Gulf of California. *Proc. Calif. Acad. Sci.*, 4th ser., 12: 695–949. 77 pls.

———. 1924b. Phycological contributions, VII. *Univ. Calif. Publ. Bot.* 13: 1–13.

———. 1925. The marine algae of the Pacific coast of North America. III. Melanophyceae. *Ibid.*, 8: 383–739. 73 pls.

———. 1926a. [Membranoptera weeksiae] in N. L. Gardner, New Rhodophyceae from the Pacific coast of North America, I. *Ibid.*, 13: 209.

———. 1926b. [Erythroglossum divaricata] in *ibid.*, p. 207.

———. 1927. [Asymmetria expansa] in N. L. Gardner, New Rhodophyceae from the Pacific coast of North America, III. *Ibid.*, p. 341.

———. 1930. Marine algae of the Revillagigedo Islands Expedition in 1925. *Proc. Calif. Acad. Sci.*, 4th ser., 19: 109–215. 12 pls.

———. 1933. A preliminary survey of Gigartina, with special reference to its Pacific North American species. *Univ. Calif. Publ. Bot.*, 17: 255–340. 20 pls.

———. 1937a. A preliminary report on the algae. The Templeton Crocker Expedition of the California Academy of Sciences, 1932. *Proc. Calif. Acad. Sci.*, 4th ser., 22: 65–98. 1 fig. 23 pls.

————. 1937b. Iridophycus in the Northern Hemisphere. *Proc. Calif. Acad. Sci.*, 4th ser., 23: 169–74.

Setchell, W. A., and J. Guernsey. 1914. [Janczewskia], in W. A. Setchell, Parasitic Florideae, I. *Univ. Calif. Publ. Bot.*, 6: 9, 12.

Setchell, W. A., and H. T. A. Hus. 1900. [Porphyra] in H. T. A. Hus, Preliminary notes on west-coast Porphyras, *Zoe*, 5: 64, 65, 69.

Setchell, W. A., and A. A. Lawson. 1905. [Peyssonneliopsis epiphytica] in W. A. Setchell, Parasitic Florideae of California. *Nuova notarisia*, 16: 63.

Setchell, W. A., and M. E. McFadden. 1911. [Colacodasya verrucaeformis] in M. E. McFadden, On a Colacodasya from Southern California. *Univ. Calif. Publ. Bot.*, 4: 149.

Setchell, W. A., and L. R. Mason. 1943. New or little known crustaceous corallines of Pacific North America. *Proc. Natl. Acad. Sci.*, 29: 92–97.

Setchell, W. A., and DeA. Saunders. 1900. [Pleurophycus gardneri], in J. E. Tilden, *American Algae* (Exsiccatae). 1900: No. 346.

Setchell, W. A., and O. Swezy. 1923. [Callophyllis pinnata], in W. A. Setchell, A revision of the west North American species of Callophyllis. *Univ. Calif. Publ. Bot.*, 10: 400.

Setchell, W. A., and H. L. Wilson. 1910. [Gracilariophila oryzoides] in H. L. Wilson, Gracilariophila, a new parasite in Gracilaria confervoides. *Ibid.*, 4: 81.

Silva, P. C. 1951. The genus Codium in California. *Univ. Calif. Publ. Bot.*, 25: 79–114. 32 figs. 6 pls.

————. 1952. A review of nomenclatural conservation in the algae from the point of view of the type method. *Ibid.*, 25: 241–324.

————. 1953. The identity of certain Fuci of Esper. *Wasmann Jour. Biol.*, 11: 221–32.

————. 1957a. [Iridaea flaccida], in L. Stoloff and P. Silva, An attempt to determine possible taxonomic significance of the properties of water-extractable polysaccharides in red algae. *Econ. Bot.*, 11: 328.

————. 1957b. Notes on Pacific marine algae. *Madroño*, 14: 41–51.

Silva, P. C., and A. P. Cleary. 1954. The structure and reproduction of the red alga Platysiphonia. *Amer. Jour. Bot.*, 41: 251–60. 37 figs.

Skottsberg, C. 1922. Notes on Pacific coast algae. II. On the Californian Delesseria quercifolia. *Univ. Calif. Publ. Bot.*, 7: 427–36. 1 pl.

Smith, G. M. 1942. Notes on some brown algae from the Monterey Peninsula, California. *Amer. Jour. Bot.*, 29: 645–53. 13 figs.

————. 1943. [various Rhodophyceae], in G. M. Smith and G. J. Hollenberg, On Some Rhodophyceae from the Monterey Peninsula, California. *Amer. Jour. Bot.*, 30: 211–22. 30 figs.

————. 1944. *Marine Algae of the Monterey Peninsula, California.* Stanford, Calif. 622 pp. 98 pls.

————. 1955. *Cryptogamic Botany.* New York. 2d ed. Vol. 1. 546 pp. 311 figs.

————. 1969. *Marine Algae of the Monterey Peninsula.* Stanford, Calif. 2d ed., incorporating the 1966 Supplement by G. J. Hollenberg and I. A. Abbott. 752 pp. 53 figs. 98 pls.

Smith, G. M., and G. J. Hollenberg. 1943. On some Rhodophyceae from the Monterey Peninsula, California. *Amer. Jour. Bot.*, 30: 211–22. 30 figs.

Smith, J. E. 1814. [Conferva granulosa] in Smith, J. E., and J. Sowerby. 1814. *English Botany* . . . London. 1st ed. 1790–1814. Vol. 36, pl. 2351.

Solier, A. J. J. 1846. Sur deux algues zoosporées formant le nouveau genre Derbesia. *Duchatres Rev. Bot.*, 3: 452–54.

Sparling, S. R. 1957. The structure and reproduction of some members of the Rhodymeniaceae. *Univ. Calif. Publ. Bot.*, 29: 319–96. 15 figs. 12 pls.

Sprengel, C. 1827. *Caroli Linnaei . . . Systema vegetabilium*. Gottingen. 16th ed. Vol. 4(1). 592 pp.

Stackhouse, J. 1801. *Nereis Britannica* . . . Bath. Fasc. 3. xi + pp. 71–112.

Stewart, J. 1968. Morphological variation in Pterocladia pyramidale. *Jour. Phycol.*, 4: 76–84.

Sturch, H. H. 1926. Choreocolax polysiphoniae Reinsch. *Ann. Bot.*, 40: 585–605. 15 figs.

Suneson, S. 1937. Studien über die Entwicklungsgeschichte der Corallinaceen. *Lunds Univ. Årsskr.*, n.f., 33(2): 1–101. 42 figs. 4 pls.

Suringar, W. F. R. 1867. Algarum japonicarum Musei botanici L. B. Index praecursorius. *Ann. Mus. Bot. Lugduno-Batavi.*, 3: 256–59.

———. 1870. *Algae japonicae Musei botanici Lugduno-Batavi* (Edidit Societas Sci. Hollandicae quae Harlem). 39 pp. 25 pls.

———. 1872–74 (1873). Illustration des algues du Japon. *Musée Botanique de Leide*, 1: 85.

Tatewaki, M. 1969. Culture studies on the life history of some species of the genus Monostroma. *Sci. Pap., Inst. Algol. Res., Hokkaido Univ. Fac. Sci.*, 6: 1–56. 22 figs. 18 pls.

Taylor, W. R. 1928. The marine algae of Florida with special reference to the Dry Tortugas. *Pap. Tortugas Lab., Carneg. Inst.*, 25: 1–219. 37 pls.

———. 1939. Algae collected on the Presidential Cruise of 1938. *Smithsonian Misc. Coll.*, 98(9): 1–18. 2 pls.

———. 1945. Pacific marine algae of the Allan Hancock Expeditions to the Galapagos Islands. *Allan Hancock Pacific Expeditions*, 12: 1–528. 100 pls.

———. 1952. Notes on Vaucheria longicaulis Hoppaugh. *Madroño*, 11: 274–77. 11 figs.

———. 1955. Notes on algae from the tropical Atlantic Ocean, IV. *Pap. Mich. Acad. Sci., Arts & Letters*, 40: 67–76.

———. 1960. *Marine Algae of the Eastern Tropical and Subtropical Coasts of the Americas*. Ann Arbor, Mich. 870 pp. 80 pls.

Thuret, G. 1854. Note sur la synonymie des Ulva lactuca et latissima L. suivie de quelques remarques sur la tribu des Ulvacées. *Mém. Imp. Soc. Sci. Nat. Cherbourg*, 2: 17–32.

———. 1863. [Ulothrix flacca], in A. Le Jolis, Liste des algues marines de Cherbourg. *Ibid.*, 10: 56.

Thwaites, G. 1849. [Homospora ramosa], in W. H. Harvey, *Phycologia Britannica*. London. Vol. 2: pl. 213.

Tilden, J. E. 1894–1909. *American Algae* (Exsiccatae). Centuries 1–7. Fasc. 1. Minneapolis.

Trevisan, V. B. A. 1845. *Nomenclator algarum* . . . Padua. 80 pp.

Tseng, C. K. 1945. New and unrecorded marine algae of Hong Kong. *Pap. Mich. Acad. Sci., Arts & Letters*, 30: 157–72.

Turner, D. 1808. *Fuci* . . . London. Vol. 1. 164 pp. 71 pls.
———. 1809. *Ibid.* Vol. 2. 162 pp. 63 pls.
———. 1811. *Ibid.* Vol. 3. 148 pp. 62 pls.
———. 1819. *Ibid.* Vol. 4. 153 pp. 62 pls.
Velley, T. 1792. [Fucus elminthoides], in W. Withering, ed., *An Arrangement of British Plants* . . . Birmingham, Eng. 2d ed. P. 255.
———. 1795. *Coloured Figures of Marine Plants, Found on the Southern Coast of England* . . . Bath. (Unnumbered.)
Vickers, A. 1905. Liste des algues marines de la Barbade. *Ann. Sci. Nat. Bot.,* sér. 9, 1: 45–66.
Waern, M. 1952. Rocky-shore algae in the Öregrund Archipelago. *Acta Phytogeogr. Suecica.,* 30: 1–298. 100 figs. 32 pls.
Wagner, F. S. 1954. Contributions to the morphology of the Delesseriaceae. *Univ. Calif. Publ. Bot.,* 27: 279–346. 290 figs.
Weber-van Bosse, A. 1904. Corallineae verae of the Malay Archipelago, in A. Weber-van Bosse and M. Foslie, The Corallines of the Siboga Expedition. *Siboga-Expeditie.* Monogr. 61. Leiden. Pp. 78–110. 3 pls.
———. 1913–28. Liste des algues du Siboga. *Ibid.* Monogr. 59b. Leiden. 533 pp. 213 figs. 16 pls.
West, J. A. 1968. Morphology and reproduction of the red alga Acrochaetium pectinatum in culture. *Jour. Phycol.,* 4: 88–99. 27 figs.
———. 1969. The life histories of Rhodochorton purpureum and R. tenue in culture. *Ibid.,* 5: 12–21. 22 figs.
———. 1970. The conspecificity of Heterosiphonia asymmetria and H. densiuscula and their life histories in culture. *Madroño,* 20: 313–19. 6 figs.
———. 1972. The life history of Petrocelis franciscana. *Br. Phycol. Jour.,* 7: 299–308. 20 figs.
Widdowson, T. B. 1971. A taxonomic revision of the genus Alaria Greville. *Syesis,* 4: 11–49. 22 figs.
Wilce, R. T., E. E. Webber, and J. R. Sears. 1970. Petroderma and Porterinema in the New World. *Mar. Biol.,* 5: 119–35. 5 figs.
Wille, N. 1901. Studien über Chlorophyceen. VII. Über einige neue marine Arten von Ulothrix. *Vidensk. Selsk. Skr. Christiana (Mat.-Nat. Kl.),* 1900(6): 22–44. 1 pl.
Wittrock, V. B. 1866. *Försök till en monographie öfver algslägtet Monostroma.* Stockholm. 66 pp. 4 pls.
Wollaston, E. M. 1971. Antithamnion and related genera occurring on the Pacific coast of North America. *Syesis,* 4: 73–92. 45 figs.
———. 1972. The genus Platythamnion J. Ag. (Ceramiaceae, Rhodophyta) on the Pacific coast of North America between Vancouver, British Columbia, and Southern California. *Ibid.,* 5: 43–53.
Wollny, R. 1881. Die Meeresalgen von Helgoland. *Hedwigia,* 20: 1–32. 2 pls.
Womersley, H. B. S. 1965. The Helminthocladiaceae (Rhodophyta) of Southern Australia. *Austr. Jour. Bot.,* 13: 451–87. 75 figs. 7 pls.
Womersley, H. B. S., and A. Bailey. 1970. Marine algae of the Solomon Islands. *Phil. Trans. Roy. Soc. London,* 259B: 257–352. 4 pls.
Woodward, T. J. 1794. Description of Fucus dasyphyllus. *Trans. Linn. Soc.,* 2: 239. 3 figs. 1 pl.

Wulfun, F. X. 1803–5. [Fucus filamentosus], in Cryptogama acquatica, K. Roemer's Archiv Botanik, 3: 64. 1 pl.

Wynne, M. J. 1969a. Life history and systematic studies of some Pacific North American Phaeophyceae (brown algae). Univ. Calif. Publ. Bot., 50: 1–88. 12 figs. 24 pls.

———. 1969b. Platysiphonia decumbens sp. nov., a new member of the Sarcomenia group (Rhodophyta) from Washington. Jour. Phycol., 5: 190–202. 36 figs.

——— 1971. Concerning the phaeophycean genera Analipus and Heterochordaria. Phycologia, 10: 169–74. 9 figs.

———. 1972. Studies on the life forms in nature and in culture of selected brown algae, in I. Abbott and M. Kurogi, eds. Contributions to the Systematics of the Benthic Marine Algae of the North Pacific. Japanese Soc. Phycol. Kobe. Pp. 133–46.

Wynne, M. J., and K. Daniels. 1966. Cumathamnion, a new genus of the Delesseriaceae (Rhodophyta). Phycologia, 6: 13–28. 19 figs.

Wynne, M. J., and W. R. Taylor. 1973. The status of Agardhiella tenera and Agardhiella baileyi (Rhodophyta, Gigartinales). Hydrobiologia, 43: 93–107. 11 figs.

Yamada, Y. 1935. Marine algae from Urup, the Middle Kuriles, especially from the vicinity of Iema Bay. Sci. Pap., Inst. Algol. Res., Hokkaido Univ. Fac. Sci., 1: 1–26. 10 pls.

———. 1944. Notes on some Japanese algae, X. Ibid., 3: 11–25.

———. 1948. Marine Algae No. 1 [Colpomenia bullosa], in Y. Yamada and S. Kinoshita, Icones of the Marine Animals and Plants of Hokkaido. Hokkaido Fisheries Scientific Institution. Yoichi, Hokkaido. P. 6.

Yamada, Y., and T. Tanaka. 1934. Three new red algae from Formosa. Trans. Nat. Hist. Soc. Formosa, 24: 342–49. 5 figs.

Yendo, K. 1902a. Corallinae verae of Port Renfrew. Minn. Bot. Studies, 2: 711–22. 6 pls.

———. 1902b. Corallinae verae Japonicae. Jour. Coll. Sci. Imp. Univ. Tokyo, 16 (art. 3): 1–36. 7 pls.

———. 1905. A revised list of Corallinae. Ibid., 20 (art. 12): 1–46.

———. 1907. The Fucaceae of Japan. Ibid., 21: 1–174. 8 pls.

———. 1914. Notes on algae new to Japan, II. Bot. Mag. Tokyo, 23: 263–81.

———. 1916. Notes on algae new to Japan, IV. Ibid., 30: 47–65. 4 figs.

———. 1920. Novae algae Japonicae, I-III. Ibid., 34: 1–12.

Zanardini, G. A. M. 1839. Biblioteca italiana, 96: 134–37.

———. 1843. Saggio di classificazione naturale delle Ficee. Venice. 65 pp. 1 pl.

Zeh, W. 1912. Neue Arten der Gattung Liagora. Berlin Bot. Gart. Notizblatt, 5: 268–73.

Sources of Illustrations

As discussed in the Preface, more than two-thirds of the illustrations in this book are from three sources: Gilbert Morgan Smith, *Marine Algae of the Monterey Peninsula, California* (Stanford, Calif.: Stanford University Press, 1944); Gilbert Morgan Smith, *Marine Algae of the Monterey Peninsula, California*, Second Edition, Incorporating the 1966 *Supplement*, by George J. Hollenberg and Isabella A. Abbott (Stanford, Calif.: Stanford University Press, 1969); and material prepared expressly for this book, under the supervision of the authors and published here for the first time. The figure legends identify the artists, by initials; their names are listed in the Preface. Illustrations from other sources and used here with permission (or from the public domain) are as follows:

Allan Hancock Foundation Occasional Papers, 9: Fig. 448.
Allan Hancock Pacific Expeditions, 12 (1945): Fig. 74; 17 (1953): Fig. 420.
American Journal of Botany, 27 (1940): Figs. 512 (cross section), 513 (polysporangium); 29 (1942): Figs. 633–34, 636–38; 30 (1943): Figs. 423, 457–58; 31 (1944): Figs. 648, 650–51; 32 (1945): Figs. 563, 598, 672; 37 (1950): Fig. 2.
Botanical Gazette, 101 (1939): Fig. 670.
Bulletin of the National Museum of Canada, 207 (1966): Figs. 46–47, 50–51, 53–54, 57, 59, 62–63, 65, 67.
Bulletin of the Southern California Academy of Sciences, 44 (1945): Figs. 88, 490–91; 66 (1967): Figs. 424, 426, 430 (cross section).
Bulletin of the Torrey Botanical Club, 69 (1942): Fig. 136; 86 (1959): Fig. 55.
Canadian Journal of Botany, 46 (1968): Figs. 22, 100, 104, 235–36, 308–9, 316, 400, 418, 484.
Dansk Botanik Arkiv, 3 (1917): Fig. 558; 3 (1920): Fig. 551.
Farlowia, 3 (1947): Fig. 48.
Fisher, Lydia, ed., *Memoir of W. H. Harvey* . . . (London, 1869): portrait of W. H. Harvey.
Greville, R. K., *Scottish Cryptogamic Flora* (Edinburgh, 1822–28), Vol. 1: Fig. 182.
Harrison, E. S., *History of Santa Cruz County, California* (San Francisco, 1892): portrait of C. L. Anderson.
Institute of Algological Research, Faculty of Science, Hokkaido University, Scientific Papers, 5 (1964): Figs. 66, 68.
Journal of Phycology, 3 (1967): Figs. 378 (habit), 379–81, 387, 390 (habit); 4 (1968): Figs. 299–300 (cystocarpic nemathecia), 302–3, 305, 307, 401–3; 5 (1969): Figs. 131–33, 135, 138–41.
Lunds Universitets Årsskrift, n.f. 21 (1925): Fig. 560.
Madroño, 11 (1952): Fig. 1; 20 (1969): Figs. 459 (male leaflets, cystocarpic papillae), 460, 601, 616.
Memoirs of the Torrey Botanical Club, 11 (1902): Fig. 4.
Nova Hedwigia, 6 (1963): Figs. 663, 665; 12 (1966): Figs. 506, 569, 671.
Nova Hedwigia Beihefte, 15 (1964): Fig. 111.
Okamura, K., *Icones of Japanese Algae* (Tokyo, 1907–42), Vol. 4: Figs. 69, 149, 279.

————, *Nippon Kaiso-shi* (Tokyo, 1936): Fig. 571.

Pacific Naturalist, 1 (1960): Figs. 168, 691; 2 (1961): Figs. 419, 427, 631, 639, 643, 649.

Pacific Science, 25 (1971): Figs. 566–67, 570.

Phycologia, 9 (1970): Figs. 99, 315, 317, 629, 658, 666; 10 (1971): Figs. 7, 36, 93, 101–2, 359, 632; 11 (1972): Fig. 470.

Proceedings of the California Academy of Sciences, 1 (1898): Figs. 85, 89, 95, 98, 107, 169; 12 (1924): Fig. 87.

Syesis, 3 (1970): Fig. 464; 4 (1971): Figs. 474–78, 480–81, 517–18, 520–26.

University of California Publications in Botany, 1914 (6): Figs. 273, 276, 277 (cross section), 694–97; 1920 (7, 8): Figs. 6, 11–13, 15, 81; 1922 (7): Figs. 84, 86, 91–92, 106, 110, 112, 114, 116, 118–20, 123, 126–27, 129; 1925 (8): Figs. 117, 174, 181; 1928 (14): Figs. 251, 253, 262, 264.

Wittrock, V. B., *Catalogus Illustratus Iconothecae Botanicae Horti Bergiani Stockholmiensis* (Stockholm, 1903): portrait of J. G. Agardh.

Index

Every botanical name used in the text, at whatever level, is given here. Page numbers for major discussions are given in roman type, those for secondary mentions in italic type. Families and higher taxa are given in small capitals. Recognized genera, species, and infraspecific taxa (including a few valid algal taxa occurring only outside the range of the book) are given in roman type. Synonyms are given in italic type. (That the genus for a synonymous species is also given in italic type does not necessarily imply the invalidity of the genus name.)

9 780804 721523